D

Fundamentals of visualization, modeling, and graphics for Engineering Design

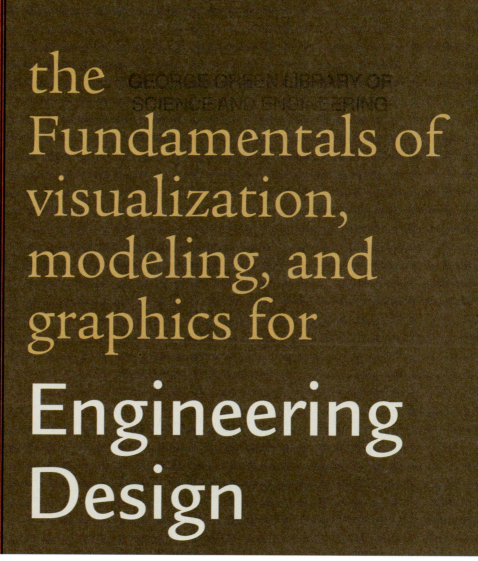

the Fundamentals of visualization, modeling, and graphics for Engineering Design

Dennis K. Lieu   ■   **Sheryl Sorby**

Professor of Mechanical
Engineering
University of California,
Berkeley

Professor of Civil and
Environmental Engineering
Michigan Technological
University

DELMAR
CENGAGE Learning™

Australia • Brazil • Japan • Korea • Mexico • Singapore • Spain • United Kingdom • United States

The Fundamentals of Visualization, Modeling, and Graphics for Engineering Design, 1st Edition
Dennis K. Lieu and Sheryl Sorby

Vice President, Technology and Trades ABU: David Garza

Director of Learning Solutions: Sandy Clark

Managing Editor: Larry Main

Senior Acquisitions Editor: James DeVoe

Development: Ohlinger Publishing Services

Marketing Director: Deborah S. Yarnell

Marketing Manager: Kevin Rivenburg

Marketing Specialist: Mark Pierro

Director of Production: Patty Stephan

Production Manager: Stacy Masucci

Content Project Manager: Michael Tubbert

Art Director: Bethany Casey

Editorial Assistant: Tom Best

Cover Image: Solid model courtesy of Hoyt USA
Background courtesy of Getty Images, Inc.

For product information and technology assistance, contact us at
Cengage Learning Customer & Sales Support, 1-800-354-9706
For permission to use material from this text or product, submit all requests online at
www.cengage.com/permissions Further permissions questions can be emailed to
permissionrequest@cengage.com

Library of Congress Control Number: 2007942952

ISBN-13: 978-1-4018-4250-5

ISBN-10: 1-4018-4250-X

1005 379946 T

Delmar
Executive Woods
5 Maxwell Drive, PO Box 8007,
Clifton Park, NY 12065-8007
USA

Cengage Learning is a leading provider of customized learning solutions with office locations around the globe, including Singapore, the United Kingdom, Australia, Mexico, Brazil, and Japan. Locate your local office at:

international.cengage.com/region

Cengage Learning products are represented in Canada by Nelson -Education, Ltd.

For your lifelong learning solutions, visit **delmar.cengage.com**

Visit our corporate website at **cengage.com**

Notice to the Reader

Publisher does not warrant or guarantee any of the products described herein or perform any independent analysis in connection with any of the product information contained herein. Publisher does not assume, and expressly disclaims, any obligation to obtain and include information other than that provided to it by the manufacturer. The reader is expressly warned to consider and adopt all safety precautions that might be indicated by the activities described herein and to avoid all potential hazards. By following the instructions -contained herein, the reader willingly assumes all risks in connection with such instructions. The publisher makes no representations or warranties of any kind, including but not limited to, the warranties of fitness for particular purpose or merchantability, nor are any such representations implied with respect to the material set forth herein, and the publisher takes no responsibility with respect to such material. The publisher shall not be liable for any special, consequential, or exemplary damages resulting, in whole or part, from the readers' use of, or reliance upon, this material.

Printed in the United States of America
1 2 3 4 5 6 7 11 10 09 08

brief contents

The Fundamentals of Visualization, Modeling, and Graphics for Engineering Design

The following chapters are on the CD in the back of this textbook:

contents

sectionone
Laying the Foundation

sectiontwo
Modern Design Practice and Tools

sectionthree
Setting Up an Engineering Drawing

sectionfour
Drawing Annotation and Design Implementation

preface

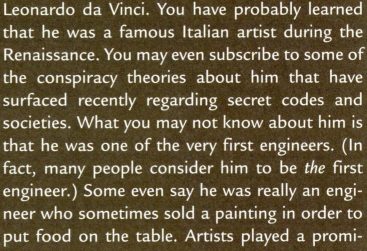

Leonardo da Vinci. You have probably learned that he was a famous Italian artist during the Renaissance. You may even subscribe to some of the conspiracy theories about him that have surfaced recently regarding secret codes and societies. What you may not know about him is that he was one of the very first engineers. (In fact, many people consider him to be *the* first engineer.) Some even say he was really an engineer who sometimes sold a painting in order to put food on the table. Artists played a prominent role at the birth of modern-day engineering, and some of the first artist-engineers included Francesco di Giorgio, Georg Agricola, and Mariano Taccola. These were the individuals who could visualize new devices that advanced the human condition. Their creativity and their willingness to try seemingly "crazy" ideas propelled technology forward at a much faster pace than had occurred in the previous thousand years.

This marriage between art and engineering has diminished somewhat since the early beginnings of the profession; however, creativity in engineering is still of paramount importance. Would the Apollo spaceship have landed on the moon without the creative thinking of hundreds of engineers who designed and tested the various systems necessary for space travel? Would we be able to instantaneously retrieve information and communicate with one another via the World Wide Web without the vision of the engineers and scientists who turned a crazy idea into reality? Would the modern-day devices that enrich and simplify our lives such as washing machines, televisions, telephones, and automobiles exist without the analytical skills of the engineers and technologists who developed and made successive improvements to these devices? The answer to all of these questions is "no." The ability to think of systems that never were and to design devices to meet the changing needs of the human population is the purview of the engineering profession.

Graphical communication has always played a central role in engineering, perhaps due to engineering's genesis within the arts or perhaps because graphical forms of communication convey design ideas more effectively than do written words. Maybe a picture really *is* worth a thousand words. As you might expect, the face of engineering graphics has evolved dramatically since the time of da Vinci. Traditional engineering graphics focused on 2-dimensional graphical mathematics, drawing, and design; knowledge of graphics was considered a key skill for engineers. Early engineering programs included graphics as an integral topic of instruction, and hand-drawn engineering graphics from 50 years ago are works of art in their own right.

However, in the recent past, the ability to create a 2-D engineering drawing by hand has become de-emphasized due to improvements and advances in computer hardware and software. More recently, as computer-based tools have advanced even further, the demand for skills in 3-D geometric modeling, assembly modeling, animation, and data management has defined a new engineering graphics curriculum. Moreover, three-dimensional geometric models have become the foundation for advanced numerical analysis methods, including kinematic analysis, kinetic analysis, and finite-element methods for stress, fluid, magnetic, and thermal systems.

The engineering graphics curriculum has also evolved over time to include a focus on developing 3-D spatial visualization ability since this particular skill has been documented as important to the success of engineers in the classroom and in the field. Spatial visualization is also strongly linked to the creative process. Would da Vinci have imagined his various flying machines without well-honed visualization skills?

We have come full circle in engineering education through the inclusion of topics such as creativity, teamwork, and design in the modern-day graphics curriculum. The strong link between creativity, design, and graphics cannot be overstated. Gone is the need for engineers and technicians who robotically reproduce drawings with little thought involved. With modern-day computational tools, we can devise creative solutions to problems without concern about whether a line should be lightly penciled in or drawn thickly and displayed prominently on the page.

It is for this new, back-to-the-future graphics curriculum that *The Fundamentals of Visualization, Modeling, and Graphics for Engineering Design* has been designed. This text is a mixture of traditional as well as modern-day topics, a mixture of analytical and creative thinking, a mixture of exacting drawing technique and freeform sketching. Enjoy.

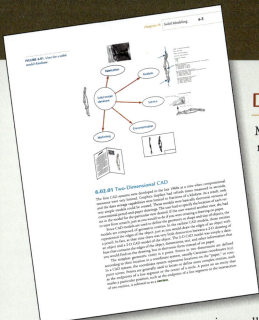

FIGURE 6.01. Uses for a solid model database.

6.02.01 Two-Dimensional CAD

The first CAD systems were developed in the late 1960s at a time when computational resources were limited. Graphics displays had refresh times measured in seconds, and the data storage capabilities were limited to fractions of a kilobyte. As a result, only very simple models could be created. Those models were basically electronic versions of conventional pencil-and-paper drawings. The user had to specify the location of each vertex in the model for the particular view desired. If the user wanted another view, she had to start from scratch, just as you would do if you were creating a drawing on paper.

Since CAD models are used to define the geometry of objects, the models are composed of geometric entities. In the earliest CAD models, those entities represented the edges of the object. Just as you would draw the edges of an object with a pencil. In fact, at that time there was very little distinction between a 2-D drawing of an object and a 2-D CAD model of the object. The 2-D CAD model was simply a database that contains the edges of the object, dimensions, text, and other information that you would find on the drawing, but in electronic form instead of on paper.

The simplest geometric entity is a point. Points in two dimensions are defined according to their location in a coordinate system, usually Cartesian coordinates (x,y). In a CAD system, the coordinate system represents locations on the "paper," or computer screen. Points are generally used to locate or define more complex entities, such as the endpoints of a line segment or the center of a circle. A point on an entity that marks a particular position, such as the endpoint of a line segment or the intersection of two entities, is referred to as a **vertex**.

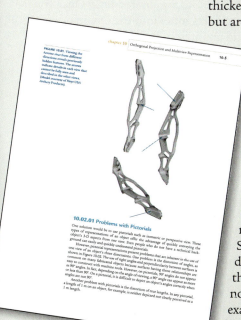

FIGURE 10.01. Viewing the *hacesaw viece* from different directions reveals previously hidden features. The arrows indicate details in each view that cannot be fully seen and described in the other views. (Model courtesy of Hoyt USA Archery Products.)

chapter 10 Orthogonal Projection and Multiview Representation 10-5

10.02.01 Problems with Pictorials

One solution would be to use pictorials such as isometric or perspective view. These types of representations of an object offer the advantage of quickly conveying the object's 3-D aspects from one view. Even people who do not have a technical background can easily and quickly understand pictorials.

However, pictorial representations present problems that are inherent in the use of one view of an object's three dimensions. One problem is the distortion of angles, as shown in Figure 10.02. The use of right angles and perpendicular surfaces is common on many fabricated objects because surfaces having those relationships are easy to construct with machine tools. However, on pictorials, 90° angles do not appear as 90° angles. In fact, depending on the angle of viewing, a 90° angle can appear as more or less than 90°. On a pictorial, it is difficult to depict an object's angles correctly when angles are not 90°.

Another problem with pictorials is the distortion of true lengths. A line of 1 m length on an object, for example, is neither depicted nor clearly perceived as a 1 m length.

Development of the Text

Many of the current graphics textbooks were written several years ago with modern-day topics such as feature-based solid modeling included as a separate add-on to the existing material. In these texts, modern-day computer-based techniques are more of an afterthought: "Oh, by the way, you can also use the computer to help you accomplish some of these common tasks." In fact, some of the more popular texts were written nearly a century ago when computer workstations and CAD software were figments of some forward-looking engineer's imagination. Texts from that era focused on drawing technique and not on graphic communication within the larger context of engineering design and creativity.

Modern engineering graphics curricula—and texts—must follow what is happening in the field. Modern product development techniques allow engineers to use computer hardware and software to examine the proper fit and function of a device. Engineers can "virtually" develop and test a device before producing an actual physical model, which greatly increases the speed and efficiency of the design process. The virtual, computer-based model then facilitates the creation of the engineering drawings used in manufacturing and production—an activity that required many hours of hand drafting just a few short decades ago.

In the real world, modern CAD practices have also allowed us more time to focus on other important aspects of the engineering design cycle, including creative thinking, product ideation, and advanced analysis techniques. Some might argue that these aspects are, in fact, the *most* important aspects of the design process. The engineering graphics curricula at many colleges and universities have evolved to reflect this shift in the design process. However, most engineering graphics textbooks have simply added CAD sections to cover the new topics. Thick textbooks have gotten even thicker. As a wise person has said, "Engineering faculty are really good at addition, but are miserable at subtraction."

When we sat down to plan this text, we wanted to produce the engineering graphics benchmark of the future—an engineering graphics approach that teaches design and design communication rather than a vocational text focused on drafting techniques and standards. We wanted to integrate modern-day design techniques throughout the text, not treat these topics as an afterthought.

A strength of this textbook is its focus not only on "what" to do, but also on "why" you do it (or do not do it) that way—concepts as well as details. This text is intended to be a learning aid as well as a reference book. Step-by-step software-specific tutorials, which are too focused on techniques, are very poor training for students who need to understand the modeling strategies rather than just which buttons to push for a particular task. In fact, we believe that mere *training* should be abandoned in favor of an *education* in the fundamentals. Students need to learn CAD strategy as well as technique. Students need to develop their creative skills and not have these skills stifled through a focus on the minutiae. In order to prepare for a lifelong career in this fast-changing technological world, students will need to understand fundamental concepts. For example, in the current methods-based approach to graphics training, students learn about geometric dimensioning and tolerancing. Many texts describe what the symbols mean, but do not explain how, why, and when they should be used. Yet, these questions of how, why and when are the questions with which most young engineers struggle and the questions that are directly addressed in this text. They are also the *important* questions—if a student knows the answer to these questions, s/he will understand the fundamental concepts in geometric dimensioning and tolerancing. This fundamental understanding will serve far into the future where techniques, and possibly the symbols themselves, are likely to change.

Organization of the Text

This textbook is organized into eighteen main chapters; a number of additional chapters are available to create custom versions for specific course needs. In organizing the chapters of this text, we were careful to group topics in a way that reflects the modern engineering design process. We purposely did not mimic the decades-old graphics classical texts, which were written in an era when the design process was based on drawings and not computer models, an era when physical and not virtual models were analyzed for structural integrity. For this reason, the order of topics in this text will not match that of the traditional graphics texts where Lettering was often the first topic of instruction.

In this text, we start with foundational topics such as sketching and visualization since these are useful in the initial or "brainstorming" step in the design process and since these are fundamental topics on which many other topics hinge. Also included are chapters on creativity, design, and teamwork—you cannot work in the real world without being a good team member. From there we move to 3-D modeling because, in the real world, design typically begins with the production of a computer model. In the next stage of the modern design process, a computer model is analyzed either virtually or sometimes physically, and these topics are covered next. Once your model is complete and thoroughly analyzed, engineers move into the design documentation stage where drawings are created and annotated. Topics such as manufacturing processes are included in the text in order to help you answer the "why" questions surrounding drawing creation and annotation. For example, many dimensions are placed on drawings based on *how* the part will eventually be manufactured, so it makes sense to understand manufacturing processes before moving on to dimensioning practice. The text is organized into four major sections as described subsequently. Additional chapters cover topics in traditional graphics instruction as well as some modern-day, "not quite ready for primetime" topics such as HTML and Web Utilization.

Section 1 – Laying the Foundation

The materials presented here focus on the needs of today's first-year engineering students who might have well-developed math and computer skills and less-developed hands-on mechanical skills. Incoming engineering students likely no longer work on their cars or bikes in the garage or may not have taken shop and drafting classes in high school. Hands-on tinkering is probably an activity of the past replaced by hands-on web page design and text messaging. (Engineering students of today in all probability have much greater dexterity in their thumbs from "texting" than do the authors of this text!) Although many engineering students enter college having spent time in a virtual computer environment, the lack of hands-on experiences that involve more than just the thumbs and that also involve real-life physical objects, often results in poorly developed three-dimensional visualization skills. In this section, these skills are explored and developed. This section also includes a project-oriented approach with inclusion of topics in design, creativity, and teamwork to prepare students for a lifetime of professional engineering practice.

Section 2 – Modern Design Practice and Tools

The modern topics found in this section reflect the current state of design in industry. Solid modeling has revolutionized engineering graphics. The widespread availability of computers has made 3-dimensional modeling the preferred tool for engineering design in nearly all disciplines. Solid modeling allows engineers to easily create mathematical models parts and assemblies, visualize and manipulate these models in real time, calculate physical properties, and inspect how they mate with other parts. The modern-day design process is characterized by computer methods that take advantage of the efficiency and advantages offered by workstations and feature-based modeling software. These new technologies have revolutionized the design process and have enabled around-the-clock engineering. By this model, engineers in Europe hand-off (via the Internet) a design project when they leave work to American engineers. The Americans, in turn, hand-off the design as they leave the office at the end of the day to Asian engineers. The Asian engineers complete the cycle by passing the design back to the Europeans at the end of their day. The sun never sets on an engineering design. Over your lifelong engineering career, the details of the design process may change again in ways that are unimagined today, but the fundamentals, as described in this section, will migrate from system to system with each advance in technology.

Section 3 – Setting up an Engineering Drawing

This section contains material found in most conventional textbooks on engineering graphics; however, the content is presented in novel ways and with a fresh approach to problem solving. The topics and techniques in this section are in wide use in engineering graphics classrooms today and are likely to continue to be invaluable into the foreseeable future. These traditional graphics topics continue to be important for several reasons. First, many legacy designs out there were produced prior to the feature-based solid modeling revolution. You may be asked to examine these designs, so it is important that you thoroughly understand drawings. Second, not every company has the capability to go directly from a computer model to a manufactured part, and drawings are still important in these environments. Finally, while the computer can usually automatically generate a drawing for you, certain conventions and dimensioning practices do not translate well. You will need to be able to verify the integrity of drawings that the computer generates for you and make changes where needed. For all of these reasons, no matter how sophisticated the computer design, hardware and software, or the manufacturing system, engineers must still be able to visualize a three-dimensional object from a set of two-dimensional drawings and vice versa.

Section 4 – Drawing Annotation and Design Implementation

The ultimate goal of the engineering design process is to develop devices where everything fits together and functions properly. The sizes of the features that define an object are crucial to the overall functionality of the system. The chapters in this section describe how sizes and geometries of entities are specified. Since no part can be made to an "exact" size even with the best in computer technologies, the allowable errors for part sizes are also described in this section. The final drawings produced in preparation for fabrication must meet exacting criteria to ensure that they are properly cataloged and interpreted for clear communication among all parties. If your drawing includes non-standard annotations, the machinist or contractor who uses those drawings to produce an engineered system may unknowingly misinterpret the drawing, resulting in higher project costs or even failure. The chapters in this section detail standard practice in drawing annotation to help you avoid making blunders. The ability to make proper, 100% correct drawing annotations will likely take you several years to develop. Be patient and keep learning.

Optional Chapters

These additional chapters are available for purchase at www.iChapters.com:
Chapter 19. Technical and Engineering Animation
Chapter 20. Topological Maps and GIS
Chapter 21. Presentation of Engineering Data
Supplemental Chapter 1. 2-D Instrument Drawing Techniques
Supplemental Chapter 2. More Working Drawings and Drawing Formats
Supplemental Chapter 3. Linkages, Cams, Gears, Springs, and Bearings.
Supplemental Chapter 4. Welding
Supplemental Chapter 5. Descriptive Geometry
Supplemental Chapter 6. HTML and Web Utilization

Chapter Structure

With a few exceptions, each chapter is organized along similar lines. The material is presented with the following outline:

1. Objectives

   Chapter-opening objectives alert students to the chapter's fundamental concepts.

2. Introduction

   This section provides an overview of the material that will be presented in the chapter, and discusses why it is important.

3. The Problem

   Each chapter directly addresses a certain need or problem in graphical communication. That problem is presented here as if the student had to face such a problem in the field. The presence of a real problem that needs to be solved gives a student added incentive to learn the material in the chapter to solve that problem.

4. Explanation and Justification of Methods

   Engineering graphics has evolved and continues to evolve at an increasingly fast pace due to advances in computer hardware and software. Although new methods associated with new technologies exist, these modern methods must remain compatible with conventional graphics practices. This consistency is required to eliminate possible confusion in the interpretation of drawings, maintain sufficient flexibility to create designs unencumbered by the tools available to document them, and reduce the time and effort required to create the drawings.

5. Guide to Problem Solving

   Sample problems of various types are presented throughout each chapter with detailed, step-by-step solutions.

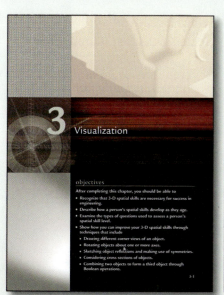

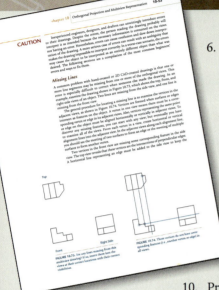

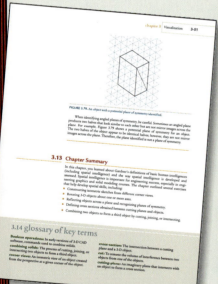

6. Caution

Often, people learn more from their mistakes than from their successes. The most common errors—and their potential consequences—are presented as examples in this section.

7. Summary

This section distills the most important information contained in the chapter.

8. Glossary of Key Terms

Formal definitions of the most important terms or phrases for the chapter are provided. Each term or phrase is highlighted the first time it is used in the chapter.

9. Questions for Review

These questions test the student's understanding of the chapter's main concepts.

10. Problems

A variety of problems and exercises help to develop skill and proficiency of the material covered in the chapter.

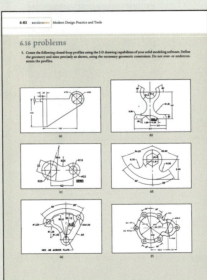

Key Features of the Presentation

We believe that this text will have a broad appeal to engineering graphics students across a wide spectrum of institutions. The following are key features of the text:

- *A focus on learning and fundamental skill development, not only on definitions, tools, and techniques.* This approach prepares students to apply the material to unfamiliar problems and situations rather than simply to regurgitate previously memorized material. In the fast-changing world we live in, an understanding of the fundamentals is a key to further learning and the ability to keep pace with new technologies.

- *Formal development of visualization skills as a key element at an early stage of the curriculum.* Development of these skills is important for students who have not had the opportunity to be exposed to a large number of engineering models and physical devices. Further, the link between visualization and creativity is strong—tools for success over a lifelong career.

- *Use of a problem-based approach.* This approach presents the student with real problems at the beginning of each chapter, shows the graphics solutions, then generalizes the solutions.

- *A casual tone and student-friendly approach.* It is a proven fact that students learn the material better if they are not fast asleep!

- *Examples of poor practices and the potential consequences of these errors.* People seem to learn the most from mistakes. By showing students the mistakes of others, perhaps they can minimize their own.

- *Several common example threads and a common project that are presented in most chapters.* The text shows how the material contained in each chapter was actually applied in the context of product development.

   One of the case studies to be presented, for example, is the Hoyt AeroTec™ Olympic style recurve target bow. The unique geometry of this bow was brought to prominence after it was part of the equipment package used to win the Gold Medal in target archery at the summer Olympic Games in Sydney, Australia. The design history of the development of this product is traced starting from its ideation as an improvement to all other target bows on the market at that time. As a student moves through the chapters of this book, the progress of the development of this product will be documented. This product was selected as an example for these reasons:

1. It was a very successful design that accomplished all of the goals set forth by its engineers.

2. It was also a product that is relatively unencumbered by the complexity of mechanisms or electronics, which are not the focus of this book.

3. The design is mature, having made it to the consumer market; this means it offers an opportunity to study some of the non-technical issues that play an important role in engineering design.

Final Remarks

This textbook contains a "core" of material covered in a traditional engineering graphics course and also a number of other chapters on modern graphics techniques. The collected material represents over 50 combined years of personal experience in the learning, application, and teaching of engineering graphics. The result is a text that should appeal to both traditional and contemporary graphics curricula. We, the authors, would like to thank you for considering this text.

Dennis K. Lieu
Professor of Mechanical Engineering
University of California, Berkeley

Sheryl Sorby
Professor of Civil and Environmental Engineering
Michigan Technological University

Contributors
Holly K. Ault, Worcester Polytechnic Institute (Chapter 6 and 17 and
      Supplemental Chapter 4)
Ron Barr, The University of Texas at Austin (Chapters 5 and 8)
Judy Birchman, Purdue University (Chapter 21)
Ted Branoff, North Carolina State University (Chapters 16 and 17)
Pat Connolly, Purdue University (Supplemental Chapter 6)
Frank Croft, The Ohio State University (Chapter 12)
La Verne Abe Harris, Purdue University (Chapter 21)
Kathy Holliday-Darr, Penn State Erie (Chapter 14 and Supplemental Chapter 5)
Tom Krueger, The University of Texas at Austin (Chapters 5 and 8)
Ann Maclean, Michigan Technological University (Chapter 20)
Kellen Maicher, Purdue University (Supplemental Chapter 6)
Jim Morgan, Texas A&M University (Chapter 4)
Bill Ross, Purdue University (Chapter 19)
Mary Sadowski, Purdue University (Chapter 21)
Tom Singer, Sinclair Community College (Supplemental Chapter 1)
Kevin Standiford, Consultant (Supplemental Chapters 2 and 3)

What is iChapters.com?

iChapters.com, Cengage Learning's online store, is a single destination for more than 10,000 new print textbooks, e-books, single e-chapters, and print, digital and audio supplements.

Choice. Students have the freedom to purchase *a-la-carte* exactly what they need when they need it.

Savings. Students can save 50% for the electronic textbook, and can pay as little as $1.99 for an individual chapter.

Get the best grade in the shortest time possible! Visit www.iChapters.com to receive 25% off over 10,000 print, digital and audio study tools.

acknowledgments

The authors would like to thank the following people for their contributions to the textbook and their support during its development:

George Tekmitchov and Darrin Cooper of Hoyt USA for their assistance and cooperation with the AeroTEC case study material; David Madsen for problem materials; David Goetsch for problem materials; George Kiiskila of U.P. Engineers and Architects for construction drawings; Fritz Meyers for use of his blueprint drawing; Mark Sturges of Autodesk for graphics materials; Rosanne Kramer of Solidworks for graphics materials; Marie Planchard of Solidworks for graphics materials; James DeVoe of Delmar Cengage Learning for his support, confidence, persistence, good humor, and patience in seeing this project through to the end; Monica Ohlinger, editor extraordinaire, of Ohlinger Publishing Services for her organization skills that resulted in the completion of this project.

In addition, the authors are grateful to the following reviewers for their candid comments and criticisms throughout the development of the text:

Tom Bledsaw, ITT
Ted Branoff, North Carolina State University
Patrick Connolly, Purdue University
Richard F. Devon, Penn State University
Kathy Holliday-Darr, Penn State University, Erie
Tamara W. Knott, Virginia Tech
Kellen Maicher, Purdue University
Jim Morgan, Texas A & M
Kellen Maicher, Purdue University
William A. Ross, Purdue University
Mary Sadowski, Purdue University
James Shahan, Iowa State University
Michael Stewart, Georgia Tech

Dennis K. Lieu

Prof. Dennis K. Lieu was born in 1957 in San Francisco, where he attended the public schools, including Lowell High School. He pursued his higher education at the University of California at Berkeley, where he received his BSME in 1977, MSME in 1978, and D.Eng. in mechanical engineering in 1982. His major field of study was dynamics and control. His graduate work, under the direction of Prof. C.D. Mote, Jr., involved the study skier/ski mechanics and ski binding function. After graduate studies, Dr. Lieu worked as an advisory engineer with IBM in San Jose CA, where he directed the specification, design, and development of mechanisms and components in the head-disk-assemblies of disk files. In 1988, Dr. Lieu joined the Mechanical Engineering faculty at UC Berkeley.   His research laboratory is engaged in research on the mechanics of high-speed electro-mechanical devices and magnetically generated noise and vibration. His laboratory also studies the design of devices to prevent blunt trauma injuries in sports, medical, and law enforcement applications. Prof. Lieu teaches courses in Engineering Graphics and Design of Electro-mechanical Devices. He was the recipient of a National Science Foundation Presidential Young Investigator Award in 1989, the Pi Tau Sigma Award for Excellence in Teaching in 1990, and the Berkeley Distinguished Teaching Award (which is the highest honor for teaching excellence on the U.C. Berkeley campus) in 1992. He is a member of Pi Tau Sigma, Tau Beta Pi, and Phi Beta Kappa. His professional affiliations include ASEE and ASME. Prof. Lieu's hobbies include Taekwondo (in which he holds a 4th degree black belt) and Olympic style archery.

Sheryl Sorby

Professor Sheryl Sorby is not willing to divulge the year in which she was born but will state that she is younger than Dennis Lieu. She pursued her higher education at Michigan Technological University receiving a BS in Civil Engineering in 1982, an MS in Engineering Mechanics in 1985, and a PhD in Mechanical Engineering-Engineering Mechanics in 1991. She was a graduate exchange student to the Eidgenoessiche Technische Hochshule in Zurich, Switzerland studying advanced courses in solid mechanics and civil engineering. She is currently a Professor of Civil and Environmental Engineering at Michigan Technological University. Dr. Sorby is the former Associate Dean for Academic Programs and the former Department Chair of Engineering Fundamentals at Michigan Tech. She has also served as a Program Director in the Division of Undergraduate Education at the National Science Foundation. Her research interests include various topics in engineering education, with emphasis on graphics and visualization. She was the recipient of the Betty Vetter research award through the Women in Engineering Program Advocates Network (WEPAN) for her work in improving the success of women engineering students through the development of a spatial skills course. She has also received the Engineering Design Graphics Distinguished Service Award, the Distinguished Teaching Award, and the Dow Outstanding New Faculty Award, all from ASEE.

Dr. Sorby currently serves as an Associate Editor for ASEE's new online journal, Advances in Engineering Education. She is a member of the Michigan Tech Council of Alumnae. She has been a leader in developing first-year engineering and the Enterprise program at Michigan Tech and is the author of numerous publications and several textbooks. Dr. Sorby's hobbies include golf and knitting.

sectionone

Laying the Foundation

The materials presented in this section focus on the needs of today's beginning engineering students, who typically have well-developed math and computer skills but less-developed hands-on mechanical skills compared to students of earlier generations. Incoming engineering students may no longer be people who work on their cars or bikes in the garage and who took shop and drafting classes in high school. Although many engineering students enter college having spent time in a virtual computer environment, the lack of hands-on experience often results in a lack of three-dimensional visualization skills. So in addition to the classical material on standard engineering graphics practices, these students need to enhance their visualization skills. Prior to the advent of CAD, the graphics classroom featured large tables topped with mechanical drafting machines and drawers full of mechanical drawing instruments. Engineering students and engineers now have additional time to focus on aspects of the engineering design cycle that are more worthy of their talents as engineers. These aspects include creative thinking, product ideation, and advanced analysis techniques to ensure a manufacturable and robust product. Formalization of the design process allows designers to focus their energies on certain areas in the process and gain more meaningful results.

1

An Introduction to Graphical Communication in Engineering

objectives

After completing this chapter, you should be able to

- Explain and illustrate how engineering graphics is one of the special tools available to an engineer

- Define how engineering visualization, modeling, and graphics are used by engineers in their work

- Provide a short history of how engineering graphics, as a perspective on how it is used today, was used in the past

1.01

introduction

Because engineering graphics is one of the first skills formally taught to most engineering students, you are probably a new student enrolled in an engineering program. Welcome!

You may be wondering why you are studying this subject and what it will do for you as an engineering student and, soon, as a professional engineer. This chapter will explain what engineering is, how it has progressed over the years, and how graphics is a tool for engineers.

What exactly is engineering? What does an engineer do? The term *engineer* comes from the Latin *ingenerare*, which means "to create." You may be better able to appreciate what an engineer does if you consider that *ingenious* also is a derivative of *ingenerare*. The following serves well enough as a formal definition of engineering:

The profession in which knowledge of mathematical and natural sciences, gained by study, experience, and practice is applied with judgment to develop and utilize economically the materials and forces of nature for the benefit of humanity.

A modern and informal definition of engineering is "the art of making things work." An engineered part or an engineering system does not occur naturally. It is something that has required knowledge, planning, and effort to create.

So where and how does graphics fit in? Engineering graphics has played three roles through its history:

1. Communication
2. Record keeping
3. Analysis

First, engineering graphics has served as a means of communication. It has been used to convey concepts and ideas quickly and accurately from one person to another without the use of words. As more people became involved in the development of products, accurate and efficient communication became increasingly necessary. Second, engineering graphics has served as a means of recording the history of an idea and its development over time. As designs became more complex, it became necessary to record the ideas or features that worked well in a design so they could be repeated in future applications. And third, engineering graphics has served as a tool for analysis to determine critical shapes and sizes, as well as other variables needed in an engineered system.

These three roles are still vital today, more so than in the past, because of the technical complexity required to make modern products. Computers, three-dimensional modeling, and graphics software have made it increasingly effective to use engineering graphics as an aid in design, visualization, and optimization.

1.02 A Short History

The way things are done today evolved from the way things were done in the past. You can understand the way engineering graphics is used today by examining how it was used in the past. Graphical communications has supported **engineering** throughout history. The nature of engineering graphics has changed with the development of new graphics tools and techniques.

1.02.01 Ancient History

The earliest documented forms of graphical communication are cave paintings, such as the one shown in Figure 1.01, which showed human beings depicting organized social

FIGURE 1.01. Undated cave painting showing hunting and the use of tools.
Source: ACE STOCK LIMITED/ Alamy

behavior, such as living and hunting in groups. The use of tools and other fabricated items for living comfort and convenience were also communicated in cave paintings. However, these paintings typically depicted a lifestyle, rather than any instructions for the fabrication of tools, products, or structures. How the items were made is still left to conjecture.

The earliest large structures of significance were the Egyptian pyramids and Native American pyramids. Some surviving examples are shown in Figure 1.02. The Egyptian pyramids were constructed as tombs for the Pharaohs. The Native American pyramids were built for religious ceremonies or scientific use, such as observatories. Making these large structures, with precision in the fitting of their parts and with the tools that were available at the time, required much time, effort, and planning. Even with modern tools and construction techniques, these structures would be difficult to re-create today. The method of construction for the pyramids is largely unknown—records of the construction have never been found—although there have been several theories over the years.

FIGURE 1.02. Mayan pyramid, Yucatan, Mexico (left), and Pharaoh Knufu and Pharaoh Khafre Pyramids, Giza, Egypt (right).
Sources: Brand X Pictures/Alamy, above; DIOMEDIA/Alamy, below.

FIGURE 1.03. Ancient Egyptian hieroglyphics describing a life story.
Source: © Bettmann/CORBIS

Egyptian hieroglyphics, which were a form of written record, included the documentation of a few occupational skills, such as papermaking and farming, although, for the most part, they documented lifestyle. An example of a surviving record is shown in Figure 1.03. As a result of those records, papermaking and farming skills could be maintained and improved over time. Even people who were not formally trained in those skills could develop them by consulting the written records.

Two engineering construction methods helped the Roman Empire expand to include much of the civilized European world. These methods were used to create the Roman arch and the Roman road.

The Roman arch, shown in Figure 1.04, was composed of stone that was precut to prescribed dimensions and assembled into an archway. The installation of the keystone at the top of the arch transferred the weight of the arch and the load it carried into the

FIGURE 1.04. Pont-du-Gard Roman aqueduct (left) built in 19 BC to carry water across the Gardon Valley to Nimes. Spans of the first- and second-level arches are 53–80 feet. The Ponte Fabricio bridge in Rome (right) built in 64 BC spans the bank of the River Tiber and Tiber Island.
Photos by William G. Godden. Reprinted with permission from EERC Library, Univ. of California, Berkeley.

FIGURE 1.05. An engraving showing the operation of an Archimedes screw to lift water. *Courtesy of Time Life Pictures/Getty Images*

remaining stones that were locked together with friction. This structure took advantage of the compressive strength of stone, leading to the creation of large structures that used much less material. The Roman arch architecture was used to create many large buildings and bridges. Roman era aqueducts, which still exist today in Spain and other countries in Europe, are evidence of the robustness of this **design**.

The method used to construct Roman roads prescribed successive layers of sand, gravel, and stone (instead of a single layer of the native earth), forming paths wide enough for commercial and military use. In addition to the layered construction methods, these roads were also crowned to shed rain and had gutters to carry away water. This construction method increased the probability that the roads would not become overgrown with vegetation and would remain passable even in adverse weather. As a result, Roman armies had reliable access to all corners of the empire.

The Roman Empire is long gone, but the techniques used for the construction of the Roman arch and the Roman road are still in use today. The reason for the pervasiveness of those designs was probably due to Marcus Vitruvius, who, during the Roman Empire, took the trouble to carefully document how the structures were made.

The Archimedes screw, used to raise water, is an example of a mechanical invention developed during the time of the Greek Empire. Variations of the device were used for many centuries because diagrams depicting its use were (and still are) widely available. One of those diagrams is shown in Figure 1.05. These early documents were precursors to modern engineering **drawings**. Because the documents graphically communicated how to build special devices and structures, neither language nor language translation was necessary.

1.02.02 The Medieval Period

Large building construction helped define the medieval period in Europe. Its architecture was more complicated than the basic architecture used for the designs of ancient buildings. The flying buttress, a modification of the Roman arch, made it possible to construct larger and taller buildings with cavernous interiors. This type of structure was especially popular in Europe for building cathedrals, such as the one shown in Figure 1.06. The walls of fortresses and castles became higher and thicker. Towers were included as an integral part of the walls, as shown in Figure 1.07, to defend the inhabitants from many directions, even when attackers had reached the base of a wall.

FIGURE 1.06. Flying buttress construction used to support the exterior walls of Notre Dame Cathedral in Paris.
Courtesy of Getty Images

FIGURE 1.07. Warwick Castle, England, circa 1350, is an example of a medieval style fortification.
Source: © Royalty-Free/CORBIS

FIGURE 1.08. The Great Wall of China, built during the medieval period, used simple engineering principles despite the large scale of the project. *Source: Nigel Hicks/Alamy*

In Asia, large fortifications, shrines, and temples, as shown in Figure 1.08, were built to last hundreds of years. The complexity of techniques to build those structures required planning and documentation, especially when raw materials had to be transported from long distances. Building structures of such sizes required an understanding of the transmission of forces among the supporting members and the amount of force those members could withstand. That knowledge was especially important when wood was the primary building material.

Large-scale civil engineering **projects** were begun during the medieval era. Those projects were designated by a civilian government to benefit large groups or the general population, as opposed to projects constructed for private or military use. The windmills of Holland, shown in Figure 1.09, are an example of a civil engineering project. The windmills harvested natural wind energy to pump large amounts of water out of vast swampland, making the land suitable for farming and habitation.

Windmills and waterwheels were used for a variety of tasks, such as milling grain and pumping water for irrigation. Both inventions were popular throughout Europe and Asia; a fact that is known because diagrams showing their construction and use have been widely available.

1.02.03 The Renaissance

The beginning of the Renaissance in the 1400s saw the rise of physical scientific thinking, which was used to predict the behavior physical **systems** based on empirical observation and mathematical relationships. The most prominent person among the scientific physical thinkers at that time was Leonardo da Vinci, who documented his ideas in drawings. Some of those drawings, which are well known today, are shown in Figure 1.10. Many of his proposed devices would not have worked in their original form, but his drawings conveyed new ideas and proposals as well as known facts.

Prior to the Renaissance, nearly all art and diagrams of structures and devices were records of something already in existence or were easily extrapolated from something already tried and known to work. When inventors applied physical science to

FIGURE 1.09. The network of windmills in Holland, used to drain water from flooded land, is an example of an early large-scale civil engineering project. *Source: © PaulAlmasy/CORBIS*

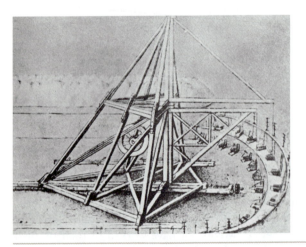
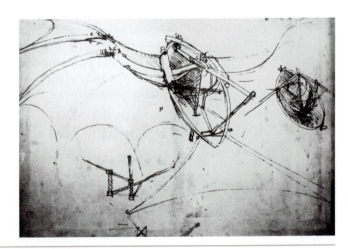

FIGURE 1.10. Images of original da Vinci drawings: a machine used for canal excavation (left) and a flying ship (right). Codex Atlanticus, folio 860; drawing from Il Codice Atlantico di Leonardo da Vinci nella biblioteca Ambrosiana di Milano, Editore Milano Hoepli 1894–1904; the original drawing is kept in the Biblioteca Ambrosiana in Milan. *Sources: © Bettmann/CORBIS*

FIGURE 1.11. A perpetual motion machine by medieval inventors: an Archimedes screw driven by a waterwheel is used to mill grain.
Source: Timewatch Images/Alamy

engineering, they could conceive things that theoretically should have worked without having been previously built. When inventors did not understand the science behind the proposed devices, the devices usually did not work. The many perpetual motion machines proposed at that time, as shown in Figure 1.11, are evidence of inventors' lack of understanding of the physical science and their resultant failed attempts to build the machines.

Engineers began to realize that accurate sizing was an element of the function of a structure or device. Diagrams made during the Renaissance paid more attention to accurate depth and perspective than in earlier times. As a result, drawings of proposed and existing devices looked more realistic than earlier drawings.

Gunpowder was introduced during the Renaissance, as was the cannon. The cannon made obsolete most of the fortresses built during the medieval era. The walls could not withstand impact from cannon projectiles. Consequently, fortresses needed to be redesigned to survive cannon fire. In France, a new, stronger style of fortification was designed. The fortification was constructed with angled walls that helped to deflect cannon fire and did not crumble as flat vertical walls did when struck head on. The new fortresses were geometrically more complicated to build than their predecessors with vertical walls. Further, the perimeter of the fortress had evolved from a simple rectangular shape to a pentagonal shape with a prominent extension at each apex. That perimeter shape, coupled with the angled walls, resulted in walls that intersected at odd angles that could not be seen and measured easily or directly. Following is a list of questions that builders of earlier fortresses could easily answer but that builders of the angled wall fortresses could not:

- What is the surface area of a wall?
- What is the fill volume?

- What are the specific lengths of timbers and beams needed to construct and brace the walls?
- What are the true angles of intersection between certain surfaces?
- What are the distances between lines and other lines, between points and lines, and between points and surfaces?

Fortunately, the French had Gaspard Monge, who developed a graphical analysis technique called **descriptive geometry**. Analytical techniques using mathematics were not very sophisticated at that time, nor were machines available to do mathematical calculations. But mechanical **instruments**, such as compasses, protractors, and rulers, together with the graphical method, were used to analyze problems without the need to do burdensome math. Descriptive geometry techniques enabled engineers to create any view of a geometric object from two existing views. By creating the proper view, engineers could see and measure an object's attributes, such as the true length of its lines, the true shape of planes, and true angles of intersection. Such skills were necessary, especially for the construction of fortifications, as shown in Figure 1.12. The complex geometry, odd angles of intersection, and height of walls were intended to maximize the cross fire on an approaching enemy, while not revealing the interior of the fortress. Another objective was to construct the ramparts and walls by moving the minimum amount of material for maximum economy.

The astuteness of the French at building fortifications kept France the prime military power in Europe until the 1700s. At that time, descriptive geometry was considered a French state secret; divulging it was a crime punishable by death. As a result of the alliance between France and the newly constituted United States, many U.S. fortifications used French designs. An example is Fort McHenry (shown in Figure 1.13), which was built in 1806 and is exquisitely preserved in Baltimore, Maryland. Fort McHenry survived bombardment by the British during the War of 1812 and is significant because it inspired Francis Scott Key to write "The Star Spangled Banner."

By the 1800s, most engineering was either civil engineering or military engineering. Civil engineering specialized in the construction of buildings, bridges, roads, commerce ships, and other structures, primarily for civilian and trade use. Military engineering specialized in the construction of fortifications, warships, cannons, and other items for military use. In both fields of engineering, as projects became more complicated, more people skilled in various subspecialties were needed. Clear, simple,

FIGURE 1.12. Using descriptive geometry to find the area of a plane.
Courtesy of D. K. Lieu

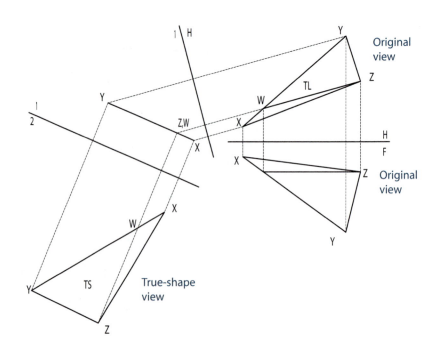

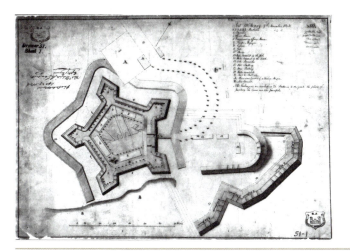 

FIGURE 1.13. French fortification design principles (left) and Fort McHenry (right) in Baltimore, Maryland, whose design was based on those principles.
Courtesy of The National Archives, College Park, Maryland, left; Courtesy of The National Park Service, Fort McHenry NMHS, right.

and universal communication was necessary to coordinate and control the efforts of specialists interacting on the same project. Different people needed to know what other people were doing in order for various **parts** and **assemblies** to fit together and function properly. To fill that need, early forms of scaled drawings began to be used as the medium for communications in constructing a building or device.

1.02.04 The Industrial Revolution

The industrial revolution began in the early 1800s with the new field of mechanical engineering. This revolution was, in part, a result of the need for new military weapons. Before the 1800s, ships and guns were fabricated one at a time by skilled craftsmen. No original plans of any ships from the Age of Discovery exist, because shipwrights did not use plans drawn on paper or parchment. The only plans were in the master shipwright's mind, and ships were built by eye. As the demand for ships grew, production methods changed. It was far more economical to build many ships using a single design of common parts than to use a custom design for each ship. Constructing from a common design required accurate specifications of the parts that went into the design.

The hardware products that general and military consumers needed then were no longer produced by skilled craftsmen but were mass-produced according the techniques and machines specified by engineers. Mass production meant that each product had to be identical to all other products, had to be fabricated within predictable and short production times, had to be made from parts that were interchangeable, and had to be produced economically in volumes much larger than in the past. The consistent and repetitive motions of machines required efficient, large-scale production which replaced manufacturing operations that had needed the skilled motions of craftsmen. Also, engines, boilers, and pressure vessels were required to provide power to machines. An early manufacturing facility with machine tools and an early steam engine are shown in Figure 1.14 and Figure 1.15, respectively.

Creating not only a product but also the machines to produce it was beyond the abilities of individual craftsmen—each likely to have a different set of skills needed for the production of a single product. The high demand for creating machines as well as products meant that the existing master-apprentice relationship could no longer supply the demand for these skills. To meet the growing demand, engineering schools had to teach courses in basic physics, machine-tool design, physical motion, and energy transfer.

FIGURE 1.14. A photo showing early factory conditions during the industrial revolution.
Courtesy of Time Life Pictures/Getty Images

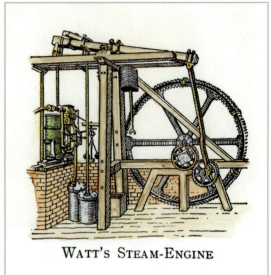

WATT'S STEAM-ENGINE

FIGURE 1.15. A schematic drawing of a James Watt steam engine; the type commonly used to power production machinery during the early years of the industrial revolution.
Source: North Wind Picture Archives/Alamy

Communication was necessary to coordinate and control the efforts of different people with different skills. Each craftsman, as well as each worker, on a project needed to know what others were doing so the various pieces, devices, structures, and/or systems would fit together and function properly. The ideas of the master designer had to be transferred without misinterpretation to those who worked at all levels of supporting roles. In the design stage, before things were actually built, the pictorial diagrams once used were soon found to be insufficient and inaccurate when new structures with new techniques were being built. More accurate representations, which would provide exact sizes, were needed. That need eventually led to the modern engineering drawing, with its multiple-view presentation, identification of sizes, and specification of allowable errors.

Around the time of the industrial revolution, patents started to become important. As a method of stimulating innovation in an industrialized society, many governments offered patents to inventors. The owners of patents were guaranteed exclusive manufacturing rights for the device represented in the patent for a prescribed number of years in exchange for full disclosure of how the device operated. Since a single successful patented invention could make its owner rich, many people were inspired to create new products. From the start, the difference between patent drawings and engineering has been that engineering drawings are made to be viewed by those formally trained in engineering skills and to show precise sizes and locations. Patent drawings, on the other hand, are made to teach others how and why a device operates. Consequently, patent drawings often do not show the actual or scaled sizes of the parts. In fact, sizes are commonly distorted to make the device more difficult for potential competitors to copy. An example of a patent drawing is shown in Figure 1.16.

1.02.05 More Recent History

As technology advanced over time, additional engineering specialties were born. In the late 1800s, as electric power became more popular and more available, electrical engineering was born. Electrical engineering at that time was concerned with the production, distribution, and use of electrical energy. The information derived from the study of

FIGURE 1.16. A U.S. patent drawing showing function but not necessarily the true sizes of the parts.
Courtesy of D. K. Lieu

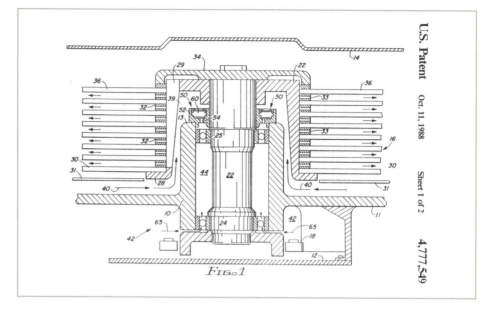

electric motors, generators (shown in Figure 1.17), power conversion, and transmission lines needed for their design was more than other engineers—not specifically electrical engineers—could be expected to know and use. Chemical engineering, as a special engineering discipline, emerged at the beginning of the twentieth century with the need for large-scale production of petroleum products in refineries, as shown in Figure 1.18, and the production of synthetic chemicals.

During the 1950s, industrial engineering and manufacturing engineering emerged from the necessity to improve production quality, control, and efficiency. Nuclear engineering emerged as a result of the nuclear energy and nuclear weapons programs.

Some of the more recent engineering disciplines include bioengineering, information and computational sciences, micro-electro-mechanical systems (MEMS), and nano-engineering. The design of a MEMS device (for example, the valve shown in

FIGURE 1.17. Later during the industrial revolution, steam engines were replaced by electric power supplied, for example, by these generators at the Long Island Railway shown (circa 1907). Electrical engineering was born.
Courtesy of SMITHSONIAN INSTITUTION Neg.#44191D

FIGURE 1.18. The demand for chemical and petroleum products led to the construction of sophisticated plants and refineries and the disciplines of chemical and petroleum engineering.
Courtesy of DOE/NREL, photo by David Parsons.

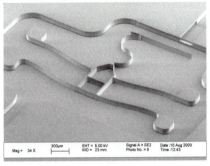

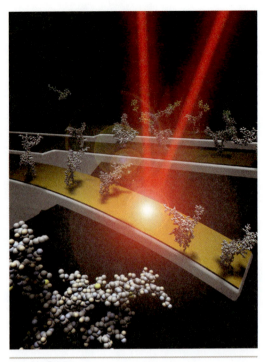

FIGURE 1.19. This MEMS valve was designed with a solid modeler and was fabricated using semiconductor processing techniques.
Courtesy of the Berkeley Sensor and Actuator Center, University of California.

FIGURE 1.20. This nano-device for sorting molecules does not actually appear as shown, but the use of graphics aids in understanding its operating principles.
Courtesy of Kenneth Hsu.

Figure 1.19) requires skills from both electrical and mechanical engineering. A nano-engineered device cannot be seen with conventional optics. Its presumed appearance, such as that shown in Figure 1.20, and function are based on conjecture using engineering graphics tools.

With the emergence of a new discipline comes formal intensive training, specifically in the specific discipline, as opposed to subspecialty training within an existing discipline. Most complex engineering projects today require the combined skills of engineers from a variety of disciplines. Engineers from any single discipline cannot accomplish landing an astronaut on the moon or putting a robotic rover, shown in Figure 1.21, on Mars.

FIGURE 1.21. Complex engineering projects, such as interplanetary space missions, require interdisciplinary engineering skills.
Source: NASA/JPL/Cornell University

1.03 The People and Their Skills

Today few engineering projects exist where a single person or a small group of people is responsible for all aspects of the project from beginning to end. Many people with many different types of technical and nontechnical skills participate in the development and production phases of a project. Whether that person is the engineer who conceives the overall idea or the fabricator who makes the individual pieces or the technician who assembles the parts to make the system operate, they all have common questions:

- What is this part, device, or structure supposed to do?
- What is it supposed to look like?
- What are the precise geometries and sizes of its features?
- What is it made of?
- How is it made?
- How does it fit into other parts, devices, or structures?
- How do I know if everything is made the way it was supposed to be made?

To answer these questions, a clear, unbroken, and unambiguous flow of communication must take place, as depicted in each diagram in Figure 1.22.

The object envisioned by the engineer must be the same object produced by the fabricator and the same object assembled into the working system by the technician. Graphical communication that follows universally accepted standards for representing shapes and sizes makes that happen.

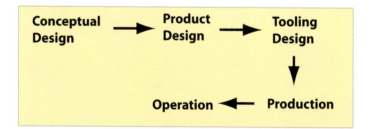

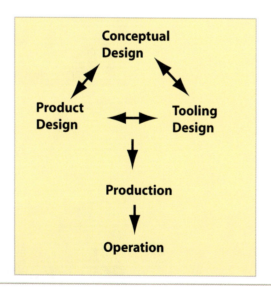

FIGURE 1.22. In conventional product design (above), phases of the development cycle occur sequentially. Concurrent engineering (below) combines two or more phases to accelerate the cycle.
Courtesy of D. K. Lieu

1.03.01 Organization of Project Life Phases

An engineering project may be as simple as a one-piece can opener or as complex as an interplanetary space mission—or anything in between. The number of people involved can be as few as one to as many as several thousand. Regardless of the complexity of the project or the number of people involved, any project can be broken into several phases over its lifetime. These life phases are as follows:

- Concept
- Design
- Fabrication
- Installation
- Operation
- Disposal

For example, consider the wind-powered electric generation facility located at California's Altamont Pass approximately 100 km east of San Francisco. This facility, composed of about 7,200 large wind turbines covering an area of several hundred square kilometers, is one of the largest wind-energy-producing facilities in the world. Many of the lessons learned in the construction and operation of the Altamont facility were incorporated into the plans to build a new wind-powered generation facility for the Solano County area, which is near the Sacramento River delta in California. A small portion of the Solano facility is shown in Figure 1.23.

FIGURE 1.23. A few of the 100-meter-high wind turbines at the Solano Wind Power Generation Facility (above) and one of the turbines on the ground before mounting (below).
Courtesy of D. K. Lieu

Before the facility could be built, it had to be decided during the Concept Phase whether the project would be economically viable and socially and environmentally acceptable. Other decisions involved the size and density of turbines.

During the Design Phase, the types of turbines were selected. The shapes and sizes of the various parts for the turbines and their supporting structures were developed, as was a scheme for collecting, controlling, and distributing the electric power that would be produced.

Parts that could not be purchased as finished units were custom-built during the Fabrication Phase. During the fabrication of custom parts for any project, the appropriate manufacturing processes are selected to reduce costs as much as possible. For large projects, it is just as common to use foreign and domestic suppliers for prefabricated as well as custom-fabricated parts.

The Installation Phase involves taking the individual parts and putting them together to create individually operating turbines. The individual turbines needed to be networked to supply power as an entire system.

The Operation Phase includes not only the control of the system but also the maintenance of the individual turbines and the networked power grid. When parts wear out or are damaged, they must be repaired or replaced.

Finally, when the entire facility has reached the end of its useful life, there must be a plan in the Disposal Phase for removal of the facility, disposal or recycling of its components, and return of the land to other uses.

1.03.02 Organization of Functional Groups

The larger the number people within each phase and between phases, the greater the need for effective communication. In a complex engineering project, the work needs to be divided into many subspecialties that are usually performed by different organizations. The personnel involved in each phase of a project generally can be organized into the following functions:

- Research and Development
- Design
- Manufacturing
- Sales and/or Buying
- Service
- Subcontractors

Depending on the complexity of the project, the same person can handle each function as a specialty, or an entire group of people may be responsible for each function. In the case of the Solano wind-powered generation facility, let's examine one of the project phases. During the Design Phase, certain people were responsible for seeking and evaluating new materials, devices, and technologies that would be of immediate use in designing and building the turbines. Those people filled the Research and Development function. Other people in the Design function were responsible for specifying the shapes and sizes of premade and custom parts so the parts would fit and function as intended. The people who actually made the parts or assembled them into operating prototypes filled the Manufacturing function. Any premade items or raw materials that had to be purchased were done by people in the Buying function. People in the Service function were responsible for operating the prototypes and, if necessary, to collect data needed to evaluate the design for improvement. Subcontractors supplied items or services that could be produced more quickly and efficiently by third parties. For the project life phases, similar responsibilities could be identified for each of the functions just mentioned.

1.03.03 Organization of Skills

Within each function of each project phase, the responsibility of the engineering personnel can be further subdivided as follows:

- Engineers
- Designers
- Drafters
- Fabricators
- Inspectors
- Technicians

Engineers are responsible for ensuring that systems and devices are specified to operate within their theoretical limits, specifying the materials and sizes of parts and assemblies so that failures do not occur, specifying the methods in which the devices are maintained and operated, and evaluating and preparing the environment in which large projects are to be placed.

Designers are responsible for the project's fit and finish, that is, specifying the geometry and sizes of components so they properly mate with each other and are ergonomically and aesthetically acceptable within the operating environment.

Drafters are responsible for documentation—the formal graphical records of parts and assemblies that are required not only for record keeping but also for unambiguous communication between people working on the project.

Fabricators are responsible for making the parts according to the specifications of the engineers and designers, using the documentation provided by the drafter as a guide.

Inspectors are responsible for checking; they take parts made by the fabricators and compare the actual sizes of the parts' features to the desired sizes. This is done to ensure that the parts are properly made and will fit and function as intended. Some projects are installed over large pieces of land. In those cases, inspectors ensure that the land has been properly prepared and that the various elements that compose the project have been made and installed according to the specifications of the engineers.

Technicians are responsible for operation and maintenance; they typically assemble various components to create working devices or structures, operate them, and maintain them.

Depending on the particular phase of the project and the particular function group within that phase, a group will have different combinations of engineering personnel. For example, the Design Phase of the wind-powered generation facility had many engineers and designers but few technicians. However, during the Operation Phase, the facility had mostly technicians, with only a few engineers. One interesting problem that engineers faced during the Operation Phase of the Altamont Pass wind power facility was how to reduce the number of birds, including large raptors, that were killed every year by the spinning turbine blades. No one foresaw this problem during the earlier phases of the project. Special avian experts had to be consulted during the Operation Phase to assist the engineers with possible solutions. Those same avian experts were consulted during the Design Phase of the Solano wind power facility. The new turbines at Solano were designed to have slower blade rotation speeds, and their towers were designed to make bird nesting difficult.

Regardless of the makeup of the engineering group, whenever a number of people participate in any aspect of a project, such as designing and constructing a part, everyone must know what that part is supposed to look like, what the part is made of, and what the part is supposed to do.

1.03.04 Concurrent Engineering

You should not finish the preceding section thinking that the only way engineering projects are done, or even the preferred way to get them done, is through formally

FIGURE 1.24. Entering the maintenance hatch in one of the Solano wind turbine towers. *Courtesy of D. K. Lieu*

organizing functional groups and separating skills. Separate functions and skills may be the classic way to do things, but many modern products use concurrent engineering to reduce the time needed for the product design and production cycles. Concurrent engineering is a process where the design and certain aspects of the fabrication phases are combined. The engineers responsible for the design of a product and the engineers responsible for manufacturing or construction work together closely. Thus, as a part is being conceived, its method of fabrication and assembly into other parts is being given careful consideration. The design of the part is then altered to facilitate its fabrication and, when economically feasible, its assembly into a larger system. Concurrent engineering also considers the method of disposal once the part has reached the end of its useful life.

As an example of concurrent engineering, consider the support tower for one of the wind turbines at the Solano facility. This structure supports the turbine, transmission gearbox, and generator; it also provides access (via a very long stairway) to these devices for maintenance, shown in Figure 1.24. Assume you are the designer of this structure and you want it made from steel.

Since several thousands of these structures may be needed for the many wind power installations around the world, you need to consider the economics of fabricating them. Using a conventional engineering timeline approach, you need to determine the required material and geometry, then make the drawings for the structure. You would have a prototype fabricated, installed on a prototype wind turbine, and tested to prove that it will do what you designed it to do and, especially, that it will not fail. Then you would turn the part and its drawings over to a manufacturing engineer to figure out the best and most economical way to **fabricate** large numbers of the structures to satisfy the worldwide needs of wind-power generation installations. For example, one way of fabricating the tower would be to make it from many small curved plates of steel, with all of the plates welded together. But it may be more economical to produce large sections of complete tubes that are the diameter of the tower and then connect those sections. However, the cost of any special tooling required to make and transport the large tubes would need to be included in the final cost of the structure. Different fabrication processes are possible for making the sections. Each process has different advantages and disadvantages in terms of cost and efficiency.

With concurrent engineering, engineers from all phases of the project work together. Engineers who ordinarily become involved later in the process get together with the designer at the early stages of the design. For example, the manufacturing engineer would advise the part designer on how to change a part so it would be easier to fabricate or handle. At nearly the same time, the manufacturing engineer would begin to design any specialized tools that would be needed to fabricate, handle, and assemble the part in large numbers. As the part prototype is being fabricated, these special tools also would be built. The advantage of concurrent engineering is that product development is reduced. The disadvantage is that large errors in design are expensive because any changes in design also require changes to the production tooling.

1.04 Engineering Graphics Technology

Mechanical drawing instruments have been a tremendous aid for the creation of engineering graphics. These instruments greatly improve the precision with which graphics can be produced and reproduced, reducing any distortion and making analyses easier and more accurate. The improvement of engineering graphics technology over the years has been a major factor in the improvement of engineering design and communication.

1.04.01 Early Years

Up until the time of the Renaissance, most drawing was done by hand without mechanical devices, because none were available. As a result, many of the drawings that were made to depict some sort of engineering device were distorted. The amount of distortion depended on the skill of the person making the drawing. **Two-dimensional (2-D) drawings** were common because they were easy to make. Attempts at drawing objects showing depth had mixed results. Leonardo da Vinci was one of few people who was good at it, but he was also a skillful artist. In general, though, handmade drawings were good for conveying ideas and some rough sizing. They were poor when precision was necessary, mostly because it was not possible to determine exact sizes from them. In fact, the inch and foot as units of measurement in Europe were not standardized until the twelfth century, and the meter was not defined until the eighteenth century. As a result, when different craftsmen built the same item, the sizes of the parts would be slightly different. Those differences made part interchangeability, and thus mass production, extremely difficult.

1.04.02 Instrument Drawing

Early instruments used to make drawings included straightedges with graduated scales, compasses and dividers, and protractors. They were generally custom-made items for the convenience of those who could afford them. Mechanical instruments for drawing did not become widely available until the industrial revolution, when, for a reasonable cost, machines could produce accurate instruments for both drawing and measuring. Both standardized units and accurate drawings made it possible for different fabricators to make the same part. With careful specifications, those parts would be interchangeable between the devices in which they functioned. Now that engineering drawing made it possible to fabricate the same part at different manufacturers, engineering drawing became a valuable means of communication.

From the industrial revolution to the late twentieth century, drawing instruments slowly improved in quality and became less expensive. Drawing instrument technology reached its most effective and highest level of use during the 1970s. Some companies and individuals today still retain, and even prefer, to use mechanical instruments for making engineering drawings. Classic drawing instruments, some of which are shown

FIGURE 1.25. Tools for instrument drawing (left) and a drafting machine (right).
Sources: © Peter Harholdt/SuperStock, left; © Françoise Gervais/CORBIS, right.

in Figure 1.25, are available from architecture, art, and engineering supply shops; these instruments include the following:

- Drafting board—a large, flat table with straight, square edges for alignment of drawing instruments
- Drawing vellum—a tough, dimensionally stable, and age-resistant paper on which drawings are made when placed on the drafting board
- T square—an instrument used to make horizontal and vertical lines by using the edges of the drafting board for reference
- Triangle—an instrument used to make lines at common angles
- Protractor—an instrument used to measure angles or make lines at arbitrary angles
- Scale—an instrument used to measure linear distances
- Drafting machine—a special machine used to hold scales at arbitrary angles while the scales are allowed to translate across the drawing, thus replacing many of the previously listed instruments
- Compass—an instrument used to make circles and arcs
- French curve—an instrument used to make curves
- Template—an instrument used to make common shapes

Using pencil or ink, engineers use instruments to draw directly on the desired sized vellum sheet. Large drawings are reproduced on special copy machines. Up until the 1980s, engineering students often were burdened with having to learn how to use the drawing instruments.

1.04.03 The Computer Revolution

During the 1970s, many large companies, particularly those in the automotive and aerospace industries, recognized the advantages of computer-based drawing and graphics: ease of storage and transmission of data, precise drawing data, and ease of data manipulation when drawings needed to be changed. Several large companies began developing computer-aided drawing (**CAD**) tools for their own use. Mainframe computers were just reaching the point where their cost, computation power, and storage capability would support computer-based drawing. The CAD systems consisted of computer terminals connected to a mainframe computer. However, the conversion to computer-based drawing was slow. Mainframe computers were expensive, the user had to have some computer skills, computer hardware and software were not very reliable, and special input and output devices were necessary. Thus, the average engineer or drafter still had a difficult time making the transition from mechanical tools to computer-based tools.

In the late 1970s and early 1980s, several companies specializing in CAD developed freestanding computer-drawing stations based on small independent computers called workstations. Those companies marketed the computer hardware and software as a complete, ready-to-operate unit known as a turnkey system. The workstation approach to CAD made the software more affordable for smaller companies. Also, CAD software became more sophisticated and easier to use. It began to grow in popularity. As personal computers (PCs) began to proliferate in the 1980s, CAD software made specifically to run on PCs became popular. One company that became a leader in this application was Autodesk, with its AutoCAD software. Companies that formerly supplied mainframe computer-based or turnkey CAD systems either quickly adapted their products for PC use or went out of business. As PCs became more powerful, cheaper, easier to use, and more prolific, CAD software did the same. Drafting boards were quickly replaced by PCs. An example of a PC-based CAD system is shown in Figure 1.26.

FIGURE 1.26. Computer graphics stations have replaced mechanical drawing instruments in most applications. A CAD drawing can be created by itself or extracted from a solid model.
Courtesy of D. K. Lieu

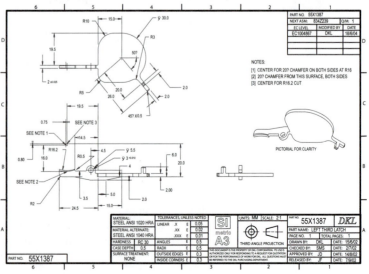

1.04.04 Graphics as a Design Tool

Computer-based **three-dimensional (3-D) modeling** as an engineering design tool began in the 1980s. CAD was a great convenience, but it produced only drawings. In this sense, CAD was just a very accurate instrument for making drawings. A drawing's representation of an object in three dimensions had to be visualized by the person reading the drawing. It was the same for any fit or function of an assembly—the person reading the drawing had to visualize it. One problem was that not all readers visualized a drawing the same way. Three-dimensional modeling addressed those problems directly. Unlike a 2-D CAD drawing, which was a collection of 2-D objects used to represent specified views of an object, computer-based solid models had 3-D properties.

The field of mechanical engineering quickly adopted 3-D modeling, calling it **solid modeling**, for the design and analysis of mechanical parts and assemblies. Extrusion or revolution of 2-D shapes created simple 3-D geometries. More complex geometries were created by Boolean operation with simple geometries. The computer calculated a 3-D pictorial **image** of the part, which the engineer could see on a computer monitor. The biggest advantage of solid modeling over CAD was that it permitted viewing a 3-D object from different perspectives, greatly easing the **visualization** of a proposed object. Multiple parts could be viewed together as an assembly and examined for proper fitting. With solid modeling, graphics became more of an engineering design tool, rather than merely a drawing tool. An example of a solid **model** for a single part is shown in Figure 1.27. An assembly model is shown in Figure 1.28.

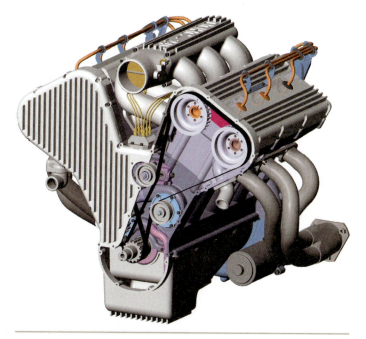

FIGURE 1.27. Solid modeling allows a proposed part to be easily visualized in a variety of orientations.
Courtesy of D. K. Lieu

FIGURE 1.28. An assembly model of an Omnica 3.2-liter V-6 engine made from a collection of solid model parts.
Courtesy of SolidWorks Corporation

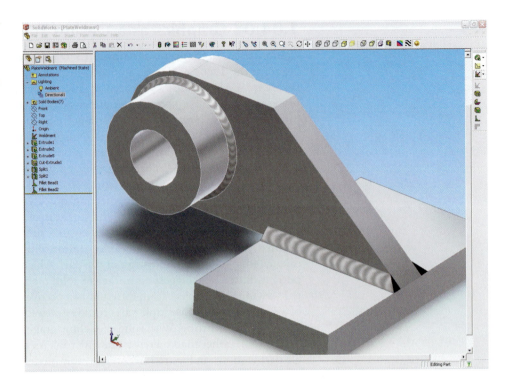

FIGURE 1.29. The graphical user interface of a solid modeling software program. *Courtesy of SolidWorks Corporation*

As you may have realized, solid modeling required more computation power and memory to process files than CAD did. That is why solid modeling was originally introduced on computer workstations using UNIX operating systems, which were relatively costly at the time. In the late 1980s, a new software algorithm increased the utility of solid modeling by making it possible to link the sizes and locations of features on an object to variables that could be input and changed easily. The process was known as parametric design. Those products made it easy for an engineer to add, delete, or change the geometry and sizes of features on a part and see the results almost immediately. Dynamic viewing, which enabled the engineer to twist and turn the part image in real time, was also a powerful software feature. A particular facility of that software—the quick and easy extraction of engineering drawings from the 3-D model—made the total software package a valuable drawing tool as well as a modeling tool.

As PCs continued to become more powerful, in the 1990s solid modeling was introduced as a PC software product. The migration of solid modeling from expensive workstations to less expensive PCs made the software popular among small companies and individuals. The later development of new graphical user interfaces, such as the one shown in Figure 1.29, as opposed to the text menus prevalent at the time, made solid modeling easy to use, even for casual users. PC-based solid modeling with graphical user interfaces soon became a standard.

1.04.05 Graphics as an Analysis Tool

Prior to the 1970s, before the days of inexpensive digital computers and handheld calculators, many types of mathematical problems were solved using graphical techniques. Those types of problems included graphical vector analysis, roots and intersections of nonlinear functions, and graphical calculus. Numerical techniques now solve these problems more quickly and easily than graphical techniques, so graphical techniques are not used much anymore. Although solid modeling has decreased the usefulness of descriptive geometry as an analytical tool in many mechanical engineering applications, descriptive geometry still has useful applications in some large-scale civil, architectural, and mining projects. For the most part, drafting boards have been replaced with computers and CAD software, considerably improving accuracy as

FIGURE 1.30. Design of many large structures, such as the Forth Road Bridge, Scotland, shown here, still requires the use of classical two-dimensional drawing and analysis techniques.
Sources: Photo by William G. Godden. Reprinted with permission from EERC Library, Univ. of California, Berkeley, above. Photo by Fredrick T. Godden. Reprinted with permission from EERC Library, Univ. of California, Berkeley, below.

well as ease of use. However, the classical methods of finding distances, areas, inclines, and intersections used for land characterization and modifications are still used. Many recent large-scale construction and landscaping projects, such as the one shown in Figure 1.30, used classical 2-D graphical analysis and presentation methods.

Using solid modeling, the calculation of important mechanical properties of parts and assemblies can be done easily. The volume that a part or assembly occupies usually can be calculated with a single command after the computer model has been built. Properties of volume, such as mass, center of mass, moments of inertia, products of inertia, and principal axes, can also be calculated. Without a solid modeler, the calculation of these properties would be laborious, especially for complex geometries.

The analysis capability of 3-D modeling also has made it popular for certain types of analyses in civil engineering applications. Two-dimensional topographic maps, such

FIGURE 1.31. Classical two-dimensional presentation of land-height contours, natural landscape (top) and development for roads and housing (bottom). *Images courtesy of Autodesk Corporation.*

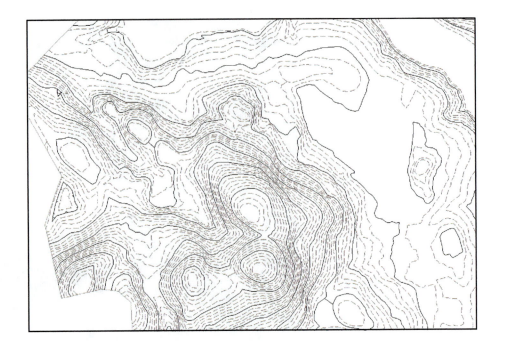

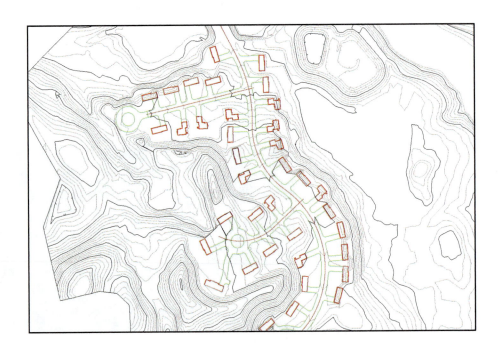

as the one shown in Figure 1.31, shows land elevations at development sites for proposed residential areas before and after the addition of roads and building pads. The elevation contours of the land change, because certain locations are excavated while other locations are filled with earth to accommodate the roads and pads.

The use of 3-D land models, shown in Figure 1.32, generated from surveying data has made it easier for both engineers and nonengineers to visualize the appearance of a landscape before and after a proposed development. Further, the analytical capability of 3-D modeling in civil engineering applications has made it possible to quickly calculate the volumes of earth that must be removed or added to accommodate the development. It is even possible to match the total addition to the total removal of earth to minimize the volume changed from the site.

FIGURE 1.32. Three-dimensional images of the contours shown in the previous figure, original land (top) and modifications to accommodate roads and buildings (bottom). *Courtesy of Autodesk ® Civil 3D © 2005 software.*

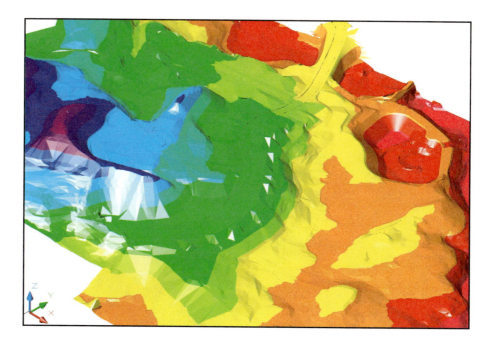

1.04.06 Graphics as a Presentation Tool

An engineer must be able to communicate not only ideas and designs but also precise engineering data. Whether this data is empirical, as collected from experiments, or analytical, as calculated from mathematical models, they must be presented so other people can understand them quickly and easily. Traditional methods of data presentation are in the form of charts and graphs. Charts include familiar items such as pie charts and bar charts commonly used for presenting data to the general public. Graphs, which are usually more technical, show data trends when the relationship between two or more variables is plotted on orthogonal axes. Examples of these types of data presentation are shown in Figure 1.33.

FIGURE 1.33. Data presentation and analysis is a vital part of engineering.
Courtesy of D. K. Lieu

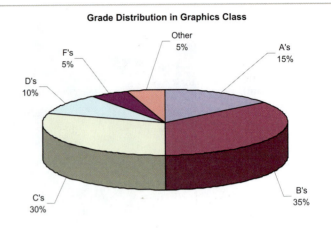

Grade Distribution in Graphics Class

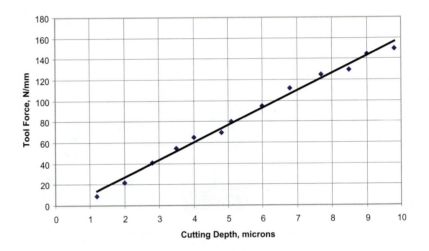

Tool Force for Cutting Ceramic

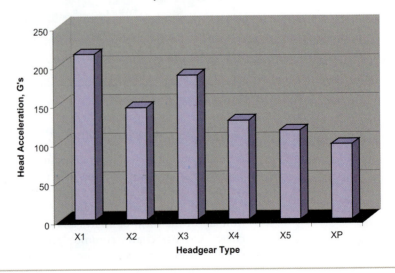

Sports Helmet Protection

Three-dimensional modeling software is also used to build geometric models that can be exported for finite element analysis (FEA). FEA is a numerical analysis method used to calculate results such as stress distribution, temperature distribution, or deformation in a part. Although FEA usually is not considered a formal part of engineering graphics, one of the most efficient and effective methods of presenting FEA results is to show the predicted contours of variables such as stress, deflection, or temperature atop a pictorial of the object. Different colors are used to represent different magnitudes of a variable. For example, Figure 1.34 shows how a solid model of the teeth, steel, and magnets of a small electric motor are created for geometry analysis. The same model is then used to generate a FEA mesh in preparation for an analysis of the magnetic-flux density distribution in the structure. The flux densities are calculated and their contours are plotted directly atop the original solid model image to show the location and magnitude of the flux densities in the motor.

A popular and effective data presentation method is to show the stress distribution in a part by plotting stress contours directly on the part image, as shown in Figure 1.35. In this way, the location and level of the highest stress in the part can be located easily. The same technique can be used for plotting the temperature distribution and magnetic flux densities in a part.

FIGURE 1.34. A three-dimensional model (top) of an electric motor is used to create a FEM mesh (center) from which magnetic flux densities can be calculated and presented (bottom).
Courtesy of D. K. Lieu

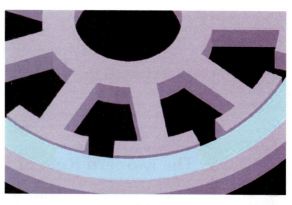

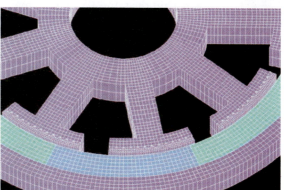

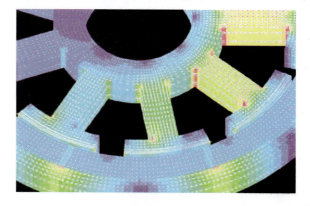

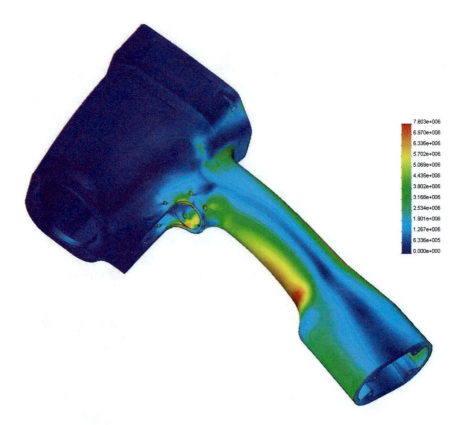

7.603e+006
6.970e+006
6.336e+006
5.702e+006
5.069e+006
4.435e+006
3.802e+006
3.168e+006
2.534e+006
1.901e+006
1.267e+006
6.336e+005
0.000e+000

1.05 The Modern Role of Engineering Graphics

Although the role of engineering graphics has evolved over the years, many aspects remain the same. Graphics remains the medium for communicating ideas and technical information. The best way to communicate an idea for a part or device is to show a picture of it. In the past, the pictures were crude handmade drawings, which required time and skill to create. Now pictures are computer-generated images of 3-D models that can be turned and rotated so they are viewable from any direction, providing more accurate depictions. Because the models are easy to create, many variations can be created and viewed in a short time. This advantage makes 3-D modeling useful not only as a means of communication but also as a means of design.

Recording the history of a design also remains an important role of engineering graphics. In the past, recording the history of designs usually meant saving master hard-copy drawings in cabinets in some sort of vault. The smallest change in a design meant changing the master copy and then sending updated copies of the master to whoever needed them. Copies commonly suffered distortion or reduced resolution due to the machines that made the copies. While hard-copy drawings are still necessary, most drawing data are now stored as electronic files. There are enormous advantages in the cataloging, retrieval, and transmission of data stored in this manner. Today model and drawing data and their updates can be sent across the world in a fraction of a second with no loss in resolution.

Engineering graphics remains an analysis tool, but the type of analysis has changed. Graphical means are no longer used to solve vector algebra, mathematics, or calculus problems. Instead, graphical models are now used to do things like examine the proper fit and function of parts within assemblies. Using 3-D models, engineers can examine parts in their final assembled state for proper motion and location. Engineers can extract the volumetric and inertial properties of the parts and assemblies, to ensure that they fit as specified. Based on externally applied forces, the stresses and deflections in the material also can be examined to ensure that failure of the device does not occur.

Formal engineering drawing remains a part of the overall design process. The traditional role of formal engineering drawings was to ensure that parts would be fabricated to specified sizes, that they would appear as specified, and that various parts would fit together properly. Prior to the 1990s, most engineering graphics classes concentrated on drawing technique and accuracy and on proper use of mechanical drawing instruments. Since engineering drawings can now be created easily and accurately with computers and software tools, the effort required by the formal drawing process is greatly reduced from what it was in the past. Since most computer graphics tools are easy to master, modern graphics classes concentrate mainly on visualization, analysis, function, and **optimization** of designs.

The development of visualization skills is a particular goal of modern engineering graphics courses. Developing visualization skill is necessary for envisioning, specifying, and creating complex designs with functional features in the three-spatial dimensions. Traditionally, these skills are developed through hand-eye coordination involving physical parts. Hands-on experience, such as repairing an automobile or a bicycle, constructing models, or playing with building toys, is helpful for developing visualization skills. In an engineering graphics curriculum, these skills can be developed by doing special visualization exercises and by building and working with solid models. Another method of developing visualization skills is to disassemble and reassemble engineered devices in a process known as mechanical dissection. During this process, students examine the operating concepts and their practical implementation, as shown in Figure 1.36.

Sketching also has proven to be a valuable technique for developing visualization skills, as shown in Figure 1.37. Sketches, which can be prepared quickly, provide a simple graphical representation of an idea with a great deal of information on concepts and appearance, without the need for formal drawing tools. For this reason, even though powerful computers and software are available, sketching remains a part of engineering graphics, both as a learning tool and as a practical skill, as you will see throughout this textbook.

FIGURE 1.36. The construction and function of a device can be learned from its disassembly, examination, and reassembly in a process known as mechanical dissection.
Courtesy of D. K. Lieu

1.06 Chapter Summary

The history of graphical communication has shown it to be vital in nearly all aspects of engineering. The development of technology, tools, and techniques used for engineering graphics has advanced, with all of the developments supporting each other. Technological tools have made the tasks associated with classical engineering graphics much easier. The technical sophistication and simple human interface of new tools have enabled engineers to concentrate on learning and developing the techniques offered by the tools, instead of merely operating the tools. Advances in computing, modeling, and display tools have increased the speed and accuracy with which communication, visualization, and analytical problems are performed. More complex designs can be produced more quickly with better functionality and fewer errors than in the past. Engineering drawing has become quicker and simpler; making it possible for engineers to concentrate on what they do best, which is to examine the functionality of a design and to optimize it for its intended environment. Engineers have new responsibilities associated with the new tools, including following protocols for the construction of proper computer models, the electronic transmission of data, and data management.

1.07 glossary of key terms

assembly: A collection of parts that mate together to perform a specified function or functions.

CAD: Computer-aided drawing. The use of computer hardware and software for the purpose of creating, modifying, and storing engineering drawings in an electronic format.

descriptive geometry: A two-dimensional graphical construction technique used for geometric analysis of three-dimensional objects.

design (noun): An original manifestation of a device or method created for performing one or more useful functions.

design (verb): The process of creating a design (noun).

drawing: A collection of images and other detailed graphical specifications intended to represent physical objects or processes for the purpose of accurately re-creating those objects or processes.

1.07 glossary of key terms (continued)

engineer (verb): To plan and build a device that does not occur naturally within the environment.

engineer (noun): A person who engages in the art of engineering.

engineering: The profession in which knowledge of mathematical and natural sciences gained by study, experience, and practice is applied with judgment to develop and utilize economically the materials and forces of nature for the benefit of humanity.

fabricate: To make something from existing materials.

image: A collection of printed, displayed, or imagined patterns intended to represent real objects, data, or processes.

instruments: In engineering drawing, mechanical devices used to aid in creating accurate and precise images.

model: A mathematical representation of an object or a device from which information about its function, appearance, or physical properties can be extracted.

optimization: Modification of shapes, sizes, and other variables to achieve the best performance based on pre-defined criteria.

part: A single object fabricated to perform one or more functions.

project: In engineering, a collection of tasks that must be performed to create, operate, or retire a system or device.

solid modeling: Three-dimensional modeling of parts and assemblies originally developed for mechanical engineering use but presently used in all engineering disciplines.

system: A collection of parts, assemblies, structures, and processes that work together to perform one or more prescribed functions.

three-dimensional (3-D) modeling: Mathematical modeling where the appearance, volumetric, and inertial properties of parts, assemblies, or structures are created with the assistance of computers and display devices.

two-dimensional (2-D) drawing: Mathematical modeling or drawing where the appearance of parts, assemblies, or structures are represented by a collection of two-dimensional geometric shapes.

visualization: The ability to create and manipulate mental images of devices or processes.

1.08 questions for review

1. Why are most cave drawings and hieroglyphics not considered to be engineering drawings?

2. In what ways did the design of military fortifications change after the discovery of gunpowder and the invention of the cannon?

3. Why did engineering drawings need to become more precise during the industrial revolution?

4. What were the three traditional roles of engineering graphics?

5. What are some of the new roles of engineering graphics created by computer graphics?

6. What are some of the advantages and disadvantages of using mechanical drawing instruments, as opposed to mathematical tools, for problem solving?

7. What are some of the advantages and disadvantages of using mechanical drawing instruments, as opposed to computational tools, for problem solving?

8. How is solid modeling different from CAD?

9. What is visualization?

10. In what ways can visualization skills be developed?

1.09 problems

Graphical communications makes the lives of engineers easier in many ways. The following exercises are intended to give you a feeling of what communication and analysis would be like without the tools and techniques used in engineering graphics. Do not become discouraged if you find these exercises to be difficult or cumbersome or if you find that the results are not accurate, which is the point of these exercises. In the chapters that follow, you will be introduced to methods of addressing the difficulties you encounter here.

1a. Do this exercise with one of your classmates. Select one or more of the objects shown in Figure P1.1, but

1.09 problems (continued)

do not show the object(s) to your partner. Using only words, give your partner a complete description of the objects you selected. Then have your partner sketch a picture of the objects based on your verbal descriptions. Reverse roles using different objects. What errors occurred between the objects that were being described and the objects that were envisioned? What can be done to reduce these errors?

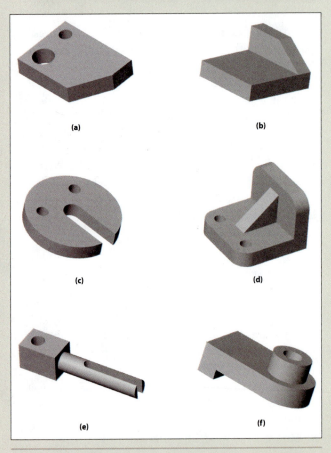

FIGURE P1.1. (a)–(f) Verbally describe these objects to your partner; then have your partner sketch a picture of the object. *Courtesy of D. K. Lieu*

1b. Give a third classmate the sketches made in Part A of this exercise. Without revealing what the original objects in the figure look like, give a complete description of the errors in the sketches and have this person make corrections to the sketches. Reverse roles using different objects. How much closer are the sketches to representing the objects shown in the figure? What additional problems occur when a third person is involved?

2. Do this exercise with a group of classmates. Select one or more of the objects shown in Figure P1.2

but do not show the figure(s) to the rest of the group. Make sketches of the object(s) you have selected, give them to the first person in the group, have that person examine them carefully, and then retrieve your sketches. Have that person use the memory of your sketches to make new sketches. Then give the new person's sketches to the second person in the group. Do not show the previous sketches to the new person. Repeat for all of the classmates in the group. When the last person is done, compare the final set of sketches to the objects selected by the first person in the original figure. What errors occurred between the final sketches and the objects that were selected? What happens to the sketches with each revision? What can be done to reduce these errors?

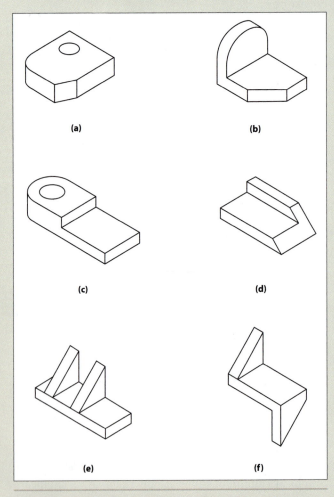

FIGURE P1.2. (a)–(f) Show one or more of these objects to your partner and have your partner sketch the object(s) from memory. Repeat the process with the newly created sketch. Compare the sketch to the original object. *Courtesy of D. K. Lieu*

1.09 problems (continued)

3. For the geometric elements shown in each of the three panels of Figure P1.3, develop formulas for finding the length, angle, area, or volume, whichever is required in each panel, using analytical methods. Generalize the solution in terms of x-, y-, and z-coordinates of the points given. What problems do you envision if the person making the calculations has no access to computers, calculators, or any other computational aids? What happens to the solution formulas as the geometries become more complicated or are rotated and translated in space?

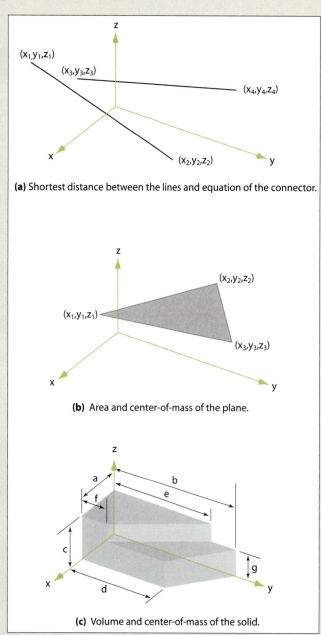

(a) Shortest distance between the lines and equation of the connector.

(b) Area and center-of-mass of the plane.

(c) Volume and center-of-mass of the solid.

FIGURE P1.3. (a)–(c) Find the specified geometric properties of the objects.
Courtesy of D. K. Lieu

2

Sketching

objectives

After completing this chapter, you should be able to

- Explain the importance of sketching in the engineering design process
- Make simple sketches of basic shapes such as lines, circles, and ellipses
- Use 3-D coordinate systems, particularly right-handed systems
- Draw simple isometric sketches from coded plans
- Make simple oblique pictorial sketches
- Use advanced sketching skills for complex objects

2.01

introduction

Sketching is one of the primary modes of communication in the initial stages of the design process. Sketching also is a means to creative thinking. It has been shown that your mind works more creatively when your hand is sketching as you are engaged in thinking about a problem.

This chapter focuses on one of the fundamental skills required of engineers and technologists—freehand sketching. The importance of sketching in the initial phases of the design process is presented, as are some techniques to help you create sketches that correctly convey your design ideas. The definition of 3-D coordinate systems and the way they are portrayed on a 2-D sheet of paper will be covered, along with the difference between right-handed and left-handed coordinate systems. The chapter will investigate how to create simple pictorial sketches. Finally, the advanced sketching techniques of shading and cartooning will be presented with a framework for creating sketches of complex objects. You will begin to explore these topics in this chapter and will further refine your sketching abilities as you progress through your graphics course.

2.02   Sketching in the Engineering Design Process

As you may remember from Chapter 1, engineers communicate with one another primarily through graphical means. Those graphical communications take several forms, ranging from precise, complex drawings to simple sketches on the back of an envelope. Most of this text is focused on complex drawings; however, this chapter focuses on simple sketches.

Technically speaking, a sketch is any drawing made without the use of drawing instruments such as triangles and T squares. Some computer graphics packages allow you to create sketches; however, you will probably be more creative (and thus more effective) if you stick to hand sketching, particularly in the initial stages of the design process. In fact, carefully constructed, exact drawings often serve as a hindrance to creativity when they are employed in the initial stages of the design process. Typically, all you need for sketching are a pencil, paper, an eraser, and your imagination.

Your initial sketches may be based on rough ideas. But as you refine your ideas, you will want to refine your sketches, including details that you left out of the originals. For example, suppose you were remodeling the bathroom in your house. Figure 2.01 shows two sketches that define the layout of the bathroom, with details added as ideas evolve. Once you have completed the layout to your satisfaction, you can create an official engineering drawing showing exact dimensions and features that you can give to the contractor who will perform the remodeling work for you.

When engineers sit down to brainstorm solutions to problems, before long, one of them usually takes out a sheet of paper and sketches an idea on it. The others in the

FIGURE 2.01. Sketches for a bathroom remodel.

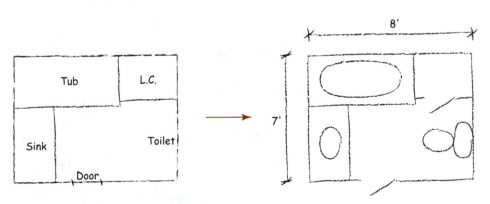

discussion may add to the original sketch, or they may create sketches of their own. The paper-and-pencil sketches then become media for the effective exchange of ideas. Although few "rules" regulate the creation of sketches, you should follow some general guidelines to ensure clarity.

2.03   Sketching Lines

Most of your sketches will involve basic shapes made from lines and circles. Although you are not expected to make perfect sketches, a few simple techniques will enable you to create understandable sketches.

    When drawing **lines**, the key is to make them as straight as possible. If you are right-handed, you should sketch your vertical lines from top to bottom and your horizontal lines from left to right. If you are sketching an angled line, choose a direction that matches the general inclination of the line—for angled lines that are mostly vertical, sketch them from top to bottom; for angled lines that are mostly horizontal, sketch them from left to right. If you are left-handed, you should sketch your vertical lines from top to bottom, but your horizontal lines from right to left. For angled lines, left-handed people should sketch from either right to left or top to bottom, again depending on the inclination of the line. To keep your lines straight, focus on the endpoint as you sketch. The best practices for sketching straight lines are illustrated in Figure 2.02.

FIGURE 2.02. Techniques for sketching straight lines.

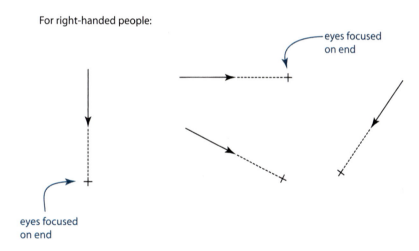

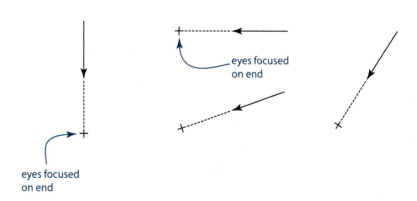

FIGURE 2.03. Rotating the paper to draw an angled line.

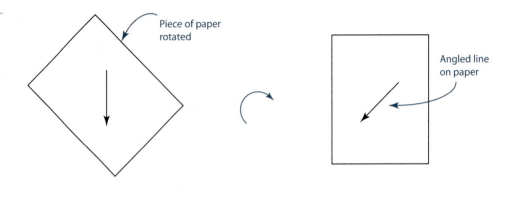

Piece of paper rotated

Angled line on paper

FIGURE 2.04. Sketching long lines in segments.

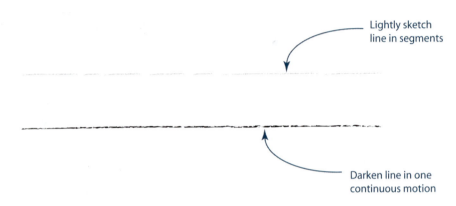

Lightly sketch line in segments

Darken line in one continuous motion

You also can try rotating the paper on the desk to suit your preferences. For example, if you find that drawing vertical lines is easiest for you and you are confronted with an angled line to sketch, rotate the paper on the desk so you can sketch a "vertical" line. Or you can rotate the paper 90 degrees to sketch a horizontal line. Figure 2.03 illustrates rotation of the paper to create an angled line.

One last point to consider when sketching lines is that you initially may have to create "long" lines as a series of connected segments. Then you can sketch over the segments in a continuous motion to make sure the line appears to be one entity and not several joined end to end. Using segments to define long lines is illustrated in Figure 2.04.

2.04 Sketching Curved Entities

Arcs and **circles** are other types of geometric entities you often will be required to sketch. When sketching arcs and circles, use lightly sketched square **bounding boxes** to define the limits of the curved entities and then construct the curved entities as tangent to the edges of the bounding box. For example, to sketch a circle, you first lightly sketch a square (with straight lines). Note that the length of the sides of the bounding box is equal to the diameter of the circle you are attempting to sketch. At the centers of each edge of the box, you can make a short **tick mark** to establish the point of tangency for the circle, then draw the four arcs that make up the circle. Initially, you may find it easier to sketch one arc at a time to complete the circle; but as you gain experience, you may be able to sketch the entire circle all at once. Figure 2.05 shows the procedure used to sketch a circle by creating a bounding box first.

One problem you may have when using a bounding box to sketch a circle occurs when the radius of the circle is relatively large. In that case, the arcs you create may be too flat or too curved, as shown in Figure 2.06. To avoid this type of error, you might try marking the radius at points halfway between the tick marks included on the bounding box. Using simple geometry, when you draw a line between the center of the circle and the corner of the bounding box, the radius is about two-thirds of the distance (technically, the radius is 0.707, but that number is close enough to two-thirds for your purposes). Then you can include some additional tick marks around the circle to guide your sketching and to improve the appearance of your circles. This technique is illustrated in Figure 2.07.

Sketching an arc follows the same general procedure as sketching a circle, except that your curved entity is only a portion of a circle. Sketching an **ellipse** follows the same general rules as sketching a circle, except that your bounding box is a rectangle and not a square. Sketching arcs and ellipses is illustrated in Figure 2.08.

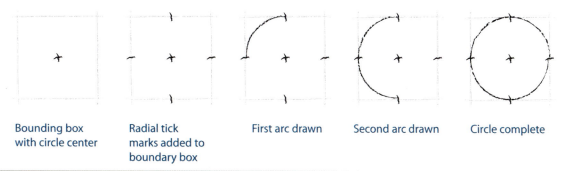

| Bounding box with circle center | Radial tick marks added to boundary box | First arc drawn | Second arc drawn | Circle complete |

FIGURE 2.05. Sketching a circle using a bounding box.

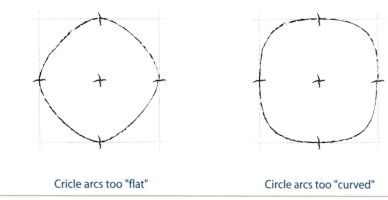

Cricle arcs too "flat"   Circle arcs too "curved"

FIGURE 2.06. Circles sketched either too flat or too curved.

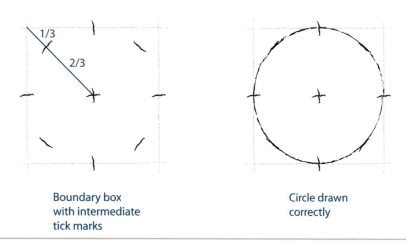

Boundary box with intermediate tick marks   Circle drawn correctly

FIGURE 2.07. Using intermediate radial tick marks for large circles.

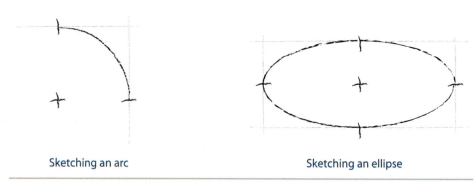

Sketching an arc   Sketching an ellipse

FIGURE 2.08. Using boundary boxes to sketch arcs and ellipses.

2.05 Construction Lines

Similar to the way you used bounding boxes to create circles and ellipses, other construction lines help with your sketching. Using construction lines, you outline the shape of the object you are trying to sketch. Then you fill in the details of the sketch using the construction lines as a guide. Figure 2.09 shows the front view of an object you need to sketch. To create the sketch, you lightly draw the construction lines that outline the main body of the object and then create the construction lines that define the prominent features of it. One rule of thumb is that construction lines should be drawn so lightly on the page that when it is held at arm's length, the lines are nearly impossible to see. The creation of the relevant construction lines is illustrated in Figure 2.10.

Using construction lines as a guide, you can fill in the details of the front view of the object until it is complete. The final result is shown in Figure 2.11.

Another way you can use construction lines is to locate the center of a square or rectangle. Recall from your geometry class that the diagonals of a box (either a rectangle or a square) intersect at its center. After you create construction lines for the edges of the box, you sketch the two diagonals that intersect at the center. Once you find the center of the box, you can use it to create a new centered box of smaller dimensions—a kind of concentric box. Locating the center of a box and creating construction lines for a newly centered box within the original box are illustrated in Figure 2.12.

Once you have created your centered box within a box, you can sketch a circle using the smaller box as a bounding box, resulting in a circle that is centered within the larger box as shown in Figure 2.13. Or you can use these techniques to create a square with four holes located in the corners of the box as illustrated in Figure 2.14.

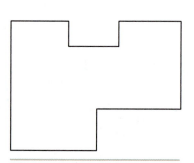

FIGURE 2.09. The front view of an object to sketch.

FIGURE 2.10. Construction lines used to create a sketch.

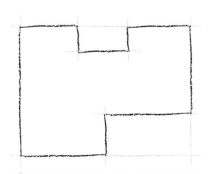

FIGURE 2.11. Completed sketch using construction lines as a guide.

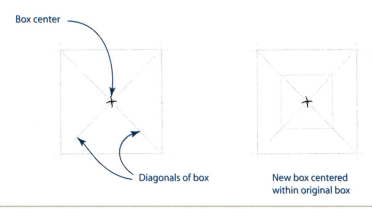

FIGURE 2.12. Creating concentric bounding boxes.

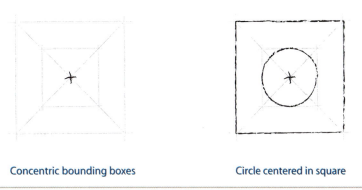

FIGURE 2.13. Sketching a circle in a box.

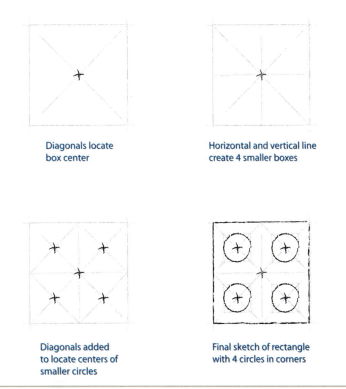

FIGURE 2.14. Using diagonal construction lines to locate centers.

2.06 **Coordinate Systems**

When sketching, you often have to portray 3-D objects on a flat 2-D sheet of paper. As is usually the case with graphical communication, a few conventions have evolved over time for representing 3-D space on a 2-D sheet of paper. One convention, called the **3-D coordinate system**, is that space can be represented by three mutually perpendicular coordinate axes, typically the x-, y-, and z-axes. To visualize those three axes, look at the bottom corner of the room. Notice the lines that are formed by the intersection of each of the two walls with the floor and the line that is formed where the two walls intersect. You can think of these lines of intersection as the x-, y-, and z-coordinate axes. You can define all locations in the room with respect to this corner, just as all points in 3-D space can be defined from an origin where the three axes intersect.

You are probably familiar with the concept of the three coordinate axes from your math classes. In Figure 2.15, a set of coordinate axes, notice the positive and negative directions for each of the axes. Typically, arrows at the ends of the axes denote the positive direction along the axes.

For engineering, the axes usually define a right-handed coordinate system. Since most engineering analysis techniques are defined by a right-handed system, you should learn what this means and how to recognize such a system when you see it. A **right-handed system** means that if you point the fingers of your right hand down the positive x-axis and curl them in the direction of the positive y-axis, your thumb will point in the direction of the positive z-axis, as illustrated in Figure 2.16. This procedure is sometimes referred to as the **right-hand rule**.

Another way to think about the right-hand rule is to point your thumb down the positive x-axis and your index finger down the positive y-axis; your middle finger will then automatically point down the positive z-axis. This technique is illustrated in Figure 2.17. Either method for illustrating the right-hand rule results in the same set of coordinate axes; choose the method that is easiest for you to use.

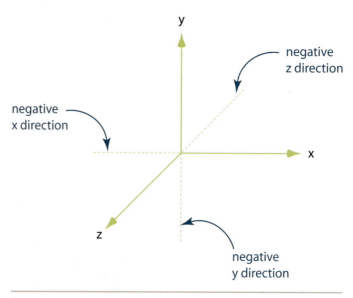

FIGURE 2.15. The x-, y-, and z- coordinate axes.

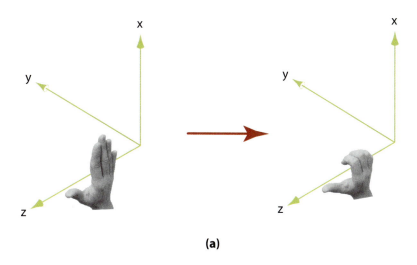

(a)

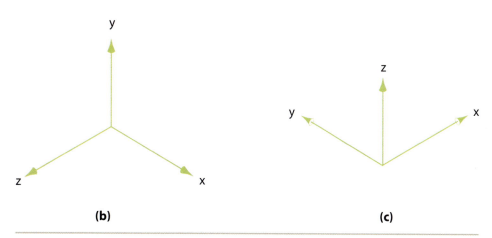

(b)   **(c)**

FIGURE 2.16. Curling the fingers to check for a right-handed coordinate system in (a) and alternative presentations of right-handed coordinate systems in (b) and (c).

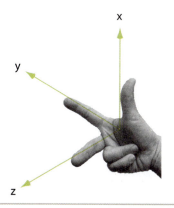

FIGURE 2.17. An alternative method to check for a right-handed coordinate system.

FIGURE 2.18. The result of using the left hand to test for a right-handed coordinate system.

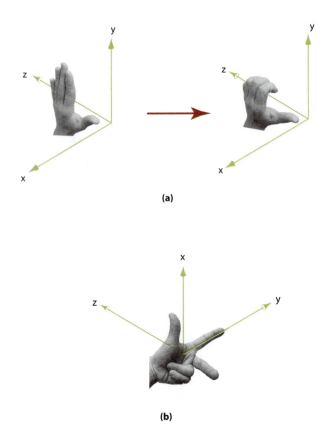

(a)

(b)

FIGURE 2.18. The result of using the left hand to test for a right-handed coordinate system.

Notice that if you try either technique with your left hand, your thumb (or middle finger) will point down the negative z-axis, as illustrated in Figure 2.18.

A **left-handed system** is defined similarly to a right-handed system, except that you use your left hand to show the positive directions of the coordinate axes. Left-handed systems are typically used in engineering applications that are geologically based—positive z is defined as going down into the earth. Figure 2.19 illustrates left-handed coordinate systems. (Use the left-hand rule to verify that these are left-handed coordinate systems.)

The question remains about how to represent 3-D space on a 2-D sheet of paper when sketching. The answer is that the three coordinate axes are typically represented as oblique or isometric, depending on the preferences of the person making the sketch. You are probably most familiar with oblique representation of the coordinate axes, which seems to be the preferred method of many individuals. With this method, two axes are sketched perpendicular to each other and the third is drawn at an angle, usually 45 degrees to both axes. The angle of the inclined line does not have to be 45 degrees, but

FIGURE 2.19. Left-handed coordinate systems.

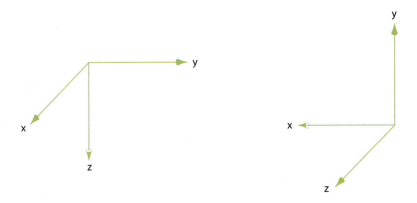

FIGURE 2.20. An oblique representation of right-handed coordinate systems.

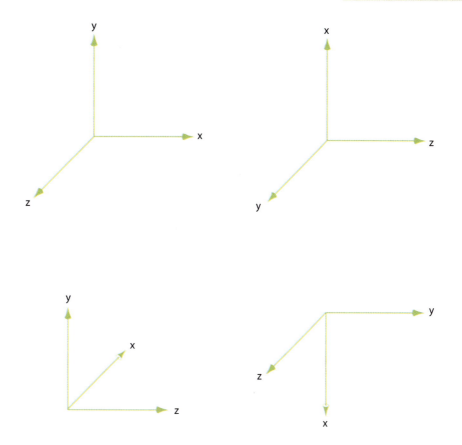

it is usually sketched that way. Your math teachers probably sketched the three coordinate axes that way in their classes. Figure 2.20 shows multiple sets of coordinate axes drawn as oblique axes. Notice that all of the coordinate systems are right-handed systems. (Verify this for yourself by using the right-hand rule.)

Another way of portraying the 3-D coordinate axes on a 2-D sheet of paper is through isometric representation. With this method, the axes are projected onto the paper as if you were looking down the diagonal of a cube. When you do this, the axes appear to be 120 degrees apart, as shown in Figure 2.21. In fact, the term *isometric* comes from the Greek *iso* (meaning "the same") and *metric* (meaning "measure"). Notice that for **isometric axes** representations, the right-hand rule still applies.

Isometric axes also can be sketched with one of the axes extending in the "opposite" direction. This results in angles other than 120 degrees, depending on the orientation of the axes with respect to the paper, as shown in Figure 2.22.

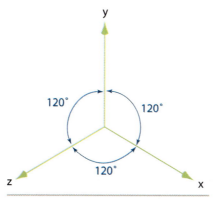

FIGURE 2.21. An isometric representation of a right-handed coordinate system.

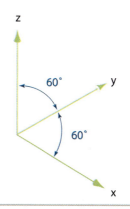

FIGURE 2.22. An isometric representation of axes with angles less than 120 degrees.

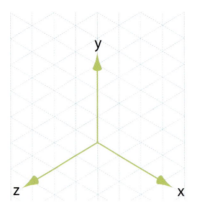 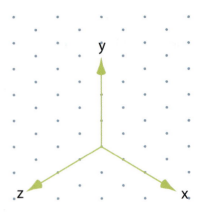

Grid or dot paper can help you make isometric sketches. With **isometric dot paper**, the dots are oriented such that when you sketch lines through the dots, you end up with standard 120 degree axes. With grid paper, the lines are already drawn at an angle of 120 degrees with respect to one another. Isometric grid paper and isometric dot paper are illustrated in Figure 2.23.

2.07 Isometric Sketches of Simple Objects

Creating isometric drawings and sketches of complex objects will be covered in more detail in a later chapter; however, this section serves as an introduction to the topic for simple objects. Mastering the techniques used to create isometric sketches of simple objects may help as you branch out to tackle increasingly complex objects. Figure 2.24 shows how **isometric grid paper** is used to sketch a $3 \times 3 \times 3$ block. Notice that there is more than one orientation from which the block can be sketched on the same sheet of grid paper. Ultimately, the orientation you choose depends on your needs or preferences.

Coded plans can be used to define simple objects that are constructed entirely out of blocks. The numerical values in the coded plan represent the height of the stack of blocks at that location. The object then "grows" up from the plan according to the numbers specified. Figure 2.25 shows a coded plan on isometric grid paper and the object that results from it.

The object shown in Figure 2.25 clearly outlines all of the blocks used to create it. When isometric sketches of an object are made, however, standard practice dictates that lines appear only where two surfaces intersect—lines between blocks on the same surface are not shown. Figure 2.26 shows the object from Figure 2.25 after the unwanted lines have been removed. Notice that the only lines on the sketch are those formed from the intersection of two surfaces. Also notice that object edges hidden from view on the back side are not shown in the sketch. Not showing hidden edges on an **isometric pictorial** also is standard practice in technical sketching.

FIGURE 2.24. Using isometric
grid paper to sketch a block.

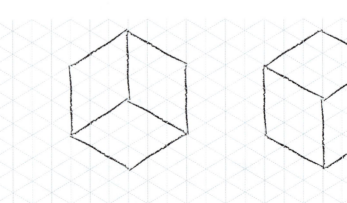

Sometimes when you are creating an isometric sketch of a simple object, part of one surface is obscured by one of the more prominent features of the object. When creating the sketch, make sure you show only the visible part of the surface in question, as illustrated in Figure 2.27.

Figure 2.28 shows several coded plans and the corresponding isometric sketches. Look at each isometric sketch carefully to verify that it matches the defining coded plan: those lines are shown only at the edges between surfaces (not to define each block), that no hidden edges are shown, and that only the visible portions of partially obscured surfaces are shown.

FIGURE 2.25. A coded plan and the resulting object.

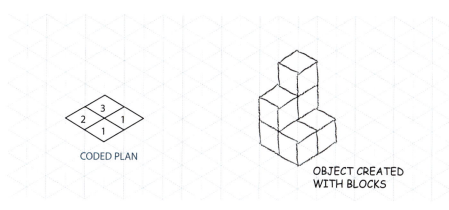

CODED PLAN

OBJECT CREATED WITH BLOCKS

FIGURE 2.26. A properly drawn isometric sketch of the object from the coded plan.

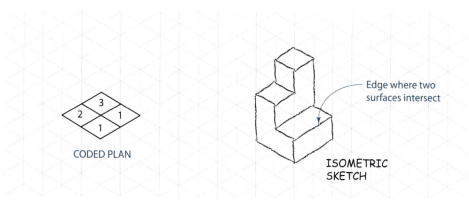

CODED PLAN

Edge where two surfaces intersect

ISOMETRIC SKETCH

FIGURE 2.27. The partially obscured surface on an isometric sketch.

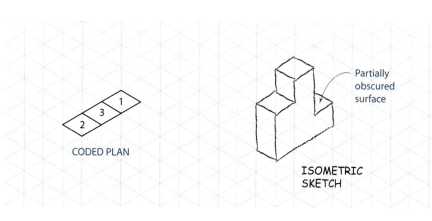

CODED PLAN

Partially obscured surface

ISOMETRIC SKETCH

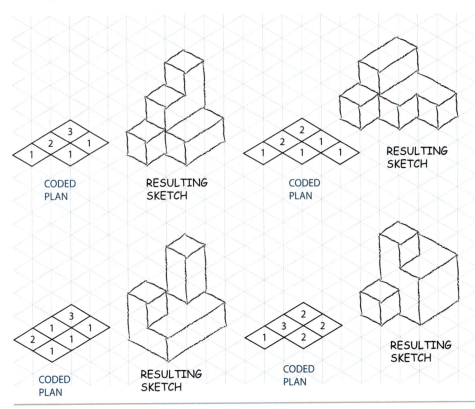

FIGURE 2.28. Four coded plans and the resulting isometric sketches.

2.07.01 Circles in Isometric Sketches

Look back at the 3 × 3 × 3 block shown in Figure 2.24. In reality, you know that all of the surfaces of the block are 3 × 3 squares; yet in the isometric sketch, each surface is shown as a parallelogram. The distortion of planar surfaces is one disadvantage of creating isometric sketches. The isometric portrayal of circles and arcs is particularly difficult. Circles appear as ellipses in isometric sketches; however, you will not be able to create a rectangular bounding box to sketch the ellipse in isometric as described earlier in this chapter. To create an ellipse that represents a circle in an isometric sketch, you first create a square bounding box as before; however, the bounding box will appear as a parallelogram in the isometric sketch. To create your bounding box, locate the center of the circle first. From the center, locate the four radial points. The direction you move on the grid corresponds to the lines that define the surface. If you are sketching the circle on a rectangular surface, look at the sides of the rectangle as they appear in isometric and move that same direction on the grid. Figure 2.29 shows a 4 × 4 × 4 cube with a circle center and four radial points located on one of the sides.

Once you have located the center of the circle and the four radial points, the next step is to create the bounding box through the radial points. The edges of the bounding box should correspond to the lines that define this particular surface. The edges will be parallel to the edges of the parallelogram that define the surface if that surface is square or rectangular. Figure 2.30 shows the cube with the circle center and the bounding box located on its side.

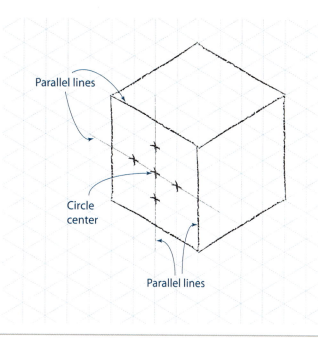

FIGURE 2.29. A cube with a circle center and radial points located.

Four arcs that go through the radial points define the ellipse, just like an ellipse drawn in a regular rectangular bounding box. The difference is that for the isometric ellipse, the arcs are of varying curvatures—two long arcs and two short arcs in this case. The arcs are tangent to the bounding box at the radial points, as before. It is usually

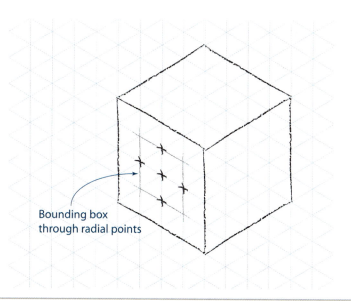

FIGURE 2.30. A cube with the circle center and bounding box on the side.

best if you start by sketching the long arcs, and then add the short arcs to complete the ellipse. Sketching the arcs that form the ellipse is illustrated in Figure 2.31.

Creating ellipses that represent circles on the other faces of the cube is accomplished in a similar manner, as illustrated in Figure 2.32 and Figure 2.33.

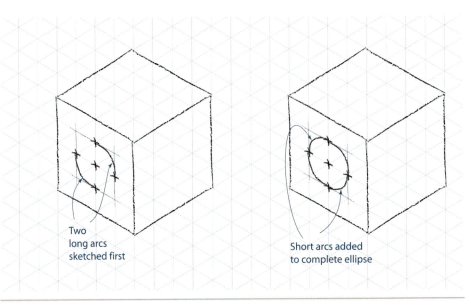

Two
long arcs
sketched first

Short arcs added
to complete ellipse

FIGURE 2.31. Sketching arcs to form an ellipse.

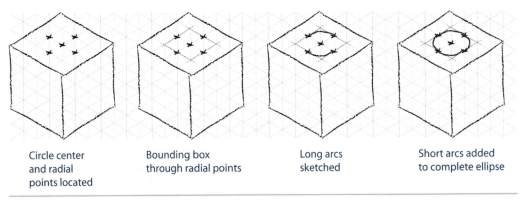

Circle center
and radial
points located

Bounding box
through radial points

Long arcs
sketched

Short arcs added
to complete ellipse

FIGURE 2.32. Sketching an ellipse on the top surface of a cube.

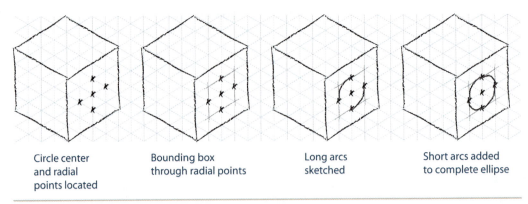

Circle center
and radial
points located

Bounding box
through radial points

Long arcs
sketched

Short arcs added
to complete ellipse

FIGURE 2.33. Sketching an ellipse on the side face of a cube.

2.07.02 Circular Holes in Isometric Sketches

One of the most common occurrences that produces a circular feature in an isometric sketch is a hole in the object. You will learn more about circular holes and object features in a later chapter, but a short introduction follows here. A circular hole usually extends all the way through an object. In an isometric pictorial, a portion of the "back" edge of a circular hole is often visible through the hole and should be included in your sketch. As a rule of thumb, the back edge of a hole is partially visible when the object is relatively thin or the hole is relatively large; when the object is thick or the diameter of the hole is small, the back edge of the hole is not visible. Figure 2.34 shows two blocks with circular holes going through them. Notice in the "thin" block that you can see a portion of the back edge of the hole; in the thicker block, though, the back edge is not visible.

To determine whether a part of the back edge of a hole is visible in an isometric sketch, you first need to locate the center of the back hole. To locate the back center, start from the center of the hole on the front surface and move in a direction perpendicular to the front surface toward the back of the object a distance equal to the object's dimension in that direction. Figure 2.35 shows the location of the center of the two back circles for the objects in Figure 2.34.

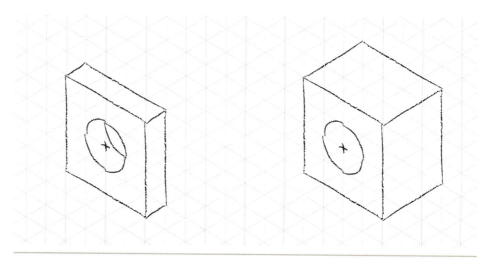

FIGURE 2.34. Blocks with circular holes in them.

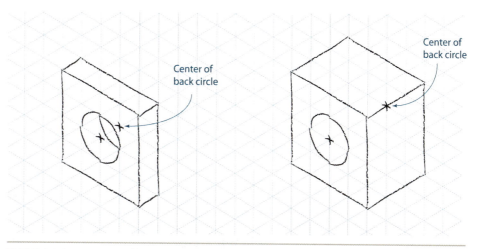

FIGURE 2.35. Centers of back circles located.

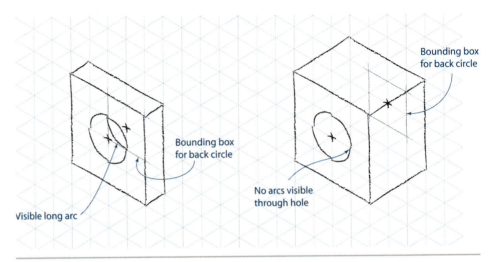

FIGURE 2.36. Determining visibility of back circles.

Starting from the back center point, lightly sketch the radial points and the bounding box for the back circle similar to the way you did for the front circle. Then add the long arc that is visible through the hole. (Note that only *one* of the long arcs is typically visible through the hole.) Add segments of the short arcs as needed to complete the visible portion of the back edge of the hole. Conversely, if after you sketch the back bounding box you notice that no portion of the ellipse will be visible on the sketch, do not include any arcs within the hole on the sketch and erase any lines associated with the bounding box. Figure 2.36 illustrates the inclusion and noninclusion of segments of the back edges of holes for the objects in Figure 2.34 and Figure 2.35.

2.08 Oblique Pictorials

Oblique pictorials are another type of sketch you can create to show a 3-D object. Oblique pictorials are usually preferred for freehand sketching because a specialized grid is not required. With oblique pictorials, as with **oblique axes**, the three dimensions of the object are shown with the height and width of the object in the plane of the paper and the third dimension (the depth) receding off at an angle from the others. Although the angle is usually 45 degrees, it can be any value.

The advantage that oblique pictorials have over isometric pictorials is that when one face of the object is placed in the plane of the paper, the object will appear in its true shape and size in that plane—it will be undistorted. This means that squares remain squares, rectangles remain rectangles, and circles remain circles. Figure 2.37 shows two pictorial representations of simple objects—one in isometric and one in

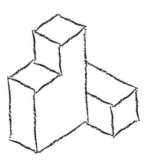

ISOMETRIC PICTORIAL

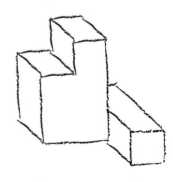

OBLIQUE PICTORIAL

FIGURE 2.37. A comparison of isometric and oblique pictorials.

oblique. Notice that the rules established for isometric pictorial sketches also hold true for oblique pictorial sketches—you do not show the hidden back edges, you show lines only where two surfaces intersect to form an edge, and you show only the visible parts of partially obstructed surfaces.

When making oblique sketches, the length of the **receding dimension** is not too important. In fact, oblique pictorials typically look better when the true length of the receding dimension is not shown. When the true length of an object's receding dimension is sketched, the object often appears distorted and unrealistic. Figure 2.38a shows the true length of a cube's receding dimension (use a ruler to make sure), and Figure 2.38b shows the same cube with the receding dimension drawn at about one-half to three-fourths its true length. Notice that the sketch in Figure 2.38a appears distorted—it does not look very much like a cube—whereas the sketch in Figure 2.38b looks like a cube.

Other conventions pertain to the way the receding dimension is portrayed in an oblique sketch; you will learn about them in a later chapter. For now, you will concentrate on trying to make a sketch that looks proportionally correct.

When creating oblique pictorials, you can choose to have the receding dimension going back and to the left or back and to the right. The direction you choose should be the one that produces the fewest obstructed surfaces in the resulting sketch. Figure 2.39 shows two possible sketches of the same object—one with the receding dimension to the left and one with the receding dimension to the right. Notice that the first sketch (Figure 2.39a) is preferable since none of the surfaces are obscured as they are with the second sketch (Figure 2.39b).

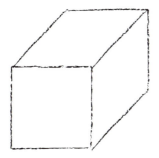

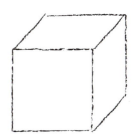

(a) Receding dimension drawn true length.

(b) Receding dimension drawn less than true length.

FIGURE 2.38. Oblique pictorials of a cube.

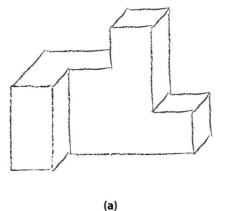

(a)   **(b)**

FIGURE 2.39. Two possible orientations for an oblique pictorial.

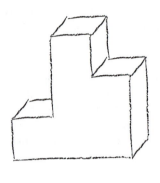

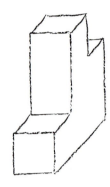

(a) Irregular surface in plane of paper.

(b) Irregular surface in receding direction.

FIGURE 2.40. Two possible orientations for an oblique pictorial.

When creating an oblique pictorial, you should put the most irregular surface in the plane of the paper. This is particularly true about any surface that has a circular feature on it. Figure 2.40 shows two different oblique pictorials of the same object. In the first sketch (Figure 2.40a), the most irregular surface is placed in the plane of the paper as it should be; in the second sketch (Figure 2.40b), the irregular surface is shown in the receding dimension. Notice that the first sketch shows the features of the object more clearly than the second sketch does.

2.08.01 Circular Holes in Oblique Pictorial Sketches

When circular holes appear in an oblique pictorial sketch, as with isometric sketches, you show the partial edges of the back circle where they are visible through the hole. Once again, partial circles are visible when the object is relatively thin or when the hole has a relatively large diameter; otherwise, partial edges are not shown. Figure 2.41 shows two oblique sketches—one in which a portion of the back edge of the hole is visible and the other in which it is not.

The procedure you use to determine whether a portion of the back circle edge is visible and, if so, which portion is visible follows the procedure outlined for isometric sketches. You start by locating the center of the back edge of the hole and marking off the four radial points. You then lightly sketch the bounding box that defines the circle. Finally, as needed, you sketch the visible portions of arcs within the circular hole. Figure 2.42 shows the procedure used to sketch the visible back edges of a circular hole in an oblique pictorial.

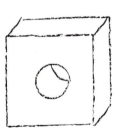

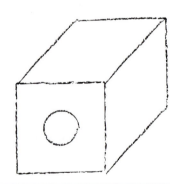

FIGURE 2.41. Oblique pictorials with circular holes in objects.

FIGURE 2.42. Determining visible back arcs in a hole.

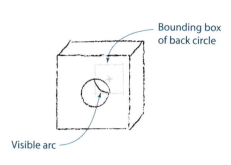

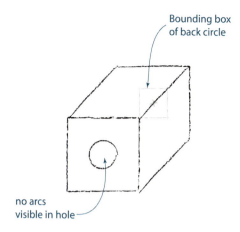

FIGURE 2.42. Determining visible back arcs in a hole.

Bounding box of back circle

Visible arc

Bounding box of back circle

no arcs visible in hole

2.09 Shading and Other Special Effects

One thing you can do to improve the quality of your pictorial sketches is to include **shading** on selected surfaces to make them stand out from other surfaces or to provide clarity for the viewer. Figure 2.43 shows an isometric sketch with all of the top surfaces shaded. Notice that the shading better defines the object for the viewer. When including shading on a pictorial sketch, try not to overdo the shading. Too much shading can be confusing or irritating to the viewer—two things you should avoid in effective graphical communication.

Another common use of shading is to show curvature of a surface. For example, the visible portion of a hole's curved surface might be shaded in a pictorial sketch. A curved surface on an exterior corner also might be shaded to highlight its curvature. Figure 2.44 shows a pictorial sketch of a simple object with curved surfaces that are shaded.

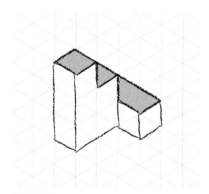

FIGURE 2.43. An object with the top surface shaded.

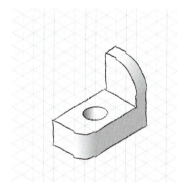

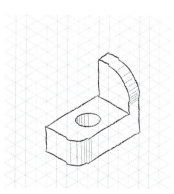

FIGURE 2.44. A simple object with two possible types of progressive shading used to emphasize the curvature of surfaces.

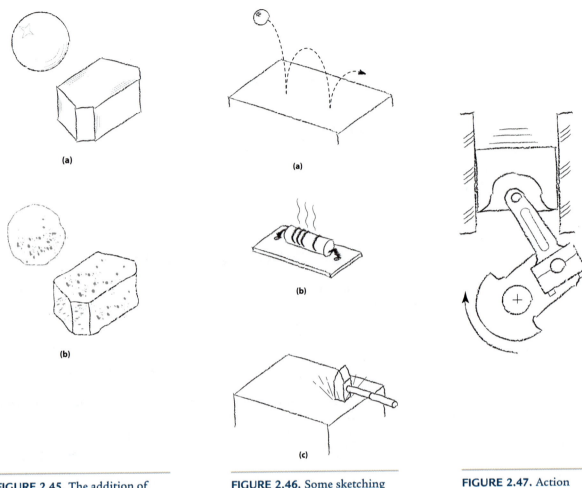

FIGURE 2.45. The addition of surface treatments to convey smooth surfaces (a) and rough surfaces (b).

FIGURE 2.46. Some sketching techniques that can be used to convey motion (a), temperature (b), and sound (c).

FIGURE 2.47. Action lines used to convey the motion of linkages.

Other sketching techniques can be used to convey features such as smooth or rough surfaces. Figure 2.45 shows different types of surface treatments that are possible for sketched objects.

You are probably familiar with techniques used in cartoons to convey ideas such as motion, temperature, and sound. Figure 2.46 shows typical cartooning lines that convey concepts not easily incorporated in a static sketch. Many of these same markings can be used in technical sketches. For example, Figure 2.47 uses action lines to convey motion for the sketch of linkages.

2.10 Sketching Complex Objects

As you refine your sketching skills, you will be able to tackle increasingly complex objects. Figures 2.48, 2.49, and 2.50 show pictorial sketches of small electronic devices. These sketches were not made to any particular scale, but were constructed so the object features appear proportionally correct with respect to one another. Notice the use of shading to enhance object appearance and to make the objects look

FIGURE 2.48. A sketch of a cell telephone.

FIGURE 2.49. A sketch of a set of headphones.

FIGURE 2.50. A sketch of a camera.

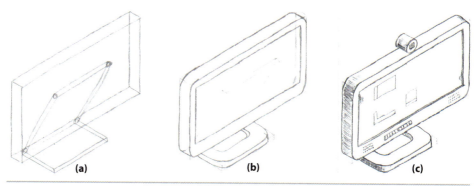

FIGURE 2.51. A sketch of a computer monitor using the method of "foundation (a), frame (b), finish (c)."

more realistic. Being able to sketch relatively complicated objects such as these will improve your ability to communicate with colleagues throughout your career. To develop this important skill, you should practice often. Do not be afraid to make mistakes—just keep trying until you get the results you want.

One way to tackle sketching a complex object is to think about it in the same way that a house is constructed—namely, "foundation, frame, finish." Using this method, you start with the "foundation" of the sketch, which usually consists of multiple guidelines and construction lines. When creating the sketch foundation, think about outlining the volume taken up by the entire object. You next "frame" the object by darkening some of the construction lines to define the basic shape of the object and its features. Once the basic frame is complete, you "finish" the sketch by adding necessary details and special features such as shading, especially on curved surfaces. Figure 2.51 shows a sketch of a flat panel computer monitor by the "foundation, frame, finish" method. Several of the exercises at the end of this chapter ask you to use this technique to develop your skills in sketching complex objects.

2.11   Strategies for Simple Pictorial Sketches

In this chapter, you learned two different ways to construct pictorial views of objects. This section outlines strategies for each type.

2.11.01 Simple Isometric Sketches

When creating an isometric pictorial from a coded plan, remember that the object "grows" up from the base according to the specified heights. You should start your sketch by drawing the visible V at the base of the object, as shown in Figure 2.52. You can determine the length of each side of the V from the coded plan. For the object defined by the coded plan in Figure 2.52, the left leg of the V is 2 units long and the right leg is 3 units long. The remaining bottom edges of the coded plans are hidden from view in the sketch and, therefore, are not included in the first drawing stages.

After you have created the base V, sketch the corner of the object at the correct height of the apex. Note that this corner will be the edge that is closest to you, the viewer. For the object shown in Figure 2.52, the height of this corner is 2 units as defined by the coded plan. The start of the isometric sketch including this corner is shown in Figure 2.53.

Starting at the "top" of this corner, move back and to the left the number of squares that are at this same height. If a change in object height is specified in the coded plan, move up or down (as shown in the coded plan) where the change occurs. When you reach the back corner, draw a vertical line back to the tip of the V you first sketched. This procedure is illustrated in Figure 2.54.

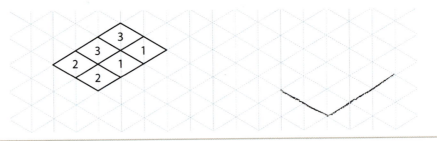

FIGURE 2.52. A coded plan with the V for the isometric sketch drawn.

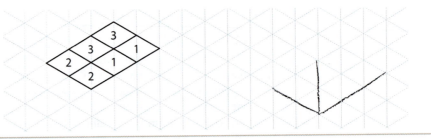

FIGURE 2.53. An isometric sketch with the nearest corner included.

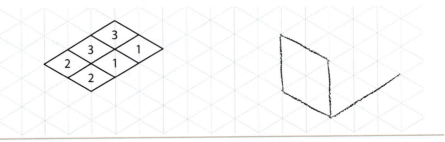

FIGURE 2.54. An isometric sketch with the first side of the surface drawn.

Follow the same procedure for the surface going off to the right from the apex of the *V*, as shown in Figure 2.55.

Complete the sketch by drawing the missing top and side surfaces of the object as shown in Figure 2.56. When adding these final features, make sure you do not include lines on surfaces—only *between* surfaces. Also, include only the visible portions of surfaces that are partially obscured.

Some of the objects you sketch may not form a simple *V* at a point nearest the viewer; instead, they will have a jagged edge along the bottom. You can use a similar procedure to sketch these objects, again starting at the bottom and outlining the shape of the object from the coded plan, as shown in Figure 2.57.

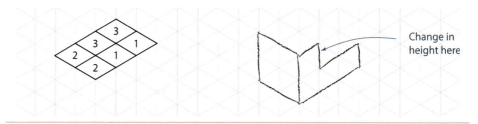

FIGURE 2.55. An isometric sketch with two side surfaces drawn.

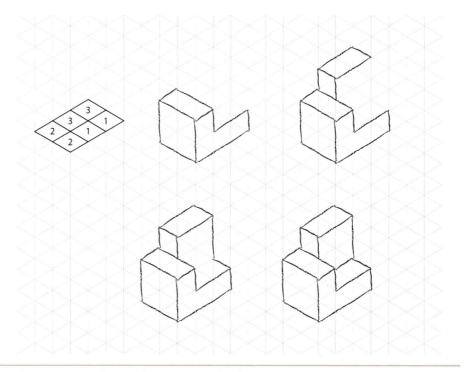

FIGURE 2.56. Completion of an isometric sketch.

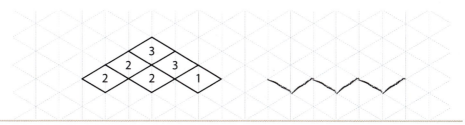

FIGURE 2.57. A jagged *V* from a coded plan.

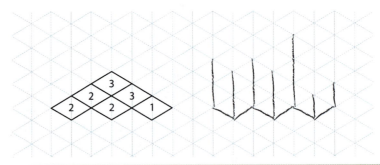

FIGURE 2.58. Heights at each corner included.

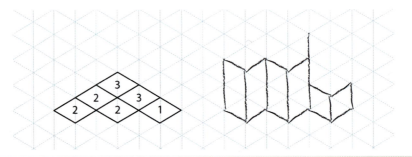

FIGURE 2.59. Side surfaces sketched.

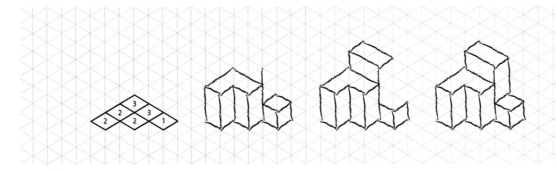

FIGURE 2.60. A completed isometric sketch.

Then you can sketch the lines that represent the height at each corner, similar to the way you sketched the height from the apex of the single *V* (see Figure 2.58). Complete the sketch by including the side and top surfaces of the object as illustrated in Figure 2.59 and Figure 2.60. For the object shown in the example, note that the final step involves erasing a portion of one of the first lines drawn (the corner at a height of 3). You need to remove part of this line so a line does not appear on the jagged side of the object.

2.11.02 Oblique Sketches

To begin your oblique sketch, you need to determine which surface on the object is closest to the viewer. Figure 2.61 shows an isometric sketch of an object with an arrow denoting the direction of the desired oblique pictorial. For this object and viewing direction, the surface labeled *A* is the one closest to the viewer in the oblique pictorial sketch.

Sketch the closest surface (in this case, surface A) in its true shape and size and decide whether you want the third dimension on the object receding back and to the left or back and to the right. Draw the visible edges receding back from each corner of

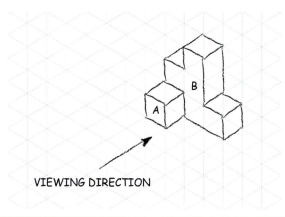

FIGURE 2.61. An isometric pictorial of an object and a viewing direction for an oblique sketch.

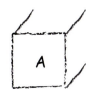

FIGURE 2.62. Surface A with receding dimensions sketched.

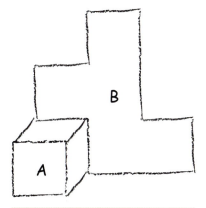

FIGURE 2.63. Surface B included in pictorial.

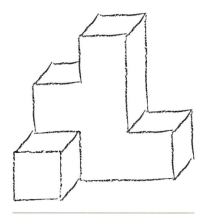

FIGURE 2.64. A completed oblique pictorial sketch.

the surface. Note that at least one corner on the surface will not have a receding line extending back from it—the receding edge will not be visible in the sketch. Figure 2.62 shows surface A with the receding edges sketched in place.

Now sketch the next surface that is parallel to the plane of the paper. For the object shown in Figure 2.63, the next closest surface is the one labeled *B*. Notice that by sketching this surface, you are connecting the endpoints of the lines drawn receding from the corners of the initial surface and, thus, are defining the side and top surfaces of the object in the pictorial. Figure 2.63 shows the result from including surface B in the oblique pictorial sketch.

Repeat these steps as often as necessary until the pictorial sketch is finished. Note that the final step is to include the back edges of the object (connecting the ends of the last set of receding lines drawn) to complete the sketch as shown in Figure 2.64.

CAUTION When creating isometric pictorials of simple objects, remember the general rules presented earlier in this chapter—that lines are included only at the intersection between surfaces, that no hidden lines are shown, and that only the visible portion of partially obscured surfaces are sketched. One common error novices are prone to make is to include extra lines on a single surface of an object, especially when there are several changes in the object's height. Figure 2.65a shows an improper isometric pictorial sketch. Notice the extra lines included on the sketch. Figure 2.65b shows the sketch after it has been cleaned up to remove the unnecessary lines.

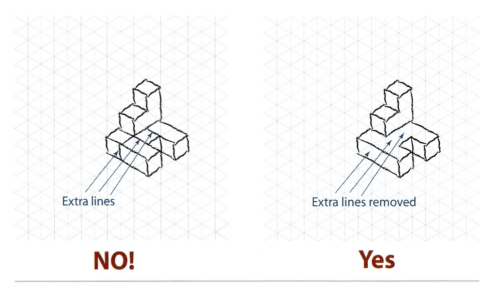

Extra lines

Extra lines removed

NO!   **Yes**

FIGURE 2.65. Isometric pictorials with and without extra lines.

Students make other common mistakes when sketching holes in isometric pictorials. One of those mistakes involves including the "back" edges of holes, even when they are not visible. Figure 2.66 shows an isometric pictorial with a hole in the object that goes all the way through. An arc representing the back edge of the hole is shown improperly in the visible part of the hole. Including the arc implies that the hole does not go all the way through the object, but stops part-way back. (Such holes are referred to as blind holes.) To avoid confusion in your isometric pictorial sketches, show only the back edge if it is visible—do not include a back edge every time you sketch an object with a hole.

Sometimes students use grid points improperly to mark off the bounding box for an isometric circular hole. Those novices fail to remember that in order to set the radial points, they need to move in the directions of the edges of the face of the object. Consider a simple box in which you want to include a circular hole emanating from the top surface. Figure 2.67a shows the four radial points incorrectly located from the center of the circle, and Figure 2.67b shows the resulting incorrect hole. Figure 2.68a shows the radial points located properly, and Figure 2.68b shows the resulting correct circular hole.

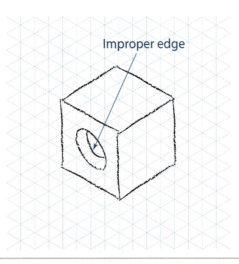

Improper edge

FIGURE 2.66. An isometric pictorial improperly showing the back edge of a hole.

One final error that students commonly make involves the creation of oblique pictorials. Novices sometimes forget to put the most complicated surface in the plane of the paper and show it in the receding direction instead. Figure 2.69a shows an oblique pictorial with a complex surface in the receding dimension, and Figure 2.69b shows the same object with the complex surface in the plane of the paper. Observe how the object is more understandable when you are viewing the complex surface "straight on." Also note that putting the complex surface in the plane of the paper actually makes your job easier; it is far easier to sketch the complex surface in its true size and shape than it is to sketch it as a distorted receding surface.

FIGURE 2.67. Improperly locating radial points on an isometric pictorial.

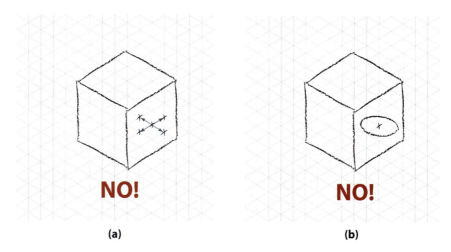

(a)   (b)

FIGURE 2.68. Properly locating radial points on an isometric pictorial.

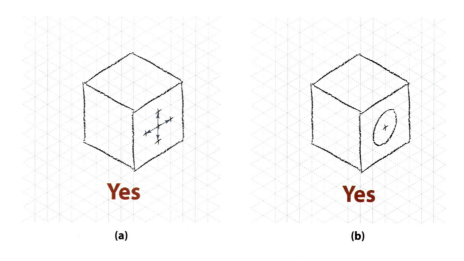

(a)   (b)

FIGURE 2.69. Oblique pictorials showing improper and proper placement of a complex surface.

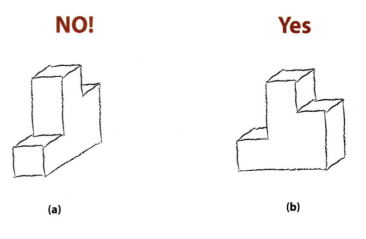

(a)   (b)

2.12 Chapter Summary

In this chapter, you learned about technical sketching and about some techniques to help you master this important form of communication. Specifically, you:

- Learned about the importance of sketching for engineering professionals and the link between creativity and freehand sketching.
- Developed techniques for successfully sketching basic shapes such as lines, arcs, circles, and ellipses.
- Learned about the right-hand rule and the way it is used to define 3-D coordinate systems in space. The axes can be portrayed on paper in either isometric or oblique format.
- Discovered how to make basic isometric sketches of objects from coded plans and about some of the rules that govern the creation of these sketches. You also learned about creating ellipses in isometric to represent circular holes in objects.
- Developed techniques for creating oblique pictorials. You also learned that for this type of pictorial, you should not show the receding dimension of the object true to size in order to avoid a distorted image.

2.13 glossary of key terms

arc: A curved entity that represents a portion of a circle.

bounding box: A square box used to sketch circles or ellipses.

circle: A closed curved figure where all points on it are equidistant from its center point.

construction line: A faint line used in sketching to align items and define shapes.

ellipse: A closed curve figure where the sum of the distance between any point on the figure and its two foci is constant.

isometric axes: A set of three coordinate axes that are portrayed on the paper at 120 degrees relative to one another.

isometric dot paper: Paper used for sketching purposes that includes dots located along lines that meet at 120 degrees.

isometric grid paper: Paper used for sketching purposes that includes grid lines at 120 degrees relative to one another.

isometric pictorial: A sketch of an object that shows its three dimensions where isometric axes were used as the basis for defining the edges of the object.

left-handed system: Any 3-D coordinate system that is defined by the left-hand rule.

line: Shortest distance between two points.

oblique axes: A set of three coordinate axes that are portrayed on the paper as two perpendicular lines, with the third axis meeting them at an angle, typically 45 degrees.

oblique pictorial: A sketch of an object that shows one face in the plane of the paper and the third dimension receding off at an angle relative to the face.

receding dimension: The portion of the object that appears to go back from the plane of the paper in an oblique pictorial.

right-hand rule: Used to define a 3-D coordinate system whereby by pointing the fingers of the right hand down the x-axis and curling them in the direction of the y-axis, the thumb will point down the z-axis.

right-handed system: Any 3-D coordinate system that is defined by the right-hand-rule.

shading: Marks added to surfaces and features of a sketch to highlight 3-D effects.

3-D coordinate system: A set of three mutually perpendicular axes used to define 3-D space.

tick mark: A short dash used in sketching to locate points on the paper.

2.14 questions for review

1. What is the role of sketching in engineering design? In creativity?

2. Describe which procedure you should use to sketch straight lines. (Are you right- or left-handed?)

3. How do circles appear on an isometric pictorial? On an oblique pictorial?

4. What is a bounding box?

5. How are construction lines used in sketching?

6. Why is it important to know the right-hand rule?

2.15 problems

1. For each of the coordinate axes shown below, indicate whether they are isometric or oblique and whether they represent right-handed or left-handed systems.

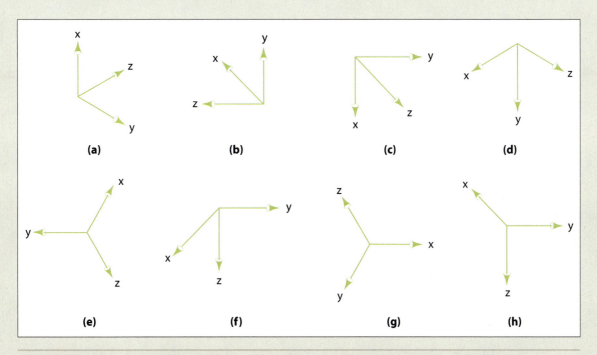

FIGURE P2.1.

2.15 problems (continued)

2. Label the third axis in each of the following figures to define a right-handed system.

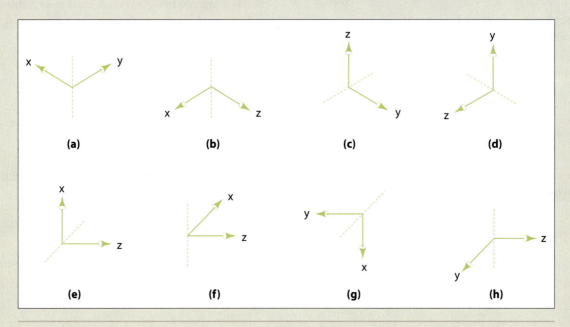

FIGURE P2.2.

2.15 problems (continued)

3. Create isometric sketches from the coded plans shown below.

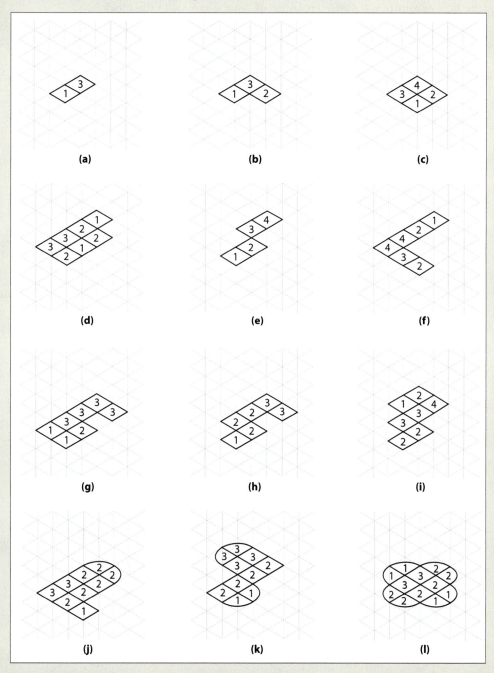

FIGURE P2.3.

4. Sketch a 6 × 6 × 2 block in isometric. On the 6 × 6 side, sketch a hole of diameter 4, making sure you include back edges of the hole as appropriate. Also create an oblique pictorial of the block.

5. Sketch a 6 × 6 × 2 block in isometric. On the 6 × 6 side, sketch a hole of diameter 2, making sure you include back edges of the hole as appropriate. Also create an oblique pictorial of the block.

2.15 problems (continued)

6. Sketch a 6 × 6 × 4 block in isometric. On the 6 × 6 side, sketch a hole of diameter 2, making sure you include back edges of the hole as appropriate. Also create an oblique pictorial of the block.

7. From the isometric pictorials and viewing directions defined in the following sketches, create oblique pictorial sketches that look proportionally correct.

8. Use the "foundation, frame, finish" method to create sketches of the following:

   a. stapler   c. coffee mug   e. calculator

   b. speedboat   d. bicycle   f. laptop computer

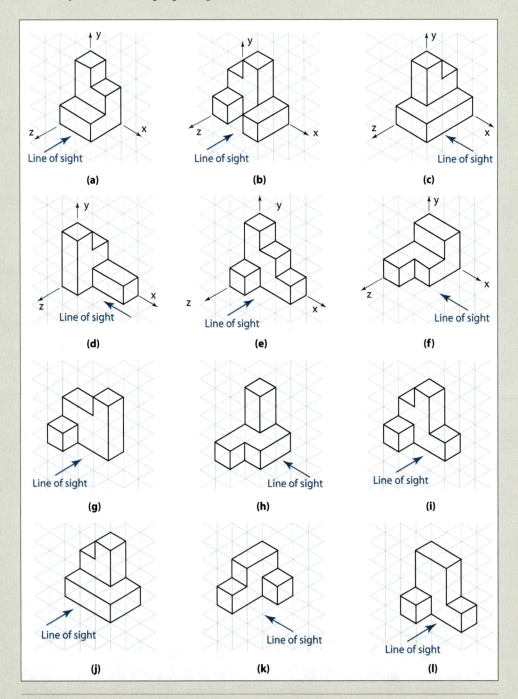

FIGURE P2.4.

3

Visualization

objectives

After completing this chapter, you should be able to

- Recognize that 3-D spatial skills are necessary for success in engineering

- Describe how a person's spatial skills develop as they age

- Examine the types of questions used to assess a person's spatial skill level

- Show how you can improve your 3-D spatial skills through techniques that include

  - Drawing different corner views of an object.

  - Rotating objects about one or more axes.

  - Sketching object reflections and making use of symmetries.

  - Considering cross sections of objects.

  - Combining two objects to form a third object through Boolean operations.

3.01
introduction

When you start your first job in the real world, an engineer or a technologist is likely to hand you a drawing and expect you to understand what is on the page. Imagine your embarrassment if you have no clue what all of the lines and symbols on the drawing mean. One of the fundamental skills you need to understand that drawing is the ability to visualize in three dimensions. The ability to visualize in three dimensions is also linked to creativity in design. People who think creatively are able to "see" things in their minds that others cannot. Their imaginations are not confined by traditional boundaries.

In this chapter, you will learn about the different types of three-dimensional (3-D) spatial skills and ways they can be developed through practice. The chapter will begin with an introduction to the background research conducted in education and to 3-D spatial skills. Then the chapter will take you through several types of visualization activities to further develop your 3-D skills through practice.

3.02 Background

Beginning in the early part of the twentieth century, IQ testing was developed to categorize a person based on his or her intelligence quotient. Anyone who took the IQ test was defined by a number that identified a level of intelligence. IQ scores over 140 identified geniuses; scores below 100 identified slow thinkers. Beginning in the 1970s, scholars began to perceive problems with this one-number categorization of a person's ability to think. One scholar in particular, Howard Gardner, theorized that there were multiple human intelligences and the one-number-fits-all theory did not accurately reflect the scope of human thought processes. Although some of his theories might be subject to scrutiny, they have gained acceptance within the scientific and educational communities. Gardner theorized that there are eight distinct human intelligences; he identified them as:

- Linguistic—the ability to use words effectively in speaking or in writing.
- Logical-Mathematical—the ability to use numbers effectively and to reason well.
- Spatial—the ability to perceive the visual-spatial world accurately and to perform transformations on those perceptions.
- Bodily-Kinesthetic—the capacity of a person to use the whole body to express ideas or feelings and the facility to use the hands to produce or transform things.
- Musical—the capacity to perceive, discriminate, transform, and express musical forms.
- Interpersonal—the ability to perceive and make distinctions in the moods, intentions, motivations, and feelings of other people.
- Intrapersonal—self-knowledge and the ability to act adaptively on the basis of that knowledge.
- Naturalist—the ability to recognize plant or animal species within the environment.

You may be acquainted with someone who has a high level of linguistic intelligence but a low level of musical intelligence. Or you might know someone who has a high level of logical-mathematical intelligence but who lacks interpersonal intelligence relationships. You may even have a friend who is generally smart but who lacks intrapersonal intelligence and attempts stunts that are beyond his or her limitations.

Most people are born with one or more of the intelligences listed. As a child, Tiger Woods was gifted with a natural ability in bodily-kinesthetic intelligence. Mozart was born with a high level of musical intelligence. However, just because a person naturally has a high level of intelligence in one area, does not mean that he or she cannot learn and improve his or her abilities in weaker areas. A person might naturally have strength in linguistics and musical intelligences, but he or she can still learn and improve in logical-mathematical endeavors. The goal of this chapter is to help those of you not born with a high level of spatial intelligence as defined by Gardner.

Learning in general and spatial skills in particular have been the subject of education research studies over the past several decades. The following are a few important questions that the research raised in the area of spatial intelligence:

- How does a person develop spatial skills?
- Why does a person need well-developed spatial skills?
- How are spatial skills measured?

The next few sections will examine researchers' answers to these questions.

3.03 Development of Spatial Skills

As a child grows, the brain develops in ways that enable the child to learn. If you think of each of the eight intelligences described by Gardner, you can understand how these skills and abilities develop as a child grows to maturity. Consider kinesthetic intelligence. Newborn infants cannot move on their own during the first few weeks of life. Within a few months, they can hold up their heads without support. By the age of four months, they can roll over; by six months, they can crawl; by one year, they can walk. Children learn to run, skip, and jump within the next year or so. Eventually, they usually develop all sorts of kinesthetic abilities that enable them to enjoy physical activities such as basketball, swimming, ballet, and bike riding. Nearly every child goes through this natural progression. However, some children develop more quickly than others; some even skip a step and go directly from rolling over to walking without ever crawling. As with most types of intelligence, some individuals—such as professional athletes—have exceptional kinesthetic skills, while others have poorly developed skills and struggle to perform the simplest tasks. However, even people who naturally have little kinesthetic ability can improve through practice and perseverance.

The remaining intelligences (mathematics, verbal, etc.) also have a natural progression; for example, to develop your mathematical intelligence, you have to learn addition before you can learn algebra. Children also acquire spatial skills through a natural progression; however, you may not be as aware of that progression of development as you are of the progressions for the other intelligences. Educational psychologists theorize that there are three distinct stages of development for spatial skills.

The first stage of development involves 2-D spatial skills. As children develop these skills, they are able to recognize 2-D shapes and eventually are able to recognize that a 2-D shape has a certain orientation in space. If you watched Sesame Street as a child, you may remember the game where four pictures of 2-D objects are shown on the TV screen—three objects are identical; the fourth is different in some way. A song urges you to pick out the object that does not belong with the other three. A child who can accomplish this task has developed some of the spatial skills at the first stage. You also may remember playing with a toy similar to a Tupperware ShapeSorter, shown in Figure 3.01. The toy is a ball that is half red and half blue with ten holes in it, each hole a different shape. A child playing this game not only has to recognize that the star-shaped piece corresponds to the star-shaped hole but also has to turn the piece to the correct orientation to fit the piece through the hole. This game challenges different 2-D skills found at the first stage of development of spatial intelligence—a child must recognize the 2-D shape of the object and then must be able to recognize its orientation in 2-D space to complete the task.

Three-dimensional spatial skills are acquired during the second stage of development. Children at this stage can imagine what a 3-D object looks like when it is rotated in space. They can imagine what an object looks like from a different point of view, or they can imagine what an object would look like when folded up from a 2-D pattern.

FIGURE 3.01. A Tupperware
ShapeSorter toy.

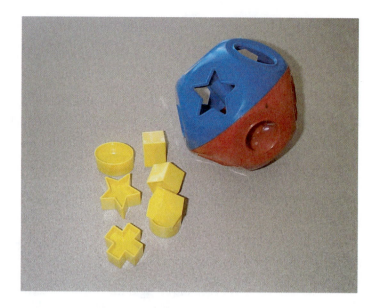

People who are adept at solving the Rubik's Cube puzzle have well-developed 3-D spatial skills. Computer games such as 3D Tetris require well-developed 3-D spatial skills to perform the manipulations required to remain "alive." Soccer players who can imagine the trajectory that puts the ball in the goal from any angle on the playing field typically have well-developed 3-D spatial skills. Children have usually acquired 3-D spatial skills by the time they are in middle school. For some children, it may take a few more years, depending on their natural predisposition toward spatial intelligence and their childhood experiences.

People at an advanced stage of the development of spatial intelligence can combine their 3-D skills with concepts of measurement. Assume you are buying sand for a turtle-shaped sandbox. You go to the local gravel pit where an employee loads the sand in the back of your pickup using a big "scoop." How many scoopfuls will you need? If you can successfully visualize the volume of sand as it is transformed from the 3-D volume of one full scoop to the 3-D volume of the turtle-shaped sandbox, you have acquired this advanced 3-D visualization skill.

Many people never develop the advanced level in spatial intelligence, just like the many people who never achieve advanced skill levels in mathematics or kinesthetic intelligence. Not achieving advanced levels in some of the intelligence areas is not likely to hamper your ability to become a productive, well-adjusted member of society. However, just as a lack of basic development in verbal intelligence is likely to hurt your chances professionally, a lack of basic skills in spatial intelligence may limit your ability to be successful, especially in engineering or a technical field.

Schools help students develop most of the intelligence types, although schools do not usually provide formal training to develop spatial intelligence. You began learning mathematics in kindergarten and are likely continuing your education in math at the present time. If you get a graduate degree in a technical area, you will probably be developing your mathematical intelligence for many years thereafter. The focus on developing spatial skills from an early age, continuing through high school and beyond, is typically absent in the U.S. educational system. Developing spatial intelligence is largely ignored in schools for a variety of reasons; however, those reasons are not the subject of this text.

The lack of prior spatial training may not be a problem for you—you developed your spatial skills informally through everyday experiences or you naturally have a high level of ability in spatial intelligence. However, poorly developed 3-D spatial skills may hinder your success in fields such as engineering and technology. This is especially true as you embark on a journey through an engineering graphics course. Poorly developed spatial skills will leave you frustrated and possibly discouraged about engineering graphics. The good news if you do not have a natural ability in 3-D spatial skills is that you can develop them through practice and exercise.

3.04 Types of Spatial Skills

According to McGee (1979), spatial ability is "the ability to mentally manipulate, rotate, twist, or invert pictorially presented visual stimuli." McGee identifies five components of spatial skills:

- **Spatial perception**—the ability to identify horizontal and vertical directions.
- **Spatial visualization**—the ability to mentally transform (rotate, translate, or mirror) or to mentally alter (twist, fold, or invert) 2-D figures and/or 3-D objects.
- **Mental rotations**—the ability to mentally turn a 3-D object in space and then be able to mentally rotate a different 3-D object in the same way.
- **Spatial relations**—the ability to visualize the relationship between two objects in space, i.e., overlapping or nonoverlapping.
- **Spatial orientation**—the ability of a person to mentally determine his or her own location and orientation within a given environment.

A different researcher (Tartre, 1990) proposed a classification scheme for spatial skills based on the mental processes that are expected to be used in performing a given task. She believes that there are two distinct categories of 3-D spatial skills—spatial visualization and spatial orientation. Spatial visualization is mentally moving an object. Spatial orientation is mentally shifting the point from which you view the object while it remains fixed in space.

Regardless of the classification scheme you choose to believe, it is clear that more than one component skill makes up the broad category of human intelligence known as spatial visualization. Thus, you cannot do just one type of activity and expect to develop equally all of the components of spatial skills. You need to do a variety of tasks to develop your spatial intelligence, just as developing linguistic intelligence requires you to speak, read, write, and listen.

3.05 Assessing Spatial Skills

As with the seven other intelligence types, standardized tests have been developed to determine your level of achievements in spatial intelligence. There are many different tests—some are for 2-D shapes, and some are for 3-D objects. Some evaluate mental rotation skills, and others measure spatial relations skills. The standardized tests usually measure only one specific component of visualization skill. If you were to take a number of different visualization tests, you might find that you have a high level of ability in one component (perhaps paper folding) relative to a low ability in a different component, such as 3-D object rotations. That is normal. Many educators and psychologists believe there is no one-size-fits-all measure of spatial intelligence, just as a single IQ number does not give a clear indication of a person's overall intelligence.

One of the tests designed to measure your level of 2-D spatial skills is the Minnesota Paper Form Board (MPFB) test. Figure 3.02 shows a visualization problem similar to those found on the MPFB test. This problem tests a person's ability to determine which set of five 2-D shapes, A through E, is the composite of the 2-D fragments given in the upper left corner of the figure. The way to solve this test is to mentally rotate or move the three pieces to visualize how to put them together to coincide with the combined shape that contains the pieces. The test may seem easy, but you should have fully developed the 2-D spatial skills needed to solve this test when you were four or five years old. During the years since then, you should have developed more advanced 2-D visualization skills. For example, you should now be able to follow a map and determine whether to make a right or a left turn without turning the map.

FIGURE 3.02. A problem similar to that found on the Minnesota Paper Form Board Test.

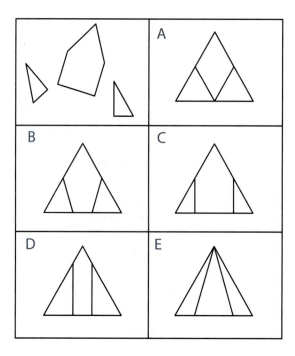

Figure 3.03 shows a visualization problem similar to what is found on the Differential Aptitude Test: Space Relations. This test is designed to measure your ability to move from the 2-D to the 3-D world. The objective is to mentally fold the 2-D pattern along the solid lines, which designate the fold lines, so the object will result in the 3-D shape. You then choose the correct 3-D object from the four possibilities shown in the figure. In your previous math classes, these 2-D patterns may have been referred to as *nets*. In engineering, the 2-D figures are called *flat patterns* or *developments*.

Mental rotations—the ability to visualize the rotation of 3-D objects—is a necessary component skill in engineering graphics and in the use of 3-D modeling software. Figure 3.4 and Figure 3.5 show problems similar to those found on two widely used 3-D spatial tests for rotations.

In the Purdue Spatial Visualization Test: Rotations, an object such as shown in Figure 3.04 is given on the top line before and after it has been rotated in 3-D space. You then have to mentally rotate a different object on the second line by the same amount and select the correct result from the choices given on the third line.

In the Mental Rotation Test, you are given an object such as shown in Figure 3.05 on the left. Of the four choices given, you pick the *two* that show correct possible rotations in space of the original object. (Note that two choices are the same object and two choices are different objects.)

FIGURE 3.03. A problem similar to that found on the Differential Aptitude Test: Space Relations.

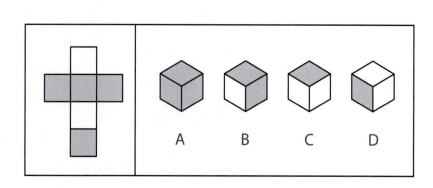

FIGURE 3.04. A problem similar to that found on the Purdue Spatial Visualization Test: Rotations.

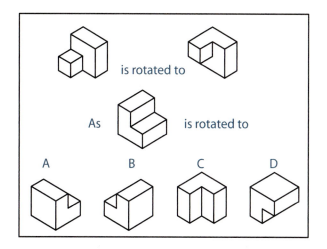

FIGURE 3.05. A problem similar to that found on the Mental Rotation Test.

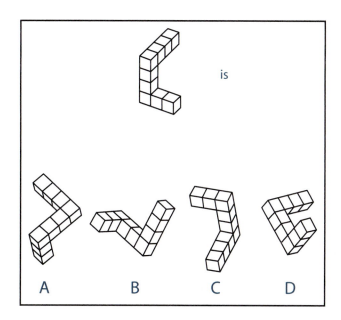

Another type of spatial skill that is often tested is the ability to visualize the **cross section** that results from "slicing" a 3-D object with a **cutting plane**. One popular test of this type is the Mental Cutting Test. Figure 3.06 shows the type of problem found on this test, which challenges you to imagine the 2-D shape that is the intersection between the cutting plane and the 3-D object.

Engineers and technologists communicate with each other largely through graphical means. They use drawings, sketches, charts, graphs, and CAD models to convey ideas. Design solutions commonly have a graphical component that is backed up by pages of calculations and analysis. Your designs will not be complete without graphics. Even chemical and electrical engineers use drawings for the processes and circuits they design.

So to communicate as an engineer, you must be able to visualize and interpret the images represented in the drawings. Besides satisfying the need for effective communication, a side benefit to having well-developed 3-D spatial skills is that your brain works better when *all* parts of it are focused on solving a problem. Sketching and visualization have been shown to improve the creative process. Well-developed spatial skills contribute to your ability to work innovatively, as well as to learn to use 3-D modeling software.

FIGURE 3.06. A problem similar to that found on the Mental Cutting Test.

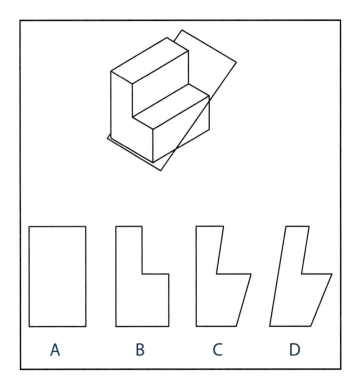

The remaining sections of this chapter will provide exercises for your brain—exercises that develop your 3-D spatial skills; exercises that help you think differently from the way you are thinking in your math and science courses; exercises that will help you improve your sketching skills.

3.06 Isometric Corner Views of Simple Objects

In Chapter 2, you learned how to create a simple isometric sketch of an object made out of blocks as specified by a coded plan. The coded plan is a 2-D portrayal of the object, using numbers to specify the height of the stack of blocks at a given location. Figure 3.07 illustrates the relationship between the coded plan, the object constructed out of blocks, and the resulting isometric sketch of the object—remember, you show edges only between surfaces on the isometric sketch.

The coded plans you viewed in Chapter 2 were constructed on isometric grid paper. The building "grew up" from the plan into the isometric grid. In the previous exercises,

FIGURE 3.07. A coded plan and its resulting isometric sketch.

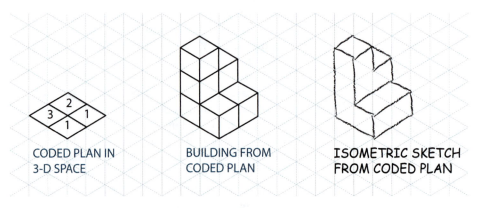

CODED PLAN IN
3-D SPACE

BUILDING FROM
CODED PLAN

ISOMETRIC SKETCH
FROM CODED PLAN

the coded plan was oriented in 3-D space on the isometric grid, which represents 3-D space. Now think about laying the coded plan flat on a 2-D sheet of paper. Figure 3.08 shows the coded plan for the object shown in Figure 3.07 laid flat in a 2-D orientation. Figure 3.9 shows the relationship between a coded plan in 2-D space, the coded plan in 3-D space, the object made of blocks, and the resulting isometric sketch.

When you orient the coded plan in 2-D space, everything you learned about these plans still applies: you "build up" from the plan. The numbers represent the height of the stack of blocks at a given location, and you show lines only where two surfaces intersect. However, now one more consideration has been introduced into the isometric sketching equation—the orientation of your "eye" with respect to the object itself. (Note that *the orientation of your eye* is often referred to as *your viewpoint*.)

Examine again the coded plans in 2-D space. Figure 3.10 shows a simple coded plan with its four corners labeled as *W*, *X*, *Y*, and *Z*. A **corner view** of the object represented by the coded plan in Figure 3.10 is the view from a given corner when the viewpoint is

FIGURE 3.08. The relationship between a coded plan in 2-D space and 3-D space.

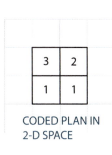

CODED PLAN IN
2-D SPACE

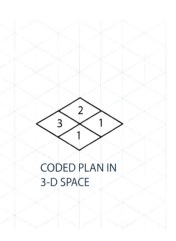

CODED PLAN IN
3-D SPACE

FIGURE 3.09. The relationship between coded plans, a building, and an isometric sketch.

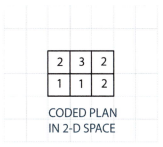

CODED PLAN
IN 2-D SPACE

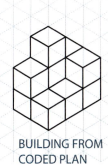

CODED PLAN
ON 3-D SPACE

BUILDING FROM
CODED PLAN

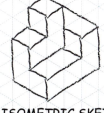

ISOMETRIC SKETCH
FROM CODED PLAN

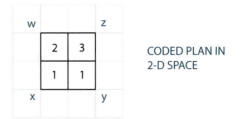

CODED PLAN IN
2-D SPACE

FIGURE 3.10. A simple coded plan with corners labeled.

above the object in question. This view is sometimes referred to as *the bird's-eye view*, because the viewpoint is above the object. A worm's-eye view is the viewpoint from *below* the object. Figure 3.11 shows the four corner views for the coded plan from Figure 3.10.

When the four corner views of the object are created, the object does not change—just your viewpoint of the object. The importance of viewpoint in visualization is readily apparent when you think about a complex system such as an automobile. When you are looking at a car from the front, you may have an entirely different mental image of the car than if you look at it from the side or rear. What you "see" depends largely on where your eye is located relative to the object.

With more practice, you will find it easier to make corner views from coded plans. At first, you may need to turn the paper to visualize what an object will look like from a given corner. With continued practice, however, you should be able to mentally turn the paper to sketch the object from the vantage point of any corner.

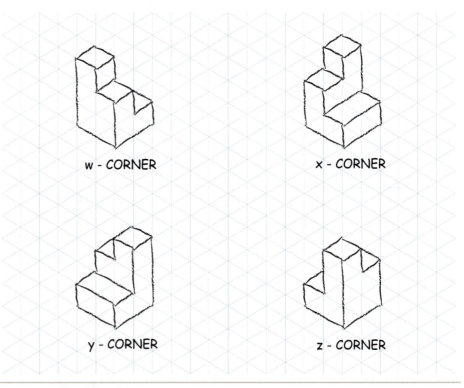

FIGURE 3.11. Sketched isometric views from the corners of the coded plan in Figure 3.10.

3.07 Object Rotations about a Single Axis

Being able to mentally visualize an object as it rotates in space is an important skill for you to acquire as an engineer or a technologist. You already have had limited exposure to the concept of rotating objects through your work with mentally rotating coded plans to obtain different corner views. In the preceding section, you started with the Y-corner view to draw the isometric. Having done that, you should be able to imagine what the object will look like from the X-corner view. If you can see in your mind what the object looks like from the X-corner view, you are mentally rotating the object in space. In this section, you will continue to work with object rotations, tackling increasingly complex objects and using increasingly complex manipulations.

You probably learned in your math classes how 2-D shapes are rotated in 2-D space about a pivot point, as illustrated in Figure 3.12. In this figure, the shape has been rotated 90 degrees counterclockwise (CCW) about the pivot point, which is the origin of the 2-D xy coordinate system. After rotation, the newly oriented shape is referred to as the "image" of the original shape. Notice that when the 2-D shape is rotated about the pivot point, each line on the shape is rotated by the same amount—in this case, 90 degrees CCW about the pivot point. Also notice that the point on the shape that was originally located at the pivot point, the origin, remains at that same location after rotation.

In Chapter 2, you learned about 3-D coordinate systems and how three axes (the x-, y-, and z-axes) can be used to describe 3-D space. When you rotate an object in 3-D space, the same principles apply as for 2-D rotations. In fact, you can reexamine the rotation of the shape in Figure 3.12 from a 3-D perspective. Figure 3.13 shows the 2-D shape drawn in 3-D space before and after it was rotated 90 degrees CCW about the pivot point, which is the origin of the xyz coordinate system.

Observe and understand how each line on the shape is rotated the same amount—90 degrees CCW about the origin—and that the point on the shape originally in contact with the origin remains at the origin after rotation. One other thing you may notice is that the pivot point is the point view of the z-axis. The point view of a line is what you see as you look down the length of the axis. To illustrate this principle, take a pen or pencil and rotate it so you are looking directly at its point; notice that the length of the pen "vanishes" and only the "point" remains visible, as shown in Figure 3.14. As such, the original rotation of the 2-D shape, as shown in Figure 3.12, could be considered a 90 degree CCW rotation about the z-axis in 3-D space.

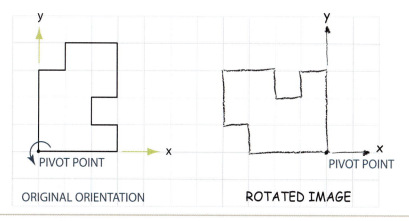

FIGURE 3.12. A shape rotated about a pivot point in 2-D space.

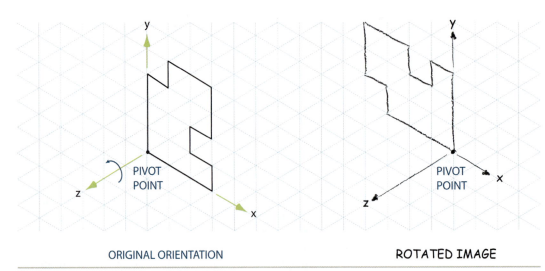

ORIGINAL ORIENTATION   ROTATED IMAGE

FIGURE 3.13. A 2-D shape rotated in 3-D space.

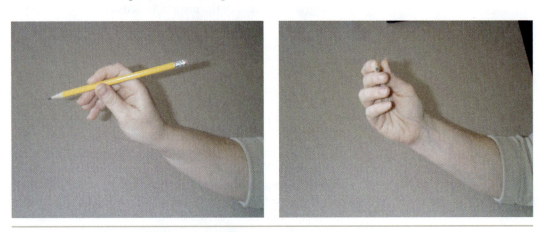

FIGURE 3.14. Looking down the end of a pencil.

Think back to what you learned in Chapter 2 about the right-hand rule. If you point the thumb of your right hand in the positive direction of the z-axis and curl your fingers, you will see that the 90 degree CCW rotation mimics the direction that your fingers curl, as illustrated in Figure 3.15. This CCW rotation of the 2-D shape represents a *positive* 90 degree rotation about the z-axis. The CCW rotation is positive because the thumb of your right hand was pointing in the positive direction of the z-axis as the shape was rotated. If you point the thumb of your right hand in the negative direction of the z-axis and the shape is rotated in the direction the fingers of your right hand curl, your fingers indicate a clockwise (CW) rotation of the shape about the z-axis, as shown in Figure 3.16. A CW rotation about an axis is defined as a negative rotation. Remember that the thumb of your right hand is pointing in the negative z-direction. Also remember that the pivot point of the shape remains at a fixed location in space as it is rotated in the negative z-direction.

You should now be ready to tackle rotations of 3-D objects in 3-D space. Imagine the 2-D shape from the past several figures is a surface view of a 3-D object. Assume you can extend the surface you have been seeing in the xy plane into the z-dimension. The result of extending that surface in the third dimension is a solid object. The terminology of 3-D CAD software says that the shape was extruded. You will learn more about extrusion later in this text. If this shape is "extruded" 3 units into the z-direction, the object will appear as shown in Figure 3.17. In this figure, notice that instead of a single point located on the axis of rotation (the z-axis in the figure), an entire edge of the object is located on that axis. The edge is hidden from sight in this view, but you can imagine it

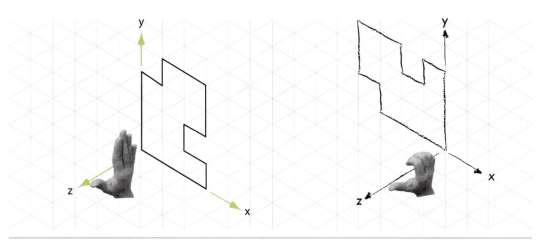

FIGURE 3.15. Positive rotation of a 2-D shape about the z-axis.

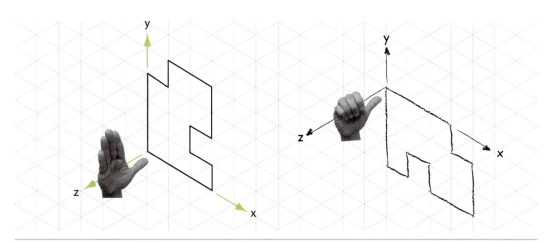

FIGURE 3.16. Negative rotation of a 2-D shape about the z-axis.

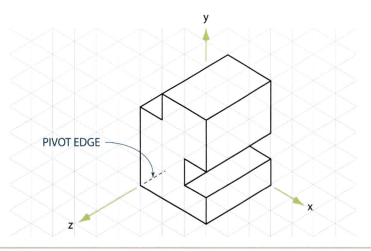

FIGURE 3.17. A 2-D shape from Figure 3.12 extruded three units in the z-direction.

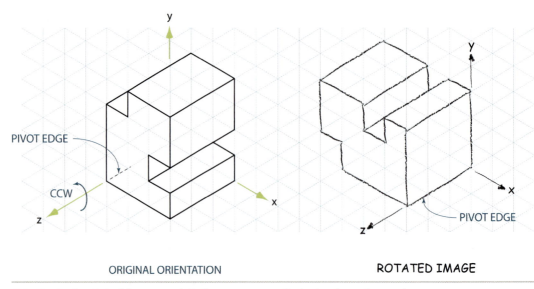

ORIGINAL ORIENTATION   ROTATED IMAGE

FIGURE 3.18. A 3-D object rotated 90 degrees counterclockwise about the z-axis.

nonetheless. Now think about rotating the entire object about the z-axis in a positive direction (or CCW) 90 degrees from its original position. When this happens, the image shown in Figure 3.18 appears. Instead of a single pivot point, the 3-D rotation has a pivot edge. Throughout the rotation, the edge remained in contact with the axis of rotation. All parts of the object also rotated by the same amount (90 degrees CCW about z) just as all parts of the surface were rotated when you were considering 2-D shapes.

Just as 2-D shapes can be rotated positively (CCW) or negatively (CW) about the z-axis, 3-D objects can be rotated in either direction. Figure 3.19 shows the same object after it has been rotated negative 90 degrees (CW) about the z-axis. This figure also makes clear that the pivot edge of the object remains in contact with the axis of rotation as the object is rotated.

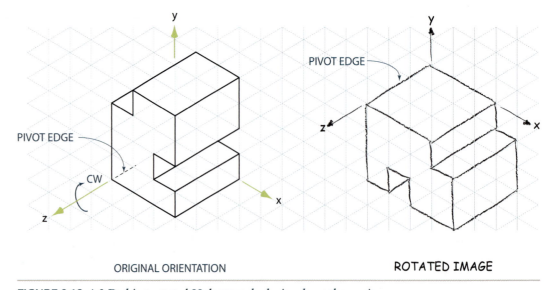

ORIGINAL ORIENTATION   ROTATED IMAGE

FIGURE 3.19. A 3-D object rotated 90 degrees clockwise about the z-axis.

Any object can be rotated about the x- or y-axis by following the same simple rules established for rotation about the z-axis:

1. The edge of the object originally in contact with the axis of rotation remains in contact after the rotation. This edge is called the pivot edge.

2. Each point, edge, and surface on the object is rotated by exactly the same amount.

3. The rotation is positive when it is CCW about an axis and negative when it is CW about an axis. The direction is determined by looking directly down the positive end of the axis of rotation.

4. An alternative method for determining the direction of the rotation is the right-hand rule. Point the thumb of your right hand into the axis of rotation—into either the positive or negative end of the axis of rotation—and curl your fingers in the direction the object is rotated. The direction you obtain from the right-hand rule is the same as the direction defined in number 3 above, positive is CCW and negative is CW.

Figure 3.20 and Figure 3.21 illustrate the positive and negative 90 degree rotations obtained about the x-axis and the y-axis, respectively.

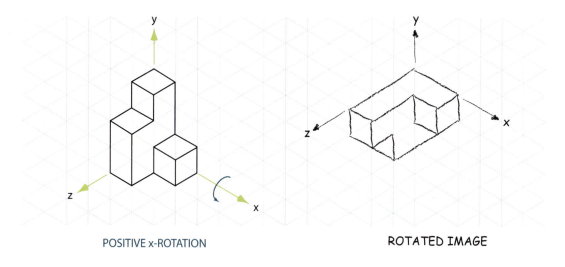

POSITIVE x-ROTATION   ROTATED IMAGE

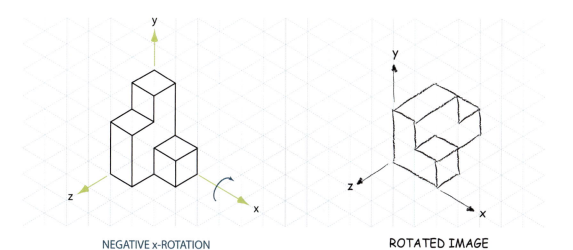

NEGATIVE x-ROTATION   ROTATED IMAGE

FIGURE 3.20. Positive and negative rotations about the x-axis.

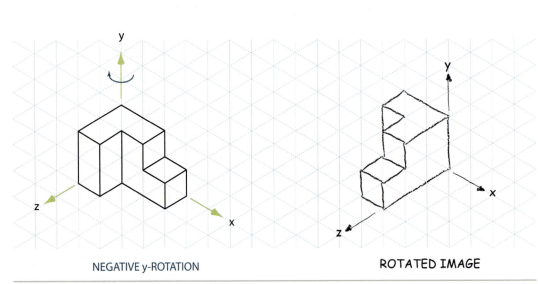

FIGURE 3.21. Positive and negative rotations about the y-axis.

3.07.01 Notation

Specifying in writing a positive, or CCW, rotation about any axis is cumbersome and time-consuming. For this reason, the following notations will be used to describe object rotations in this text:

■ To denote positive rotations of an object about the indicated axis.

■ To denote negative rotations of an object about the indicated axis.

■ Also, for simplicity in sketching, this text will always rotate an object in increments of 90 degrees about the indicated axis. Figure 3.22 illustrates the result when you rotate the object according to the notation given.

3.07.02 Rotation of Objects by More Than 90 Degrees about a Single Axis

In all examples and figures in the preceding sections, objects were rotated exactly 90 degrees about a single axis. In reality, you can rotate objects by any number of degrees. If you rotate an object in two increments of 90 degrees about the same axis, the total rotation will be 180 degrees. Similarly, if you rotate an object in three increments,

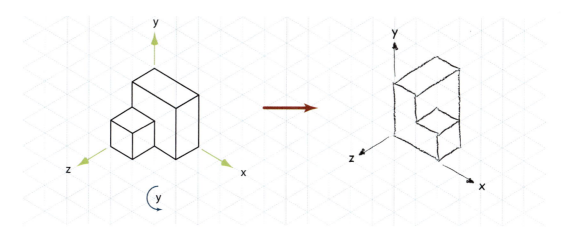

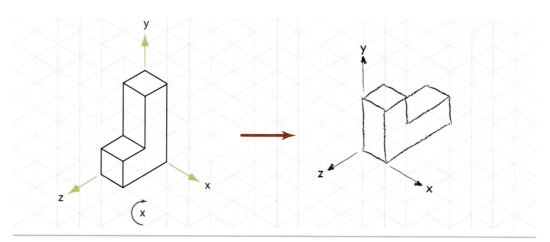

FIGURE 3.22. Object rotations specified by notation.

the total rotation will be 270 degrees. Figure 3.23 shows an object that has been rotated 180 degrees about a single axis, along with the symbol denoting the amount and direction of rotation. Notice that the two 90 degree positive x-axis rotations indicate the total 180 degree rotation achieved.

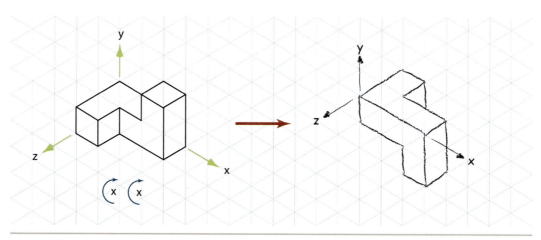

FIGURE 3.23. An object rotated 180 degrees about an axis.

Once you are free to rotate objects in multiple increments of 90 degrees, you can achieve several equivalent rotations. The term *equivalent rotations* means that two different sets of rotations produce the same result.

3.07.03 Equivalencies for Rotations about a Single Axis

When an object is rotated in multiple increments about an axis, the following equivalencies can be observed:

- A positive 180 degree rotation is equivalent to a negative 180 degree rotation.
- A negative 90 degree rotation is equivalent to a positive 270 degree rotation.
- A positive 90 degree rotation is equivalent to a negative 270 degree rotation.

These equivalencies are illustrated in Figures 3.24, 3.25, and 3.26, respectively.

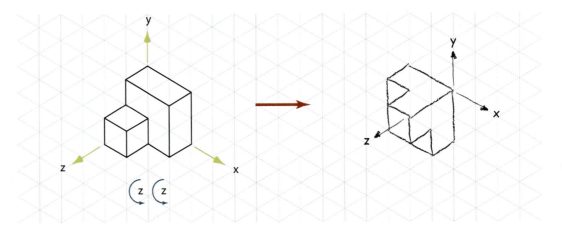

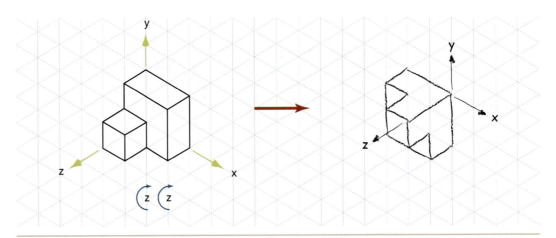

FIGURE 3.24. A positive 180 degree rotation is equivalent to a negative 180 degree rotation.

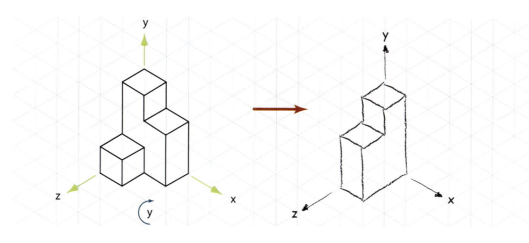

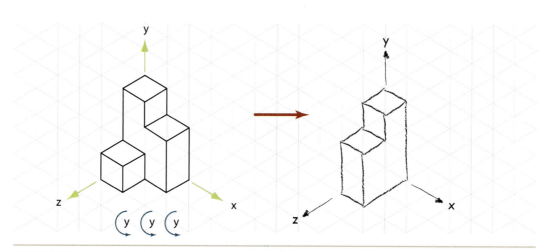

FIGURE 3.25. A negative 90 degree rotation is equivalent to a positive 270 degree rotation.

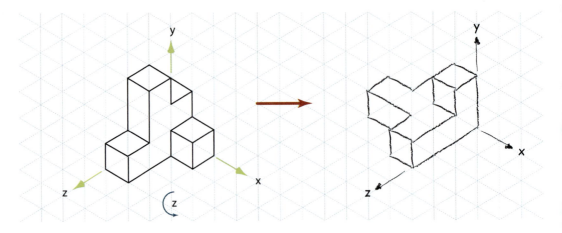

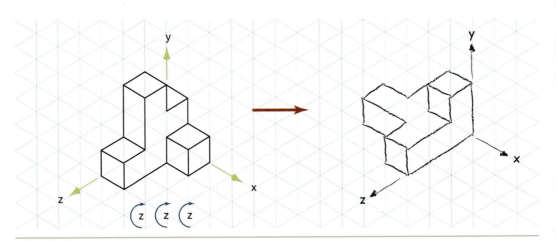

FIGURE 3.26. A positive 90 degree rotation is equivalent to a negative 270 degree rotation.

3.08 Rotation about Two or More Axes

In the same way you rotated an object about a single axis, you can also rotate the object about more than one axis in a series of steps. Figure 3.27 shows an object that has been rotated in the positive direction about the x-axis and then rotated in the negative direction about the y-axis. The rotation notation used in the figure indicates the specified two-step rotation. Figure 3.28 shows the same set of rotations, only this time they are shown in two single steps to achieve the final result. Notice that when an object is rotated about two different axes, a single edge no longer remains in contact with the axis of rotation (since there are now two of them). For rotations about two axes, only a single point remains in its original location, as shown in Figure 3.27 and Figure 3.28.

When rotating an object about two or more axes, you must be careful to perform the rotations in the exact order specified. If the rotations are listed such that you rotate the object CCW in the positive direction about the x-axis and then rotate it CW in the negative direction about the z-axis, you must perform the rotations in that order. Object rotations are not commutative. (Remember that the commutative property in math states that 2 + 3 = 3 + 2.) For object rotations, rotating about the x-axis and then rotating about the y-axis is *not* the same as rotating about the y-axis and then rotating about the x-axis.

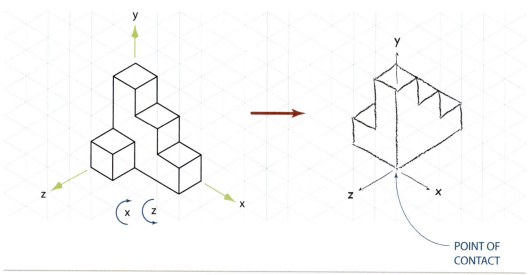

FIGURE 3.27. An object rotated about two axes.

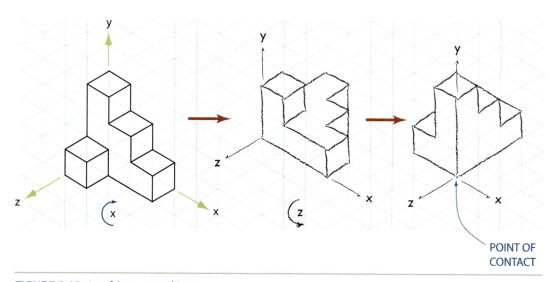

FIGURE 3.28. An object rotated in two steps.

In the top portion of Figure 3.29, the object has been rotated about positive y and then rotated about negative z to obtain its image. In the bottom portion of the figure, the object has been rotated about negative z and then rotated about positive y to obtain a new image of the rotated object. The second image is obtained by reversing the order of the rotations. The resulting images are not the same when the order of rotation is changed. Why? Because with the first set of rotations, the edge of the object on the y-axis serves as the pivot line for the first rotation, which is about positive y. For the second set of rotations, the edge of the object on the z-axis serves as the pivot edge for the first of the two rotations. When you rotate first about negative z, you are using an entirely different object edge than the initial pivot line; hence, the difference in rotated images.

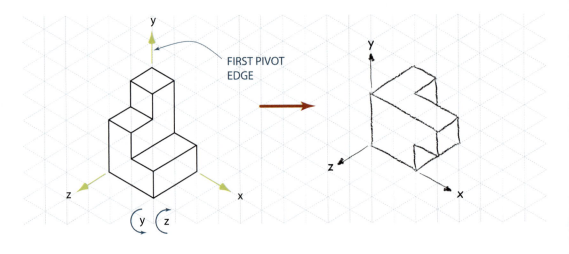

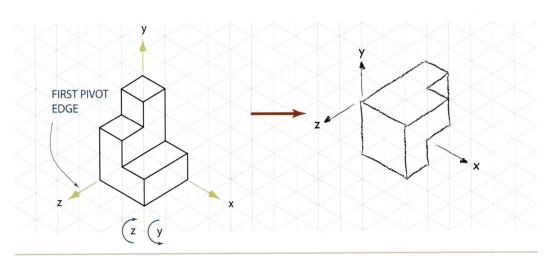

FIGURE 3.29. Object rotations about two axes—order not commutative.

3.08.01 Equivalencies for Object Rotations about Two or More Axes

Just as there are equivalencies for rotations of an object about a single axis, there are equivalencies for object rotations about two axes. Figure 3.30 shows one pair of rotational equivalencies. Can you find another set? How about positive x and then negative z? No! Or positive y and then positive z? Yes! There are several possibilities for each pair of rotations. But it is impossible to come up with simple rules for equivalency, as in the previous discussion of equivalent rotations about a single axis. Equivalent rotations for objects about two or more axes are likely to be determined through trial and error and a great deal of practice.

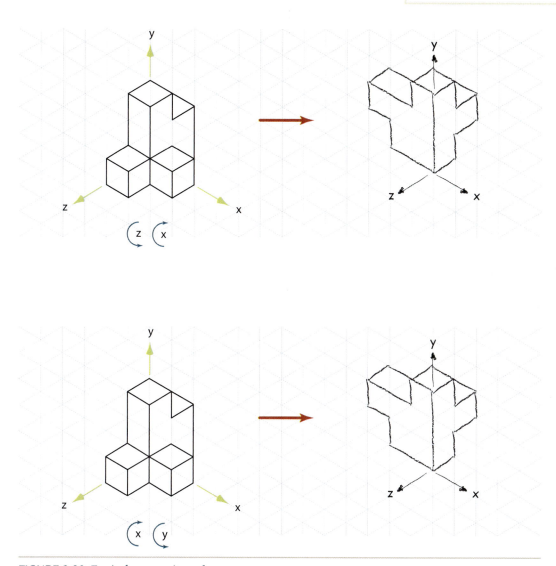

FIGURE 3.30. Equivalent rotations about two axes.

3.09 Reflections and Symmetry

Now that you know the basics of how to visualize an object rotated about an axis, you are ready to move on to visualizing reflections and symmetry. Visualizing planes of symmetry, for example, could save you a great deal of computation time when you are using analysis tools such as FEA. You will learn more about FEA in later chapters of this text.

You are probably familiar with the concept of **reflections** because you are used to looking at your image reflected back to you from a mirror. With a mirror, you see a reflected 2-D image of your face. If you have a mole on your right cheek, you will see the mole on the right cheek of the reflection. Even though your face is three-dimensional, your face in the mirror is a 2-D reflection—as if your face were projected onto a 2-D plane with your line of sight perpendicular to the plane. You may be able to see somewhat in the third dimension from this mirror plane; however, your depth perception will be a bit off because the image is only two-dimensional. Three-dimensional reflection of objects is different from 2-D reflections with mirrors. For one thing, you reflect a 3-D object *across* the plane so that a 3-D image ends up on the other side of the reflection plane.

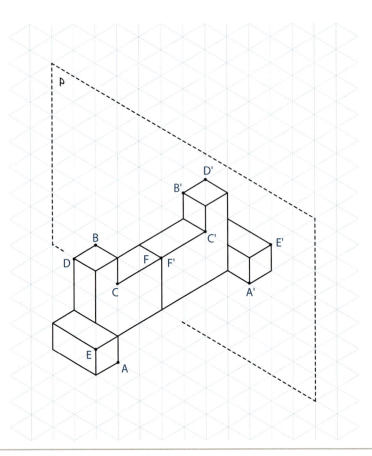

FIGURE 3.31. An object and its 3-D reflection.

Formally stated, in the case of 3-D object reflections, such as shown in Figure 3.31, each point A of the object is associated with an image point A´ in the reflection such that the plane of reflection is a perpendicular bisector of the line segment AA´. What this means is the distance between a point on an object and the reflection plane is equal to the distance between the corresponding point on the image and the reflection plane. The distances are measured along a line perpendicular to the plane of reflection. Figure 3.31 shows a simple object and its reflection across a reflection plane.

In this figure, several points on the original object are labeled, as well as their corresponding points on the reflected image. In this case, the plane of reflection coincides with one planar end of the original object; therefore, the corresponding planar end of the reflected image also coincides with the reflection plane. If you measure the distance between point A on the object and the reflection plane, you will find that it is 3 units. Then if you measure the distance between A´ and the reflection plane, you will find the distance to be 3 units again. It is also possible to reflect an object across a plane when the object is located some distance from the reflection plane, as illustrated in Figure 3.32.

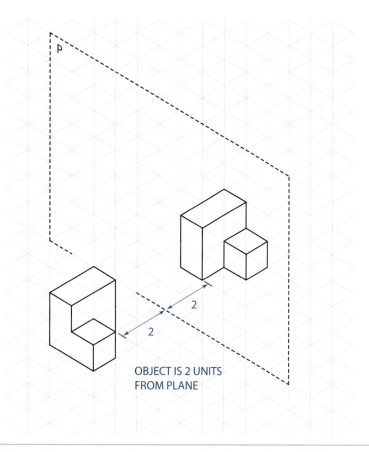

FIGURE 3.32. An object located at a distance from the plane and its reflection.

3.09.01 Symmetry

Your job as an engineer may be easier if you can recognize planes of symmetry within an object. A plane of **symmetry** is an imaginary plane that cuts through an object such that the two parts, one on either side of the plane, are reflections of each other. Not all objects have inherent symmetry. The human body is roughly symmetrical and has one plane of symmetry—a vertical plane through the tip of the nose and the belly button. The left side is a reflection of the right side. Some objects contain no planes of symmetry, some contain only one plane of symmetry, and still others contain an infinite number of planes of symmetry. Figure 3.33 shows several objects and their planes of symmetry: one object contains no planes of symmetry, one object has just one plane of symmetry, one object has two planes of symmetry, and the last object contains an infinite number of planes of symmetry.

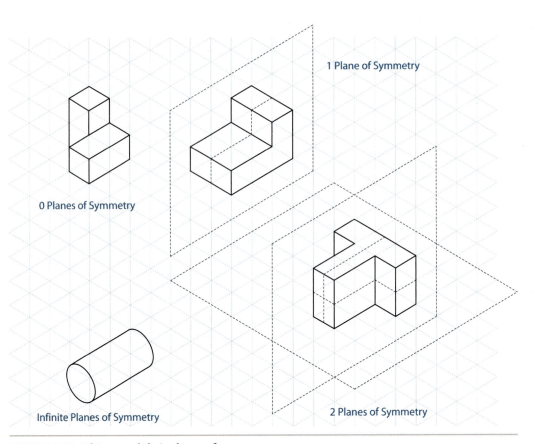

FIGURE 3.33. Objects and their planes of symmetry.

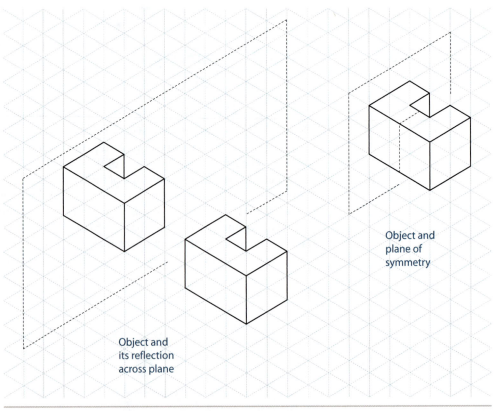

FIGURE 3.34. A comparison of object reflection and symmetry.

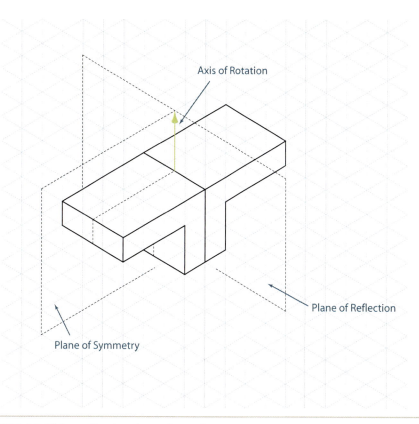

FIGURE 3.35. Object reflection through rotation.

There is one major difference between object reflection and object symmetry. With reflections, you end up with two separate objects (the original and its reflected image); with symmetry, you have a single object that you imagine is being sliced by a plane to form two symmetrical halves. Figure 3.34 illustrates the difference between the two.

For an object that is symmetrical about a plane, you can sometimes obtain its reflection by rotating the object 180 degrees. To do this, the axis of rotation must be the intersection between the plane of reflection and the plane of symmetry (two planes intersect to form a line). This concept is illustrated in Figure 3.35. Note that a reflection of an object that is not symmetrical cannot be achieved through a simple 180 degree rotation of the object. Hold up your hands in front of you to obtain an object (left hand) and its reflected image (right hand). Note that because your hands have no planes of symmetry, it is impossible to rotate one of them in space to obtain the other one.

3.10 Cross Sections of Solids

Visualizing cross-sections enables an engineer to figure out how a building or a mechanical device is put together. Visualizing cross-sections enables an electrical engineer to think about how circuit boards stack together within the housing that contains them. Chemical engineers and materials engineers think about the cross sections of molecules and the way those molecules combine with other molecules. Geological engineers and mining engineers visualize cross sections of the earth to determine where veins of rock and ore may be located. Most of the skills described in these examples are at an advanced level; in this section, you will learn about cross-sections of solids from a fundamental level. Then you can apply the principles to the visualization of more complex parts and systems in later courses and, of course, in your professional work.

Simply stated, a cross-section is defined as "the intersection between a solid object and a cutting plane." Because a plane is infinitely thin, the resulting intersection of the two planes is a planar section. The limits of the cross-sectional plane are the edges and

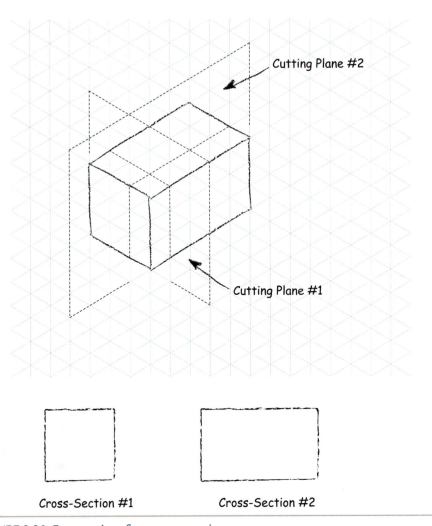

Cross-Section #1   Cross-Section #2

FIGURE 3.36. Cross sections from a square prism.

the surfaces where the plane cuts through the object. Consider a loaf of bread. Imagine a single slice of infinitely thin bread. One slice of bread would represent the cross section obtained by slicing a vertical plane through the loaf. Because most loaves of bread are not "constant" in shape along their lengths, the cross section changes as you go along the loaf. You know from experience that the cross sections, or slices, on the ends of the loaf are typically smaller than the slices in the middle.

The cross-section obtained by intersecting a cutting plane with an object depends on two things: (1) the orientation of the cutting plane with respect to the object and (2) the shape of the original object.

Consider the square prism shown in Figure 3.36. It is cut first by a vertical cutting plane perpendicular to its long axis to obtain the square cross section shown. If the cutting plane is rotated 90 degrees about a vertical axis, the result is the rectangular cross section shown in the figure. The two cross-sections are obtained from the same object. The difference in the resulting cross-sections is determined by changing the orientation of the cutting plane with respect to the object.

Now consider the cylinder shown in Figure 3.37. If a cutting plane is oriented perpendicular to the axis of the cylinder, a circular cross-section results; if the plane is located along the axis of the cylinder, a rectangular cross-section is obtained. Observe that this rectangular cross section through the cylinder is identical to the cross section obtained by slicing the rectangular prism along its long axis in Figure 3.36.

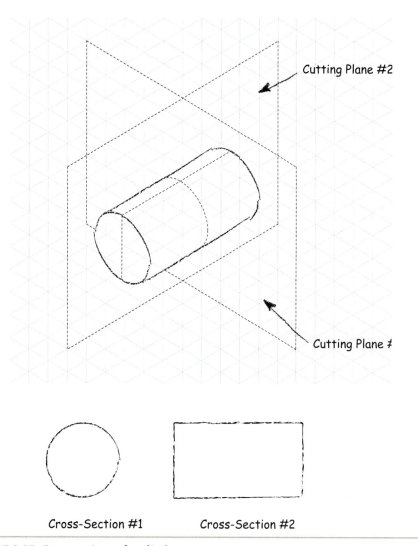

Cutting Plane #2

Cutting Plane #

Cross-Section #1   Cross-Section #2

FIGURE 3.37. Cross sections of a cylinder.

Because a resulting cross-section through an object depends on the orientation of the cutting plane with respect to the object, most objects have several cross-sections associated with them. Figure 3.38 shows a cylinder with four possible cross sections. Can you imagine the orientation of the cutting plane with respect to the cylinder for each cross-section?

You already know that the first two cross-sections, rectangle and circle, were obtained by orienting the cutting plane perpendicular to and along the long axis of the cylinder, respectively.

What about the third cross-section? It was obtained by orienting the cutting-plane at an angle with respect to the axis of the cylinder.

The fourth cross-section was also obtained by angling the cutting plane with respect to the cylinder axis, but the angle was such that a portion of the cutting plane went through the flat circular end surface of the cylinder.

Figure 3.39 shows several cross-sections obtained by slicing a cube with cutting planes at different orientations.

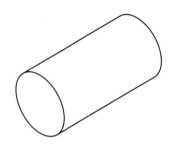

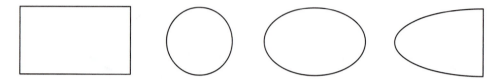

FIGURE 3.38. Various cross sections of a cylinder.

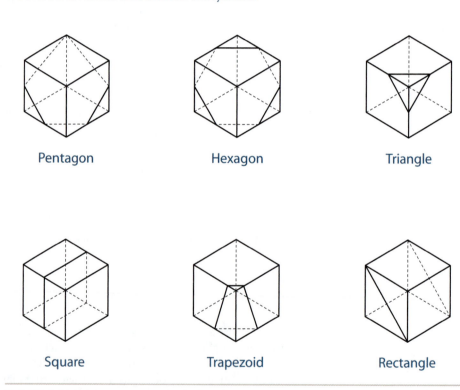

Pentagon   Hexagon   Triangle

Square   Trapezoid   Rectangle

FIGURE 3.39. Various cross sections of a cube.

3.11 Combining Solids

Another skill that will be helpful to you as an engineer is the ability to visualize how two solids combine to form a third solid. The ability to visualize **combining solids** will be helpful as you learn how to use solid modeling software. In early versions of 3-D CAD software, commands used to combine solids were sometimes known as **Boolean operations**. This terminology was borrowed from mathematics set theory operations, called Booleans, where basic operations include unions, intersections, and complements between sets of numbers. Boolean logic is now the foundation of many modern innovations. In fact, if you have performed a search on the Web using an AND or an OR operator, you have used Boolean logic to help you narrow or expand your search. In terms of 3-D CAD, the Boolean set operations typically correspond to software commands of Join, Intersect, and Cut. To help you become familiar with the terminology since you probably will be building 3-D computer models, this section will use the same terminology.

Two overlapping objects can be combined to form a third object with characteristics of each original object apparent in the final result. To perform any Cut, Join, and Intersect operations to combine objects, the objects must be overlapping initially. What is meant by overlapping is that they share a common volume in 3-D space—called the **volume of interference**. Figure 3.40a shows two objects that overlap; Figure 3.40b shows the volume of interference between the two objects. Notice that the volume of interference takes shape and size characteristics from each of the two initial objects.

When two objects are **joined**, the volume of interference is absorbed into the combined object. The result is a single object that does not have "double" volume in the region of interference. The Boolean Join operation is illustrated in Figure 3.41.

When two objects are combined by **intersecting**, the combined object that results from the intersection is the volume of interference between them, as shown in Figure 3.42.

In the **cutting** of two objects, the combined object that results from the cutting depends on which object serves as the cutting tool and which object is cut by the other object. The result of a cutting operation is that the volume of interference is removed from the object that is cut, as illustrated in Figure 3.43.

FIGURE 3.40. Overlapping objects and volume of interference.

Overlapping
Objects

Volume of
Interference

FIGURE 3.41. Result of two objects joined.

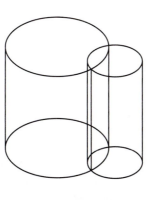
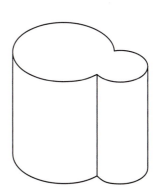

Overlapping Objects

Objects Joined

FIGURE 3.42. Result of two objects intersected.

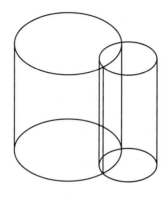

Overlapping Objects   Objects Intersected

FIGURE 3.43. Result of two objects cutting.

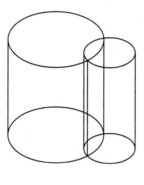

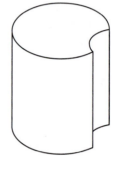

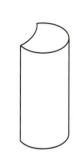

Overlapping Objects   Small Cylinder Cuts   Large Cylinder Cuts
                             Large Cylinder   Small Cylinder

3.12 Strategies for Developing 3-D Visualization Skills

In this chapter, you learned about several different types of exercises you might want to tackle in order to develop your 3-D visualization skills. In this section, you will learn how to get started with those exercises.

3.12.01 The Sketching of Corner Views

When you have a large task, such as constructing the isometric view of an object, the easiest way to get started is to complete an isometric view of a small piece first and then move on successively to other pieces of the object. You should follow this same process for almost every task in this chapter and throughout this textbook. In terms of objects, the basic building blocks are points, edges, and surfaces. To get started, you need to break down the object into its elements.

When sketching an isometric view of an object from its coded plan, you should look first at the corner from which you are sketching to determine which side of the object is on the left and which is on the right. The corner will be defined by a vertical line on the grid paper with the left and right sides emanating from there. Figure 3.44 shows a coded plan with the corners identified. If you want to sketch the Y-corner view, the arrows indicated show the left and right directions emanating from this corner. If you cannot immediately see this, turn the page so that *Y* is directly in front of you.

Next, you should sketch the height of the object from the corner you selected. As you move to the left from this corner, how does the height of the object change? For the coded plan in Figure 3.44, the height at the corner is 2; and as you go to the left, the object goes back two squares at the same height and then switches to a height of 3 for

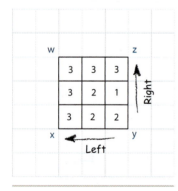

FIGURE 3.44. A coded plan.

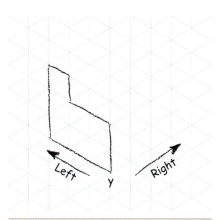

FIGURE 3.45. The left side surface of a building.

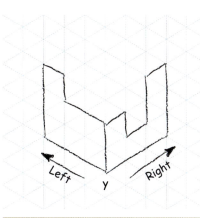

FIGURE 3.46. Left and right side surfaces of an object.

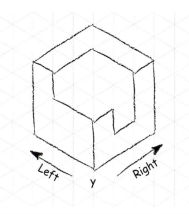

FIGURE 3.47. The top surface of the object.

one additional square. With this height information, you should be able to sketch the left-side surface of the object as shown in Figure 3.45.

You have now completed drawing one surface on the object. Go back to the original corner and think about moving in the other direction to define the surface on the object's right side. The height of the object goes from 2 to 1 to 3, one square deep at each height just indicated. Using this information, you can sketch the right-side surface going to the right from your chosen corner as illustrated in Figure 3.46.

Look again at the coded plan. You will see that the maximum height of the object is 3 units. At the same height, five blocks form an L going from the left side to the right side. You can sketch the L-shaped top surface at the given height as shown in Figure 3.47.

Look at the coded place once more, this time to determine where the object's height is 2. You will see three blocks at this height; together these three blocks trace another smaller L, as shown in Figure 3.48.

Finally, examine the sketch to determine where you need to add lines to complete the isometric. Add those lines. For the object you have been working with, you need to add lines to define the top surface that has a height of 1 and add whatever lines are needed at vertical corners to complete the isometric sketch. Figure 3.49 shows the completed sketch of the object.

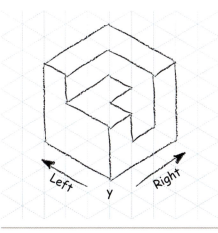

FIGURE 3.48. The second top surface of the object.

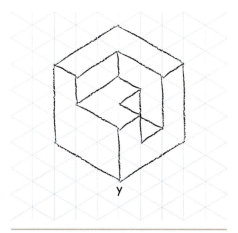

FIGURE 3.49. The completed isometric sketch.

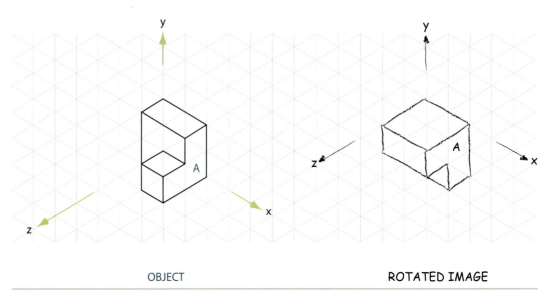

OBJECT   ROTATED IMAGE

FIGURE 3.50. An object in original position and its rotated image.

3.12.02 Object Rotations about One Axis

Once again, the simplest way to visualize rotation of an object is to focus on just one of the object's surfaces. You should look at the object and mentally select one surface to serve as your focal point for the rotation. This surface is called the key surface for purposes of this exercise. When you mentally rotate the key surface, you usually can let the remainder of the object follow. Figure 3.50 shows an object and its rotated image. What axis was the object rotated about, and was it a positive or a negative rotation?

This exercise will use the L-shaped surface, labeled *A*, as the key surface. By focusing on surface A on the original object, you can see that the surface was rotated positive 90 degrees about the x-axis. By definition, when an object is rotated in space, all points, edges, and surfaces on the object are rotated the same amount. This means that the entire object from Figure 3.50 is rotated 90 degrees about the x-axis, because surface A was rotated by that amount. Note that the surface you choose as the key surface could be any of those on the object, and this surface is likely to change from problem to problem.

Just as focusing on one surface can help you *identify* an object rotation, starting with one surface also can assist you in the task of *performing* an object rotation. Look at the object in Figure 3.51. Your task is to rotate the object negative 90 degrees about the z-axis and sketch the image that results.

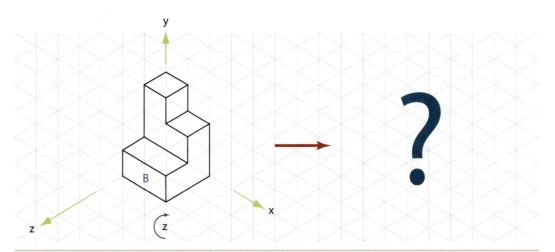

FIGURE 3.51. An object for rotation.

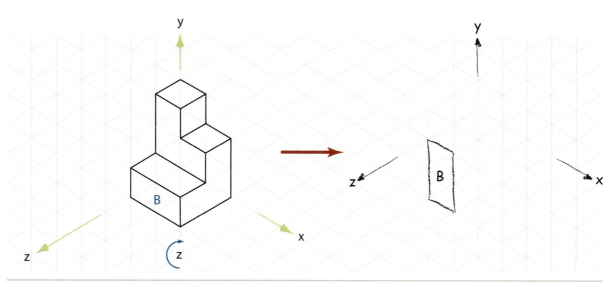

FIGURE 3.52. Rotation of surface B only.

To perform the task, you can select any surface; however, because the surfaces on the right side of the object will be hidden from view after the rotation is performed, they are probably not the best choice with which to begin this exercise. The surface of the object labeled B is probably the easiest one to begin with. Imagine surface B is rotated by the specified amount and sketch the surface only where it will appear after rotation. Figure 3.52 illustrates the rotation of the surface 90 degrees negatively (CW) about the z-axis.

Knowing the location of surface B after rotation, you can think about rotating surfaces C and D by the same amount as surface B and sketching the rotated surfaces C and D as shown in Figure 3.53. Notice that with 3-D rotations of an object, surfaces visible in the original view of the object are not always visible in the rotated image and surfaces that were hidden in the original view often become visible in the image of the rotated object.

Now you need to imagine what the "back" left-side surface of the object looks like after rotation so you can include it in your sketch. The back surface is hidden from view in the original orientation; but after rotation, it is the "top" surface of the object and is clearly visible. Going back to the object in its original position, you can see that this back surface starts at a height of 1 and jumps to a height of 3, meaning that it will

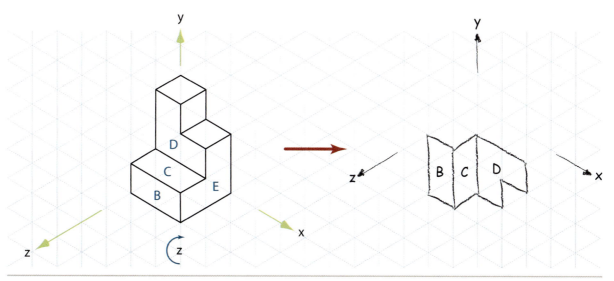

FIGURE 3.53. Rotation of surfaces C and D.

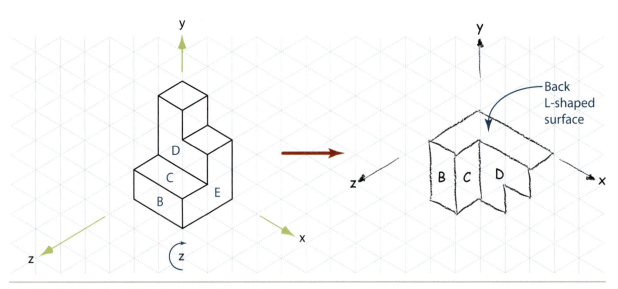

FIGURE 3.54. The back left-side surface after rotation.

appear L-shaped when rotated into position in the new view. This L-shaped surface can be sketched along with the other surfaces of the object as shown in Figure 3.54.

The last step is to clean up the drawing, adding lines as needed to complete the isometric sketch and to define the remaining surfaces on the image of the rotated object as illustrated in Figure 3.55.

3.12.03 Object Rotations about Two or More Axes

A step-by-step procedure for visualizing rotations about two or more axes is not as simple as the procedure for visualizing rotations of an object about one axis. The procedure is not simple because when rotating about two or more axes, many of the surfaces on the object that start out visible before rotation will be hidden after the first or successive rotations. One way to deal with the difficulty of visualizing rotation about two axes is to sketch the object as it appears after the first rotation and before the second rotation. For example, if an object is to be rotated positively about the x-axis and then negatively about the y-axis, sketch the intermediate step—the rotation about the x-axis—and complete the task from there. This way, the complex rotation is divided into a series of two single axis rotations. Figure 3.56 illustrates the method.

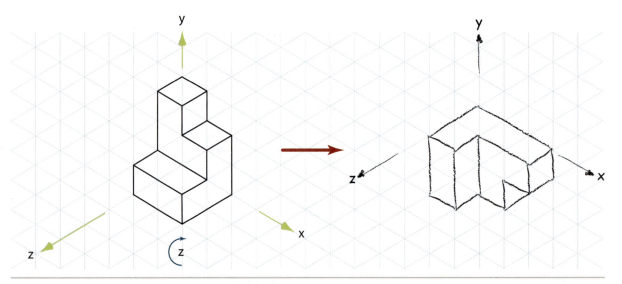

FIGURE 3.55. The completed sketch of the object after rotation.

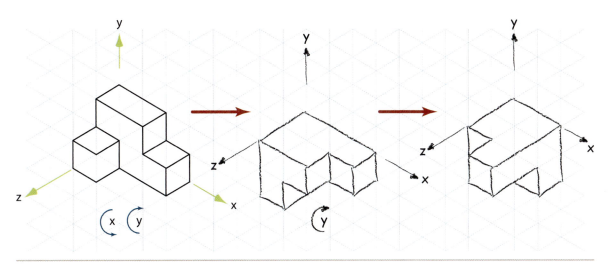

FIGURE 3.56. Intermediate rotation sketched.

You might be able to visualize the rotation about two or more axes by concentrating on a single key surface. But if you do try to do this, take care to assure that the key surface you choose to rotate remains visible after the rotation has been performed. Figure 3.57 shows an object and its rotated image. Note that surfaces A and B are no longer visible after the specified two rotations, so you would have to concentrate on surface C when mentally rotating the object.

One technique that may help you visualize rotations about two or more axes is to focus on a single surface that was not visible originally (instead of a visible one) and imagine its orientation after rotation. For example, the object shown in Figure 3.58 has a U-shaped surface on its back left side. Although you cannot see the surface in the original object orientation, the overall shape of the objects tells you that it is there. The surface of the object in its original position is hidden from view, but you should be able to visualize the surface nevertheless. Whenever you are asked to perform the rotations specified, you can think about rotating just the one surface and sketching it in its new position, similar to the way you started with the rotations about one axis. Figure 3.59 shows just the back surface rotated about both axes specified and into its final position.

When sketching the surface in the position after being rotated as specified, you can fill in the lines composing the entire object that corresponds to the same specified series of rotations as the single key surface. The complete sketch of the object after both specified rotations have been performed is shown in Figure 3.60.

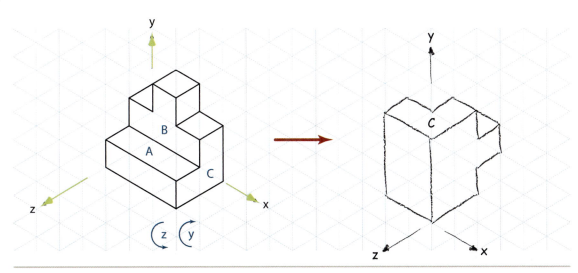

FIGURE 3.57. Visible surfaces before and after rotation.

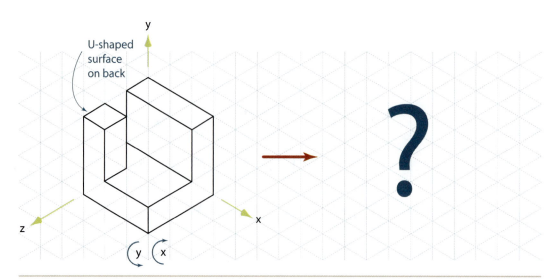

FIGURE 3.58. Original object position.

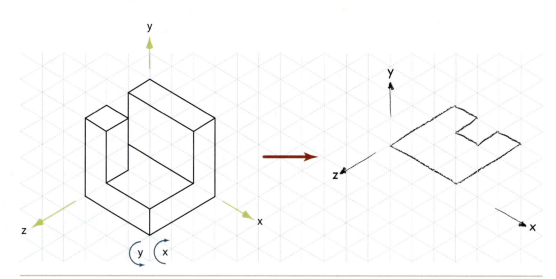

FIGURE 3.59. Rotation of U-shaped surface.

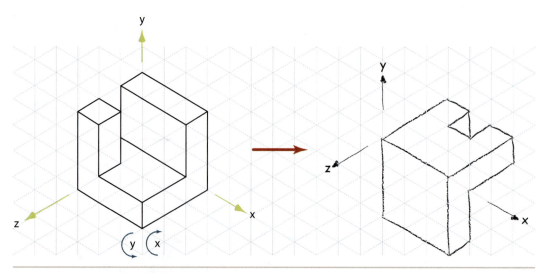

FIGURE 3.60. Completed object rotation.

3.12.04 Reflections

When creating a sketch of the image of an object as it appears on the other side of a reflection plane, you must remember that points on the object that are invisible in the original object may become visible in its reflection. The easiest way to create an object reflection is to start with one surface that is fully visible in the original view and will be fully visible in the reflection; then transfer that surface, point by point, from your view of the object to your view of the reflected image. (Usually, at least one surface on the object meets this criterion.) Since the object is the same distance from the reflection plane on either side of the plane, each point on the object and its corresponding reflected point on the image must be the same distance from the reflection plane on either side of the plane.

Figure 3.61 shows an object located 2 units from the plane of reflection. For this object, each of A, B, C, D, and E are located 2 units from the plane; another point hidden from view directly below A is the same distance from the plane of reflection. If each of these six points is reflected across the plane, they will be a total of 4 units—2 units to the plane and 2 units across the plane on the other side of it—from their location on the original object in a direction "perpendicular" to the plane. By "connecting the dots," you can sketch the surface they define, as in Figure 3.62, which shows the reflection of these points and the surface they form (the reflected surface) on the other side of the reflection plane.

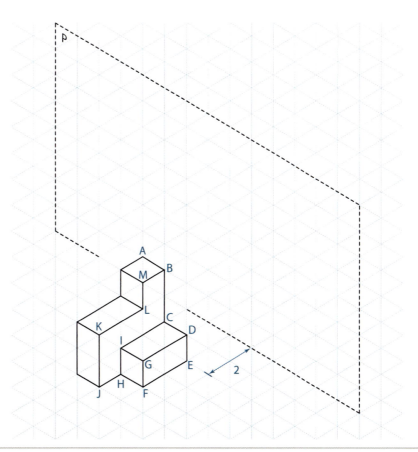

FIGURE 3.61. An object located 2 units from reflection plane.

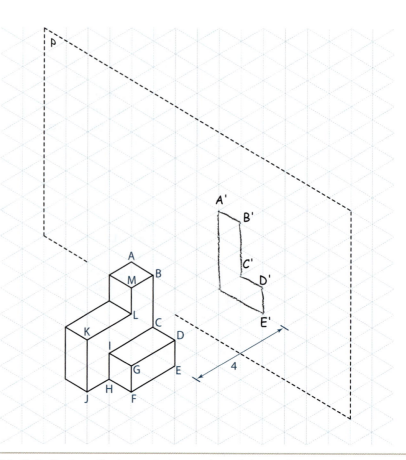

FIGURE 3.62. Reflection of points A, B, C, D, and E and the resulting surface.

For the object in this example, you can focus next on two surfaces that also will be visible in the reflection. One surface is defined by points D, E, F, and G; and the other surface is defined by points C, D, G, and I. Points C, D, and E have already been reflected, so the only points you have to consider to define the two surfaces are G, F, and I. Each point G, F, and I is located 4 units from the plane of reflection; thus, the reflection of each of these three points will be 8 units from its location on the original object in a direction perpendicular to the plane of reflection. The reflection of these points and the surfaces that result from connecting the reflected points are shown in Figure 3.63.

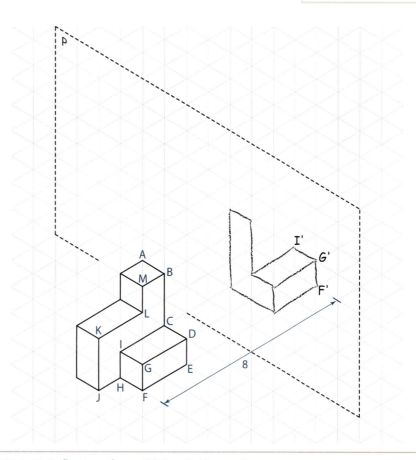

FIGURE 3.63. Reflection of two additional object surfaces.

Now consider the Z-shaped surface defined by points B, C, I, H, J, K, L, and M. You can reflect each of these points one at a time, realizing that H will be obscured from view in the reflected Z-shaped surface. Figure 3.64 illustrates the result of defining the reflected Z-shaped surface.

You now have enough of the object reflected that you can easily complete the sketch by adding the missing surfaces. The completed sketch is shown in Figure 3.65. Point labels have been excluded from the figure for clarity's sake. You could have reflected the object's other two surfaces point by point, but you can probably manage without that step.

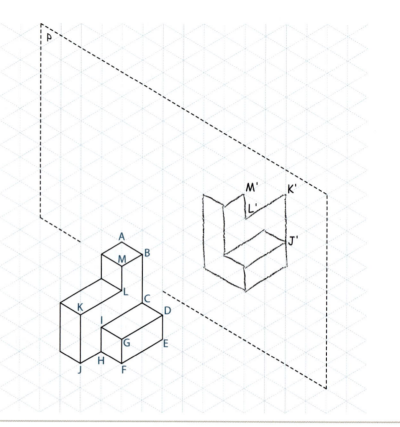

FIGURE 3.64. Reflection of Z-shaped surface.

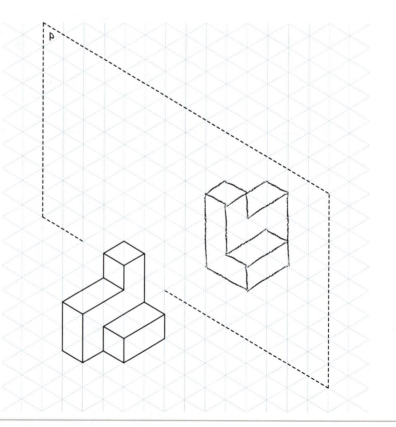

FIGURE 3.65. Completed object reflection.

3.12.05 Object Symmetry

To determine planes of symmetry for an object, you might want to sketch a dashed line on a surface of the object where you *think* a plane "splits" the object into two symmetrical parts. If you extend this line in your imagination, you can visualize an infinite plane. Look on each side of the dashed line you sketched on the object. Is one side a mirror image of the other side? If the sides are mirror images, then the dashed line you drew lies within a plane of symmetry; if the sides are not mirror images, then your dashed line does not lie within a plane of symmetry for the object.

You probably want to start with horizontal or vertical planes to identify an object's potential planes of symmetry. Figure 3.66 shows an object repeated three times with three potential planes of symmetry identified; two of the planes are vertical, and one is horizontal.

Notice that for the three planes identified, only planes 1 and 3, the vertical planes, are actually planes of symmetry. The part of the object on one side of plane 2 is not a mirror image of the part on the other side, so this plane cannot be a plane of symmetry. The object, then, has two planes of symmetry.

What about planes that are neither vertical nor horizontal? Those planes can be handled as was just shown for vertical and horizontal planes. First, sketch a line on the object that you think will lie within an imagined plane of symmetry; then examine each side of the object to see if the two halves sliced by the plane are mirror images of each other. This time the lines you sketch will be at an angle and not horizontal or vertical. Figure 3.67 shows an object with several planes of symmetry identified.

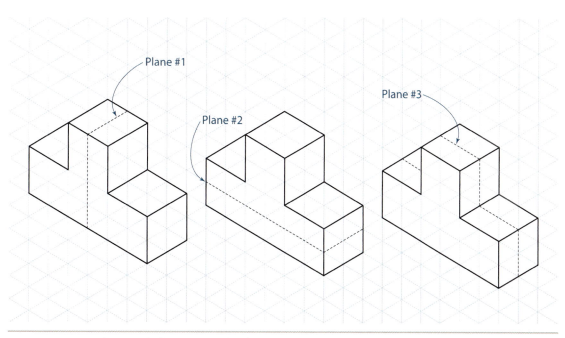

FIGURE 3.66. An object with three potential planes of symmetry identified.

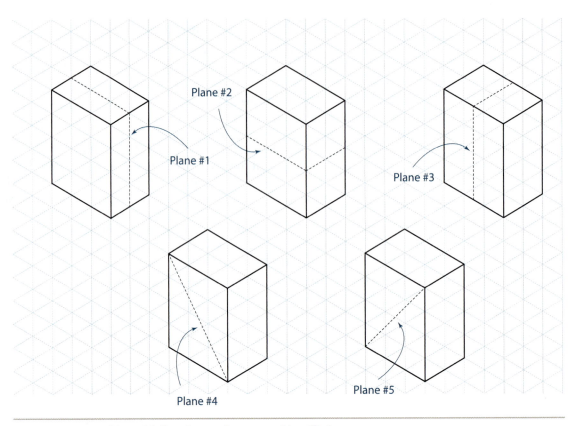

FIGURE 3.67. An object with five planes of symmetry identified.

3.12.06 Cross-Sections

When cutting planes are angled, the dimensions of the resulting cross-section "along" the plane becomes "stretched" compared with the dimensions of the cross section that results when the cutting plane is not angled. Figure 3.68 shows a simple object that has two cutting planes passing through it and the cross section corresponding to each cutting plane. For the vertical cutting plane, the shape of the cross section corresponds to the overall height and width of the object. When the cutting plane is angled, its height is stretched but its width remains the same. This point was illustrated previously in Figure 3.38. is, when the cylinder is cut along a plane perpendicular to its axis, a circular cross section results; however, when the cutting plane is angled, the diameter is stretched in one direction but remains the same in the other direction, resulting in the elliptical cross section.

To visualize the cross section that is obtained with an "angled" cutting-plane, think in terms of edges, either existing or imagined, on the object that are parallel to or perpendicular to the edges of the cutting plane. As the cutting-plane slices through the object, it intersects with the edges and surfaces of the object; the boundaries of

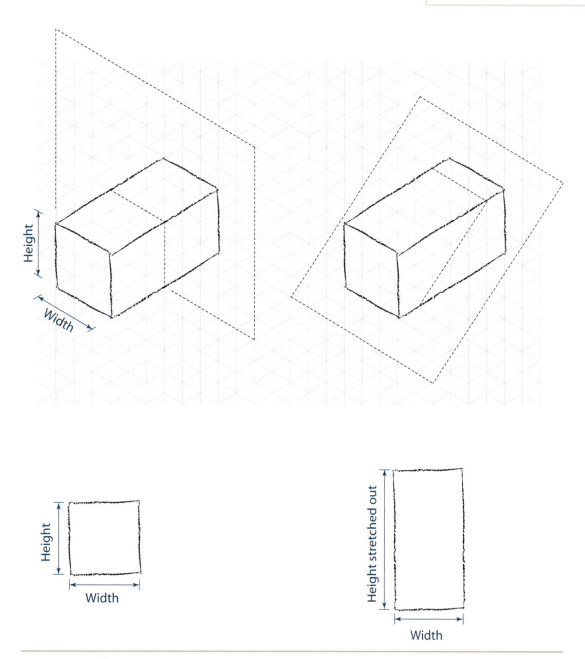

FIGURE 3.68. Effect of angling the cutting plane.

the cross-section will generally be parallel to the "edges" of the plane. (In reality, the cutting plane is infinite in size and, therefore, has no edges; but it is usually sketched as a finite size with edges so its orientation is shown.) After you have defined the edges of the cross section, you should mentally rotate the resulting planar cross section so it is perpendicular to your view direction. This allows you to "see" the cross section in its true shape and size. Obtaining the cross section for an object with an angled cutting plane is illustrated in Figure 3.69.

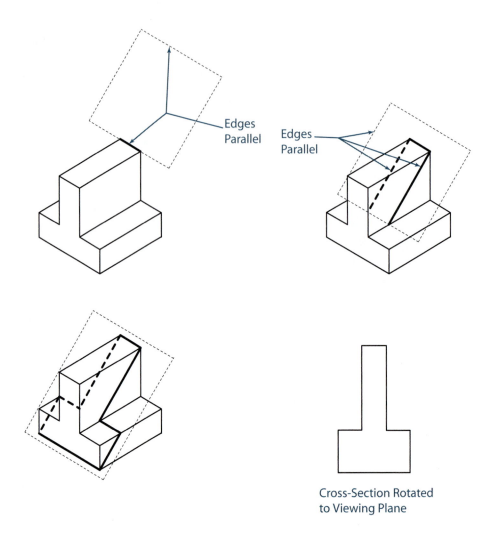

FIGURE 3.69. Cross section of an object with an angled cutting plane.

3.12.07 Combining Solids

The examples in the preceding three figures resulted from Boolean operations between simple objects combining to form a third, slightly more complicated object. The Cut, Join, and Intersect operations can be used in a series of steps to create a more complex object, which is often the case in 3-D modeling software applications. When creating a complex object using these methods, you first need to examine the final object you want to end up with; then visualize the steps needed to get to that final object. Figure 3.70 shows an object that you need to create using Boolean operations.

How would you create the object shown in Figure 3.70? One method you might employ is based on the concept of "material removal," which simply means cutting. Using this method, you need to create a block that is the overall size of the final object. Then you must create a smaller block that you will use to cut the larger block to form the basic staircase shape of the object. The first operation is illustrated in Figure 3.71.

Working with the newly formed object, now you need to create a cylinder and use it to cut the hole in the object, as shown in Figure 3.72. The final step is to create a block small enough to cut the "slot" in the top of the object, as illustrated in Figure 3.73. After completing all of these steps, you end up with the final object you set out to create.

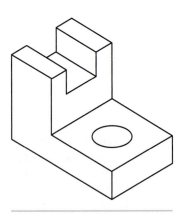

FIGURE 3.70. An object to be created through Boolean operations.

Most objects in 3-D modeling software can be created using many different methods. Can you think of a different series of steps you could use to create the object shown in Figure 3.70? Based primarily on joining operations, Figure 3.74 shows another method that uses Boolean operations to create the object.

Can you think of any other methods? As you gain familiarity with the use of 3-D modeling software, you will develop your own preferred methods for creating parts. Sometimes the method you use will depend on the object's final design characteristics. Other chapters in this text as well as texts devoted entirely to 3-D modeling software cover 3-D computer-aided modeling in greater detail and provide information about other types of operations that can be used to create solid models effectively and efficiently.

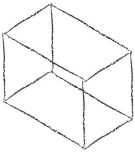

Create block

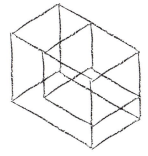

Create small block

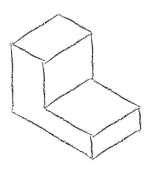

Small block cuts larger block

FIGURE 3.71. Creation of a stepped shape through cutting.

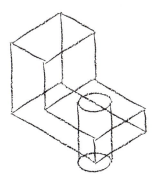

Create cylinder

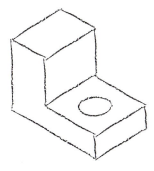

Cylinder cuts object

FIGURE 3.72. Cutting a hole in an object.

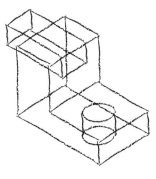

Create small block

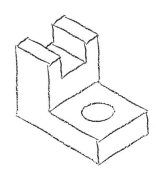

Small block cuts object -- desired result

FIGURE 3.73. Cutting an upper slot to achieve the desired result.

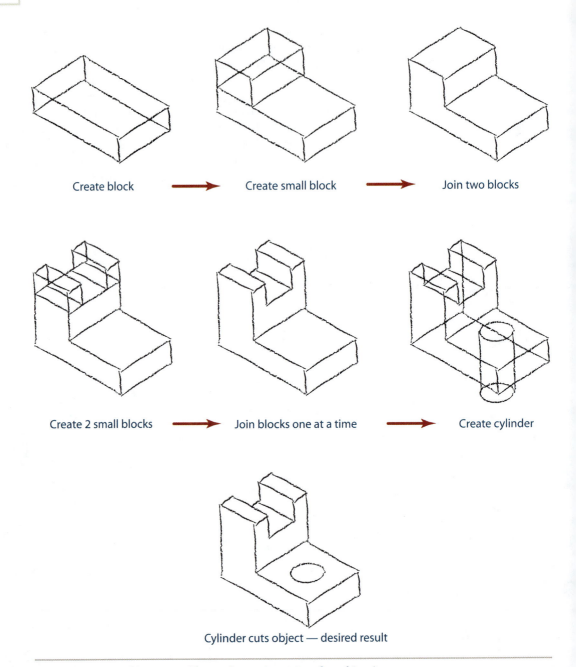

Create block → Create small block → Join two blocks

Create 2 small blocks → Join blocks one at a time → Create cylinder

Cylinder cuts object — desired result

FIGURE 3.74. An object created by an alternative series of combinations.

CAUTION When creating corner views of simple objects, remember the general rules for creating simple isometric sketches described in an earlier chapter—that lines are included only at the intersection between surfaces, that no hidden lines are shown, and that only the visible portion of partially obscured surfaces are sketched. One common error novices often make is to include extra lines on a single surface of an object, especially when there are several changes in the height of the object. Figure 3.75 shows an improper sketch from a coded plan. Can you detect the "extra" unnecessary line?

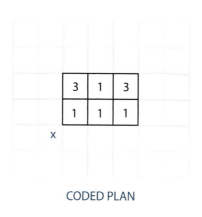

CODED PLAN

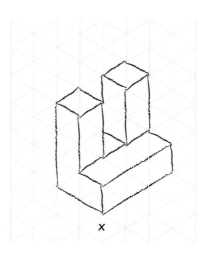

ISOMETRIC SKETCH

FIGURE 3.75. A common error: a sketch containing an extra line.

The extra line for the surface exists on the lower surface between the two "towers" on the object, as highlighted in Figure 3.76. Since a line defines the intersection of each of the towers with the lower surface, novices often extend the line through the gap even though there is no intersection between surfaces there. Figure 3.77 shows the correctly drawn object from the coded plan.

When rotating an object about an axis using the right-hand rule, right-handed people often forget that they must put down their pencil in order to rotate the object correctly in space. Right-handed people often use their left hand to define the rotation as they sketch with their right hand. If you forget to put down your pencil to define the direction of your rotation, you will end up with a rotation in the opposite direction of what you intended, as shown in Figure 3.78.

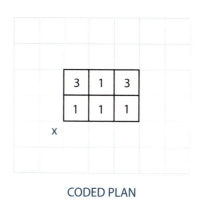

CODED PLAN

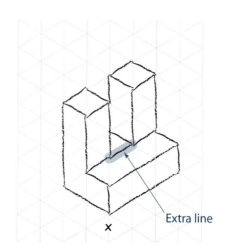

Extra line

ISOMETRIC SKETCH

FIGURE 3.76. A sketch with the extra line highlighted.

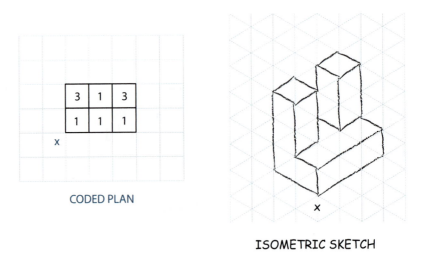

CODED PLAN

ISOMETRIC SKETCH

FIGURE 3.77. A correctly sketched object from a coded plan.

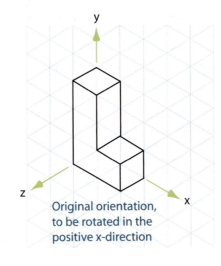

Original orientation,
to be rotated in the
positive x-direction

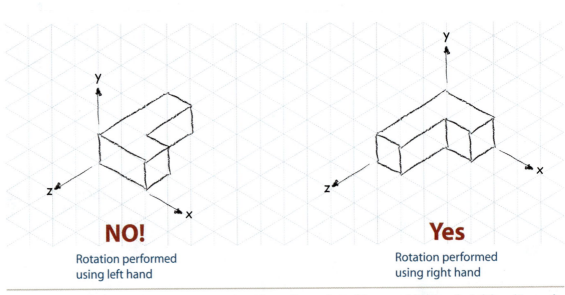

NO!

Rotation performed
using left hand

Yes

Rotation performed
using right hand

FIGURE 3.78. A common error: positive rotation (about the x-axis in this example) using the left hand instead of the right hand.

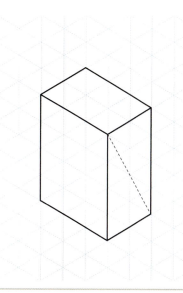

FIGURE 3.79. An object with a potential plane of symmetry identified.

When identifying angled planes of symmetry, be careful. Sometimes an angled plane produces two halves that look similar to each other but are not mirror images across the plane. For example, Figure 3.79 shows a potential plane of symmetry for an object. The two halves of the object appear to be identical halves; however, they are not mirror images across the plane. Therefore, the plane identified is not a plane of symmetry.

3.13 Chapter Summary

In this chapter, you learned about Gardner's definitions of basic human intelligences (including spatial intelligence) and the way spatial intelligence is developed and assessed. Spatial intelligence is important for engineering success, especially in engineering graphics and solid modeling courses. The chapter outlined several exercises that help develop spatial skills, including:

- Constructing isometric sketches from different corner views.
- Rotating 3-D objects about one or more axes.
- Reflecting objects across a plane and recognizing planes of symmetry.
- Defining cross sections obtained between cutting planes and objects.
- Combining two objects to form a third object by cutting, joining, or intersecting.

3.14 glossary of key terms

Boolean operations: In early versions of 3-D CAD software, commands used to combine solids.

combining solids: The process of cutting, joining, or intersecting two objects to form a third object.

corner views: An isometric view of an object created from the perspective at a given corner of the object.

cross-section: The intersection between a cutting plane and a 3-D object.

cut: To remove the volume of interference between two objects from one of the objects.

cutting-plane: An imaginary plane that intersects with an object to form a cross section.

3.14 glossary of key terms (continued)

intersect: To create a new object that consists of the volume of interference between two objects.

join: To absorb the volume of interference between two objects to form a third object.

mental rotations: The ability to mentally turn an object in space.

reflection: The process of obtaining a mirror image of an object from a plane of reflection.

spatial orientation: The ability of a person to mentally determine his own location and orientation within a given environment.

spatial perception: The ability to identify horizontal and vertical directions.

spatial relations: The ability to visualize the relationship between two objects in space, i.e., overlapping or nonoverlapping.

spatial visualization: The ability to mentally transform (rotate, translate, or mirror) or to mentally alter (twist, fold, or invert) 2-D figures and/or 3-D objects.

symmetry: The characteristic of an object in which one half of the object is a mirror image of the other half.

volume of interference: The volume that is common between two overlapping objects.

3.15 questions for review

1. What are some of the basic human intelligences as defined by Gardner?

2. What are the stages of development for spatial intelligence?

3. What are some of the basic spatial skill types?

4. What do the numbers on a coded plan represent?

5. What are some general rules to follow when creating isometric sketches from coded plans?

6. When a person is looking down a coordinate axis, are positive rotations CW or CCW?

7. Describe the right-hand rule in your own words.

8. Are object rotations about two or more axes commutative? Why or why not?

9. What is one difference between object reflection and object symmetry?

10. Are all objects symmetrical about at least one plane? Explain.

11. The shape of a cross section depends on two things. Name them.

12. What is the effect on the resulting cross section of a cutting plane that is tilted?

13. What are the three basic ways to combine solids?

14. In the cutting of two objects, does it matter which object is doing the cutting?

3.16 problems

1. For the following objects, sketch a coded plan, labeling the corner marked with an *x* properly.
2. Indicate the coded plan corner view that corresponds to the isometric sketch provided.

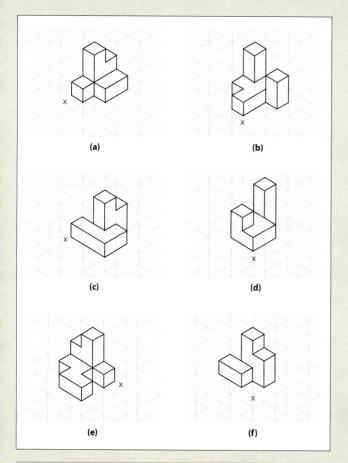

(a)　(b)

(c)　(d)

(e)　(f)

FIGURE P3.1.

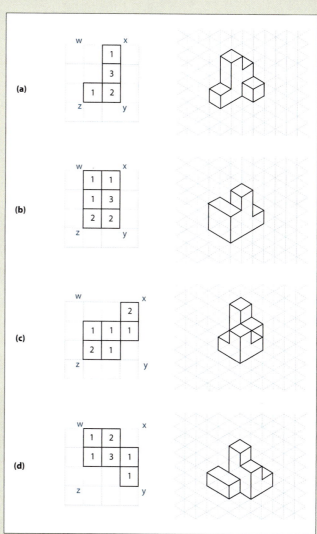

FIGURE P3.2.

3.16 problems (continued)

3. Use isometric grid paper to sketch the indicated corner view (marked with an *x*) for the coded plan.

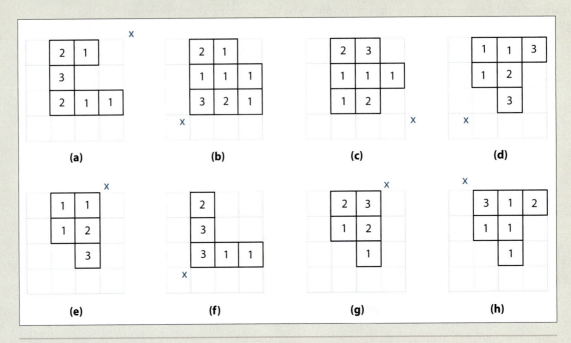

FIGURE P3.3.

3.16 problems (continued)

4. Using the notation developed in this chapter, indicate the rotation the following objects have experienced.

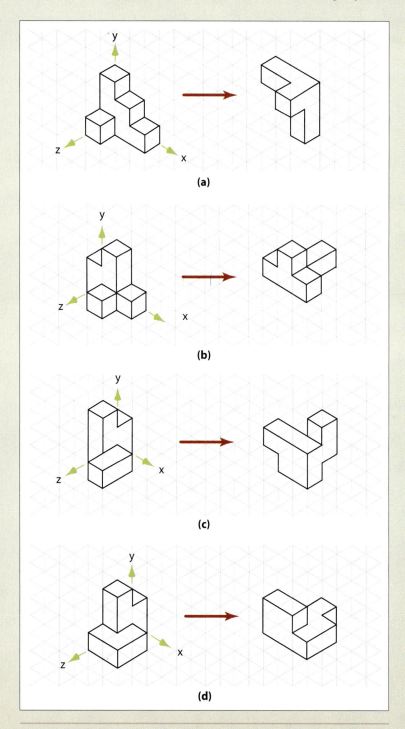

FIGURE P3.4.

3.16 problems (continued)

5. Rotate the following objects by the indicated amount and sketch the results on isometric grid paper.

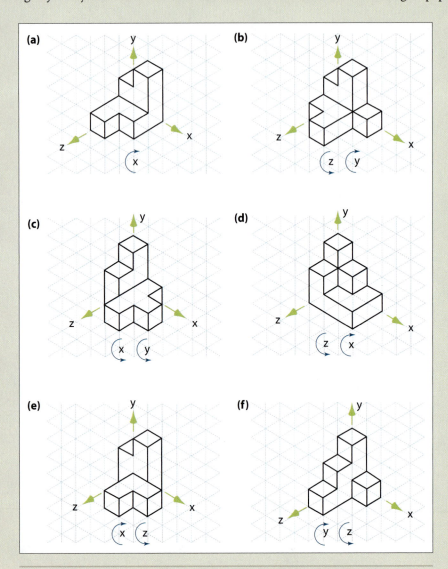

FIGURE P3.5.

3.16 problems (continued)

6. Copy the following object on isometric grid paper and sketch its reflection across the indicated plane. Note that the sketch of the reflection has been started for you.

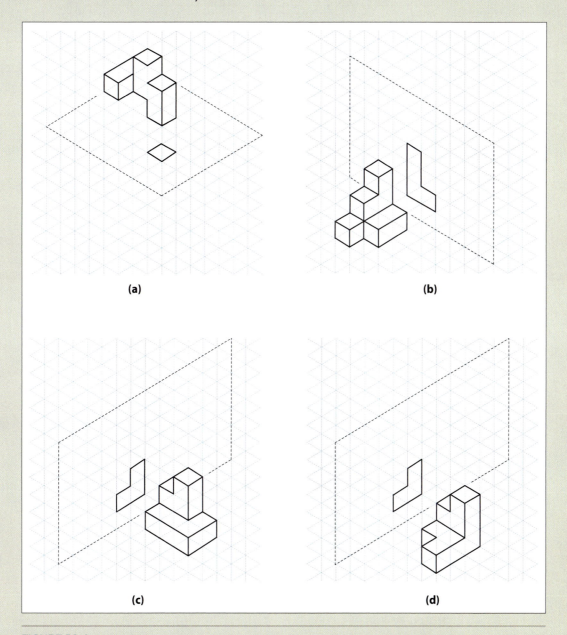

(a)

(b)

(c)

(d)

FIGURE P3.6.

3.16 problems (continued)

7. How many planes of symmetry do each of the following objects have?

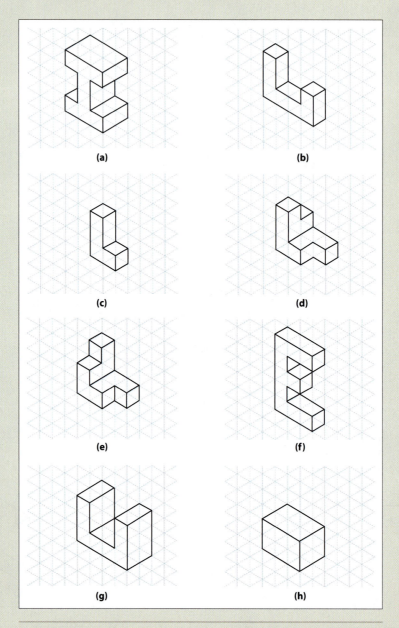

FIGURE P3.7.

3.16 problems (continued)

8. Sketch the cross section obtained between the intersection of the object and the cutting plane.

9. Sketch the result of combining the following objects by the indicated method.

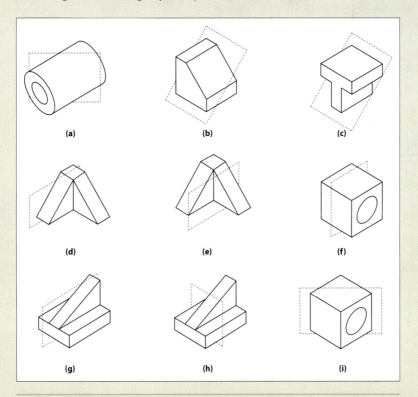

(a)   **(b)**   **(c)**

(d)   **(e)**   **(f)**

(g)   **(h)**   **(i)**

FIGURE P3.8.

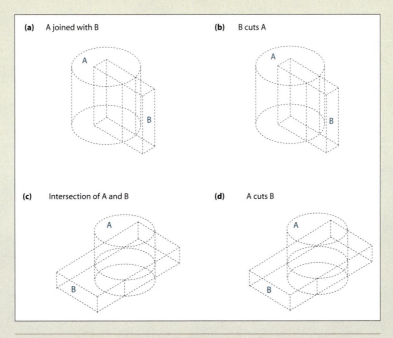

(a)   A joined with B

(b)   B cuts A

(c)   Intersection of A and B

(d)   A cuts B

FIGURE P3.9.

3.16 problems (continued)

10. Describe by words and sketches how you would create the following objects by combining basic 3-D shapes.

11. Create isometric sketches from these coded plans using the corner view that is circled or the corner prescribed by your instructor.

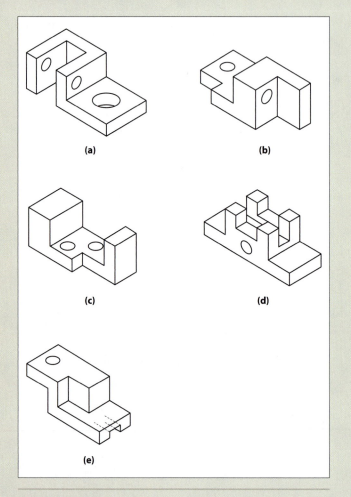

FIGURE P3.10.

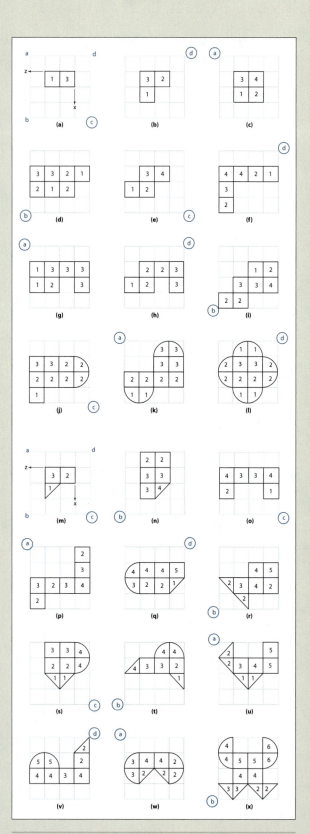

FIGURE P3.11.

3.16 problems (continued)

12. Add the reflected images or redraw these objects with symmetry using the xy, yz, or xz planes as indicated or the planes prescribed by your instructor.

13. The object shown in (a) is show again in (b) rotated by −90 degrees about the y-axis to reveal more detail. Rotate the object sequentially about the axes indicated or about the axes prescribed by your instructor.

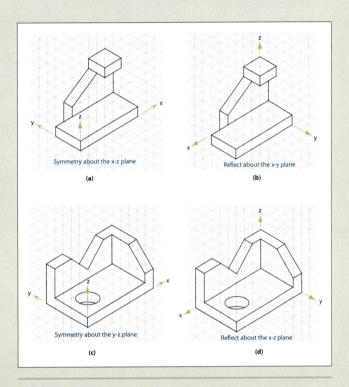

FIGURE P3.12.

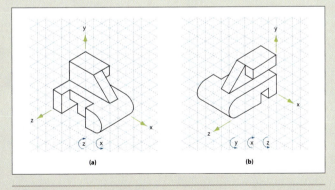

FIGURE P3.13.

3.16 problems (continued)

14. Create an isometric sketch of the objects created from coded plans A and B. Rotate object A sequentially about the axes indicated or about the axes prescribed by your instructor. Show the new object created by the indicated Boolean combination of object A and object B or the Boolean operation prescribed by your instructor when the coordinate axes of A and B are aligned.

15. Triangular volume A, triangular volume B, and rectangular volume C are shown intersecting in space. On the dashed outline drawings, darken and add edges to show all visible edges of the final volume created by the indicated Boolean operations.

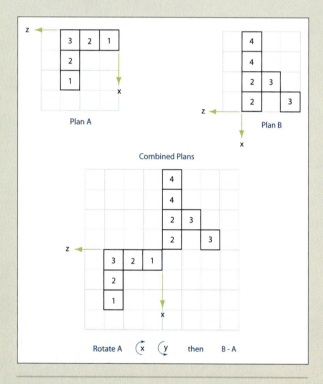

Plan A

Plan B

Combined Plans

Rotate A   ⟲x   ⟳y   then   B - A

FIGURE P3.14.

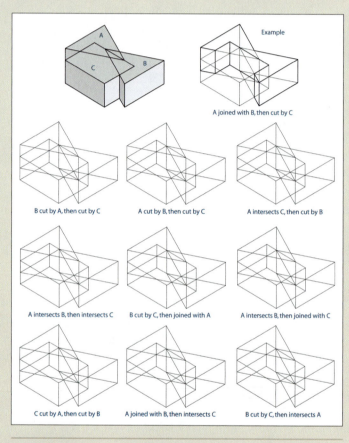

Example

A joined with B, then cut by C

B cut by A, then cut by C

A cut by B, then cut by C

A intersects C, then cut by B

A intersects B, then intersects C

B cut by C, then joined with A

A intersects B, then joined with C

C cut by A, then cut by B

A joined with B, then intersects C

B cut by C, then intersects A

FIGURE P3.15.

4

Working in a Team Environment

objectives

After completing this chapter, you should be able to

- Understand the benefits of working in a team
- Organize team projects and team member responsibilities
- Communicate and work in a team
- Assess the strengths and weaknesses of your team
- Apply strategies for improving team performance
- Solve problems with team members
- Work in a team effectively

4.01
introduction

Assume your instructor has assigned a "design" project for the semester. Moreover, your instructor has assigned everyone in the class to work in teams. You probably have worked on team projects in the past. Perhaps your experiences were positive; however, you may have been frustrated at times because not everyone on the team put forth the same amount of effort in completing the project. You may be a bit skeptical about the viability and value of working on a team.

In real life, most engineering projects are accomplished by teams. Think of the space missions. The astronauts would not be able to travel into space without the effort of thousands of engineers, technicians, ground crew, and other support staff. As another example, think of the last movie you watched. The movie "team" included the director, producer, cast, camera operator, costumer, sound effects crew, and many other people who pulled together to develop the successful final product. Automobiles, computers, iPods, cell phones, and countless other everyday products were designed, produced, packaged, marketed, and distributed through a team effort. So if you really want to have a successful career as an engineer or a technologist, you have no choice but to learn to work in a team environment. In fact, you may even find that you enjoy the team atmosphere as you begin to appreciate the way individuals pull together and complement one another to create products that are innovative and timely.

While working on team projects in high school or college classes, you may have experienced less than satisfactory results. In those often dysfunctional teams, task assignments may not have been distributed evenly or team members may not have delivered on promises to complete their tasks. While being on a dysfunctional team is an experience you would like to avoid, being a part of a good team is a rewarding experience. This chapter will outline reasons for working on a team, ways to get started on the road to a successful team, strategies for making your team effective and efficient, and tools for dealing with problems that will inevitably confront you and your team. As with most activities, being prepared for the challenges and opportunities, and having a plan for how to deal with them will improve your chances of success and lead to an enjoyable and productive team experience. Further, learning how to develop successful team environments will prepare you for a future technological career. As you read this chapter, pay particular attention to the sections on organization, time management, and communication, because they are the keys to a successful project—as an individual or as a team.

4.02   Why Work in a Group?

In a later chapter, you will learn about creativity and the design process. Many of the projects suggested in that chapter are difficult, if not impossible, to do by yourself. Amost all of them can be enhanced by the interpersonal interaction that comes from being part of a good team and the environment that comes with being a member of a team. Obviously, more team members means more bodies at work (and the likelihood of a better final product); but a team is more than a group of people who divide up a task into manageable parts. The diversity of different life and professional experiences of team members leads to a larger group of ideas and a variety of approaches in solving problems. Team discussions can generate ideas, expand options, and improve the final product. Even questions from naysayers are helpful—both in clarifying ideas and identifying fatal flaws—before a great deal of time and effort is spent on an idea that ultimately leads to a dead end.

In addition, employers value employees (whether for summer jobs, internships, co-op positions, or full-time jobs after graduation) who can work as a member of a team and who are team players. An engineer rarely works alone. Most projects require a range of training and skills beyond what the most skilled engineer can be expected to do. Even if a project is small enough for a single person to design it from start to finish, engineers rely on information collected or generated by others and need others to fabricate or construct what they design; often engineers rely on others to make sure the project has been built correctly and to identify problems.

Working on a team does not involve personal relationships; a team is based on professional relationships that require you to respect and value the skill sets that other members of the team bring to the project. A team is a group of colleagues who work together to complete an assigned task.

4.03 What Does It Mean to Be a Team Player?

Being a good team member takes work. Most people are used to working on their own—making decisions, prioritizing tasks, and being accountable for their own work. Working with others requires a different approach than working alone. To be a successful part of a team, you need to consider several issues. You should be prepared *not* to be in charge of everything. For some people, this requires a great deal of effort; for other people, it is less taxing. At times, you will be the supervisor; other times you will be supervised. You need to be flexible and understand that a team consisting only of leaders (or only of followers) is not likely to perform well.

Also be prepared to have some interesting (and some frustrating) encounters with your new work mates. Be prepared to exchange points of view and to learn from those around you. Everyone on a team is responsible for success and is accountable for failure.

Most importantly, prepare to learn how to be a team member. Share your strengths with the team and be willing to contribute. Remember, the combined efforts of all team members should yield a better outcome than the efforts of one individual. Learn new team skills and be adaptable.

Many teams have problems when everyone tries to be in charge or when no one tries to be in charge. The result can be the same: uneven distribution of work, incomplete work, missed deadlines, subpar performance, and frustration. Even though a team is a united effort, each individual is accountable for the overall performance of the team.

Individuals generally react differently in groups than they do on their own. If you miss deadlines or produce inferior work as an individual, you can expect to be held accountable if your work habits are the same when you are part of a team. Conversely, if you produce high-quality work on your own and do the same as part of a team, you will be rewarded accordingly. Remember that team members are accountable first for their individual performance and second for the group's performance. Keep everyone informed of your progress.

4.04 Differences between Teaming in the Classroom and Teaming in the Real World

Selecting personnel and identifying skills are the most important tasks of assembling a team to work on a real-world project. While it may be advantageous to pick people who have worked together previously and who have established a good working relationship, you need to make sure that all of the skills required for the completion of the project are represented by at least one person on the team. For example, if the team is designing a building, the team must have a member who understands, among other things, the:

- Design of a foundation.
- Design of the structure.
- Design of elevator and/or escalator systems.
- Design of air-conditioning systems.

Additional skills are likely to be on the list if the building is to be made of reinforced concrete, or if it is to be constructed in Alaska or California or Louisiana.

In the real world and in the classroom, the goal is to complete a successful project on time and within budget. However, the skills and training of potential team members in the classroom are all virtually the same (unlike the real world). Furthermore, the *primary* goal in the classroom is for each member of the team to learn about each task required in the project. Whereas a mechanical engineer is not expected to teach other members of a building-design team how the air-conditioning system works or why the particular components were selected, each member of a classroom team is expected to explain her part of the team project. Unless the team members complete all of the tasks together, each member must teach the rest of the team what she did on her part of the project.

4.05   Team Roles

For your team to operate like a "well-oiled machine," you need to understand that the members must fill specific **team roles**, if effective collaborative work is to result. Typically, well-functioning teams have a leader, a timekeeper, and a note taker at a minimum. If there are additional team members, assigning someone to the role of devil's advocate is also a good idea. The **team leader** does just that—she leads. This does not mean the team leader dictates or makes all of the decisions for the group. The team leader sets the meeting time, sets the agenda for the meeting, and generally keeps the meeting moving. The team leader also makes sure the team stays on target and remains focused on the task at hand.

The **note taker** keeps a written record of the team's progress. He or she records what tasks have been assigned to whom and records the expected completion dates of the tasks. The note taker is responsible for sending the minutes of the meeting to all team members. The minutes are a written record of what transpired during the meeting and serve as a reminder of who is responsible for completing what task(s).

The **timekeeper** makes sure the schedule is maintained and that meetings do not run over the allotted time. If meetings routinely last longer than planned, team members may skip them or resent coming to them—either of which leads to less productive team encounters.

Finally, the role of the **devil's advocate** is to challenge ideas without being too overbearing or unpleasant. The devil's advocate makes sure that all options are considered and that ideas are sound. However, a devil's advocate should not challenge ideas just for the sake of the challenge; doing this can annoy teammates and detract from the overall effectiveness of the team's operation.

Depending on your personality, you might be naturally inclined toward one role over another. For example, you may naturally be a critic who performs the role of devil's advocate very well. In the classroom setting, you should try out other team roles, so you can develop additional team skills. You may need to hone your note taking skills, and filling that role on the team may help your personal development. In classroom projects, team members can rotate roles so everyone has a chance to experience each role. By performing roles that are unfamiliar to you, you learn to appreciate people who work in these roles. Developing an appreciation for and respecting the skills of the other members of your team are the first steps toward your becoming an effective team member.

4.06 Characteristics of an Effective Team

Most successful teams either knowingly or unknowingly operate by certain ground rules that contribute to overall team productivity. Some of these ground rules are described in subsequent sections of this chapter, giving you the opportunity to learn the rules and adapt them to your particular setting and project.

4.06.01 Decisions Made by Consensus

For a diverse team, it will be nearly impossible to get 100 percent agreement on all of the decisions. Trying to achieve this unreachable goal will lead to frustration and poor productivity for the team. **Consensus** means finding an option that all team members will support. It does not mean that all team members would select that option as their first choice, although some of them probably would. When making decisions by consensus, everyone on the team is invited to voice an opinion. Some people may be naturally shy and unwilling to speak up. The team leader should note when a team member has not voiced an opinion and invite the individual to speak up. Silence should not be interpreted as agreement—often it is not. Another important aspect in making team decisions is to consider the data carefully. Decisions made based on feelings, where data is ignored, are usually not optimal.

4.06.02 Everyone Participates

No member of a team should be allowed to sit back and watch others do the work. As mentioned previously, it is important for every member of the team to voice an opinion during meetings. It is just as important for every member of the team to participate in the work of the team. Tasks should be assigned to members based on talents/skills, and no one should be allowed to choose not to do something. The leader is responsible for making sure that every member participates equally in the work of the team. This does not mean that every task needs to be divided into equal parts, but it does mean the *overall* work should be divided evenly.

4.06.03 Professional Meetings

Team meetings should be productive and engaging. If they are ineffective and a waste of time, team members will likely skip them or not participate fully. This, in turn, will lead to poor-quality work from the team. Team meetings work best when a procedure has been established for the conduct of the meeting and an agenda has been created in advance. The agenda should be prepared by the team leader and e-mailed to participants in advance of team meetings. As a rule of thumb, the **agenda** should include (1) a review of progress to date, (2) a review and possible revision of the project schedule, and (3) new task assignments for team members as needed. This list is not exhaustive—your agenda will be dictated somewhat by the project you are working on. The timekeeper is responsible for making sure the team follows the agenda so that all items on the agenda are completed within the time allotted for the meeting. Punctuality at team meetings is a necessary ingredient for success. A person who shows up late for a meeting is not being fair to or respectful of her teammates. Attendees should do their best to be on time.

4.07 Project Organization—Defining Tasks and Deliverables

Your team was likely formed to work on a project, maybe even a design project. In a subsequent chapter, you will learn about the design process and its various stages. For now, understand that design is an iterative process that proceeds from stage to stage until completion. At some point in the process, you may have to return to an earlier stage to redo something the team thought was completed. Redoing earlier work is a normal part of the design process, especially in cases where you are trying something new and do not know if your idea or solution will actually work. During each stage, you meet as a team to review your progress and to determine what needs to be done next. In the early stages of the project, you probably made a list of all of the tasks that needed to be done. You should review this list at each meeting because the list will likely change as the project evolves.

Once you are sure you have a complete list of what needs to be done before the next meeting, it is time to make a list of tasks and assign responsibility to the person who will be completing each task. When reviewing the items on the list of tasks to accomplish, you need to determine which tasks depend on the outcome of other tasks. For example, if you are going to machine a part, you need to create a drawing of the part. To create the drawing, you may need to create a solid model of it in your CAD system. Thus, it would be unreasonable to expect someone to do the machining *before* the modeling work has been done—the task of machining depends on the outcome of the modeling task.

Another consideration when assigning tasks is to determine which tasks can be done by an individual and which require a group effort. If someone on your team has a difficult schedule to work around, you may want to assign individual tasks to that person most of the time (but not always) to accommodate their schedule. In addition to looking at the individual/group effort required of each task, you also should try to estimate the amount of time each task will require for completion. If one group member ends up completing a task that requires ten hours while two other members complete tasks that require one hour each, resentment is likely to build, thereby hindering group progress. However, as stated earlier in this chapter, division of labor for the project should be balanced overall. So if the person assigned the ten-hour task has been a slacker on previous assignments, perhaps that person should complete a significant task the next time one is assigned.

When assigning tasks, you should try to match the talents and abilities of each team member to the requirements of the task. However, you want to rotate duties so one person is not burdened with all of the writing or all of the modeling or all of the calculating (similar to the way the team roles are rotated so everyone has the opportunity to experience each role). As you are thinking about the assignment of tasks, ask yourself the following questions:

- Which team member is *best qualified* to do the task?
- Who is *able* to do the task in terms of either time or skill?
- Who is *willing* to do the task?

You need to make choices between assigning a task to a person who can accomplish it and assigning the task to the best person. The best outcome may not result from the best person being assigned to a task. If the person is overloaded as a result of the task assignment, she may not do a very good job or may not be able to complete the task. Balancing task assignments is key to producing the best possible project. A project may not be the best one a team can produce; rather, it is the best project a team can produce within the limits of available resources.

4.08 Time Management—Project Scheduling

Once you have organized your team and started work on your project, you should begin developing a plan for completing the project. Think of the plan as being dynamic, not static, since you are likely to be making changes to the plan as the project unfolds. Think of the initial plan as a flowchart of activities or a calendar of events that should include items such as who was assigned to work on each task and where each task fits in the overall project. When examining the various tasks that make up the final project, think about the interrelationships between tasks. What task precedes/follows each task? How does information flow from one task to another? In the previous example, modeling was the first task to be accomplished, which was followed by the creation of a drawing, culminating in a fabricated part. The task that precedes and follows each task is well-defined, but the method of communicating information between tasks may not be as straightforward. Ideally, this information flows seamlessly through the CAD software; however, you may need to run file translator routines to move information between tasks.

Perhaps the most important activity in project scheduling is to determine how each task fits within the overall project and when each task should be completed for your project to end successfully. Usually when you are organizing your project, you can begin at the beginning or you can begin at the end. If you start at the beginning, the organization of the project can be done in a cyclic manner. As a team, ask the following questions:

1. What needs to be done first?
2. What do we need to know before we can do WhatNeedsToBeDoneFirst (WNTBDF)?
3. Now that we have a new WNTBDF, repeat step 2.

Sometimes it is more efficient to begin by considering the *final* deliverable for the project (product, design, prototype, sketch, etc.). You then work backward through the process, identifying what needs to happen before a specific task can be accomplished. Another way of thinking about this is to consider all of the other tasks that must be completed before a specific task can be accomplished. When organizing your project from the beginning or the end, you also need to consider who will be completing each task—you cannot establish a timeline without considering the realities of everyone's schedule.

Sometimes the result of this activity is to discover that the timeline for your well-planned project does not match the deadlines established by your client (or instructor).

In this case, you must revisit the task list and eliminate tasks or compress the time to complete each task. In others words, you must determine how good of a project you can deliver in the time allowed. Even in the real world, you do not always have enough time to produce the best design (or the client does not have enough money to build the best product). The goal is to produce the best product within the constraints given. These constraints are usually time, money, materials, and talent.

By now, you may have realized that the schedule for your project and the organization of tasks in your project (presented in the previous section) are linked. The following sections include information on two tools you may find useful as you organize and plan your project: the Gantt chart and the Critical Path Method.

4.08.01 Gantt Charts

When working in teams, it is essential to establish a well-thought-out plan for completing the project. If you are working on a project as an individual, the planning stage is not nearly as critical, since no one else is depending on you to complete a task to an acceptable level of quality within a certain time frame. One useful tool to help you organize your project, assigning a timeline for the completion of various tasks, is a Gantt chart. A **Gantt chart** is a table that lists the tasks in the leftmost column that must be completed, and it identifies the dates across the top row by which each task must be completed. Shading indicates the times for working on each of the project tasks. Without a detailed plan that lays out due dates and establishes a timeline, most projects get bogged down in trivial details and important tasks are delayed. This will often put the success of the project in jeopardy of being completed on time. Figure 4.01 shows a simple Gantt chart for a student project in reverse engineering, a topic that will be discussed in more detail in a subsequent chapter.

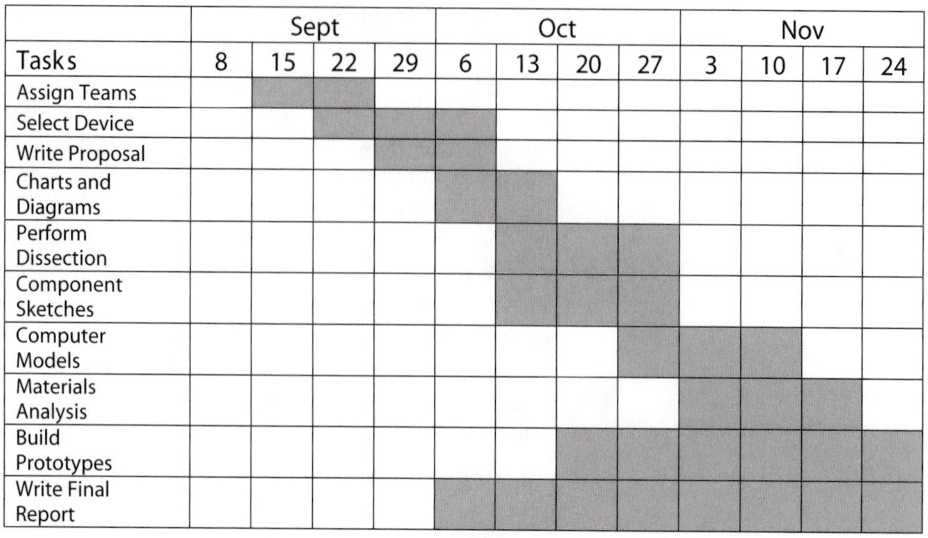

Tasks	Sept				Oct				Nov			
	8	15	22	29	6	13	20	27	3	10	17	24
Assign Teams		▓	▓									
Select Device				▓	▓							
Write Proposal					▓							
Charts and Diagrams					▓	▓						
Perform Dissection						▓	▓	▓				
Component Sketches							▓	▓				
Computer Models								▓	▓	▓		
Materials Analysis									▓	▓	▓	
Build Prototypes							▓	▓	▓	▓	▓	▓
Write Final Report					▓	▓	▓	▓	▓	▓	▓	▓

FIGURE 4.01. Gantt chart for reverse engineering project.

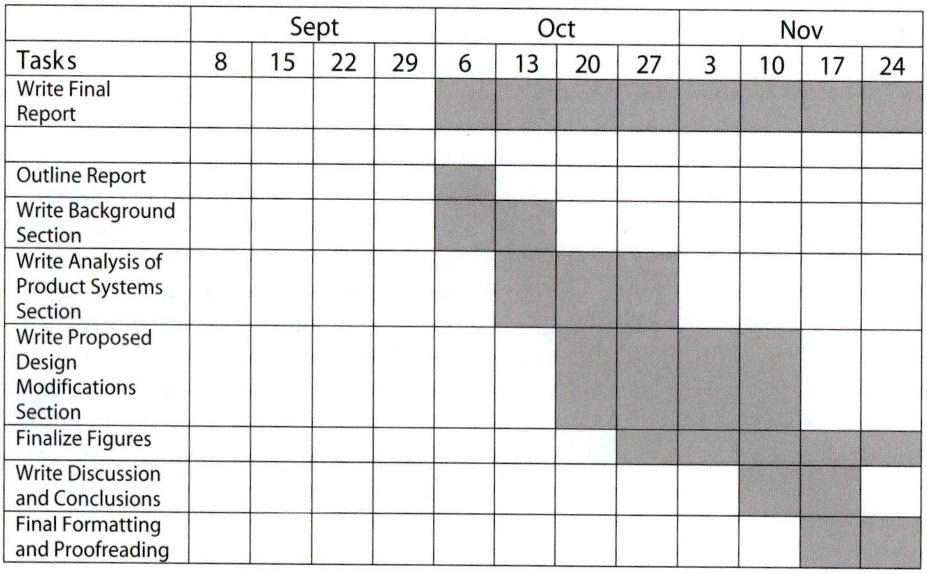

Tasks	Sept				Oct				Nov			
	8	15	22	29	6	13	20	27	3	10	17	24
Write Final Report					▓	▓	▓	▓	▓	▓	▓	▓
Outline Report					▓							
Write Background Section					▓	▓						
Write Analysis of Product Systems Section						▓	▓					
Write Proposed Design Modifications Section							▓	▓	▓			
Finalize Figures									▓	▓	▓	
Write Discussion and Conclusions										▓	▓	
Final Formatting and Proofreading											▓	▓

FIGURE 4.02. Gantt chart for report writing task.

Note that each of the major task headings can be broken down further and a Gantt chart created for each major task. For example, Figure 4.02 shows a new Gantt chart created just for the last task listed in the previous Gantt chart.

4.08.02 Critical Path Method

The **critical path method (CPM)** is used in project scheduling to determine the least amount of time needed to complete a given project. CPM is also used to determine which activities are most critical to the on-time completion of the project (hence, the name) and which activities are not as critical to the overall project schedule. The **critical path** includes the sequence of the activities that have the longest duration. When these activities are strung together on the critical path, you can determine the shortest possible duration of the project. Activities that are not on the critical path can be allowed to "float" with regard to schedule, which will not impact completion of the overall project.

The CPM is like a flowchart for the project. It helps everyone visualize what has been accomplished, in what stage of the project the team is (what percentage is complete, whether the project is ahead or behind schedule, etc.), and what needs to be done next at each step of the project. If names are associated with each task, the CPM can also serve as a reminder of who is waiting for a finished task before that person can begin the next task. The information required to construct a CPM diagram is:

- A list of all activities required to complete the project.
- The amount of time each activity will take to complete.
- The dependencies between tasks (i.e., what task relies on the completion of another task before it can be started).

As an example of a CPM, consider a project broken down into six major activities. The task breakdown is characterized as follows:

Activity	Duration	Depends on
1	2 days	—
2	4 days	1
3	6 days	2
4	5 days	1
5	2 days	2, 4
6	8 days	3, 5

The critical path diagram for this project can be constructed as shown in Figure 4.03. Note that arrows are used to show forward progress through the project, and dependencies between tasks are shown graphically. For this example, there are three possible "paths" through the project from start to finish, as shown in Figure 4.04. The first path includes Activities 1, 2, 3, and 6. This path has a total duration of 20 days. The second path includes Activities 1, 2, 5, and 6 with a duration of 16 days. The final path includes Activities 1, 4, 5, and 6 with a duration of 17 days. The critical path is the first one (with Activities 1, 2, 3, and 6) because this path has the longest duration. Based on the critical path, the *shortest* possible completion schedule for the project is 20 days; anything less than this is impossible.

Each activity on the critical path is now a critical activity; that is, if any of the activities are not completed on time, the overall time needed to complete the project will be lengthened. This means that Activity 4 and Activity 5 (the only activities not on the critical path) have some float time—if they are not completed on time, the overall project will still be on schedule. If you are a project manager, this information shows you where to concentrate your efforts. If it looks as though Activity 4 is beginning to interfere with Activity 3, you can suspend work on Activity 4 for a short while to make sure Activity 3 proceeds unhindered. Or if Activity 2 starts to flounder, you can shift resources away from Activity 4 to make sure the critical activity (Activity 2) is completed on schedule.

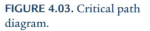

FIGURE 4.03. Critical path diagram.

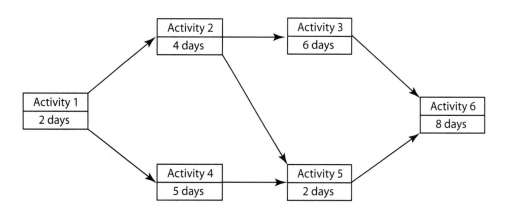

FIGURE 4.04. Three possible paths through project completion.

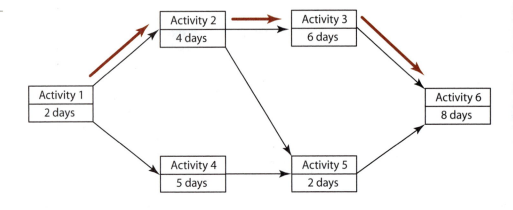

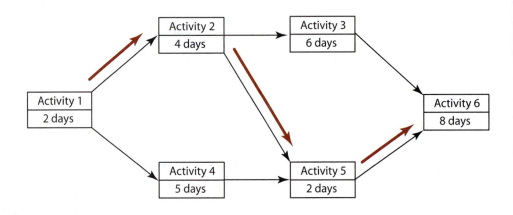

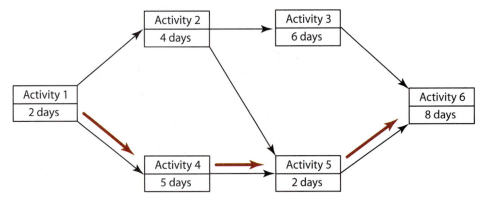

Because completion of any project is dynamic rather than static, as with Gantt charts, you should review your critical path diagram periodically to ensure that it still accurately reflects the realities of your project.

4.09 Communication

Following is a discussion about communication between the team and the outside world and communication among members of the team. Each member of the team must communicate openly and honestly during task assignments (Do I agree with the plan? Do I have other commitments that will interfere with the timeline? Can I commit enough time to the task? Do I have the skills required to complete the task alone?). A team member's silence is usually interpreted as agreement with the plan, or at least acceptance of the plan. Most problems between team members result from a lack of open and honest communication.

4.09.01 Agreeing How to Communicate

E-mailing, chatting, and text messaging are useful modes of communication, and you have probably used all of them; however, regular face-to-face meetings are essential for a team's successful communication. Electronic modes of communication can be used, but nothing takes the place of face-to-face meetings. Because no one has time for an unnecessary or poorly conducted meeting, when team members do meet, they want the meetings to be productive. Decide early in the project how often the team needs to convene to conduct its business. Team members have other obligations, and the team project will get only a portion of their time. Having a regularly scheduled meeting time enables all team members to plan their activities around project time.

Not all information is shared in meetings; and the options of e-mail, notes, memos, and voice mail are appropriate under the right circumstances. All team members need to agree on how to communicate.

In your initial team meeting, find out how the team wants to handle communication. How much time do members have for meetings? Does everyone have e-mail? How often do they check it? Is voice mail reliable? The value of written records of team meetings and decisions cannot be overemphasized. By investing whatever time is needed to agree on how to communicate, team members will save time during the course of the project.

Documentation is essential after meetings. The note taker's responsibility is to summarize the conduct of the meeting and communicate his or her notes to the rest of the group. Sometimes the team keeps a bound notebook of team-meeting notes, with the notebook passed from one note taker to the next (when team roles are being rotated). Usually the note taker will convert meeting notes to an electronic document and e-mail them to all team members, summarizing the following:

- Did we say what needed to be said?
- Did we begin and end on time?
- What was decided?
- What tasks were assigned?
- Who is responsible for each task?
- What questions still need to be answered?
- What must be done before the next meeting?

After the note taker e-mails the document out for review, you should look over the notes carefully, making comments if any points conflict with how you remember the meeting. If you do not take the time to review the notes or if you do not bother to raise objections with the notes as written, the notes will become the permanent record of what transpired at the meeting. Once again, silence is interpreted as agreement. You cannot come back later to say that you did not know you were responsible for completing a task when that assignment was included in the meeting notes that you agreed (either actively or passively) were complete and accurate.

4.09.02 Communicating Outside Meetings

As the team's level of trust increases, the need for face-to-face meetings may decline. Alternative modes of communication can be used to keep members informed of progress, changes, and the need for team meetings. The team should also decide how members unable to meet with the entire group will communicate with the team, as well as how the team will communicate meeting results to absent members. Progress reports should include information about what is or is not happening. Sometimes what is not happening is as important as what is actually taking place. Other points to include in a progress report include pointing out tasks that need immediate attention or changes that need to be made. One of the most important parts of a progress report is recommendations for what needs to be done to get a task or an activity back on track.

A **process check**, another way of communicating in face-to-face meetings or in electronic forms of communication, can help move the team (or individuals) toward improved performance. Usually, a process check is a reflection of:

- What the team members did well that they want to continue doing.
- What did not go as well as the team would have liked.
- What the team can do to improve the things they want to improve and not detract from the things they think are going well.

The team should periodically conduct a process check at the beginning or end of a team meeting; doing so will help resolve people's differences and reinforce good feelings. Either way, a process check facilitates a professional, functioning team.

4.09.03 Communicating with the Outside World

A key to effective teamwork is effective communication. In the previous sections, you learned how to communicate with other team members. In reality, teams also need to be able to communicate with the outside world. In a university setting, the outside world may include instructors and classmates. In the working world, the outside world may consist of bosses, coworkers, clients, and/or the general public. The modes with which your team might need to communicate include progress reports, final reports, final design documentation, and project presentations.

When preparing a progress report, focus on the status of the project and any obstacles the team has encountered. Typically, a progress report is prepared for an instructor or boss, so the points must be clear and direct. Progress reports are meant to give an overview of the progress since the last report, so shorter is probably better as long as all necessary details are included. If your team has identified additional resources that are necessary for the successful completion of the project, these resources should be identified in the progress report as well. In final reports to the client or boss, you should provide the details of your design. Final reports are typically several pages in length (depending on the project) and outline the choices you made, the analysis you performed, and any test results you obtained. Design documentation usually includes drawings, specifications, and the details of any analysis you performed. Your design documentation should show how you arrived at the answer(s) you did. Presentations should show the highlights of your project. Usually, you will be given separate instructions from your professor about her expectations for a group presentation. Make sure you adhere to any guidelines you are given. If you are going to make a group presentation, you should practice at least once as a team to make sure you stay within the time limit and cover all necessary points.

For classroom projects, you will probably need to convey to your instructor how each person on the team contributed to the overall project. Specifically, you may be asked to address the following:

- Who contributed?
- What did each person contribute?
- How much credit does each person deserve?

Be honest and fair in your appraisal of your teammates' work on the project, as well as in your appraisal of your own work. Giving someone a pass that does not deserve one is not fair to the rest of the team, nor is judging someone too harshly.

4.10 Tools for Dealing with Personnel Issues

The following sections will give you tools for avoiding problems usually attributed to team members who become difficult to work with because of a personality trait that does not adapt well to team work or because the team member is not properly committed to completing the assignment. Keeping everyone motivated is best done proactively rather than reactively.

4.10.01 Team Contract

A **team contract** (also known as a code of conduct, an agreement to cooperate, or rules of engagement) is a formal written document, which should be readily available during all team meetings. The document should be established only after careful, thorough, and honest discussion by all team members. The contract lists the rules the team agrees to live by. Often team members need to revise a contract once they spend enough time working together that they (and others on the team) discover pet peeves. A good method for establishing a contract is to ask each member to bring to a team meeting a list of two or three of the biggest problems they encountered while members of previous teams. At the meeting, the team should consider each item in a round-robin fashion, until all members' lists have been exhausted. The resulting contract is a list of rules/agreements that, if followed, will incorporate the items on each person's list. This means that if everyone follows the rules in the contract, no one will violate any pet peeve or cause any of the previously experienced team-related problems.

Changes to the contract may be required as the team progresses, because well-meaning members, in spite of their best intentions, revert to inappropriate behavior. However, this does not have to mean an end to the team. Although revised contracts often include rewards and penalties to assist members in bringing about the desired behavior, you want to avoid coercing a member into accepting a contract. It is imperative that everyone on the team be treated with respect and that all disagreements are viewed as legitimate. Benjamin Franklin, upon the signing of the Declaration of Independence, said, "We must . . . all hang together or, most assuredly, we shall all hang separately." This quote applies to teams as well.

4.10.02 Publication of the Rules

Once you have determined how to operate as a team, write down the agreed-upon rules. Figure 4.05 shows a sample team contract established by a student project group.

Once you have created the contract, make sure every team member receives a copy and use the ground rules as a tool for effective communication and teamwork. Establishing, revisiting, and revising (if necessary) a contract can start the team out on the right foot or keep the team on track as crunch time materializes. Make sure all members support the rules. If you find that the rules are not working, change them, making sure everyone agrees to the new set of rules.

4.10.03 Signature Sheet and Task Credit Matrix

You should include a signature sheet with every assignment, whether it is requested by your instructor or not. On the sheet, explain that the individuals' signatures mean that the individuals participated in the assignment, have a general understanding of the entire submission, and are deserving of the credit indicated on the task credit matrix.

The **task credit matrix** is a table that lists each member on the team, each task, and indicates how the credit should be shared among the members of the team. If a team member does not deserve credit for the project, he or she should not be allowed to

FIGURE 4.05. Sample team contract.

Sample Contract

1. All members will attend meetings or notify the team by e-mail or phone in advance of anticipated absences.

2. All members will be fully engaged in team meetings and will not work on other assignments during meetings.

3. All members will complete assigned tasks by agreed-upon deadlines.

4. Major decisions will be subject to group discussion and consensus or majority vote.

5. The roles of recorder and timekeeper will rotate on a weekly rotational basis (all members will take their turn — NO EXCEPTIONS).

6. The team meetings will occur only at the regularly scheduled (weekly) time or with at least a two-day notice.

sign the signature sheet and a zero should be entered in the task credit matrix. It is not expected that each member will have contributed to each task. What is expected is that the team split the work fairly and that each individual deserves credit. Note that the credit does not have to be equal, although ideally (and usually) it is. Your team may decide to weight different tasks, based on the amount of effort or contribution required. You also may decide to split the tables, one for the task breakdown (to aid the instructor in knowing to whom to direct questions) and a second expressing the team's desire for distribution of points or credit. Often instructors will allow the team this privilege.

4.11 Chapter Summary

In this chapter, you learned about the importance of teams in engineering and in working on student projects. You learned about organizing a team and assigning roles for efficient, effective team meetings. The need to rotate roles among members was also discussed. You learned about the keys to successful team meetings and about organizing projects for optimal teamwork. The critical path method and Gantt charts were introduced as tools to help keep a project moving forward and for maintaining a reasonable project schedule. Finally, you learned about the importance of communication when working on a team. Two types of communication were discussed: internal communication among team members and external communication for working with other people, such as bosses and instructors.

4.12 glossary of key terms

agenda: The list of topics for discussion/action at a team meeting.

consensus: A process of decision making where an option is chosen that everyone supports.

critical path: The sequence of activities in a project that have the longest duration.

critical path method (CPM): A tool for determining the least amount of time in which a project can be completed.

devil's advocate: The team member who challenges ideas to ensure that all options are considered by the group.

Gantt chart: A tool for scheduling a project timeline.

note taker: The person who records the actions discussed and taken at team meetings and then prepares the formal written notes for the meeting.

process check: A method for resolving differences and making adjustments in team performance.

task credit matrix: A table that lists all team members and their efforts on project tasks.

team contract: The rules under which a team agrees to operate (also known as a code of conduct, an agreement to cooperate, or rules of engagement).

team leader: The person who calls the meetings, sets the agenda, and maintains the focus of team meetings.

team roles: The roles that team members fill to ensure maximum effectiveness for a team.

timekeeper: The person who keeps track of the meeting agenda, keeping the team on track to complete all necessary items within the alotted time frame.

4.13 questions for review

1. What are some of the advantages of working in teams?

2. What are some disadvantages of working in teams?

3. What are some important member roles that all teams should include, and what are the responsibilites of each role?

4. How should tasks be assigned in a team project?

5. In what ways is the Critical Path Method useful to the successful completion of a project?

6. What can be done to ensure that the workload in a project is fairly distributed among the team members?

7. What are some methods that can be used to ensure that all team members contribute equally to the team effort?

8. What can be done to solve the problem of a nonperforming team member?

9. What can be done to ensure effective communication within a team?

4.14 problems

1. In a memo to your instructor, describe the team roles outlined in this chapter.

2. In a memo to your instructor, describe an unsatisfactory team experience you encountered previously.

3. Meet with your team and develop a code of conduct for future team meetings.

4. Create a Gantt chart for the steps (and timeline) you would take to secure a summer job. Some of the tasks might include writing a resume, looking online for opportunities, and preparing letters.

5. If you have been assigned a project by your instrutor, create a Gantt chart for its completion.

6. Create a critical path diagram for planning and hosting a surprise birthday party for your mother or another family member.

5

Creativity and the Design Process

objectives

After completing this chapter, you should be able to

- Describe the steps in the design process
- Identify engineering tasks at the various stages of the design process
- Explain the role of creativity in the engineering design process
- Use several techniques that aid in the development of creative thinking

5.01
introduction

What distinguishes engineers from scientists is that engineers design and create solutions to the many technical problems of the world, whereas scientists study problems and report their findings in literature. For example, environmental scientists studied Earth's atmosphere several years ago and found that, as a society, the world is creating too many air pollutants—especially from automobile emissions. Certain engineers acted on the results of those studies and designed cars that were more fuel-efficient and that produced less pollution. Other engineers worked on changing the composition of fuel so that fewer pollutants were produced during combustion. Engineers are now working on solar-powered and electrically powered vehicles and on hydrogen fuel cells to further reduce the quantity of air pollutants emitted into the atmosphere. Other engineers are working on ways to use biofuels for cars.

Engineering is a design-oriented profession. Engineers design devices or systems and figure out how to mass-produce them, how to package them, and how to ship them to their intended destination. As an engineer, the type of device or system you design may depend on your discipline, but the design procedure you use will be essentially the same regardless of the industry in which you work.

As described in a previous chapter, engineers function in many capacities in business enterprises. Some engineers are involved in manufacturing, some in marketing, some in management, and some in testing. However, design is the central function of engineers. Typically, as you move through an engineering career, you will be responsible for different aspects of the enterprise at different times; but at some point, you will probably play a primary role in the design of new products. At other times, you may oversee the design of new products as a supervisor of a team of engineers. You also may be responsible for making sure the entire enterprise—including designers, manufacturers, and marketers—is running smoothly and the various team members are communicating with one another and working together to achieve the optimal end product. Because design is such a central portion of the engineering endeavor, you need to understand the design process thoroughly. The remainder of this chapter describes the **engineering design** process and the role creativity plays in the process.

5.02   What Is Design?

Design is a goal-oriented, problem-solving activity that typically takes many iterations—teams rarely come up with the "optimal" design the first time around. As an example, think of the minivan. When Chrysler developed the first minivan, it was considered a revolutionary concept in design and other car manufacturers quickly developed competing models. With each model, improvements were made to the original design such that the minivans of today are much improved compared to the initial product. The key activity in the design process is the development and testing of a descriptive model of the finished product before the product is finally manufactured or constructed. In the case of a manufactured product, the descriptive model usually includes solid 3-D computer models, engineering sketches and drawings, and possibly rapid prototypes. For a civil engineering project, the descriptive model includes drawings, specifications, and sometimes a scale model made of wood or plastic. Three-dimensional CAD models also are becoming more prevalent in the civil/construction industry. Engineering design includes a systematic approach to product definition, conceptualization, development, testing, documentation, and production/construction. Design is usually accomplished in a group environment with many people contributing ideas and skills to complete the finished design. Hence, creativity and interpersonal skills are important attributes of the design engineer.

When designing a device or a system, the engineer must keep certain factors in mind. These factors are usually related to the function and cost of the resulting system. For **sustainable design**, life cycle analysis and environmental impact are especially important factors. In most cases, engineers must make a number of choices during the design process. For example, consider the automobile. Engineers may choose from metals, composites, or plastics for the materials that make up the car's body. Each type of material has advantages and disadvantages. Although steel is strong and ductile, it is prone to corrosion (rust) and is relatively heavy, reducing the fuel efficiency of the car, which, in turn, leads to increased pollution. Composites are strong and can absorb a great deal of energy during crashes, but can be brittle and may be more expensive than steel. Plastics are readily formed in almost any shape and are resistant to corrosion, but are relatively weak, making safety a significant concern. Although plastics are widely used in car bumper systems, they typically are not used in the car's body. These are just a few of the factors that engineers must consider as they design an automobile body.

Engineers make choices by weighing the often competing factors associated with function, cost, and environmental impact. A car could be built that causes virtually no pollution and that is perfectly safe; however, the average person may not be able to afford it. So engineers make trade-offs between cost, safety, and environmental impact and design a car that is reasonably safe, is relatively inexpensive, and has minimal emissions.

Design is an aspect in virtually every discipline of engineering; however, chemical engineers typically view design differently than do mechanical engineers. For a chemical engineer, design includes determining the correct chemicals/materials to combine in the correct quantities and in the correct order to achieve the desired final product. Chemical engineers determine when to stir the mixture, heat it, or cool it. Electrical engineers may design computer chips or wiring for a building or the antenna system for a car or a satellite receiver. Civil engineers, like most chemical engineers, typically design one-off systems with features and/or specifications that are unique to a single application or location. They may design a single bridge or roadway, or a water distribution system or sewage system. Because civil and chemical engineering designs usually are not mass-produced, it is often impossible to create a **prototype** for testing before construction begins. Imagine the cost of building a "practice" bridge for every bridge that a civil engineering team designs and constructs.

Mechanical engineers typically design products that will be mass-produced for consumer use—cars, bicycles, washing machines, etc. Therefore, prototyping is an important part of the design process for mechanical engineers. Prototypes are the initial design concepts that are often created so that further design analysis can be performed before machines are retooled to produce, say, 10 million copies of a product. The process of creating and testing prototypes often saves a company money because engineers can work out the kinks or discover flaws early in the design stage. In the past, the design process included the production of several prototypes for testing and analysis. Today much of the testing can be accomplished using computer software tools, greatly reducing the need for prototypes. However, most manufacturing companies still produce at least one prototype before going into the production of a new product. The foundation for many computer-based testing and analyses in the design process is a 3-D solid model, which is a focal topic of this text.

Although modifications exist in the design process for engineers of different disciplines, there are similarities too. Almost all designs require drawings, sketches, models, and analysis (calculations). The remainder of this chapter will focus on design in a manufacturing arena—the type of design most familiar to mechanical engineers. This type of design results in products that are mass-produced. Where appropriate, variations to the design process for one-off designs (as in civil and chemical engineering) will be described.

5.02.01 Computers in Design

Computers have been used in engineering design for several decades. In the early years, computers were used primarily for their number-crunching capabilities. In other words, computers were employed to perform the tedious calculations involved in engineering design. Over the years, the role of computers in the design process changed significantly. Graphical computer workstations evolved and with them the ability of engineers to see their designs before building them. Engineers also can do much of the testing of design iterations on-screen, eliminating the need for numerous prototypes. Numerical methods such as **finite element analysis (FEA)**, modal analysis, and thermal analysis have enabled engineers to design systems in a fraction of the time that traditional design methods require. Modern design software often is easily incorporated into the manufacturing process, enabling the designer to establish cutting tool paths on-screen for the efficient manufacture of computer-generated models. Even other manufacturing capabilities, such as rapid prototyping, in which physical prototypes can be created within a matter of hours rather than days, have been a direct result of computer-aided design capabilities.

Today **computer-aided design (CAD)** is an efficient design method. The basis for CAD is the construction of a graphical 3-D model on the computer. This model can then be tested by any of the available numerical methods. Design modifications can be accomplished on-screen and the modified 3-D models tested again. When the engineer is satisfied that the design will meet or exceed all of the design criteria, a 2-D drawing can be created using the 3-D model as its basis. From this drawing, a physical prototype can be created and tested by traditional means to ensure compliance with the design criteria. Then the drawing is usually handed over to the manufacturing division for mass production. Alternatively, when a **computer-aided manufacturing (CAM)** system is available, the part or parts are produced directly from the 3-D computer models.

5.02.02 Classification of Engineering Designers

Engineering design is a broad concept with many integrated stages, competing alternatives, and diverse requirements for success. As such, many design teams in industry have specialty engineers who are responsible for certain aspects of the design process. Most design projects have a team leader, or **chief designer**, who oversees the work of the individual team members. In the early stages of the process for mass-produced designs, **industrial designers** lend their creative skills to develop the product concept and style. For civil engineering projects, **architects** may be employed for their creative talents in **conceptual design**. Specialists in CAD, **CAD designers**, develop the computer geometry for the new design. **Design analysts**, specialists in FEA and other software tools, check the new design for stress and load distribution, fluid flow, heat transfer, and a host of other simulated mechanical properties. **Model builders** are engineers who make physical mock-ups of the design using modern rapid prototyping and CAM equipment. **Detail designers** complete the final design requirements by making engineering drawings and other forms of **design documentation**. Before you begin reading about the **design process**, you should consider the role that creativity plays in the process.

5.03 Creativity in Design

Creativity is an important feature of the design process. Often engineers get hung up on the "rules" and "constraints" of design and forget to think creatively. Historically, there is a strong link between engineering and art. One of the earliest recognized engineers is Leonardo da Vinci. In fact, some people say that da Vinci was an engineer who sometimes sold a painting so he could make a living. If you examine the sketches and drawings of da Vinci, you will see that he was interested in the development of products

FIGURE 5.01. Visual thinking model.

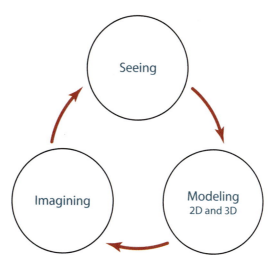

to help improve people's lives, including some of the first recorded conceptual designs of airplanes and helicopters. Creativity is at the heart of innovation, and it can be enhanced with both individual and group activities. Some of the more common activities used to facilitate creative thinking are described subsequently. However, psychologists have found that the brain works most creatively when the hands are engaged in completing a mindless task. You probably have experienced occasions when you try to remember something but cannot, then find that the thought pops into your head when you begin another task. So to free your mind for creative thoughts, it may be best to take a break and wash the dishes, do some yard work, lift weights, or do laundry. Performing any of these mindless tasks will help free your mind for creative thoughts.

5.03.01 Visual Thinking

Visual thinking is the process of expanding one's creative ideas using visual cues and feedback. The visual cues can take the form of sketches or computer models; however, in the *initial* stages of design, sketching is often viewed as a necessary ingredient for creative thought. Visual thinking can be thought of as a circular feedback loop, as illustrated in Figure 5.01. The visual thinking process can start at any place in the feedback loop; but for the sake of simplicity, start with the step labeled "Imagining." You first imagine an idea for a new design or product and then sketch the idea in some graphical mode (2-D sketch, 3-D sketch, or computer image). Seeing the idea adds to your understanding of it, which can be extrapolated more deeply. You get a better mental image with a visual cue, which allows you to take the preliminary idea and refine it, sketch it again, etc. You can continue the process until you have a well-defined sketch or idea of the product for formal analysis and design.

5.03.02 Brainstorming

Brainstorming is the most common form of group ideation and concept generation and is a process used for generating as many ideas as possible. Brainstorming is typically done as early in the design process as possible (i.e., before you start solving the problem, before breaking the project into tasks, and before deciding who will do each task). Once individuals begin to focus on specific tasks, it may be too late to consider alternatives. Too often, teams will tackle a problem by taking the first idea presented and pursuing that idea without considering any alternatives. Teams also can use brainstorming to generate a list of the tasks that need to be done before the project can be completed. Five simple steps define a brainstorming session:

Step 1: Assemble your project team and make sure you allow ample time for the session. Diversity in the group will enhance the quality and breadth of the ideas generated. Select a group leader to run the session and select a group recorder to take notes.

Step 2: Define the idea of the design project to be discussed. Write down the idea and make sure everyone in the group understands it.

Step 3: Discuss the rules about brainstorming and make sure everyone in the group agrees to abide by them. If necessary, keep the rules on display as a reminder to members who may stray. Rules for brainstorming are intended to help the team generate more ideas. Comments about an idea, whether positive or negative, can stifle the brainstorming process. Although the following rules are simple, you may find them difficult to follow:

1. Everyone participates.
2. Every idea is recorded.
3. Judgment is suspended—there is no such thing as a bad idea.
4. No criticism is permitted.
5. No commentary is permitted.
6. No one dominates the process.

Step 4: Start the brainstorming session by asking everyone in the group to offer an idea. If possible, the recorder writes down all responses so everyone can see them. Alternatively, each member of the team can keep a list of his own ideas. These lists will prevent ideas from being lost and will help restart the process if there is a pause in the flow of ideas.

Step 5: At the end of the session, spend time going through all of the ideas. Combine, categorize, and eliminate the ideas to narrow the list. Once team members are sure that all possible ideas have been included, the team should discuss the advantages and disadvantages of each idea. After talking about the pros and cons, each member should rank the best three ideas on the list. The team should keep the ideas with the highest rank and decide as a group which approach to use.

Instead of conducting the entire brainstorming session by the process outlined previously, you may consider using the following steps for a brainstorming session:

1. Individually spend ten minutes writing down ideas for tackling the assigned project.
2. Combine the individual lists onto a flip-chart, blackboard, or whiteboard. Team members may ask questions to clarify an idea, but no other comments are allowed during this process.
3. Continue brainstorming as a group until all ideas have been exhausted.

5.03.03 Brainwriting (6-3-5 Method)

Brainwriting is an alternative to brainstorming. In brainwriting, each member of the group focuses on sketching his ideas rather than verbalizing them. With brainwriting, you typically start with a team of six people. Each person sketches three ideas on a sheet of paper, leaving ample room for additional graphics and annotations. The idea sketches are then passed around the table so fellow members can add their own comments and ideas, as shown in Figure 5.02. Usually the idea sketches are passed around the group five times. The expectation is that by the fifth time around, a favorable design idea will have emerged. Brainwriting also is called the *6-3-5 method* (six people, three ideas each, five times around the table).

FIGURE 5.02. Brainwriting.

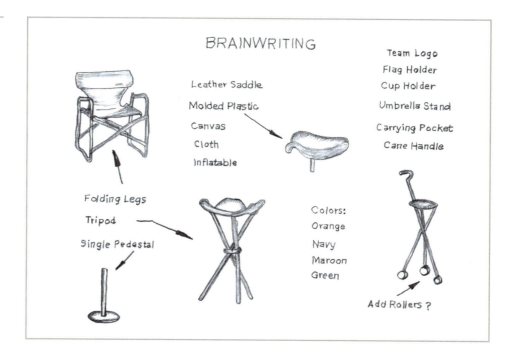

5.03.04 Morphological Charts

Morphology refers to the study of form and structure. A **morphological chart** can be used to generate ideas about a new design concept. The chart has a leading column that lists the various desirable functions of the proposed design. Along each row, various options for each function are listed, as shown in Figure 5.03. You can use brainstorming techniques to list as many options as possible for each function. The group then reviews and decides on a priority pathway through the options to address each desired function.

5.03.05 Concept Mapping

Concept mapping is a technique used to network various ideas together, as shown in Figure 5.04. During idea generation, the main design concept is placed in the center of the map with the various options linked outward in a brainstorming-like session. Each option then serves as a node for other choices. In that manner, the team can explore a big picture of all ideas and see a strong visual image of the connectivity of the different ideas.

FIGURE 5.03. A morphological chart.

MORPHOLOGICAL CHART					
Function	**Options**				
Seat Style	Saddle	Molded Dish	Strap	Inflatable	
Seat Materials	Leather	Plastic	Canvas	Rubber	Cloth
Number of Legs	One	Three	Four		
Leg Material	Wood	Aluminum	Plastic	Wrought Iron	
Leg Assembly	Pin	Hinge	Force Fit	Folding	
Accessories	Cup Holder	Beverage Cooler	Umbrella Holder	Flag Post	
Aesthetic Offerings	Team Logo	Choice of Colors			
Carrying Style	Carrying Handle	Handle Like Cane	Strap On Back	Roller Wheels	

FIGURE 5.04. A concept map.

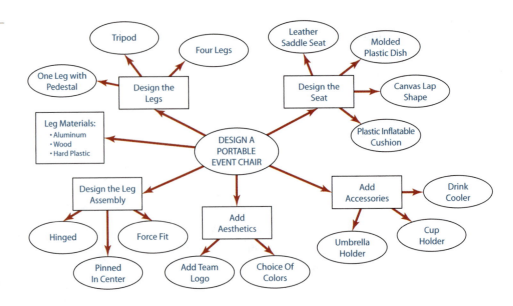

5.04 The Engineering Design Process

Design is a multistep process. However, there is considerable disagreement on exactly how many steps are involved in the process. Figure 5.05 shows one example of the sequence of steps in the design process. The design team often starts with stage 1 (Identify) and continues to stage 7 (Produce) and then begins again. Often the process does not proceed sequentially from stage 1 to stage 7; the design team might discover a serious problem in stage 4 (Analyze) and then return to stage 2 (Conceptualize) before moving on to stage 5 (Prototype).

Textbooks and writings on the design process include many different versions of and names for the stages in the design process. The stages presented previously are just one way to look at the design process; therefore, you should not think of them as the definitive word on the subject. However, the stages described in the remainder of this chapter are related to the graphical tools you will study in this textbook. For this reason, they have been adopted here. Knowing the number of stages and the labels for each stage is not nearly as important as understanding the overall process.

FIGURE 5.05. An engineering design process.

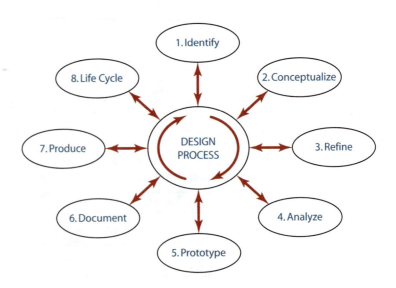

5.04.01 Stage 1: Problem Identification

Good design practice starts with a clearly defined need for a new product or system. Alternatively, a revised, improved design for an existing product may be required. A market survey may demonstrate that the new product or system is useful, has market appeal, is producible with today's technology, and will make a profit for the company supporting the design effort. Sometimes a new design idea is simply an alternate solution to a problem answered by an existing competing product. Indeed, many of today's highly successful products are the evolutionary result of free market enterprise. In the civil engineering world, a design project is typically the result of a client requesting a specific structure or system. For example, a governmental agency such as a county may request that a civil engineering firm design a new water distribution system to serve the needs of the county's residents. In this case, the client may have already defined the problem.

In the **problem identification** stage, the design engineer must address questions and answers from the customer's/client's perspective and from the engineer's perspective. For example, Table 5.01 shows two different perspectives for a new urban bicycle design.

When a product is designed, one of the design considerations is how long will it be used before it is no longer effective, which is called the **life cycle** of the product. Some products are designed for replacement on a regular basis. Thus, environmental considerations and disposal of a product must be considered throughout the design process. In the design process called **green engineering**, environmental concerns are considered throughout the process, not just at the end, because an engineer's choices in the problem definition stage often influence the overall environmental impact and life cycle of the product. Once the functional requirements have been identified, the design team can start the design process by generating some concepts.

5.04.02 Stage 2: Concept Generation

Concept generation is the most creative phase of the design process. You learned about some methods for creative thinking and concept generation earlier in this chapter. Typically, concept generation starts with brainstorming, brainwriting, or a similar team meeting where ideas are tossed around and discussed. Criticism is usually limited

TABLE 5.01. Functional Requirements for a New Urban Bicycle.

CUSTOMER'S PERSPECTIVE	ENGINEER'S PERSPECTIVE
MODERATE STREET SPEED	SUSTAINABLE SPEED OF XX MPH MAXIMUM SPEED OF YY MPH
COMPACT SIZE	DIMENSIONS NOT TO EXCEED A X B X C
SAFE	STRUCTURAL STRENGTH CONTROLLABILITY BRAKING CAPABILITY TIRE PUNCTURE RESISTANCE
COMFORTABLE	ERGONOMICS OF SEAT EFFICIENCY OF POWER TRAIN POSITION(S) OF HANDLEBAR
ATTRACTIVE	CHOICES OF PAINT COLOR LIGHTS AND REFLECTORS
AFFORDABLE	SELECTION OF MATERIALS MANUFACTURING PROCESS NUMBER OF PARTS SALES VOLUME

FIGURE 5.06. Concept sketches.

in the concept generation stage, since maximizing the number of good ideas is desirable. At the end of the concept generation stage, the team should have selected a few main ideas that it will focus on for future refinement and analysis. Sketching is an integral way to develop concepts for a new design. Figure 5.06 shows some examples of concept sketches.

5.04.03 Stage 3: Concept Selection and Refinement

Once a few quality concepts have been identified, the design team must converge on one or two final concepts to further explore in the design process. A common technique for selecting the final concept(s) is to use a **weighted decision table**, as shown in Figure 5.07. With a weighted decision table, all of the common attributes and desirable features of each concept are list in the first column. A weighting factor for each feature/attribute is then established (e.g., using a scale of 0 to 10). The various design options are listed in subsequent columns in a parallel fashion to the listed features/attributes. The team then conscientiously scores each option for every feature/attribute, each time applying the weighting factor to the score, as illustrated in Figure 5.07. Adding all of the scores for individual attributes yields a final "bottom-line" number that can be used to select the highest-ranked option.

Sometimes the initial concepts may need to be refined before a final decision can be made. Refinement will likely include the development of 3-D computer models for defining geometry not accurately expressed in the concept sketches. For example, as shown in Figure 5.08, different computer models of a new product can assist the members of the design team in visualizing the specific model that has the marketing appeal they are seeking and in making the final decision.

FEATURE ATTRIBUTE			DESIGN 1		DESIGN 2		DESIGN 3		DESIGN 4	
			OPEN BOTTOM		**THUMB GRIP**		**DOUBLE LOOP**		**SINGLE LOOP W/FINGER SUPPORT**	
		WEIGHT 0–10	**S**CORE -- TOTAL 0–10 **S * W**		**S**CORE -- TOTAL 0–10 **S * W**		**S**CORE -- TOTAL 0–10 **S * W**		**S**CORE -- TOTAL 0–10 **S * W**	
AESTHETICS	Color	5	3	**15**	3	**15**	3	**15**	3	**15**
	Form	8	7	**56**	7	**56**	7	**56**	7	**56**
ERGONOMICS	Grip ability	8	7	**56**	6	48	2	16	9	72
	Drinking Ease	6	3	**18**	5	30	4	24	8	48
FUNCTIONALITY	Adapts to Hand	8	5	**40**	7	56	2	16	8	56
STABILITY	Base size	6	7	**42**	9	54	3	18	9	54
	Height	8	7	**56**	8	64	3	24	8	64
MANUFACTURABILITY	Injection Molding	5	3	**15**	3	15	3	15	3	15
	Slip Molding	5	3	**15**	3	15	3	15	3	15
			Weighted Total	**313**	**Weighted Total**	**353**	**Weighted Total**	**199**	**Weighted Total**	**395**

FIGURE 5.07. A weighted decision table for concept selection.

FIGURE 5.08. Computer models for concept selection.

FIGURE 5.09. Object mass properties.

MASS PROPERTIES REPORT (Concept 4)

Mass Properties of ergonomic cup (Material – ABS Plastic)

Output coordinate System: -- default --

Density = 0.037 pounds per cubic inch

Mass = 0.948 pounds

Volume = 25.722 cubic inches

Surface area = 134.693 inches^2

Center of mass: (inches)
          X = 0.071
          Y = 1.648
          Z = 0.000

Principal axes of inertia and principal moments of inertia: (pounds * square inches)
Taken at the center of mass.
          Ix = (0.033, 0.999, 0.000)   Px = 2.401
          Iy = (-0.999, 0.033, 0.000)   Py = 3.770
          Iz = (0.000, 0.000, 1.000)   Pz = 3.921

Moments of inertia: (pounds * square inches)
Taken at the center of mass and aligned with the output coordinate system.
          Lxx = 3.768   Lxy = 0.046   Lxz = -0.000
          Lyx = 0.046   Lyy = 2.402   Lyz = -0.000
          Lzx = -0.000   Lzy = -0.000   Lzz = 3.921

Moments of inertia: (pounds * square inches)
Taken at the output coordinate system.
          Ixx = 6.341   Ixy = 0.157   Ixz = -0.000
          Iyx = 0.157   Iyy = 2.407   Iyz = -0.000
          Izx = -0.000   Izy = -0.000   Izz = 6.498

5.04.04 Stage 4: Design Evaluation and Analysis

In this stage, the selected concept is further analyzed by any number of numerical methods. Before the advent of CAD and analysis tools, this stage of the design process involved building and testing physical models. Now the building and testing can be done on the computer, saving companies a great deal of time and money. The tests are conducted to determine mechanical properties of objects or systems and their performance during simulated conditions.

One of the simpler types of analysis that can be performed with a computer model is the computation of the object's mechanical properties, such as mass, center of gravity, and moment of inertia. All of these properties may not be meaningful to you right now; however, they are key quantities used in performing most types of static and dynamic analyses. A **mass properties analysis** report, shown in Figure 5.09, is a useful document for evaluating and presenting the static mechanical conditions of the design.

Further analysis of the design might include an FEA of stress contours and deformation. Heat transfer and aerodynamic flow also can be simulated using modern computational software (Figure 5.10). These numerical methods will be discussed in more detail in a later chapter and are themselves topics of entire texts.

FIGURE 5.10. A numerical analysis model.

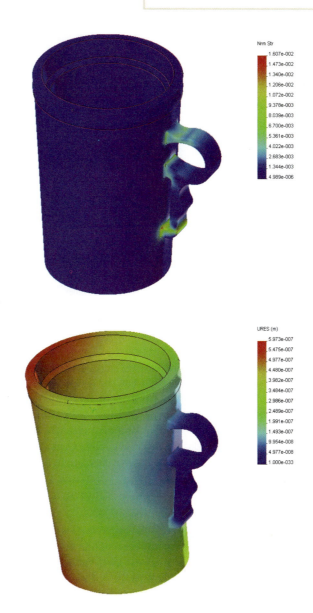

5.04.05 Stage 5: Physical Prototyping

Most designers and clients would like to see a physical model of the design—they want to look at it, hold it in their hand, and show it to other interested parties. Several different types of physical models can be developed during this stage of the design process. Engineers can have the shop people build a *scale model*, an actual *true-size model*, or just a simple *mock-up concept model* that shows the general physical appearance of the design.

In recent years, modern technology has accelerated the production of prototype models in the design process. CAM systems can take data from a 3-D solid model and cut the pattern using computer numerical control (CNC) machines. Figure 5.11 shows a part that was created through CNC machining. CNC machining will be covered in more detail in a later chapter of this text.

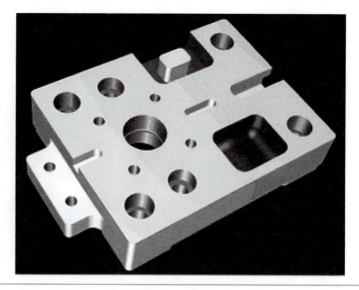

FIGURE 5.11. A part created through CNC machining.

Today rapid prototyping systems can perform some of the same functions of traditional machining tools, except that they require far less time (hence, the term *rapid*) and fewer resources than traditional methods. Some of the modern rapid prototyping methods include stereolithography (SLA), Selective Laser Sintering (SLS), Laminated Object Manufacturing (LOM), Fused Deposition Modeling (FDM), Solid Ground Curing (SGC), and inkjet printing techniques. Most recently, 3-D printers have become affordable prototyping alternatives for the office environment. Reasonable 3-D models can be printed using 3-D printers, as shown in Figure 5.12.

5.04.06 Stage 6: Design Documentation

There are many forms of design documentation, but the most common form is a finished detailed drawing, as shown in Figure 5.13. A detailed drawing shows the information needed to manufacture the final part. A good portion of the rest of this text discusses detailed engineering drawings. You will learn how to create drawings and how

FIGURE 5.12. Models created with a 3-D printer.

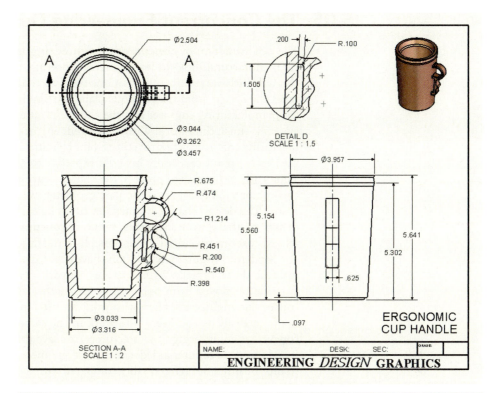

FIGURE 5.13. A detailed design drawing.

to interpret them correctly. You will learn about dimensioning and tolerancing for annotation of the drawing. You will learn about conventions developed over the years that provide everyone with the same understanding of what is on the drawing. You also will learn how drawings are created from 3-D computer models of parts and systems.

5.04.07 Stage 7: Production

Once the design documentation is complete, it is time to begin the production stage. For a civil engineering design, production is called the construction stage; for mechanical engineering, production is the manufacturing stage. Many engineers claim that the design process ends with the documentation stage. However, the way the product is designed may impact its production and distribution processes later on. For mechanical engineering projects in this stage, the goal is usually (but not always) to produce the product in large quantities, to meet performance standards, and to keep manufacturing costs low. Many different methods for manufacturing parts have been developed and are widely used, including machining, casting, rolling, and sheet metal cutting processes. These methods will be covered in more detail in a later chapter.

For civil engineering projects, because no prototypes have been built, design modifications are common in the production stage. Contractors may find, for example, that ductwork does not fit within the space provided on the drawings, or they may find that piping needs to be rerouted around an obstruction. For this reason, in civil engineering projects, it is important to continue to document the design by making notations on the drawings where changes are made. These drawings are called **as-built drawings**, and they reflect the way the project was actually constructed—not just the way it was designed. As-built drawings are an important part of the design process because if piping is rerouted through a building, for example, someone will need to know exactly where the plumbing is in case leaks or other problems arise.

5.05 The Concurrent Engineering Design Process

The new paradigm of **concurrent engineering** is sometimes referred to as "design for manufacturability." In traditional engineering, the part being designed progresses through each stage, moving from one team to the next. At each new stage, the team takes the design from the previous team and applies its own expertise. The first time the manufacturing engineer sees the part for production is when the design and analysis teams finish their work. With concurrent engineering, designers, analysts, and manufacturing engineers work together from the initial stages of the design process. In this way, each person can apply his own expertise to the problem at hand *from the start*. Thus, early in the design process, the manufacturing engineer might say, "If you made this minor modification to the part geometry, we would save $100,000 in retooling costs." The design change in question could be easily implemented during the initial phases of the process; whereas, without concurrent engineering, the change (and related cost savings) might be impossible in the final stage.

Modern computer workstations have enabled concurrent engineering to become a reality in the workplace. With local area networks, wide area networks, and the Internet, data can be moved from one desktop to another almost effortlessly. Members of the concurrent engineering team who work in different countries can share design ideas nearly as easily as engineers who work in the same building. Using the principles of concurrent engineering, manufacturers can save thousands, even millions, of dollars. In addition, computer-aided concurrent engineering design is more efficient and results are often of higher quality compared with designs produced in the past.

5.06 Patents

Patents are a way to protect the intellectual property of a new design. In the United States, a patent generally gives the inventor sole claim to intellectual property rights for twenty years. Application for a patent is made to the director of the United States Patent and Trademark Office and typically includes three components:

1. A written document made up of a specification (description and claims) and an oath or a declaration.

2. A patent drawing (Figure 5.14) in those cases in which a drawing is necessary.

3. Filing, search, and examination fees.

Frequently, inventors hire a lawyer to do the legal work and to make sure the idea has not already been patented. In some cases, the individual inventor conducts a patent search using a for-hire Internet site.

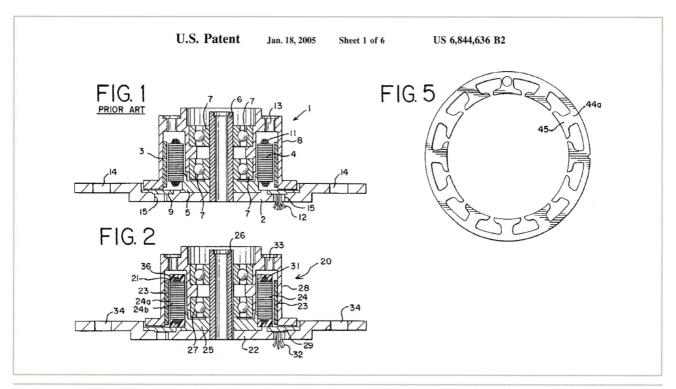

FIGURE 5.14. A patent drawing.

5.07 Chapter Summary

In this chapter, you learned that design is an iterative process and that the process has several stages; however, there is no general agreement about the exact number of stages. You learned that engineers must weigh competing factors such as cost, function, and environmental impact when making design decisions. You learned about design in the information age and ways computers are used throughout the process, greatly reducing costs. You learned about the importance of creativity in engineering design and about several techniques you can use, either as an individual or as part of a team, to foster creative thinking. Finally, you learned about concurrent engineering, which is enabled by computer technologies and can be used to reduce product costs and improve product quality.

casestudy
Integrated Project: Conception of the Hoyt AeroTec Target Bow

Hoyt USA was founded in 1931 by sportsman and bow maker Earl Hoyt. The company is located in Salt Lake City, Utah, where it has both engineering and fabrication facilities. Hoyt USA has a long-standing reputation as a high-quality maker of bows for sports, recreation, and competition. In 1972, the company revolutionized competition archery when it introduced its first metal-handled collapsible recurve bow at the 1972 Olympic games. Given only to the U.S. team, the metal-handle design offered significant advantages over other bows, which were mostly made of wood at that time. A metal handle was relatively immune to the effects of changing temperature, humidity, and time, which affected the geometry, stiffness, and vibration properties of the bow. These variations made it difficult to use a wooden bow to land arrows shot after shot in the same place on a target. In addition, lighter arrows produce higher stresses in a wooden bow, sometimes causing the bow to break. By contrast, the strength of the handle's metal enabled an archer to use lighter arrows, which reached more distant targets in shorter times. Soon after Hoyt USA introduced the metal-handled bow, other bow manufacturers followed suit.

In the early 1990s, Hoyt USA wanted to improve the design of its metal bow handle to improve its share of the target archery equipment market. This market was very competitive, with a typical recurve bow lasting only three or four years before it needed to be replaced. In the development of a new design, Hoyt USA had to consider several things. For product performance, the main considerations were strength, weight, and vibration. A new product had to be stronger than previous ones. Super-light carbon arrows, light composite bow limbs, and synthetic strings produced increasing levels of stress in bows, to the point where even metal handles were breaking. But additional strength could not be gained at the expense of weight. Many archers already complained about the excessive weight of metal-handled bows. Any new product also needed to be less flexible and have less vibration than existing bows. Target archers considered excessive flexibility, noise, and vibration to be undesirable characteristics.

The new product needed to be developed with analysis and production in mind. Detailed stress analysis was necessary to ensure that there would be no breakage problems, as had been the case with other products. Also, the new product had to be designed so it could be easily produced in state-of-the-art fabrication facilities consisting mostly of computer-controlled four-axis milling machines, which Hoyt USA was expanding at the time.

A conventional metal-handled recurve target bow.

The new concept that Hoyt USA developed was a structural support member located behind the grip of the bow handle. The support member was designed with an outward bend so it would not touch the arm of the user. Touching the bow at any location except the grip was a violation of the rules in target archery. The new design allowed the forces in the bow handle to be more widely distributed, resulting in reduced stresses without a significant increase in weight.

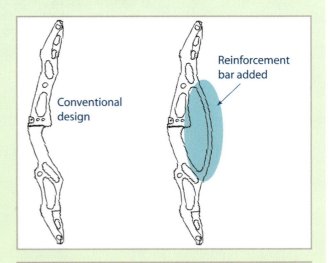

Conventional design

Reinforcement bar added

Concept sketches showing the addition of a structural reinforcement bar to a conventional bow handle to reduce its flexibility and vibration.

Conceptual sketches for the new design were developed first. The sketches enabled Hoyt USA engineers to communicate ideas among themselves, as well as to engineers outside their group, managers, production specialists, and potential customers. About twenty design variations were examined on the drawing board, after which three or four were selected for further development. A solid model was built using a computer, because the model could be used for stress analysis with finite element methods and because the model could be exported directly to the fabrication machines on Hoyt USA's production floor.

A solid model of the new design used for analysis and fabrication.

After final selection and refinement of the design as a result of the stress analyses and field testing, Hoyt USA began full production of the new product less than one year after the concept was first discussed. To protect the innovative design from being copied by competitors, Hoyt USA applied for and was granted a U.S. patent. Patents also were secured in foreign countries. The design concept was trademarked TEC for "Total Engineering Concept." Because of the relatively radical appearance of the product (and the rather conservative nature of archers), at

The production version of the structurally reinforced handle.

first, the product was slowly and cautiously adopted by the market. But after a few years (and some outstanding performances at the Olympics), TEC products by Hoyt USA were eagerly embraced by the market as a superior technology.

DISCUSSION QUESTIONS/ACTIVITIES

1. Explain the design process that the developers at Hoyt used to engineer the TEC bow.

2. Create a weighted decision table based on the questions and answers that might have been generated during the design process of the TEC bow.

3. In what ways were concurrent engineering techniques used throughout the development of the TEC bow?

5.08 glossary of key terms

architects: Professionals who complete conceptual designs for civil engineering projects.

as-built drawings: The marked-up drawings from a civil engineering project that show any modifications implemented in the field during construction.

brainstorming: The process of group creative thinking used to generate as many ideas as possible for consideration.

brainwriting: A process of group creative thinking where sketching is the primary mode of communication between team members.

CAD designers: Designers who create 3-D computer models for analysis and detailing.

chief designer: The individual who oversees other members of the design team and manages the overall project.

computer-aided design (CAD): The process by which computers are used to model and analyze designed products.

computer-aided manufacturing (CAM): The process by which parts are manufactured directly from 3-D computer models.

concept mapping: The creative process by which the central idea is placed in the middle of a page and related concepts radiate out from that central idea.

conceptual design: The initial idea for a design before analysis has been performed.

concurrent engineering: The process by which designers, analysts, and manufacturers work together from the start to design a product.

design analysts: Individuals who analyze design concepts by computer methods to determine their structural, thermal, or vibration characteristics.

design documentation: The set of drawings and specifications that illustrate and thoroughly describe a designed product.

design process: The multistep, iterative process by which products are conceived and produced.

detail designers: The individuals who create engineering drawings, complete with annotation, from 3-D computer models or from engineering sketches.

engineering design: The process by which many competing factors of a product are weighed to select the best alternative in terms of cost, sustainability, and function.

finite element analysis: A numerical method used to analyze a product in terms of it structural, thermal, and vibrational performance.

green engineering: The process by which environmental and life cycle considerations are examined from the outset in design.

industrial designers: The individuals who use their creative abilities to develop conceptual designs of potential products.

life cycle: The amount of time a product will be used before it is no longer effective.

mass properties analysis: A computer-generated document that gives the mechanical properties of a 3-D solid model.

model builders: Engineers who make physical mockups of designs using modern rapid prototyping and CAM equipment.

morphological chart: A chart used to generate ideas about the desirable qualities of a product and all of the possible options for achieving them.

patents: A formal way to protect intellectual property rights for a new product.

problem identification: The first stage in the design process where the need for a product or a product modification is clearly defined.

prototype: The initial creation of a product for testing and analysis before it is mass-produced.

sustainable design: A paradigm for making design decisions based on environmental considerations and life cycle analysis.

visual thinking: A method for creative thinking, usually through sketching, where visual feedback assists in the development of creative ideas.

weighted decision table: A matrix used to weigh design options to determine the best possible design characteristics.

5.09 questions for review

1. What are the main stages in the design process?
2. Why is creativity important in the engineering design process?
3. How does engineering design differ from the type of design artists perform?
4. What is meant by concurrent engineering?
5. How is a computer used in the modern-day design process?
6. What are some of the differences in design for a civil engineering project versus a mechanical engineering project?

5.10 design projects

The following sections will outline specifications for design projects. These projects were tested with students at the University of California at Berkeley over the years and are suitable for use in a first- or second-year design course at a university. The projects are designed for completion by a team of four or five students.

5.10.01 STANDARD PROJECT MATERIALS

Use the following standard list of materials, in addition to any special items listed in the specific design rules for the project, to construct the device assigned by your instructor. No other materials are permitted.

- Paper, 30# (maximum): 2 square meters maximum; 2 layers maximum
- Poster board, single-ply, medium weight: 1 square meter maximum
- Foam core modeling board, 3/16″ nominal thickness: 1 square meter maximum
- Twine, 60 lbs. (maximum) labeled breaking strength: 3 meter length maximum
- Wood dowel, 1/4″ nominal diameter: 1 meter length maximum
- Mailing tube, 2″ nominal diameter, medium-weight cardboard: 1 meter length maximum; no endcaps
- Rubber bands (sample to be supplied), #62 or #64: 10 maximum
- Elmer's Glue-All glue: 30 cc maximum
- Hot melt adhesive (polyolefin): 30 cc maximum
- Scotch brand transparent cellophane tape: 1 meter maximum

All of the materials can be purchased at local art supply or convenience stores. Equivalent material may be substituted only with the instructor's permission. Paints, markers, flags, and other decorative items not on the list may be used as long as they are purely decorative; for example, paint cannot be used as weight or ballast.

5.10.02 STANDARD PROJECT DELIVERABLES

The following list provides the standard deliverables for your project. Your instructor may assign additional deliverables and will let you know the due date for each deliverable. When you are organizing your team effort, you can use these deliverables as the milestones to produce a Gantt chart or critical path diagram to help you stay on track and complete the project on time.

Required Drawing Deliverables:

1. Conceptual sketches—alternative and final designs
2. Outline assembly drawings
   Multiview of assembled project
   Isometric or pictorial view of assembled project
   Cutaway views as required for clarity
   Sectional views as required for clarity
   Overall dimensions only
   Balloons to identify subassembly or part numbers and names
3. Detail drawings
   One multiview drawing per part (isometric or pictorial)
   All dimensions, datums, and tolerances
   Quantity
   Material
   Sectional views as required for clarity
   Isometric views as required for clarity

5.10 design projects (continued)

4. Exploded assembly drawings
   Blow-apart pictorial view of all assemblies
   Blow-apart pictorial view of all subassemblies
   Balloons to identify part numbers and part and subassembly names
   Subassemblies as required (highly recommended)
5. Bill of materials
   List of all parts by PN, showing name, quantity, and material
   List of all materials needed for assembly (e.g., tape and glue)

Use millimeter dimensions and proper title blocks and borders for your engineering drawings. It is recommended that all drawings be cross-checked by different people. Alternate the functions of drafter, designer, and checker. The team leader must give final approval.

Final Demonstration and Oral Presentation:

Each team is expected to give an informative final presentation of its design, as well as a demonstration during the distance contest. Use descriptive graphics slides to complement the presentation. Keep the presentation short and direct.

Written Report:

In your written final report detailing the project results, describe the alternatives your team considered, describe which ones you selected, and explain why you selected them. Use drawings to illustrate key points in the design process. Include the results of your product testing and include a section on what you would have done differently.

DESIGN PROJECT #1: ESCAPE!

NASA is once again looking for a few good engineers. This time the agency is seeking conceptual ideas for an escape device that would allow launchpad crews and astronauts to leave the area quickly in case of a potentially explosive, toxic, or otherwise harmful situation. The device is to be remotely launched and should be designed to place personnel as far away from the launch point as possible. The device must land safely, leaving the personnel unharmed. However, for this project, you will demonstrate the concept of your device using a hard-boiled egg.

The Mission:

Your mission is to design and build a device that will launch a hard-boiled egg (USDA Grade A Large, which your team must provide) into the air and have it land as far from the launch point as possible. The device must land the egg totally intact (no cracks in the shell). The design, for example, may be composed of a mechanism for launching the egg and a device attached to the egg for lowering it slowly (like a parachute). You may surround the egg with a protective covering. However, the covering cannot penetrate the shell or be bonded to it. The function of your device will be graded on the distance from the point of launch relative to that of the other teams in the class. A stiff distance penalty will be assessed if the egg is damaged.

Design Rules:

The device must be constructed out of the standard project materials listed previously, in addition to *only* the following item: one hard-boiled egg (USDA Grade A Large).

Equivalent material may be substituted only with the permission of the instructor. Any design deemed by the instructor as unsafe will not be acceptable (e.g., no sharp flying objects, no explosive devices, and no raw or rotten eggs).

More Contest Rules:

- The device, including the launcher, must initially fit within a 1.0 m x 1.0 m x 1.0 m volume without external support, except for the triggering means.
- Once the egg has been launched, the device may expand to any size.
- The device must be freestanding and may not be taped, glued, or in any other way affixed to the ground.
- The device must be remotely triggered (e.g., by a string or rod). Team members (all parts of the body) must remain a minimum of 1 meter away from the device when the egg is launched.
- The device must be set up within 3 minutes; otherwise, a 3-meter distance penalty on the total distance will be assessed for every 10 seconds of overtime.
- Human power may be used to trigger the device, but not to impart motion to the egg. However, human power may be used to store energy (e.g., into the rubber bands) for any use.
- Distance is measured from the point the egg is completely clear of the launch structure (e.g., completely airborne with no attachment to the launch structure) to the point stops where any part of the device containing the egg touches a solid object connected to the ground.
- The egg must attain a distance of at least 3 meters from the launch position.
- The egg must survive the landing without cracking or sustaining any other visible signs of damage.
- Surviving eggs will be peeled and eaten by the team leader to ensure that they have not been altered in any way. If the team leader does not survive, the entire team will fail the project.
- The egg must be removed from its protective covering within 30 seconds.

5.10 design projects (continued)

- If the egg does not survive, the total distance will be recorded as zero.
- If the egg is damaged, it must be replaced in time for the next launch.
- The maximum distance from three trials will be recorded. A misfire will count as one trial.
- Spare parts are recommended and do not count in the materials inventory of the final assembly.

DESIGN PROJECT #2: FAST FOOD!

Food Service in the dormitories is experimenting with a new method of feeding students in the morning. Instead of going to the dining commons for breakfast, students will open a window before a prescribed time. Food Service will then deliver breakfast by launching it into the dorm room. That way, students do not need to wake up early just to get breakfast and can sleep late if they so choose. All students need to do is leave the window open in the evening to ensure that breakfast will be delivered in the morning. The dining service has asked your team to demonstrate a conceptual prototype of a device that will perform this function.

The Mission:

Your mission is to design and build a device that will launch a bagel with cream cheese into the air and have it land as far from the launch point as possible. The device must land the bagel totally intact and unsoiled. The design, for example, may be composed of a mechanism for launching the bagel and a box around the bagel to help protect it. However, the covering cannot pierce the bagel and cannot be bonded to it. The function of your device will be partially graded on the distance from the point of launch to the point of landing relative to that of the other teams in the class. A stiff distance penalty will be assessed if the bagel is damaged or soiled: you will have to eat it.

Design Rules:

The device must be constructed out of the standard project materials listed previously, in addition to *only* the following item: one plain bagel sliced horizontally and smeared with plain soft cream cheese to 1/4″ average thickness. The bagel cannot be more than 24-hours old at the time of launch.

All of the materials (except the bagel with cream cheese) can be purchased at local art supply stores. Any design deemed by the instructor as unsafe will not be acceptable (e.g., no sharp flying objects, no explosive devices, and no spoiled food).

More Contest Rules:

- Once the bagel has been launched, the device may expand to any size.

- The device, including the launcher, must initially fit within a 1.0 m x 1.0 m x 1.0 m volume without external support, except for the triggering means.
- The device must be freestanding and may not be taped, glued, or in any other way affixed to the ground.
- The device must be remotely triggered (e.g., by a string or rod). Team members (all parts of the body) must remain a minimum of 1 meter away from the device when the bagel is launched.
- The device must be set up within 3 minutes; otherwise, a 3-meter distance penalty on the total distance will be assessed for every 10 seconds of overtime.
- Human power may be used to trigger the device, but not to impart motion to the bagel. However, human power may be used to store energy (e.g., into the rubber bands) for any use.
- Distance is measured from the point the bagel is completely clear of the launch structure (e.g., completely airborne with no attachment to the launch structure) to the point stops where any part of the device containing the bagel touches a solid object connected to the ground.
- The bagel must survive the landing without cracking, opening, soiling, or sustaining any other visible signs of damage.
- The surviving bagel with the longest distance will be eaten by the team leader to ensure that it has not been altered in any way. If the team leader does not survive, the entire team will fail the project.
- The bagel must be removed from its protective covering within 15 seconds.
- If the bagel does not survive, the total distance will be recorded as zero.
- If the bagel is damaged, it must be replaced in time for the next launch.
- The average of the three longest distances from as many launches as can be accomplished within a single 60-second period will be recorded. Thus, you should have multiple bagels and containers ready to launch. A misfire will be considered as zero distance.
- Spare parts are recommended and do not count in the materials inventory of the final assembly.

DESIGN PROJECT #3: REWARD!

Several problems are on the horizon for engineering graphics classes of the future. First, the classes are getting larger, requiring that lectures be held in larger rooms. This trend makes it difficult for the instructor to toss candy rewards to specific students in the class, because the instructor's throwing range is limited. Second, the course CD is apparently a flop in the market and customers from

5.10 design projects (continued)

all over the country are returning their disks to the publisher. To solve both problems at the same time, someone recommended that candies be strapped to CDs and thrown together. The aerodynamic properties of the CD can be used to increase the range of the candy. This idea was immediately adopted, so here is your project.

The Mission:

Your mission is to design and build a device that will launch a Hershey's chocolate Nugget (with almonds) taped to a CD so it passes through an 8′ x 8′ target frame placed as far from the launch point as possible. The target frame will be placed such that the opening faces the launcher, with the bottom of the target frame on the ground. The launching field will be relatively flat. Each team will have 3 minutes to hit the target at least once. The target distance from the launcher will be specified by the team. The single longest distance at which the target is hit will be recorded for each team. If the target is not hit on any of the tries, the final recorded distance will be recorded as zero.

Design Rules:

The device must be constructed out of the standard project materials listed previously, in addition to *only* the following items:

- One genuine Hershey's chocolate Nugget (with almonds) at room temperature, still in the wrapper
- One standard 120 mm diameter optical CD

    Any design deemed by the instructor as unsafe will not be acceptable (e.g., no sharp flying objects, no explosive devices, and no spoiled food).

More Contest Rules:

- The device, including the launcher, must initially fit within a 1.0 m x 1.0 m x 1.0 m volume without external support, except for the triggering means.
- The device must be freestanding and may not be taped, glued, or in any other way affixed to the ground.
- The device must be remotely triggered (e.g., by a string or rod). Team members (all parts of the body) must remain a minimum of 1 meter away from the device when the Nugget is launched.
- When the launcher is armed and ready to be triggered, it must be entirely self-supporting and stable (e.g., it does not require any external support from team members).
- The device must be set up within 3 minutes; otherwise, a 1-meter distance penalty on the total distance will be assessed for every 10 seconds of overtime.
- Human power may be used to trigger the device, but not to impart motion to the Nugget. However, human power may be used to store energy (e.g., into the rubber bands) for any use.
- Distance is measured from the point the CD is completely clear of the launch structure (e.g., completely airborne with no attachment to the launch structure) to the 8′ x 8′ target frame.
- The Nugget with the longest distance will be eaten by the team leader to ensure that it has not been altered in any way. If the team leader does not survive, the entire team will fail the project.
- The single longest distance at which you hit the target from as many launches as can be accomplished within a single 3-minute period will be recorded. Thus, you should have multiple Nuggets on CDs ready to launch. A misfire will be considered as zero distance.
- Spare parts are recommended and do not count in the materials inventory of the final assembly.

DESIGN PROJECT #4: DEPLOY IT!

NASA is once again looking for a few good engineers. This time the agency is seeking conceptual ideas for the deployment of structures such as antennas and solar panels in spacecraft. The device is to be self-deploying and should be designed to extend as far from the base point as possible while still remaining connected to the base point. It is possible that deployment will occur in various environments—from gravity-free space to high gravity on planets and moons. However, you will demonstrate the concept for your device in earth gravity.

The Mission:

Your mission is to design and build a device that will deploy a structure to reach as far from the origin point as possible. The device must remain physically connected from the origin point to the point of furthest extension. Deployment must be automatic upon activation of a trigger mechanism, and the base structure is to be fixed to the ground. The design, for example, may be composed of a mechanism for extending a boom, cantilevered structure, or suspended structure from the origin point.

Design Rules:

The device must be constructed out of the standard project materials listed previously, in addition to *only* the following item: 4 meters of 3M #2090 Long-Mask masking tape, 2″ wide, to fix the base to the ground.

    Any design deemed by the instructor as unsafe will not be acceptable (e.g., no sharp flying objects and no explosive devices).

5.10 design projects (continued)

More Contest Rules:

- The device, without external support, must initially fit within a 1.0 m x 1.0 m base area on the ground, except for the triggering means.
- The device must initially be less than 0.5 m in height, except for the triggering means.
- Once triggered, the device must deploy automatically to its final state without further assistance and may expand to any size.
- The device must be remotely triggered (e.g., by a string or rod). The trigger may be used only to release energy from the system. The trigger cannot add energy to the system. Team members (all parts of the body) must remain a minimum of 1 meter away from the device when the device is deployed. The means of triggering must be contained on the allowable materials list but will not be counted in the final materials inventory.
- The device may be fixed to the ground only in the original base area, using only the 3M tape specified for this purpose.
- The device must be set up within 3 minutes; otherwise, a 0.1-meter distance penalty on the total distance will be assessed for every 10 seconds of overtime.
- Human power may be used to trigger the device, but not to impart motion to the structure. However, human power may be used during setup to store energy (e.g., into the rubber bands) for any use.
- Distance is measured from the forward-most point of the base prior to deployment to the forward-most connected point of the structure after deployment (in a predefined direction).
- Except within the original base area, no part of the structure may touch the ground in the final deployed position. Incidental (accidental) contact with the floor is permitted during deployment. However, prolonged contact (e.g., using the ground for support, using a wheeled carriage, or bouncing along the ground) is not permitted. No external structures (e.g., wall, ceiling, or pipes) can be used for guidance or support at any time.
- The maximum distance from three trials will be recorded. A misfire will count as one trial.
- Spare parts are recommended and do not count in the materials inventory of the final assembly.

DESIGN PROJECT #5: VERTICAL LIMIT

Your local fire department is looking for conceptual ideas for rescuing people in high-rise buildings. The firefighters have asked you to develop and build a test model for their review. The device is to be self-deploying and is to be designed to extend as high as possible while still remaining connected to the ground. The structure is to be freestanding in its original and deployed states.

The Mission:

Your mission is to design and build a device that will deploy from a prescribed initial size to a freestanding structure that reaches as high as possible. Deployment must be automatic upon activation of a trigger mechanism. The base structure is to be fixed to the ground. The design, for example, may be composed of a mechanism for extending a boom or truss structure.

Design Rules:

The device must be constructed out of the standard project materials listed previously, in addition to *only* the following item: 4 meters of duct tape, 2″ wide, to fix the base to the ground.

Any design deemed by the instructor as unsafe will not be acceptable (e.g., no sharp flying objects and no explosive devices).

More Contest Rules:

- The device, without external support, must initially fit within a 1.0 m x 1.0 m base area on the ground, except for the triggering means.
- The device must initially be less than 0.5 m in height, except for the triggering means.
- Once triggered, the device must deploy automatically to its final state without further assistance and may expand to any size.
- The device must be remotely triggered (e.g., by a string or rod). The trigger may be used only to release energy from the system. The trigger cannot add energy to the system. Team members (all parts of the body) must remain a minimum of 1 meter away from the device when the device is deployed. The means of triggering must be contained on the allowable materials list but will not be counted in the final materials inventory.
- The device may be fixed to the ground only in the original base area, using only duct tape.
- The device must be set up within 3 minutes; otherwise, a 0.1-meter distance penalty on the total height will be assessed for every 10 seconds of overtime.
- Human power may be used to trigger the device, but not to impart motion to the structure. However, human power may be used during setup to store energy (e.g., into the rubber bands) for any use.
- Distance is measured from the ground to the highest point of the structure when it is fully deployed.
- No external structures (e.g., wall, ceiling, or pipes) can be used for guidance or support at any time.

5.10 design projects (continued)

- The maximum height from three trials will be recorded. The structure height must be maintained for the time it takes to measure the height (approximately 2 minutes). A misfire will count as one trial.
- Spare parts are recommended and do not count in the materials inventory of the final assembly.

DESIGN PROJECT #6: THERE 'N BACK

The problem of air pollution caused by automobiles has plagued cities worldwide for decades. Several solutions have been proposed over the years, including public mass transportation systems, electric vehicles, hybrid vehicles, low-emission fuels, human-powered vehicles, and solar- or wind-powered vehicles. None of these options have been very successful to date. Consequently, it is time to develop new concepts in powered vehicles. Recently, your instructor received an anonymous e-mail stating that energy in a vehicle might be stored in elastic elements. This idea was immediately adopted, and a study was commissioned to investigate the possibility of using a large number of surplus rubber bands to power a commuter vehicle.

The Mission:

Your mission is to design and build a small-scale concept vehicle that travels in a linear trajectory as far as possible and then automatically returns along the same trajectory. The device is to be powered by two rubber bands—either #62 or #64. On the day of testing, a travel line will be taped on the floor. Travel distances will be measured in the direction of the line only. Each team will have three launches of their vehicle. *The travel distance to be recorded will be the distance the vehicle travels backward along the trajectory line after the vehicle stops its forward travel.* The backward travel distance cannot exceed the forward travel distance. The single longest distance the vehicle travels backward in three attempts will be recorded for each team. If the vehicle has no forward or backward travel, the final distance will be recorded as zero.

Design Rules:

The device must be constructed out of the standard project materials listed previously (and *only* the materials listed).

Any design deemed by the instructor as unsafe will not be acceptable (e.g., no sharp flying objects, no explosive devices, and no burning or combustible materials).

More Contest Rules:

- The vehicle must be entirely self-contained (e.g., no external launching or guidance devices).
- The entire vehicle must initially fit within a 1.0 m x 1.0 m x 1.0 m volume without external support. After launching, the vehicle can expand to any size.
- The vehicle can be released by hand or remotely triggered. Any number of team members can be involved with the release. Once released, the vehicle cannot be touched.
- The device must be set up within 3 minutes for each launch; otherwise, a 0.5-meter distance penalty on the total distance will be assessed for every 10 seconds of overtime.
- The vehicle or a part of the vehicle must remain in contact with the ground at all times.
- Human power may be used to trigger the vehicle, but not to impart motion to the vehicle (i.e., no pushing or pulling the vehicle) However, human power may be used to store energy in the rubber bands for any use.
- Gravity cannot be used to produce motion (e.g., no launching from a ramp).
- Travel distance is measured in the direction parallel to the length of the path. If the vehicle hits the side wall of the hallway and stops, all vehicle motion is considered finished.
- The final recorded travel distance will be the distance from the closest point of forward travel (from the starting line) on the vehicle to the closest point of return travel (from the forward mark) on the vehicle.
- Objects expelled from the vehicle are still considered a part of the vehicle for measurement of travel distance.
- Spare parts are recommended and do not count in the materials inventory of the final assembly.

sectiontwo

Modern Design Practice and Tools

The widespread availability of computers has made three-dimensional modeling the preferred tool for engineering design in nearly all disciplines. Solid modeling allows engineers to easily create mathematical models parts and assemblies, visualize and manipulate these models in real time, inspect how they mate with other parts, and calculate their physical properties. The geometry of a part, as well as allowable errors, is determined to a large degree on the fabrication method used to make it. Thus, a basic understanding of fabrication processes is important for creating three-dimensional designs. There are many available manufacturing processes, and all have advantages and disadvantages. Three-dimensional modeling also is used as the foundation for many sophisticated computational analysis techniques such as mass properties, interference checking, and finite-element analysis. The ease with which this can be done has moved what was formerly a complicated analytical process performed by specialists into the realm of standard practice by many design engineers as part of the design process.

6

Solid Modeling

objectives

After completing this chapter, you should be able to

- Introduce solid modeling as an engineering design graphics tool
- Explain how solid models are created
- Show how parts and models can be decomposed into features
- Develop strategies for creating a solid model
- Explain how solid models support the entire product life cycle

6.01
introduction

Solid modeling is a computer-based simulation that produces a visual display of an object as if it existed in three dimensions. **Solid models** aid in forming a foundation for the product development process by providing an accurate description of a product's geometry and are used in many phases of the design process and life cycle of the product. This chapter will focus on methods for creating robust solid models of mechanical parts; however, these methods can be applied to other domains as well.

Solid models are created with specialized software that generates files for individual as well as assembled parts. These models are then used in a variety of applications throughout the design and manufacturing processes, as shown schematically in Figure 6.01. During the product concept stage, solid models are used to visualize the design. As the product is refined, engineers use solid models to determine physical properties such as the strength of the parts, to study how mechanisms move, and to evaluate how various parts fit together. Manufacturing engineers use solid models to create manufacturing process plans and any special tools or machines needed to fabricate or assemble parts. Solid models also can be used to generate formal engineering drawings to document the design and communicate details of the design to others. People responsible for the product life cycle may depend on solid models to help create images for service manuals and disposal documentation. Even sales and marketing functions use graphics generated from solid models for business presentations and advertising. Thus, it is very important not only to learn how to create solid models but also to understand how others will use the models. Solid models must be built with sound modeling practices if they are to be useful in downstream applications. In this chapter, you will learn how to create robust solid models that not only look like the real thing but also support the entire product life cycle. You also will learn about the history of CAD tools and the importance of solid modeling as part of an engineering design graphics system.

6.02 Tools for Developing Your Idea

Many tools have been developed for creating accurate images of an object as an aid in analyzing its function, recording its history, or visualizing its appearance. One of the simplest tools is a pencil, which is used to make sketches of an object on paper. More formal tools include rulers, protractors, compasses, and various types of manually operated drafting machines. These tools are used to make more accurate, standardized drawings according to precise rules and conventions, as discussed in a previous chapter.

CAD systems are among the most sophisticated graphics and design tools available to engineers and designers. Many types of CAD systems are on the market today. The simplest systems are general purpose drawing or drafting packages that can be used to create 2-D images, similar to the way pencil images are created on paper (except faster and easier). More complex packages allow you to create simulations of 3-D models that can be used not only to generate conventional 2-D drawings of a design but also to create 3-D images for visualization. The core of a CAD model is a geometric **database**. The database includes information about the geometry and other engineering properties of an object. The CAD software uses the database to display the model and to conduct further engineering analysis. A short discussion of CAD history will demonstrate how these systems evolved and provide some insight into the modeling processes used by designers with various CAD systems.

FIGURE 6.01. Uses for a solid model database.

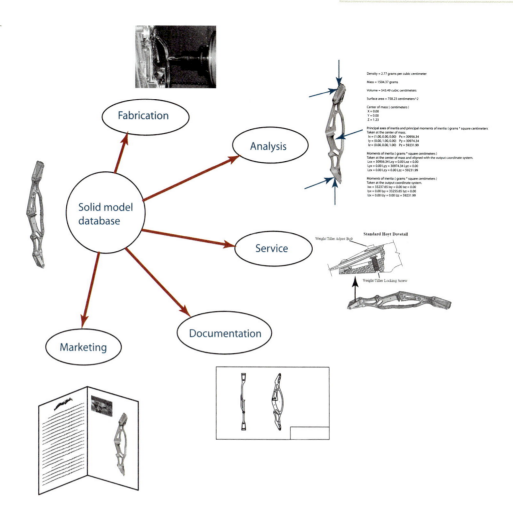

6.02.01 Two-Dimensional CAD

The first CAD systems were developed in the late 1960s at a time when computational resources were very limited. Graphics displays had refresh times measured in seconds, and the data storage capabilities were limited to fractions of a kilobyte. As a result, only very simple models could be created. Those models were basically electronic versions of conventional pencil-and-paper drawings. The user had to specify the location of each vertex in the model for the particular view desired. If the user wanted another view, she had to start from scratch, just as you would to do if you were creating a drawing on paper.

Since CAD models are used to define the geometry or shape and size of objects, the models are composed of geometric entities. In the earliest CAD models, those entities represented the edges of the object, just as you would draw the edges of an object with a pencil. In fact, at that time there was very little distinction between a 2-D drawing of an object and a 2-D CAD model of the object. The 2-D CAD model was simply a database that contains the edges of the object, dimensions, text, and other information that you would find on the drawing, but in electronic form instead of on paper.

The simplest geometric entity is a point. Points in two dimensions are defined according to their location in a coordinate system, usually Cartesian coordinates (x,y). In a CAD system, the coordinate system represents locations on the "paper," or computer screen. Points are generally used to locate or define more complex entities, such as the endpoints of a line segment or the center of a circle. A point on an entity that marks a particular position, such as the endpoint of a line segment or the intersection of two entities, is referred to as a **vertex**.

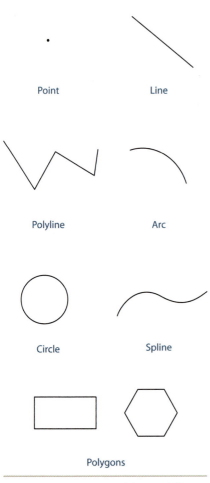

Point

Line

Polyline

Arc

Circle

Spline

Polygons

FIGURE 6.02. Some entities used for 2-D CAD.

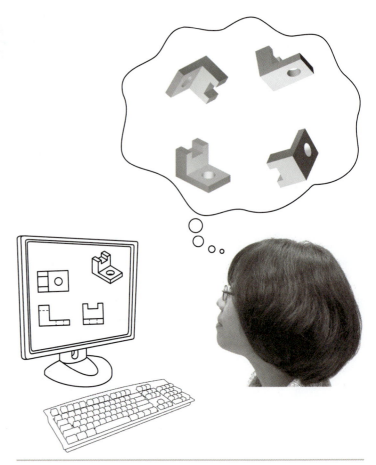

FIGURE 6.03. With 2-D models, visualization of a 3-D object must be done mentally.

Two-dimensional geometric entities are those that can be created as a path or curve on a plane. Those entities include lines, circles, **splines**, arcs, polygons, and conic sections, which are shown in Figure 6.02. The entities can be assembled to create images of a desired object as it would be seen from different viewing directions, as shown in Figure 6.03.

One weakness of 2-D CAD systems is that to visualize and manipulate a 3-D model of the object, you must mentally assemble and reform the 2-D views. Another weakness of 2-D CAD (and pencil drawings) is that it is possible to create images of objects that are physically impossible to build, such as the three-pronged fork and the triangle shown in Figure 6.04.

6.02.02 Wireframe Modeling

In the early 1980s, 2-D CAD drafting packages evolved into 3-D modeling systems. In these newer systems, 3-D information could be included for the model. The computer could then perform the calculations needed to create views of an object as if it was seen from different directions. These systems were still limited to using entities such as

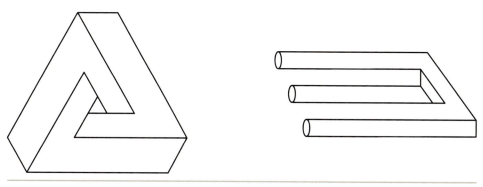

FIGURE 6.04. Impossible 3-D objects can be drawn with 2-D elements.

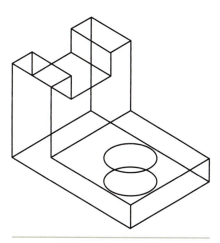

FIGURE 6.05. A wireframe model of a 3-D object.

lines, circles, and arcs; but the assemblage of the entities was no longer restricted to being on a single plane. The geometric entities were represented in a 3-D database within a 3-D coordinate system with x-, y-, and z-coordinates. Since simple curve or path entities were used to define the edges of an object, such models were called **wireframe models**. Think of a wireframe as being similar to a box kite. A wireframe model of a bracket is shown in Figure 6.05.

Wireframe models are still very limited in their representation of parts. The same wireframe model can represent an object from two different viewing directions, as demonstrated in Figure 6.06. Thus, the models were sometimes difficult to visualize as solids. Some models were ambiguous, being interpreted by viewers as different objects. Look at Figure 6.07 and try to imagine the solid object represented by the wireframe model in (a). Can you visualize the shape of the object? Does this figure represent more than one object? When the hidden edges are removed and the surfaces shaded, as in (b) and (c), it is much easier to see the desired shape.

FIGURE 6.06. Two possible view interpretations of the same wireframe model.

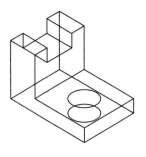

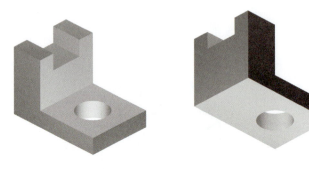

FIGURE 6.07. The wireframe model in (a) can represent the object in (b) or the object in (c).

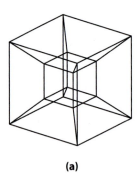

(a)

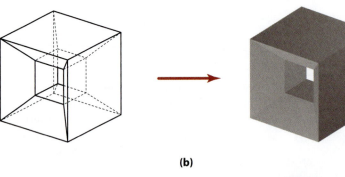

(b)

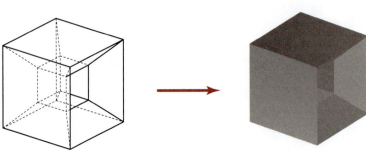

(c)

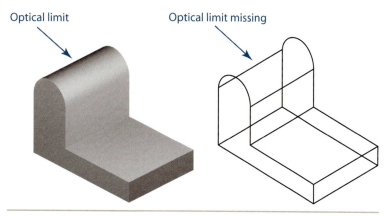

Optical limit

Optical limit missing

FIGURE 6.08. Wireframe models do not show the optical limit of a curved surface.

Another problem with wireframe CAD systems was that the geometry was limited to shapes with simple planar and cylindrical surfaces. Also, parts with cylindrical features, such as the one shown in Figure 6.08, generated wireframe models that did not show the optical limit or silhouette of the cylindrical surface. Even so, wireframe modeling represented a tremendous advance in technology compared to the drafting board. It is estimated that more than 75 percent of all common machined parts can be accurately represented using 3-D wireframe models.

6.02.03 Surface Modeling

As computers became more powerful and data storage capabilities increased, surface modeling techniques were developed. With a **surface model**, the designer could display the surfaces of a part, such as those shown in Figure 6.09, and use the model to perform engineering analyses such as calculating the part's mass properties. Such models also could be used to generate computer programs that controlled the fabrication of parts, for example, on a computer-controlled cutting machine called a mill.

Surface models evolved from wireframe models by mathematically describing and then displaying surfaces between the edges of the wireframe model. Thus, a surface model is a collection of the individual surfaces of the object. This modeling method is called **boundary representation**, or **b-rep**, because the surfaces "bound" the shape. The bounding entities of a simple part created using boundary representation are shown in Figure 6.10. The bounding entities can be planes, cylinders, and other surfaces in three dimensions. These surfaces are in turn bounded by simpler curve entities such as lines and arcs.

FIGURE 6.09. A surface model with semitransparent surfaces to reveal detail.

FIGURE 6.10. A surface model exploded to show individual surfaces.

The use of surface models eliminates most of the problems with visual ambiguity encountered with wireframe models.

6.02.04 Solid Modeling

Solid models are visually similar to surface models, so it is sometimes difficult to distinguish between them. With a solid model, however, the software can distinguish between the inside and outside of a part and the objects can have thickness. Thus, the information stored in the 3-D database is sufficient to distinguish between an empty shoe box and a brick. The software also easily computes information such as the object's volume, mass, center of mass, and other inertial properties. Early solid models, developed in the late 1980s, were made using a technique known as **constructive solid geometry (CSG)**. CSG models are composed of standard building blocks in the form of simple solids such as rectangular prisms (bricks), cylinders, and spheres, called **primitives**. The shapes are easy to define using a small number of dimensions. Figure 6.11 shows some of these basic solids. To create more complex solids, the primitives are assembled using Boolean operations such as addition (union), subtraction (difference), and interference. Examples of these operations are shown in Figure 6.12.

Surface and CSG models were very powerful tools for design, but their early versions were rather cumbersome to use. As computational resources improved, so did the capabilities of modeling software. Increasingly more sophisticated modeling methods, such as creating a solid model by moving or rotating a closed 2-D outline on a path

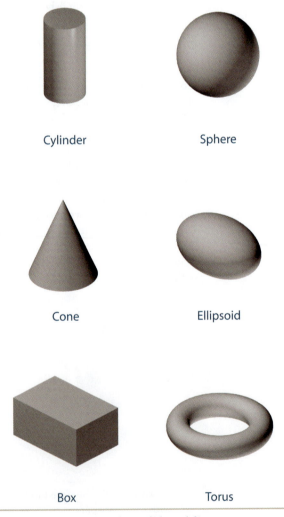

Cylinder   Sphere

Cone   Ellipsoid

Box   Torus

FIGURE 6.11. Some 3-D primitives used in solid modeling.

FIGURE 6.12. Steps in using solid primitives to build a more complicated solid model using Boolean operations.

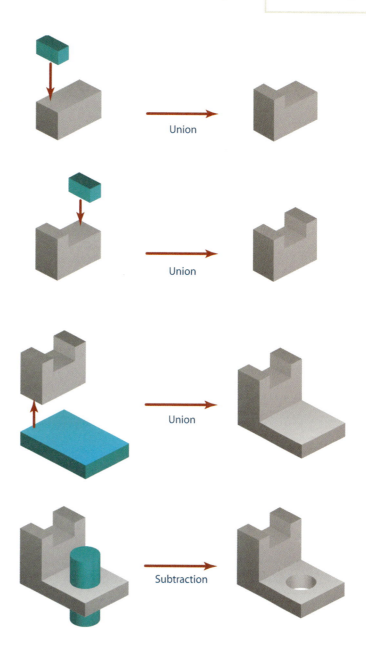

through space, as shown in Figure 6.13, were developed. Further developments included software tools for taking many individual solid model parts and simulating their assembly into a larger structure, as explained in Chapter 1, and for easily creating formal engineering drawings for parts and assemblies from their solid models.

A more accurate and efficient modeling tool called **feature-based solid modeling** was developed in the mid-1990s. This modeling method permitted engineers and designers to create a more complex part model quickly by adding common features to the basic model. Features are 3-D geometric entities that exist to serve some function. One common and easily recognizable feature is a hole. Holes in a part exist to serve some function, whether it is to accommodate a shaft or to make the part lighter. Other features, such as bosses, fillets, and chamfers, will be defined later in this chapter.

Parametric solid modeling is a form of feature-based modeling that allows the designer to change the dimensions of a part or an assembly quickly and easily. Since parametric feature-based solid modeling is currently considered the most powerful 3-D CAD tool for engineers and designers, the remainder of this chapter will be devoted to this modeling method.

FIGURE 6.13. Solids created by (a) moving and (b) revolving a 2-D outline through space.

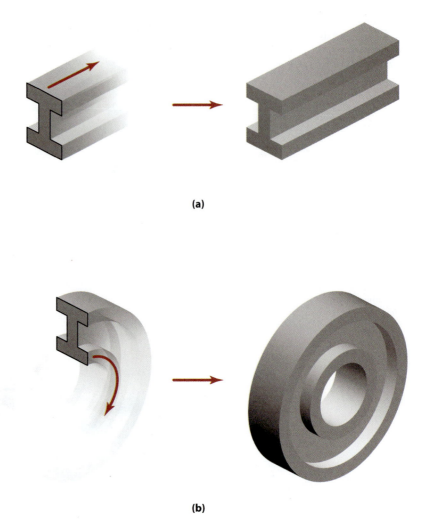

(a)

(b)

6.03 A Parametric Solid Model

So how does one go about creating a parametric feature-based solid model? In this section, a very simple model will be created to demonstrate basic concepts. More detail and sophistication will be presented in subsequent sections of this chapter. The tools that you need to create a parametric model are solid modeling software and a computer that is powerful enough to run the software. As you create the model, the software will display an image of the object which can be turned and viewed from any direction as if it actually existed in three dimensions.

Using the mouse and keyboard, you will interact with the software through a **graphical user interface (GUI)** on the computer's display device (i.e., the computer monitor). The GUI gives you access to various tools for creating and editing your models. GUIs differ slightly in different solid modeling software. However, most of the packages share some common approaches. When creating a new model (i.e., with nothing yet existing), you will probably be presented with a display of 3-D Cartesian coordinate x-, y-, and z-axes and the three **primary modeling planes**, which are sometimes called the **principal viewing planes** or **datum planes**. These planes help you visualize the xy, yz, and xz planes and are usually displayed from a viewing direction from which all three planes can be seen, as shown in Figure 6.14.

Nearly all solid modelers use 2-D **sketches** as a basis for creating solid features. Sketches are made on one of the planes of the model with a 2-D sketching editor similar to a drawing editor found on most 2-D CAD drafting software. When you begin a new model, you often make a sketch on the one of the basic modeling planes. When the sketching plane is chosen, some modelers will reorient the view so you are looking straight at the 2-D sketching plane. You can then begin sketching.

Line segments are usually inserted using mouse clicks, as shown in Figure 6.15(a). A sketch is initially created without much attention being paid to precise dimensions and exact orientations of the different segments. For convenience, the **sketching editor** in most solid models automatically corrects sloppy sketches by making assumptions about the intended geometry. For example, if a line segment is sketched almost vertically or almost horizontally, the sketching editor will force the line into a vertical or horizontal orientation. Figure 6.15(a) shows a sketch of a rectangle created by clicking the four corners, or vertices; Figure 6.15(b) shows the cleaned-up sketch after the sketching editor corrects the user input and reorients the line segments.

6.03.01 Valid Profiles

Before a solid feature can be created by extrusion or rotation, the final profile of the shape must be a closed loop. Extra line segments, gaps between the line segments, or overlapping lines create problems because the software cannot determine the boundaries of the solid in the model. Samples of proper and improper profiles are shown in Figure 6.16.

6.03.02 Creation of the Solid

A completed sketch that is used to create a solid is called a **profile**. A simple solid model can be created from the profile by a process known as **extrusion**, as shown in

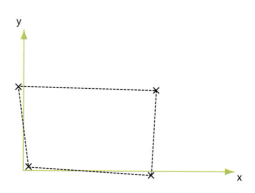

(a) corners of rectangle specified by user

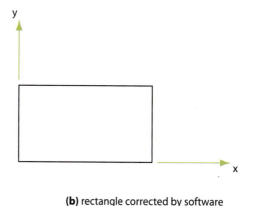

(b) rectangle corrected by software

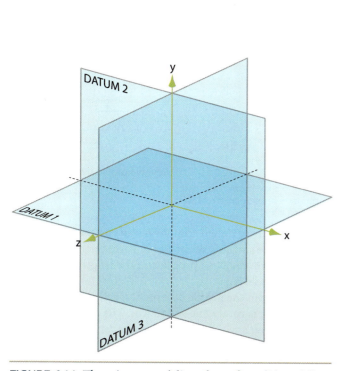

FIGURE 6.14. The primary modeling planes for solid modeling.

FIGURE 6.15. 2-D sketching.

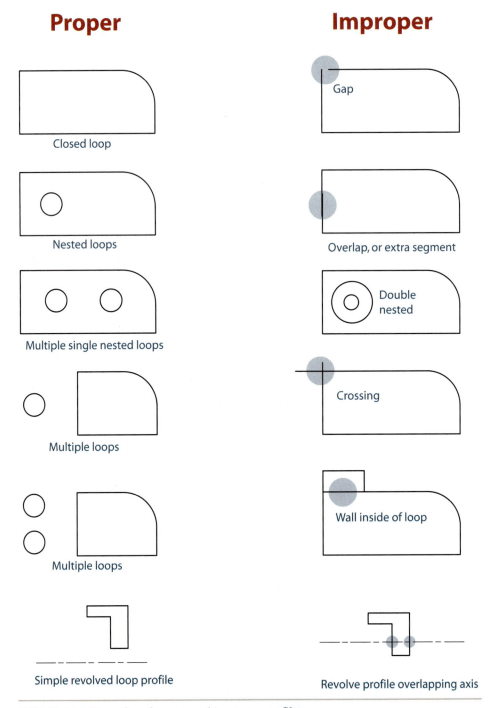

FIGURE 6.16. Examples of proper and improper profiles.

Figure 6.17. Imagine the profile curve being pulled straight out of the sketching plane. The solid that is formed is bound by the surfaces swept out in space by the profile as it is pulled along the path. Both the geometry of the profile and the length of the extrusion must be specified to define the model fully.

A different model can be created from the profile by a process called revolution. To create a **revolved solid**, a profile curve is rotated about an axis. The process is similar to creating a clay vase or bowl on a potter's wheel. The profile of a revolved part is also planar, and the axis of revolution lies in the profile plane (sketching plane). One edge of

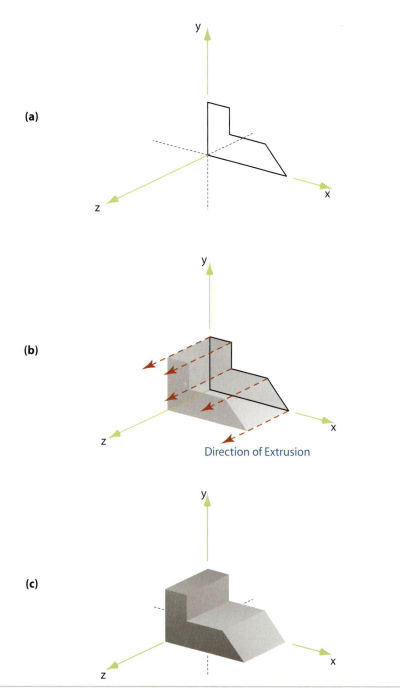

FIGURE 6.17. A solid created by extrusion of a 2-D profile.

the sketch may lie along the axis of revolution, as shown in Figure 6.18(a); or the sketch may be offset from the axis, as shown in Figure 6.18(b). It is important to make sure that the profile does not cross over the axis of revolution. This would create a self-intersecting model (i.e., a solid created inside another solid), which most solid modeling software interpret to be a geometric error. The geometry of the profile and the angle of rotation must be specified to define the model fully. The models shown in Figure 6.18 are revolved through a full 360 degrees.

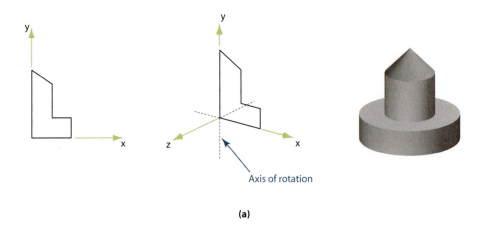

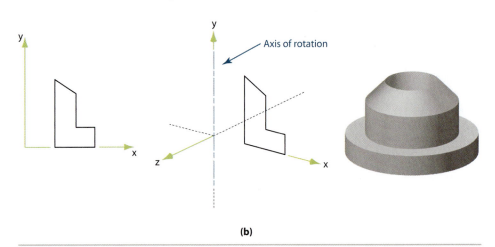

FIGURE 6.18. A solid created by rotation of a 2-D profile, with the axis on the profile in (a) and with the axis off the profile in (b).

6.04 Making It Precise

Before a part can be submitted for analysis or fabrication, the sizes and locations of all of its features must be completely specified. To see how this is done, let's back up a few steps in our discussion of the creation of the model.

6.04.01 Orientation of the Sketch

Before you begin to create the first extrusion or revolution, you must decide where to place the part in the space relative to the xyz coordinate system. With the model shown in Figure 6.17, the initial sketch was placed on one of the basic modeling planes. If the sketch was placed on one of the other basic modeling planes instead, the model would have the same geometry but with a different orientation in space, as shown in Figure 6.19.

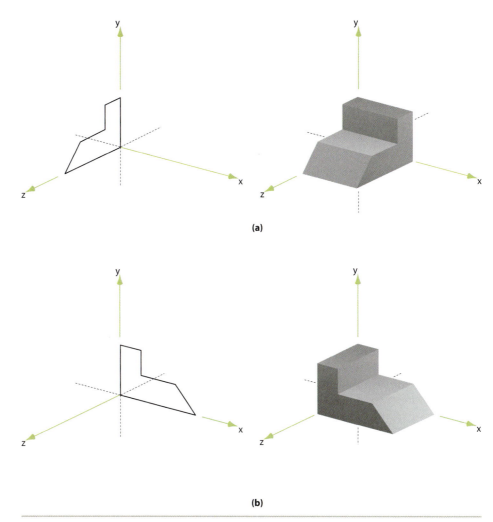

(a)

(b)

FIGURE 6.19. The same profile made in different sketching planes produces the same object but in different orientations. In (a), the profile is made in the yz plane; and in (b), the profile is made in the xy plane.

6.04.02 Geometric Constraints

Formally, **constraints** are the geometric relationships, dimensions, or equations that control the size, shape, and/or orientation of entities in the profile sketch and include the assumptions that the CAD sketcher makes about your sloppy sketching. Constraints that define the size of features will be discussed in the following section. The previous section provided a few examples of **geometric constraints** that were applied to a simple sketch: lines that were drawn as nearly horizontal were assumed to be horizontal, and lines that were drawn as nearly vertical were assumed to be vertical. Those assumptions reduce the number of coordinates needed to specify the location of the endpoints. Some solid modelers require you to constrain the profile fully and specify the sizes and locations of all of its elements before allowing the creation of a solid feature; others allow more free-form sketching. Geometric constraints may be either implicitly defined (hidden from the designer) or explicitly displayed so you can modify them. A set of geometric constraints is not unique, as demonstrated in Figure 6.20. In

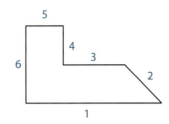

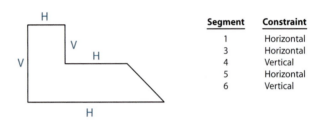

Segment	Constraint
1	Horizontal
3	Horizontal
4	Vertical
5	Horizontal
6	Vertical

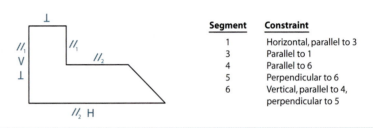

Segment	Constraint
1	Horizontal, parallel to 3
3	Parallel to 1
4	Parallel to 6
5	Perpendicular to 6
6	Vertical, parallel to 4, perpendicular to 5

FIGURE 6.20. The line segments in a profile are numbered in (a). The implied geometric constraints for each segment are shown in (b), and an equivalent set of applied constraints is shown in (c). A letter or symbol beside a segment signifies the type of geometric constraint applied to it.

this example, a set of geometric constraints that restricts some lines to being horizontal or vertical is equivalent to another set of constraints that restricts some lines to being either parallel or perpendicular to each other.

Geometric constraints specify relationships between points, lines, circles, arcs, or other planar curves. The following is a list of typical geometric constraints. The results of applying the constraints are shown graphically in Figure 6.21.

- Coincident—forces two points to coincide
- Concentric—makes the centers of arcs or circles coincident
- Point on Line—forces a point to lie on a line
- Horizontal/Vertical—forces a line to be horizontal/vertical
- Tangent—makes a line, a circle, or an arc tangent to another curve
- Colinear—forces a line to be colinear to another line
- Parallel—forces a line to be parallel to another line
- Perpendicular—forces a line to be perpendicular to another line
- Symmetric—makes two points symmetric across a centerline

Before constraint   **After constraint**

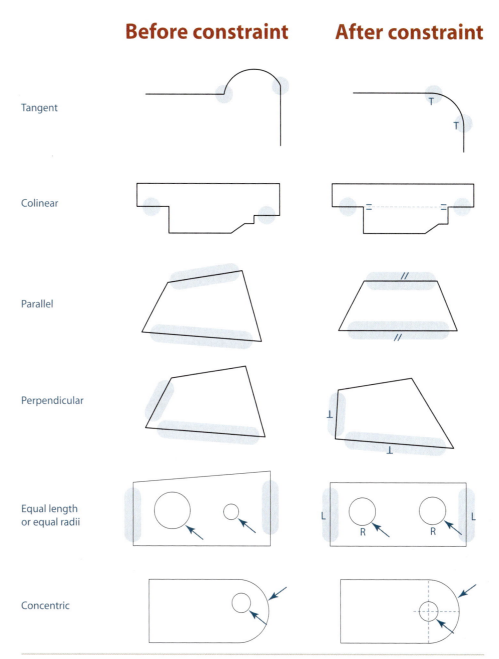

FIGURE 6.21. Geometric constraints commonly found in sketching editors.

The sketching editors in most solid modeling software are usually configured to try to interpret the user's sketching intent such that certain constraints are created automatically. In addition to adjusting nearly horizontal or vertical lines into true horizontal or vertical lines, if two lines are nearly perpendicular or parallel or an arc and a line are nearly tangent at the common endpoint, the sketching editor will impose the assumed geometric relationship. These automatically applied geometric constraints can be changed at a later time if desired.

6.04.03 Dimensional Constraints

Each of the 2-D entities in the profile must have size and position. **Dimensional constraints** are the measurements used to control the size and position of entities in your sketch. Dimensional constraints are expressed in units of length, such as

FIGURE 6.22. A profile fully constrained with geometric and dimensional constraints.

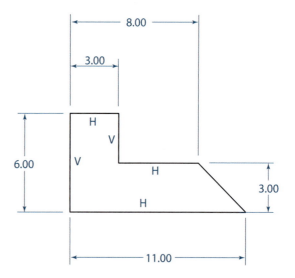

millimeters, meters, inches, or feet. For example, look at the profile in Figure 6.22, which shows dimensional constraints that define its size. If you, the designer, do not fully specify all of the necessary information, the software will default to some value that you may not want. It is better if you control the model, rather than have the software assign assumed parameters and conditions to the model.

Dimensional constraints can be created interactively while you are sketching, but also automatically as a result of a feature operation, an extrusion, or a revolution. There are three principal types of dimensional constraints:

- Linear dimensional constraints define the distance between two points, the length of a line segment, or the distance between a point and a line. Linear dimensions can be measured horizontally or vertically or aligned with the distance being measured.

- Radial and diametral dimensional constraints specify the radius or diameter of an arc or a circle.

- Angular dimensional constraints measure the angle between two lines. The lines do not need to intersect, but they cannot be parallel.

6.04.04 Uniqueness of Constraints

A set of dimensional constraints is not unique. It is possible to apply a different set of dimensional constraints on a profile to produce exactly the same geometry, as shown in Figure 6.23.

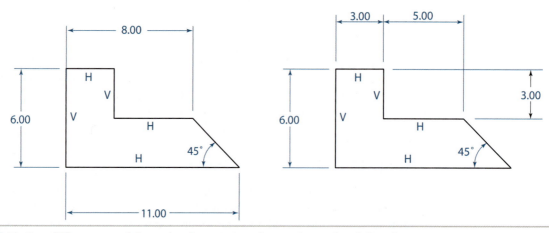

FIGURE 6.23. Two different sets of dimensional constraints that can be used to define the same geometry.

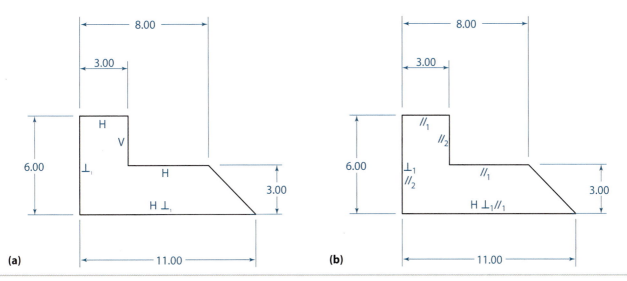

FIGURE 6.24. Two different sets of geometric constraints that define the same geometry.

Combinations of dimensional and geometric constraints also are not unique. It is possible to have different combinations of geometric constraints and dimensional constraints define exactly the same geometry, as shown in Figure 6.24.

The natural question then becomes, which set of constraints is correct or preferred? The answer depends on what the function of the part and the design intent is or how the designer wants to be able to change the model. You also should consider how the solid model will be used for analysis, manufacturing, and documentation when applying sets of constraints. One of the greatest advantages of a parametric solid model is that the model can be changed easily as the design changes. However, the constraints limit the ways in which the model can be changed.

6.04.05 Associative and Algebraic Constraints

Associative constraints, sometimes called **algebraic constraints**, can be used to relate one dimensional constraint to another. The dimensional constraints on a profile are expressed in terms of variables. Each dimensional constraint is identifiable by a unique variable name, as shown in Figure 6.25. Algebraic constraints can be used to control the values of selected variables as the result of algebraic expressions. Algebraic expressions consist of constants and variables related to each other through the use of arithmetic functions (+, −, *, absolute value, exponent, logarithm, power, square root, and sometimes minimum and maximum); trigonometric functions; and conditional expressions (if, else, or when) including inequalities comparisons (if A > B then . . .).

There are two different methods for solving sets of algebraic constraint equations. Software that uses **variational techniques** solves the equations simultaneously. A compatible solution for all of the variables can be calculated when there are a sufficient number of equations. In a system using **parametric techniques**, the equations are usually solved in sequential order. The equations will have only one unknown variable. All other variables in the algebraic expression must be known for the value of the unknown variable to be calculated, which is called the dependent or **driven dimension**. The known variables are called the **driving** dimensions. As shown in Figure 6.25, when the value of a driving dimensional constraint is changed, the value of its driven dimensional constraints are automatically changed too.

FIGURE 6.25. Dimensional constraints are shown in term of variables and a set of algebraic constraints in (a). Dimensions D3, D4, and D5 are automatically specified by specifying dimensions D1 and D2 in (b). Dimensions D3, D4, and D5 change automatically when D1 and D2 are changed in (c).

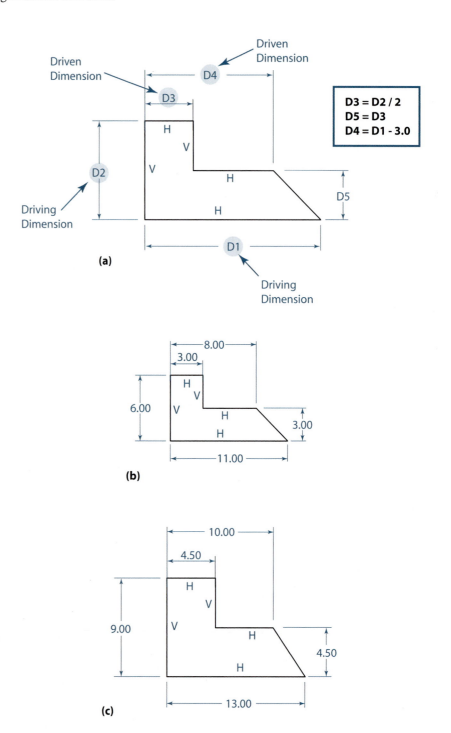

6.05 Strategies for Combining Profile Constraints

A completed profile is constrained using a combination of geometric and dimensional constraints and may include algebraic constraints as well. The constraint set must be complete for the geometry to be fully defined. If a profile is overconstrained or underconstrained, it may not be possible to create a solid feature from the profile. Some solid modeling software automatically applies constraints, but these constraints usually need to be changed to reflect the design intent. Furthermore, most software systems expect the user to apply constraints in addition to the automatically generated constraints; in particular, variational modelers allow underconstrained sketches and do not require user-applied constraints, but these systems can yield unpredictable results when the dimension values are changed. By gaining a thorough understanding of constraints (and how and when to apply them), you will be able to control the behavior of your models and capture your design intent. A strategy for applying geometric and dimensional constraints to a profile is demonstrated next.

The first constraint usually applied to a new sketch is a **ground constraint**. Ground constraints serve as anchors to fix the geometry in space. Ground constraints may have various forms. The most common type of ground constraint is a geometric entity such as a line or point on the profile having been made coincident with one of the basic modeling planes or with the origin of the coordinate system. For example, if the first feature of a model is created by extrusion, it may be convenient to place a corner of the profile on the origin of the coordinate system. This is usually done by placing one of the vertices of the sketch exactly at the origin. If the first feature is created by rotation, it may be convenient to place one of the endpoints of the center axis at the origin.

When the profile is closed and the automatically generated constraints have been applied the interactive constraint definition phase begins. Some software creates a fully constrained sketch, including both geometric and dimensional constraints; but the constraint set chosen by the software is usually not exactly what you want. Other software does not fully constrain the sketch, but leaves this task to the designer. Ground constraints should be specified if this was not already done when the profile was sketched. Next, geometric and dimensional constraints should be added and/or changed until the profile is fully constrained. Typically, your solid modeling software will alert you when the profile is fully constrained or when you try to overconstrain the profile. In particular, you should take care to delete any unwanted geometric and dimensional constraints that may have been automatically added. Finally, the profile should be changed to reflect the design intent and the dimensional constraints adjusted to the desired values. Some sketching editors automatically readjust the profile after each constraint is added; others wait until all of the constraints have been specified before readjusting the profile. Updating the profile to show its new shape after constraints are added or changed is called **regeneration**.

The way dimensional constraints are added depends largely on what the intended function of the part is, how it is to be fabricated, and how the geometry of the part may change in the future. What would a simple L-bracket look like if some of the dimensions were changed, as shown in Figure 6.26? In this case, d1 was changed from 30 to 40 and d2 was changed from 3 to 8. The result is shown in the figure. But if you want to make the bracket by bending a piece of sheet metal, the part should have a uniform thickness throughout. One way to do this is to force the length of line segments that define the thickness of both legs of the *L* to be equal. The geometric constraint shown in Figure 6.27 has this effect. The equal length geometric constraint replaces the dimensional constraint for the thickness of the vertical leg of the bracket. If you tried to apply the equal length constraint and the dimensional constraints on both line segments, the sketch would be overconstrained, a situation the software would not accept. In addition, an associative constraint needs to be added between the radius of the inside corner of the bracket and the radius of the outside corner to ensure that the thickness of the part is constant around the corner.

This constraint strategy demonstrates how to make your parts more robust. Through this simple example, you can see the importance of fully understanding the behavior of your model and the effects of your selection of dimensions and constraints. Your choices for geometric, dimensional, and algebraic constraints are not unique; but the decisions you make in selecting a set of constraints will have a big impact on the behavior of your model if you make changes to it. You must choose a modeling strategy that will reflect your design intent.

FIGURE 6.26. Changing the values of the dimensions changes the geometry of the model, without the need to reconstruct the model.

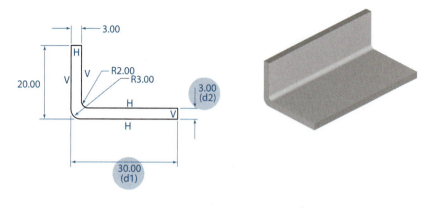

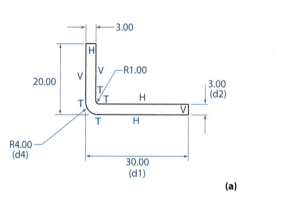

FIGURE 6.27. When compared to the original model in (a), the addition of the equality and associative constraints in (b) ensures a constant material thickness even if the dimensions are changed, thus adding functionality to the model if that is the intent.

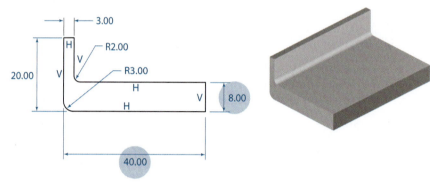

(a)

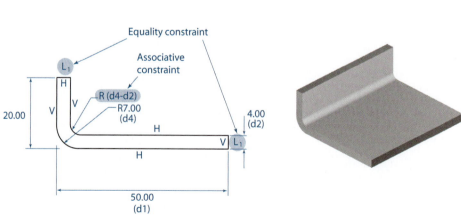

(b)

As an exercise for developing skill in the application of constraints, consider the rough sketch and the finished profile shown in Figure 6.28. For the profile to be fully constrained using only the dimensional constraints shown, certain geometric constraints are needed. Segment 1, for example, needs to be horizontal and tangent to Segment 2. Segment 2 needs to be tangent to Segment 1 as well as to Segment 3. Segment 3 needs to be tangent to Segment 2, perpendicular to Segment 4, and parallel to Segment 5. Segment 4 needs to be perpendicular to Segment 3 and equal in length to Segment 14. These constraints and the required geometric constraints on the remaining segments are shown in Figure 6.29. Keep in mind that a set of geometric constraints may not be unique. Can you specify another set of geometric constraints for this example that would create the same profile with the same dimensional constraints?

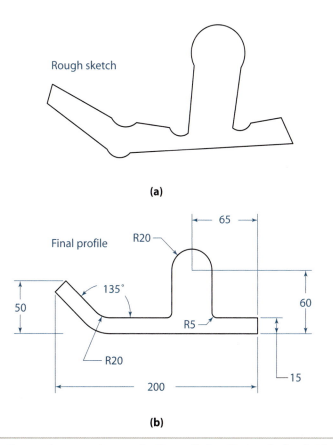

FIGURE 6.28. Geometric constraints need to be applied to the rough sketch (a) to produce the desired, fully constrained profile (b).

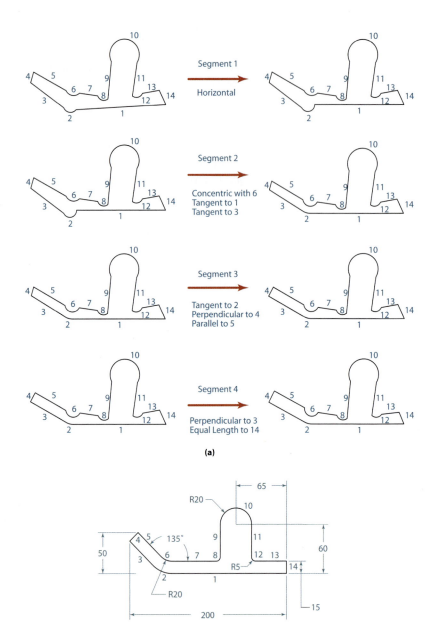

(a)

(b)

Segment	Constraint
1	Horizontal, Tangent to 2
2	Concentric with 6, Tangent to 1, Tangent to 3
3	Perpendicular to 4, Parallel to 5, Tangent to 2
4	Equal Length to 14, Perpendicular to 3
5	Parallel to 3, Tangent to 6
6	Concentric with 2, Tangent to 5, Tangent to 7
7	Horizontal, Tangent to 6, Tangent to 8
8	Tangent to 7, Tangent to 9
9	Vertical, Tangent to 8, Tangent to 10
10	Tangent to 9, Tangent to 11
11	Vertical, Tangent to 10, Tangent to 12
12	Tangent to 11, Tangent to 13
13	Horizontal, Tangent to 12
14	Vertical, Equal Length to 4

6.06 More Complexity Using Constructive Solids

You have seen how to create solid models by sketching a 2-D profile on one of the basic modeling planes and then using a single extrusion or a single rotation to create a 3-D model. Adding material to or removing material from the original model can create a more complex model. When material is added, a **protrusion** feature is created. When material is removed, a **cut** feature is created. Both protrusions and cuts begin with sketched profiles that are then extruded or revolved to form solid shapes that are added to or removed from the existing body of the model. For an extruded feature, the profile lies in the sketch plane and is extruded in a direction perpendicular to the sketching plane. For a revolved feature, the profile and the axis of revolution must be coplanar so both will lie on the sketch plane.

When protrusions or cuts are made on an existing model, sketches and profiles are no longer restricted to be located on one of the basic modeling planes. Instead, any planar surface on the model can be selected and used as a **sketching plane** on which sketches and profiles can be created. Once a sketching plane has been selected, any 2-D element that is created will be forced to lie on that plane. After a sketching plane is selected, the model can be reoriented to look directly into the sketching plane. Although you can sketch when not looking directly into the sketching plane, you need to be very careful when viewing from a different orientation. Edges of your sketch may not be shown in their true shape, and angles may appear distorted. Most people find it easier to create 2-D profiles when they are looking directly into the sketching plane, just as it is easier for someone to draw straight lines and angles with correct measurements when the paper is oriented straight in front of her.

Examples of profiles on various sketching planes on a model and resulting extruded protrusions are shown in Figure 6.30; examples of extruded cuts are shown in Figure 6.31. Examples of revolved protrusions are shown in Figure 6.32, and examples of revolved cuts are shown in Figure 6.33.

As with the first extrusion or revolution that created the main body of the model, the profiles for the added protrusions or cuts must be fully defined by geometric, dimensional, and algebraic constraints before they can be extruded or revolved. A common geometric constraint for protrusions or cut features is to make one or more edges or vertices of the new profile coincident with edges of the surface used as the sketching plane. In Figure 6.30(a), notice that one surface of the original object has been selected as a sketching plane and a rectangular profile has been sketched on the selected plane. The top and bottom edges of the sketched profile are coincident with edges of the sketching surface. The direction of extrusion is, by default, perpendicular to the selected sketching plane.

The length of the extrusion or angle of rotation also must be specified. There are several options for defining the length of the extrusion, as shown in Figure 6.30 and Figure 6.31. The simplest is to specify a **blind extrusion**. A blind extrusion is one that is made to a specified length in the selected direction, analogous to specifying a dimensional constraint, as shown in Figure 6.30(b). If your extrusion is the first feature used to create your initial model, it will be a blind extrusion. For a cut such as a hole, a blind extrusion creates a hole of a specified depth, as shown in Figure 6.31(b).

Another way to determine the length of the extrusion is to use existing geometry. One option for specifying an extrusion length is to **extrude to the next surface**. With this option, the extrusion begins at the profile and the protrusion or cut stops when it intersects the next surface encountered, as shown in Figure 6.30(c) and Figure 6.31(c). Another option is to **extrude to a selected surface**, where the protrusion or cut begins at the profile and stops when it intersects a selected surface, which may not necessarily be the first one encountered. See Figure 6.30(d) and Figure 6.31(d). For extruded cuts, there is an option to **extrude through all**. This option creates a cut or protrusion that starts at the profile and extends in the selected direction through all solid features, as shown in Figure 6.31(e). A **double-sided extrusion** permits the

FIGURE 6.30. Different ways to terminate an extruded protrusion from the profile in the sketching plane in (a). Blind extrusion in (b), extrude to next surface in (c), extrude to a selected surface in (d).

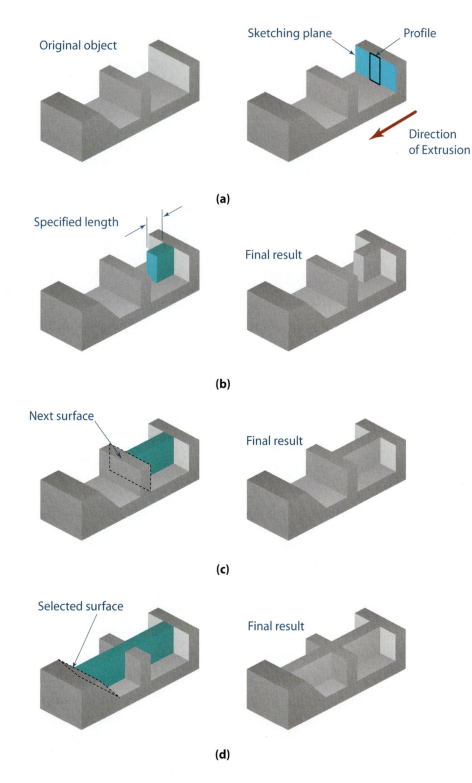

protrusion or cut to extend in both directions from a profile. The method of termination in each direction can then be specified independently. Other methods of terminating the extrusion length may be available depending on the specific solid modeling software used. You also can specify the angle of rotation of a revolved protrusion or cut in a similar manner by using a specified angle (blind revolution) or by revolving up to next or selected surfaces.

FIGURE 6.31. Different ways to terminate an extruded cut from the profile in the sketching plane in (a). Blind cut in (b); cut to next surface in (c). Cutting to a selected surface (d) and cutting through all (e).

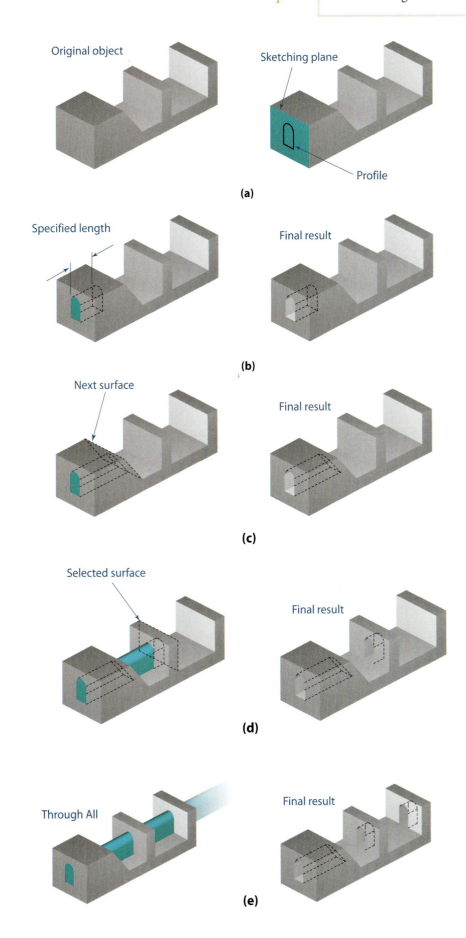

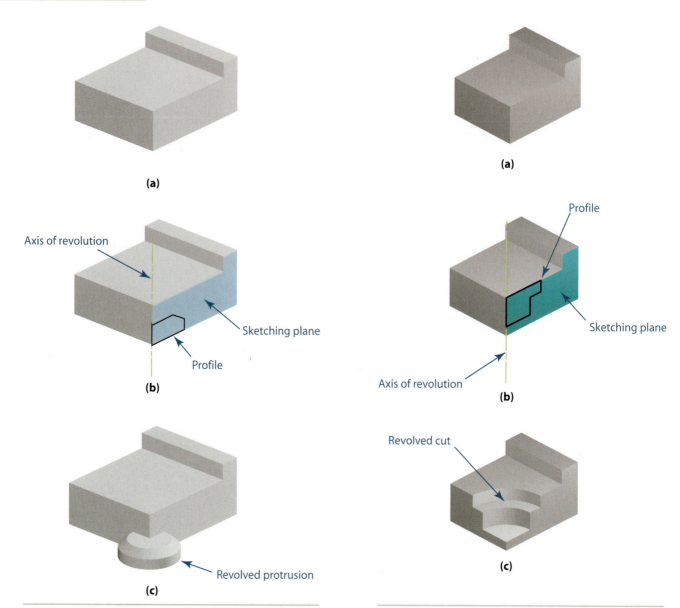

FIGURE 6.32. The addition of a revolved protrusion to an existing base in (a) by using one of its surfaces as a sketching plane to create a centerline and profile in (b) and revolving it to produce the final result in (c).

FIGURE 6.33. The addition of a revolved cut to an existing base in (a) by using one of its surfaces as a sketching plane to create a centerline and profile in (b) and revolving it to produce the final result in (c).

6.07 Breaking It Down into Features

When you build a solid model, you need to decide how to create the various shapes that compose the part. Very few parts can be modeled as a single extrusion or revolution. The various protrusions and cuts on the main body of a model are called **features**. What are features? If you consider your face, you might say that its features are your eyes, nose, lips, and cheeks. It is not much different on a manufactured part; a feature can be any combination of geometric shapes that make up the part and are distinctive in shape, size, or location. Features are characteristic elements of a particular object, things that stand out or make the object unique. Features often have characteristic geometric shapes and specific functions. A simple hole, for example, is a cylindrical cut that is often used as a receptacle for a fastener such as a bolt or screw. A manufactured part may have many different types of features. Since these features are the foundation of contemporary solid modeling systems, you must be able to recognize them.

Engineered parts also have features that are composed of repeated combinations of shapes. Most feature-based modelers have a collection of standard built-in features and may also allow you to define your own features. This can be handy when your products are designed with a particular shape that varies in size, such as gear teeth, airfoils, or turbine blades. The challenge for designers is to identify part features and build solid models that reflect the function of the part and design intent.

6.07.01 The Base Feature

All of your parts will be created from a collection of features, but you need to start your model with a basic shape that represents the general shape of the object. Your first step should be to study the part and identify the shape that you will use as the **base feature**. The base feature should be something that describes the overall shape of the part or something that gives you the greatest amount of functional detail that can be created with a single extrusion or rotation. Figure 6.34 shows several parts with the base features used to create the solid models.

After the base feature is created, you can modify the shape by adding or subtracting material to it to create form features. A **form feature** is a recognizable region or area on the part geometry that may have a specific function and/or method of manufacture. The geometric components or shapes within the feature usually have some geometric relationships or constraints. Different CAD systems use various names for these features, but you should become familiar with some of the common terms. The following section discusses common feature types.

FIGURE 6.34. Parts and their base features.

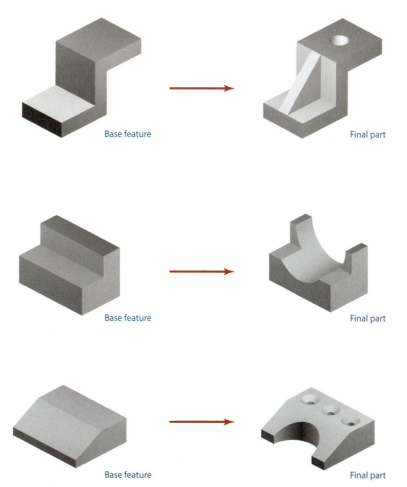

Base feature   Final part

Base feature   Final part

Base feature   Final part

FIGURE 6.34. (CONTINUED)
Parts and their base features.

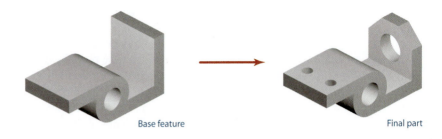

Base feature   Final part

Base feature   Final part

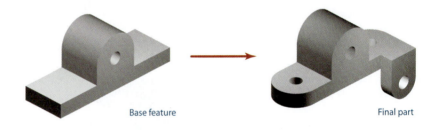

Base feature   Final part

Base feature   Final Part

Base feature   Final Part

6.07.02 Chamfers, Rounds, and Fillets

Unless otherwise specified, adjoining surfaces on a virtual part can intersect to form sharp corners and edges, but real parts often have smooth transitions along the edges of these surfaces. The most common edge transitions are **rounds**, **fillets**, and **chamfers**. A round is a smooth radius transition of the external edge created by two intersecting surfaces. A fillet is a smooth transition of the internal edge created by two intersecting surfaces. Geometrically, the rounds and fillets are tangent to both intersecting surfaces. Examples of rounds and fillets are shown in Figure 6.35. Fillets and rounds are specified by the size of their radii and the edge(s) that are rounded.

Chamfers also provide a transition between two intersecting surfaces, but the transition is an angled cut instead of a radius. Examples of chamfers are shown in Figure 6.36. Chamfers can be specified by the distance along each intersecting surface to the original edge or by the distance along one of the original surfaces and the angle made with that surface.

Functionally, on an inside edge, a fillet may be necessary to facilitate fabrication or to reduce stresses at the corner so the part does not break as easily. On an outside edge, rounds and chamfers are usually used to eliminate sharp edges that can be easily damaged or that can cause injury or damage when the part is handled. Rounds, fillets, and chamfers are generally small when compared to the overall size of the associated base or parent feature.

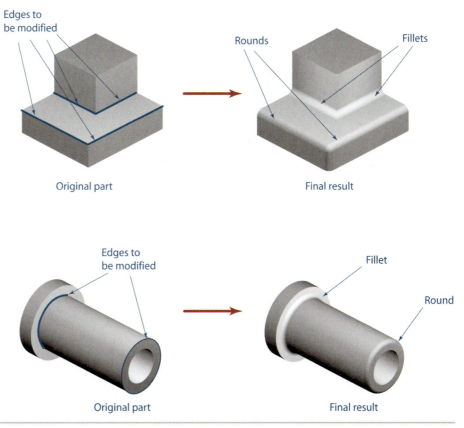

FIGURE 6.35. Examples of rounds and fillets applied to the edges of a part.

FIGURE 6.36. Examples of chamfering applied to the edges of a part.

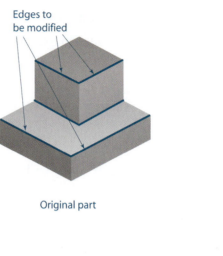

Original part
Final result

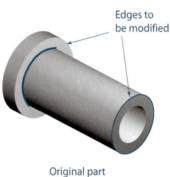

Original part
Final result

6.07.03 Holes

Holes are ubiquitous in nearly all manufactured parts and, therefore, can be inserted into a model as features by most solid modeling software. Holes are often used with bolts or screws to fasten parts together. Many different types of holes can be used with specific fasteners or can be created using different manufacturing processes. Some special types include holes that are blind, through, tapped, counterbored, or countersunk, as shown in Figure 6.37. Each type of hole has a particular geometry to suit a specific function. You should study the hole types so you recognize them when you model your parts.

Many solid modeling software packages include standard or built-in features to help you with your modeling task. When you use a standard hole feature, the solid modeling software makes certain assumptions about the geometry of a hole so you do not need to specify all of the dimensions and constraints that make up the feature. A countersunk hole, for example, can be made as a revolved cut. What do you need to do to create this feature? You begin by selecting a sketching plane, then create and constrain the sketch and revolve the sketch about a specified axis. Many things can go wrong if you are not careful. There might not be a plane on which to sketch. Your sketch might not have the proper shape, or the axis might not be perpendicular to the desired surface. However, a countersunk hole feature can often be created from a standard feature by selecting the location of the axis of the hole on the desired surface, the diameter of the hole, the diameter and angle of the countersink, and the depth of the hole. The shape of the profile, axis of revolution, and angle of revolution are included automatically in the feature definition. No sketching plane is needed.

FIGURE 6.37. Cross sections of various types of holes to reveal their geometry.

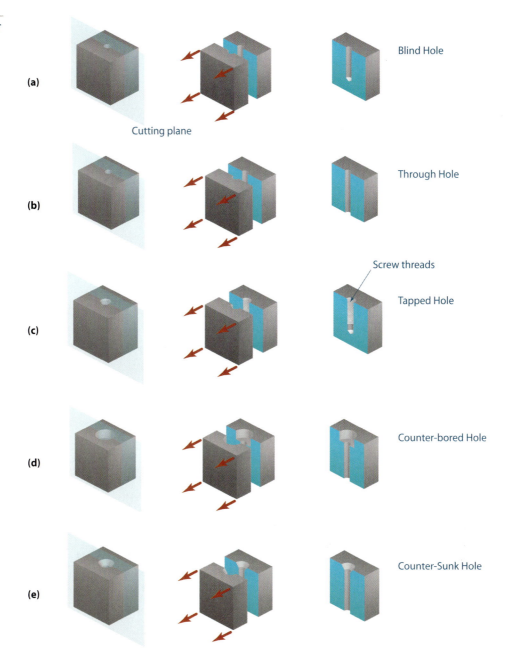

(a)

Cutting plane

Blind Hole

(b)

Through Hole

Screw threads

(c)

Tapped Hole

(d)

Counter-bored Hole

(e)

Counter-Sunk Hole

Figure 6.38 illustrates the use of a cut feature compared to the use of a built-in hole feature to create a countersunk hole. In most cases, it is more desirable to create a hole using the built-in hole feature instead of a general purpose cut feature. Besides being a more natural way to place a hole in a model, you avoid potential errors in creating the desired geometry. Furthermore, using a general cut feature does not incorporate the specific geometry and function of a "hole" in the knowledge base of the model, which may be useful in downstream applications such as process planning for manufacturing the hole.

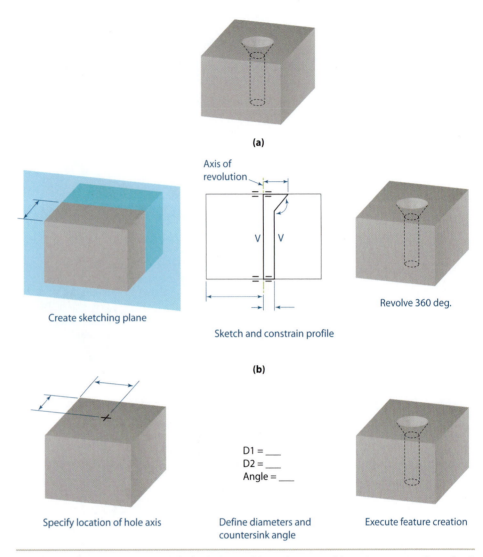

(a)

Axis of revolution

Create sketching plane

Sketch and constrain profile

Revolve 360 deg.

(b)

Specify location of hole axis

D1 = ___
D2 = ___
Angle = ___

Define diameters and countersink angle

Execute feature creation

FIGURE 6.38. The countersunk hole shown in (a) can be created by using a general revolved cut, as shown in (b), or by specifying the hole as a built-in feature, as shown in (c).

6.07.04 Shells

The process of creating a shell, or **shelling**, removes most of the interior volume of a solid model, leaving a relatively thin wall of material that closely conforms to the outer surfaces of the original model. Shelled objects are often used to make cases and containers. For example, a soda bottle is a shell, as are cases for electronic products such as cell phones and video displays. The walls of a shell are generally of constant thickness, and at least one of the surfaces of the original object is removed so the interior of the shell is accessible. Figure 6.39 shows examples of a model that has been shelled.

Shelling is sometimes considered an operation rather than a feature. It is usually performed on the entire model, including all of its features, by selecting the surfaces to be removed and the thickness of the shell wall. Any feature not to be shelled should be added to the model after the shelling operation is complete. The order of feature creation and shelling operations may have a dramatic effect on the shape of the part, as will be shown later in this chapter.

FIGURE 6.39. Examples of shelling.

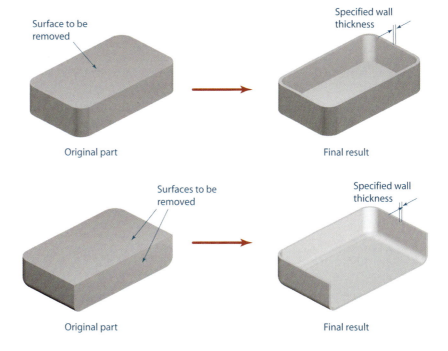

6.07.05 Ribs and Webs

Ribs are small, thin, protrusions of constant thickness that extend predominantly from the surface of a part. Ribs are typically added to provide support or to stiffen a part. Sometimes they are added to improve a part's heat transfer ability. **Webs** are areas of thin material that connect two or more heavier areas on the part. Examples of ribs and webs are shown in Figure 6.40. These features are usually specified by their flat geometry, thickness, and location.

FIGURE 6.40. Ribs (a) added to parts to reinforce them, and webs (b) connect thicker sections on parts.

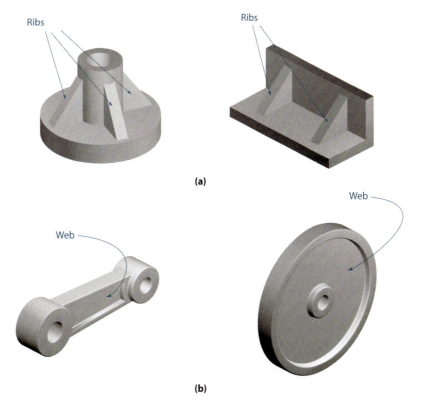

6.07.06 Other Feature Types

The features that follow (and that are shown in Figure 6.41) are less commonly found in solid modelers. When available, they should be used as needed. When such features are not available in the solid modeler, the geometric shapes can still be created from sketched profiles as protrusions or cuts. Note that special feature types usually imply a particular shape, function, manufacturing process, or other feature attribute.

- Boss—a slightly raised circular area, usually used to provide a small, flat, clean surface
- Draft—a slight angle in the otherwise straight walls of a part, usually used to facilitate its removal from a mold
- Groove—a long, shallow cut or annulus
- Island—an elongated or irregularly shaped raised area, usually used to provide a flat, clean surface
- Keyseat—an axially oriented slot of finite length on the outside of a shaft
- Keyway—an axially oriented slot that extends the entire length of a hole
- Slot—a straight, long cut with deep vertical walls
- Spot face—a shallow circular depression that has been cut, usually used to provide a small, flat, clean surface
- Taper—a slight angle in the otherwise cylindrical walls of a part, usually used to facilitate its insertion or removal into another part

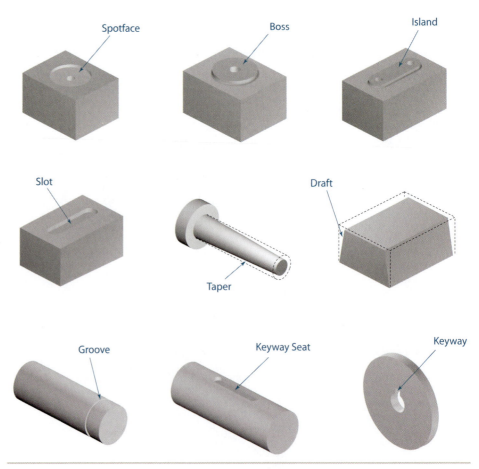

FIGURE 6.41. Various features with specific functions that may be added to a solid model.

Thread

Knurl

FIGURE 6.42. Some cosmetic features.

6.07.07 Cosmetic Features

Parts can be modified by altering their surfaces characteristics. These characteristics are called **cosmetic features** because they generally modify the appearance of the surface but do not alter the size or shape of the object, just like lipstick or hair coloring. Cosmetic features are necessary to the function of the part and may be included in the model so they can be used in later applications, such as fabrication. Some common cosmetic features include threads and knurls. Since the geometric changes are small and detailed, the cosmetic features usually are not modeled in their exact geometric form in the database of the object, but are included as notes or with a simplified geometric representation. You will learn more about simplified representations on drawings in later chapters. Some cosmetic features are shown in Figure 6.42.

6.07.08 An Understanding of Features and Functions

As a design engineer, you need to become familiar with the different types of features on various parts. Doing so will help you communicate with other engineers as well as imbed more of a part's engineering function into your models. For example, if you look at Figure 6.43, you will notice a rectangular cut on the edge of the hole. This cut is a geometric feature called a keyway. Why is it there? What purpose does it serve? In Figure 6.44, the gear is mounted to a shaft, which also has a rectangular cut. A small part called a key is used to line up the shaft and the gear and transmits torque from the shaft to the gear. If you were to create a feature-based solid model of the gear, you could identify the rectangular cut as a keyway feature. If the model parts were to be assembled with assembly modeling software (which is explained in detail in a subsequent chapter), the computer and software would recognize the models as mating parts and orient the gear, key, and shaft automatically.

FIGURE 6.43. A gear with teeth, a bore, and a keyway as functional features.

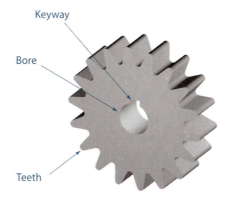

Keyway

Bore

Teeth

FIGURE 6.44. A gear and shaft assembly. The key functions to transmit torque. The keyseat receives the key in the shaft, and the keyway receives the key in the gear.

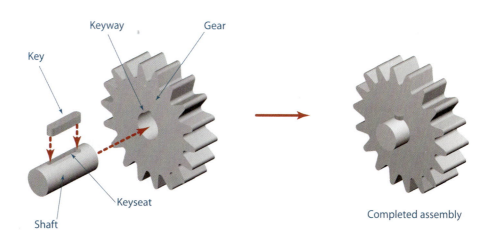

Keyway   Gear

Key

Keyseat

Shaft

Completed assembly

6.08 More Ways to Create Sophisticated Geometry

Creating protrusions and cuts by extending the sketch profiles made on either the basic modeling plane or one of the existing surfaces of the model results in a wide variety of possible models. Even more sophisticated models, however, can be created by using reference geometries called datums, which can be added to the model, displayed, and used to create features. Generally, solid modelers offer at least three types of **datum geometries** that can be placed into a model: datum points, datum axes, and datum planes. These datum geometries do not actually exist on the real part (i.e., they cannot be seen or felt) but are used to help locate and define features. Consider, for example, the part shown in Figure 6.45(a). The angled protrusion with the hole would be easy to create if an angled sketching plane could be defined as shown. The extrusion could be made to extend from the sketching plane to the surface of the base feature. This feature would be more difficult, although not impossible, to define using extruded protrusions and cuts that extended only from the basic modeling planes or one of the surfaces on the existing model. In Figure 6.45(b), the uniquely shaped web would be easy to create if a sketching plane could be placed between the connected features as shown. An extruded protrusion could extend from both sides of the sketching plane to the surfaces of the connected features.

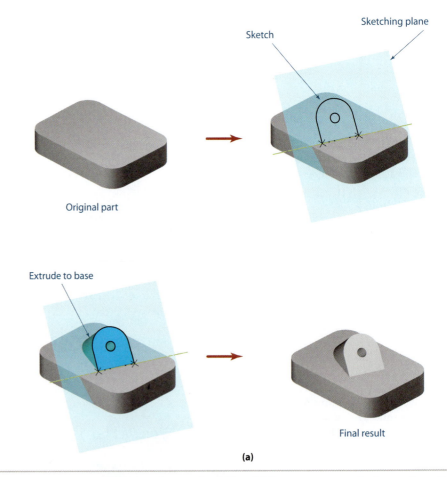

FIGURE 6.45. Using a sketching plane and profile, which are not on an existing surface of the object, to create a protrusion feature (a).

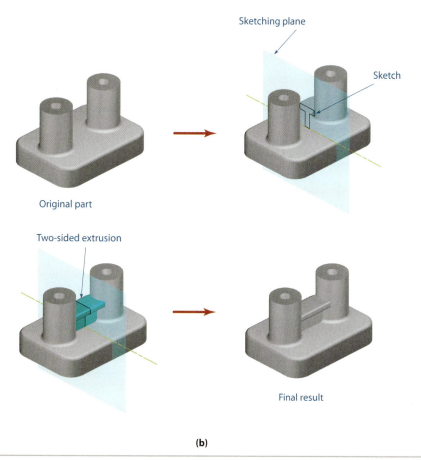

(b)

FIGURE 6.45. (CONTINUED) Using a sketching plane and profile, which are not on an existing surface of the object, to create a web feature (b).

The next few sections will describe methods in which the three different types of datums can be defined geometrically and how the datums can be used to create a variety of new types of features. Depending on the specific solid modeling software being used, some of the methods described here for datum definition may or may not be available.

6.08.01 Defining Datum Points

Following are some of the different ways a datum point can be defined and created. The definitions are shown graphically in Figure 6.46.

- At a vertex
- On a planar surface at specified perpendicular distances from two edges
- At the intersection of a line or an axis and a surface that does not contain the line

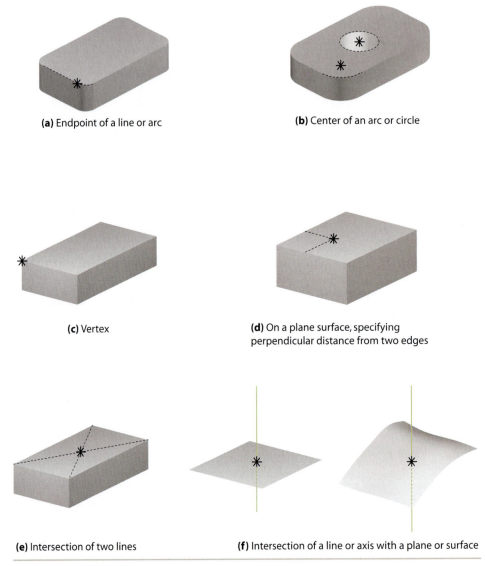

(a) Endpoint of a line or arc

(b) Center of an arc or circle

(c) Vertex

(d) On a plane surface, specifying perpendicular distance from two edges

(e) Intersection of two lines

(f) Intersection of a line or axis with a plane or surface

FIGURE 6.46. Various ways to define a datum point.

6.08.02 Defining Datum Axes

Following are some of the different ways a datum axis can be defined and created. The definitions are shown graphically in Figure 6.47.

- Between two points (or vertices)
- Along a linear edge
- At the intersection of two planar surfaces
- At the intersection of a cylinder and a plane through its axis
- Along the centerline of a cylinder or cylindrical surface

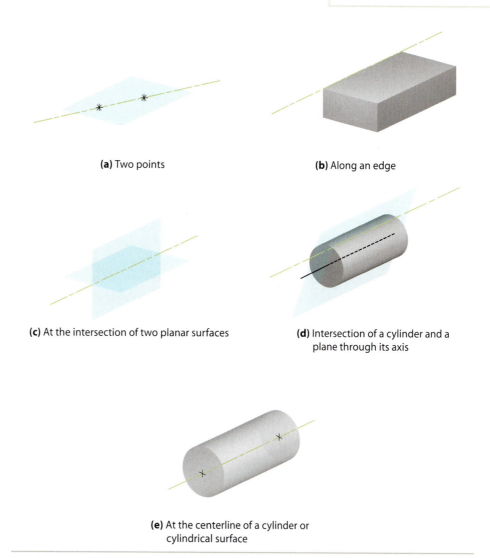

(a) Two points

(b) Along an edge

(c) At the intersection of two planar surfaces

(d) Intersection of a cylinder and a plane through its axis

(e) At the centerline of a cylinder or cylindrical surface

FIGURE 6.47. Various ways to define a datum axis.

6.08.03 Defining Datum Planes

Following are some of the different ways a datum plane can be defined and created. The definitions are shown graphically in Figure 6.48.

- Through three noncolinear points
- Through two intersecting lines
- Through a line and a noncolinear point
- Offset from an existing flat surface at a specified distance
- Through an edge or axis on a flat surface at an angle from that surface
- Tangent to a surface at a point on that surface
- Perpendicular to a flat surface and through a line parallel to that surface
- Perpendicular to a flat or cylindrical surface through a line on that surface
- Tangent to a cylindrical surface at a line on that surface

FIGURE 6.48. Various ways to define a datum plane. More ways to define a datum plane.

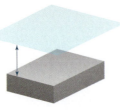

(a) Offset from Existing flat surface, and a distance

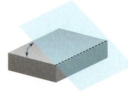

(b) A flat surface, through an edge or axis on that surface, and an angle from that surface

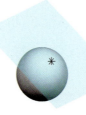

(c) Tangent to a surface at a point on that surface

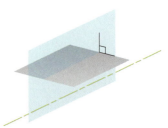

(d) Perpendicular to a flat surface, and through a line parallel to that surface

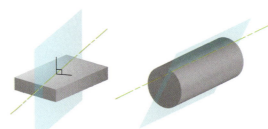

(e) Perpendicular to a flat or cylindrical surface and through a line on that surface

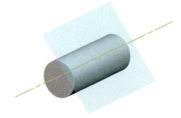

(f) Tangent to a cylindrical surface at a line on that surface

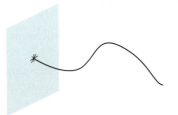

(g) Perpendicular to a curve at a point on that curve

(h) Three points

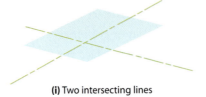

(i) Two intersecting lines

(j) A Line and a point

6.08.04 Chaining Datums

Series of simply defined datums are often used for creating more complex datums. In the example shown in Figure 6.45(a), the angled protrusion was created in this manner: On the top surface of the base extrusion, two datum points were created by defining each of their locations from the edges of the base extrusion. A datum axis was then created using the two datum points as the endpoints of the axis. Finally, the desired datum plane was defined using the top surface of the base extrusion, using the datum axis created in that plane, and specifying the angle that the new datum plane makes with the top surface.

Another example is shown in Figure 6.49, where a datum plane is created to be tangent to the surface of a cylindrical extrusion. An intermediate datum plane is defined by one of the basic planes; the axis of the cylinder, which lies on that basic plane; and the angle the intermediate datum plane makes with the basic plane. The final datum plane is then created to be tangent to the surface of the cylindrical extrusion at its intersection with the intermediate datum plane. A datum plane tangent to the surface of a cylinder is commonly used to create cuts that extend radially into a cylindrical surface, such as holes or slots, and protrusions that extend radially from the cylindrical surface, such as spokes or vanes. Note that with protrusion from a tangent datum plane, the extrusion must be specified to extend in both directions from that datum; otherwise, there will be a gap between the extrusion and the curved surface.

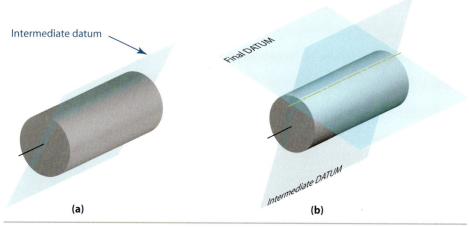

Intermediate datum

Final DATUM

Intermediate DATUM

(a)   (b)

FIGURE 6.49. To create a datum plane that is tangent to a cylindrical surface at a specific location, an intermediate datum plane, shown in (a), can be created through the centerline of the cylinder. The intersection of the intermediate datum with the cylinder creates a datums axis that is used to locate the final datum plane.

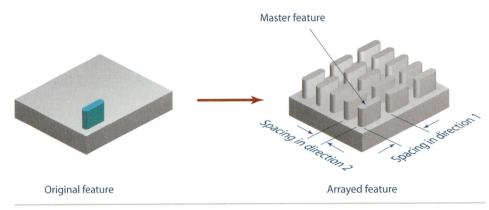

Master feature

Spacing in direction 2

Spacing in direction 1

Original feature   Arrayed feature

FIGURE 6.50. A rectangular array of protrusions created from a master feature.

6.08.05 Using Arrays (Rectangular and Circular)

One method of creating multiple identical copies of a feature in a model is to create a **feature array**, which is sometimes called a **feature pattern**. A feature array takes one feature, called the **master feature**, and places copies of it on the model at a specified spacing. The copied features are identical to the master feature, and changing the geometry of the master at a later time also changes the geometry of the copies at that time. Including features in this manner can save time and effort in creating the entire model, especially when the features are rather complex. An example of a model with a rectangular array of features is shown in Figure 6.50. An array of rectangular cuts is shown in Figure 6.51. As shown, rectangular arrays can generate copied features in two directions. These directions must be specified, as well as the spacing of the copied features in each direction. Finally, the number of copies in each direction must be specified. Care must be taken to ensure that there is enough room on the model to accommodate all of the copied features.

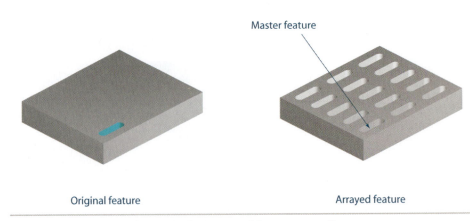

Master feature

Original feature   Arrayed feature

FIGURE 6.51. A rectangular array of cuts created from a master feature.

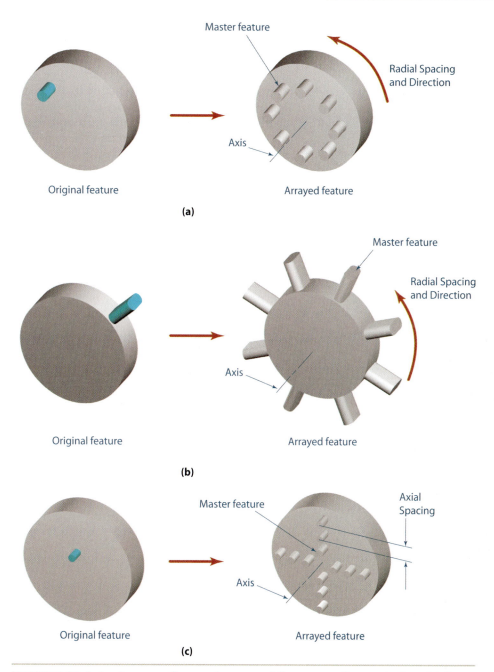

FIGURE 6.52. A circular array of protrusions created from master features in the axial direction (a) and (b), both in the axial and radial direction (c).

Examples of models with radial arrays of protrusions are shown in Figure 6.52. Radial arrays can extend radially or axially. For radial arrays, in addition to the master feature being selected, the axis of revolution for the array must be selected. If such an axis does not already exist on the model, one must be created from an added datum axis. The number of copies, the direction of the array, and the radial and axial spacing of the copies must be specified.

6.08.06 Using Mirrored Features

Another method of creating a feature, when applicable, is to create its mirrored image. To create a **mirrored feature**, you must first identify a mirror plane. You can use an existing plane or define a new datum plane to use as the mirror plane, as shown in Figure 6.53. A mirrored duplicate of the master feature can then be created on the model on the opposite side of the mirror plane. Mirrored features can be cuts or protrusions; however, keep in mind that the copied feature will be a mirror image of the master, not an identical copy. Changing the master feature at a later time also will change the mirrored feature correspondingly. As with arrayed features, using mirrored features can save a great deal of time in model creation, especially when the mirrored feature is complex.

6.08.07 Using Blends

Not all models can be created using just extruded or revolved features. One complex feature is a **blend**. Figure 6.54 shows models with blended surfaces. A blend requires at least two profile sketches, and the model is formed by a smooth transition between

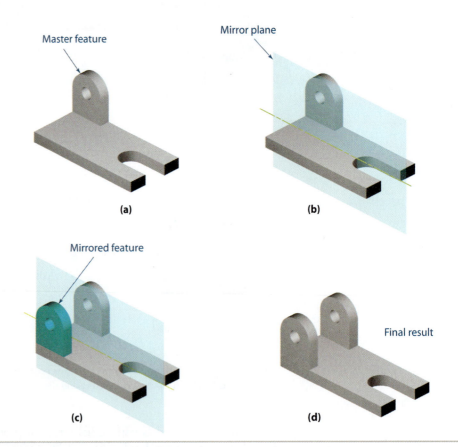

FIGURE 6.53. Creation of a mirrored feature. The master feature in (a) is mirrored by creating a datum plane as a mirror plane (b). A mirror image of the feature is produced on the opposite side of the datum plane in (c), and the final result is shown in (d).

these profiles. The profiles can be sketched on the basic modeling planes, on surfaces of an existing model, or on datum planes. In the simplest blends, the profiles are sketched on parallel planes. Many software packages require the number of vertices on each of the sketched profiles to be equal. If your profiles do not have the same number of vertices, you will have to divide one or more of the entities to create additional vertices. In some sketching editors, circles include four vertices by default. The vertices in all profiles are usually numbered sequentially, and the software usually tries to match the vertices to create an edge between vertices with the same number, as shown in (a) and (b) of Figure 6.54. Rotating the profiles or redefining the vertex numbering can control twisting of the blended transition, as shown in Figure 6.54(c). Further control on the model transition usually can be performed by specifying the slope of the transition at each vertex for each shape.

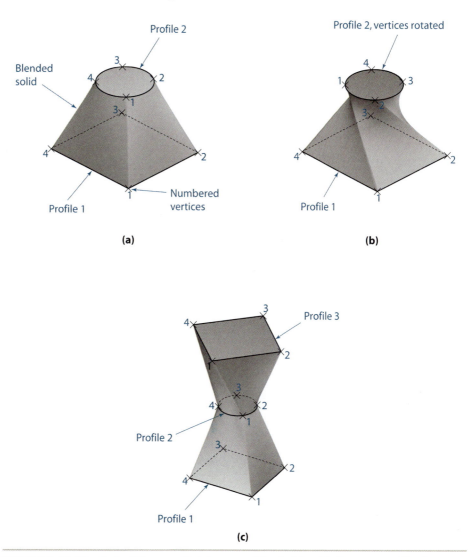

FIGURE 6.54. Blended solids created with two profiles in (a), with the same profiles but rotated vertices in (b), and with three profiles in (c).

6.08.08 Sweeps

Swept features, as with simply extruded or revolved features, are created with a single profile. The difference is that a swept feature does not need to follow a linear or circular path, but can follow a specified curve called a **path** or **trajectory**. The profile is created at an endpoint of the path on a sketching plane that is perpendicular to the path at that endpoint. In sweeping out a solid volume, the profile is imagined to travel along the path. Usually the profile is constrained to remain perpendicular to the path. A good example of a swept solid is a garden hose. The cross section or profile is a simple circle, but the path can be curved. Figure 6.55(a) shows the path and profile of a swept feature where the path is open. Figure 6.55(b) shows a swept feature where the path is closed. Care must be taken in defining the profile and path of a swept solid. Just as you cannot bend a garden hose around a sharp corner without creating a kink, if the path of your sweep contains a sharp corner or a small radius, the feature may fail by trying to create a self-intersecting solid. A special case of a swept solid is a coil spring. In this case, the path is a helix, as shown in Figure 6.56. Many solid modelers include a hel-

FIGURE 6.55. Features created by sweeps. The sketching plane is perpendicular to the path. The path in (a) is open, and the path in (b) is closed.

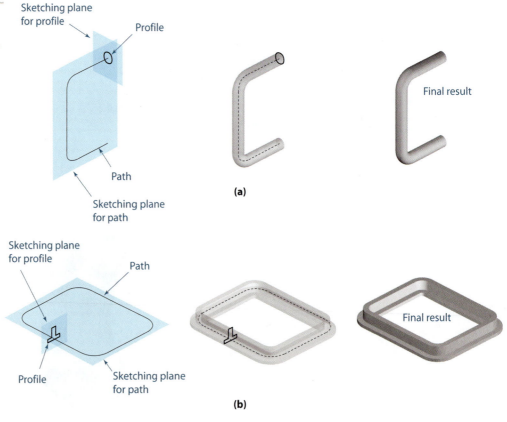

FIGURE 6.56. A tapered spring created by sweeping a circular profile on a helical path.

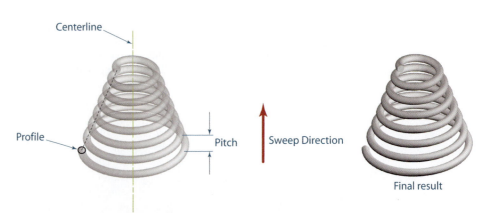

ical sweep as a special feature so you do not have to sketch the helix. In this case, you sketch the profile and specify an axis on the sketching plane. The helix is specified by a pitch dimension, which is the distance between coils, and the direction of the sweep. To avoid self-intersection, the pitch must be larger than the maximum size of the profile in the sweep direction.

6.09 The Model Tree

An extremely useful editing tool included in most solid modeling software is the **model tree**, sometimes called the **feature tree**, **design tree**, or **history tree**. The model tree lists all of the features of a solid model in the order in which they were created, providing a "history" of the sequence of feature creation. Further, any feature in the model tree can be selected individually to allow the designer to edit the feature. An example of a model tree and its associated solid model are shown in Figure 6.57. Usually new features are added at the bottom of the model tree. Some software allows the designer to "roll back" the model and insert new features in the middle of the tree. In this case, the model reverts to its appearance just before the insertion point, so any inserted feature cannot have its geometry or location based on features that will be created after it.

FIGURE 6.57. A typical model tree showing the features of a model in the order in which they were created.

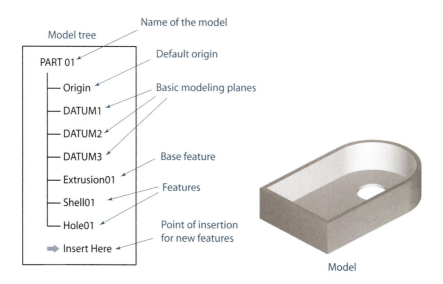

FIGURE 6.58. The result of reversing the order of creating the hole and shell features.

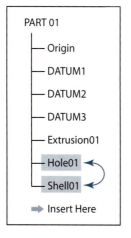

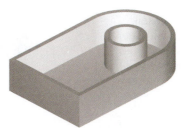

The order in which features are created may have a profound effect on the results. In the previous example, a shell feature, which has the effect of hollowing out a part, was performed with the top surface of the part removed from the feature. A hole was then added to the model after the shelling operation. If the hole was added to the block before the shelling operation, the result would be different, as shown in Figure 6.58, because the surface around the hole through the block would have been considered a part of the shell. In most solid modeling software, removing the feature from one location in the model tree and inserting it in a new location changes the order of creation of the feature.

The model tree also provides access to the editing of features. Each feature item on the model tree can be expanded. The base extrusion in the previous example is composed of a fully constrained rectangular sketch profile that has been extruded to a specified length. The feature can be expanded in the model tree, as shown in Figure 6.59, to give access to the profile so it can be selected for editing. The sketch can then be edited by restarting the sketching editor. The dimensional constraints can be changed by selecting and editing their numerical values. Access to the sketching editor and feature parameters may vary with different software, and changes made through the model tree may be one of several different ways to modify your model.

In many models, certain features are dependent upon the existence of other features. For example, consider the features shown in the model in Figure 6.60. The location of the counterbored hole is measured from the edges of the rectangular base.

FIGURE 6.59. Use of the model tree to access and edit the sketch used to create the base feature (Extrusion01).

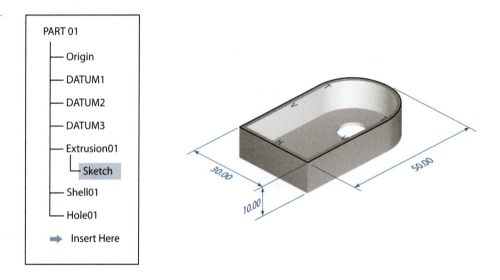

FIGURE 6.60. The holes in the model show parent-child dependencies. The existence of the straight hole depends on the existence of the countersunk hole, which depends on the existence of the counterbored hole. Elimination of a parent also eliminates its child.

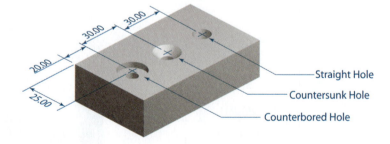

However, the location of the countersunk hole is measured from the location of the counterbored hole and the location of the straight hole is measured from the location of the countersunk hole. Imagine what would happen to the straight hole if the countersunk hole were deleted. There would be no reference for placing the straight hole; therefore, it could not be created. Similarly, if the counterbored hole was deleted, neither the countersunk hole nor the straight hole could be created. This relationship is often referred to as a parent-child relationship. The straight hole is considered the **child feature** of the countersunk hole, and the countersunk hole is considered the child of the counterbored hole. The counterbored hole is considered the **parent feature** of the countersunk hole, and the countersunk hole is considered the parent of the straight hole. Just as you would not be reading this text if your parents did not exist, neither can features in a solid model exist without their parent (or grandparent) features. On the model tree, if you try to delete a particular feature, its progeny also will be deleted. However, different software behaves differently; and while some software provides specific warnings about the deletion of features, other software does not.

Understanding parent-child relationships in solid models is important if your model needs to be flexible and robust. As a designer, you undoubtedly will want to change the model at some time. You might need to add or delete features to accommodate a new function for the part or reuse the model as the basis of a new design. If you minimize the number of dependencies in the feature tree (like a family tree), it will be easier to make changes to your model. When it is likely that some features will be deleted or suppressed in a future modification of the part, those features should not be used as parents for other features that must remain present. The most extreme example of this strategy is called **horizontal modeling**, where the feature tree is completely flat; that is, there are no parent features except the base feature. This type of modeling strategy was patented by Delphi and has been used successfully by many companies. In Figure 6.61, the locations of three holes have been redefined so they are measured from the edge of the rectangular base instead of relative to one another. The base then becomes the parent to all three holes, and deleting any one of the holes does not affect the others.

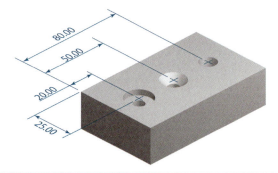

FIGURE 6.61. This model demonstrates horizontal modeling. Each hole has no parent-child dependencies except to the base feature.

6.10 Families of Parts

Groups of engineered parts often have very similar geometry. An everyday example is bolts and screws. A group of bolts may have the same head and thread geometries, but differ in their available length. Another example is the family of support brackets shown in Figure 6.62. Each bracket has a rough L-shaped base feature, holes, and a support rib (except for Version 3). Only the size and number of holes are different for each version.

When a group of parts is similar, it is possible to represent the entire group with a **family model**, with different versions of that model selected to specify particular parts. Such a model includes a **master model**, which has all of the features that are in any of the members of the group, and a **design table**, which lists all of the versions of that model and the dimensional constraints or features that may change in any of its versions. The attributes that may change are sometimes called **parameters**. The first

Version 1

Version 2

Version 3

FIGURE 6.62. A family of three parts with similar features and geometry.

step in building a family model is to identify all of the features and parameters that can be varied in the members of the family. In addition to a numerical value, every dimensional constraint in a model has a unique **dimension name**, which can be shown by selecting the appropriate display option. In Figure 6.63, all of the dimensional constraints have been changed to show their dimension names and the features have been identified by the feature names that appear on the design tree.

The next step is to select the option for the construction of a design table, which is usually an internal or external spreadsheet, in the solid modeling software. The spreadsheet table should look similar to that shown in Figure 6.64. The first column usually contains the names of the different versions of the model. In Figure 6.64, these versions are called Version 1, Version 2, and Version 3 for convenience. The first row usually contains the names of the parameters that can change with each version. The individual cells of the spreadsheet show what the corresponding numerical values are of the

FIGURE 6.63. The master model showing the numerical values of its dimensions in (a) and the names of the features and dimensions in (b).

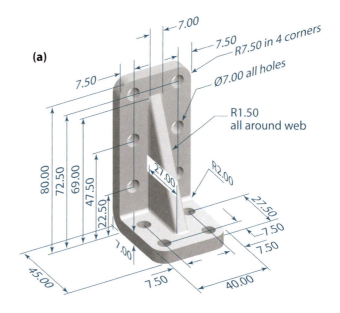

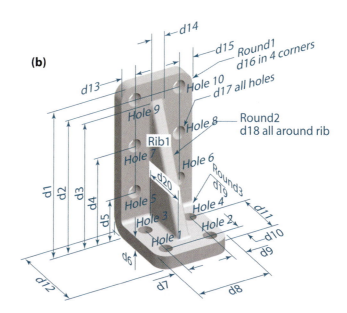

	Hole 3	Hole 4	Hole 7	Hole 8	Hole 9	Hole 10	Rib1	d1	d3	d12
Version 1	U	U	U	U	U	U	U	80.00	69.00	45.00
Version 2	U	U	U	U	S	S	U	60.00	50.00	45.00
Version 3	S	S	S	S	S	S	S	30.00	N/A	30.00

S = Suppressed
U = Unsuppressed

FIGURE 6.64. The design table for the parameters that change within the three versions of the family of parts in Figure 6.62 and Figure 6.63.

dimensional constraints for each version and whether a particular feature is present in that version. When a particular feature is present, it is specified as being **unsuppressed**. When that feature is not present, it is specified as being **suppressed**. When the version of the part to be displayed has been selected, the corresponding model with its specified parameters is shown.

With the existence of a design table, editing the values in the table can change the numerical values of those dimensional constraints for any model version. In Figure 6.65, selecting and editing the appropriate cell in the design table changed the height of the L-bracket. In Figure 6.66, the support rib is no longer present because it was suppressed in the design table. When suppressing a feature, remember to be cautious, because suppressing a feature will also suppress its entire progeny.

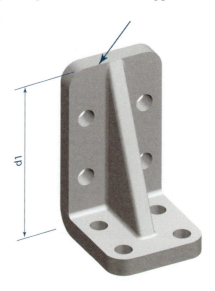

	Hole 3	Hole 4	Hole 7	Hole 8	Hole 9	Hole 10	Rib1	d1	d3	d12
Version 1	U	U	U	U	U	U	U	69.00	69.00	45.00
Version 2	U	U	U	U	S	S	U	60.00	50.00	45.00
Version 3	S	S	S	S	S	S	S	30.00	N/A	30.00

S = Suppressed
U = Unsuppressed

FIGURE 6.65. The height of the L-bracket in the model has been changed by changing the value of the cell in the design table associated with this parameter.

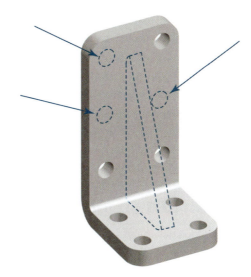

	Hole 3	Hole 4	Hole 7	Hole 8	Hole 9	Hole 10	Rib1	d1	d3	d12
Version 1	U	U	S	S	S	U	S	80.00	69.00	45.00
Version 2	U	U	U	U	S	S	U	60.00	50.00	45.00
Version 3	S	S	S	S	S	S	S	30.00	N/A	30.00

S = Suppressed
U = Unsuppressed

FIGURE 6.66. Features in the model can appear or not appear by changing their suppression states in the design table.

6.11 Strategies for Making a Model

You have a blank computer display in front of you, and your solid modeling software is running. So where do you start? The first step in modeling a solid part is to decompose it into features. Study the part and try to identify the base feature. Expanding on a previous statement, the base feature should be something that describes the overall shape of the part or something that gives you the greatest amount of functional detail that can be created with a single extrusion, rotation, sweep, or blend. Next, break the rest of the part into subsections that can be created using extruded, revolved, swept, or blended shapes. Look for standard features such as holes and slots that are manufactured using a particular process. Identify the edge features such as chamfers and rounds. Once you have studied the part, you can create the model using the following eight-step procedure:

1. Create any datum geometries or paths required to create the base geometry.

2. Sketch and constrain profiles needed for the base.

3. Extrude, rotate, sweep, or blend to create the base.

4. Create any necessary datum geometries or paths to create the next feature.

5. For sketched features, sketch and constrain or otherwise specify the feature profiles; then extrude, rotate, sweep, or blend to create the feature.

6. For standard features such as holes and edge features, specify the desired parameters and placement on the existing geometry.

7. Array or mirror the feature if necessary to create identical features.

8. Repeat steps 4–9 until the model is complete.

Once the model is complete, it can be modified to become more robust. For example, additional associative constraints may be added in place of dimensional constraints or design tables may be created for families of parts.

6.11.01 Step-by-Step Example 1—The Guide Block

Consider the guide block in Figure 6.67 as an example. How would you build a solid model for this part? What should be its base feature? What are its secondary features?

One reasonable base feature would be an extrusion made with the profile shown in Figure 6.68. This extrusion would capture many details of the part in a single operation and is representative of the general shape of the part. The sketch is made on one of the basic planes and is geometrically and dimensionally constrained. Note the use of horizontal and vertical geometry constraints, which would likely be applied automatically if the segments were sketched approximately in the orientations shown. Also note that a corner of the profile is grounded by constraining the vertex to be coincident with the origin of the coordinate system; therefore, dimensional constraints locating the profile on the plane are not needed. Note the use of the colinear constraint on the two short horizontal sketch segments, which eliminates the need to place separate dimensional constraints on the height of the segments. Once the profile is complete, it can be extruded to the width of the part, as shown in Figure 6.68(d).

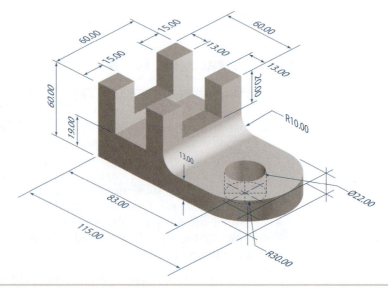

FIGURE 6.67. A solid model of this part is to be created. What operations should be performed, and in what sequence should they be made?

FIGURE 6.68. The base feature is created by sketching on one of the basic modeling planes (a), constraining the sketch to create a profile (b), and extruding the profile (c) to the required depth to obtain the desired result (d).

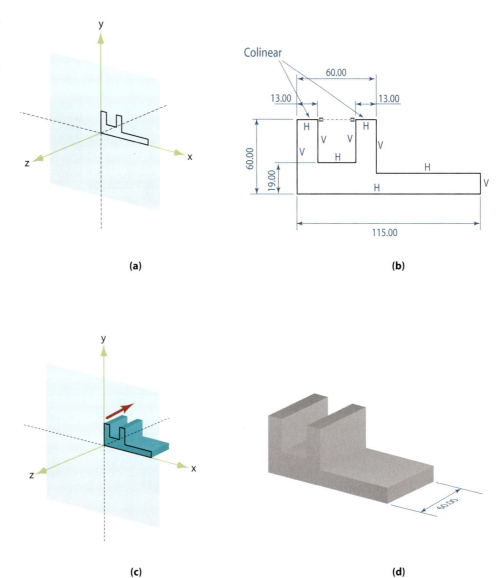

(a)

(b)

(c)

(d)

The first feature to be added is the slot across the upper portion of the part. The slot can be made on the model by an extruded cut using a rectangular profile on the sketching plane shown in Figure 6.69. Geometric and dimensional constraints are added to the sketch. One edge of the sketch is constrained to be colinear with the top edge of the base feature, guaranteeing that the slot always will be a slot (and not a square hole) if the height of the part increases. An extruded cut is then made by extruding the completed profile to the limit of the part or through the entire part. If this extruded cut was made to a specific length just beyond the limit of the part (e.g., with a blind extrusion extending past the part), the resulting model would appear identical to the desired model. However, blind extrusions like this are usually considered poor modeling practice, because if selected length constraints of the part are increased, the slot no longer extends entirely through the part.

Next, the hole is added. Simple through holes can be created as a cut feature by selecting the desired plane or face of the existing solid, sketching a circle, and extruding a cut or negative feature. However, a better way to make a hole is to use the standard

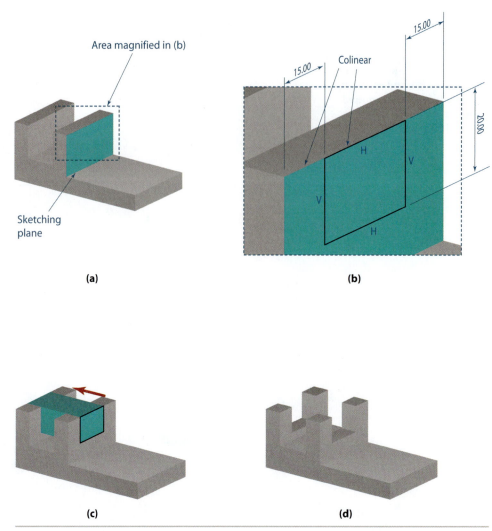

FIGURE 6.69. A slot is created by selecting a surface on the model to be the sketching plane in (a), on which a sketch is created and constrained in the magnified view in (b). The profile is extruded to the end of the part in (c), and the material is removed to create the result in (d).

hole feature, which will identify the feature as a hole in the database. Therefore, if your part is to be manufactured by an automated production system, the holes can be recognized and automatically drilled or bored. The way that the location dimensions of the hole are included in the model is also important. Why does this matter? Looking at the part, you might assume that the hole should stay centered on the width of the part. What will happen if the width of the part changes? Wouldn't you want the hole to remain centered on the width of the part? Using an associative constraint on the width location of the hole, making it always equal to one-half the part's width, ensures that the design intent will be maintained. No matter who uses your model or changes the dimensions, the intended symmetry will remain embedded in the part. Adding the hole with its associative constraint is shown in Figure 6.70.

Finally, the round and fillet features need to be added to the model geometry. Rounds and fillets are associated with particular edges, so no sketching is involved for this step. Simply pick the desired edge and apply the round or fillet feature, specifying the desired radius. The result of adding the rounds is shown in Figure 6.71, and the fillet is shown in Figure 6.72. There are no array or mirror features, so the model is now complete.

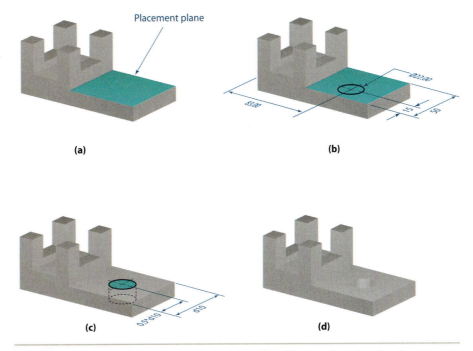

Placement plane

(a)

(b)

(c)

(d)

FIGURE 6.70. The surface to which the hole is to be added is selected as the placement plane in (a). The location and diameter of the hole are specified in (b). An associative constraint is used on the variable names to ensure that it remains centered in (c), and the final result is shown in (d).

FIGURE 6.71. The edges to be rounded are selected in (a), and the radius is specified in (b) as an associative constraint to ensure a full radius across the part. The result of the rounding operation is shown in (c).

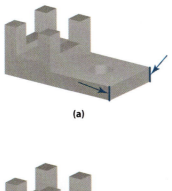

(a)

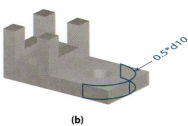

0.5*d10

(b)

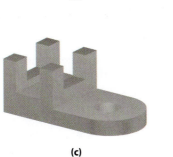

(c)

FIGURE 6.72. The edge to be filleted is selected in (a), and the radius is specified in (b). The result of the fillet operation is shown in (c).

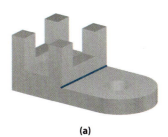

(a)

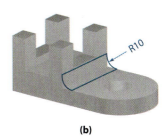

R10

(b)

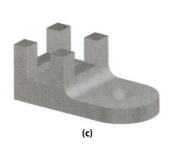

(c)

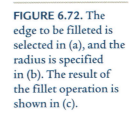

Depending on the design intent, several different modeling strategies can be used for the same part. If the designer wants the entire part to be symmetrical, another way to achieve the desired symmetry would be to create a two-sided extrusion of the base profile from the original sketching plane. By creating a two-sided extrusion, no extra datum planes are needed to ensure symmetry. Then the slot is constrained to be symmetrical across the plane, and the hole center is constrained to lie in the symmetry plane.

6.11.02 Step-by-Step Example 2—The Mounting Brace

The mounting brace shown in Figure 6.73 includes two mounting plates with holes connected by a cross-shaped web. The base feature is not as easily identifiable as in the previous example. Also, one mounting plate is set at an angle with respect to the other mounting plate, which is located on the bottom of the part. Note that no dimensions are given for the location of the web on either mounting plate because the web is assumed to be centered on both plates. The height of the web is measured at the center of the cross section. Creating the model of this part provides a good example of how datum axes and datum planes are used to help create and locate geometry.

A rectangular block that will help form the bottom mounting plate is selected as the base feature. Even though this rectangular block does not dominate the overall part shape, it is easy to locate and create; and the other features can be easily located and created from it. The sketch, profile, and extrusion of this simple base using the basic modeling planes are shown in Figure 6.74. The initial sketch is made on datum plane 1. Note that the origin is not used to constrain the location of the profile. The sketch is to be symmetrical across datum planes 2 and 3. The strategy used to achieve symmetry may vary depending on your software. Some software allows you to place symmetry constraints across the basic datum planes that appear as edges in the sketch view, such as datums 2 and 3 in this example. Otherwise, a datum axis must be used as a centerline. With some solid modeling software, datum axes are present at the intersections of the basic modeling planes and need not be created. Sometimes you may need to create datum axis 1 at the intersection of datum planes 1 and 3 and datum axis 2 at the

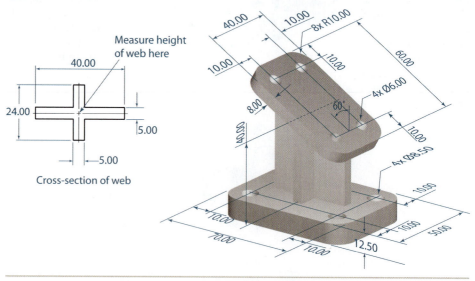

FIGURE 6.73. A solid model of this part is to be created. What operations should be performed, and in what sequence should they be made?

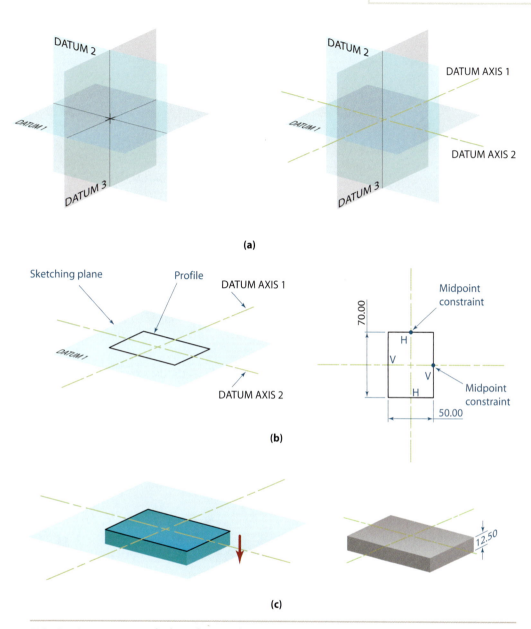

FIGURE 6.74. To create the base feature, datum axes are added to the intersections of the basic modeling planes in (a). Those axes are used to help center the base profile in (b) by constraining the midpoints of the segments to be colinear with the axes. The base feature is extruded to the required depth in (c).

intersection of datum planes 2 and 3. Now the midpoints of two adjacent legs of the rectangle can be constrained to lie on the datum axes using either coincidence or colinearity constraints. With these constraints, the rectangular profile will be centered about the axes and will remain centered about the axes if the design is modified in the future. The extrusion direction is chosen to be downward from datum plane 3, thus putting that datum plane on top of the base.

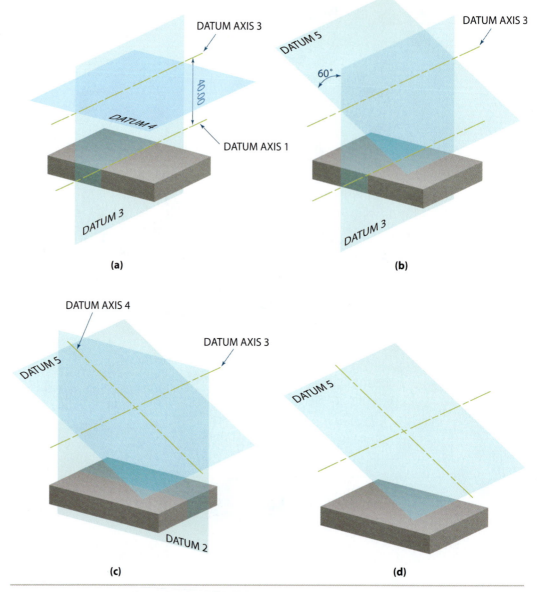

FIGURE 6.75. To create the angled feature, datum axis 3 is created in (a) above the midline of the model using intermediate datum plane 4 created above and parallel to the top of the base extrusion. Angled datum plane 5 is created through the new datum axis in (b). Datum axis 4 is created at the intersection of datum plane 5 and basic modeling plane 2 in (c) to create datum axes that can be used to locate the center of the angled feature, as shown in (d).

The next major feature to be added is the angled mounting plate. It may seem peculiar to create disconnected geometry, but you will soon see how useful this strategy can be. No sketching plane is available for creating the angled plate; so before the extrusion is created, some additional datum geometries need to be created. In addition to the datum (sketching) plane, some datum axes are needed to position the sketch at the center of the web. These datum geometries, shown in Figure 6.75, are as follows:

- Datum plane 4—located 40 mm above and parallel to datum plane 1
- Datum axis 3—at the intersection of datum plane 3 and datum plane 4

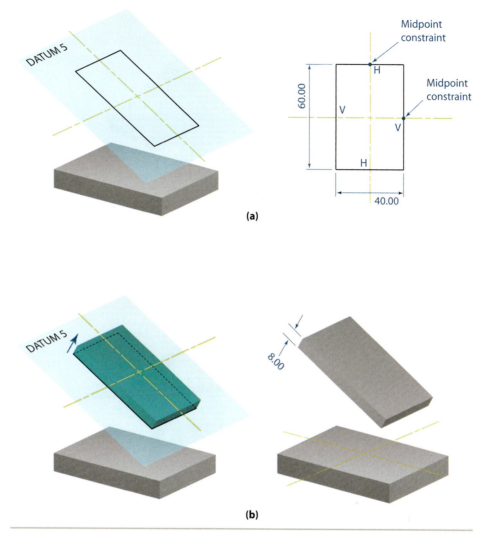

(a)

(b)

FIGURE 6.76. The profile for the angled feature is sketched and constrained on the angled datum plane in (a), constraining the midpoints of the segments to be colinear with the axes. The profile is extruded to the required depth in (b).

- Datum plane 5— through datum axis 3, 60 degrees from datum plane 3
- Datum axis 4—at the intersection of datum plane 2 and datum plane 4

Datum axes 1 and 2 mark the center of the base; and datum axes 3 and 4 mark the center of the angled mounting plate, where the web will be located.

To create the angled mounting plate, a sketch of its rectangular profile is constrained and dimensioned on datum plane 5, as shown in Figure 6.76. Note that each of the midpoints of two adjacent edges of the profile have been constrained to be colinear with datum axis 3 or 4. In this way, the rectangular profile is always centered about the axes. The sketched profile is then extruded to form the angled mounting plate.

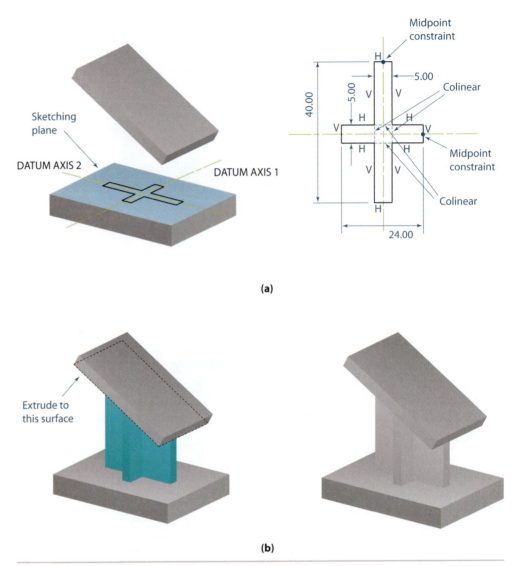

(a)

(b)

FIGURE 6.77. To create the web, the top of the base is selected as the sketching plane in (a) and the profile of the web is constrained using midpoint and colinear constraints to ensure symmetry. The profile is extruded to the underside of the angled feature in (b).

The top of the base feature is used as the sketching plane for the profile of the web. A cross-shaped profile is used, as shown in Figure 6.77. The midpoints of the line segments representing the thickness of two adjacent legs of the cross are constrained to be coincident with datum axis 1 or with datum axis 2. Note the use of colinear constraints on the edges of opposing legs of the cross. These constraints make the thickness of the legs the same on opposite sides when one of the legs is dimensioned. The profile is extruded up to the next surface that it intersects, which is the bottom of the angled mounting plate.

The rounds are added using the round and fillet feature creation tool by selecting the edges shown in Figure 6.78 and specifying the desired round radius. By grouping the rounds in one feature, their radii will be equal and will remain equal as the part is modified in the future. Adding holes to both mounting plates using a hole feature creation tool is shown in Figure 6.79. The centers of the holes are located on the upper surface of each plate, with their locations specified relative to the edges of the plates. The holes are then defined by their diameters and depth (all the way through to the next surface on each plate). Note that you would not want to use the "through all" option on these holes, as the holes from the angled plate would extend through the baseplate as well. The model is now complete.

FIGURE 6.78. The edges for rounding are selected in (a), and the desired radius of the rounds is applied in (b).

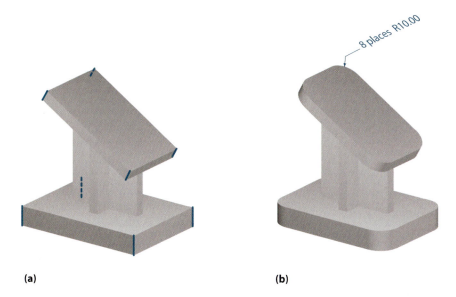

(a)

(b)

FIGURE 6.79. The holes are added to the base feature in (a) and to the angled feature in (b).

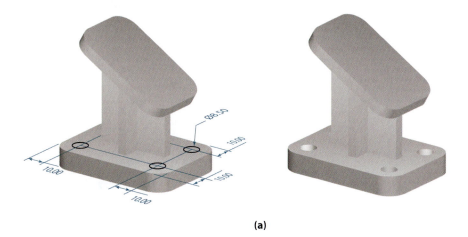

(a)

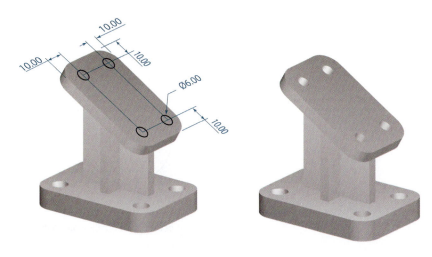

(b)

6.11.03 Step-by-Step Example 3—The Handwheel

The part shown in Figure 6.80 is a handwheel. It consists of a cylindrical hub with a D-shaped hole at its center, a circular rim with a circular cross section, and spokes that connect the hub to the rim. The handwheel is a good example of a model that should be created with a combination of rotation, sweep, and array.

A good base feature might be either the hub or the rim. Both of these features can be created by revolving a suitable profile. Some solid modelers will let you create both features in the same operation, as will be shown here. One of the basic modeling planes is selected as the sketching plane. A centerline will be required on the same sketch as the profiles, as shown in Figure 6.81. The centerline is placed coincident with the intersection of the sketching plane and another basic modeling plane. The length of the centerline is unimportant. The rectangular profile for the hub is sketched and constrained as shown in Figure 6.81(a). The circular profile of the rim is sketched and

FIGURE 6.80. The handwheel to be modeled is shown in (a). It is cut away in (b) to help reveal all of its dimensions. The D-shaped center hole and a cross section of a spoke are magnified in (c) to show their dimensions.

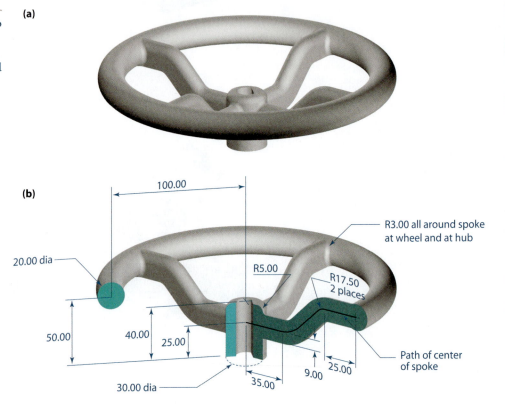

(a)

(b)

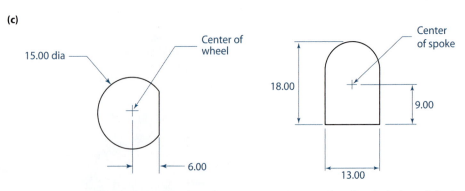

(c)

Dimensions of D-shaped hole (enlarged view)   Cross-section of spoke (enlarged view)

FIGURE 6.81. The profile for the base feature is sketched in (a), and the base feature is created by revolution in (b).

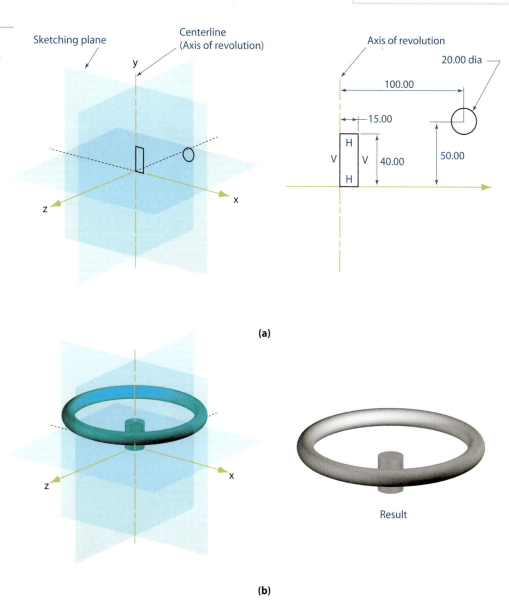

(a)

Result

(b)

constrained. Notice that the two profiles are not constrained to each other, but only to the basic planes and centerline or axis of revolution. Thus, if the shape of the hub or rim is changed, the other profile in the sketch will not change. Both profiles are rotated 360 degrees about the centerline in the same operation to produce the base feature. The base feature appears as two disconnected solids, but this is fine as long as you remember to connect them later.

The next feature to be created is one of the spokes. Because it will be created using a sweep, a path will be required in addition to the cross-section profile, as shown in Figure 6.82(a). The path can be created on the same basic modeling plane that was used for sketching the base profiles. Even though the same sketching plane is used, the profiles for the base and the path are considered separate entities. The spoke cross section must be sketched on a plane perpendicular to the path. The only planes available for sketching in this example are the basic modeling planes. Therefore, you will start the swept profile on the basic modeling plane at the center of the hub, which is perpendicular to the desired path.

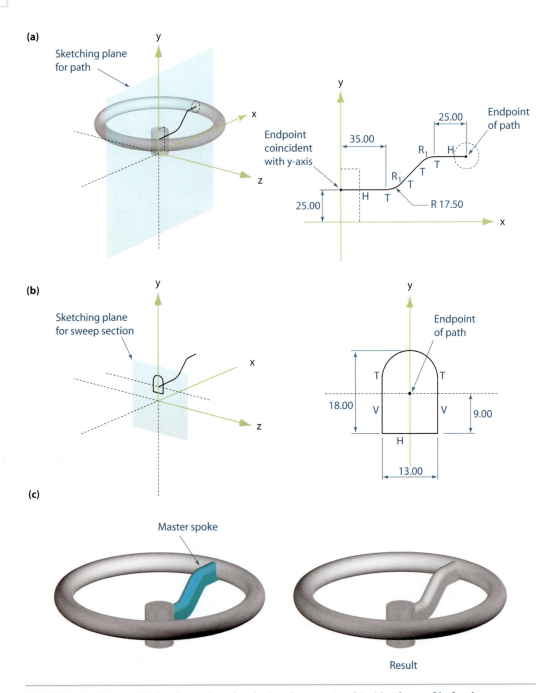

FIGURE 6.82. The profile for the path is sketched and constrained in (a). The profile for the sweep section is sketched and constrained in (b). The master spoke is created by sweeping in (c).

Some software requires the designer to create the path first, then the cross-section profile; other software reverses the order. For this example, the creation of the path will be described first. The path is comprised of three line segments and two arcs. The starting point on the hub is constrained to be coincident with the rotation axis, and the endpoint is constrained to be coincident with the center of the circle that represents the profile of the rim. Additional geometric and dimensional constraints are added, as shown in Figure 6.82(a). The profile must be created on a sketching plane that is perpendicular to the path at one of its endpoints. As previously noted, another one of the basic modeling planes satisfies this requirement, as shown in Figure 6.82; so creating a new datum plane is not necessary. The profile is sketched and constrained. Notice that the bottom of the cross-section profile does not lie on the basic modeling plane; its

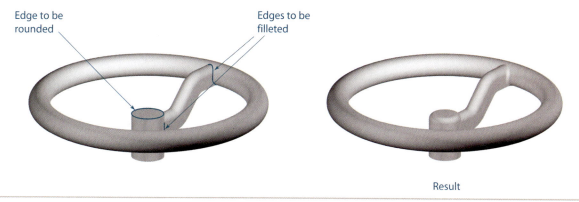

FIGURE 6.83. The edge at the top of the hub is rounded, and the intersections of the spoke are filleted to the specified dimensions.

vertical location is controlled by a dimensional constraint that is measured from the endpoint of the path curve. A sweep operation using the just-created path and profile is performed to create a solid spoke. A round is added to the top of the hub, and fillets are added at the intersections of the spoke with the hub and rim, as shown in Figure 6.83.

Multiple spokes are created with a circular array. The axis for the array is selected to be the same as the axis of rotation of the base. The features to be arrayed are selected to be the swept spoke as well as all of the fillets on it. Most solid modeling systems allow multiple features to be arrayed or patterned in a single operation. This is particularly useful when a parent feature has children that should be included in the pattern. In fact, some software systems automatically include the child features in the pattern. Some software requires a separate pattern operation for each feature. In either case, four spokes are made with equal rotational spacing, as shown in Figure 6.84.

The final feature to be added is the D-shaped hole. This feature is created as an extruded cut, as shown in Figure 6.85. The top surface of the hub is selected as the sketching plane. The profile is sketched and constrained. Note that the center of the arc for the hole is constrained to be coincident with the center axis of the hub. The profile is extruded to the next intersecting surface to create the cut and thus completes the model. Notice that the D-shaped cut for the hole was performed after the creation of the spokes. What would happen if the hole was created before the spokes were modeled? Since the path of the master spoke starts at the axis of revolution of the hub, the spoke is essentially embedded in the hub. If the hole was created first, material would be created in this region and would fill up portions of the hole when the spokes were created. A good rule of thumb to remember when modeling is to create solid geometry first (protrusions), then remove material (cuts) whenever possible.

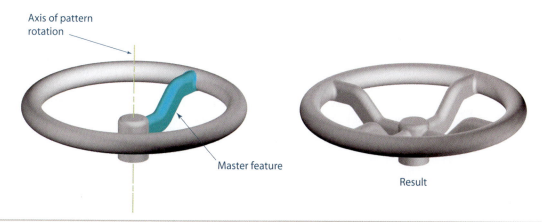

FIGURE 6.84. Multiple spokes are created by a circular pattern, or array, using the center of the wheel as the axis of rotation.

FIGURE 6.85. The D-shaped hole is created by selecting the top of the hub as the sketching plane in (a). The profile is sketched and constrained in (b). The profile cut is extruded through the hub in (c).

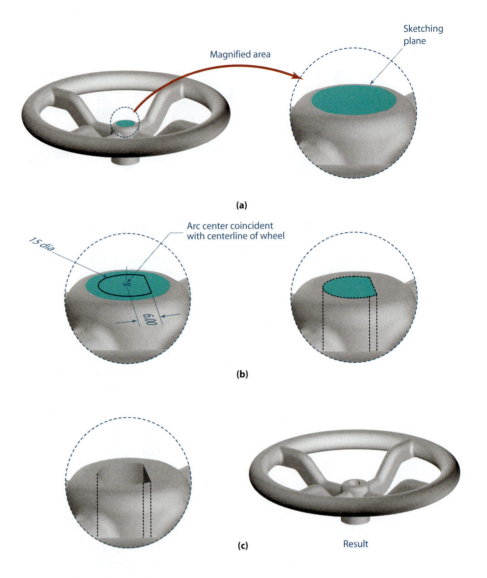

6.12 Extraction of 2-D Drawings

Nearly all solid modeling software packages have a facility for easily creating 2-D engineering drawings from solid models. Formal engineering drawing, which is covered in detail in later chapters, displays the part with all of its features in multiple predesignated views. It also displays the sizes and locations of the features. Solid modelers, which display a model from any viewpoint, can easily create the required views and display dimensions, thus greatly reducing the time and effort required to produce a drawing. Note that the dimensional constraints used in creating the solid model may be different from the dimension values that should be displayed on the engineering drawing. For example, the drawing is required to display all of the dimensions that are necessary for manufacturing the part. Some of these dimension values may be controlled by geometric constraints and are not included in the model as dimensional constraints. You will learn more about proper dimensioning practices in later chapters of this text. An example of a 2-D engineering drawing produced from a solid model is shown in Figure 6.86.

FIGURE 6.86. A typical solid model and a formal working drawing extracted from the model.

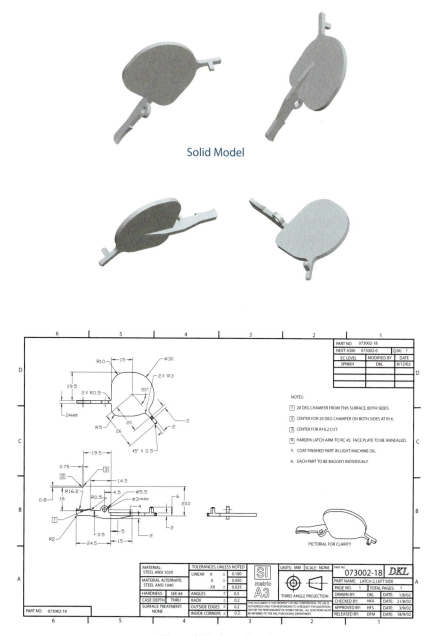

Solid Model

Formal Working Drawing

CAUTION

Inexperienced users of solid modeling software usually commit common errors in creating models. Some of these errors are merely a nuisance, such as generating extra unnecessary dimensions. Other errors do not let the user proceed with the creation process and must be resolved before the user can proceed. Still other errors let the user create a model that may appear like the one desired; however, problems manifest themselves later when the model is edited or when the part is fabricated. The following sections are a compilation of common errors made in solid modeling.

Base Profile Not Properly Positioned

A 3-D model is defined not only by its geometry but also by its location in space. When a solid model is built, the location of the model should be defined relative to the origin of the model's coordinate system. Defining this location is done by defining the location of the profile used for the base feature, most commonly by making one vertex (or point on the centerline) on the profile coincident with the origin. If this is not done, as shown in Figure 6.87, the profile will not be fully constrained and extra dimensional constraints would have to be added to define the location of the profile. These extra dimensional constraints are meaningless and add confusion to the model.

Invalid Profile

Valid profiles were discussed earlier in the chapter. Often invalid profiles are created inadvertently, usually through careless use of the computer's pointing device. Three common errors are shown in Figure 6.88. In Figure 6.88(b), the user attempted to close the sketch to make a valid profile, but missed the target endpoint and left the sketch open. Valid profiles must be closed. In Figure 6.88(c), while attempting to close the sketch, the user crossed over the first sketch element with the final sketch element. Valid profiles cannot cross, overlap, or self-intersect. A similar problem occurs when the profile contains a duplicate line segment. Overlapping lines can be very difficult to

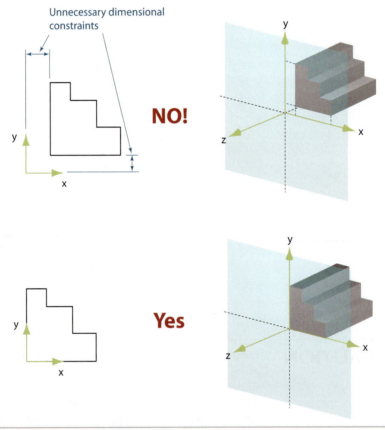

FIGURE 6.87. The base feature should be aligned with the origin whenever possible to avoid the need for extra dimensions.

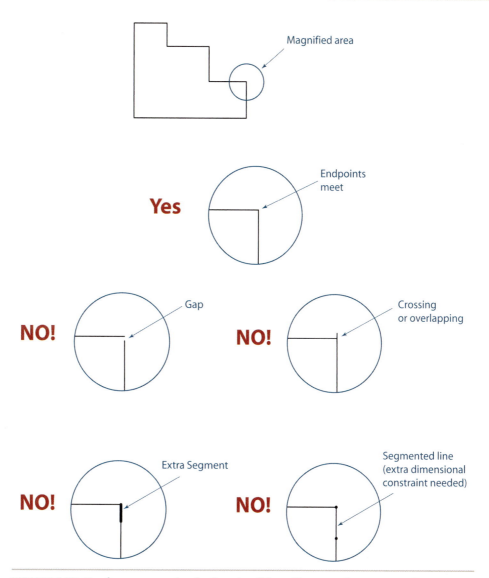

FIGURE 6.88. Careless construction leads to invalid profile errors that are sometimes difficult to find.

find; you may want to delete a line segment if you suspect that it might be duplicated, then redraw the line if no duplicate is found. In Figure 6.88(d), the user inadvertently used two line segments in place of one continuous line segment, resulting in an internal vertex along the desired edge. The profile cannot be fully constrained until the lengths of both line segments are defined. This results in an unnecessary and meaningless dimensional constraint. A profile with a segmented line like this is considered poor modeling practice. Invalid profile errors are usually difficult to see and, therefore, difficult to resolve. Some sketching editors alert the user by highlighting the location of gaps or intersections. Otherwise, you may be faced with the tedious task of searching for the source of the problem. Careful use of the computer's pointing device usually keeps these errors to a minimum.

Profile Not Constrained with Design Intent

It is sometimes tempting to use dimensional constraints instead of geometric constraints that reflect the intent of the design, because application of dimensional constraints follows traditional drafting practice and is typically easy to do in a sketching editor. The availability of geometric constraints in solid modelers offers an opportunity to include aspects of design intent that were previously unavailable in 2-D drafting. Consider the profile for the base feature of a rod clamp, shown in Figure 6.89. The profile is fully constrained, specifying the location of the center of the circular cutout. The design intent, however, is for the circular cutout to be concentric with the rounded part of the exterior of the clamp. If the overall height of the profile was changed, this design intent would not be maintained. In replacing the location dimensions of the circular cutout with a concentricity constraint (coincident arc centers), the design intent is maintained as the overall height of the profile changes. Also note that symmetry constraints on the lower edge of the part ensure that the gap is centered along the axis of the clamp and that both sides have equal thickness.

FIGURE 6.89. The desired behavior is for the outer radius of this part to be concentric with the hole. A geometric constraint rather than a dimensional constraint is preferred to guarantee this behavior, even if other dimensions are changed.

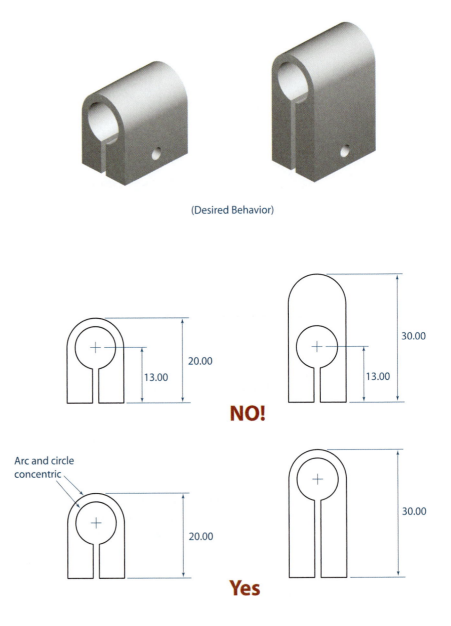

(Desired Behavior)

Profile Overconstrained by Automatic Constraints

For the most part, sketching editors that intelligently apply presumed geometric constraints are very useful. For example, a great deal of time is saved when lines that are sketched to be almost horizontal or almost vertical have these constraints applied automatically. However, automatic constraint application can sometimes lead to problems, particularly when the constraints are applied inadvertently. If your sketching editor has been set to search automatically for equal element sizes, as in Figure 6.90, line segments created to almost the same length in the sketch will automatically be constrained equal in length to each other. Arcs created to almost the same radius as other arcs also will be automatically constrained equal in radii to each other. If that is your intent, then fine. But if that is not your intent, then you must be careful not to create line segments or arcs that are too close in size to each other or you must delete the unintentional constraints when they appear and replace them with separate dimensional constraints on each entity. Figure 6.91 shows a case where an intelligent colinear constraint has been unintentionally applied. In Figure 6.92, an unintentional perpendicularity constraint has been applied. In all cases of unintentional constraints, the addition of the desired geometric constraints or dimensional constraints cause the profile to be overconstrained. In this case, you need to delete the unintentional constraints and then add constraints to capture your design intent.

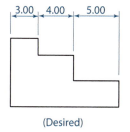

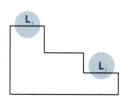

NO!

FIGURE 6.90. Automatic geometric constraints in the sketching editor sometimes cause equal length constraints to be applied inadvertently.

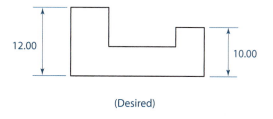

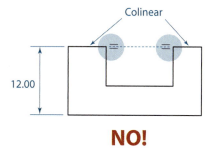

NO!

FIGURE 6.91. Automatic geometric constraints in the sketching editor sometimes cause colinear constraints to be applied inadvertently.

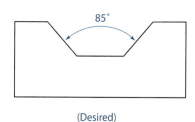

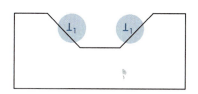

NO!

FIGURE 6.92. Automatic geometric constraints in the sketching editor sometimes cause perpendicular constraints to be applied inadvertently.

Dimensional, Instead of Geometric, Constraints

A common error in making extruded cuts and protrusions is to use a blind depth to extrude the feature a specific distance, when what is really desired is to extrude the feature to a particular surface. In the model shown in Figure 6.93, for example, the design intent is for the extruded cut to extend all the way through the base. However, the cut was created by specifying a dimensional constraint for the length of the extrusion. The specified length of the extrusion is an unnecessary, meaningless dimension, because the cut was to extend all the way through. Also, if the thickness of the base increases, as sometimes occurs when the design changes, the extruded cut may no longer extend all the way through the base. By specifying, instead, that the extruded cut is to continue until it intersects the bottom of the base, the design intent is always fulfilled.

In the model shown in Figure 6.94, the design intent is for the protrusion to extend from the angled datum plane to the top of the base. However, it was created by specifying a dimension for the length of the extrusion. The specified length of the extrusion is a meaningless dimension, because the protrusion was to extend to the top surface of the base. Also, if the thickness of the base decreases, as may occur if the design ever

FIGURE 6.93. When a cut is to extend to a specific surface or all the way through a part, cutting to a specified distance should not be done.

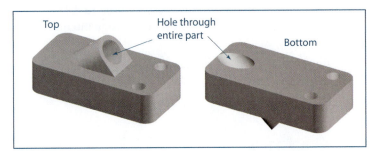

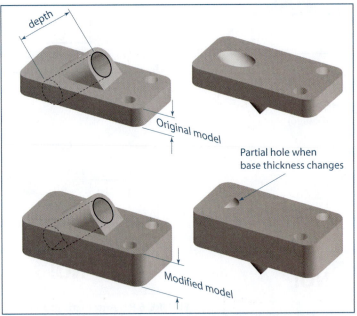

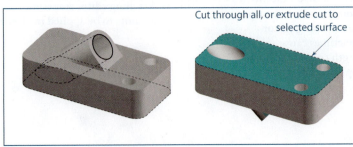

FIGURE 6.94. When a protrusion is to extend to a specific surface, extending to a specified distance should not be done.

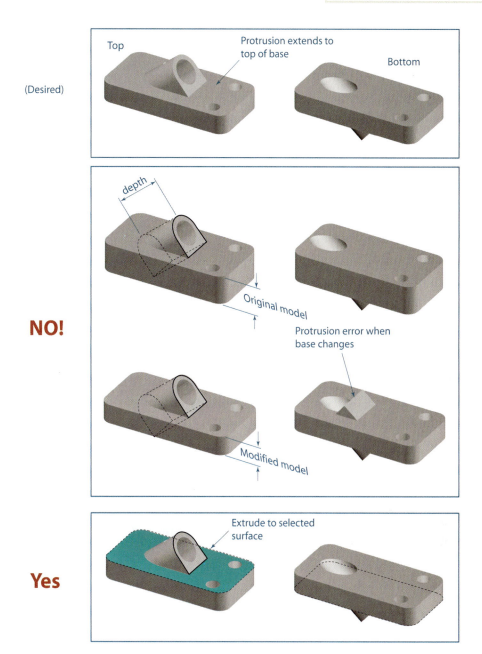

changes, the extruded protrusion may extend beyond the bottom of the base. By specifying, instead, that the extruded protrusion is to continue until it intersects the top of the base, the design intent is always fulfilled.

Using a Cut or Protrusion Instead of a Built-in Feature

A common practice for beginning designers is to use a cut feature to create a hole, for example. Figure 6.95 shows the creation of a counterbored hole using two concentric circular cut features compared to using the built-in counterbored hole feature. Even though the geometry of the resultant holes is identical, the use of cut features can result in undesired results. What if you decreased the diameter of the counterbore to a value that was smaller than the diameter of the through hole? What if the location constraints for the two cut features had different values? The resulting geometry would not represent your intent, a counterbored hole. The built-in counterbored hole feature does not permit such changes in the geometry. Another reason to use the built-in features is to capture more design information. While it may seem easy to sketch a circle on a given surface to create

FIGURE 6.95. Functional features such as this counterbored hole should be created as features and not constructed from extruded cuts or protrusions.

(Desired)

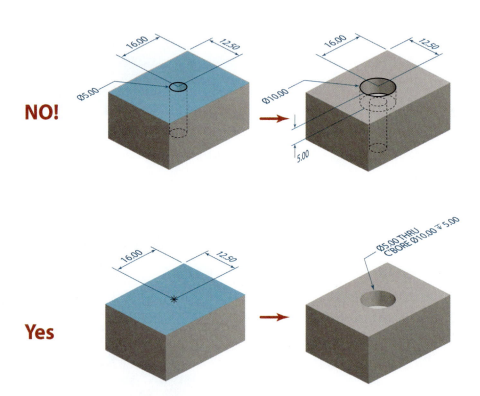

a hole, the solid modeler may be integrated with a larger, more sophisticated CAD/CAM system that might not recognize the feature as a hole. If the CAD system has intelligent assembly capabilities, it may automatically align the axis of the hole with the shaft of a bolt or screw. If you are creating manufacturing plans from your solid model, the software may recognize certain types of holes and specify the appropriate fabrication sequences automatically. But when you create the hole using a cut feature, the CAD system does not know it is a hole; and you will have to specify the assembly and fabrication sequences yourself. That leaves a great deal of room for errors such as putting bolts in upside down or omitting one of the machining operations. To save time and avoid errors in later applications, it is prudent to use the appropriate feature type that imbeds the special attributes of the feature in your solid model.

6.13 Chapter Summary

In this chapter, you learned about features and parametric solid modeling. Features are distinctive shapes that compose a solid model. Parametric models have the capability to be modified by changing the sizes and other attributes of the features in the model. The history of solid modeling shows how CAD has evolved from wireframe to solid models and provides some insight regarding the strategies used to create solid models.

Part modeling can be a very complicated process, but some general strategies make it easier to create good, robust solid models. But before you go to the computer, you need to consider how the part model will be used. Later applications such as manufacturing and documentation will be easier when the part is modeled properly. Thus, you need to plan carefully and ask yourself some questions before the first feature is created: How can you decompose that complicated widget into simpler features that are available on your solid modeler? Which feature should you create first? Which features are related to each other? How is the part used? Manufactured? Can standard features such as holes be modeled to imbed design intent and/or manufacturing information in addition to simple geometric characteristics? How does the part fit into an assembly? What will the engineering drawing look like? These are just a few questions you need to consider before you begin. For now, you may not know the answers to all of these questions; but as you gain experience, you will develop an appreciation for the importance of building a robust solid model that captures your design intent.

The solid modeling process begins with identification of the features of the part, followed by selection of the base feature. Profiles for extrusion, rotation, sweeps, and blends are created with sketches and are controlled using different types of geometric, dimensional, and associative constraints. After the base feature is created, other features are added to the model; these features are dependent upon the base feature or other previously created features. Care must be taken in the creation of solid models to make flexible models that are robust and that can be used for purposes such as analysis, manufacturing, and documentation.

As modeling systems continue to develop, designers and engineers want to include more information in the model to better simulate the physical characteristics of the parts. These models, called behavioral models, might include features such as physical properties, manufacturing tolerances, surface finish, and other characteristics of the parts. Besides a person's appearance, you would need to know something about his or her education or physical abilities to determine whether the person might be able to do a particular job. Likewise, a designer or an engineer may need to know more than just the shape of an object or assembly model to determine whether it will perform its intended function. As they become more realistic and can simulate the actual behavior of the parts and assemblies, product models of the future will contain even more characteristics and features.

6.14 glossary of key terms

algebraic constraints: Constraint that define the value of a selected variable as the result of an algebraic expression containing other variables from the solid model.

associative constraints: *See* algebraic constraints.

base feature: The first feature created for a part, usually a protrusion.

blend: A solid formed by a smooth transition between two or more profiles.

blind extrusion: An extrusion made to a specified length in a selected direction.

boundary representation (b-rep): A method used to build solid models from their bounding surfaces.

chamfers: Angled cut transitions between two intersecting surfaces.

child feature: A feature that is dependent upon the existence of a previously created feature.

constraints: Geometric relationships, dimensions, or equations that control the size, shape, and/or orientation of entities in a sketch or solid model.

constructive solid geometry (CSG): A method used to build solid models from primitive shapes based on Boolean set theory.

cosmetic features: Features that modify the appearance of the surface but do not alter the size or shape of the object.

cut: A feature created by the removal of solid volume from a model.

database: A collection of information for a computer and a method for interpretation of the information from which the original model can be re-created.

6.14 glossary of key terms (continued)

datum geometries: Geometric entities such as points, axes, and planes that do not actually exist on real parts, but are used to help locate and define other features.

datum planes: The planes used to define the locations of features and entities in the construction of a solid model.

design table: A table or spreadsheet that lists all of the versions of a family model, the dimensions or features that may change, and the values in any of its versions.

design tree: *See* model tree.

dimensional constraints: Measurements used to control the size or position of entities in a sketch.

dimension name: The unique alphanumeric designation of a variable dimension.

driven dimension: A variable connected to an algebraic constraint that can be modified only by user changes to the driving dimensions.

driving dimension: A variable used in an algebraic constraint to control the values of another (driven) dimension.

double-sided extrusion: A solid formed by the extrusion of a profile in both directions from its sketching plane.

extrude through all: An extrusion that begins on the sketching plane and protrudes or cuts through all portions of the solid model that it encounters.

extrude to selected surface: An extrusion where the protrusion or cut begins on the sketching plane and stops when it intersects a selected surface.

extrusion: A solid that is bounded by the surfaces swept out in space by a planar profile as it is pulled along a path perpendicular to the plane of the profile.

family model: A collection of different versions of a part in a single model that can display any of the versions.

feature array: A method for making additional features by placing copies of a master feature on the model at a specified equal spacing.

features: Distinctive geometric shapes on solid parts; 3-D geometric entities that exist to serve some function.

feature-based solid modeling: A solid modeling system that uses features to build models.

feature pattern: *See* feature array.

feature tree: *See* model tree.

fillets: Smooth transitions of the internal edge created by two intersecting surfaces and tangent to both intersecting surfaces.

form feature: A recognizable area on a solid model that has a specific function.

geometric constraints: Definitions used to control the shape of a profile sketch through geometric relationships.

graphical user interface (GUI): The format of information on the visual display of a computer, giving its user control of the input, output, and editing of the information.

ground constraint: A constraint usually applied to a new sketch to fix the location of the sketch in space.

history tree: *See* model tree.

holes: A cut feature added to a model that will often receive a fastener for system assembly.

horizontal modeling: A strategy for creating solid models that reduces parent-child relationships within the feature tree.

master feature: A feature or collection of features that is to be copied for placement at other locations in a model.

master model: In a collection of similar parts, the model that includes all of the features that may appear in any of the other parts.

mirrored feature: A feature that is created as a mirror image of a master feature.

model tree: A list of all of the features of a solid model in the order in which they were created, providing a "history" of the sequence of feature creation.

parameters: The attributes of features, such as dimensions, that can be modified.

parametric solid modeling: A solid modeling system that allows the user to vary the dimensions and other parameters of the model.

parametric techniques: Modeling techniques where all driven dimensions in algebraic expressions must be known for the value of the dependent variables to be calculated.

parent feature: A feature used in the creation of another feature, which is called its child feature.

path: The specified curve on which a profile is placed to create a swept solid.

primary modeling planes: The planes representing the XY-, XZ-, and YZ-planes in a Cartesian coordinate system.

primitives: The set of regular shapes, such as boxes, spheres, or cylinders that are used to build solid models with constructive solid geometry methods (CSG).

principal viewing planes: The planes in space on which the top, bottom, front, back, and right and left side views are projected.

profile: A planar sketch that is used to create a solid.

6.14 glossary of key terms (continued)

protrusion: A feature created by the addition of solid volume to a model.

regeneration: The process of updating the profile or part to show its new shape after constraints are added or changed.

revolved solid: A solid formed when a profile curve is rotated about an axis.

ribs: Constant thickness protrusions that extend from the surface of a part and are used to strengthen or stiffen the part.

rounds: Smooth radius transitions of external edges created by two intersecting surfaces and tangent to both intersecting surfaces.

shelling: Removing most of the interior volume of a solid model, leaving a relatively thin wall of material that closely conforms to the outer surfaces of the original model.

sketches: Collections of 2-D entities.

sketching editor: A software tool used to create and edit sketches.

sketching plane: A plane where 2-D sketches and profiles can be created.

solid model: A mathematical representation of a physical object that includes the surfaces and the interior material, usually including a computer-based simulation that produces a visual display of an object as if it existed in three dimensions.

splines: Polynomial curves that pass through multiple data points.

suppressed: Refers to the option for not displaying a selected feature.

surface model: A CAD-generated model created to show a part as a collection of intersecting surfaces that bound a solid.

swept feature: A solid that is bound by the surfaces swept out in space as a profile is pulled along a path.

trajectory: *See* path.

unsuppressed: Refers to the option for displaying a selected feature.

variational techniques: Modeling techniques in which algebraic expressions or equations that express relationships between a number of variables and constants, any one of which can be calculated when all of the others are known.

vertex: A point that is used to define the endpoint of an entity such as a line segment or the intersection of two geometric entities.

webs: Small, thin protrusions that connect two or more thicker regions on a part.

wireframe models: CAD models created using lines, arcs, and other 2-D entities to represent the edges of the part; surfaces or solid volumes are not defined.

6.15 questions for review

1. What are some of the uses of solid models?
2. What is a feature?
3. Why are features important in solid modeling?
4. What types of features can be used as base features for your solid models?
5. Why are wireframe models inferior to solid models?
6. What are the steps in creating a solid model?
7. What are some errors that make a sketch invalid for creating a solid?
8. Why is it necessary to constrain a 2-D sketch?
9. What are the different types of geometric constraints?
10. What are associative constraints?
11. What are dimensional constraints?
12. What does it mean when a feature is a child of another feature? A parent of another feature?
13. What are some errors that constitute poor modeling practices?
14. What are some examples of good modeling-strategies?

6.16 problems

1. Create the following closed-loop profiles using the 2-D drawing capabilities of your solid modeling software. Define the geometry and sizes precisely as shown, using the necessary geometric constraints. Do not over- or undercon-strain the profiles.

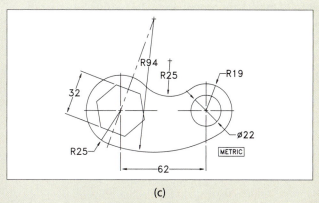

(a)

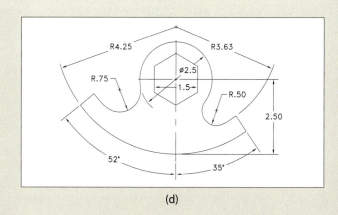

(b)

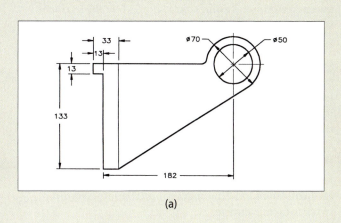

(c)

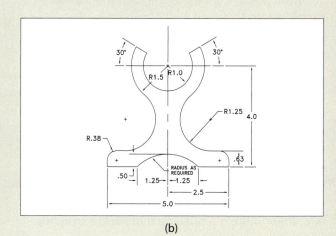

(d)

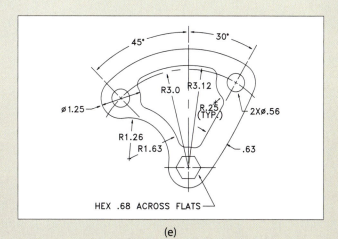

(e)

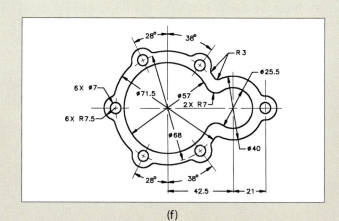

(f)

6.16 problems (continued)

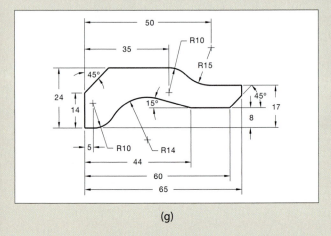

(g)

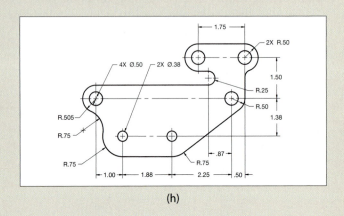

(h)

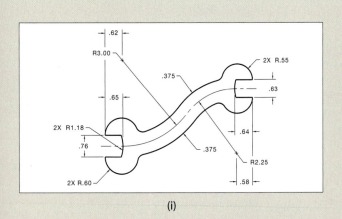

(i)

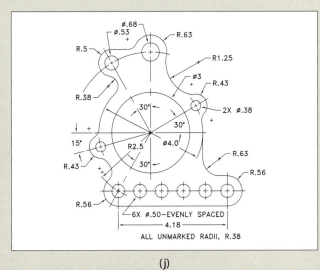

(j)

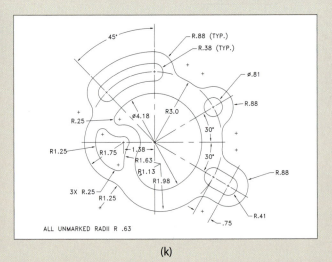

(k)

FIGURE P6.1.

6.16 problems (continued)

2. Study the following closed-loop profiles for which geometric constraints have not been added. Number each segment of the profiles and specify the necessary geometric constraints on each segment to create the final profile. Do not over- or underconstrain the profiles.

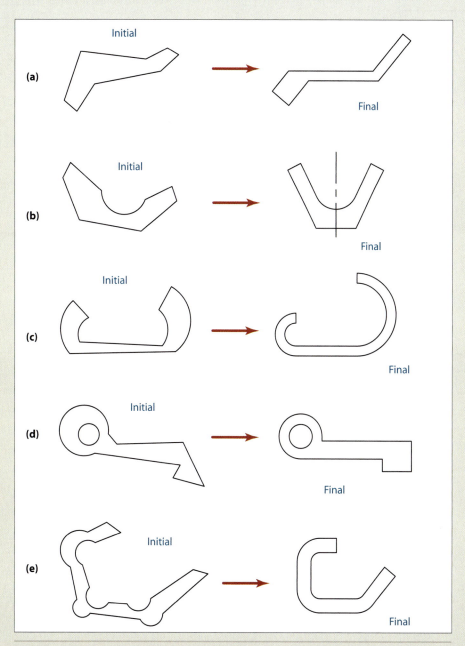

FIGURE P6.2.

6.16 problems (continued)

3. At first glance, these profiles may appear to be missing key dimensions. However, they are fully constrained by the addition of geometric constraints. Number each segment of the profiles. What were the geometric constraints used for each segment?

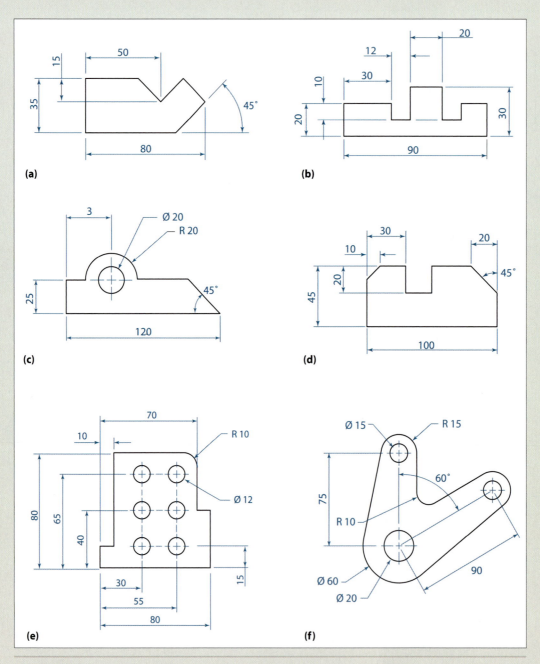

(a) **(b)**

(c) **(d)**

(e) **(f)**

FIGURE P6.3.

6.16 problems (continued)

4. Create solid models of the following parts in your CAD system. Identify what you consider to be the base geometry for each part. Are any (child) features dependent upon the existence of other (parent) features? If so, specify the hierarchy.

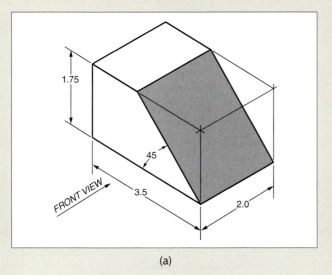

(a)

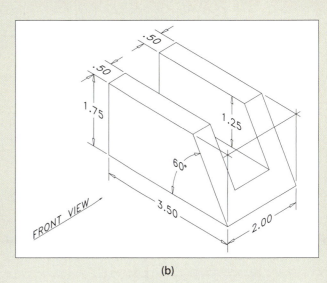

(b)

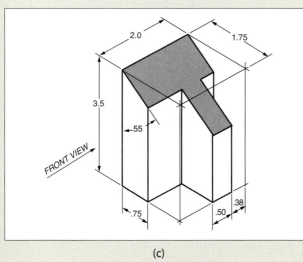

(c)

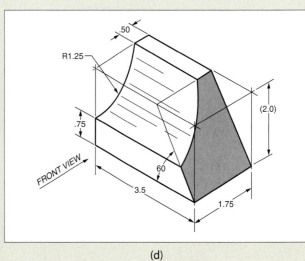

(d)

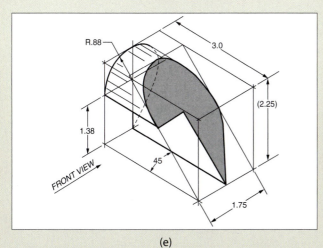

(e)

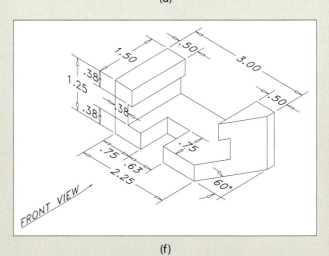

(f)

6.16 problems (continued)

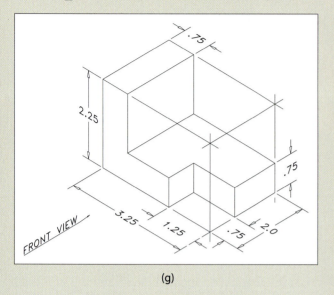

(g)

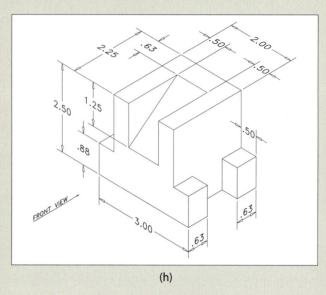

(h)

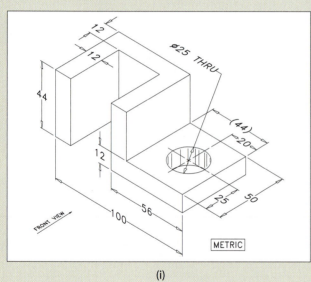

(i)

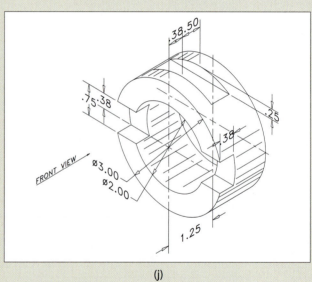

(j)

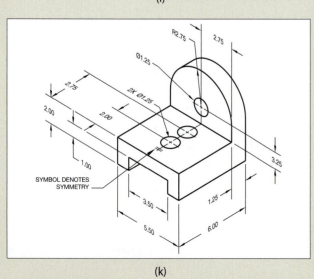

(k)

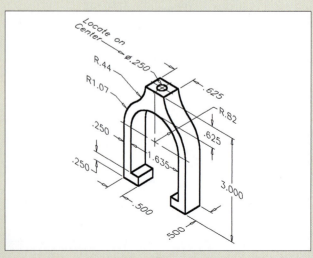

(l)

6.16 problems (continued)

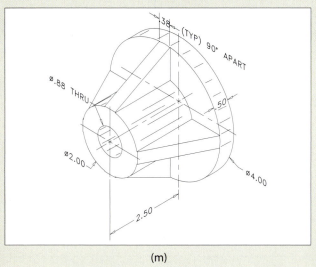

(m)

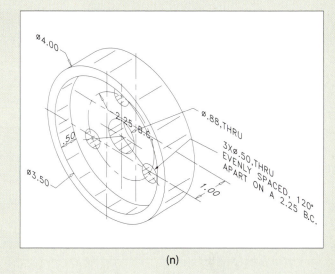

(n)

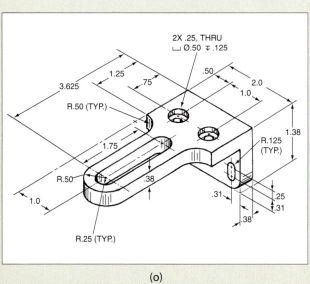

(o)

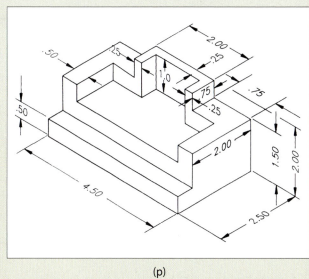

(p)

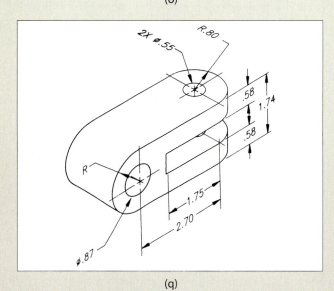

(q)

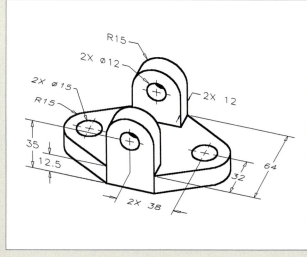

(r)

6.16 problems (continued)

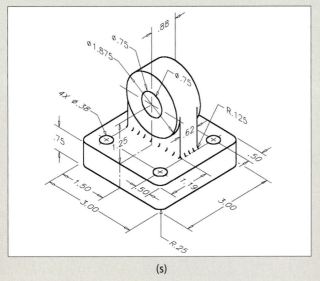

(s)

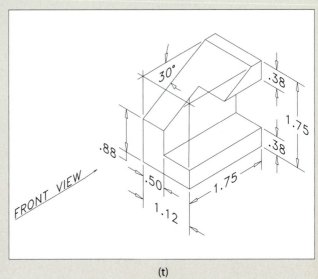

(t)

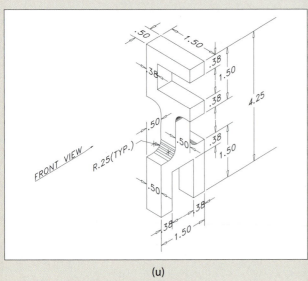

(u)

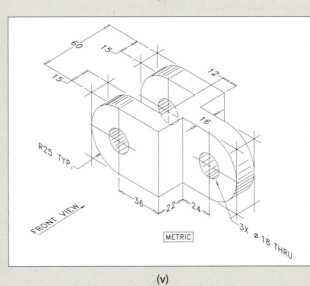

(v)

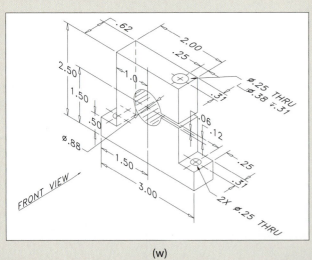

(w)

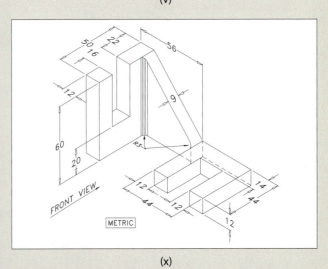

(x)

6.16 problems (continued)

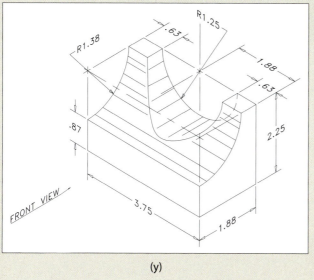

(y)

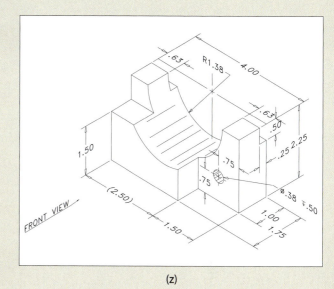

(z)

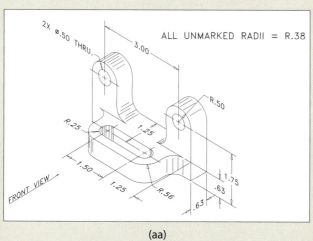

(aa)

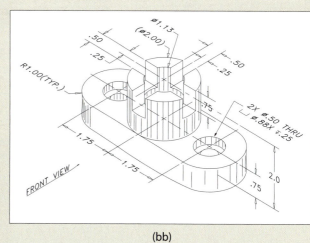

(bb)

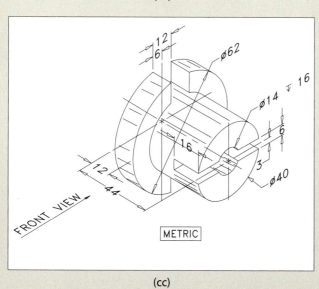

(cc)

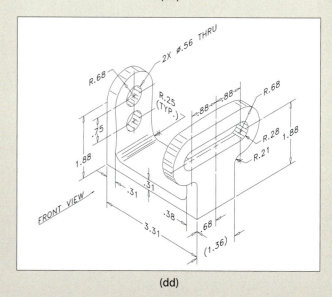

(dd)

6.16 problems (continued)

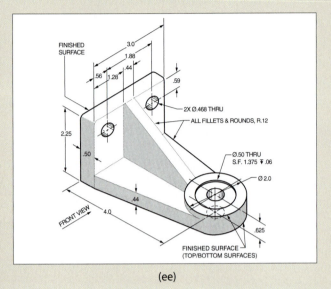

(ee)

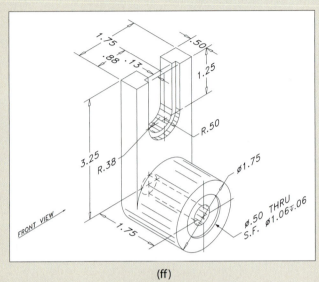

(ff)

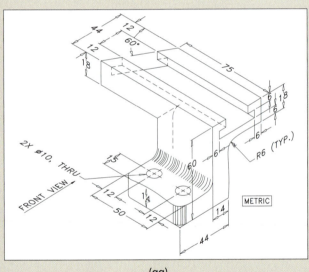

(gg)

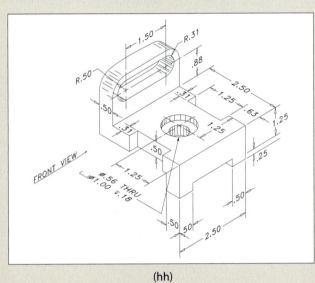

(hh)

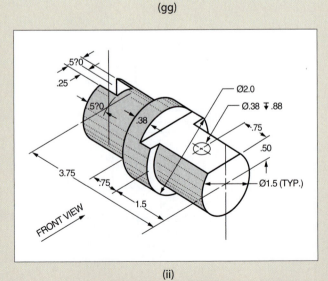

(ii)

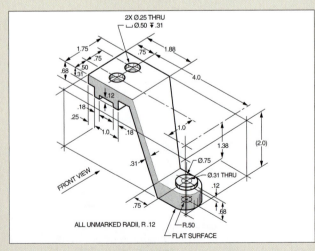

(jj)

6.16 problems (continued)

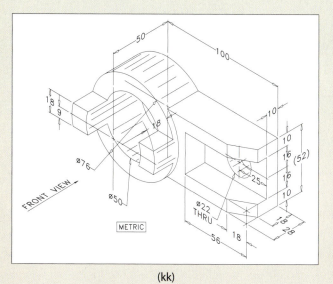

(kk)

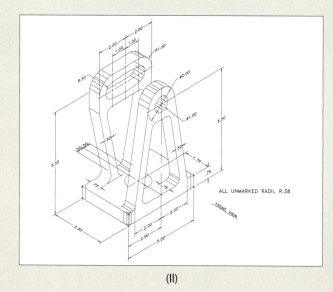

(ll)

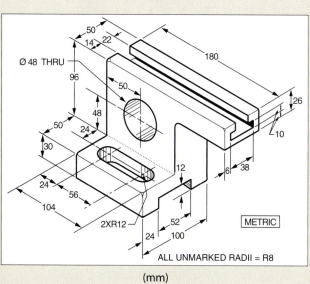

(mm)

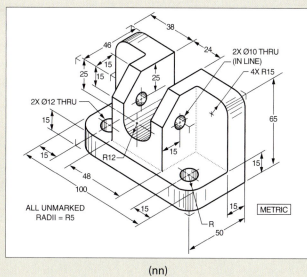

(nn)

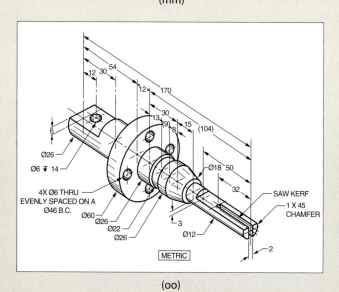

(oo)

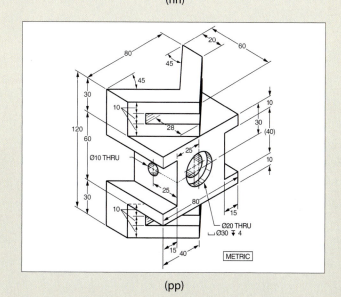

(pp)

6.16 problems (continued)

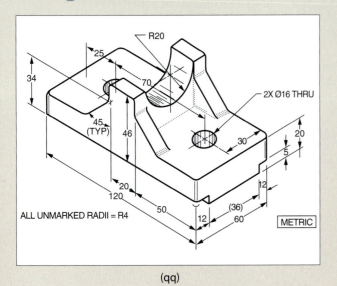

(qq)

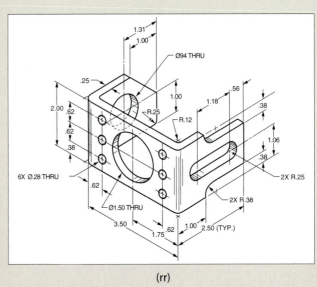

(rr)

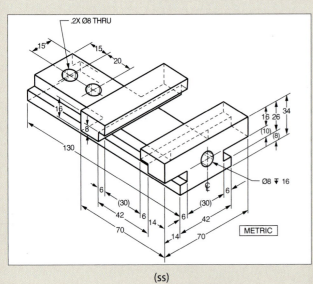

(ss)

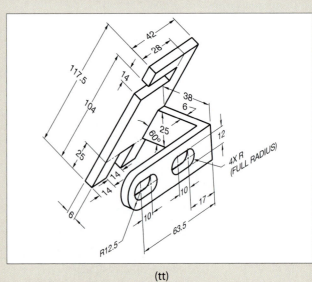

(tt)

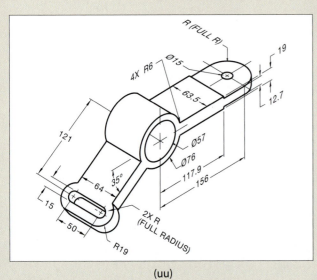

(uu)

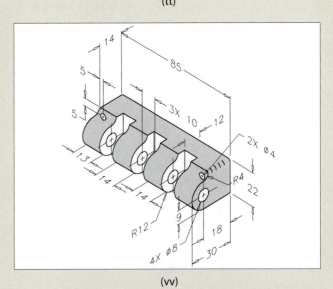

(vv)

6.16 problems (continued)

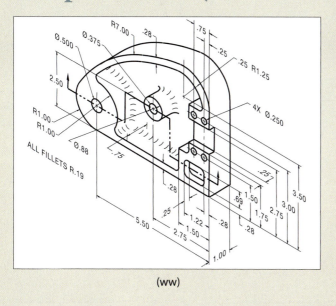

(ww)

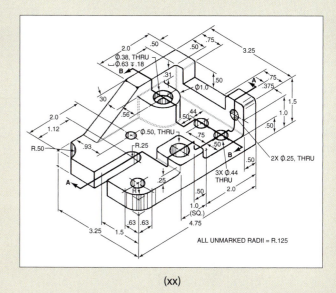

(xx)

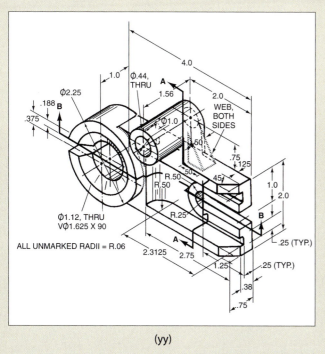

(yy)

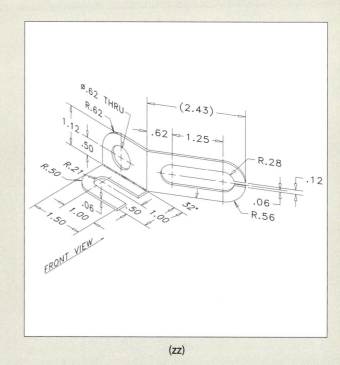

(zz)

6.16 problems (continued)

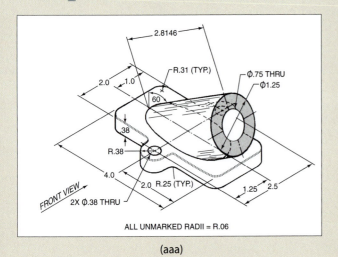

(aaa)

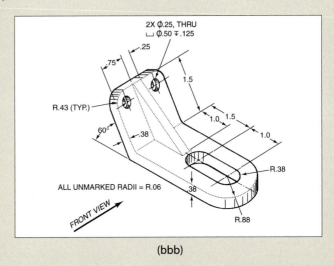

(bbb)

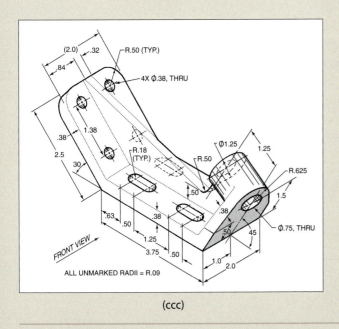

(ccc)

FIGURE P6.4.

7

Assembly Modeling

objectives

After completing this chapter, you should be able to

- Apply new terminology in the context of assembly modeling
- Create an appropriate hierarchy for effective assembly modeling
- Apply assembly constraints between instances
- Create a bill of materials and an assembly drawing
- Determine interferences and clearances between instances in an assembly

7.01
introduction

In a previous chapter, you learned about the techniques used to create solid models with computer tools. Most engineered systems, however, do not consist of a single part, but comprise multiple parts that work together and form the system or assembly. Think of a bicycle. It is used for personal transportation, and its purpose is to allow an individual to get from point A to point B through pedaling. It is composed of several subsystems, each of which serves a distinct purpose in allowing a person to operate the bicycle in a safe and consistent manner. Some of the subsystems found on a bicycle are the pedal system, the gear system, the tire system, and the frame. Each subsystem is composed of other subsystems or parts. For example, the gear system contains many single parts (e.g., the individual gears and the cables), as well as other subsystems (e.g., the derailleur).

This chapter describes how systems or assemblies are created with the use of CAD software and what type of information can be extracted from the assembly models. Because the foundation of assembly modeling is most often the initial creation of the components that make up the assembly, the assumption is that you are already familiar with the creation of solids.

7.02 Assembly Terminology

As with most categories in engineering design and analysis, one of the first things you must learn is the terminology particular to the topic. For assembly modeling, you need to learn a few new terms so you can work productively. Some of these terms will be discussed in the subsequent paragraphs.

In the creation of solid models, computer-generated geometry was referred to as parts, features, or objects; however, the objects that make up a system are referred to as **components**. A system component is identical to its referenced object geometry, and the change in terminology is to avoid confusion between two closely related modeling tasks. Think of a component as a 3-D part that has been brought into an assembly model. For example, consider the footbridge shown in Figure 7.01. Notice that the assembly model consists of abutments, approaches, a bridge deck, and handrails.

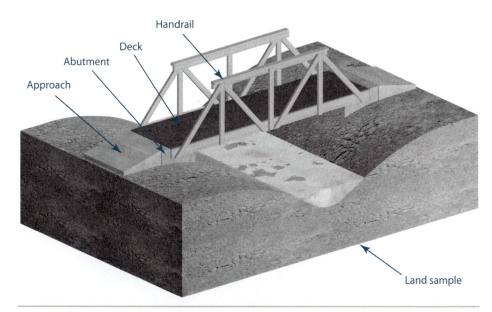

FIGURE 7.01. An assembled footbridge placed on a sample cutout section of land.

You can create a solid model of the abutment within solid modeling software. In this case, the solid model of the abutment is the desired final outcome of the modeling process. When you are working on assembly modeling where the desired final outcome is a bridge assembly, the parts brought into the model that you are working on (the assembly) are called components. When the solid model of the abutment is brought into the assembly model, it is referred to as a component and is no longer thought of as a "part."

In the case of the bridge assembly model, two abutments are in the desired final result. It would be cumbersome to bring a new abutment component into the assembly each time it is required. Further, the abutments are identical to each other (they are just oriented differently in space), so you want it to be clear that the abutments in the final assembly are copies of the same component. In assembly modeling, the copies of a component within the system are called **instances**. Thus, for the bridge, an abutment can be an overall component associated with the system with two instances of the particular component within the assembly. Figure 7.02 illustrates the concept of instances of components in an assembly model.

In reality, system components do not actually appear on the computer screen—they are stored in memory within the assembly model, waiting for you to instance them into the system, but are not displayed in the work area. It is similar to having certain vocabulary words stored in your brain. You can put the words on a sheet of paper to form sentences and paragraphs, but the words do not physically exist anywhere in space—they exist only in your memory. When you want to put a copy of a component in an assembly, you *instance* it into your workspace.

FIGURE 7.02. The concept of parts, components, and instances in assembly modeling.

Part modeling

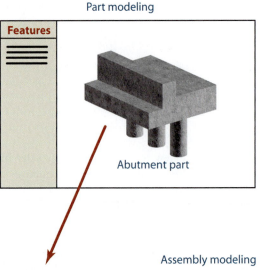

Assembly modeling

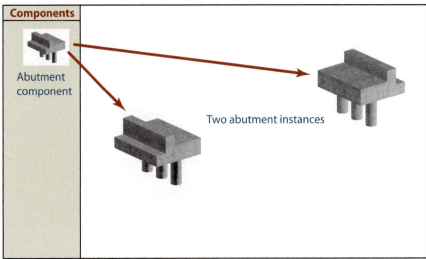

FIGURE 7.03. The handrail subassembly (with the bridge deck shown in phantom for reference only).

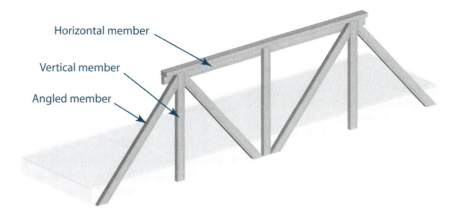

Horizontal member

Vertical member

Angled member

FIGURE 7.04. The second handrail subassembly as an instance of the first handrail subassembly.

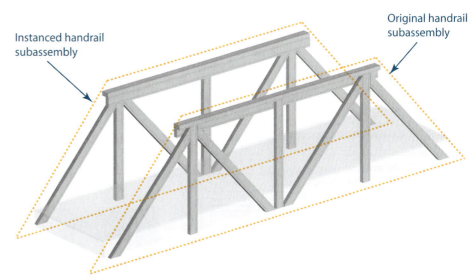

Instanced handrail subassembly

Original handrail subassembly

Another new term for you to consider in assembly modeling is the concept of a **subassembly**. A subassembly is a grouping of components that serves a single purpose within the overall assembly. Going back to the bridge example, a subassembly could be created to represent the handrail. Notice that the handrail consists of vertical members, angled members, and the horizontal crossbar at the top, as shown in Figure 7.03.

Instead of instancing each of the individual members that make up the handrail into the overall assembly, you could create one subassembly of the handrail that includes appropriate instances of all of its components—including the vertical, angled, and horizontal members—and just instance that subassembly into the overall system two times. The completed handrail subassembly is shown in Figure 7.04.

Since it is unlikely that the horizontal member of the handrail will act independently of the vertical or angled members, it makes sense to put them into a subassembly together so you can work with them as a single unit. You might think it would be easier to make the handrail a single part. But because the individual components may be made of different materials, you want to keep them as separate objects for any analysis you perform later on. Linking them in a subassembly allows you the flexibility to alter properties of components independent of one another, but also affords you the efficiency of treating them as one unit in the assembly modeling process.

Subassemblies are unique in that they are composed of instances, but subassemblies also can be instances. When you are creating the overall system, you can insert instances that are individual components or that are subassemblies made up of several individual components. Subassemblies also can contain other subassembly instances. For the handrail subassembly, you could choose to combine two angled members and one vertical member as one subassembly, as shown in Figure 7.05. The subassembly

FIGURE 7.05. The web subassembly composed of two angled components and one vertical component.

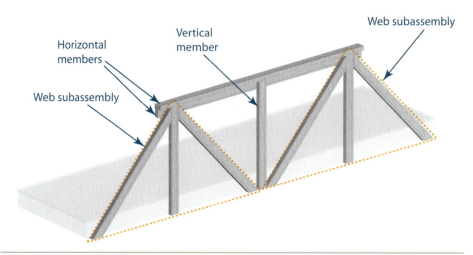

FIGURE 7.06. The complete handrail subassembly composed of one vertical member, two horizontal members, and two web subassemblies (with the bridge deck shown in phantom for reference only).

could be instanced into the overall handrail subassembly two times, along with one additional vertical member and the horizontal bar, to make up the overall handrail subassembly shown in Figure 7.06.

It may seem that the change in terminology between solid modeling and assembly modeling is picky and overly complicated. Why not just put multiple copies of the abutment part within the assembly? Why do you have to learn about components and instances? Why are subassemblies that are made up of instances and that are instances themselves important? The main reasons for the changes are to improve assembly modeling efficiency and to save computer storage space. Saving computer storage space may seem like a trivial matter; however, the less memory a model takes up, the faster your computer modeling software will work. Files that are too large (in terms of memory) are difficult and slow to work with. The improvement in assembly modeling efficiency through the use of subassemblies should be apparent. But how do you save on computer storage space or memory?

To answer that question, consider just the horizontal member of the handrail sub-assembly. Depending on the design of the member, the computer must "remember" many points to define the model. Each time there is a vertex where two or three edges intersect, the software must remember where the point is located in 3-D space. If the edges of the member are rounded, not sharp, corners, several points on the rounded edge must be remembered for the part geometry to be defined completely. Because some designs for a horizontal handrail are fairly complex, the number of geometric data points used to define it can be large; and the number of 3-D locations in space for the part the computer model must remember also is large. If you were to put the horizontal member directly into the assembly model, each time you moved the member in space within the assembly model, the computer would have to remember the new location of each multiple point that defines the horizontal member. In the case of a bridge with two handrails, the number of points to be remembered is relatively large.

By organizing with components and instances, however, the data to be stored is significantly reduced. The component is referenced to an object, and all the component has to remember is that the component looks exactly like the object. When an instance of the component is placed in the assembly (or subassembly), all that has to be remembered is location of the instance relative to that of the component. In other words, when the instance is located 6 units along the x-axis and 0 units along both the y- and z-axes and it is rotated 90 degrees about Z relative to the component definition, the only thing to be remebered is the location of the instance compared to the location of the component. Thus, for each instance, only six pieces of data must be remembered by the computer—translations in X, Y, and Z and rotations about X, Y, and Z. The amount of data stored

FIGURE 7.07. A schematic representation of a horizontal member part (a) and an instance (b).

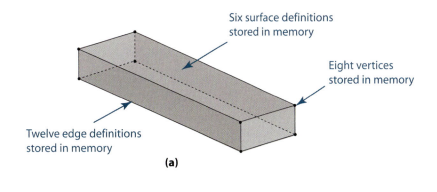

Six surface definitions stored in memory

Eight vertices stored in memory

Twelve edge definitions stored in memory

(a)

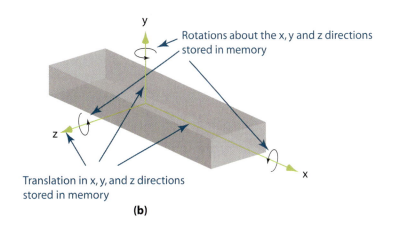

Rotations about the x, y and z directions stored in memory

Translation in x, y, and z directions stored in memory

(b)

under this scheme is significantly less than what would be stored if copies of an object were placed directly into the system. Figure 7.07 shows the object for the horizontal handrail member and an instance of it in a subassembly. Also shown in the figure are the data points defining the location of the instance that define its location within the subassembly.

7.02.01 Associativity

In the solid modeling chapter, you learned about associativity between parts and drawings. When an object changed, any associated drawings also changed in the same way. Associativity also can exist between parts, components, and assemblies. In assembly modeling, **associativity** means that if you change the geometry of a part, the component and all instances of it also will change by the same amount. With some software, when you change a component within assembly modeling, the object geometry also changes by the same amount. That also means that any drawings, finite element models, or manufacturing models associated with the part also will be updated.

Associativity greatly increases your computer modeling efficiency for complex parts and systems. Thinking back on the footbridge example, imagine you are working for a company that designs and produces footbridges for parks and recreation areas. The company might have several bridge models that use the same bridge deck. If the bridge deck design division of the company changes its design, the changes can be reflected throughout all bridge models simultaneously as long as the associative relationships between parts and assemblies have been maintained throughout the company's product line. If the associations have been broken, the engineers working with the various models must be aware of the changes in the bridge deck design and then make the required changes model by model. If the engineers forget one of the models, the overall design will be incorrect, which could have a significant impact when the production phase of the project is initiated.

7.03 Assembly Hierarchy

Assembly models are easier to work with when they are organized in a logical manner. An assembly is usually thought of as a composition of several smaller subassemblies, each of which may consist of other subassemblies or individual components. The organization or structure of a system is referred to as its **hierarchy**. The hierarchy is similar to an inverted family tree. Another way to think of an assembly hierarchy is as a corporate structure. The president of the company is at the top of the hierarchy and at the next level has several vice presidents reporting to him. Each vice president, in turn, is responsible for several managers; and structure of the organization continues until you reach the lowest level in the company hierarchy, where the laborers are located. In this analogy, the individual laborers can be thought of as subassembly instances, the managers would be the first level of subassemblies, the vice presidents would be the next level of subassemblies, etc., until the entire assembly is defined. The top-level assembly could be considered the company president.

As with objects and features, the associations between components and subassemblies are often called parent-child relationships. Going back to the handrail subassembly, the handrail would be a parent whose children are the horizontal member, the subassembly of two angled members and a vertical one, etc. The overall bridge would be the parent whose children consist of two handrails, two approaches, two abutments, the bridge deck, etc. As with objects and features, to work effectively, you must understand the various parent-child relationships defined in assembly modeling. A schematic of a possible hierarchy for the footbridge is shown in Figure 7.08.

Organizing an assembly in a logical hierarchy enables you to work more efficiently with the assembly. The technique is similar to the way you organize files on a computer, establishing folders to group files together in your work space. As stated previously, one advantage of creating an assembly hierarchy is that subassemblies can be dealt with as a whole rather than as separate components. For example, one subassembly can be moved as a single unit within the system, rather than individual components of the system being moved separately, just like the way you move folders and all of the files in them on your computer workstation.

FIGURE 7.08. The footbridge assembly hierarchy.

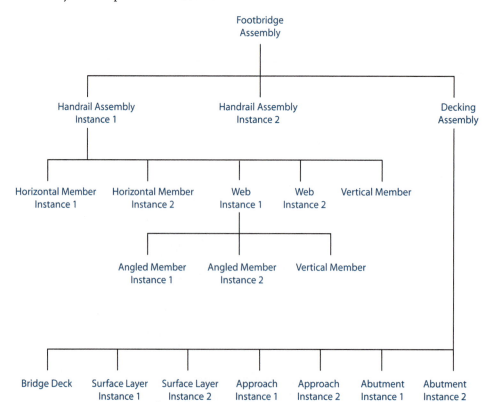

7.04 **Assembly Constraints**

The first thing you want to do after setting up your system hierarchy is to orient the instances so they are properly located in space relative to one another. To do that, you need to select one component to serve as the **base instance**. The base instance remains stationary with the other instances moving into place around it. For example, if your assembly model was a car, you might choose the chassis as the base instance. Usually the choice of a base instance is fairly obvious; however, in some cases, you may have to choose one from among several choices.

After selecting the base component, you should establish reference planes for the assembly that are connected to the base component. These reference planes will serve as the coordinate planes for the space defined by the system. For the footbridge assembly, the bridge deck was selected as the base component, with the reference planes established as shown in Figure 7.09. You will use these reference planes as you orient the various components within the system.

When creating parts, you used constraints to establish geometric and dimensional relationships between 2-D entities or features. Thus, you constrained two lines to be parallel or perpendicular to each other. Or you constrained the diameter of a circle to be a specific size and its center to be located given distances from lines on the drawing or from edges on an object. In assembly modeling, you can apply **assembly constraints** between two 3-D instances so the instances maintain dimensional or geometric relationships with respect to each other within the assembly.

Each time you bring an instance of a component into an assembly, you introduce six degrees of freedom (DOF). The instance will have a set of coordinate axes associated with it, and the six new DOFs will correspond to the three translational DOFs (distance along X, Y, and Z from the instance origin to the base component origin) and to the three rotational DOFs (rotations about X, Y, and Z) relative to the coordinate planes associated with the base component. Each time you apply a constraint to the assembly, you remove one or more DOFs. As with part modeling, you can continue to apply constraints until all of the DOFs are removed from the assembly model.

7.04.01 Concentric Constraints

One of the more useful applications of assembly constraints is to define two different instances to be concentric with each other. This is especially useful in dealing with cylindrical shafts that fit within a cylindrical hole in another part. You can constrain the centerline of one instance to coincide with the centerline of a different instance,

FIGURE 7.09. The bridge deck as the base instance, with reference planes and axes.

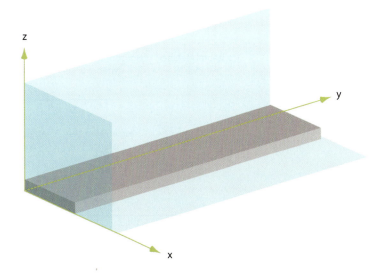

FIGURE 7.10. Concentric constraint used to locate a bolt in its hole. Before application of constraint (a) and after constraint (b).

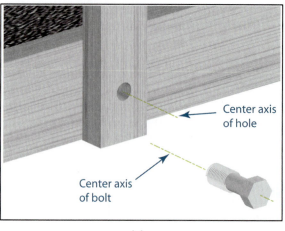

(a)

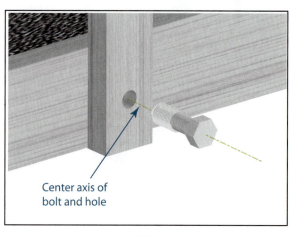

(b)

forcing the two instances to be concentric. After constraining the instances to be concentric, you only need to determine how far along each centerline the instances are located relative to each other. For the bridge assembly, nuts and bolts are used to fasten the instances of the system together. (The nuts and bolts were not shown in previous figures of the bridge assembly to simplify the discussion.) Figure 7.10 shows a concentric constraint applied between two instances in the bridge assembly. Note that when the concentric constraint is applied to the system, several DOFs are removed, since the x-, y-, and z-rotations are now fixed and one of the translation DOFs also has been removed.

In some modeling packages, an "insert" constraint is a special type of concentric contraint used to insert a fastener such as a bolt into a hole in a different part. When the bolt is inserted into the hole, the system automatically applies a concentric constraint between the two components.

7.04.02 Mating Surfaces Constraints

Another useful constraint available in most assembly modeling software is used to define two surfaces as mating surfaces. Mating surfaces coincide with each other—in other words, they line up on top of each other. Usually that type of constraint works only with flat or planar surfaces. When creating a mating surface constraint, you are essentially working with the surface normals. A surface normal is defined as "a vector that is perpendicular to the surface and points away from it." So when you are applying

a mating surface constraint, you are forcing the normals of the two surfaces to be parallel to each other. Depending on the final desired result, you may have to flip one of the instances around in order to apply the mating surface constraint to achieve the correct orientation between instances. Figure 7.11a shows a mating surface constraint applied between the inner face of the hex head at the end of the bolt and the outer surface of the handrail member for the bridge. Figure 7.11b shows the result of incorrect mating surface orientation. Figure 7.11c shows the correct surface orientation, with bolt flipped in the final result. Also notice that when the constraint is applied, several DOFs for the system are removed, including both rotational and translational DOFs.

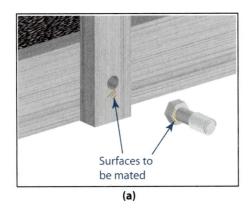

Surfaces to
be mated

(a)

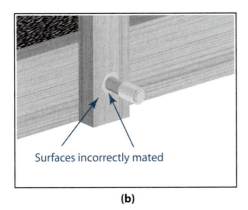

Surfaces incorrectly mated

(b)

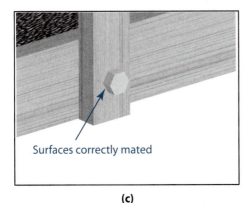

Surfaces correctly mated

(c)

FIGURE 7.11. Applying mating surfaces constraint after concentric constraint to the bolt and hole. The mating surfaces are identified in (a). The incorrect mating orientation is shown in (b); the correct orientation, in (c).

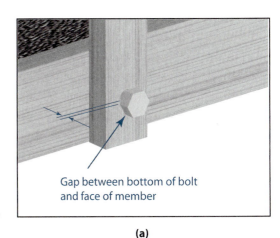

Gap between bottom of bolt
and face of member

(a)

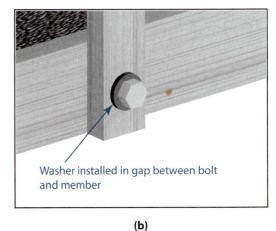

Washer installed in gap between bolt
and member

(b)

FIGURE 7.12. Applying an offset constraint between the bolt and member to permit a washer instance.

Another type of constraint that is related to the mating surfaces constraint is one whereby you include an "offset" from one surface to the other. Basically, you are saying that the surface normals for the two surfaces are still parallel to each other, but now the surfaces do not line up on top of each other. Figure 7.12a shows the bolt and handrail member after an offset constraint has been included between the instances to make room for the washer that is included in Figure 7.12b.

7.04.03 Coincident Constraints

One definition of *coincidence* is that two entities take up the same space. For that type of constraint in assembly modeling, coincidence can occur between two lines, between two points, or between a point and a line. (Note that coincidence of two planes has already been defined with the mating surfaces constraint.) In assembly modeling, defining coincident constraints often means that you select a corner or an edge on one instance to be coincident with a corner or an edge on another instance. When applying this type of constraint, you typically are required to input offset distances between points in order to achieve your final desired result. Figure 7.13a shows the effect of making one edge on a vertical handrail member coincident with an edge on the horizontal member, and Figure 7.13b shows the final result achieved after including a dimensional constraint between corresponding endpoints of the edges. Figure 7.14a and Figure 7.14b show the effect of making the edges of two members coincident and then making the two corresponding corner points coincident to achieve the desired final result.

When applying coincident constraints, you may again have to flip one instance around to achieve the desired final result. Figure 7.15a shows two instances with a coincident constraint applied, and Figure 7.15b shows the same two instances after one has been flipped to put it into its desired orientation.

FIGURE 7.13. A coincident edge constraint is used in (a) to align the edges of the horizontal and vertical members. A dimensional constraint is used in (b) to perform the final location of the horizontal member with respect to the vertical member.

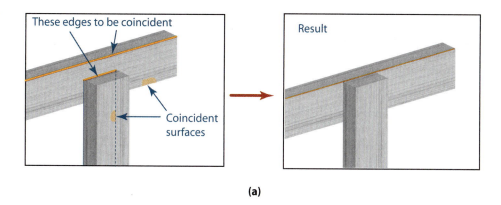

(a)

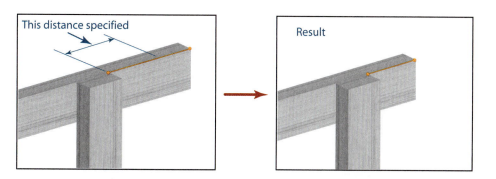

(b)

FIGURE 7.14. A coincident edge constraint is used in (a) to align the edges of the vertical and angled members. A coincident vertices constraint is used on the vertices shown in (b) to perform the final location of the angled member with respect to the vertical member.

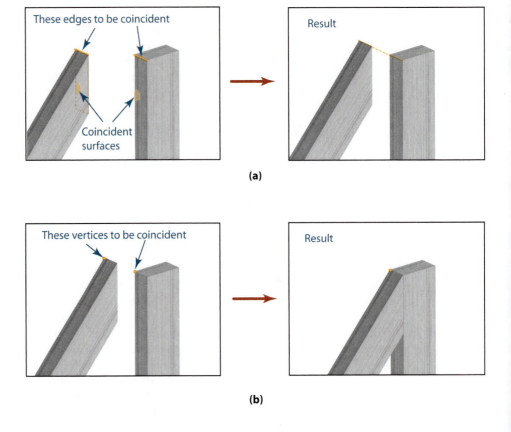

(a)

(b)

FIGURE 7.15. The effect of flipping a coincident surface constraint. Coincident surfaces, edges, and vertices are used in (a) to align the vertical and angled members. The coincident surface constraint is flipped in (b) to realign the members.

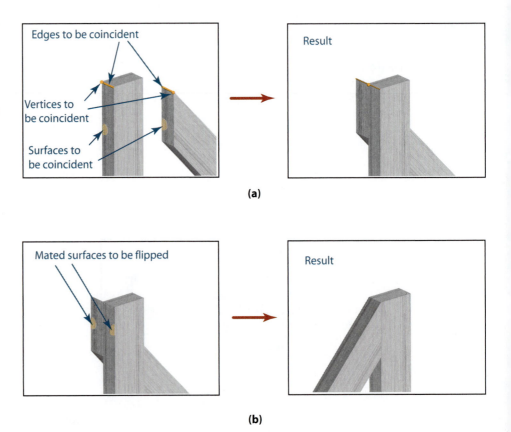

(a)

(b)

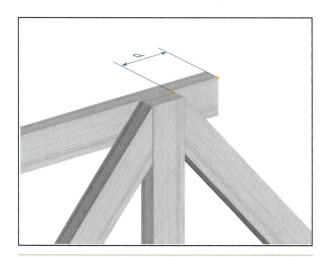

FIGURE 7.16. A distance constraint between vertices to locate the horizontal member with respecte to the angled member.

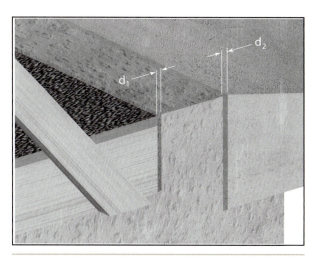

FIGURE 7.17. Parallel distance constraints used to create clearance gaps between the deck and the abutment and between the abutment and the approach.

7.04.04 Distance Constraints

When creating a 3-D solid model, you often add dimensions between features to define sizes of parts. With assembly models, you add distance constraints to define the relationship between two instances. When adding the distance constraints, you use points, edges, and surfaces to define the distances between the instances. Often a distance constraint is used in conjunction with a second or third assembly constraint. For example, for the two instances shown in Figure 7.16, a coincident constraint was applied between the point on the angled member and the edge on the horizontal crossbar. A distance constraint was then applied between the endpoint of the two instances, as shown in the figure.

A distance constraint also can be applied between parallel surfaces on two separate instances, as shown in Figure 7.17.

7.04.05 Adding Constraints to Your Assembly

In addition to putting your parts together to see how they look, certain types of analysis that are best performed on an assembled system will aid you in your engineering design tasks. One predominant type of analysis that you can perform with an assembled model is to model it as a mechanism. With this type of analysis, you add joints such that two parts are able to rotate freely about an axis or you add a joint such that two parts slide relative to each other. Consider a piston in a cylinder. The piston can slide in and out; and as is illustrated schematically in Figure 7.18, the cylinder and the connecting rod are also free to pivot about the piston pin and the connecting pin.

Mechanism analysis is important for a piston assembly. The analysis allows you to determine whether the connecting rod bumps up against the walls of the cylinder as the mechanism goes through its cycle of motion or whether the piston bumps up against the fixed end of the cylinder. For you to accomplish this type of analysis, however, your assembly must be properly constrained.

When you learned about including constraints with sketches or models as you were studying part creation, you learned that sketches could be unconstrained, fully constrained, or partially constrained. Similarly, assemblies can be unconstrained, fully constrained, or partially constrained. You should avoid unconstrained systems since all instances in the assembly are free to move relative to one another. If you have an unconstrained assembly, you need to move each instance and subassembly one at

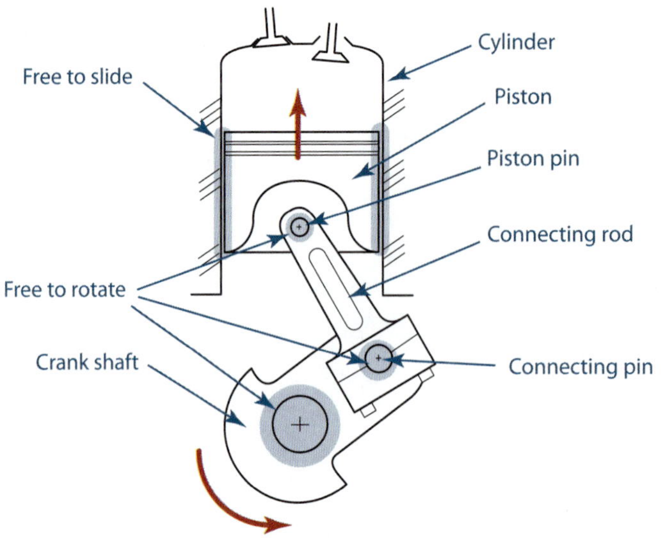

FIGURE 7.18. Schematic drawing of piston/cylinder assembly.

a time—a tedious task—to perform your analysis. An assembly that is fully constrained is rigid and unable to move. Each time you try to move one instance, the entire system reverts to its original orientation to satisfy all of the constraints on it. Like Goldilocks, you want a system that is "just right"—with just enough constraints to define the permissible motion in the assembly but not too many constraints that its ability to function properly is hindered.

Going back to the piston/cylinder assembly, you want the constraint between the piston and the pin to be rigidly defined—these two instances should not be able to move relative to each other (ever!). However, the cylinder and piston should be partially constrained such that they remain concentric to each other but the piston is still free to move up and down along the centerline of the cylinder. Likewise, the connecting rod should be partially constrained such that the pin remains concentric to the pin hole but is free to rotate about its axis as the system moves. To achieve the balance between systems that are over- or underconstrained, you should think about the types of motion your assembly will go through in computer analysis as well as in real life. If you think about the bridge assembly presented earlier in this chapter, the instances in this system should never move relative to one another—a bridge is a rigid structure that is built to remain rigid its entire life cycle. In the case of the footbridge, having a fully constrained assembly is permissible.

Some software requires assembly models be fully constrained. In this case, you are usually permitted to include "assumptions" with your constraints in order to satisfy the need for system flexibility and movement. For example, in the case of the connecting rod and piston pin, you could include the concentric constraint and then add an "assumption" that the two instances are able to rotate relative to each other. In this case, the assumption will eliminate DOFs in the system and the software will be satisfied that the assembly is "fully" constrained.

7.05 Exploded Configurations

For most mechanical systems, an assembly drawing is necessary in order to put the system together. You probably saw that kind of drawing if you ever put together a furniture kit or a model. Once you assemble your system, you can create an **exploded configuration** that will essentially be an assembly drawing for the system. Figure 7.19 shows an exploded configuration for a small flashlight assembly. Note that the configuration shows how all of the parts in the system will be put together, or assembled.

FIGURE 7.19. An exploded view of a small flashlight assembly.

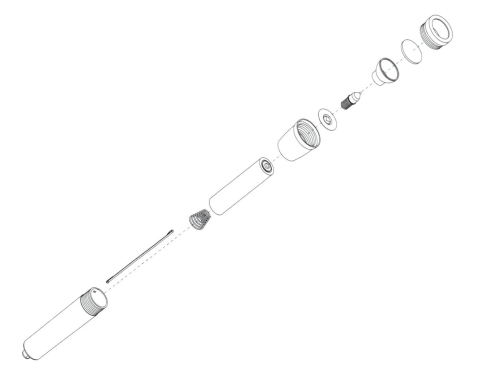

7.06 Interferences and Clearances

One of the main advantages to be gained from assembling a system of parts is that you are able to determine whether two parts overlap or interfere with each other. Since different engineering teams often work on separate parts of an assembly and since communication between groups may not always be clear, being able to determine overlap is especially important when working collaboratively on a large design. The amount that two instances overlap is referred to as the **interference** between them. Another advantage is that you can determine **clearances**, or minimum distances, between parts. You can use that information to optimize the assembly. You can change the size of features as appropriate to remove interference or to increase clearances between parts. For some systems, you may choose to perform a kinematic analysis of the parts in the assembly. You can then check each part for interference or clearance problems as the parts move through the kinematic analysis. For example, if you are designing a piston and cylinder system, in the initial position, none of the parts will overlap. However, as the crankshaft is rotated and the piston slides in or out of the cylinder, you may find that the connecting rod interferes with the end of the cylinder, as illustrated in Figure 7.20. Since that is a problem with the design, you can modify the design to alleviate the problem and re-check the interferences and clearances.

One other consideration in accomplishing interference or clearance checks for an assembly is the *number* of checks you want to perform. If you have a system with twenty-five parts, you probably will not want to check for interference between all parts in the assembly; that would require (25 – 1)! (24 factorial, or more than 6×10^{23}) checks. With most CAD software, you can specify two groups of parts. The software will perform the interference or clearance checks, checking all of the parts in one group against all of the parts in the second group.

Only one of the two, interference or clearance, is possible between two instances. If two parts overlap, there is an interference; and a clearance is not defined between the two. Conversely, if two instances have a clearance, or distance, between them, they do not overlap; and an interference is not defined.

One common place where you need to check interferences and clearances is between shafts and circular holes. When you align shafts in holes by making them concentric, you may not realize that the shaft is too large or too small for the hole. In a later chapter, you will learn about classes of fits between holes and shafts. In that chapter, you will learn that sometimes you want the shaft to be larger than the hole and other times you want the shaft to be smaller than the hole. The class of fit you need depends on your design intent. Figure 7.21 shows an end view of two shafts in holes—in one case, the shaft is larger than the hole; in the other case, the shaft is smaller than the hole. The clearance and interference between the shaft and the hole in each case are shown in the figure.

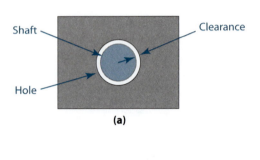

(a)

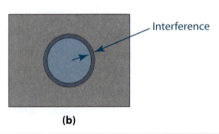

(b)

FIGURE 7.21. When the shaft has a smaller diameter than the hole, as in (a), there is clearance between the two parts. If the shaft has a larger diameter than the hole, as in (b), there is interference between the parts.

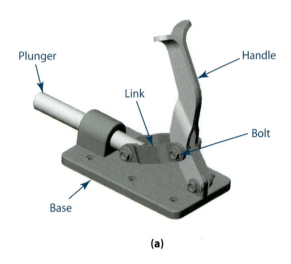

(a)

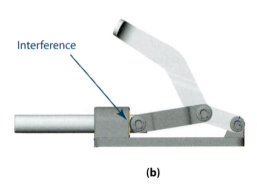

(b)

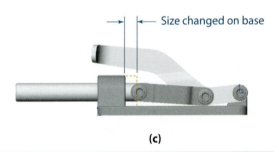

(c)

FIGURE 7.20. The completed toggle clamp assembly is shown in (a). In its original form, the link and the base interfere, as shown in (b), preventing the handle from being pushed all the way down into a closed and locked position. The base has been modified in (c) to eliminate the interference, allowing the toggle to work properly.

7.07 Bill of Materials

Most real-life assembly drawings contain a **bill of materials**, which consists of a table listing all parts in the assembly as well as the quantity of each part required to put the assembly together. If you were to begin assembling a system, you could look at the bill of materials and easily determine how many screws, how many washers, etc., you needed to gather to assemble the system. Figure 7.22 shows an assembly drawing (exploded configuration) for a system, along with the associated bill of materials.

In some cases, a bill of materials lists subassembly items, as shown in Figure 7.23. In that case, the assembly drawing for the subassembly is usually shown on a different sheet of paper and you are expected to have constructed the subassembly before you begin putting the entire assembly together.

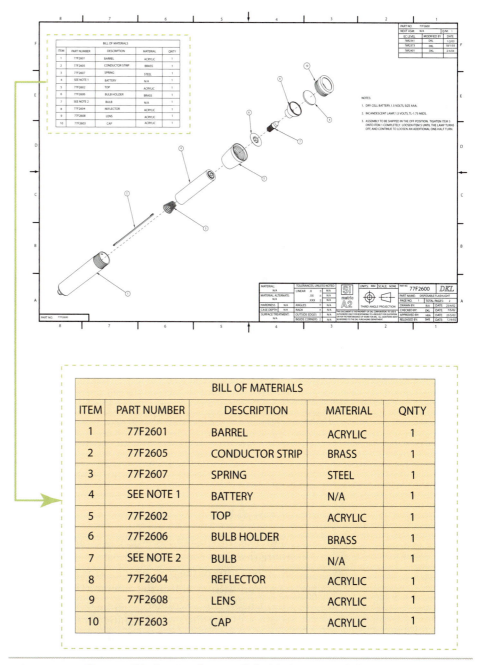

\multicolumn{5}{c}{BILL OF MATERIALS}
ITEM	PART NUMBER	DESCRIPTION	MATERIAL	QNTY
1	77F2601	BARREL	ACRYLIC	1
2	77F2605	CONDUCTOR STRIP	BRASS	1
3	77F2607	SPRING	STEEL	1
4	SEE NOTE 1	BATTERY	N/A	1
5	77F2602	TOP	ACRYLIC	1
6	77F2606	BULB HOLDER	BRASS	1
7	SEE NOTE 2	BULB	N/A	1
8	77F2604	REFLECTOR	ACRYLIC	1
9	77F2608	LENS	ACRYLIC	1
10	77F2603	CAP	ACRYLIC	1

FIGURE 7.22. The assembly drawing for a small flashlight, with its bill of materials on the drawing.

FIGURE 7.23. The bill of materials for two levels of subassembly, leading to the final assembly of the footbridge.

BILL OF MATERIALS, PART NUMBER XKZ0030, WEB ASSEMBLY				
ITEM	PART NUMBER	DESCRIPTION	MATERIAL	QNTY
1	XKZ0001	VERTICAL MEMBER	WOOD	1
2	XKZ0002	ANGLED MEMBER	WOOD	2

BILL OF MATERIALS, PART NUMBER XKZ0015, HANDRAIL ASSEMBLY				
ITEM	PART NUMBER	DESCRIPTION	MATERIAL	QNTY
1	XKZ0001	VERTICAL MEMBER	WOOD	1
2	XKZ0003	HORIZONTAL MEMBER	WOOD	2
3	XKZ0030	WEB ASSEMBLY	WOOD	2

BILL OF MATERIALS, PART NUMBER XKZ0009, DECKING ASSEMBLY				
ITEM	PART NUMBER	DESCRIPTION	MATERIAL	QNTY
1	XKZ0006	DECK	WOOD	1
2	XKZ0007	SURFACE LAYER	FIBER REINFORCED CONCRETE BOARD	2

BILL OF MATERIALS, PART NUMBER XKZ0001, FOOTBRIDGE ASSEMBLY				
ITEM	PART NUMBER	DESCRIPTION	MATERIAL	QNTY
1	XKZ0023	ABUTMENT	POURED CONCRETE	2
2	XKZ0013	APPROACH	POURED CONCRETE	2
3	XKX0009	DECKING ASSEMBY	WOOD, CONCRETE	1
4	XKZ0015	HANDRAIL ASSEMBLY	WOOD	2

7.08 Assembly Strategy

When creating an assembly, you can choose from two modeling methods—top-down or bottom-up. Bottom-up assembly modeling was used extensively in the past; however, top-down modeling is now gaining in popularity. A description of each method follows.

7.08.01 Bottom-up Assembly Modeling

With **bottom-up modeling**, you create all of the parts required for the system. When more than one copy of a given part is in an assembly, you do not have to create more than one copy of the part—you include multiple instances of the component in the system. You then establish the assembly hierarchy, making sure you organize instances into smaller subassemblies as appropriate. You orient all of the instances using constraints to establish relationships between instances. Next, you check for clearances and interferences between instances to make sure your constraints were properly applied and you achieved your design intent. If you so choose, you obtain a bill of materials for the assembly. Your final step is to create an exploded view of the system as necessary to create its assembly drawing.

7.08.02 Top-down Assembly Modeling

Top-down assembly modeling has evolved in recent years in much the same way concurrent engineering evolved—both were facilitated through the modern computer tools and CAD systems. With **top-down modeling**, the system is first defined, including its hierarchy. In some cases, physical space may be assigned to the assembly as well as to all subassemblies and components contained in it. The function of the system and its components also is articulated at this time. After the framework of the system has been established, engineering teams then work on creating the individual parts and subassemblies that go into the overall system. For example, because the engineers working on subassembly A know that their subassembly connects and interacts with subassemblies B, D, and G, they can collaborate through e-mails or other means as they create their assigned subassembly. Further, the team members working on subassembly A know that the space they occupy cannot encroach on the space occupied by subassembly C and must take that fact into account when they do their design work. Teams working on subassembly B know that they must work with teams working on subassemblies A and C as they complete their assigned task. In this way, efficiency in assembly design is achieved—problems can be solved before they arise. With the bottom-up design approach, problems between subassemblies A and D might not be apparent until the entire design is completed, meaning that all teams need to start from scratch to solve the problem. Top-down assembly modeling is especially effective when multiple teams, many times in different parts of the world, are working on a single design. Most modern-day CAD systems have the tools necessary to implement a system of top-down assembly modeling; however, as a student, you may not have the opportunity to work in this efficient environment.

7.09 Strategy for Bottom-up Assembly Modeling

Suppose you needed to model the assembly shown in Figure 7.24. The distinct components that make up the assembly are shown in Figure 7.25. Note that the bottom-up approach will be employed in the assembly modeling of this case, since all of the parts have already been created and must simply be put together to form the assembly model.

FIGURE 7.24. A fully assembled model of a bearing block.

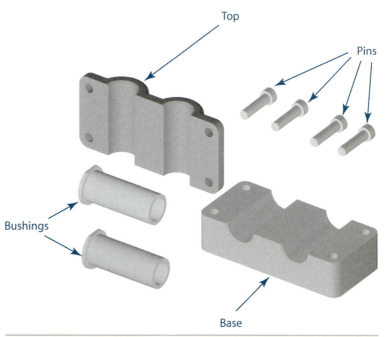

FIGURE 7.25. An exploded view of the parts that form the bearing block assembly.

FIGURE 7.26. Two possible hierarchies for the bearing block assembly, with no subassemblies (a) and with a housing block subassembly (b).

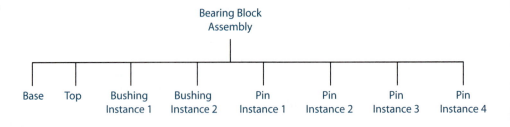

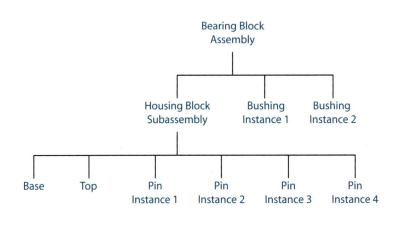

With the bottom-up assembly modeling approach, after the individual components have been modeled, you need to define a hierarchy for the system. For this particular assembly, it might make sense to link the four pins into one subassembly; however, this is probably not necessary. Since the bushings have a flat surface for alignment, you probably need to know if they should be aligned independent of each other before you decide whether they make up a subassembly. Since the base block and the cap block will likely be considered a single unit, the two components should be put into a subassembly. Two possible hierarchies for the system are shown in Figure 7.26.

For purposes of demonstration, you will work with the hierarchy established in Figure 7.26a. In this case, your first step is to assemble the base subassembly. This subassembly consists of the base block and the cap block. You should establish the base block as the base instance for the assembly, setting the coordinate planes for the system as shown in Figure 7.27.

If you bring the cap block onto the screen, you can put it in place within the system through the use of coincident constraints. For this subassembly, the edges of the half circles of the cap and base should coincide. If this constraint is applied, notice the two possible orientations for the cap where this constraint is satisfied, as shown in Figure 7.28.

To achieve the desired orientation for the cap block, you need to apply one more coincident constraint using any of the remaining edges, or you can use the endpoints of the corresponding half circles. Figure 7.29 shows the second coincident constraint applied, which leads to the desired final result.

Now you put the pin subassembly in place within the overall system. The pin subassembly consists of four instances of the pin component. You can bring the pins on-screen one at a time and locate each of them relative to the base subassembly that is

FIGURE 7.27. Definitions of coordinate axes and datum planes for the base as the first object in the assembly model.

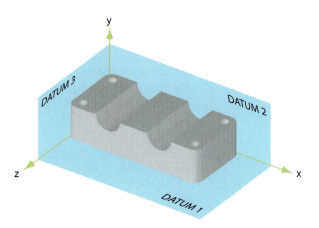

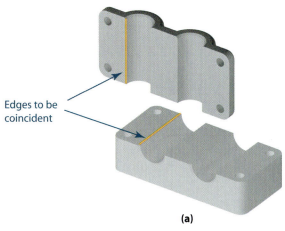

Edges to be coincident

(a)

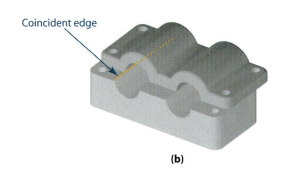

Coincident edge

(b)

Coincident edge

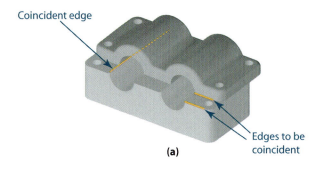

Coincident edge

Edges to be coincident

(a)

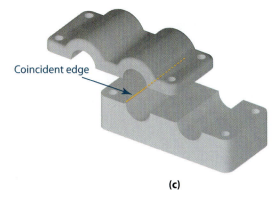

Coincident edge

(c)

Result

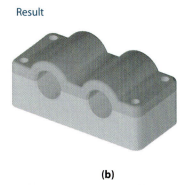

(b)

FIGURE 7.28. A coincident constraint applied to the edges shown in (a) can result in two possible orientations of the top, as shown in (b) and (c).

FIGURE 7.29. Adding a coincident edge constraint to the edges shown in (a) is one method used to create the final alignment between the base and the top, as shown in (b).

already in place. If you bring the first instance of the pin on-screen, you need to establish two constraints to put it in its final location. The two constraints to be added are a concentric constraint between the axis of the pin and the axis of the corresponding hole, and a mating surfaces constraint can be applied between the "bottom" surface of the pinhead and the top surface of the cap, block. These two constraints are shown in Figure 7.30a, with the final result shown in Figure 7.30b.

Using a similar strategy, the remaining three instances of the pin component can be oriented within the assembly, with the result shown in Figure 7.31.

Finally, the two bushings can be inserted one at a time into the assembly using a strategy similar to the one used in locating the pins—applying a concentric constraint between the centerlines of the bushing and the base-block semicircular cutout and a mating surfaces constraint between the corresponding surfaces of the bushing and the base block. Figure 7.32a shows the constraints applied to the leftmost bushing, and Figure 7.32b shows the result of applying these constraints.

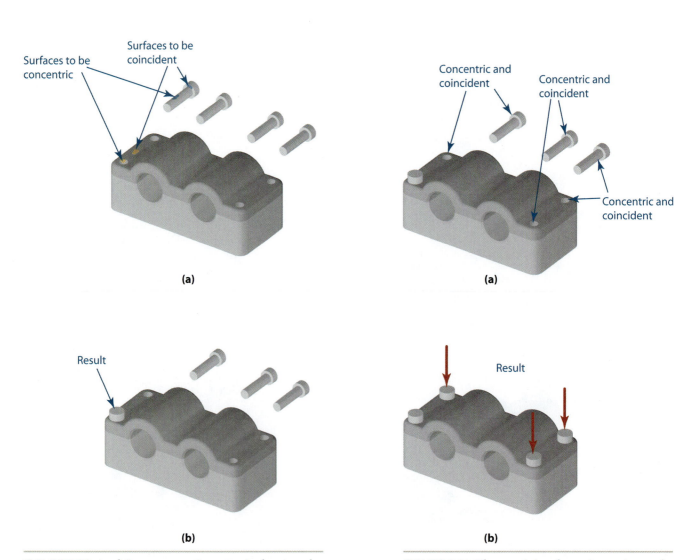

FIGURE 7.30. Applying a concentric constraint between the shaft of the pin and the hole in the top and a coincident constraint between the bottom of the pinhead and the top, as shown in (a), locates the pin in the hole, as shown in (b).

FIGURE 7.31. The remaining three pins are inserted in their holes by applying the same types of concentric and coincident constraints used for the first pin.

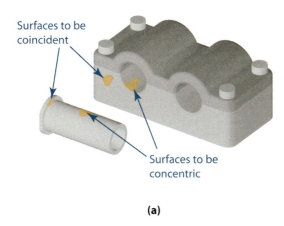

Surfaces to be
coincident

Surfaces to be
concentric

(a)

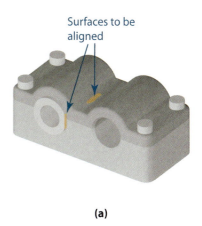

Surfaces to be
aligned

(a)

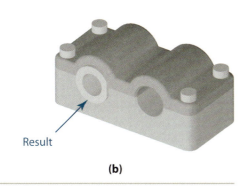

Result

(b)

FIGURE 7.32. The bushing is placed by applying the concentric and coincident constraints to the surfaces indicated in (a) to produce the result shown in (b).

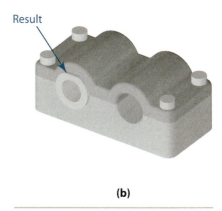

Result

(b)

FIGURE 7.33. An alignment constraint is applied between the flat on the bushing and the top surface, as shown in (a), to create the desired orientation of the bushing, as shown in (b).

When the pins were added to the assembly, it did not matter what angular orientation they had with respect to the axis of the pin; however, the bushing instances include a flat surface that can be used for alignment. To align the surface properly, you want to include a distance constraint between the flat surface on the bushing and the bottom (or top) flat surface on the base. In this case, the distance constraint forces the two surfaces to be parallel to each other. Figure 7.33a shows the constraint applied between the flat surface of the bushing and the upper flat surface of the top, and Figure 7.33b shows the result of applying this constraint.

Finally, the second instance of the bushing can be brought into the assembly and oriented with the use of appropriate constraints. The final assembly model is shown in Figure 7.34.

Once your assembly model is complete, you may want to check for interferences or clearances between instances. Figure 7.35 includes the results from an interference and clearance check for the instances in the assembly. Since there is a relatively small number of instances in the assembly, all instances were checked against all others.

Other items you might need from this model include an assembly drawing and a bill of materials. Most modern-day software can generate these automatically. Figure 7.36 shows an assembly drawing with the bill of materials for the block assembly you have been working with thus far.

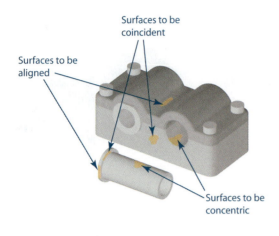

Surfaces to be coincident

Surfaces to be aligned

Surfaces to be concentric

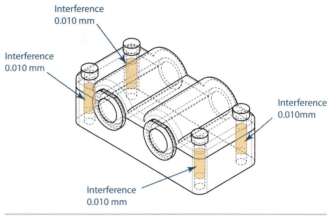

Interference 0.010 mm

Interference 0.010 mm

Interference 0.010mm

Interference 0.010 mm

FIGURE 7.35. The result of an interference check between all of the parts in the bearing block assembly, showing the interference from the forced fit between the pins and the base block.

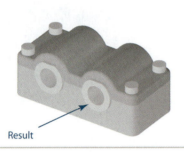

Result

FIGURE 7.34. The second bushing is placed by applying the same type of concentric, coincident, and aligned constraints as were used to place the first bushing, as shown in (a). The final position of the second bushing is shown in (b).

FIGURE 7.36. An assembly drawing of the block assembly (removed from its drawing header) identifying its parts in its bill of materials.

BILL OF MATERIALS, PART NUMBER CDX010, BEARING BLOCK ASSEMBLY				
ITEM	PART NUMBER	DESCRIPTION	MATERIAL	QNTY
1	CDX011	BASE BLOCK	ALUMINUM, 6061 T6	1
2	CDX012	CAP BLOCK	ALUMINUM, 6061 T6	1
3	CDX089	PIN	STEEL, 1060	4
4	CDX076	BUSHING	BRONZE, SINTERED	2

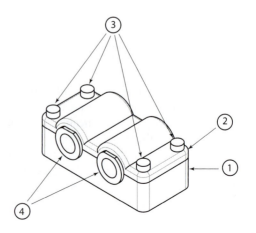

FIGURE 7.37. A fully assembled model of a table vise.

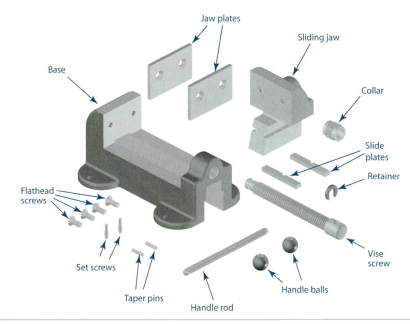

FIGURE 7.38. The various pieces of the table vise prior to assembly.

As another example, consider the vise assembly shown in Figure 7.37, with the individual components shown in Figure 7.38.

The first step is to establish the hierarchy for the system. In this case, you can consider the vise base to be the base instance. The sliding jaw is a subassembly that consists of two subassemblies—a jaw plate subassembly and a handle subassembly. A jaw plate subassembly also attaches to the base. Each subassembly includes several screws that are used for fastening. The overall hierarchy for that assembly is shown in Figure 7.39.

FIGURE 7.39. The assembly hierarchy for the table vise.

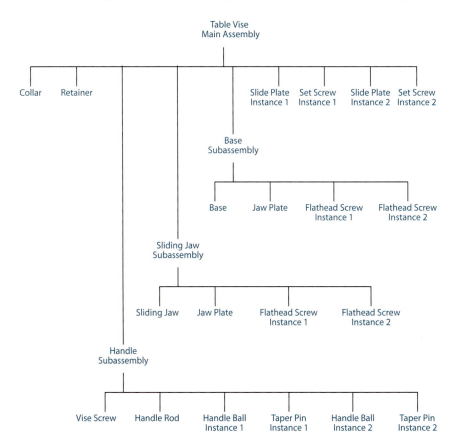

These surfaces
coincident

Conical surfaces
coincident

These holes
concentric

Shaft and hole
concentric

These holes
concentric

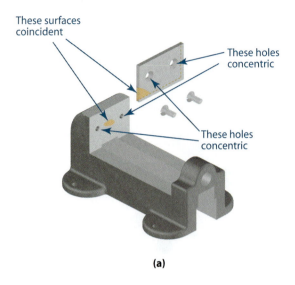

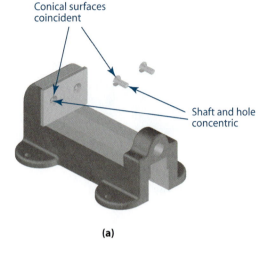

(a)

(a)

Result

Result

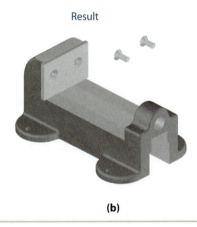

(b)

(b)

FIGURE 7.40. Adding assembly constraints between the jaw plate and the base on the base subassembly (a) to get the assembled results in (b).

FIGURE 7.41. Adding assembly constraints to one flathead screw (a) to get the assembled results in (b). The second flathead screw is constrained in a similar manner within the other hole.

To put this assembly together, you once again start with the base instance. Attached to the base instance is the jaw plate subassembly that consists of a jaw plate and two screws that hold the jaw plate in place. Using concentric and mating surfaces constraints, you can assemble the jaw plate subassembly as shown in Figure 7.40; and that subassembly can be added to the base instance through coincident constraints as shown in Figure 7.41.

At this point, it is probably easier to put the base instance (along with the attached jaw plate subassembly) away and work only with the sliding jaw subassembly. When you start with the sliding jaw as the "base" of the subassembly, you can insert another instance of the jaw plate subassembly and put it in place using coincident constraints between the appropriate elements, as shown in Figure 7.42.

The handle subassembly can be put together with a concentric constraint between the handle and the vise screw, as shown in Figure 7.43.

FIGURE 7.42. The sliding jaw subassembly is created from its parts. The jaw plate is constrained in (a); and the flathead screws are constrained, first one and then the other, in (b).

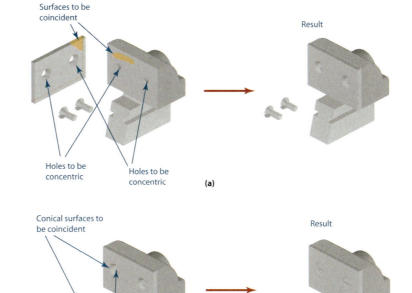

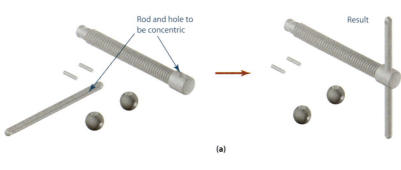

FIGURE 7.43. The handle subassembly is created by applying constraints to the rod (a), the end balls (b), and the pins that retain the end balls (c).

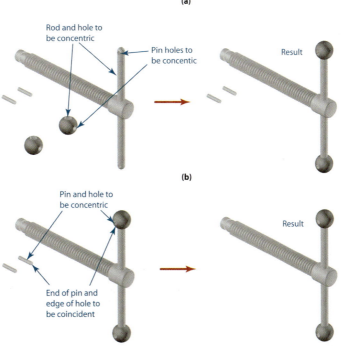

Base subassembly

Sliding jaw subassembly

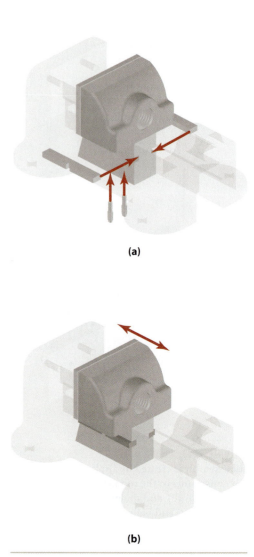

(a)

(b)

FIGURE 7.45. By applying coincident constraints between the two slide plates and the sliding jaw and concentric and coincident constraints between the two setscrews and the sliding jaw in (a), the sliding jaw can move freely in one direction only, as shown in (b).

(a)

Surfaces to be coincident

Surfaces to be coincident

(b)

Result

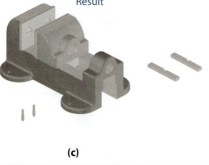

(c)

FIGURE 7.44. The base subassembly and sliding jaw subassembly are brought together in (a). The sliding jaw subassembly is rotated to expose the surfaces to be constrained in (b). The final result is shown in (c).

Next, the handle subassembly and the set screws that hold it in place can be added to the sliding jaw through use of mating surfaces and concentric constraints, as shown in Figure 7.44.

Finally, the sliding plate and associated screws can be added to the bottom of the sliding jaw through the use of coincident and distance constraints, as shown in Figure 7.45.

You now can retrieve the base component and orient the sliding jaw subassembly through the use of coincident constraints between appropriate edges on the base and

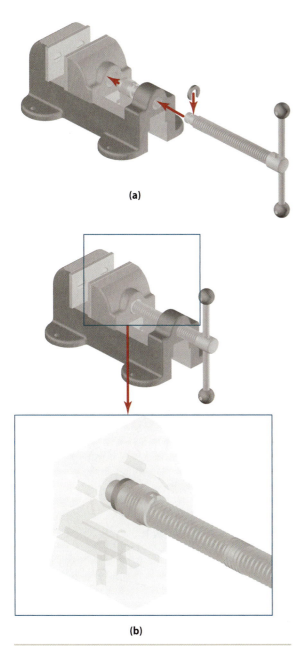

(a)

(b)

FIGURE 7.46. The retainer and collar are assembled onto the screw and sliding jaw in (a) to complete the vise, shown in (b). The detail in (b) shows the final positions of the retainer and collar in the sliding jaw.

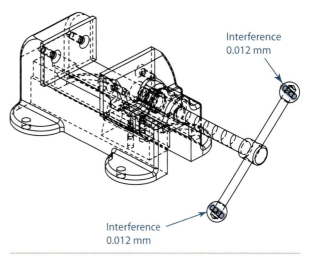

Interference
0.012 mm

Interference
0.012 mm

FIGURE 7.47. An interference check on the entire vise assembly warns that there will be interference between the taper pins and the pin holes on the end balls on the handle rod. The interference, in this case, is desirable for the purpose of keeping the end balls on the handle rod.

the sliding jaw. Note that you do not want to include any distance constraints at this time, enabling the sliding jaw to be located anywhere along its "track." Figure 7.46 shows the completed assembly of the vise model.

If you like, you can check for clearances and interferences (Figure 7.47) or create an assembly drawing and a bill of materials for the assembly (Figure 7.48).

	BILL OF MATERIALS, PART NUMBER RNP1000, TABLE VISE ASSEMBLY			
ITEM	PART NUMBER	DESCRIPTION	MATERIAL	QNTY
1	RNP050	BASE	STEEL, CAST	1
2	RNP051	SLIDING JAW	STEEL, CAST	1
3	RPN010	JAW PLATE	STEEL, AISI 1020	2
4	RPN015	SCREW	STEEL, AISI 1060	1
5	RPN017	COLLAR	BRASS	1
6	RPN018	RETAINER	STEEL, AISI 1060	1
7	RPN020	HANDLE ROD	STEEL, AISI 1040	1
8	RPN022	HANDLE BALL	STEEL, AISI 1040	2
9	RPN012	SLIDE PLATE	STEEL, AISI 1060	2
10	RPN008	TAPER PIN	STEEL, AISI 1060	2
11	RPN009	FLATHEAD SCREW	STEEL, AISI 1060	4
12	RPN007	SETSCREW	STEEL, ASIS 1060	2

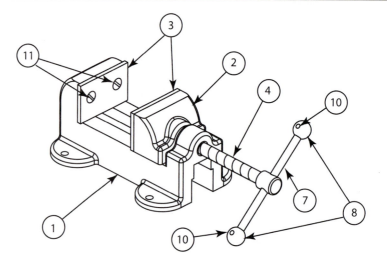

CAUTION

One thing to avoid when working with assembly models is the creation of a flat hierarchy. With a flat hierarchy, the main assembly is the parent and every instance is a child of the main assembly—there are no subassemblies, just individual instances. With a flat hierarchy, you cannot treat several instances as one, but must deal with each one individually. Creating a flat hierarchy may save you time in the planning stages, but it is likely to cost you a great deal of time when you work with the assembly. To avoid a flat hierarchy, make sure you logically group instances together in subassemblies to save time later. Figure 7.49 shows an undesirable flat hierarchy for the footbridge assembly, in contrast to Figure 7.08 at the beginning of this chapter which shows a hierarchy that advantageously employs subassemblies to link instances together logically.

As discussed previously, another situation you should avoid in assembly modeling is the creation of a system that is too rigid. A rigid system consists of instances that are fixed in space and not fixed relative to one another. To avoid an overly rigid system, make sure you establish relationships between instances with constraints rather than merely locate instances by moving them in 3-D space. Using constraints will add flexibility to your assembly model. For example, consider the bolt used to hold together two

FIGURE 7.49. A "flat" assembly hierarchy, such as this one for the footbridge model, should be avoided. A more reasonable hierarchy is shown in Figure 7.08.

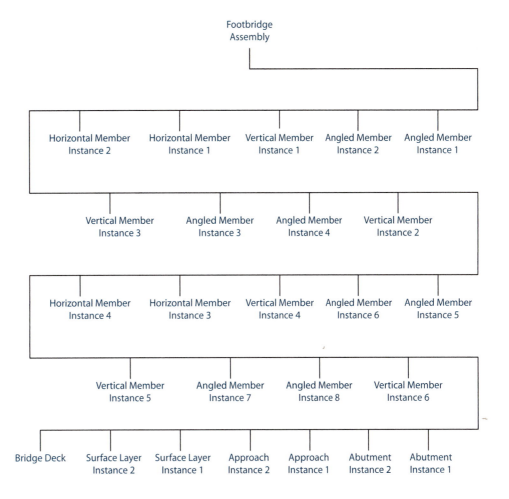

NO!

members of the handrail for the footbridge. If you use a concentric constraint between the bolt and the holes in the members, as you move the members around, the bolt will move with them. However, if you put the bolt in the hole without a constraint, when you move the members, the bolt will stay where you put it and will not move with the members. Without a constraint, you would have to move the members and then move the bolt in an additional step.

7.10 glossary of key terms

assembly constraints: Used to establish relationships between instances in the development of a flexible assembly model.

associativity: The situation whereby parts can be modified and the components referenced to the parts will be modified accordingly.

base instance: The one fixed instance within an assembly.

bill of materials: A tabular list of the components, with quantities of each for the parts, that make up an assembly.

bottom-up modeling: The process of creating individual parts and then creating an assembly from them.

clearances: The minimum distances between two instances in an assembly.

components: References of object geometry used in assembly models.

exploded configuration: A configuration of an assembly that shows instances separated from one another. An exploded configuration is used as the basis for an assembly drawing.

7.10 glossary of key terms (continued)

hierarchy: The parent-child relationships between instances in an assembly.

instances: Copies of components that are included within an assembly model.

interference: The amount of overlap between two instances in an assembly.

subassembly: A logical grouping of assembly instances that is treated as a single entity within the overall assembly model.

top-down modeling: The process of establishing the assembly and hierarchy before individual components are created.

7.11 questions for review

1. Describe the differences between an object and a component.

2. What is an instance?

3. What does the term *associativity* mean in the context of assembly modeling?

4. What type of relationships are made when an assembly model is established?

5. Name and describe three types of assembly constraints.

6. Define *interference* and *clearance* in the context of assembly modeling.

7. What is a bill of materials?

8. What is the primary difference between bottom-up assembly modeling and top-down assembly modeling?

7.12 problems

1. The two parts in Figure P7.1 are to be mated together as shown. Using only the features labeled, apply assembly constraints to mate the two pieces so that the top part is fully constrained and assembled correctly with the bottom part. Assume the bottom part is already fixed in position. Specify five ways of doing that using only coincident and concentric constraints. An example follows:

   Constraint set 1:   hole 1 concentric with hole 2

   surface 1 coincident with surface 3

   surface 2 coincident with surface 4

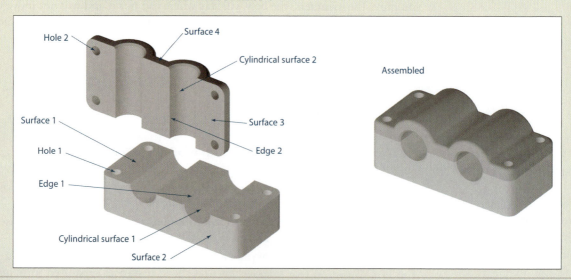

FIGURE P7.1. Constrain the edges and surfaces of (a) to create the assembled position (b).

7.12 problems (continued)

2. The parts shown in Figure P7.2 are to be assembled into a screw clamp. Create a solid model for each part and apply assembly constraints to create an assembly model of the clamp. All parts are made of steel. The notation "M10" designates a standard metric screw thread with a 10 mm outer diameter. (You will learn more about this in a later chapter.)

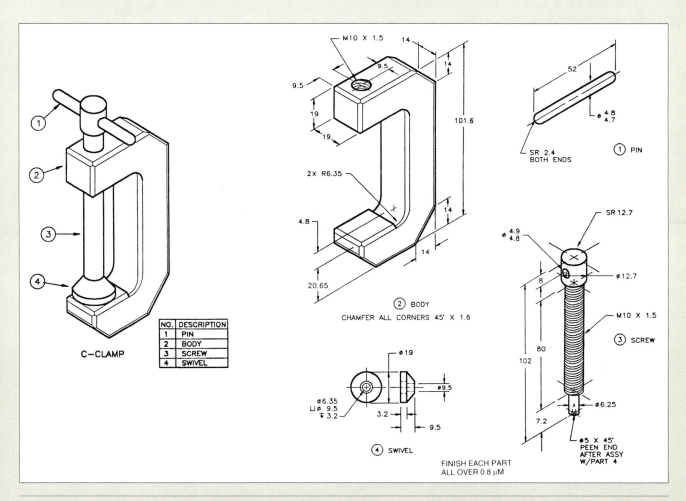

FIGURE P7.2. Create an assembly model for the screw clamp.

7.12 problems (continued)

3. Figure P7.3 shows a conceptual model for a pen-type eraser in whole and in cutaway view. Using reasonable materials and dimensions of your choice, expand the concept to create solid models of the individual parts. Using assembly constraints, create an assembly model of the eraser.

FIGURE P7.3.

4. Figure P7.4 shows a conceptual model for a garden hose nozzle in whole, cutaway, and exploded views. Using reasonable materials and dimensions of your choice, expand the concept to create solid models of the individual parts. Using assembly constraints, create an assembly model of the nozzle.

FIGURE P7.4.

5. Conceptual sketches (top view, front view, and side view) for a toggle clamp are shown in Figure P7.5. Using reasonable materials and dimensions of your choice, expand the concept to create solid models of the individual parts. Using assembly constraints, create an assembly model of the toggle clamp.

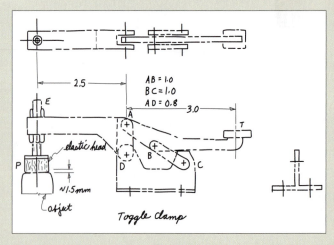

FIGURE P7.5.

7.12 problems (continued)

6. Conceptual sketches (top view, front view, and side view) for a wheelbarrow are shown in Figure P7.6. Using reasonable materials and dimensions of your choice, expand the concept to create solid models of the individual parts. Using assembly constraints, create an assembly model of the wheelbarrow.

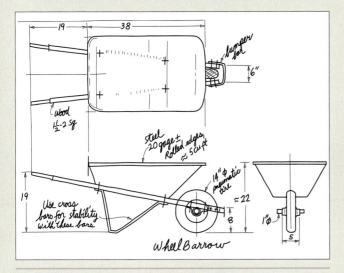

FIGURE P7.6.

7. Conceptual sketches (top view, front view, and side view) for a caster are shown in Figure P7.7. Using reasonable materials and dimensions of your choice, expand the concept to create solid models of the individual parts. Using assembly constraints, create an assembly model of the caster.

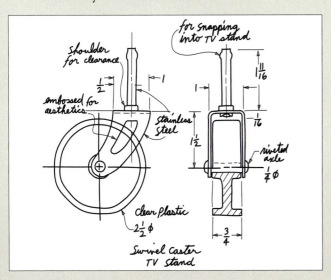

FIGURE P7.7.

8. Conceptual sketches (top view, front view, and side view) for a gear puller are shown in Figure P7.8. Using reasonable materials and dimensions of your choice, expand the concept to create solid models of the individual parts. Using assembly constraints, create an assembly model of the gear puller.

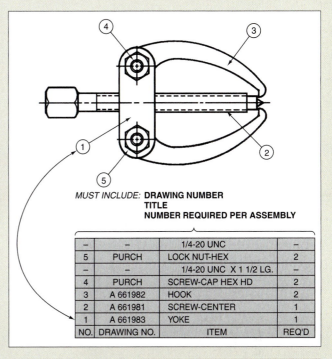

MUST INCLUDE: DRAWING NUMBER
TITLE
NUMBER REQUIRED PER ASSEMBLY

–	–	1/4-20 UNC	–
5	PURCH	LOCK NUT-HEX	2
–	–	1/4-20 UNC X 1 1/2 LG.	–
4	PURCH	SCREW-CAP HEX HD	2
3	A 661982	HOOK	2
2	A 661981	SCREW-CENTER	1
1	A 661983	YOKE	1
NO.	DRAWING NO.	ITEM	REQ'D

FIGURE P7.8.

8

Design Analysis

objectives

After completing this chapter, you should be able to

- Explain the importance of analysis in the design process
- Apply the reverse engineering process to a simple device
- Perform a mass properties analysis for a computer-generated model
- Describe the fundamental principles in performing a Finite Element Analysis

8.01 introduction

In an earlier chapter, you learned about the stages in the design process and about the role of the computer in the process. You also learned that some engineers focus on the design stage, some on the manufacturing stage, and some on the analysis stage. Most of the computer-based analysis techniques described in this chapter require a high level of mathematical understanding and are beyond the scope of this text. In fact, for some analysis techniques, such as the finite element method, entire books are written on the topic. In this chapter, you will learn about analysis techniques that are typically found in an introductory design graphics course. You should understand that many other types of analysis techniques are available to engineers as they design devices and systems. You will probably learn about many of these techniques as you make your way through your engineering curriculum. This chapter introduces three analysis techniques: reverse engineering, mass properties analysis, and the finite element method.

Analysis is a broad term that involves the study of the behavior of a physical system under certain imposed conditions. Analysis usually involves quantitative reasoning and techniques, but it might not. As you learned previously, engineering analysis characterizes a significant part of the design process, since all design ideas have to be verified for feasibility.

Analysis can be conducted in many different ways, but generally an underlying physical principle (equation) is involved. Although modern analysis is typically done using special computer programs, conducting an experiment in a laboratory setting or reverse-engineering a product also is a feasible analysis method. Many of the numerical analysis techniques are based on advanced-level mathematical descriptions and are highly computerized. A few numerical techniques are as follows: structures can be analyzed for stresses and deformations; bodies, for motion; engines, for temperature distributions; and dynamic systems, for their vibrations. In performing lab experiments, engineers can determine the strength of a new design component through tensile testing. Through reverse engineering, systems and parts can be analyzed for their function and possible modification. These are just a few of the analysis tools available to engineers as they design products or infrastructure for the general public.

8.02 Reverse Engineering

Reverse engineering is a systematic methodology for analyzing the design of an existing device or system, either as an approach to study the design or as a prerequisite for redesign. Reverse engineering essentially is a process used to gain information about the functionality and sizes of existing design components. For instance, you might have an idea or a concept concerning a unique addition to an existing product or system. It would make sense that you first analyze the products or systems already on the market to see how they work and what features they have that could be used or improved for the new design. A careful dissection and study of several similar designs could contribute much toward the new process or design. In industry, the most common reasons for reverse engineering are these:

- Existing prototypes are available but no design data, CAD computer files, or drawings exist.

- The original manufacturer of the product or part is hesitant, unwilling, or unable to provide replacement parts for a system in use or the data to reproduce them.

- Old, worn, or broken parts for a system exist for which there are no known data sources.
- Existing geometry data needs to be modified to improve a part's functionality, dimensionality, or appearance.

It should be noted that reverse engineering is a legitimate activity in industry and is not the same as industrial espionage. Determining "how something works" is not stealing someone's ideas and manufacturing counterfeit products. Indeed, some engineering companies offer reverse engineering services using sophisticated measuring systems.

Reverse engineering is a technique within the practice of engineering design that can be useful in several ways. Reverse engineering can save time because there is no need to "reinvent the wheel" when you can start from existing geometric data. The reverse engineering technique also can help an engineer develop a systematic approach to thinking about and improving the design of devices and systems. Seeing and holding the existing parts, noting their dimensions, and understanding their mating relationships improve an engineer's design and visualization abilities. By repeated practice of the reverse engineering process, you can gradually acquire a mental data bank of many design solutions for use in future products.

Reverse engineering is sometimes called **mechanical dissection** because it involves taking apart, or dissecting, a mechanical system. As you dissect the system, you carefully sketch the parts and show how they fit and work together so the system can be reassembled at a later date. You also need to measure all of the features on each part carefully during the dissection process so you can create solid models of them at a later time. Since accurate measurements are a significant part of the reverse engineering process, this chapter will discuss some of the more common measurement tools available. In engineering, the practice of measuring parts is called **metrology**.

8.03 Metrology Tools for Reverse Engineering

Reverse engineering is essentially a measurement and documentation process. As you take systems apart, you *document* how the parts work together and *measure* the sizes of the features of the parts. Tools used to do this measuring vary from simple handheld devices to the most sophisticated computer-driven machines. Accuracy will usually be a direct function of the cost of the metrology device, with less expensive tools providing less accurate measurements. Traditional inexpensive **engineering scales** or a ruler can be used for linear measurements with 0.25 mm "eyeball" accuracy; however, this is not always accurate enough for reverse engineering purposes. The following sections outline some common reverse engineering metrology tools.

8.03.01 Handheld Calipers

Other than engineering scales or rulers, the most economical tool for reverse engineering is a handheld **caliper**. Fairly accurate measurements can be obtained using a set of calipers such as those shown in Figure 8.01. With calipers such as these, the jaw, shown on the left, expands to fit around the part being measured. The distance between the flat surfaces of the caliper—and thus the distance you are trying to measure—can be determined by reading the scale on the long side of the device in combination with the digital readout. Calipers also can be used to determine the inside diameters of holes and other features through use of the opposite ends of the jaws. Typical calibers have a jaw-gap range of up to 100 mm, with a resolution accuracy of 0.025 to 0.010 mm. The numerical readout of calipers can be either a dial gauge needle or a digital LCD display, and they come in both English and metric units.

FIGURE 8.01. Handheld calipers.

FIGURE 8.02. A single probe coordinate measuring machine. *Courtesy Hexagon Metrology, Inc.*

8.03.02 Coordinate Measuring Machine (CMM)

Coordinate measuring machines, such as the one shown in Figure 8.02, use a probe to touch the part. The computer system built into the probe senses the x-, y-, and z-coordinates of the location of the probe and records that information into memory. The probe then lifts and touches the next point on the part, once again recording the coordinates of the endpoint of the probe. The system repeats the process until the whole part has been systematically scanned. This process is sometimes called digitizing the model since it converts the 3-D continuous part geometry to digital data. The points determined by the probe can then be used to build a 3-D computer model of the part. High-density scans, where the probe moves a small distance each time, may take a long time; but the result is a data set with accuracy in the 0.001-mm range. Typical volume capacities (measurement space) for CMM systems are 0.6 m x 0.6 m x 0.6 m and larger, with larger volumes greatly increasing the cost of the system.

8.03.03 3-D Laser Scanner

Laser scanning systems use a laser projector and cameras to determine part dimensional coordinates to a high degree of accuracy. During a scan, the projector shines a series of white-light stripes on the object. The two cameras pick up the light and use the principle of triangulation to determine the point's x-, y-, and z-coordinates. As many as 1.3 million coordinate points can be obtained in one scan session in less than 24 hours with an accuracy of 0.025 mm per point. Similar to working with a CMM system, the points can then be downloaded to a CAD system for solid modeling creation. A two-camera laser scanning system is shown in Figure 8.03. Due to the high cost of both CMM and laser scanning systems, you will likely be using rulers, scales, and calipers for any mechanical dissection projects you are involved in at the university.

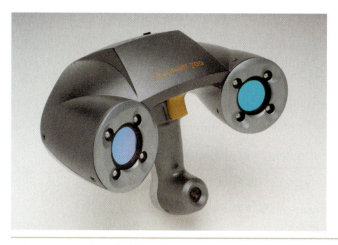

FIGURE 8.03. A handheld 3-D laser scanner used to extract geometry. *Courtesy of Z Corporation*

8.04   The Reverse Engineering Process

Like many technical activities, reverse engineering can be viewed as a process. In the following sections, the reverse engineering process is illustrated through a simple example—a common kitchen scale such as the one shown in Figure 8.04.

8.04.01   Defining the Reverse Engineering Project

Just like design projects, the first step in a reverse engineering project is to clearly define the project and determine the relevant factors and parameters. Careful planning of the project begins with assessing the customer needs and the engineering specifications. The overall functionality of the system, such as input and output relationships, also must be defined. Finally, subsystems and individual components should be estimated and outlined, perhaps through the use of a product dissection approach. After you have determined the relevant parameters and factors, you should summarize your findings in a problem statement so you have a clear picture of the reverse engineering project—what you hope to discover and what you hope to show. If your goal is to make improvements to a product, you should include that goal in the problem statement.

The problem statement can be in narrative or bulleted format, but you may find it easier to list the items rather than write a narrative statement. The problem statement should include a clear definition of the project/problem, a look at what a customer might be expecting in the product, a view of what an engineer will have to consider, as well as a definition of how the product must function. Figure 8.05 illustrates a problem statement for the kitchen scale project.

FIGURE 8.04. A kitchen scale.

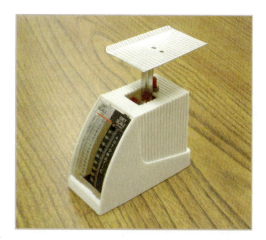

The Small Kitchen Scale

The Problem Statement

The Kitchen Scale is a useful device to measure out small amounts of food that are to be prepared. There are many people who need to be careful of the quantity of food that they take in. Diabetics must be careful of the amount of certain foods that they eat in order to maintain a proper blood sugar level. People on weight loss diets must also be careful with the number of calories that they consume at every meal. The Kitchen Scale is a device that can help to control the amount of food consumed by an individual.

Customer's Perspective

Some of the things that the customer might look for in a kitchen scale are the cost of the item and its appearance. Some other factors he/she would be concerned about would be how sturdy the device is, the accuracy of the weighing process, and its size. The customer may even ask if he/she is getting quality for the price.

Engineer's Perspective

The engineer has a totally different perspective from which she operates. The engineer will be looking at things like the material that the object is made of, the mechanisms that are required, the strength of the materials used, etc. In relation to the kitchen scale: Is the riveting process of the plate to the stem adequate? Are the materials used strong enough to hold 1 pound of weight? Does the scale have any sharp edges that could cause injury? What mechanism is necessary to give an accurate reading on the scale?

Functional Requirements

The kitchen scale must be able to accommodate 1 pound of food.

The plate must be large enough to hold 1 pound of food.

The stem must be strong enough to support the plate and the food.

The linkage must be designed so that the reading is accurate to the nearest .5 ounce.

The read-out must be easily read.

The design must lend itself for easy cleaning and sanitizing.

FIGURE 8.05. A problem statement for the kitchen scale reverse engineering project.

BLACK BOX DIAGRAM
(Example: Small Kitchen Scale)

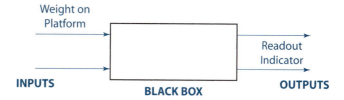

Weight on
Platform

Readout
Indicator

INPUTS　　　　　　　**BLACK BOX**　　　　　　　**OUTPUTS**

FIGURE 8.06. A black box diagram for the small kitchen scale.

As part of this initial stage in the reverse engineering project, you may find it helpful to create a black box diagram. With a **black box diagram**, you specify the overall inputs and outputs on the system, ignoring the inner workings of the device. Sometimes this black box diagram is referred to as the "view from 10,000 feet." A black box diagram of the small kitchen scale is shown in Figure 8.06.

At this time, you also might want to develop a sketch of the device for reference later. Figure 8.07 shows a sketch of the kitchen scale.

8.04.02 Dissecting a System

Once the preliminary examination of the project is completed, the problem statement is completed, and the black box diagram has been created, it is time to dissect the device. As you take apart the device, you need to document how the mechanism was dissected. Write down detailed notes for every step in the dissection process, because you may need the directions to reassemble the device at a later date. Figure 8.08 shows the dissection notes obtained as the kitchen scale is dissected. As part of the documentation of the dissection, you also should create a fishbone diagram. A **fishbone diagram** shows each of the subassemblies as a major "stem" from the "backbone" with

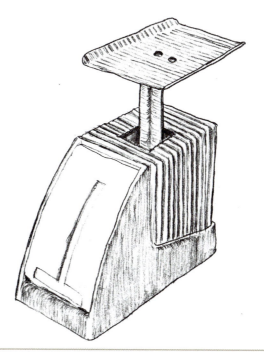

FIGURE 8.07. A sketch of the kitchen scale.

Dissection Notes for a Small Kitchen Scale

1. Remove the small Phillips-Head screw from the bottom of the scale.
2. Remove the cover plate from the bottom of the scale.
3. Remove the small brass knurled nut located under the weighing plate.
4. Release the spring from the lower end of the metal stem inside the scale, and remove the spring and threaded stem subassembly.
5. The four-bar linkage will now release from the back wall of the scale body.
6. Gently remove the metal stem from the plastic four-bar linkage.
7. The weighing plate and the stem can now be removed from the scale.
8. The read-out scale can now be removed from the front of the scale body.
9. If deemed necessary, the four-bar linkage can be taken apart and the spring can be removed from the threaded shaft.

FIGURE 8.08. Dissection notes.

all of the individual parts within a subassembly shown as branches from the corresponding stem. Figure 8.09 shows a fishbone diagram for the kitchen scale. Notice that for this diagram, four major subassemblies are identified: the housing, the platform, the four-bar mechanism, and the spring mechanism.

You also should sketch the subassemblies as you take them apart so you can refer to them later as needed. Figure 8.10 shows a sketch of the housing subassembly; Figure 8.11 shows the platform and spring mechanism subassemblies; Figure 8.12 shows the sketch of the four-bar linkage for the kitchen scale. It may not seem important to document each small step of dissection on a simple product; however, on a more complicated mechanism, it is essential to document each step for reassembly at a later time.

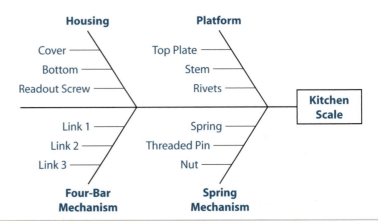

FIGURE 8.09. A fishbone diagram for the dissection of the kitchen scale.

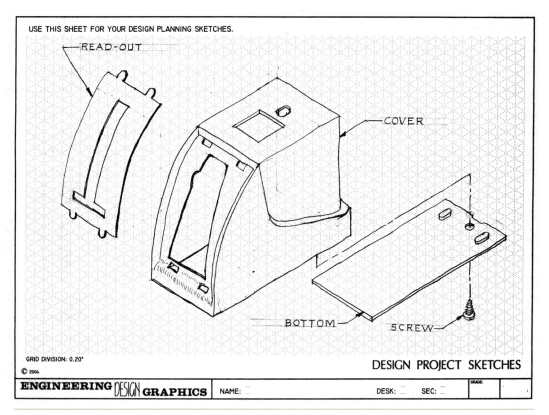

FIGURE 8.10. A sketch of a housing subassembly.

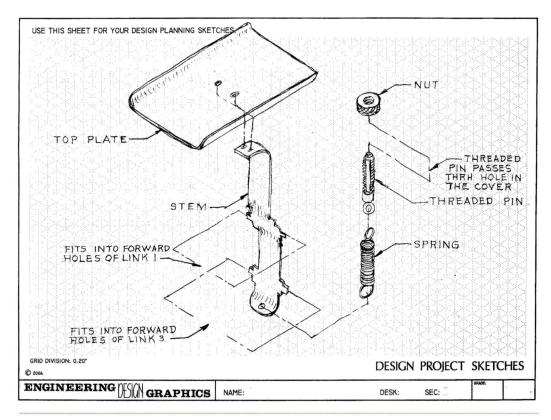

FIGURE 8.11. A sketch of platform and spring subassemblies.

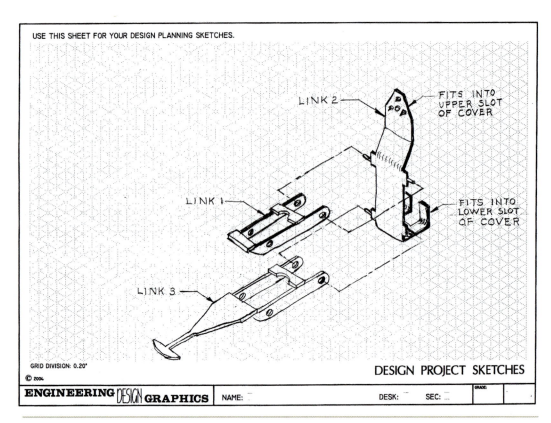

USE THIS SHEET FOR YOUR DESIGN PLANNING SKETCHES.

LINK 2

FITS INTO
UPPER SLOT
OF COVER

LINK 1

FITS INTO
LOWER SLOT
OF COVER

LINK 3

GRID DIVISION: 0.20"

© 2004

DESIGN PROJECT SKETCHES

ENGINEERING DESIGN **GRAPHICS**   NAME: DESK: SEC: GRADE:

FIGURE 8.12. A sketch of a four-bar linkage subassembly.

8.04.03 Obtaining Part Sizes

Through the use of metrology equipment, the individual parts are now studied for size, shape, and position data, which is usually done with simple handheld devices such as a set of scales or calipers. Or you can use more accurate and sophisticated measurement systems such as a coordinate measuring machine if you have one available or if greater accuracy is needed.

Sometimes this dimensional data can be written down in a set of notes or in an inspection report. Making sketches of the part and applying dimensions directly to the sketch also can be used to document the part. Note that the assembly sketches you made in the previous step may not be suitable for this part of the exercise, since they are essentially assembly pictorials. You may need to make new sketches of individual parts for this step.

8.04.04 Developing a 3-D CAD Model

Some believe that the most important phase of the reverse engineering process is to obtain an accurate 3-D computer model of the devices being dissected. You have learned about 3-D solid modeling and assembly modeling in previous chapters of this text. To create accurate 3-D solid models, you may need to refine the CAD data once the preliminary model is completed, adjusting the data as needed. If you used a coordinate measuring machine for the part measurements, the computer system might be able to download the data automatically to a CAD file and create the parts for you. If not, you will need to build the 3-D CAD parts and systems individually using the available software commands. Figure 8.13 shows a CAD model of the kitchen scale. In this figure, the front plate of the model is not shown, exposing the interior of the assembly.

FIGURE 8.13. A computer model of the kitchen scale assembly.

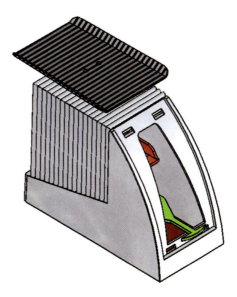

Once you have created each of the 3-D CAD models, you can generate engineering drawings from them. These drawings will contain dimensions and could be used by a machinist to create the part. Figure 8.14 shows an engineering drawing of the kitchen scale body that was created from its 3-D CAD model. You will learn more about engineering drawings and the machining process in later chapters of this text.

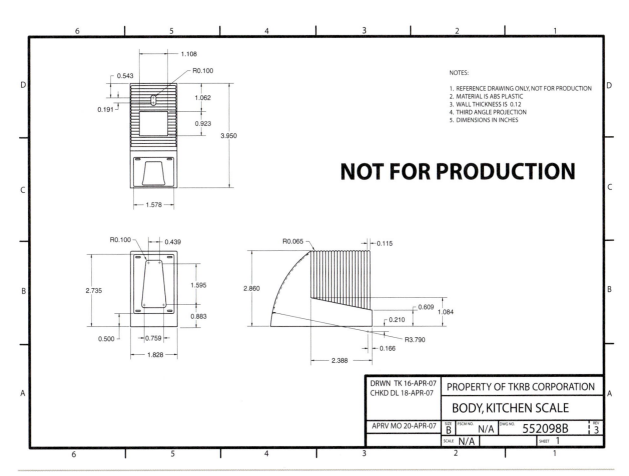

FIGURE 8.14. An engineering drawing of the kitchen scale body.

8.04.05 Considering Potential Redesign

Since one purpose of the reverse engineering process is to study an existing design and suggest improvements to it, the final task of the dissection project should be to consider potential modifications to the product. During your thorough study of the design, some weaknesses may have become apparent. For example, the hole on the top of the body of the kitchen scale may be too large, causing the stem and weighing plate to be unstable. For the model studied here, the weighing plate and the stem are made of stamped metal and have relatively sharp edges. A different process of manufacturing or a different selection of materials might be desirable. The materials used in the spring may be prone to creep, meaning that the springs "stretch out" slightly but permanently with repeated use and thus will not accurately weigh food over the long term. You also may decide that the capacity of the scale should be increased from 0.5 kg to 1 kg. Some of these modifications would require that the entire system be redesigned; other modifications would merely require a different specification for materials. In any case, if you are asked to suggest design modifications for the product you dissected, you should outline them in a memo or report, including new drawings or 3-D CAD models as needed to illustrate your new product.

8.05 Geometric Properties Analysis

In the preceding sections of this chapter, you learned about one type of analysis—reverse engineering. Although 3-D CAD played a role in the reverse engineering process, solid modeling was not an absolute necessity in performing reverse engineering. You could have stopped with the creation of sketches and drawings of the physical parts in the device and not created 3-D CAD models. Your analysis of the device would have been complete at that time. For the remainder of this chapter, you will learn about two types of analysis for which 3-D CAD modeling is absolutely essential. For these analysis activities, the *basis* for the analysis is a 3-D solid model. In other words, to perform the analysis, you must create the solid model and then proceed from there.

The physical properties of a part can be computed about any set of axes—the global, the local, or some other set of user-defined axes. Properties can be input by material type for computational purposes. For example, in specifying that a given part is composed of steel, the software will automatically use the standard density of steel in its calculation of physical properties.

In some instances, you may want to compute the physical properties of your 3-D models. For example, you may want to minimize an object's volume or weight. Or you may need to know where a system's **center-of-mass** is so you can increase the device's stability. If you check the physical properties of the object as you make changes to it, you should be able to optimize your design through an iterative process. The types of physical properties that are typically calculated from a 3-D part definition include the surface area, volume, mass, density, and center-of-mass. Distances and dimensions on the parts also can be measured or computed. In addition, inertial properties (such as **radii-of-gyration**, **moments-of-inertia**, principal-axes of rotation, and products-of-inertia) also can be calculated from the models. You learn more about these types of properties in courses such as Statics and Dynamics; but for now, you merely need to understand that these are calculated from the geometry, mass, and mass distribution of the part.

8.05.01 Measurement Analysis

With measurement analysis, you can determine the dimensions and other geometric parameters of a CAD model. Most modern 3-D modeling software can display the dimensions of the object one way or the other, especially the dimensions you input to create the model. In addition, the software can compute dimensions that are not shown. For example, most software has a Measure function that can be applied to a 2-D sketch or to a 3-D solid model. When using this type of command in the software, you query the solid model on-screen and the software returns the value you desire.

- *Measure Point:* This command returns the coordinates (x-, y-, and z-) of a specific point on a sketch or model. Usually, this point has to be a pickable point, such as a corner or circle center. You could, however, interpolate between two pickable points to return the coordinates of a point that is halfway between two defined points. Figure 8.15a illustrates the Measure Point command.

- *Measure Line Length:* This measure command returns the length of a line, as shown in Figure 8.15b. The line can but does not have to be a specific edge on the model. In other words, the software can be used to determine the distance between two points on the model.

- *Measure Line Distance:* This measure command returns the shortest (perpendicular) distance between a line and another identified entity, such as a point. The Measure Line Distance command is illustrated in Figure 8.15c.

- *Measure Circle:* This measure command returns the center and diameter of a circle, such as a hole diameter. Figure 8.15d illustrates the use of the Measure Circle command.

- *Measure Arc:* The Measure Arc command returns the center and radius of an arc, as shown in Figure 8.15e.

- *Measure Surface:* Measure Surface returns the area of a specified surface and the length of the perimeter surrounding that surface. This command is illustrated in Figure 8.15f.

8.05.02 Mass Properties Analysis

Mass properties are the static properties of a solid body. Mass properties depend on two things: the geometry of the part and its density, where **density** is defined as the mass per unit volume. Table 8.01 includes the fundamental mass properties that can typically be computed with 3-D CAD software. Your computer system may include other properties; however, the properties provided here are the "basics." For now, you may not understand all of the terminology found in the table; you will likely learn about some of these mass properties in more advanced-level engineering and science courses.

The mass properties of an object are related to one another through various formulas. For example, the density is the mass divided by the volume. Likewise, the radius-of-gyration about a given axis is derived directly from the object's moment-of-inertia about that same axis. For most CAD systems, densities of standard materials are internally stored and are available for assignment to the parts you create. So for an assembly model, you could assign some parts to be made from steel, others from aluminum, and still others from plastic. The software would automatically insert the correct density

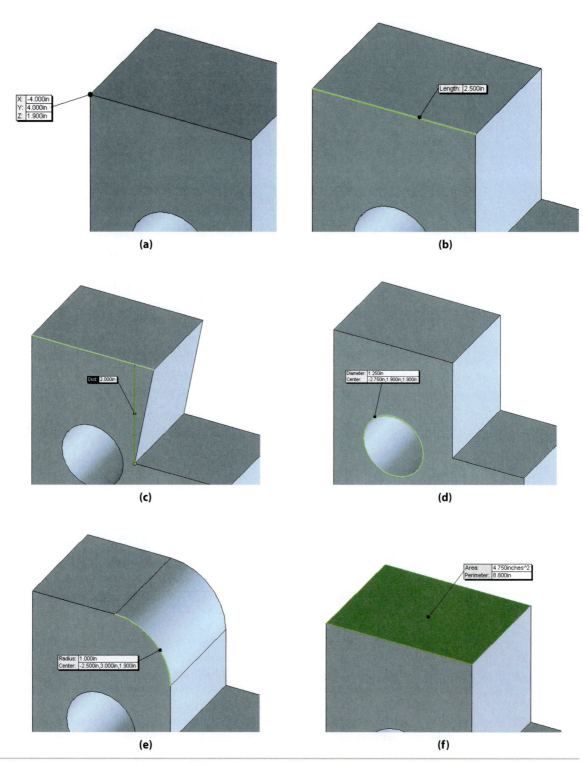

FIGURE 8.15. Various measure commands.

TABLE 8.01. Mass Properties of a Solid Body.

PROPERTY	DEFINITION
DENSITY	DENSITY IS THE WEIGHT PER UNIT VOLUME FOR THE MATERIAL FROM WHICH THE PART IS MADE.
MASS	THE MASS OF A BODY IS THE MEASURE OF THE BODY'S PROPERTY TO RESIST CHANGE IN ITS STEADY MOTION. THE MASS DEPENDS ON THE VOLUME OF THE BODY AND THE DENSITY OF THE MATERIAL FROM WHICH THE BODY IS MADE.
VOLUME	THE VOLUME OF A BODY IS THE TOTAL VOLUME OF SPACE ENCLOSED BY THE BODY'S BOUNDARY SURFACES.
SURFACE AREA	THE SURFACE AREA IS THE TOTAL AREA OF THE BOUNDARY SURFACES DEFINING THE SOLID.
CENTER-OF-MASS OR CENTROID	CENTER-OF-MASS (OR CENTROID) OF A VOLUME IS THE ORIGIN OF COORDINATE AXES FOR WHICH FIRST MOMENTS ARE ZERO. IT IS CONSIDERED THE CENTER OF A VOLUME. FOR A PARALLEL GRAVITY FIELD, THE CENTER OF GRAVITY COINCIDES WITH THE CENTROID.
PRINCIPAL AXES-OF-INERTIA AND PRINCIPAL MOMENTS-OF-INERTIA	PRINCIPAL MOMENTS-OF-INERTIA ARE EXTREME (MAXIMAL, MINIMAL) MOMENTS OF INERTIA FOR A BODY. THEY ARE ASSOCIATED WITH PRINCIPAL AXES-OF-INERTIA THAT HAVE THEIR ORIGIN AT THE CENTROID, AND THE DIRECTION OF EACH IS USUALLY GIVEN BY THE THREE UNIT VECTOR COMPONENTS.
MOMENTS-OF-INERTIA	A MOMENT-OF-INERTIA IS THE SECOND MOMENT OF MASS OF A BODY RELATIVE TO AN AXIS, USUALLY X-, Y-, OR Z-. IT IS A MEASURE OF THE BODY'S PROPERTY TO RESIST CHANGE IN ITS STEADY ROTATION ABOUT THE AXIS. IT DEPENDS ON THE BODY'S MASS AND ITS DISTRIBUTION AROUND THE AXIS.
RADIUS-OF-GYRATION	THE RADIUS-OF-GYRATION IS THE DISTANCE FROM THE AXIS OF INTEREST WHERE ALL OF THE MASS CAN BE CONCENTRATED WHILE STILL YIELDING THE SAME MOMENT-OF-INERTIA.

value in its calculations. After computing the volume of the object, the software would simply multiply the volume by the density and return the object's mass for your further calculations.

Many engineering calculations require data on mass properties of the component parts of the mechanical system for further analysis. The data required might include the mass, the centroid, moments-of-inertia, products-of-inertia, or radii-of-gyration. The moments-of-inertia, products-of-inertia, or radii-of-gyration may be calculated for any set of axes of rotation. The most commonly used sets of axes for these calculations are either the Cartesian axes (standard x-, y-, or z-axes) or the centroidal axes, with the centroidal axes being the preferred set for this type of calculation. The origin of the centroidal axes is located at the centroid of the object and is the "mass center" of the object. For simple objects such as a sphere, the centroid coincides with the exact center of the sphere. For an object such as a cube, the centroid would be at a point in the cube's interior that is halfway between all six surfaces that make up the cube. For more complex objects, the software can compute the exact centroid (and the various inertial properties) for you.

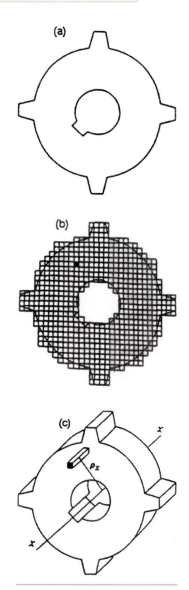

(a)

(b)

(c)

ρ_x

x

x

FIGURE 8.16. Numerical computation of the moment-of-inertia.

As stated previously, you will learn about most of these mass properties in other engineering courses; but consider just one example now. To spin a wheel about an axis, you have to apply a moment or torque to it. The level of resistance of the wheel to rotate faster and faster under the action of the moment is governed by its moment-of-inertia. If the moment-of-inertia is large, the wheel will require a higher torque to accelerate to a desired rotation rate; conversely, if the moment-of-inertia is small, the wheel will easily accelerate to the same rotation rate with a smaller torque applied to it. In essence, the moment-of-inertia is the weight of the mass multiplied by a distance the mass is away from the rotation axis. Thus, a wheel with a large diameter will be harder to accelerate to a desired rotation rate than a wheel with a small diameter.

Moments-of-inertia are easily computed for standard shapes by simple formulae; however, most objects are not made up of simple shapes and the computer is an effective tool for computing moments-of-inertia for complex shapes. The computer software determines an object's moment-of-inertia by dividing the body into small pieces. The software then multiplies the mass of each piece by the distance the small piece is away from the axis of interest. Adding up all of the individual moments-of-inertia, you obtain the moment-of-inertia for the entire body. An example of this procedure is shown in Figure 8.16. In this figure, a solid model of a cam wheel is shown in 8.16a and the wheel has been subdivided into small pieces in Figure 8.16b. The moment-of-inertia of each small piece (one of which is shown in Figure 8.16c) is calculated by multiplying the mass of the piece by the distance from the axis; in this case, ρx. Summing all of the individual moments-of-inertia from each piece produces the overall moment-of-inertia for the cam wheel. You should realize that trying to calculate by hand the moment-of-inertia about X for the cam wheel would be a tedious, painstaking, and error-prone procedure. With the use of computer software tools, quantities such as the moment-of-inertia can be calculated easily with a few mouse clicks, provided you already created the solid model of the part.

To see how an engineer might use a property such as a moment-of-inertia in the design process, consider the two flutter plates shown in Figure 8.17 and Figure 8.18 (A and B). The mass properties for design A are given in Table 8.02, and the mass properties of design B are given in Table 8.03.

Flutter Plate A

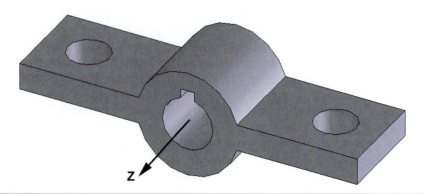

Z

FIGURE 8.17. Flutter plate design A.

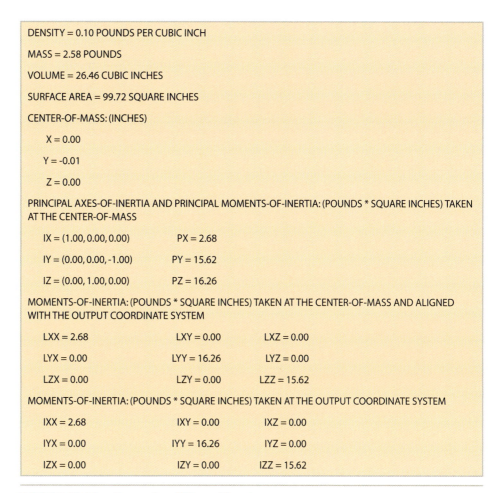

DENSITY = 0.10 POUNDS PER CUBIC INCH

MASS = 2.58 POUNDS

VOLUME = 26.46 CUBIC INCHES

SURFACE AREA = 99.72 SQUARE INCHES

CENTER-OF-MASS: (INCHES)

X = 0.00

Y = -0.01

Z = 0.00

PRINCIPAL AXES-OF-INERTIA AND PRINCIPAL MOMENTS-OF-INERTIA: (POUNDS * SQUARE INCHES) TAKEN AT THE CENTER-OF-MASS

IX = (1.00, 0.00, 0.00)   PX = 2.68

IY = (0.00, 0.00, -1.00)   PY = 15.62

IZ = (0.00, 1.00, 0.00)   PZ = 16.26

MOMENTS-OF-INERTIA: (POUNDS * SQUARE INCHES) TAKEN AT THE CENTER-OF-MASS AND ALIGNED WITH THE OUTPUT COORDINATE SYSTEM

LXX = 2.68	LXY = 0.00	LXZ = 0.00
LYX = 0.00	LYY = 16.26	LYZ = 0.00
LZX = 0.00	LZY = 0.00	LZZ = 15.62

MOMENTS-OF-INERTIA: (POUNDS * SQUARE INCHES) TAKEN AT THE OUTPUT COORDINATE SYSTEM

IXX = 2.68	IXY = 0.00	IXZ = 0.00
IYX = 0.00	IYY = 16.26	IYZ = 0.00
IZX = 0.00	IZY = 0.00	IZZ = 15.62

TABLE 8.02. Mass Properties of Flutter Plate A.

Flutter Plate B

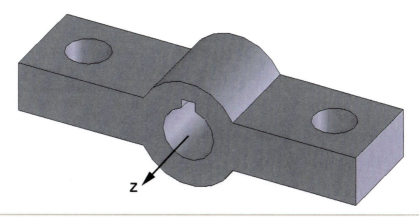

FIGURE 8.18. Flutter plate design B.

TABLE 8.03. Mass Properties of Flutter Plate B.

DENSITY = 0.10 POUNDS PER CUBIC INCH

MASS = 3.66 POUNDS

VOLUME = 37.51 CUBIC INCHES

SURFACE AREA = 115.29 SQUARE INCHES

CENTER-OF-MASS: (INCHES)

   X = 0.00

   Y = -0.01

   Z = 0.00

PRINCIPAL AXES-OF-INERTIA AND PRINCIPAL MOMENTS-OF-INERTIA: (POUNDS * SQUARE INCHES) TAKEN AT THE CENTER-OF-MASS

IX = (1.00, 0.00, 0.00)	PX = 3.74
IY = (0.00, 0.00, -1.00)	PY = 27.81
IZ = (0.00, 1.00, 0.00)	PZ = 28.86

MOMENTS-OF-INERTIA: (POUNDS * SQUARE INCHES) TAKEN AT THE CENTER-OF-MASS AND ALIGNED WITH THE OUTPUT COORDINATE SYSTEM

LXX = 3.74	LXY = 0.00	LXZ = 0.00
LYX = 0.00	LYY = 28.86	LYZ = 0.00
LZX = 0.00	LZY = 0.00	LZZ = 27.81

MOMENTS-OF-INERTIA: (POUNDS * SQUARE INCHES) TAKEN AT THE OUTPUT COORDINATE SYSTEM

IXX = 3.74	IXY = 0.00	IXZ = 0.00
IYX = 0.00	IYY = 28.86	IYZ = 0.00
IZX = 0.00	IZY = 0.00	IZZ = 27.81

The primary motion of the flutter plate is rotation about the given z-axis. In this case, the thickness of the arm plates is the key difference in the two designs, with a larger thickness used for design B. A check of the moment-of-inertia about the z-axis (Izz) for design A has a value of 15.62 lbs-in^2, and the moment-of-inertia about the z-axis for design B has a value of 27.81 lbs-in^2. This means that design B would require more torque to accelerate the flutter plate to a desired rotation rate, making A the better choice if required torque were the only design consideration. In some cases, however, a strength requirement for the flutter plate arms might supersede the torque requirement; and design B, with its thicker and thus stronger arms, might be the better choice overall.

Mass properties analysis offers fast insight into the performance of a design concept without requiring significant additional effort. Once you have created the CAD model, you can obtain the mass properties with a few simple mouse clicks. Hence, computing mass properties is an initial design analysis tool that allows you to identify possibilities that can be ruled out or possibilities that require further exploration. More advanced analysis tools, such as the finite element method described in the next section of this chapter, require more sophisticated input parameters and take more time to complete, but they also yield a better in-depth insight into the mechanical behavior of the designed object.

8.06 Finite Element Analysis

Finite Element Analysis (FEA) is an advanced design analysis technique that has been made possible through the development of sophisticated 3-D CAD solid modeling tools. In engineering practice, there are some cases where desired quantities can be calculated by simple mathematical expressions. For example, for a simple circular rod that is pulled on either end, the stress in the member can be computed as the pulling force applied to the end of the shaft divided by the cross-sectional area of the member. Armed with knowledge of the physical properties of the rod's material, an engineer can calculate whether the stress applied to the rod is too large, resulting in permanent deformation. Design changes can then be made to reduce the stress in the rod if desired. One simple method for reducing the stress is to increase the cross-sectional area of the rod.

Unfortunately, most real-life objects are not simple circular rods and the determination of stresses within these objects is a complicated process. Thus, determining whether an object will be permanently deformed through applied loads would be a nearly impossible proposition without the use of tools such as FEA. For complex objects, engineers can still model their behavior mathematically under various loading conditions; however, for these objects, the equations are virtually impossible to solve using simple mathematics. In these cases, the behavior of the object is governed by a set of differential equations for which no simple solution exists. Similar to the way the computer solved for the moment-of-inertia of a complex object, the FEA method is founded on dividing the part into small pieces, which are together called a **mesh**. The governing differential equation can be easily solved for each of these individual pieces, called elements. The results are then compiled by the computer across all of the elements to produce an accurate picture of the stresses throughout the object.

In essence, FEA is in the category of engineering analysis tools that apply computational approximations to classic field problems. By dividing an object into small elements, solving for stresses within the individual elements, and then compiling the results across the entire object, you have obtained an approximate solution to the problem. Generally speaking, this approximate solution is "close enough." In fact, since an exact solution is virtually impossible to achieve, the FEA approximation is far better than no solution at all. The types of field problems for which the FEA method is a viable analysis technique are presented in the following paragraphs.

Mechanical stress and displacement fields. Objects that are subjected to loads experience **mechanical stresses**. On a microscopic level, these stresses tend to pull molecules apart or push them closer together. A shear stress, for example, tries to slide molecules apart in a motion similar to when you slide your hands against each other. The stresses experienced by an object will result in slight changes in shape as the molecules move relative to one another. If you are pulling on an object, it will tend to elongate or stretch. If you are pushing on the object, it will compress or squash together. If you are applying a shear stress to the object, it will shear and become distorted. The FEA method can be used to solve for all of these types of stresses and deformations for objects subjected to various types of loads.

Figure 8.19 shows the stresses experienced by a simple cantilever beam that has been loaded by pushing down on one end. In this case, the lower stresses are found at the end of the beam where the load is applied, with higher stresses found at the end of the beam where it is attached to the wall.

Fluid flow and pressure fields. Another application of the finite element method is in the computation of fluid flow and pressure fields, and one of the primary uses is in analyzing airflow. (In engineering analysis, air is considered to be a fluid.) This type of analysis is used to compute the airflow around airplane wings, enabling engineers to design bigger and better jets that can develop sufficient lift with bigger payloads. Air pressure fields also can be determined through this type of analysis. Figure 8.20 shows the movement of particles suspended in a moving fluid. The particles, for example, could be blood cells or nano-particles in a blood vessel. By predicting and altering the way the fluid flows, or altering the geometry of the particles, a desired motion of the particles can be produced.

Thermal flow and temperature fields. Another class of problems that can be solved using the finite element method is the computation of thermal flow and temperature fields. For a satellite system, one side of the system is typically exposed to the sun

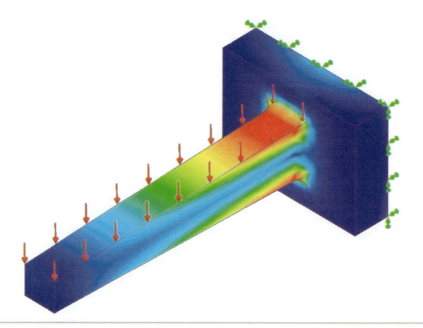

FIGURE 8.19. Stress analysis of a cantilever beam.

and the other is not. As you can imagine, the side exposed to the sun is much warmer than the other side. It is important to determine how the temperature flows from one side to the other, because temperature may affect the performance of the satellite. For example, large variations in temperature can cause thermal stresses, because the warmer side expands and the colder side contracts. If the stresses are too large, permanent deformation

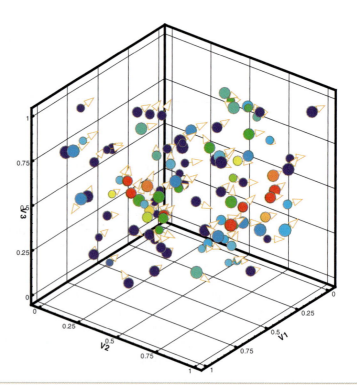

FIGURE 8.20. Particle motion is a moving flow field. *Courtesy of Tarek Zohdi, University of California.*

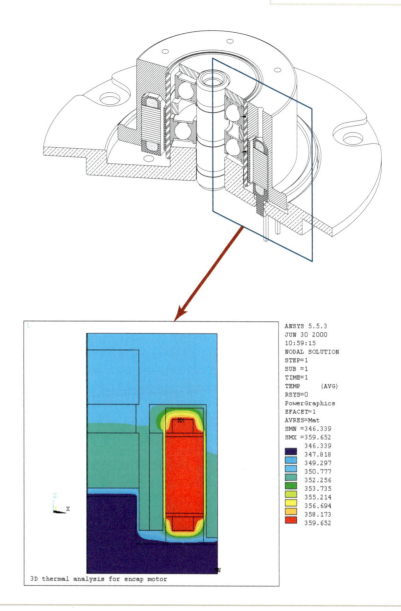

FIGURE 8.21. Thermal analysis of the inside of a disk drive spindle motor.

or even fracture can result. Figure 8.21 shows a spindle motor for a disk drive that has been analyzed for its temperature distribution.

Electromagnetic fields. When electricity flows through wire, an electromagnetic field is created. Electromagnetic fields are also a significant consideration in the design of antennas. The size and shape of the electromagnetic field depends on the type of antenna used. If the steel in the body of the automobile interferes with the electromagnetic field from the antenna, radio reception will be poor and owners will not be satisfied with their cars. Due to poor electromagnetic performance, the simple vertical mast antenna has been virtually eliminated in new car models—new antennas are more sophisticated and are less likely to have poor performance issues.

The finite element method can be used to solve for electromagnetic fields so that antennas for automobiles can be optimized. Figure 8.22 shows the output from a finite element model that was used to determine an electromagnetic field for a linear step-motor.

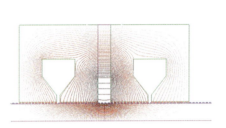 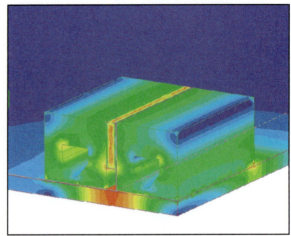

FIGURE 8.22. Electromagnetic analysis of a linear step-motor.

The key to successful numerical solution of engineering field problems is the division of the solid volume into small, finite-sized elements. The process of subdividing the solid into elements is referred to as meshing. In general, the smaller the elements, the more accurate the solution; however, if the elements are too small, you may introduce error into your model due to computational approximations. Further, smaller element sizes mean more elements and computation time will significantly increase. Element sizes should be reduced wherever there is curvature in the model to compute desired quantities accurately in those regions. In general, you can use larger elements in areas where stresses are relatively static and you need small elements in areas where the stresses are rapidly changing. Knowing what sizes to use for elements will take practice.

Fortunately, most of the tedium in meshing has been eliminated or greatly reduced with modern FEA software applications. In most cases, you merely specify a nominal element size and the software will automatically create the elements for you; the software also will automatically reduce the element sizes when it encounters curved regions. As you gain experience with the finite element method, you will get a better feel for reasonable nominal element sizes; but for now, you should rely on your instructor to provide insight into the modeling procedure. Figure 8.23 shows a simple part for which a finite element mesh has been created.

In the FEA meshing scheme, the finite elements are connected through their edges and nodal points. It is at these nodal points and along the element edges that the **boundary conditions** are specified. In the case of stress analysis, boundary conditions might be forces or defined displacements; for thermal analysis, boundary conditions might be input as a temperature. In the case of objects subjected to external forces, you must fix some of the nodal points in space so the object is not able to move at those points.

During the analysis, each finite element is treated as a simple solid body with limited properties, interacting with its surrounding region only through the nodal points and element edges. Although this approximation seems very rough, proper selection of the element properties and sizes will ensure an accurate solution to the problem. When the analysis results are displayed, deformations are usually given to an enlarged scale and colored contours are used to show the stress distribution, as shown in Figure 8.24. Thermal gradients and fluid pressure fields also can be shown as colored contours on the output plots.

FIGURE 8.23. A finite element mesh applied to a 3-D CAD model.

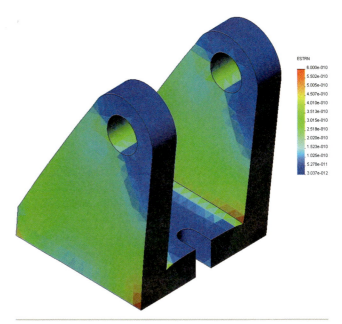

FIGURE 8.24. Stress distribution contours obtained through FEA.

8.07 The Finite Element Analysis Process

The following example details a step-by-step approach toward the use of FEA in solving an engineering problem. This example introduces the power of the FEA technique in solving numerical problems and illustrates that it is possible to solve problems such as these with a limited mathematical background. However, it is important to realize that in engineering practice, you should not perform this type of analysis without a proper mathematical foundation. Solving complex problems without a thorough understanding of the scientific and mathematical principles involved could lead to significant mistakes, sometimes having catastrophic results.

8.07.01 Creating the Model Geometry

The first step in the process is to create the model geometry—you cannot perform an analysis unless there are objects or systems to analyze. For this example, you will consider a pillow block and a shaft positioned at its center. Figure 8.25 shows a solid model of the pillow block, and Figure 8.26 shows the solid model of the shaft. As you create the parts, make sure you input the correct material characteristics so the software will use the appropriate material properties as it solves for stresses and displacements.

After the two parts are created, an assembly model is defined in which the shaft is centered in the hole of the pillow block. The assembly will look like the model in Figure 8.27.

The complete analysis of the pillow block could be a complex procedure, since analysis of the pillow block might have to include the effects of the rotating shaft on the pillow block or the thermal stresses developed in the assembly. For this example, only a simple static analysis of the pillow block will be carried out.

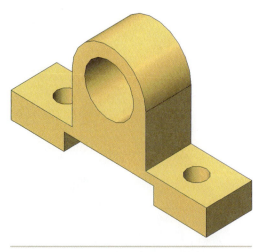

FIGURE 8.25. A solid model of a pillow block.

FIGURE 8.26. A solid model of a shaft.

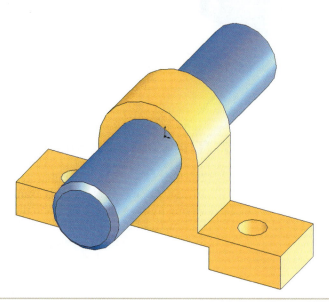

FIGURE 8.27. A pillow block assembly.

8.07.02 Applying Constraints

If a body is subjected to an external force, it will move unless it is constrained. Imagine an eraser sitting on your desk as you read this book. If you push the eraser, it will merely slide across the desk until you stop pushing it. If, however, you hold the eraser in place with one hand while pushing with the other hand, it will deform in response to the applied force. For a finite element study, your object or assembly must be fixed in space. If you do not constrain the object, it will merely translate in space. Stresses and **displacements** will develop in response to the combination of forces and **constraints**. For the pillow block assembly under consideration here, you can imagine that it is sitting on a flat surface. As you push the shaft downward, the flat surface resists the force, preventing the assembly from moving in that direction. Therefore, putting constraints along the bottom surface, fixing that surface in space, would realistically model the assembly. Figure 8.28 shows constraints applied to the pillow block assembly that fix these points in space, mimicking the action of the flat surface on which the assembly rests.

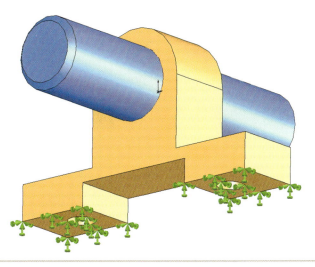

FIGURE 8.28. The pillow block assembly with restraints applied.

8.07.03 Applying Loads

After the constraints are specified, loads (in the form of forces, pressures, and/or torques) can be applied. For this example, forces applied to the ends of the shaft will test the integrity of the pillow block. The forces could be applied in any direction—in a direction parallel to the axis of the shaft or in a direction perpendicular to the axis. Alternatively, a torque could be applied to twist the shaft and produce shearing stresses in it. If you wanted to apply a torque to the shaft, you would have to make sure that the shaft was constrained from rotating within the pillow block hole; otherwise, the shaft would merely spin in the hole and no stresses would be generated. For this example, the force will be applied downward perpendicular to the axis of the shaft, as shown in Figure 8.29.

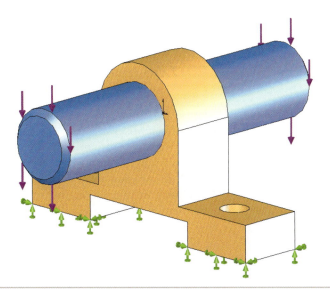

FIGURE 8.29. The pillow block assembly with applied forces.

FIGURE 8.30. The pillow block assembly with the finite element mesh applied.

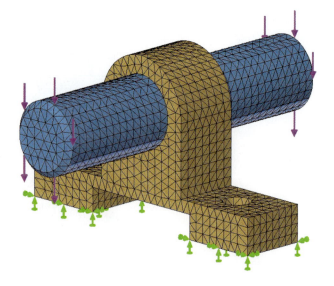

8.07.04 Meshing

The final preparatory step before the model can be submitted for analysis is to create the mesh. The nominal mesh size is carefully chosen to minimize computation time without compromising accuracy. Most modern-day software will automatically generate the mesh for you; and at the end of the meshing process, your model will look similar to that shown in Figure 8.30. For some software, you may have to mesh the objects before you apply the constraints and loads, because the boundary conditions can be applied only to existing nodes and element edges.

8.07.05 Computing the Results

The finite element solver computes various quantities for the model based on the boundary conditions—constraints and loads—that you applied. For stress analysis, the two primary quantities of interest are the stresses and displacements. Other quantities can be computed, such as strains; but since strains are directly related to stresses, they are of secondary importance in the analysis. After the analysis is complete, the results are retained in the computer's memory. The results can be accessed and displayed in various ways as you investigate the model performance.

8.07.06 Investigating the Results

At this point, the software has all needed calculations and you can display the results. Figure 8.31 shows the von Mises stress (or a combined stress) distribution. An explanation of von Mises stresses is beyond the scope of this text; however, you should understand that red contours typically signify high stresses (hot areas) and blue contours typically signify low stresses (cool areas). For the pillow block assembly, the von Mises stress distribution display shows high stresses in the shaft where the shaft and the pillow block are in contact. High stresses also are found in the base of the pillow block in the vicinity directly below the shaft near the corners.

Figure 8.32 shows the displacements for the model. The displacements are typically shown exaggerated in the display—the true displacements are usually too small to be seen with the naked eye, and the exaggeration helps you visualize the response of the model. From the displacements shown in Figure 8.32, it is apparent that the ends of the shaft experience the largest displacements, as expected.

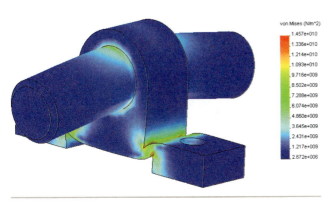

FIGURE 8.31. Stress distribution on the pillow block assembly.

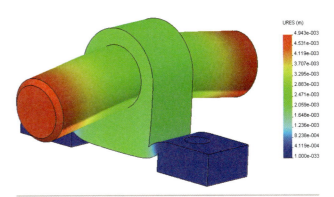

FIGURE 8.32. Displacement distribution for the pillow block assembly.

Some software may have the capability to perform a Design Check. With a Design Check, the software identifies areas of the model that could be modified to improve performance. Typically, the Design Check identifies areas of high stress or large displacements. Figure 8.33 shows the results from a Design Check performed for the pillow block assembly. In this case, the sharp corners on the bottom thru slot of the pillow block have been identified as a potential area for modification.

8.07.07 Modifying the Design

One of the purposes of analysis is to identify design flaws or areas for improvement prior to production. For the pillow block example, some fundamental design changes can be made to improve the performance of the original model. If the internal edges of the pillow block were rounded, stresses in the model would be greatly reduced. Sharp internal corners and edges produce stress concentrations in a model (as you will learn in later courses), so rounding the corners should relieve the stress concentrations. Another design modification would be to reduce the size of the slot on the lower part of the pillow block. By reducing the size of the slot, more material would be located beneath the portion of the pillow block where the shaft is pressing against it. Another design modification would be to increase the wall thickness around the shaft hole—again providing greater resistance to the force and reducing the stress. A modified pillow block is shown in Figure 8.34.

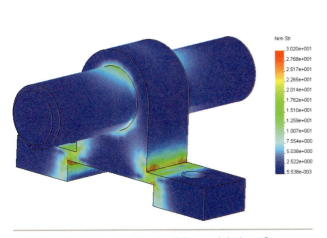

FIGURE 8.33. A Design Check of the model identifies areas for potential redesign.

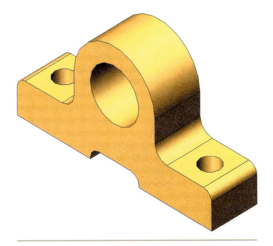

FIGURE 8.34. A modified pillow block design.

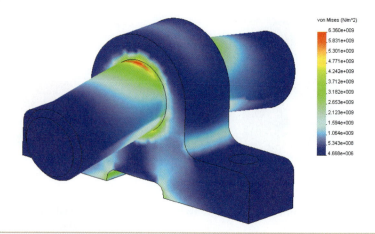

FIGURE 8.35. Stress distribution for the modified pillow block assembly.

After these modifications have been made, the constraints and forces are reapplied to the model and a new mesh is created so the system can be analyzed again. In this way, you can verify that the new design has better design performance in terms of stress concentration and displacement when compared to the original design. With this redesign of the pillow block, its performance was significantly improved, as shown by the von Mises stress plot shown in Figure 8.35.

8.08 Chapter Summary

In this chapter, you learned about the importance of analysis in the engineering design process. You learned that many of the analysis methods are founded in mathematical and scientific principles. You learned that reverse engineering is a type of analysis where devices are dissected and the functions of the parts that make up the devices are studied. The dissection process is documented through notes and sketches so the device can be reassembled at a later time. Parts in the device are measured, and computer models are created and assembled. You also learned about mass properties analysis as a technique for computing various quantities from part geometry. Computing mass properties provides a quick look at the part geometry and is an analysis technique that helps rule out possible designs or identify potential solutions that are worth pursuing in the initial stages of the design process. Finally, you learned about the FEA technique and the way it is used to solve complex engineering problems. You learned about the classes of problems that can be solved through FEA, including stress analysis, fluid flow, thermal flow, and electromagnetic fields.

8.09 glossary of key terms

analysis: The study of the behavior of a physical system under certain imposed conditions.

black box diagram: A diagram that shows the major inputs and outputs from a system.

boundary conditions: The constraints and loads added to the boundaries of a finite element model.

caliper: A handheld device used to measure objects with a fair degree of accuracy.

center-of-mass (centroid): The origin of the coordinate axes for which the first moments are zero.

constraint: A boundary condition applied to a finite element model to prevent it from moving through space.

coordinate measuring machine: A computer-based tool used to digitize object geometry for direct input to a 3-D CAD system.

density: The mass per unit volume for a given material.

displacement: A change in the location of points on an object after it has been subjected to external loads.

engineering scale: A device used to make measurements in much the same way a ruler is used.

Finite Element Analysis: An advanced computer-based design analysis technique that involves subdividing an object into several small elements to determine stresses, displacements, pressure fields, thermal distributions, or electromagnetic fields.

fishbone diagram: A diagram that shows the various subsystems in a device and the parts that make up each subsystem.

laser scanning (three-dimensional): A process where cameras and lasers are used to digitize an object based on the principle of triangulation.

mass: A property of an object's ability to resist a change in acceleration.

mechanical dissection: The process of taking apart a device to determine the function of each part.

mechanical stress: Developed force applied per unit area that tries to deform an object.

mesh: The series of elements and nodal points on a finite element model.

metrology: The practice of measuring parts.

moment-of-inertia: The measure of an object's ability to resist rotational acceleration about an axis.

radius-of-gyration: The distance from an axis where all of the mass can be concentrated and still produce the same moment-of-inertia.

reverse engineering: A systematic methodology for analyzing the design of an existing device.

surface area: The total area of the surfaces that bound an object.

volume: The quantity of space enclosed within an object's boundary surfaces.

8.10 questions for review

1. How does reverse engineering differ from industrial espionage?

2. Describe the following metrology tools: scales, calipers, coordinate measuring machine, and 3-D laser scanner. Which is least expensive? Which is most accurate?

3. How are volume, density, and mass for an object related to one another?

4. What are some of the mass properties that can be computed for a CAD model?

5. What is a moment-of-inertia? Is it easier to spin a wheel with a large diameter or a small diameter?

6. What are constraints in an FEA?

7. What kinds of boundary conditions can be applied to a finite element model?

8. Describe a finite element mesh. In general, will accuracy improve as element size gets smaller or as it gets larger?

9. What is one disadvantage in using a small element size for a finite element model?

8.11 problems

1. For the objects shown in Figure P8.1, the given coordinate system is to be placed at the lower rear corner of each object. Create a solid model for each object and calculate its volume using the units shown. Relative to the given coordinate system, find the x-, y-, and z-coordinates of the centroid for each object.

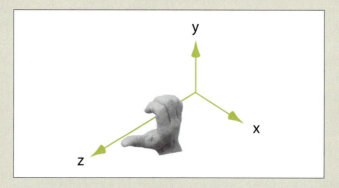

Coordinate system used from Problem 1.

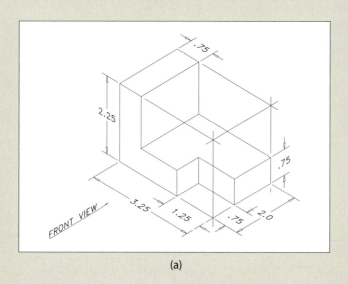

(a)

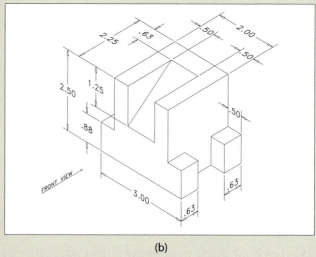

(b)

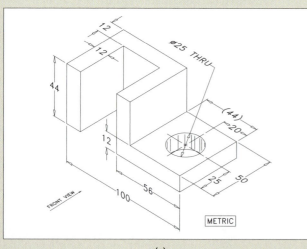

(c)

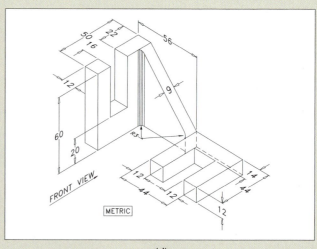

(d)

8.11 problems (continued)

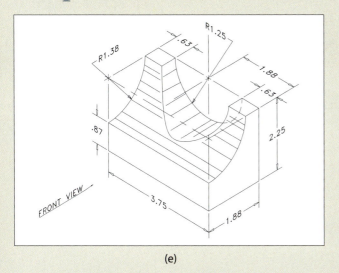

(e)

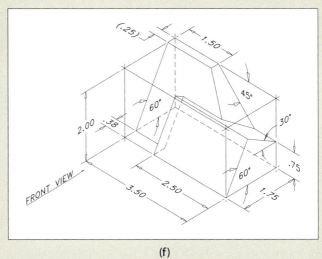

(f)

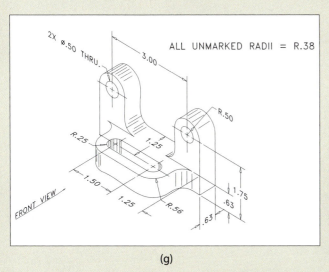

(g)

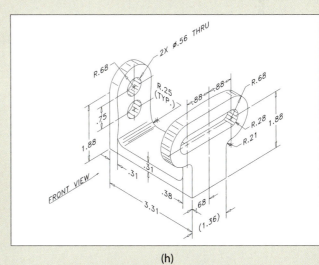

(h)

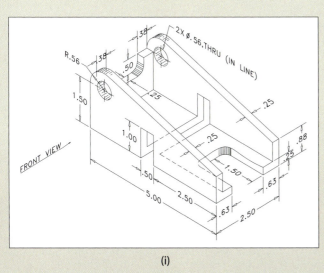

(i)

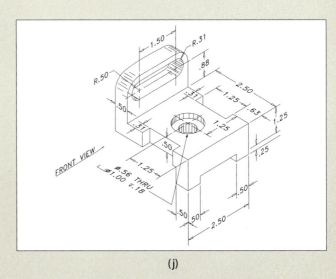

(j)

8.11 problems (continued)

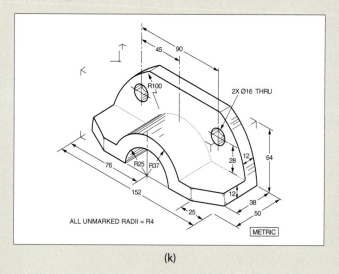

(k)

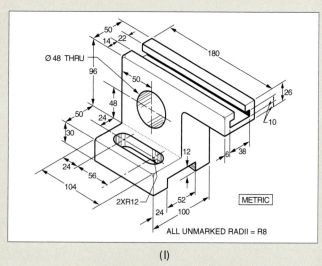

(l)

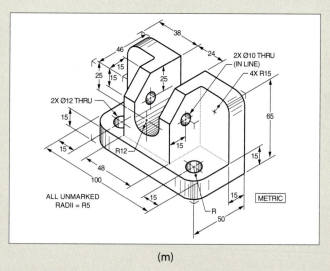

(m)

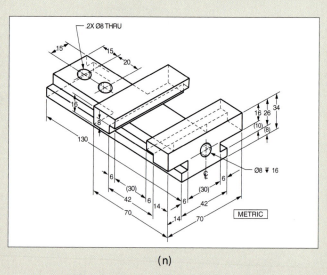

(n)

FIGURE P8.1.

9

Fabrication Processes

objectives

After completing this chapter, you should be able to

- List common production processes available for part fabrication

- Visualize how an engineered part can be made to ensure that it can indeed be made

- Discuss the relative difficulty, time, and expense involved in using specific fabrication methods

- Demonstrate how a design should change in order to match the fabrication method most suitable for its production volume

9.01
introduction

You might be wondering why an engineering graphics textbook contains a chapter on fabrication processes. The fact is that most engineers and designers who deal with modeling and drawing are the same people who specify the geometries of the parts that go into the devices they design. Thus, it would be useful to have some knowledge of the different types of fabrication processes that are available, their relative precisions, and their relative costs. As you can imagine, many fabrication processes exist. Some of the processes are highly specialized for the production of a specific type of geometry. Others are flexible in their application and can be used to produce parts of all sorts of different geometries. When a part is being designed, its fabrication process must be considered to ensure that the part can indeed be fabricated within the desired cost constraints. The design may need to change to suit a fabrication process that is appropriate for the part's intended production volume. Some geometries that are easily produced using high-volume processes may be difficult to produce using low-volume processes and vice versa. In this sense, an engineer or a designer who is responsible for the specification of part geometries must be familiar with the fabrication processes used to produce these geometries.

9.02 Making Sure It Can Be Made

As an engineer or a designer, you ultimately will be asked to specify the precise geometry of a part that will fit into a device or structure of some kind. In response, you probably would deliver a computer model or drawing that shows what the part will look like, with the exact sizes of various geometries and features that go into the part. Perhaps your part looks like the object shown in Figure 9.01.

When the various people who are involved with the function, fabrication, installation, or use of this part see your specifications, they are likely to have the following questions, which you must be prepared to answer before the part is made:

1. Is the geometry of the part producible?
2. Can the part be produced within existing time and budget constraints?
3. Can the geometry be changed to reduce production costs without sacrificing function?
4. How must the geometry be changed as production volumes increase, forcing a change in the production process?

To answer those questions, you must have some familiarity with common fabrication processes. You must be familiar not only with the processes themselves but also with the relative time, costs, and precisions associated with each process. While detailed explanations of the various fabrication processes are outside the scope of this book, it will present an introduction to the more common fabrication methods.

The fabrication process used to make a part is often dependent upon its production volume; that is, the fabrication process is dependent upon the number of parts that are to be made. Fabrication processes that are ideal for making one or two identical parts may be unsuitable in terms of cost and time when trying to make ten thousand or even a million of the same parts. Those who are familiar with machine shops, for example, are well aware of the relative differences between a production shop and a prototyping shop. The machines in production shops have tools and fixtures made specifically for handling large numbers of parts of a specific design. The tools and fixtures in a production shop cannot be easily changed to fabricate a part with a different geometry. On the other hand, machines in prototype shops have tools and fixtures that can be easily changed or

FIGURE 9.01. This part needs to be fabricated. What processes will be needed, and in what order should they be performed?

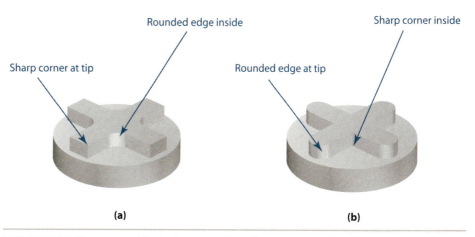

Sharp corner at tip

Rounded edge inside

Sharp corner inside

Rounded edge at tip

(a)

(b)

FIGURE 9.02. Two parts with similar cross-shaped protrusions. The sharp corner at the tip of (a) may create problems for a high-volume fabrication process. The sharp inside corner of (b) may create problems for a low-volume fabrication process.

rearranged to build a variety of different part geometries. However, a prototype shop can produce parts only in small numbers.

The intended fabrication process affects the way a part should be designed. For example, consider the part shown in Figure 9.02(a), which has a cross-shaped protrusion atop a simple base. The squared corners on the tips of the cross and the rounded edges on the inside of the cross would be considered simple to produce using a low-volume production process such as milling (which will be explained in a later section). However, to produce the sharp corner at the tip using a high-volume process such as die casting (again, to be explained in a later section) would present some problems because the mold would be difficult to make. Modifying the protrusion cross geometry to squared inside corners, as shown in Figure 9.02(b), would make it easier to fabricate using a high-volume process but more difficult to fabricate using a low-volume process.

Now consider a similar part that has a cross-shaped cavity instead of a protrusion, as shown in Figure 9.03(a). The squared corners at the tips of the cross would be difficult to produce using a low-volume production process. However, to produce the same feature geometry using a high-volume process would be simple. Modifying the cut-out cross geometry to have squared inside corners, as shown in Figure 9.03(b), would create some problems for fabrication using a high-volume process but simplify fabrication using a low-volume process.

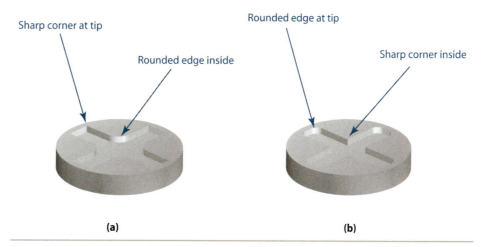

Sharp corner at tip

Rounded edge inside

Rounded edge at tip

Sharp corner inside

(a)

(b)

FIGURE 9.03. Two parts with similar cross-shaped cavities. The sharp corner at the tip of (a) may create problems for fabrication using a low-volume process. The sharp corner on the inside of (b) may create some difficulty for production with a high-volume fabrication process.

FIGURE 9.04. Elimination of sharp corners, as shown in these redesigned parts, eliminates potential problems that may occur with either low-volume or high-volume fabrication methods.

(a)   **(b)**

By modifying the cross geometry to have both rounded outside edges and rounded inside corners, as shown in Figure 9.04, the part is well suited for low-volume and high-volume fabrication methods. This modification allows the same design to be used for both prototyping and high-volume production.

Why do some feature geometries create difficulties when one production method is used compared to another? The reasons will be explained later in this chapter. But first, you should have a basic understanding of the different ways geometries and features on parts can be made. In addition to the processes of milling and die casting briefly mentioned, many other fabrication processes exist. For convenience, the individual processes can be loosely grouped into families of similar processes. These families can be based on the relative speed of process, precision of the formed features, cost of any secondary equipment required to support the production, or the ease with which the fabricated geometry can be changed once it is in production. Bear in mind that the processes outlined in the following sections include some of the most common fabrication processes, but not every process in existence. A complete listing and description of all of the different types of fabrication processes that currently exist would be too large to present in a single book chapter. Detailed information about the processes to be presented here can be found in comprehensive reference books such as *Machinery's Handbook*, published by Industrial Press, at a library, or on the Internet by searching on the keywords *machinery* and *handbook*.

9.03   Processes for Low-Volume Production

Parts that are produced at a rate of a few per day (or less) will likely be fabricated using a low-volume process. These processes are usually appropriate for one-of-a-kind parts and for prototypes of parts that may be fabricated later at higher rates of production. Low-volume processes are performed by machines that are commonly found in well-equipped fabrication shops. These machines are designed for general-duty applications and are capable of making a wide variety of different part geometries with commonly available **fixtures** and few or no special tools. Fixtures are mechanical devices, such as clamps and brackets, which are used to hold and align the **workpiece** as it is being shaped. A workpiece is a common name for a part while it is still in the fabrication process, that is, before it is a finished part. A workpiece usually begins as a piece of material that has been cut from a length of standard, or stock, shape that is commercially available. The workpiece is then modified by additional cutting operations that remove more material or by joining operations that add more material until the desired part geometry is produced. A fixed or movable, replaceable cutting implement with one or more sharpened edges used for removing material from a workpiece is known as a **tool bit**.

9.03.01 Standard Commercial Shapes

Common raw materials for engineered parts include metals such as steels, aluminums, and brasses and plastics such as polycarbonate, nylon, and Delrin. Materials manufacturers, suppliers, and distributors of raw materials deliver these products in the form of **standard commercial shapes**. These shapes are usually long lengths of the material that have a uniform cross-section shape and plates or sheets of uniform thickness.

Some of the common shapes are shown in Figure 9.05. These shapes include rounds, bars, tubes, channels, tees, angles, and a variety of beam geometries for making structures. Each shape comes in a variety of standard sizes, as detailed in the Appendix. Not all shapes or sizes will be available for less common materials such as titanium, magnesium, copper, and plastics; so the material suppliers' catalogs must be checked. When the raw material shape used to fabricate a part is selected, it is prudent to start the workpiece with a standard commercial shape and size that is close to, but slightly larger than, the final external size of the part. In this way, the amount of material that needs to be removed (and usually the time and effort required to make the final product) is minimized.

The standard commercial shapes are produced by high-volume processes, usually extrusion and/or rolling, which are discussed later in this chapter.

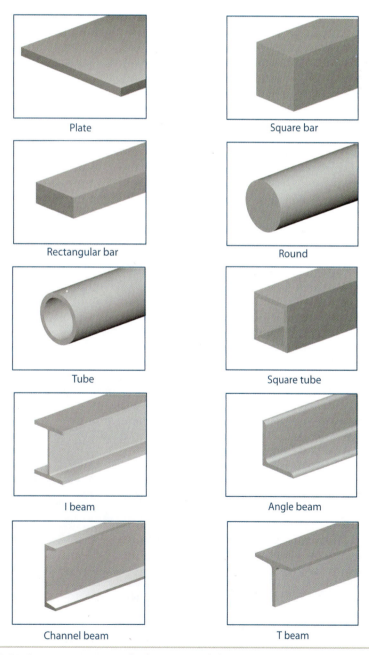

Plate

Square bar

Rectangular bar

Round

Tube

Square tube

I beam

Angle beam

Channel beam

T beam

FIGURE 9.05. Some common standard commercial shapes.

9.03.02 Sawing

A workpiece is usually cut to a rough size from a standard commercial shape by a **sawing** operation. In a sawing operation, a multitoothed cutting blade is pulled through the material, as shown in Figure 9.06. The direction of rapid tool motion and material removal is along the length of the blade. To extend the cut, either the workpiece or the blade must be moved in a direction across the length of the blade. The blade then cuts a path through the part.

Sawing is a relatively quick operation, but its precision and surface finish quality are relatively poor when compared to other machining processes. For sizes up to 1 m in length, reliably creating sizes that are accurate to within 1 mm is usually considered easy with a sawing operation, whereas accuracy to within 0.1 mm is considered difficult. In addition, paths cut by a saw on a workpiece usually are limited to straight lines and large-radius curves.

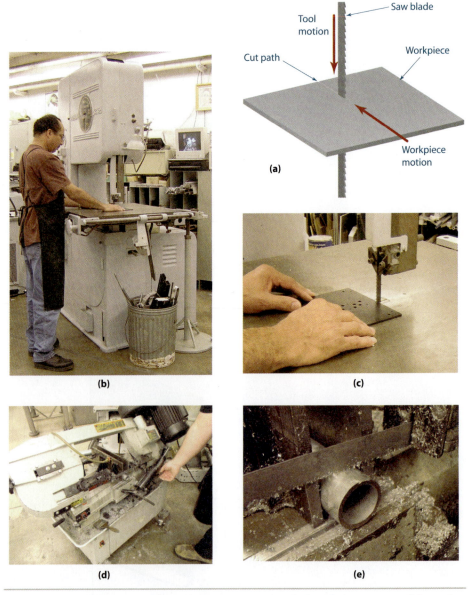

FIGURE 9.06. A basic sawing operation is shown in (a) with the relative motion of the tool and the workpiece. A vertical band saw is shown in (b) with the workpiece being hand-fed in (c). The horizontal band saw in (d) automatically feeds the blade into the workpiece to cut off pieces of stock material shown in (e).

FIGURE 9.07. A basic turning operation is shown in (a) and a basic boring operation is shown in (b) with the relative motion of the tool and workpiece. A manually controlled lathe is shown in (c), with a close-up of the tool and workpiece in (d).

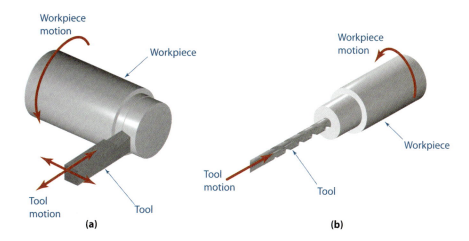

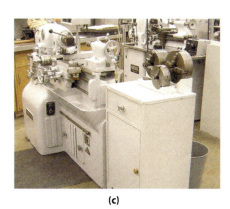

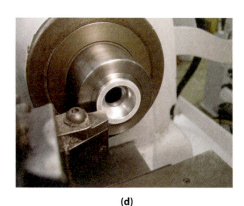

9.03.03 Turning

Axially symmetrical features can be made on a machine tool called a **lathe**, on which the workpiece is rapidly rotated about a single axis. A cutting tool is then moved slowly in the radial and/or axial direction relative to the workpiece, removing material in these directions. The process of shaping the workpiece in this manner is known as **turning**. The process is similar to shaping clay on a potter's wheel. Turning can be used to create concentric features on the outside, inside, or face of the part, as shown in Figure 9.07. Circumferential grooves and screw threads are made using this process, as are stepped, tapered, and rounded shafts. Removal of material on the inside of the workpiece is known as **boring**. Deep holes along the rotation axis can be made using a boring tool that is shaped like a drill bit.

For hand-sized parts, turning can be used to easily create feature sizes that are radially accurate to within 0.01 mm, whereas accuracy to within 0.001 mm is considered difficult. Surface finishes of excellent quality can be created. Features can easily be located axially within an accuracy of 0.05 mm, whereas axial location within 0.005 mm is considered difficult.

9.03.04 Drilling

Round holes are most commonly made by **drilling**. In this operation, a rapidly spinning cutting tool called a **drill bit** is slowly pushed into a stationary workpiece, as shown in Figure 9.08. Drill bits are usually long, with a sharpened tip and spiral grooves along their length to help remove the material that is cut away. Material is cut and removed from the workpiece in the axial direction of the drill bit and the motorized spindle that turns it.

A machine called a **drill press**, also shown in Figure 9.08, can be used to spin a drill bit and plunge it into the workpiece. Holes can easily be located on the face of the workpiece within an accuracy of 0.2 mm, whereas axial location within 0.02 mm is considered difficult. This location accuracy can be improved by performing the drilling operation using a milling machine, as described in the next section. Holes that go completely through the workpiece are called **through holes**; whereas holes that go part of the way through the workpiece are known as **blind holes**. Screw threads can be made inside a hole using a special cutting tool called a **tap**. A hole with screw threads inside it is known as a **tapped hole**. For holes up to 20 mm in diameter, drilling can easily create diameters that are accurate to within 0.2 mm, whereas accuracy to within 0.02 mm is considered difficult. The depth of a hole can easily be controlled to within an accuracy of 0.2 mm, but controlling the depth to within 0.02 mm is considered difficult. The

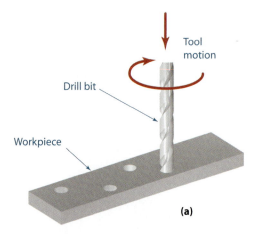

(a)

(b)

(c)

FIGURE 9.08. The basic drilling operation is shown in (a) with the motion of the tool. A manually controlled drill press is shown in (b), with a close-up of the tool (drill bit) and the workpiece in (c).

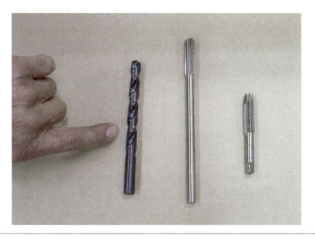

FIGURE 9.09. Some tools for making holes. A drill bit (left) is used to make a basic hole. The reamer (center) is used to slightly enlarge a hole to a very precise diameter. The tap (right) is used to create screw threads inside a hole.

accuracy of hole diameters can be improved with a process called **reaming**, whereby a special drill bit with a very precise diameter that is just slightly larger than an existing hole is used to open the hole slowly to the final desired diameter. With a reaming operation, hole diameters can be made that are accurate to within 0.02 mm, whereas accuracy to within 0.002 mm is considered difficult. The differences between a drill bit, a reaming tool, and a tap are shown in Figure 9.09.

9.03.05 Milling

Milling is performed with **milling machines**, which are some of the most versatile machines found in a shop. Various operations that can be performed on a mill are shown in Figure 9.10. The **spindle** of a milling machine can hold and rapidly spin a variety of different cutting tools, including drill bits. However, the most common cutting tool used in a milling machine is an **end mill**. Although sometimes shaped like a drill bit, an end mill is unique because it can cut and remove material from the workpiece in the axial and transverse directions with respect to the spindle axis. Whereas only the tip of a drill bit is sharpened, both the bottom and sides of the flutes on a milling tool are sharpened.

By shaping the end mill, special geometries can be created easily. Ball and round end mills, for example, can be used for rounding the inside and outside edges of a workpiece. Angle end mills can be used for beveling edges. Fly cutters can be used for removing small amounts of material over a relatively large area to create a clean and flat surface. Examples of these types of cutting tools are shown in Figure 9.11.

Although drilling can be done on a milling machine for improved accuracy in hole location, a drill bit can cut and remove material in the axial direction only. Through the use of end mills, geometric features in the workpiece can be created by slowly moving the workpiece, which is mounted to a **stage** that can move in both the axial and transverse directions with respect to the spindle axis, as shown in Figure 9.12. A milling machine whose stage can be moved in this manner is called a **three-axis mill** because its stage can move in three different linear directions.

FIGURE 9.10. Some standard milling operations showing the relative motions of the tool and workpiece. Standard milling is shown in (a). Internal features such as slots can be made by plunging the tool and then milling (b). Specially shaped tools can be used to create chamfered or rounded edges on inside (c) or outside (d) edges. Making a chamfered edge (e). Using a fly cutter for facing to create a smooth, flat surface is shown in (f). Large, round internal features are created by boring (g). Large, round external features can be created by moving the workpiece on a circular or semicircular path (h).

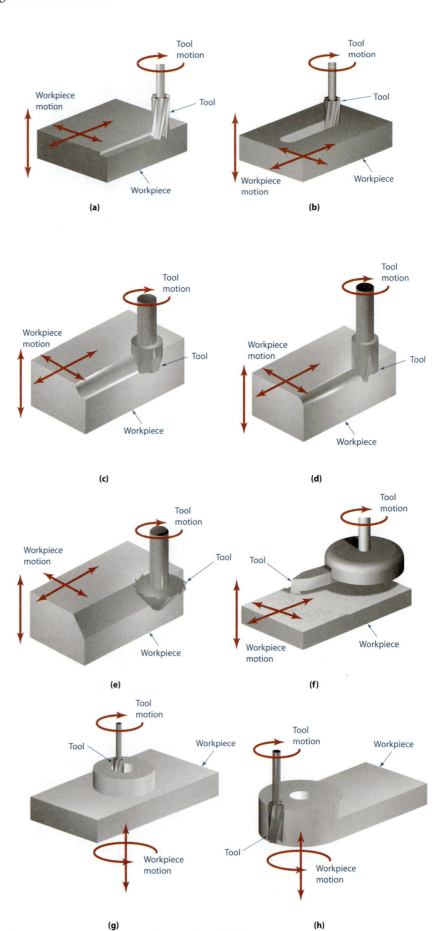

FIGURE 9.11. Some tools for milling special geometries. From left to right, a straight end mill, a ball end mill, a corner rounding end mill, and a dovetail end mill.

(a)

(b)

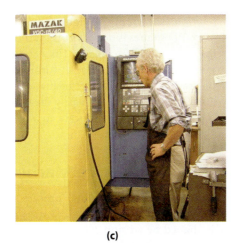

(c)

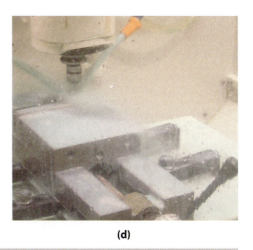

(d)

FIGURE 9.12. A three-axis manually controlled milling machine is shown in (a) with a close-up of the tool and workpiece in (b). The numerically controlled three-axis milling machine shown in (c) can coordinate the motion between all three axes, making it possible to create more complex geometries. Through the window of the enclosed work area, the action of the tool, workpiece, and lubricating coolant can be seen in (d).

FIGURE 9.13. A four-axis milling machine allows the spindle head, which holds the rotating tool, to be turned on an independent axis. Thus, the tool can cut from different directions, allowing the creation of even more complex geometries.

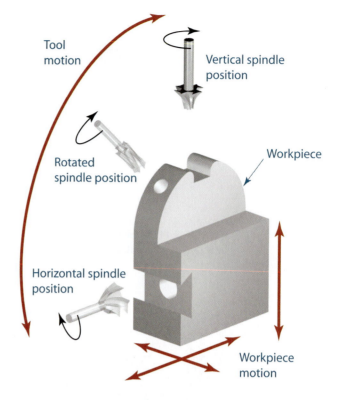

A manually controlled milling machine relies on a human operator to control the stage motion in one direction at a time. A numerically controlled milling machine uses an electronic controller that can be programmed to move the stage in all three directions simultaneously, enabling the production of more complex shapes. For example, when the stage is programmed to move in a circular fashion, axially symmetric features, such as those shown in Figure 9.10(c) and Figure 9.10(d), can be created. On a manually operated mill, axially symmetric features can be created by securing the workpiece to a special rotating stage called a rotary table. Even more complex shapes can be created with a four-axis milling machine. With this machine, the direction of the spindle axis can be changed, as shown in Figure 9.13. This flexibility enables the mill to cut and remove material in many different directions without the workpiece having to be removed and reoriented, which would be time-consuming and expensive. In the case of the Hoyt AeroTec riser (shown in the highlighted fabrication case study), the production of the complex 3-D shapes is done with a minimum of operation intervention that would have been extremely difficult and time-consuming on a three-axis mill. Even more complex geometries, such as those of turbine blades, require the use of four- or five-axis mills for production and cannot be made on three-axis mills.

For hand-sized parts, milling can easily create feature geometries in both transverse and axial directions that are accurate to within 0.10 mm, whereas accuracy to within 0.01 mm is considered difficult.

9.03.06 Electric Discharge Machining

The electric discharge machining (**EDM**) process uses a charged electrode through which a high-level electric current is passed to the workpiece through a dielectric fluid. The workpiece is shaped by erosion caused by the electric current. The most common EDM configuration is wire EDM, by which a thin (0.05 to 0.25 mm in diameter) wire is used to erode a path through the thickness of the workpiece, as shown in Figure 9.14. The wire and workpiece are charged to opposite electric potentials. The workpiece is moved slowly in a direction perpendicular across the wire to create a path through the workpiece. The electric discharge, similar to a small spark, erodes both the wire and

FIGURE 9.14. The wire EDM process is shown in (a) with the relative motion of the wire and the workpiece. A wire EDM machine is shown in (b) with a close-up of the work area in (c).

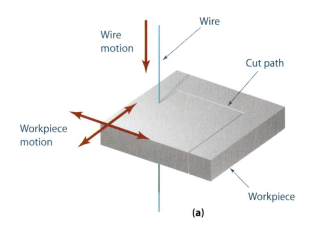

(a)

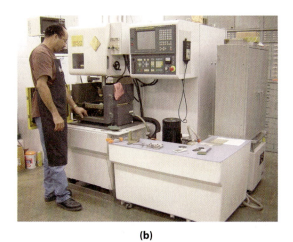 

(b) **(c)**

the workpiece. To prevent the wire from completely eroding at the workpiece, the wire is moved slowly along its length from a feed spool to a take-up spool. Note that the cutting path is two-dimensional and that the axial orientation of the wire (and therefore the erosion path in the direction along the wire) is unchangeable in most wire EDM machines.

For the EDM process to work, the workpiece must be electrically conductive; therefore, insulating materials such as wood and most plastics cannot be shaped by EDM. The EDM process is capable of creating complex paths, with very little deformation force applied to the workpiece, thereby creating accurate features of complex geometries. An example of this work is the lamination shown in Figure 9.15 from a small electric motor.

FIGURE 9.15. This prototype lamination for a small electric motor is an example of the fine detail and precision that can be achieved with the wire EDM process.

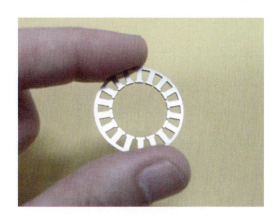

However, the EDM process is relatively slow. The path creation rate is dependent upon the thickness of the workpiece, with material removal rates of approximately 0.20 to 400 cubic cm per hour. Thus for relatively thick workpieces, such as those made of metal plate or bar, the process is much slower than it would be for thin workpieces, such as those made of sheet metal. For hand-sized parts, EDM can easily create feature geometries in the transverse direction that are accurate to within 0.10 mm, whereas accuracy to within 0.01 mm is considered difficult.

9.03.07 Broaching

Some holes, slots, and cutouts with odd geometries can be made with an operation called **broaching**. In a broaching operation, rapidly sliding a tool called a **broach** through the workpiece forms a hole or slot, as shown in Figure 9.16. The broach is a long cutting tool with teeth that are shaped to cut away the workpiece and form the final geometry of the cutout as the tool travels along the length of the workpiece.

Broaching is a relatively fast operation that is suitable for prototyping and high-volume production. However, a custom broach usually is required for each feature that is to be formed by this operation. For hand-sized parts, broaching can easily create feature geometries in the transverse direction that are accurate to within 0.20 mm, whereas accuracy to within 0.02 mm is considered difficult.

FIGURE 9.16. A broaching process used in making a slot. The starting workpiece with a hole is shown in (a). A stationary guide is placed in the hole, and a broach slides down the guide in (b) to produce a slot (c). The broach shown in (d) is part of a set of slotting broaches that gradually increase in depth of cut.

Workpiece

(a)

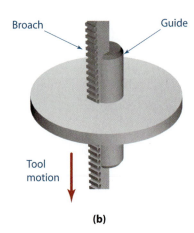

Broach   Guide

Tool motion

(b)

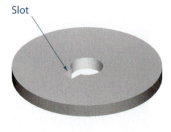

Slot

(c)

(d)

9.03.08 Rapid Prototyping

Many types of processes fall into the category of **rapid prototyping**. Rapid prototyping use various methods to create parts of complex geometries quickly by selective hardening of a bed of powdered raw material or a pool of liquid raw material at room temperature. Rapid prototyping processes are constructive processes whereby the desired part is built up by the selective addition of material as opposed to destructive material removal or reforming used in more conventional manufacturing methods. Rapid prototyping is often used to make preliminary models for visualization, fit, form, and function studies. At the present time, these processes are considered slow and expensive and the available materials are limited; so these processes are typically not used for high-production volumes. However, rapid prototyping can be used to create patterns of complex parts, which are then used in the various **casting** and molding processes. That process will be described in the following section on high-volume production.

One common rapid prototyping process is **stereolithography** (SLA). In the SLA process, solid spots are formed in a pool of liquid polymer by focusing laser beams at a single spot in the pool, curing and hardening the liquid at that spot. Moving the lasers and focusing on successively adjacent spots can form a solid part of complex geometry. This process is shown schematically in Figure 9.17. The process can create complex part geometries more quickly than the conventional prototyping processes discussed previously; however, only a limited range of plastic materials can be used. The parts made from this process typically have slightly poorer strength and surface finish when compared to parts cut from stock raw materials or parts formed by conventional high-volume processes described later in this chapter. Metals parts cannot be formed by SLA. The **fused deposition** (FD) process is another common rapid prototyping process. In this process, a thin stream of molten plastic is deposited from a nozzle that can be moved in two or three dimensions. Depositing the plastic in the desired locations and allowing it to cool and harden after it is deposited creates complex, solid objects. This process, which is not unlike using a miniature hot melt glue gun, is shown schematically in Figure 9.18.

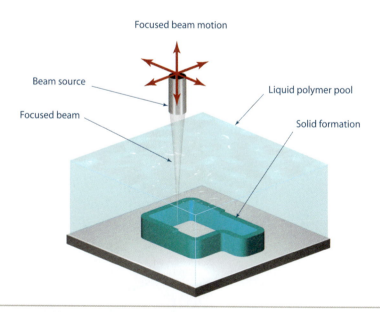

FIGURE 9.17. Rapid prototyping by stereolithography. The part is formed in a pool of special liquid polymer that cures (solidifies) when exposed to intense heat or ultraviolet light. A solid is built by focusing a laser or ultraviolet beam along sequential spots in the pool, solidifying those spots.

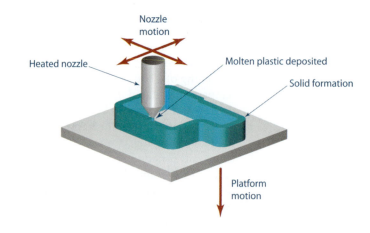

FIGURE 9.18. Rapid prototyping by fused deposition. The part is formed by depositing hot molten plastic, which cools and hardens, along sequential spots on the solid to be formed. Depending on the maker of the machine, the platform or the nozzle can move vertically or transversely.

Closely related to FD is **3-D printing**. This process fuses small plastic (or other powdered material) particles, which lie on a horizontal bed, by depositing from a nozzle a thin stream of solvent that acts as an adhesive. Moving the nozzle, similar to moving the ink head on a printer, controls the volume and position of the stream and, thus, the pattern of fused particles. When the fusing of particles is completed on one layer, the bed is moved downward slightly and a fresh layer of particles is spread over the previous layer. The solvent stream is then used to fuse the particles on this layer into a pattern; the solvent also fuses this layer to the previous layer. Successively moving the bed downward, spreading

FIGURE 9.19. Rapid prototyping by 3-D printing. The part is formed by spraying a jet of adhesive or solvent along sequential spots atop a bed of powdered plastic (a), solidifying the plastic at those spots. The platform is then lowered slightly (b), a new layer of powder is spread on top, and the jet is sprayed again (c). The process of lowering the platform (d), spreading a new layer, and solidifying at selected spots (e) is continued until the desired solid is formed (f). The excess powder is then brushed away.

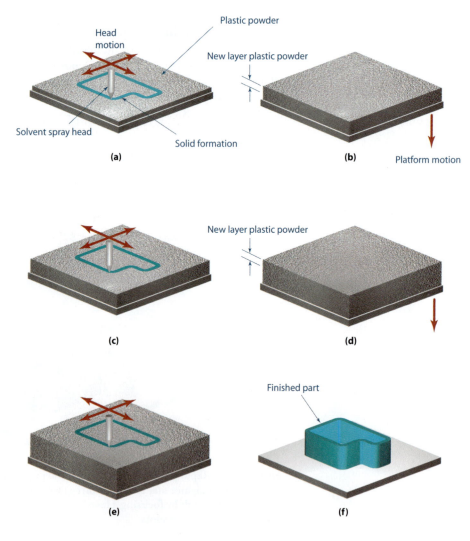

fresh particles, and then fusing the particles of each new layer to the previous layer can produce complex 3-D shapes. The 3-D printing process is shown schematically in Figure 9.19.

As with STL, only a small range of plastic materials can be used for FD and for 3-D printing. Metals parts cannot be formed by these methods. Metal parts can, however, be formed by a rapid prototyping process called **selective laser sintering** (SLS). Laser sintering is similar to 3-D printing except that a powdered metal replaces the powdered plastic. Instead of a nozzle being moved to control the position of a stream of solvent to fuse plastic powder, a focused laser beam is moved to selectively melt and fuse the metal powder.

The parts made from these processes typically have poorer strength and surface finish when compared to parts cut from stock materials or when compared to parts formed by the conventional high-volume processes described later in this chapter. Also, rapid prototyping processes are best suited for small numbers of small parts with complex geometries. Some examples of the complex geometries that are possible using rapid prototyping methods are shown in Figure 9.20. As parts become larger or more numerous, the time required to form them bit by bit, which is inherent in all of the rapid prototyping processes described, makes these processes uneconomical.

The accuracies of rapid prototyping processes are generally equivalent to many high-volume production processes. For hand-sized parts, rapid prototyping processes can easily create feature geometries in all three dimensions that are accurate to within 0.50 mm, whereas accuracy to within 0.05 mm is considered difficult. As these technologies are further developed, accuracies will improve.

FIGURE 9.20. Examples of parts that can be created by rapid prototyping methods. (Images courtesy of Z Corporation.)

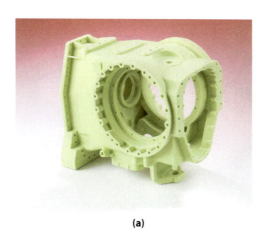

(a)

(b)

(c)

(d)

9.03.09 Welding and Brazing

Unlike the previous processes that shape a workpiece by removing material, **welding** creates shapes by adding pieces together. Welding fuses two or more separate workpieces to create a single workpiece by melting the workpieces at their joints and adding more material to build up the joints as necessary to make them stronger. For metals, a gas flame, an electric arc, or a concentrated laser beam usually generates the heat necessary to melt the material. Plastics are typically welded by a stream of hot air or by the friction heat generated by ultrasonic vibration of the workpieces. A typical welding process is shown schematically in Figure 9.21.

Welding is a relatively fast process; however, the heat necessary to fuse the workpieces has a tendency to distort them. Thus, welding is usually not used for parts that are of high precision. The heat required for welding also changes the properties of some of the materials used for workpieces. Some steels and aluminums, for example, lose much of their strength when they are heated. Other materials become hard and brittle. Finally, the welding process has a tendency to introduce flaws, such as cracks in the workpiece, greatly reducing its effective strength.

In **brazing**, two or more separate metal workpieces are bonded together with a metal alloy filler that has a lower melting point than the workpieces. Unlike welding, the workpieces are not melted during brazing and the bonding alloy is not built up around the joint to strengthen it. Instead, the workpieces are heated at the joints by a gas flame, for example, to a temperature just above the melting point of the filler. The filler is then applied to

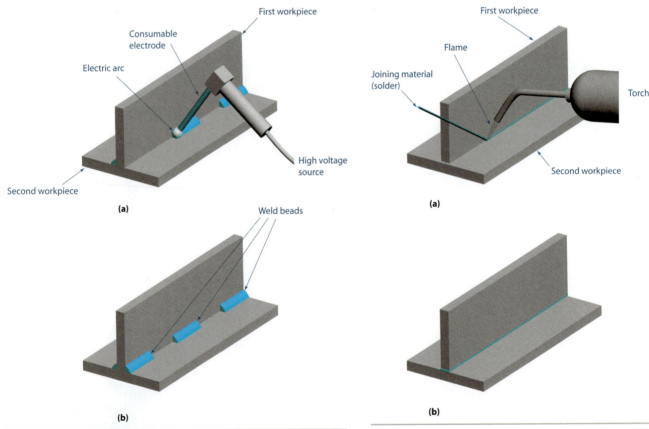

(a)

(b)

(a)

(b)

FIGURE 9.21. Two separate workpieces are joined using a welding process. An electric arc is produced between the electrode and the workpieces, melting the workpieces and the electrode at the point of contact in (a). The electrode is consumed by the process to form beads of material that fuse the workpieces together in (b).

FIGURE 9.22. Two separate workpieces are joined using a brazing process. A gas flame from a torch heats the workpieces and melts the joining metal (solder) in (a). The molten joining metal is pulled into the small space between the workpieces by capillary action. When the joining metal cools and hardens, the two workpieces are fused, as in (b).

the joint, and capillary action pulls the filler between the workpieces. A schematic of the brazing process is shown in Figure 9.22. Brazing typically distorts the workpieces less than welding because the temperatures involved are lower. Also, brazing introduces fewer flaws to the joint than welding. However, as with welding, the temperatures are high enough to cause some metals, such as heat-treated steels, to lose much of their strength.

9.03.10 Grinding

The **grinding** process uses a rapidly rotating abrasive wheel to remove material from a workpiece very slowly. Common processes include flat grinding and inside cylindrical and outside cylindrical grinding, as shown in Figure 9.23. The workpiece or the spindle of the abrasive wheel is moved slowly in the transverse or vertical direction to shape the desired surface on the workpiece. A photograph of a common flat grinder, used to make flat surfaces, is shown in Figure 9.24.

Because the rate of material removal is very slow, the forces applied to the workpiece from the tool are smaller than with the cutting processes detailed previously. The small applied forces result in less deformation of the part compared to most other cutting processes. Consequently, high precisions in flatness and cylindrical roundness can be achieved by grinding. The quality of the surface finish is very smooth.

FIGURE 9.23. Various grinding processes showing the relative motion of the workpiece and abrasive wheel (tool). Flat grinding is shown in (a). Internal cylindrical grinding is shown in (b). External cylindrical grinding is shown in (c).

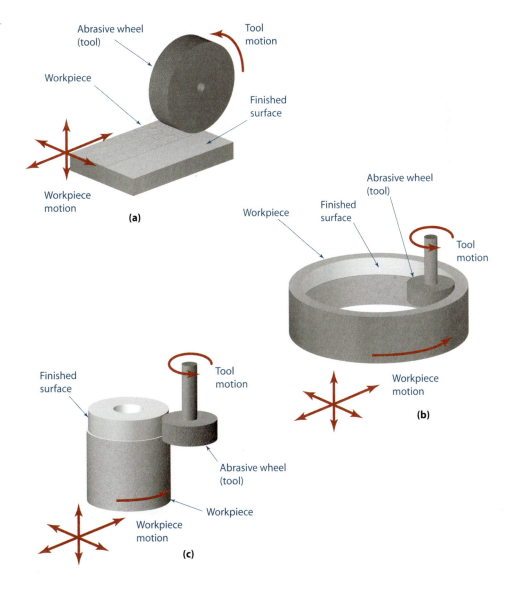

FIGURE 9.24. Grinding machines. Flat surface grinder is shown in (a), with a close-up of the abrasive wheel and workpiece in (b). A tool grinder, used for making and sharpening tools and tool bits, is shown in (c), with a close-up of the abrasive wheel and workpiece in (d).

Grinding can easily produce flatness or roundness accuracy of 0.020 mm on hand-sized parts. Accuracies of better than 0.002 mm are considered difficult with this process. Although the accuracies produced by grinding are excellent, the process is slow. Consequently, grinding is usually used as a finishing operation to create highly accurate final sizes after the workpiece has been roughly shaped by cutting operations. Grinding is most commonly performed on hard materials such as steels. Softer materials, including aluminums and plastics, are usually not ground because those materials tend to fill and clog the rough surface of the abrasive wheel, greatly reducing the wheel's ability to remove material.

SUMMARY OF COMMON LOW-VOLUME PRODUCTION PROCESSES		
PROCESS	**ADVANTAGES**	**DISADVANTAGES**
SAWING	RELATIVELY FAST	POOR ACCURACY, POOR SURFACE FINISH; CUTS ARE GENERALLY LINEAR OR LARGE-RADIUS
TURNING	EXCELLENT ACCURACY FOR AXIALLY SYMMETRIC GEOMETRY	CAN PRODUCE AXIALLY SYMMETRIC GEOMETRY ONLY
DRILLING AND REAMING	FAST AND ACCURATE FOR AXIALLY SYMMETRIC HOLES	CAN PRODUCE AXIALLY SYMMETRIC HOLES ONLY
MILLING	EXTREMELY VERSATILE	MILLING MACHINES CAN GET EXPENSIVE, ESPECIALLY FOUR- AND FIVE-AXIS VERSIONS
EDM	PRODUCES 2-D CUTS OF ODD GEOMETRY AND FINE DETAIL WITH GREAT ACCURACY	VERY SLOW; LIMITED TO CONDUCTIVE MATERIALS
BROACHING	PRODUCES 2-D CUTOUTS QUICKLY	MAY REQUIRE BUILDING A BROACH OR SERIES OF BROACHES FOR EACH APPLICATION; LIMITED GEOMETRIES
RAPID PROTOTYPING	PRODUCES COMPLEX 3-D GEOMETRIES	SLOW; MATERIALS GENERALLY LIMITED TO CERTAIN PLASTICS; LIMITED ACCURACY AND SURFACE FINISHES
WELDING AND BRAZING	PRODUCES COMPLEX GEOMETRIES MORE QUICKLY THAN CUTTING PROCESSES	LIMITED MATERIALS; HIGH-TEMPERATURE PROCESS DEFORMS THE PARTS AND OFTEN REDUCES THEIR STRENGTH
GRINDING	PRODUCES CYLINDRICAL AND FLAT SURFACES WITH GREAT ACCURACY	VERY SLOW; GEOMETRIES GENERALLY LIMITED TO FLAT AND ROUND SURFACES; CANNOT BE USED FOR SOFT MATERIALS, SUCH AS ALUMINUMS AND PLASTICS

9.04 Processes for Higher-Volume Production

Processes that depend on material removal for shaping of a workpiece are generally slow because most engineering materials are strong and cutting or abrading them takes time. It is possible to increase the speed of shaping by deforming the workpiece under a great deal of force or pressure or forming the geometry of the workpiece while its material is still in a hot liquid state. Some of these techniques are described next. While these methods can create a part relatively quickly, usually a substantial investment in **tooling** is needed before these methods can be used. Tooling includes any special tools, fixtures, or other devices needed to create, align, hold, or transport a part during its fabrication. However, for the most part, the same tooling can be reused to create many parts with the same geometry. The cost of the tooling must be included with the cost of fabricating the parts to determine the overall cost of the project. When only a few of these parts are made, the low use of the tooling may not justify the cost. However, as larger numbers of a part are made, the cost of the tooling associated with each part will decrease, helping to justify the initial cost.

Processes for high-volume production usually require a substantial investment in tooling and are not used unless the volume of production of a part can economically justify this investment. These processes often start with raw materials in their most basic form as delivered from the material manufacturer. Plastics, for example, can be delivered in the form of bulk pellets. Metals can be delivered as ingots, powder, or roughly formed bars, plates, or sheets. These materials are then remelted and/or reformed under pressure to produce the desired part. Most high-volume processes cannot achieve the accuracies that are possible with low-volume processes. Cutting and grinding operations are minimized and are used only as necessary to finish creating geometries that cannot be produced by a high-volume process alone.

9.04.01 Sand Casting

All casting processes require a **mold** of the desired part. A mold is a supported cavity shaped like the desired part into which molten raw material is poured or injected. A mold can be disassembled into two or more pieces for removing the part once the raw material has solidified. The places where the mold can be disassembled are called **split lines**. In the case of **sand casting**, the mold is created by making a master **pattern**, which has the same geometry as the desired part. The master can be made of metal, plastic, or wood because only its geometry is needed to create the mold and it does not need to assume the material properties of or function as the real part. Sand mixed with clay as a binder is packed around the master to create the mold. The mold is then split and the master removed, leaving a cavity in the mold the shape of the master. Next, the mold is fired to harden the sand and binder, similar to hardening clay brick or pottery. After the mold is reassembled, molten metal is poured into the cavity through one or more access ports. When the metal hardens, the mold is opened and the cast part is removed. With more complicated molds, it may be necessary to split the mold into three or more pieces to remove the master and the cast part. The sand casting process is shown schematically in Figure 9.25.

Sand casting is used for various alloys of aluminum, steel, zinc, bronze, lead, and magnesium. The molds for sand casting are relatively inexpensive when compared to molds for other casting processes; however, the accuracy of the geometry and the surface finish created by sand casting are considered poor. The process also requires the generation and handling of molten metal. For hand-sized parts, sand casting can easily create feature geometries that are accurate to within 1 mm, whereas accuracy to within 0.1 mm is considered difficult.

Investment casting is a form of sand casting that can be used to create relatively intricate shapes with good accuracy. This process begins with the fabrication of a wax master form that is geometrically nearly identical to the desired production part. The wax master is then repeatedly dipped in slurry that is composed of fine-grained sand with a binder such as clay or cement. Allowing the slurry to dry, or set, between each dip forms a thick-walled mold. The set mold is then fired to dramatically improve its strength, hardness, and temperature resistance. This step also melts the wax master, which runs out of the mold through a hole. Metal is then poured into the mold to replace the wax. After the metal hardens, the mold is broken to retrieve the part. Investment casting can easily create feature geometries that are accurate to within 0.25 mm, whereas accuracy to within 0.025 mm is considered difficult.

Casting and molding processes usually require the sides in the mold to be slightly tapered in the direction of the mold opening, as shown in Figure 9.26. This makes it easier for the part to be removed from the mold. This slight angling of the surfaces is known as **draft**, and the angle is usually at least 1 or 2 degrees. Larger draft angles would make it easier to remove the part from the mold.

FIGURE 9.25. The sand casting process. In (a), a master pattern that matches the geometry and size of the desired final part is made using prototype fabrication methods. The sand and binder are formed around the master in (b). The master is removed in (c) to leave the desired cavity in the sand. The mold is fired in (c) to create a hardened mold capable of withstanding high temperatures. The mold is aligned and closed in (e), and molten raw material is poured to fill the mold cavity in (f). After the raw material cools and hardens, the mold is opened in (g) and the completed part is removed in (h).

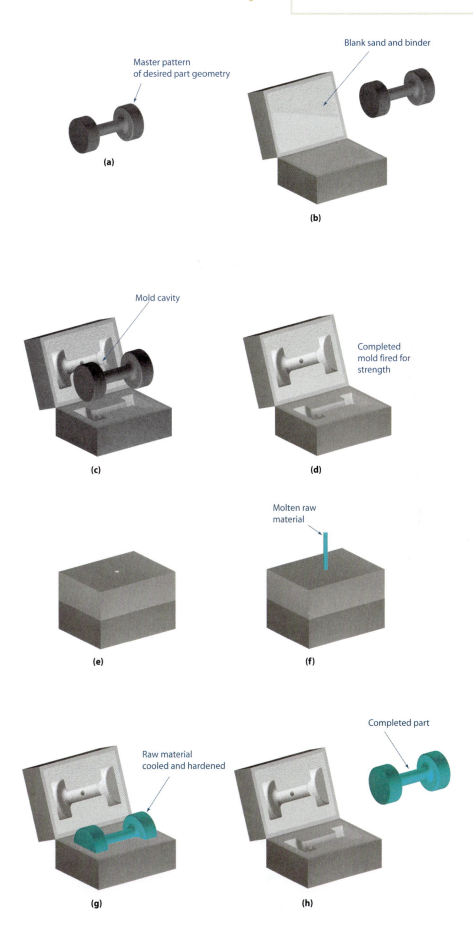

FIGURE 9.26. A cast part with straight walls, as in (a), might be difficult to remove from its mold. By adding draft to the walls, as shown in (b), the part would be easier to remove from its mold. The split line on the part coincides with the part line on the mold.

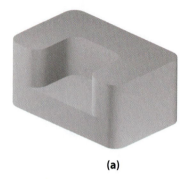

(a)

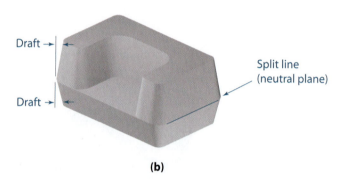

(b)

Both sand casting and investment casting are considered medium-volume production processes because of the labor required to build and handle the molds. Because these molds usually must be broken to retrieve the parts, the molds are inherently fragile and must be handled carefully to avoid unintentional damage.

9.04.02 Extrusion

Extrusion is a method for creating long lengths of material with a uniform cross-section shape. Many of the standard commercial shapes are made with this process. In this process, raw materials are placed in a chamber where elevated pressure and temperature are created. Under these conditions, the raw material fluidizes and is forced out of the chamber through an orifice shaped like the desired cross section. The special machine that creates the pressure and temperature required for this process is called an extrusion machine. These machines are typically very expensive, so their use is usually scheduled and shared between the production of different geometry extrusions. The piece of the machine that contains the orifice is called the extrusion **die**, and it is usually a replaceable item so that different dies can be used on the same machine. Once pushed outside the die orifice, the raw material solidifies, creating the desired final shape. This process is shown schematically in Figure 9.27. For hand-sized cross sections, extrusion can easily create feature geometries that are accurate to within 0.20 mm, whereas accuracy to within 0.02 mm is considered difficult.

Extrusion can be used for either metals or plastics. Extrusion also can be used to create many custom shapes, as shown in Figure 9.28.

9.04.03 Drawing

Two different processes are referred to as drawing. One is more accurately called **wire drawing**. Wires and small round bars can be reduced in diameter by pulling slightly larger stock diameters through one or more nozzles with gradually reducing orifices. This process is shown schematically in Figure 9.29.

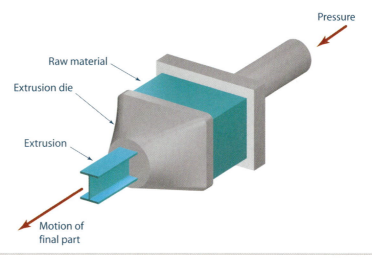

FIGURE 9.27. In the extrusion process, elevated temperature and pressure are applied to the raw material, which liquefies and is forced through a shaped orifice. The raw material hardens once it leaves the orifice.

FIGURE 9.28. By changing the shape of the orifice in the extrusion die, custom geometries such as these heat sinks can be created.

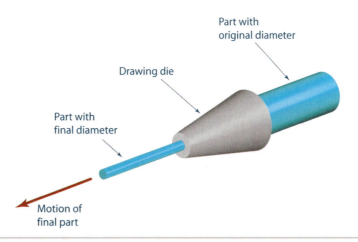

FIGURE 9.29. The diameters of a wire and small rod can be reduced to the desired size by pulling them through a nozzle with a reducing orifice. This process is called wire drawing.

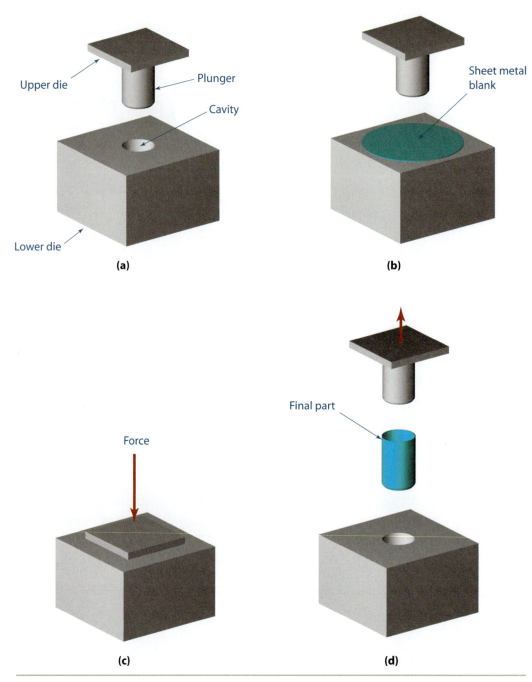

FIGURE 9.30. Thin shells can be created using a shaped cavity with a closely shaped plunger (a) in a process called deep drawing. Sheet metal is placed across the cavity (b). The plunger is inserted into the cavity, stretching the sheet metal (c). The plunger is removed, and the part is ejected (d).

This process produces circular cross-section shapes only, but it can do so with diameter accuracies within 0.10 mm. Diameter accuracy better than 0.010 mm is considered challenging.

In the process known as **deep drawing**, thin-shelled parts are created from sheets of thin, malleable metal, also known as sheet metal. In this process, a plunger shaped like the desired shell is pressed into a piece of sheet metal. The tool stretches and deforms the sheet into a shaped cavity that fits closely with the plunger. The resulting part closely

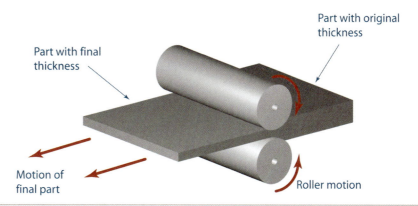

FIGURE 9.31. Plates and bars can be reduced to their desired thickness by squeezing the material through a set of large rollers. This process is called rolling.

resembles the tool and cavity. This process is shown schematically in Figure 9.30. Most aluminum beverage cans are made with this process, as are many automobile parts such as engine oil pans and valve covers. For hand-sized parts, drawing can easily create feature geometries that are accurate to within 0.25 mm, whereas accuracy to within 0.025 mm is considered difficult.

9.04.04 Rolling

Many stock commercial shapes are reduced to their final shapes by a process called **rolling**. In this process, malleable raw material is reduced in thickness by squeezing it through large rollers, as shown schematically in Figure 9.31, to produce plates and rectangular bars. The process is similar to rolling and flattening pasta dough on a pasta-making machine. Rolling can be done while the material is in a red hot, soft state (hot rolling) or while the material is at or near room temperature (cold rolling). Cold rolling requires larger rollers and more force than hot rolling, but the resulting products generally have better size accuracy, surface finish, and strength than that achieved with hot rolling. For hand-sized cross sections, cold rolling can easily create feature geometries that are accurate to within 0.10 mm, whereas accuracy to within 0.01 mm is considered difficult. Cold rolling accuracies are typically three to five times better than those achievable by hot rolling.

Round bars can be created by rolling material between two flat anvils that move in opposite parallel directions with a gradually decreasing gap between them, as shown in Figure 9.32, similar to rolling cookie dough on a table. As with plates, round bars can be hot-rolled or cold-rolled.

9.04.05 Die Casting and Molding

In **die casting**, the mold is made of steel with surfaces that have been hardened to improve its durability and wear resistance. Die casting is usually reserved for certain grades of aluminum, zinc, and magnesium. Steels cannot be die-cast because the temperatures necessary to create molten steel for the product would soften or melt the mold. **Injection molding** is very similar to die casting except that the part material is plastic instead of metal and the molten material is injected into the mold under pressure. The mold cavity is not formed from a master pattern as with sand casting; instead, it is cut to shape using the low-volume production processes described earlier. The mold includes a set of ports through which molten material can be injected into the cavity and a set of vents from which air can escape as the molten material is

FIGURE 9.32. A round bar (a) can be rolled between two dies (b), or anvils with a reducing gap, to produce a product with its diameter reduced to a final size.

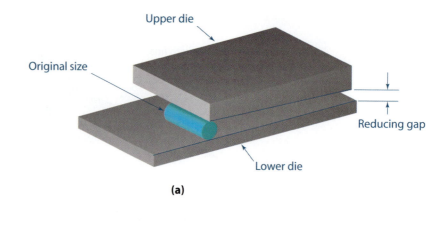

(a)

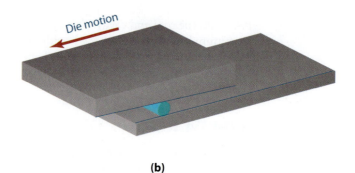

(b)

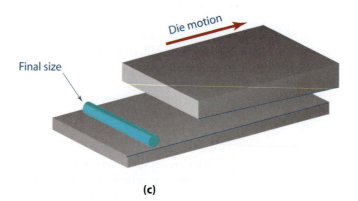

(c)

injected. Leftover bits of material at either set of ports sometimes remain with the part and are known as **sprue**. If the mold is made such that it must be opened to remove the part, leftover material may appear on the part at the split-line location. This bit of material is known as **flash**. Ejector pins inside the mold help to dislodge the cooled part. A die casting mold and a part produced from it are shown in Figure 9.33.

FIGURE 9.33. In die casting and molding, a steel mold must be created using low-volume processes (a). An actual mold for a plastic case is shown in (b), and the part produced from it is shown in (c) and (d).

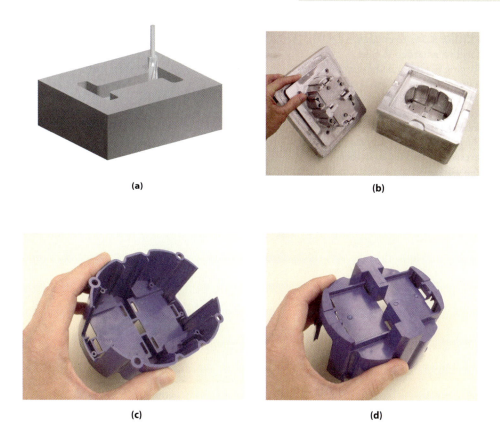

(a)

(b)

(c)

(d)

For hand-sized parts, die casting and molding can easily create feature geometries that are accurate to within 0.50 mm, whereas accuracy to within 0.05 mm is considered difficult.

9.04.06 Forging

The **forging** process deforms a workpiece by pressing it with a great deal of force between two preshaped molds, or forging dies, as shown schematically in Figure 9.34. This operation can be done with the workpiece at or near room temperature or at elevated temperatures to reduce the amount of force required and to improve the flow of material in the die. Often the workpiece is simply sawn or cut from a standard commercial shape but is otherwise featureless. The amount of feature detail that can be created by forging is not nearly as great as that achieved by casting. Since the raw material starts as a hard solid in a forging operation, the amount of material deformation cannot be large; otherwise, the workpiece would break. Therefore, the beginning workpiece must have the same rough shape and the same overall volume as the finished forged part. Forging is often used as a rough-shaping process to reduce the amount of machining required to produce a final product.

For hand-sized parts, forging can easily create feature geometries that are accurate to within 0.50 mm, whereas accuracy to within 0.05 mm is considered difficult.

FIGURE 9.34. To create a forging, a hardened steel forging die (a) must be created using prototyping processes. The die halves are aligned (b), and a featureless workpiece is inserted into the die (c). The die is closed (d), and great pressure is applied to deform the workpiece to conform to the shape of the cavity. The die is opened (e), and the finished part is ejected (f).

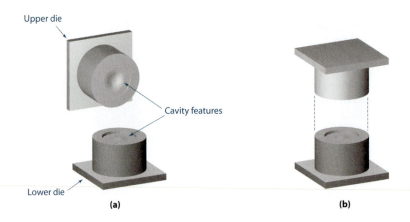

(a) **(b)**

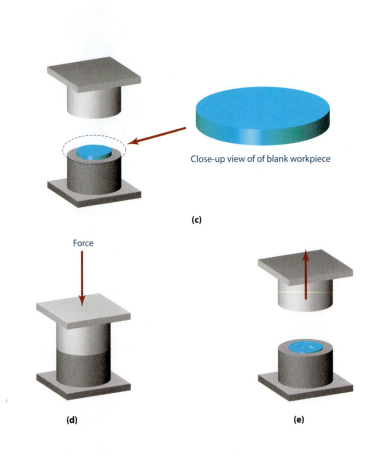

(c)

(d) **(e)**

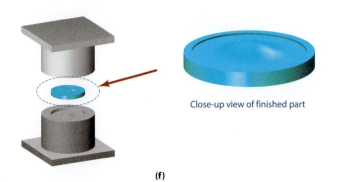

(f)

FIGURE 9.35. To create a stamping, a stamping die (a) must be created from hardened steel using prototyping processes. The die is aligned (b), and the sheet metal workpiece is inserted into the die (c). The die is closed (d), shearing and deforming the workpiece to create the final part. The die is opened, and the part is removed (e).

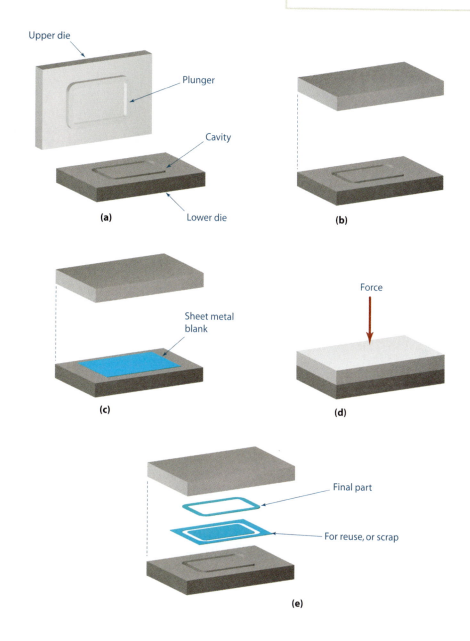

9.04.07 Stamping

Stamping is an operation that is used on sheet metals. As with deep drawing, stamping deforms the metal between a cavity and a closely fitting plunger. The edge of the plunger fits closely to the edge of the cavity, and the metal is cut with a shearing action. This tool is called a stamping die, and it is like a large version of the paper-punching tool used for cutting shaped holes in paper. This process is shown schematically in Figure 9.35. In addition to cutting the metal, stamping also can simultaneously deform the sheet metal to produce a variety of different-shaped features by using a plunger that gradually fits into the cavity. This deformation is similar to deep drawing except that the deformation is not nearly as severe; thus, the thickness of the sheet metal is not changed much. Successive stamping operations—with each operation creating a little more of the final desired part geometry as the raw material is fed sequentially into a series of dies—can create impressive 3-D features. Examples of successively stamped parts are shown in Figure 9.36.

For hand-sized parts, stamping can easily create feature geometries that are accurate to within 0.25 mm, whereas accuracy to within 0.025 mm is considered difficult.

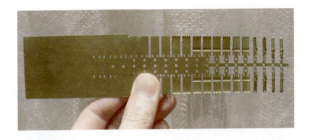

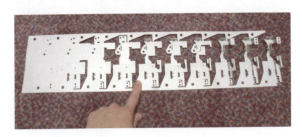

FIGURE 9.36. Examples of successively stamped parts. Each operation, moving from left to right on the sample strips, makes the part more complete. The final part geometry is on the right end of the strip.

9.04.08 Sintering

The raw material for the **sintering** process is metal in powdered form. For this reason, sintering is sometime known as powder metallurgy. Metal powder is placed in a mold; and a great deal of pressure and high temperature are applied, fusing the metal particles together to form a solid part. This process is shown schematically in Figure 9.37. Sintering can produce parts with the same detail as cast parts. However, sintering is usually restricted to small (i.e., hand-sized) parts due to the great amount of pressure required for this process. Sintered parts are porous and have slightly lower density and strength than solid metal parts. The porosity of sintered parts is sometimes an advantage because the parts can be impregnated with oil and used as a bearing surface to reduce contact friction between other parts.

For hand-sized parts, sintering can easily create feature geometries that are accurate to within 0.25 mm, whereas accuracy to within 0.025 mm is considered difficult.

SUMMARY OF COMMON HIGH-VOLUME PRODUCTION PROCESSES		
PROCESS	**ADVANTAGES**	**DISADVANTAGES**
SAND CASTING	PRODUCES 3-D GEOMETRIES; MOLDS ARE INEXPENSIVE	SLOW DUE TO MANUAL HANDLING; LOW TO MODERATE ACCURACY AND SURFACE FINISH; PARTS ARE POROUS
INVESTMENT CASTING	PRODUCES 3-D GEOMETRIES WITH GOOD DETAIL AND ACCURACY	EVEN SLOWER THAN SAND CASTING DUE TO MANUAL HANDLING; MOLDS CANNOT BE REUSED
EXTRUSION	PRODUCES COMPLEX SHAPES OF CONSTANT CROSS-SECTION GEOMETRY	EXPENSIVE TOOLING; CANNOT BE USED TO PRODUCE 3-D SHAPES
DRAWING	PRODUCES VERY ACCURATE GEOMETRY; USES SOLID RAW MATERIAL	GEOMETRIES LIMITED TO CIRCULAR OR CIRCULAR TUBES
ROLLING	PRODUCES VERY ACCURATE GEOMETRY; USES SOLID RAW MATERIAL; PROCESS USUALLY STRENGTHENS THE MATERIAL	GEOMETRIES LIMITED TO AXIALLY SYMMETRIC SHAPES
DIE CASTING AND MOLDING	PRODUCES COMPLEX 3-D GEOMETRIES	MOLDS ARE EXPENSIVE; CANNOT BE USED FOR STEELS; POROSITY IS SOMETIMES A PROBLEM
FORGING	USES SOLID RAW MATERIAL; PROCESS USUALLY STRENGTHENS THE MATERIAL	CAN PRODUCE SHALLOW 3-D DETAIL ONLY
STAMPING	FAST AND CAN PRODUCE COMPLEX 3-D SHAPES; USES SOLID RAW MATERIAL	CAN BE USED ON SHEET METAL ONLY
SINTERING	PRODUCES 2-D SHAPES ACCURATELY	3-D SHAPES LIMITED TO SHALLOW FEATURES; FINAL PART USUALLY HAS POROSITY

FIGURE 9.37. To create a sintered part, a hardened steel die (a) must be created using prototyping processes. Powdered metal is placed in the cavity (b). The die is closed (c). The die is closed (d), and elevated temperature and pressure are applied. The die is opened (e), and the finished part is ejected.

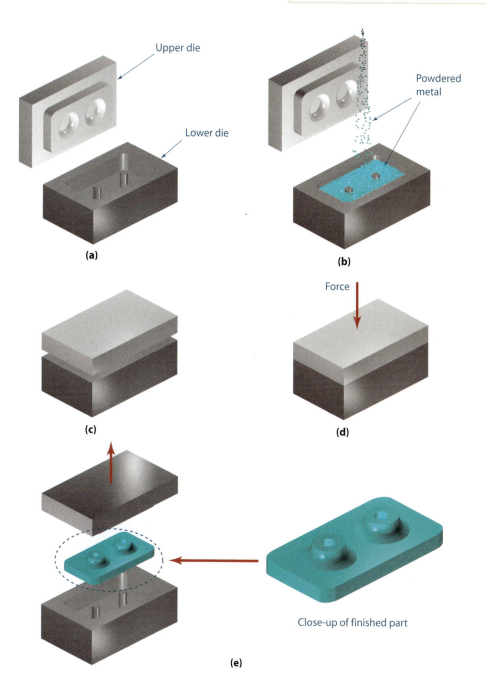

9.05 Burr Removal

Part features that have been produced by cutting processes are prone to have sharp edges and small bits of extraneous raw material attached to them. These bits of material are called burrs; and unless they are removed, they later become trapped between other parts, impairing the mating and function between the parts. Also, parts with sharp edges can cut its handlers, as well as damage surrounding parts. Unless there is a specific need to retain a sharpened edge, it should be removed for safety. For low-volume production, a skilled worker using a file and other handheld tools can remove burrs and edges. For medium- and high-level production volumes, burrs and edges can be removed through a process called **tumbling**. In this process, a part is

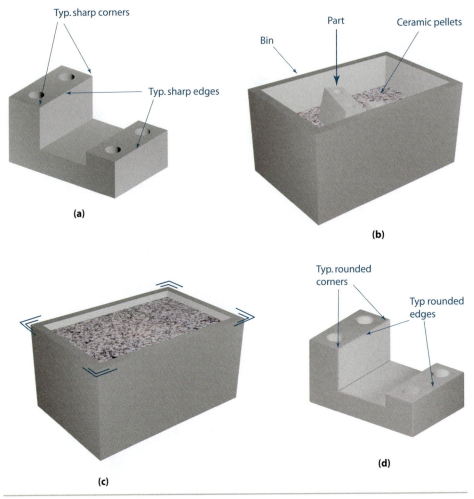

FIGURE 9.38. Tumbling to remove burrs and edges. The part in (a) may have sharp edges and burrs from machining operations. The part is placed in a bin filled with abrasive ceramic pellets (b) and is completely covered and shaken (c), causing the pellets to rub the part. The action removes the burrs and sharp external edges from the part (d).

surrounded in a bin filled with ceramic pellets. The bin is shaken or rotated, causing the part and pellets to rub against one another. Because the ceramic in the pellets is harder than the material of the part, this action abrasively removes part material most drastically at its sharpest external points. The longer the part is tumbled or the more abrasive the pellets, the more part material removed. The material from relatively flatter surfaces is removed very slowly, so they are left relatively unaltered while sharp edges and burrs are removed. Internal surfaces and edges of the part that are not in contact with the pellets, such as the inside of small holes and screw threads inside small holes, are unaffected by the process. Parts with smaller feature details require smaller pellets. The process of tumbling is shown schematically in Figure 9.38.

9.06 Combined Processes

One way to learn about using specific fabrication processes is to imagine how you would create a part using the processes. It is quite common to use a combination of processes, each one being best suited for creating specific types of geometries. Return to the part presented at the beginning of this chapter, shown again in Figure 9.39, and assume the part is about the size of your hand and is to be made of aluminum. With the wide array of fabrication methods presented so far, the part can be made many

FIGURE 9.39. This part needs to be fabricated. What processes will be needed, and in what order should they be performed?

FIGURE 9.40. A possible fabrication sequence for the block using low-volume production processes. A slightly oversized workpiece is sawn from a length of stock commercial shape in (a). The external shape of the block is roughly sawn in (b). The external surfaces are milled to their final sizes and finishes in (c).

ways. However, considering the number of parts that are to be produced, some fabrication methods will be better than others.

First, consider that only one of these parts is needed. Investing in the tooling that would be required for using high-volume processes makes no economic sense in this case. Using the low-volume processes that have been described, you would get the part much quicker with less cost.

One possible sequence of operations to create the part is shown in Figure 9.40. First, the initial workpiece is made by sawing it off the end of a stock commercial shape that is slightly larger in overall size than the final part. The basic L shape of the final part is then roughly sawn out. If good size accuracy of the final part is required (say within 0.10 mm), the next step would be a milling operation to refine the surfaces of the L shape. Next (since the workpiece is already in a milling machine), another milling operation is used to create the slot on top. Finally, a drilling operation, which can be done on the milling machine or a drill press, is used to create the hole.

Other sequences or processes are possible but not as desirable. For example, although milling alone could produce the basic L shape, an initial sawing operation removes most of the undesired material much quicker. Milling is then used for the

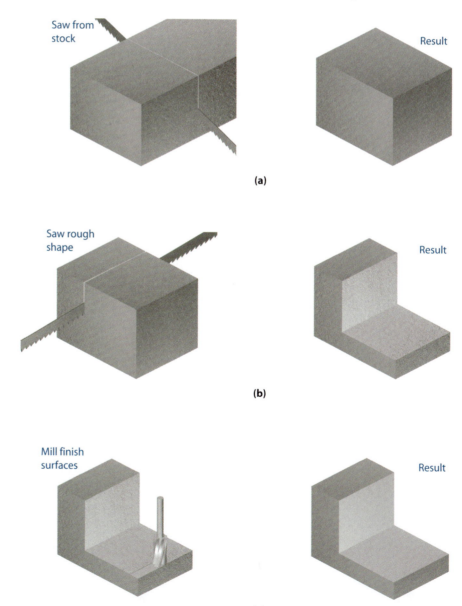

Saw from stock

Result

(a)

Saw rough shape

Result

(b)

Mill finish surfaces

Result

(c)

FIGURE 9.40. (CONTINUED)
The slot is milled in (d), and the
hole is drilled in (e).

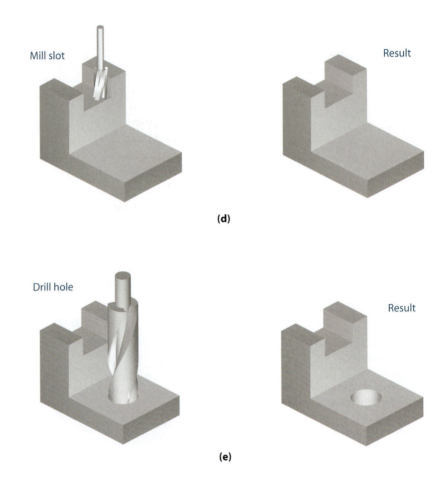

Mill slot

Result

(d)

Drill hole

Result

(e)

final finish work. The basic L shape, slot, and hole also could have been produced by EDM operations. However, EDM removes material at a much slower rate than either milling or drilling and the simple geometries of the features did not require the intricate path-cutting capability of EDM.

If you decide that a large number of parts need to be made, you can consider high-volume processes. The processes used depend on what material the part is made of and how accurately the feature sizes need to be controlled. Most metals can be cast. Die casting can create more accurate sizes and better surface finishes but cannot be used on steels. Small plastic parts can be molded. Small metal parts can be sintered or forged. Sheet metals can be stamped and drawn. Both metals and plastics can be extruded. When the feature sizes from these raw operations are not sufficiently accurate for the part to function, postprocesses such as additional machining or grinding can be added.

Presume the part in Figure 9.40 is now to be made in larger numbers. Assume about a thousand of them are needed. With a required size accuracy of 0.1 mm, die casting would certainly do the job; but even with a thousand pieces, the tooling cost for die casting may not justify its use. The required accuracy would be near the limit of what can be done with sand casting, which is much less expensive but possible. If features such as the hole or the slot need to be made with higher accuracy, they can be finished with milling or drilling operations. This process is shown in Figure 9.41. Investment casting would be more expensive than sand casting, but the part would not require any postmachining. The suitability of the material also needs to be checked. Not all alloys of aluminum or any metal can be cast successfully. Many alloys lose much of their strength after they have been melted, poured, and rehardened.

FIGURE 9.41. A fabrication sequence using medium-volume production processes. A master form is created in (a). The form is used to create the cavity in a sand mold in (b). The mold is closed, molten metal is poured, and the hardened casting is removed in (c). The slot is finished by milling in (d) if necessary. The hole is drilled in (e) to complete the part.

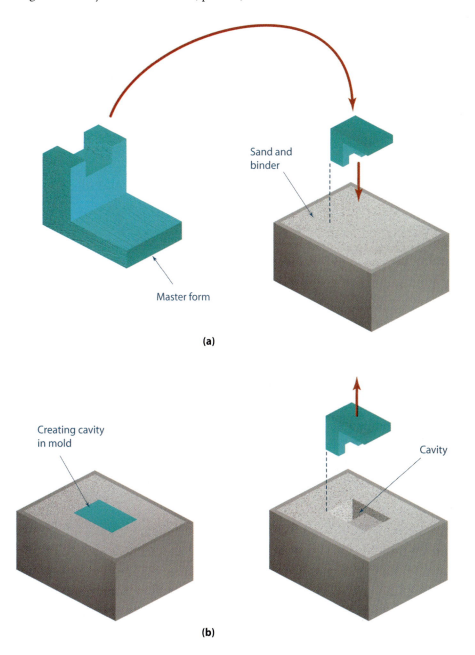

Sand and binder

Master form

(a)

Creating cavity in mold

Cavity

(b)

FIGURE 9.41. (CONTINUED)
The mold is closed, molten metal is poured, and the hardened casting is removed in (c). The slot is finished by milling in (d) if necessary. The hole is drilled in (e) to complete the part.

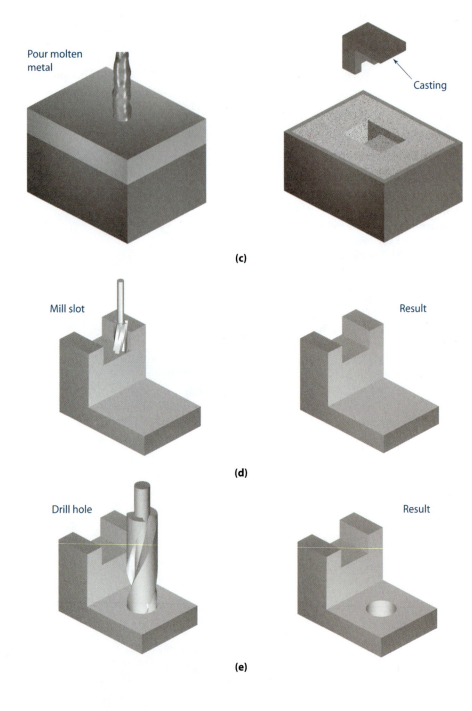

Pour molten metal

Casting

(c)

Mill slot

Result

(d)

Drill hole

Result

(e)

When the planned production volume of the part exceeds 10,000, tooling investment required for die casting and extrusion starts to become economically justifiable. Figure 9.42 shows a possible fabrication sequence that uses an extrusion to create the basic L shape. To create a custom extrusion, the extrusion die must be fabricated and fitted to an extrusion machine, which can create the temperatures and pressure needed to liquefy the raw material and press it through the extrusion die. Sections of the L shape are sawn off to create a workpiece. The slot is then milled, and the hole is drilled.

FIGURE 9.42. A fabrication sequence using a high-volume process. The basic L shape is created by extrusion in (a). The length is then sawn to the required size in (b). The slot is created by milling in (c). Broaching the slot is an alternative. The hole is created by drilling in (d).

Extrude basic shape

Result

Raw material

Extrusion die

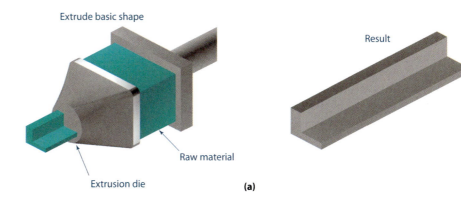

(a)

Saw to needed size

Result

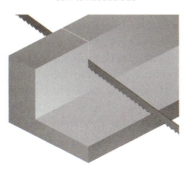

(b)

Mill slot

Result

(c)

Drill hole

Result

(d)

FIGURE 9.43. Using low-volume fabrication methods, what sequence of processes would be needed to create this part?

9.06.01 Example 1: The Retainer

A part called a retainer is shown in Figure 9.43. This part is about hand-sized and is to be made of steel. For this part to fit and function properly, its exterior size must be accurate to within 0.10 mm, but the size of the hole and slots must be accurate to within 0.025 mm. In addition, the size of the outer diameter of the long cylindrical surface must be accurate to within 0.005 mm. To create this part using low-volume fabrication processes, what sequence of operations should be used?

A possible sequence of operations is shown in Figure 9.44. Sawing off a length of standard commercial-sized steel tubing creates a workpiece. The length of tubing needed would be slightly larger than the final length of the part. The outer diameter of the tubing must be slightly larger than the maximum outer diameter of the part, and the inside diameter of the tubing must be smaller than the minimum inner diameter of the part. If tubing of an appropriate size and the proper material cannot be found, an alternative would be to start with a solid, round commercial size. Starting with a solid, round stock would not be as practical because more material would need to be removed from the center of the part, wasting time and effort as well as material. The workpiece is placed in a lathe, where material is removed by boring to form the inside diameter of the part. While the workpiece is in the lathe, the outer diameter of the lip and the outer diameter of the long cylindrical surface are cut to size by turning. Since turning cannot produce the 0.005 mm accuracy needed for the long cylindrical surface, the turning operation leaves the diameter slightly oversized so that a grinding

FIGURE 9.44. Fabrication of the retainer using low-volume production processes. A workpiece is sawn from circular stock in (a). The inside bore and chamfer are created by turning in (b). The outside geometry and chamfers are created by turning in (c). The slot is milled in (d).

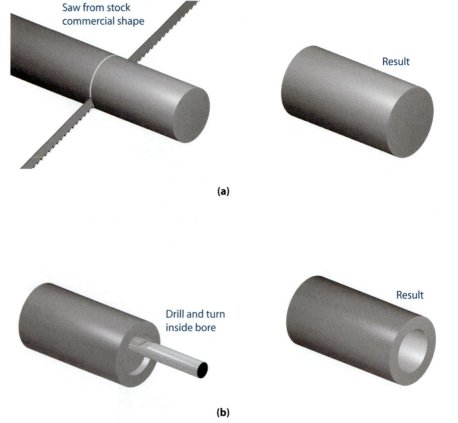

FIGURE 9.44. (CONTINUED)
The outside geometry and
chamfers are created by turning
in (c). The slot is milled in (d).

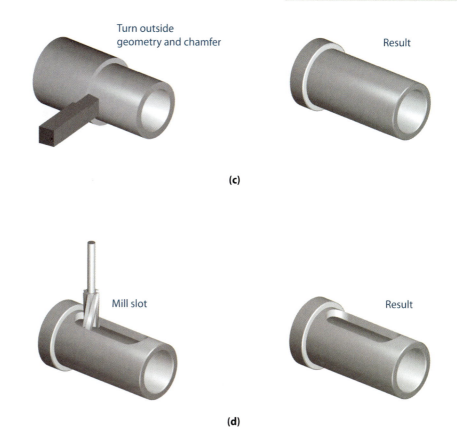

(c)

(d)

operation can be used later to create the final size. The part is cut to its desired axial length on the lathe. The workpiece is next placed in a milling machine to produce the rounded slot. Finally, a rotary grinding operation reduces the long cylindrical surface to its desired size and accuracy.

9.06.02 Example 2: The Coupling

A part called a coupling is shown in Figure 9.45. This part is about hand-sized and is to be made of steel. For this part to fit and function properly, its exterior size must be accurate to within 0.10 mm, but the size of the hole and slots must be accurate to within 0.02 mm. To create this part using medium- to high-volume fabrication processes, what sequence of operations should be used?

Investment casting can be used to create steel parts and would produce the accuracies needed for most of the part, including the cylindrical body and the axial protrusions. However, the accuracies needed for the hole and slots require a secondary machining operation after the main part is cast. Before the part can be cast, wax masters must be fabricated. The masters need to reproduce only the geometry of the final part (less the hole and slots), not its material properties. Since each wax master will ultimately be destroyed for each part made in the investment casting process, making the

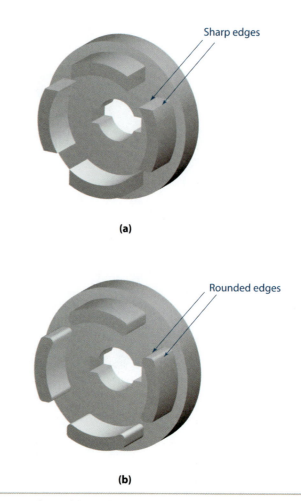

Sharp edges

(a)

Rounded edges

(b)

FIGURE 9.45. The original part (a) is redesigned with rounded edges (b), so that wax master forms can be molded more easily in preparation for making investment castings.

masters using low-volume fabrication methods would be too slow. Instead, the masters can be molded. The part in Figure 9.45 has been slightly redesigned to eliminate the sharp outside corners so that a molding die can be fabricated more easily for the wax masters. The molding die for the masters can be made of aluminum instead of steel because wax is a very soft material that will not cause much wear in the die. Although aluminum is more expensive than steel per unit volume, it also is softer and easier to machine than steel; thus, an aluminum die is less expensive to make than a steel die. A possible sequence of operations for creating the masters is shown in Figure 9.46.

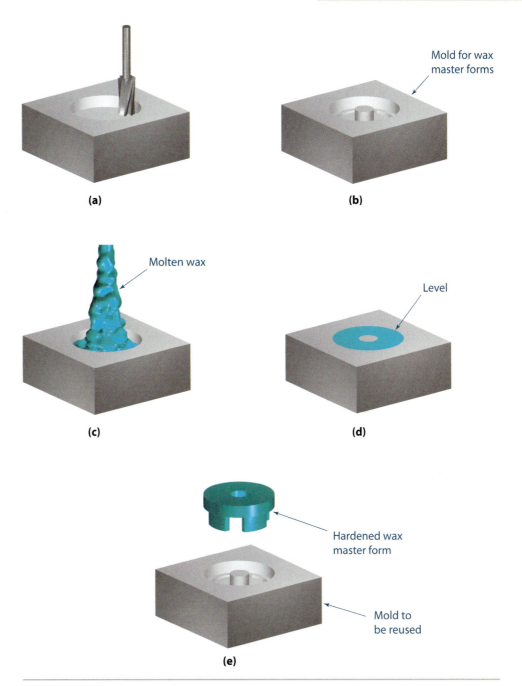

(a)

Mold for wax
master forms

(b)

Molten wax

(c)

Level

(d)

Hardened wax
master form

Mold to
be reused

(e)

FIGURE 9.46. Steps for creating wax master forms for the coupling. Aluminum (or steel) is milled (a) to create a mold (b). Wax is poured (c) and allowed to level and harden (d). The wax form is then removed (e), and the mold can be used to make additional wax forms as needed.

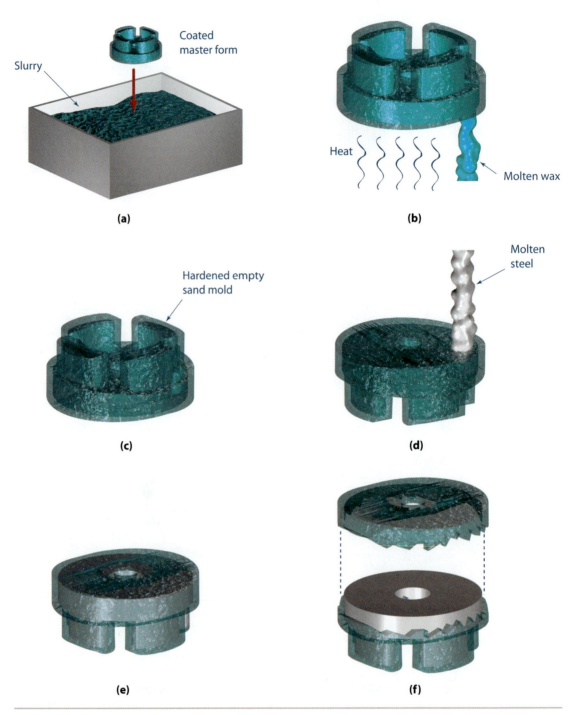

Slurry

Coated
master form

(a)

Heat

Molten wax

(b)

Hardened empty
sand mold

(c)

Molten
steel

(d)

(e)

(f)

FIGURE 9.47. Investment casting of the coupling. The wax master is repeatedly dipped in slurry (a). The coated master is heated (b) to harden the slurry and run the wax to leave an empty sand mold (c). Molten steel is poured into the sand mold (d) and allowed to harden (e). The sand mold is broken to retrieve the part (f).

Once masters have been fabricated, they can be used to create production investment casting molds. This sequence is shown in Figure 9.47. Repeated dipping of the master into liquid slurry of fine sand with a clay or cement binder—with the layer allowed to harden between each dipping operation—creates a mold. Each progressive layer can be thicker and use coarser sand. The master is then removed from the mold, and molten steel is poured into it. When the steel hardens, it is removed from the mold as a cast part.

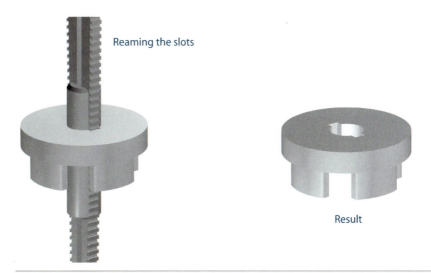

Reaming the slots

Result

FIGURE 9.48. The final hole size and the slots are created by reaming.

Drilling followed by broaching operations makes the hole and slots. This step is shown in Figure 9.48. The broaching tool can be made from hardened tool steel using an EDM process, followed by grinding if necessary.

9.06.03 Example 3: The Motor Plate

A part called a motor plate is shown in Figure 9.49. This part is about hand-sized and should be made of aluminum. Assume that in order for this part to fit and function properly, its exterior size must be accurate to within 0.30 mm, but the size of the holes and slot must be accurate to within 0.15 mm. To create this part using high-volume fabrication processes, what sequence of operations should be used?

Since the part is made of aluminum, die casting can be considered. The die casting process is capable of creating the accuracy required for the external sizes of the part and its internal features, thus eliminating the need for any secondary machining operations. A possible sequence for making the mold, which is machined out of steel, is shown in Figure 9.50. Note that for this part, the sharp outside corners on the part require sharp inside corners in the mold cavities. These corners are difficult to create by machining a single block of material. For the corners to be created, the mold must be assembled and fastened together from multiple pieces. In addition, the mold must

FIGURE 9.49. Using high-volume fabrication methods, what sequence of processes would be needed to create this part?

FIGURE 9.50. To create the sharp outside corners on the motor plate, the mold created from several pieces (a) are joined together (b) with either screws or pins.

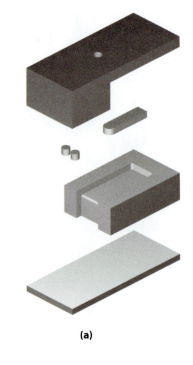

(a)

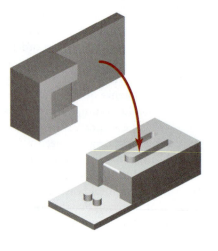

(b)

have a means of alignment, such as pins, to ensure proper positioning of its top and bottom halves, input ports from which molten metal can be injected into the cavity, and output ports from which air can escape as the molten metal is injected. After machining, the steel mold is hardened to improve its wear resistance.

After the mold has been fabricated, it can be attached to a die casting machine, which provides a source of molten aluminum. The mold halves are assembled, and the molten aluminum is injected to fill the cavity. Once the aluminum is cooled, the mold is opened and the part is ejected. This process is shown in Figure 9.51. Any flash or sprue left from the casting operation would need to be removed if it interfered with the fit, function, or handling of the part.

FIGURE 9.51. The dies for the motor plate are aligned (a) and assembled (b). Molten aluminum fills the mold (c) and is allowed to cool and harden. The mold is opened, and the part is removed (d).

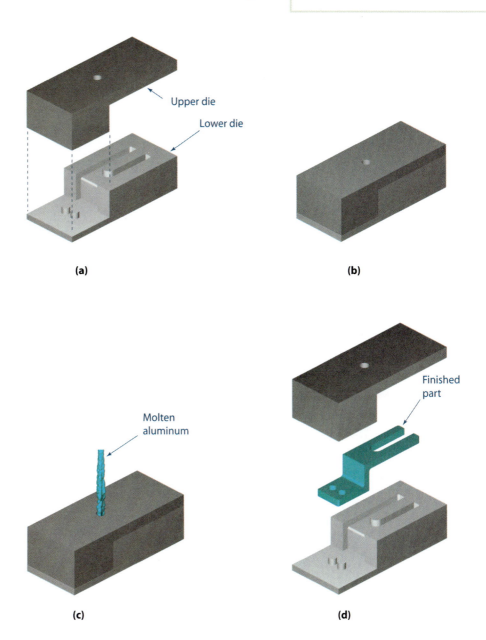

(a)

(b)

(c)

(d)

CAUTION An inexperienced engineer or designer is quickly exposed when he specifies a design that cannot be easily made without realizing that it cannot be easily made. With the vast array of special fabrication techniques, too numerous to mention here, almost any imaginable part geometry can be made. However, special fabrication techniques tend to be slow and/or expensive. Having an understanding of the basic fabrication methods described in the previous sections and avoiding geometries that are difficult to make with these methods ensures that parts can be designed to be not only functional but also simple to fabricate. The following sections describe types of geometries that engineers should avoid.

FIGURE 9.52. Long holes such as those on the original part in (a) are difficult to keep straight because drill bits bend more as they cut deeper. When possible, holes should be kept short, as on the redesigned part in (b).

Long, Skinny Holes

A hole can be considered "long and skinny" when the aspect ratio of its depth to its diameter is longer than about ten to one. A long, skinny hole would need to be made by a long, skinny drill bit. Such a drill bit has a tendency to wobble and bend under the forces generated by the cutting process, creating holes that are inaccurate in size and location. The wobble and bend also increase the chances that the drill bit will break under the cutting forces, particularly with tough materials such as stainless steels. On cast and molded parts, long, skinny holes require long, skinny pins in the mold. Such pins also have a tendency to bend and break under the forces applied during the casting and mold removal processes. Whenever possible, designs that include long, skinny holes should be modified so they are not used, as shown in Figure 9.52.

Long, Skinny Protrusions

Long, skinny protrusions, such as those shown in Figure 9.53(a), also should be avoided if possible. For a machined feature, the cutting forces from the tool would cause such a protrusion to bend as it was being created. The resulting feature would thereby be deformed. Such features can be created with EDM, which inherently applies low forces to the part; but this process is slow. On cast and molded parts, long, skinny protrusions require long, skinny cavities in the mold. Such cavities are difficult to fill during the casting process. Also, once a cavity is created as a cast feature, it is easy to

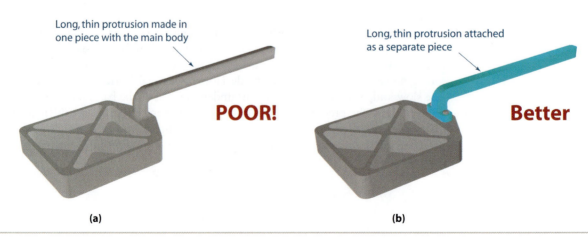

FIGURE 9.53. Parts with long, thin protrusions are more easily made as two separate pieces rather than as one continuous piece.

deform and will break under the forces applied during the casting and mold removal process. A possible solution when a long, skinny protrusion is necessary is to create it as a separate part, as shown in Figure 9.53(b).

Sharp Inside Corners Made by Cutting Processes

An inside corner is created when three concave edges meet at a point, as shown on the part in Figure 9.54(a). Such a corner would be very difficult to produce using the common cutting methods discussed earlier. On a milling machine, for example, the curvature inherent on an end mill would make it impossible to create a sharp inside corner. However, the rounded corner shown in Figure 9.54(b) can be created using an end mill, and the blunted corner shown in Figure 9.54(c) can be created using a ball end mill. However, the sharper the corner, the smaller the milling tool and the slower the material removal rate.

Whenever possible, inside corners that must be made by cutting operations should be rounded instead of sharp, as shown in the illustration.

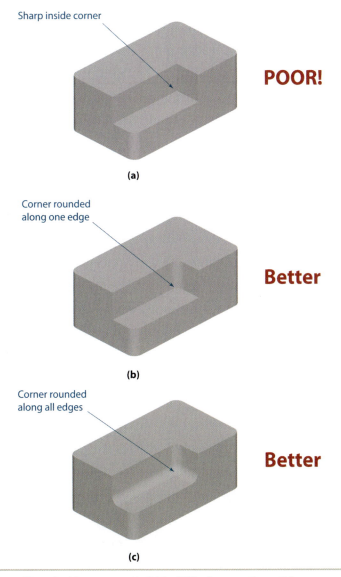

FIGURE 9.54. A sharp inside corner as in (a) is difficult to produce with common cutting processes. By rounding the corners as in (b) or (c), the feature is much easier to create.

Sharp Outside Corners with Molding and Some Casting Processes

An outside corner is created when three convex edges meet at a point, as shown on the part in Figure 9.55. Such a corner would be simple to produce using the common cutting methods discussed earlier. However, on molded parts, the existence of a sharp outside corner requires the existence of a sharp inside corner in the mold. For sand and investment casting, because packing aggregate and binder around a master pattern creates these molds, there is little or no effect on the cost of the mold. However, molds for die cast and molded parts are usually made by cutting processes, which means difficulty in cutting inside corners.

On such a mold, however, a bit more tolerance for any extra machining time usually is needed to create the mold because very few molds are made when compared to the number of parts made from the molds. A little extra time and effort invested in creating inside corners, if they are indeed necessary, would noticeably increase the cost of the mold but only slightly increase the overall cost of the high-volume part.

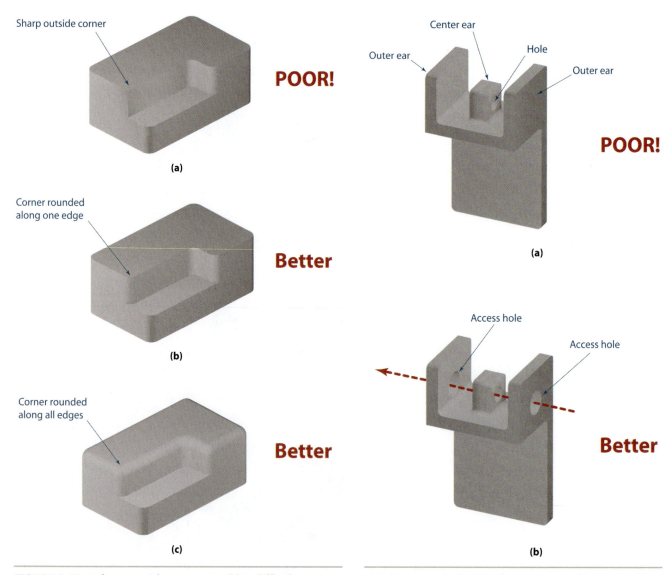

FIGURE 9.55. A sharp outside corner as in (a) is difficult to produce by processes such as die casting and molding. By rounding the corner as in (b) or (c), the feature is much easier to create.

FIGURE 9.56. The hole in the center ear of the part in (a) would be difficult to create because the outer ears interfere with tool access. One possible solution would be to create access holes in the outer ears, as in (b).

Features in Hard-to-Reach Locations

When features are created by cutting processes, consideration must be given to the access the cutting tools have to the features. Features that are obscured are difficult to make. As an example, consider the proposed part shown in Figure 9.56(a). A hole on the center ear is located at the center of the part. A hole is usually produced by drilling in a machining operation (or by a pin in a molding or casting operation). This feature seems simple enough, but how can it be produced on this particular part? Drilling cannot be used without putting access holes in the sidewalls of the part. The ears must be relocated so the holes can be created without interference occurring between the drill (or pin) and the rest of the part; or the part must be redesigned to allow access to the ears by a drill (or pin), as shown in Figure 9.56(b).

Undercuts on Cast or Molded Parts

An **undercut feature** is a concave feature in which the removed material expands outward anywhere along the depth of the feature. An example of an undercut feature is the material removed from the inside of a bore as the bore becomes deeper, as with the part in Figure 9.57. In this figure, the part is hypothetically split to show the feature. Such a feature, commonly called a relief, is sometimes used to reduce friction when a shaft is placed in the bore or to reduce the weight of the part. However, an undercut bore cannot be made by sand casting, die casting, or molding because the mold would be extremely difficult to separate from the master form or the actual part unless it was a highly flexible material, such as rubber or rubber foam. Molten material could flow into the mold; however, the hardened part would be locked around the portion of the mold used to create the part, as demonstrated in the cut-away part and mold shown in Figure 9.58. Undercut features can be made by investment casting using the lost wax method; however, the mold must be broken to remove the part, thus making this process expensive.

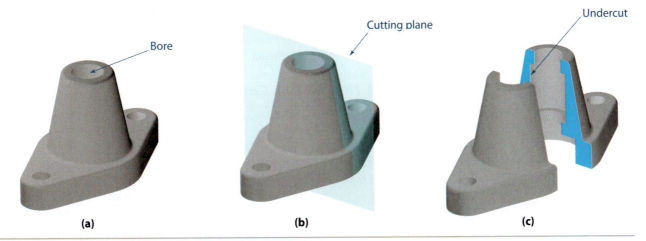

(a) (b) (c)

FIGURE 9.57. The original part (a) is hypothetically split with a cutting plane (b) to reveal the interior of the part and its undercut bore (c).

FIGURE 9.58. The part with the undercut bore (a) would be locked inside the mold. The part with the straight bore (b) can be removed from the mold without breaking it.

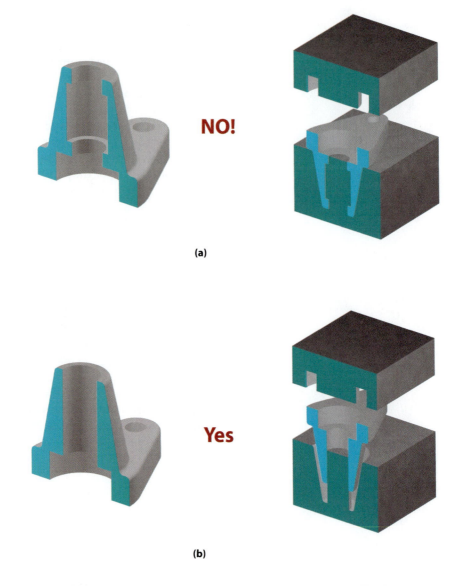

Undercut features on a cast or molded part should be avoided if possible. When necessary, these features can be made only by secondary machining processes after the casting or molding operation. In Figure 9.58, the undercut bore is replaced by a straight bore, which allows the part to be removed without destroying the mold. If the undercut is necessary, it can be created with a milling or turning operation.

A screw thread is another common example of an undercut feature. In cast or molded parts, screw threads are usually cut as a secondary machining operation.

Mixed Wall Thickness on Cast or Molded Parts

In cast and molded parts, a good design practice is to keep major wall thicknesses about the same throughout the part to reduce distortion of the part during its cooling. If the walls on a part are of vastly different thicknesses, as shown in Figure 9.59, the thicker portions would cool and harden more slowly than the thinner portions. Because almost all engineering materials shrink with decreasing temperature, thicker portions would shrink more than thinner portions, leading to distortion of the final part. Moreover, thicker wall castings take longer to cool than thinner wall castings and therefore cannot be removed from the mold as quickly, which reduces the production rate.

9.07 Considerations for 3-D Modeling

The growth in popularity and sophistication of 3-D modeling has had a tremendous effect on many fabrication methods—for both low-volume and high-volume processes. In the early days of solid modeling, the software was used as a design aid for enhanced visualization, calculation of physical properties, and generation of formal engineering drawings. The drawings would be given to fabricators, who could work directly from the printed versions to create the desired part features on manually controlled machines. Alternatively, an operator could use the drawings to enter data on a computer program to create the parts on numerically controlled machines. As 3-D modeling became more popular, software was created for taking a solid model file and generating from it the program necessary to drive the numerically controlled machines. The need for a skilled operator to specify information such as tool types and tool paths became less necessary, as this information could be derived from a solid model. Most modern fabrication shops now ask that solid model files be submitted in addition to or instead of formal engineering drawings.

Special software also was developed for creating models of the mold required to make a part for die casting and molding processes. The software not only specifies the geometry of the required mold but also aids in identifying potential problems with the design, such as large differences in part wall thickness, which can lead to distortion of the part upon cooling, and long flow paths for the molten raw material, which can lead to incomplete filling of portions of the mold.

FIGURE 9.59. A casting (a) with large differences in thickness may distort upon cooling. By making the thickness of the walls similar (b), the distortion will be minimized.

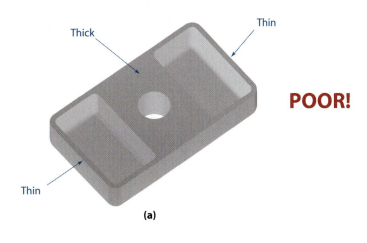

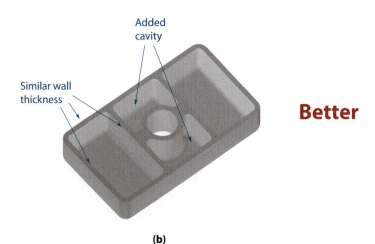

integratedproject | Fabrication of the Hoyt AeroTec Target Bow

The Hoyt AeroTec bow handle was designed with a material and fabrication method already in mind. The annual production volume of the product was anticipated to be in the thousands. This volume was high enough to consider some sort of casting process or forging process to help create at least a part of the product geometry. Earlier products of this type made by Hoyt and other manufacturers were of cast magnesium or cast aluminum. However, casting has a tendency to create voids and porosity, which can dramatically decrease the strength of the product. Structural integrity was a key consideration. The stresses created by a bow launching lightweight arrows have been known to cause some metal handle bows to crack and break. Thus, earlier products tended to be bulky due to the amount of

The workpiece starts as an 18 kg rectangular block of wrought 6061 T-6 aluminum.

The computer-generated solid model of the AeroTec bow handle. This shape is produced entirely on a numerically controlled milling machine.

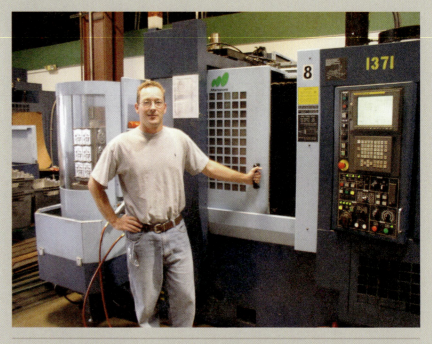

The workpiece is placed in a computer-controlled Matsuura four-axis milling machine, which reduces the workpiece into the 1.3 kg AeroTec product shape.

metal necessary to prevent failure. The dimensional accuracy desired for the AeroTec was not achievable by casting or forging alone. Those processes also produce a poor quality surface finish, which is undesirable for the high-end consumer market at which the product was targeted.

The material selected was 6061 T-6 high-strength aluminum, usually used for aircraft applications. The number 6061 specifies the alloying composition of the metal. The T-6 designation refers to a specific heat-treating process used to improve the strength of the metal. This material was lightweight, strong, not difficult to obtain, and priced such that the product could be produced within the desired cost constraints.

The production method selected was numerically controlled milling. This process, when done on a four-axis milling machine, could produce all of the features of the AeroTec, including the holes and screw threads, with the desired accuracy. Hoyt has a number of four-axis computer-controlled mills, built by Matsuura Corporation, in its production facility in Salt Lake City, Utah. These facilities are shared between several product models. The facilities could be set up to produce one particular model for several days or weeks and then reconfigured to produce a different product model.

For each part, the workpiece begins as a single block of cold-rolled, or wrought, aluminum delivered from the material supplier. Each workpiece has an initial mass of 18 kg. No formal engineering drawings are required. Skilled machine operators are not required to operate the milling machines. Hoyt engineers create a solid model

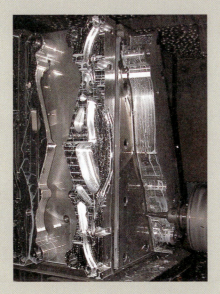 

(a)   (b)

The workpiece is held inside the Matsuura mill using special tooling that holds eight workpieces simultaneously. The mill is programmed to select the correct cutting tool for each operation.

The machined workpiece and ceramic pellets are placed in a canister and tumbled to remove burrs, sharp edges, and machining marks.

integratedproject   Fabrication of the Hoyt AeroTec Target Bow (CONTINUED)

using SolidWorks software. The engineers then convert the model into a Mastercam software program that can be used for controlling the spindle and tool path of a Matsuura four-axis mill. The program also automatically selects the proper milling, drilling, or tapping tools used to create the desired product features. An operator clamps the initial workpiece into a fixture in the numerically controlled mill, closes the door, and runs the program to begin the cutting. After the cutting is complete, the machined workpiece leaves the mill with the desired geometry: a final mass of 1.3 kg. The excess aluminum removed from the workpiece is collected and recycled.

After the machined workpiece is removed from the mill, it is sent to the tumbler to have any burrs, sharp edges, and machining marks removed. In the tumbling operation, the workpiece is placed in a canister filled with ceramic pellets and the canister is rolled, causing the pellets to rub the workpiece. This process produces a fine satin-like finish on the surface of the workpiece.

A finishing operation called anodizing is then applied to the metal. Anodizing is an electro-chemical process that adds color to the metal surface and hardens the surface, making it more durable. Finally, threaded steel inserts are pressed into the appropriate cavities in the handle and a plastic grip is added to complete the product.

Some of the ceramic pellets used in the tumbling process.

An AeroTec bow handle after the final anodized finish is applied.

9.08 Chapter Summary

A wide variety of production processes are available for creating the geometries and features contained on a part. When only a few parts (or a single part) are desired, low-volume processes can be used. These processes usually consist of various ways of removing material, such as cutting, from commonly available standard commercial shapes. The machines used for low-volume production are flexible in their configuration, allowing for a wide variety of different geometries to be produced with a minimum of setup time. However, as the required production volumes increase, different fabrication processes must be considered. High-volume processes usually involve the redistribution of raw material in molten, solid, or powdered form to produce the desired shape. High-volume processes can produce parts more economically than low-volume processes; however, the setup time and tooling expenses for high-volume production are much greater than for low-volume production. High-volume processes also tend to be less accurate than low-volume processes in producing a desired geometry.

A design is of little value if it cannot be fabricated. In the design of a part, the intended method of its fabrication must be taken into consideration when the geometry of its features is specified. For the part to be made economically, these geometries may need to change depending on whether the intended production volume of the part is high or low. These geometries also may need to change according to what production processes are available.

In revisiting the parts with the cross-shaped features presented at the beginning of the chapter, the part shown in Figure 9.02(a) has a cross-shaped protrusion atop a simple base. The squared outside corners and rounded inside edges of the cross would be considered simple to produce with an end mill on a milling machine, which is a low- to medium-volume production process. To produce the same feature geometry using a high-volume process, such as molding or die casting, would be considered difficult because the sharp outside corners on the part require sharp inside corners in the mold. Since the mold is made by machining (for example, on a milling machine), such a corner would be difficult to produce, as shown in Figure 9.60.

The axially symmetric outline of a milling tool can produce a sharp or rounded inside edge but not a sharp inside corner. Modifying the protrusion cross geometry to have rounded outside edges and squared inside corners, as shown in Figure 9.02(b), would make it easier to fabricate by die casting or molding because the mold can be made by machining. However, when the part is made by machining, the sharp inside corners cannot be produced by a milling tool.

Now revisit the part that has a cross-shaped cavity instead of a protrusion in Figure 9.03(a). Milling could easily produce the rounded inside edges of the cross. However, milling could not produce the squared outside corners. The outline cut by the curvature of the end mill would not permit it. To produce the same feature geometry using molding or die casting would be simple because the mold would have a protrusion with squared outside corners and rounded inside edges, both of which could be made with an end mill. Modifying the cavity geometry to have rounded outside edges and squared inside corners, as shown in Figure 9.02(b), would make it possible to fabricate by milling. However, such a cavity by die casting or molding would require a mold that has a protrusion with squared inside corners, which cannot be milled, as shown in Figure 9.61.

By modifying the cross geometry of the protrusion and the cavity to have rounded outside edges and rounded inside edges, as shown in Figure 9.04, the parts can be cut easily on a milling machine for low-volume production. But the molds for the part also can be cut on a milling machine in preparation for high-volume production.

FIGURE 9.60. Close inspection of the cavity reveals that the cylindrical outline of the cuts made by an end mill cannot create sharp corners at the tip of the cross.

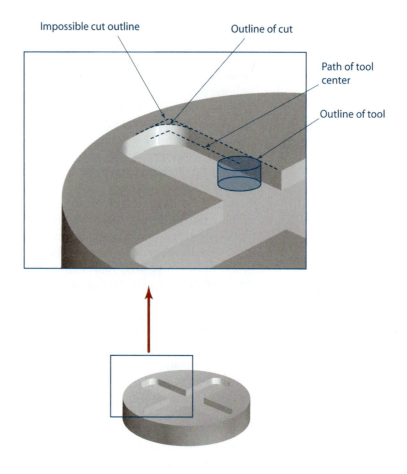

FIGURE 9.61. Close inspection of the protrusion reveals that the cylindrical outline of the cuts made by an end mill cannot create sharp corners on the inside of the cross.

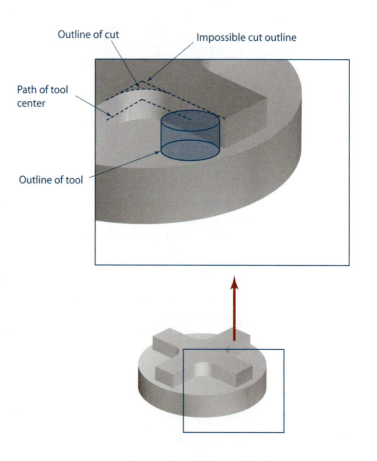

9.09 glossary of key terms

3-D printing: A process for creating solid objects from a powder material by spraying a controlled stream of a binding fluid into a bed of that powder, thus fusing the powder in the selected areas.

blind hole: A hole that does not go all the way through a part.

boring: The general process of making a hole in a part by plunging a rotating tool bit into a part, moving a rotating part into a stationary tool bit, or moving a part into a rotating tool bit.

brazing: A method for joining separate metal parts by heating the parts, flowing a molten metal (solder) between them, and allowing the unit to cool and harden.

broach: A long, shaped cutting tool that moves along the length of a part when placed against it. It is used to create uniquely shaped holes and slots.

broaching: A process of creating uniquely shaped holes and slots using a long, shaped cutting tool that moves along the length of a part in a single stroke when placed against the part.

casting: A method of creating a part by pouring or injecting molten material into the cavity of a mold, allowing it to harden, and then removing it from the mold.

deep drawing: Creating a thin-shelled part by pressing sheet metal into a deep cavity mold.

die: A special tool made specifically to deform raw or stock material into a desired outline of a part or feature in a single operation.

die casting: A method of casting where the mold is formed by cutting a cavity into steel or another hard material. *See* casting.

draft: The slight angling of the walls of a cast, forged, drawn, or stamped part to enable the part to be removed from the mold more easily.

drill bit: A long, rotating cutting tool with a sharpened tip used to make holes.

drilling: A process of making a hole by plunging a rotating tool bit into a part.

drill press: A machine that holds, spins, and plunges a rotating tool bit into a part to make holes.

EDM: Electric discharge machining; a process by which material is eroded from a part by passing an electric current between the part and an electrode (or a wire) through an electrolytic fluid.

end mill: A rotating cutting tool that, when placed in the spindle of a milling machine, can remove material in directions parallel or perpendicular to its rotation axis.

extrusion: A process for making long, solid shapes with a constant cross section by squeezing raw material under elevated temperatures and pressure through an orifice shaped with that cross section.

fixture: A mechanical device, such as a clamps or bracket, used for holding a workpiece in place while it is being modified.

flash: Bits of material that are left on a part from a casting or molding operation and found along the seams where the mold pieces separate to allow removal of the part.

forging: A process of deforming metal with a common shape at room temperature into a new but similar shape by pressing it into a mold under elevated pressure.

fused deposition: A process where parts are gradually built up by bits of molten plastic that are deposited by a heated tip at selected locations and then solidified by cooling.

grinding: A method of removing small amounts of material from a part using a rotation abrasive wheel, thus creating surfaces of very accurate planar or cylindrical geometries.

injection molding: A process for creating a plastic part by injecting molten plastic into a mold under pressure, allowing the material to solidify, and removing the part from the mold.

investment casting: A method of casting where the mold is formed by successive dipping of a master form into progressively coarser slurries, allowing each layer to harden between each dipping. *See* casting.

lathe: A machine used to make axially symmetric parts or features using a material removal process known as turning.

milling: A process of removing material from a part using a rotating tool bit that can remove material in directions parallel or perpendicular to the tool bit's rotation axis.

milling machine: A machine used to make parts through a material removal process known as milling.

mold: A supported cavity shaped like a desired part into which molten material is poured or injected.

pattern: A master part from which molds can be made for casting final parts.

rapid prototyping: Various methods for creating a part quickly by selective hardening of a powder or liquid raw material at room temperature.

reaming: A process for creating a hole with a very accurate final diameter using an accurately made cylindrical cutting tool similar to a drill bit to remove final bits of material after a smaller initial hole is created.

9.09 glossary of key terms (continued)

rolling: A process for creating long bars with flat, round, or rectangular cross sections by squeezing solid raw material between large rollers. This can be done when the material is in a hot, soft state (hot rolling) or when the material is near room temperature (cold rolling).

sand casting: A casting process where the mold is made of sand and binder material hardened around a master pattern that is subsequently removed to form the cavity. *See* casting.

sawing: A cutting process that uses a multitoothed blade that moves rapidly across and then through the part.

selective laser sintering: A process where a high-powered laser is used to selectively melt together the particles on a bed of powdered metal to form the shape of a desired part.

sintering: A process where a part is formed by placing powdered metal into a mold and then applying heat and pressure to fuse the powder into a single solid shape.

spindle: That part of a production cutting machine that spins rapidly, usually holding a cutting tool or a workpiece.

split line: The location where a mold can be disassembled for removal of a part once the molten raw material inside has solidified.

sprue: Bits of material that are left on the part from a casting or molding operation and found at the ports where the molten material is injected into the mold or at the ports where air is allowed to escape.

stage: That part of a machine that secures and slowly moves a cutting tool or workpiece in one or more directions.

stamping: A process for cutting and shaping sheet metal by shearing and bending it inside forms with closely fitting cutouts and protrusions.

standard commercial shape: A common shape for raw material as would be delivered from a material manufacturer.

stereolithography: A process for creating solid parts from a liquid resin by selectively focusing heat or ultraviolet light into a pool of the resin, causing it to harden and cure in the selected areas.

tap: A cutting tool similar to a drill bit used to create screw threads inside a hole.

tapped hole: A hole that has screw threads inside it.

three-axis mill: A milling machine whose spindle, which holds the rotating cutting tool, can be oriented along any one of three Cartesian axes.

through hole: A hole that extends all the way through a part.

tool bit: A fixed or moving replaceable cutting implement with one or more sharpened edges used to remove material from a part.

tooling: Tools and fixtures used to hold, align, create, or transport a part during its production.

tumbling: A process for removing sharp external edges and extraneous bits of material from a part by surrounding it in a pool of fine abrasive pellets and then shaking the combination.

turning: A process for making axially symmetric parts or features by rotating the part on a spindle and applying a cutting to the part.

undercut feature: A concave feature in which the removed material expands outward anywhere along its depth.

welding: A method for joining two or more separate parts by applying heat to the edges where they meet and melting the edges together along with a filler of essentially the same material composition as the parts.

wire drawing: The process of reducing the diameter of a solid wire by pulling it through a nozzle with a reducing aperture.

workpiece: A common name for a part while it is still in the fabrication process, that is, before it is a finished part.

9.10 questions for review

1. Why is it important to consider the fabrication method for a part when it is designed?
2. What processes are used to form raw materials into standard commercial shapes?
3. How is drilling similar to milling?
4. How is drilling different from milling?
5. Why are sharp inside corners difficult to produce using cutting processes?
6. Why are sharp outside corners difficult to produce using casting or molding processes?
7. What is the difference between sand casting, die casting, and investment casting?

9.10 questions for review (continued)

8. How would fixtures be different for high-volume production versus low-volume production?

9. What is the difference between a single-axis milling machine and three-axis milling machine?

10. What are some limitations to shapes that can be created using wire EDM?

11. What are some advantages and disadvantages of using rapid prototyping processes?

12. What are some disadvantages of using welding as a production process?

13. What are some disadvantages of using brazing as a production process?

14. What are burrs? Why should they be removed?

15. Why are long, skinny holes difficult to produce?

16. Why are long, skinny protrusions difficult to produce?

9.11 problems

1. For each part shown in Figure P9.1, specify practical fabrication steps that can be used to create a prototype of the part by sketching pictorial diagrams of the part after each feature is created. Start with a standard stock material geometry. Specify the best prototype fabrication process to use to create each feature. For each process, sketch the shape and orientation of the cutting tool to be used. Use arrows (with a single arrowhead for relatively slow speed and a double arrowhead for relatively fast speed) to show the motion of the tool and the workpiece during cutting processes. How would the required accuracy of feature sizes and locations affect the type processes selected?

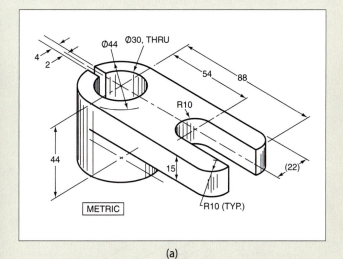

(a)

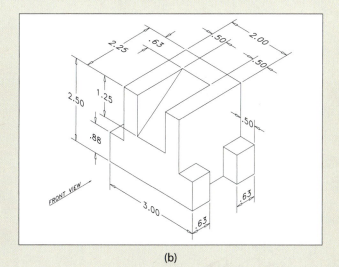

(b)

9.11 problems (continued)

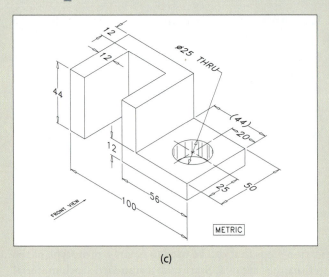

(c)

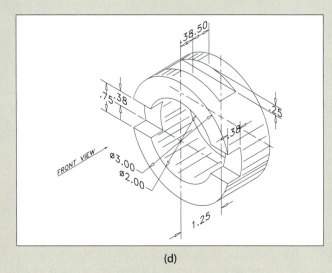

(d)

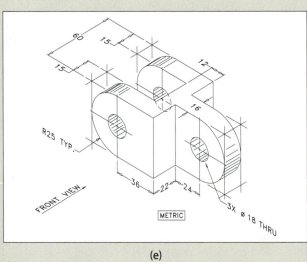

(e)

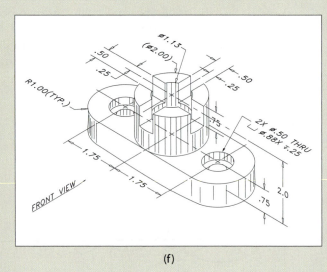

(f)

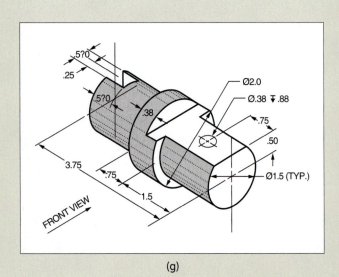

(g)

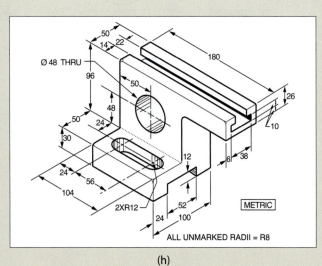

(h)

9.11 problems (continued)

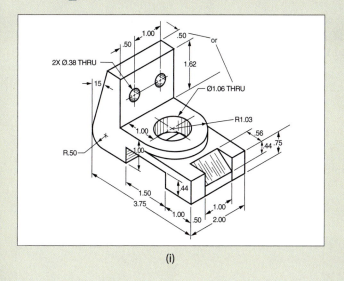

(i)

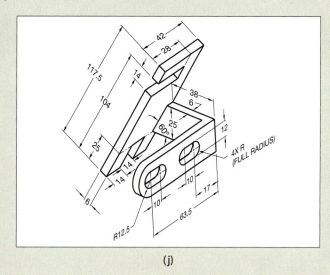

(j)

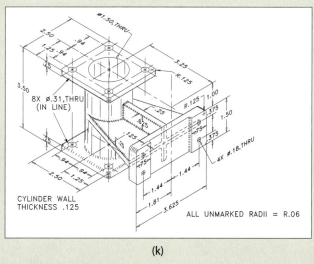

(k)

FIGURE P9.1.

9.11 problems (continued)

2. For each part shown in Figure P9.2, identify features that would be difficult to create using low-volume cutting production processes. Explain why each feature would be difficult to create. Suggest at least one method for changing each feature without altering its function so the part would be easier to make.

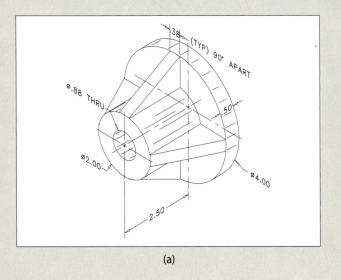

(a)

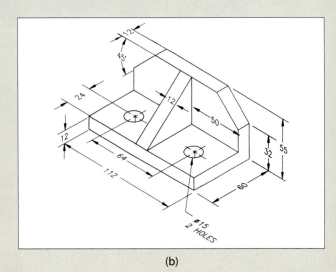

(b)

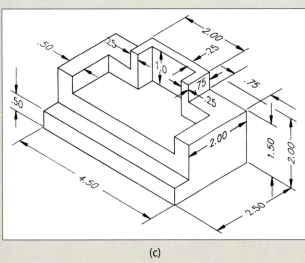

(c)

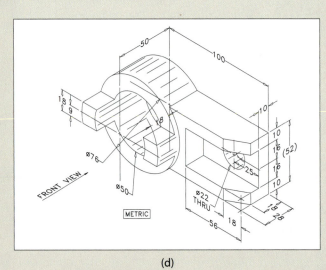

(d)

9.11 problems (continued)

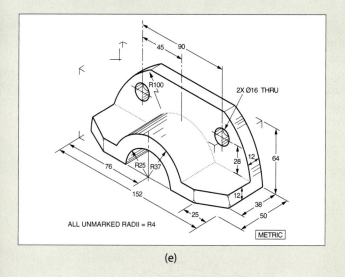

(e)

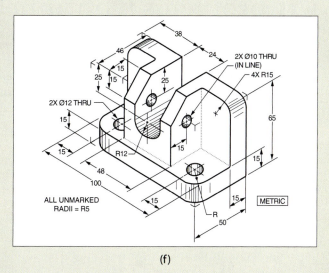

(f)

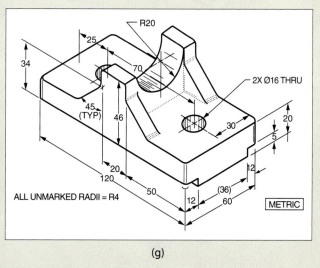

(g)

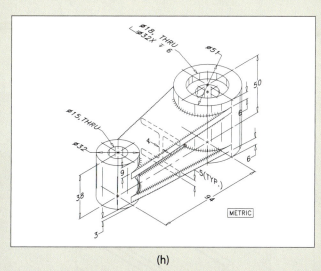

(h)

FIGURE P9.2.

9.11 problems (continued)

3. For each part shown in Figure P9.3, suggest a high-volume process that could be used to create some or all of the major features. Identify features that would be difficult to create using the high-volume production processes and explain why the features would be difficult to create. Suggest at least one method for changing each feature without altering its function so the part would be easier to make. What features would require a secondary (i.e., machining) operation to create? How would the required accuracy of feature sizes and locations affect the requirement for a secondary operation? How would the suggested processes change if the part were made of a different material (e.g., aluminum or plastic instead of steel).

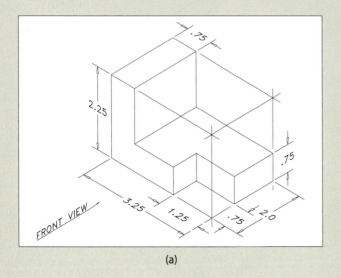

(a)

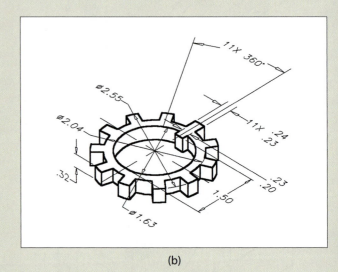

(b)

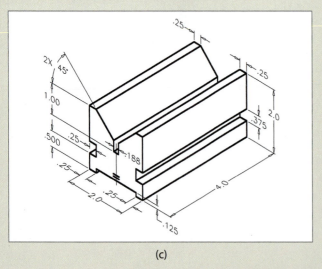

(c)

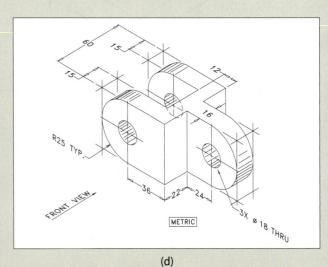

(d)

9.11 problems (continued)

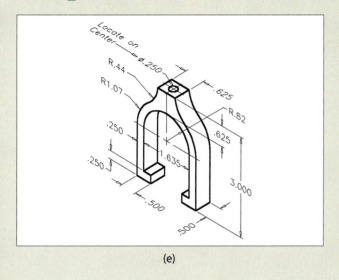

(e)

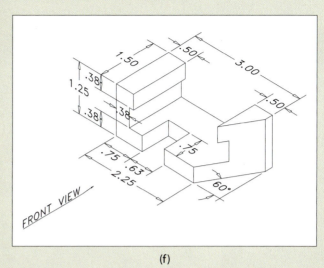

(f)

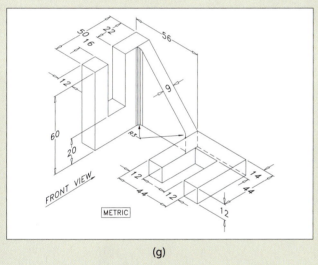

(g)

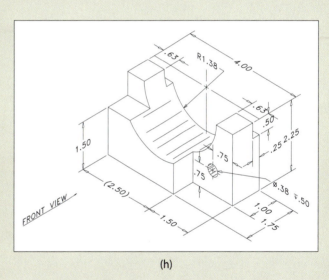

(h)

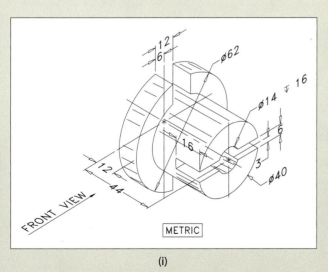

(i)

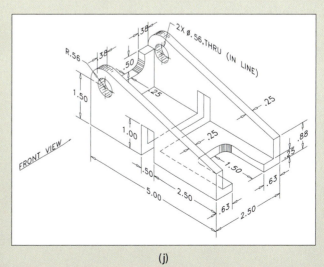

(j)

9.11 problems (continued)

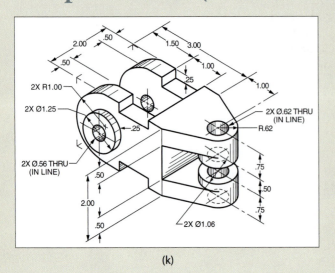

(k)

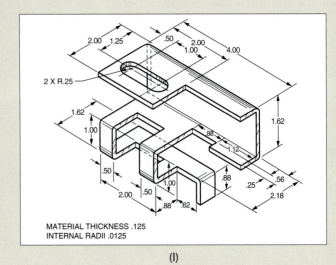

MATERIAL THICKNESS .125
INTERNAL RADII .0125

(l)

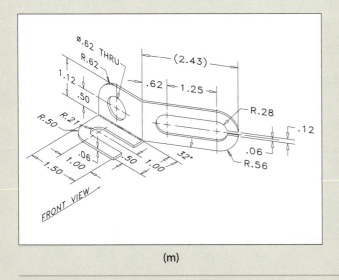

(m)

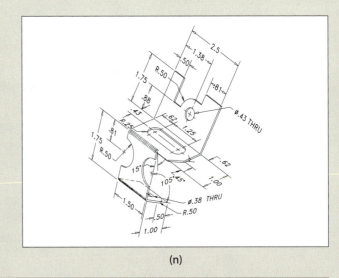

(n)

FIGURE P9.3.

sectionthree

Setting Up an Engineering Drawing

Almost all formal engineering drawings are presented in a multiview format using orthogonal projection. Most of the topics and techniques presented in this section are in wide use today and will continue to be invaluable in the foreseeable future. No matter how sophisticated the computer design hardware and software or the manufacturing system, engineers still must be able to create two-dimensional drawings of three-dimensional objects and be able to visualize three-dimensional objects from two-dimensional drawings.

10

Orthogonal Projection and Multiview Representation

objectives

After completing this chapter, you should be able to

- Discuss the principles of orthogonal projection
- Show how orthogonal projection is used to create multiple views of an object for formal engineering drawing
- Explain why orthogonal projection is necessary to represent objects in formal engineering drawing
- Create a multiview drawing from a 3-D object

10.01

introduction

The best way to communicate the appearance of an object (short of showing the object itself) is to show its image. For the purposes of the object's fabrication, analysis, or record keeping, this image must be precise. A precise description of an object begins with an accurate graphical representation of that object, which is what a formal engineering drawing is all about. It is a series of images that show the object viewed from different angles, every view accurately depicting what that object would look like from each view.

Whether you originated a drawing or you received one from the originator, the images represented in any engineering drawing must be interpreted the same way. Consistency is achieved by adhering to nationally and internationally accepted methods for creating and interpreting the images. Pictorial images, such as the isometric drawings first presented in Chapter 2 (and detailed in Chapter 12), quickly convey large amounts of qualitative information. However, pictorial images have the disadvantage of distorting the true size, configuration, and geometry of their features.

For an object to be represented without distortion or ambiguity, enough views must be provided such that all of the object's features can be clearly seen and accurately measured. In an engineering drawing, the choice of views is not arbitrary. Also, the views are carefully chosen such that the features on the object are aligned between the views and the geometries of the features are shown without distortion. With these views, size specifications can be added later to complete the description of the object.

10.02 A More Precise Way to Communicate Your Ideas

You have a wonderful idea for a new device. You believe in your idea. You want to have it fabricated. However, you must communicate to another party your thoughts about what the parts in the device will look like when they are fabricated. The other party may be another engineer who subjects your device to a more detailed analysis of what it should look like. The other party may be a fabricator who makes the device to your exact specifications. The other parties may be located in another area of the country or in another country. With the international scope of business today, design, analyses, and fabrication are commonly done in different locations around the world.

If questions arise concerning your idea, you may not be around to answer them. That is why all other parties involved in fabricating the object must envision it exactly as you do. One of your goals as the engineer or designer of a product, device, or structure is to represent it graphically in such a way (i.e., accurately) that it can be fabricated without any party misinterpreting how you want it to appear.

During the development of the Aerotec riser, the engineers at Hoyt USA faced the possibility that the product's geometry would be misinterpreted due to insufficient representation of what it would look like after fabrication. Creating a graphical image of the object in the form of a sketch or drawing as seen from only a single direction was not a good idea. The riser, which is shown in Figure 10.01, contains many features, such as cutouts and protrusions that could remain hidden when viewed from only one direction. The object had to be viewed from multiple directions to ensure that all of its features were revealed. If you were the engineer responsible for the design and manufacturing of a similar product, what would you do? How would you communicate what you want built to those who build it? How would you ensure that different people interpret and build the product the same way every time?

FIGURE 10.01. Viewing the Aerotec riser from different directions reveals previously hidden features. The arrows indicate details in each view that cannot be fully seen and described in the other views. (Model courtesy of Hoyt USA Archery Products).

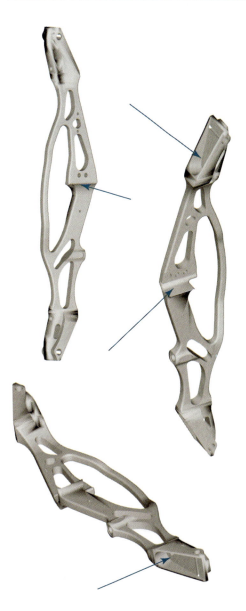

10.02.01 Problems with Pictorials

One solution would be to use pictorials such as isometric or perspective view. These types of representations of an object offer the advantage of quickly conveying the object's 3-D aspects from one view. Even people who do not have a technical background can easily and quickly understand pictorials.

However, pictorial representations present problems that are inherent in the use of one view of an object's three dimensions. One problem is the distortion of angles, as shown in Figure 10.02. The use of right angles and perpendicularity between surfaces is common on many fabricated objects because surfaces having those relationships are easy to construct with machine tools. However, on pictorials, 90° angles do not appear as 90° angles. In fact, depending on the angle of viewing, a 90° angle can appear as more or less than 90°. On a pictorial, it is difficult to depict an object's angles correctly when angles are not 90°.

Another problem with pictorials is the distortion of true lengths. In any pictorial, a length of 1 m on an object, for example, is neither depicted nor clearly perceived as a 1 m length.

FIGURE 10.02. Distortion of true lengths and angles in pictorial presentations.

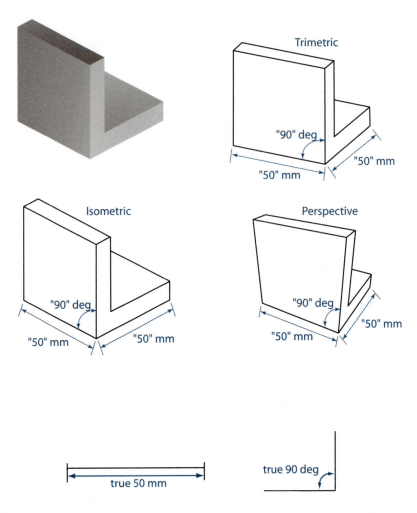

In some cases, such as an object with only rectilinear edges seen in the isometric view, this length distortion is the same in every direction. In this case, the real length can be obtained by multiplying the distorted edge length by a single correction factor. In Figure 10.02, for example, the length of each edge of the object shown in the isometric would need to be the actual edge length multiplied by a scaling factor of 0.612 if the object were drawn its full size. The formulas for getting this particular scaling factor is complicated, so do not worry about it for now. In general, however, the correction is not just a simple scaling factor. In a dimetric view, the correction factor for an edge of the object is dependent upon direction. The correction factor is more complicated for a trimetric representation, and the correction factor is even more complicated for a perspective representation.

Internal measurements also are distorted in pictorials. This distortion is dependent upon the direction of measurement, as shown in Figure 10.03. The location of the center of a hole placed at the center of a square face is, in reality, equidistant from each vertex of the square. However, in an isometric pictorial, the center of the hole must be drawn such that it is located a different distance from one vertex than from its adjacent vertices.

Figure 10.03 also shows the problem of curve distortion. The simplest curve—a circle or an arc of a circle—appears elliptical on a pictorial. On an isometric pictorial, the conversion from a circle to its representation as an ellipse is a matter of figuring out the scaling factors to calculate the major and minor axes and the orientation of the ellipse, both of which are dependent upon the circle's orientation in space. The calculation or construction is more complicated on a trimetric view because the scaling and orientation factors are different for different plane orientations on the object. On a perspective representation, the construction is more complicated

FIGURE 10.03. Distortion of internal lengths in pictorials. These different lengths on the same object represent the same length, which is the diameter of the holes in the cube.

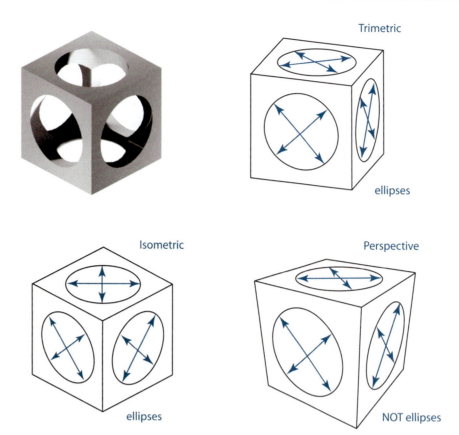

because the circle does not appear as an ellipse, but rather as an oval, or egg shape. (Remember, an egg shape is not an ellipse.)

The sum of the previous discussion is that although pictorials have the advantage of looking realistic, it is difficult or impractical to create an object with precision from them. Pictorials are subject to misinterpretation and errors in analysis and fabrication because the angles and distances are distorted. The most universally accepted solution to these problems is to use multiview representations, which are explained next.

10.02.02 Viewing Planes

A **multiview** representation depicts in one plane, such as a sheet of paper, many images of the same object, each viewed from a different direction. Pictorials can be used to enhance the clarity of the 3-D perception of an object; but the sizes of the object and its details are shown in a series of views, each view showing the sizes in their true length or shape. Any fabrication or analysis of the object's measurements can then be based on what is shown in the multiview projections, not on what is shown in the pictorial.

When you visualize an object in space, its appearance changes depending on the direction from which you view it. The lines and curves that form the graphical presentation of the object, such as the lines and curves shown in Figure 10.03, represent edges that are the intersections of surfaces. Now visualize a transparent plane, perhaps a sheet of glass, fixed in space between you and the object. This plane is called a **viewing plane**. Imagine the image of the object as seen through the plane is somehow painted onto the plane. Continuing to imagine, remove the object and look at the image painted on the viewing plane. What you see on that plane is a 2-D image of a 3-D object. The appearance of the image, however, would depend on the viewing angle of your head in front of that plane when you created the image. The simplest and most accurate view is from your head looking directly forward at the object. In general, to be accurate about the appearance of the object as seen through the plane, you would need to define the locations and orientations of the object, the viewing plane, and the viewer.

This is a great deal of information. But you would not need all of that information if you defined the image as one created by orthogonal projection, which is explained next.

10.02.03 Orthogonal Projection

In orthogonal projection, the image of an object is composed of points projected from individual points on the object onto the viewing plane such that the projection of each point is perpendicular to the viewing plane. Orthogonal projection of an object onto a transparent viewing plane is shown in Figure 10.04, where you can see the perpendicular relationship between the projection lines and the viewing plane when the plane is turned on edge.

An image created in this manner has two advantages. One advantage is that such an image is easy to create because you do not have to worry about defining the location or orientation of the viewing plane relative to the line-of-sight. The line-of-sight from a point on the object to the viewing plane is like the projection path; that is, it is always perpendicular to the viewing plane. The other advantage is that by turning the object such that an edge of the object is parallel to the viewing plane, the image of that edge shows its true length. Furthermore, the length of a projected edge is independent of its distance from the viewing plane. Both of these properties are shown in Figure 10.05.

10.02.04 A Distorted Reality

An image created by orthogonal projection is merely a convenience that allows you to analyze the image more easily when you are ready to make the object depicted. In the strictest sense, orthogonal projection does not accurately represent an image of the way a real object looks. In reality, parts of an object that are farther away appear smaller than the same-sized parts of an object that are closer. With orthogonal projection, all parts of the object appear in the same scale no matter how far the object is placed from

FIGURE 10.04. Using orthogonal projection to create an image of an object on a viewing plane. The object in (a) is in front of the viewing plane. The object in (b) is behind the viewing plane. In either case, the projection lines are perpendicular to the viewing plane, as shown in (c).

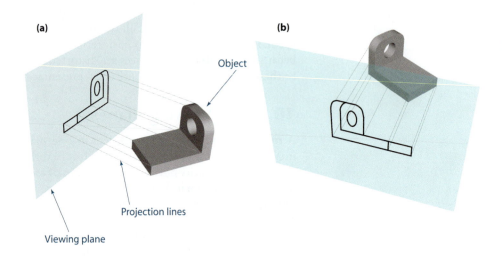

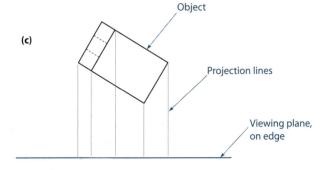

FIGURE 10.05. With orthogonal projection, the projected length of an edge is independent of its distance to the viewing plane. For this particular object orientation, the true lengths of the vertical edges are shown.

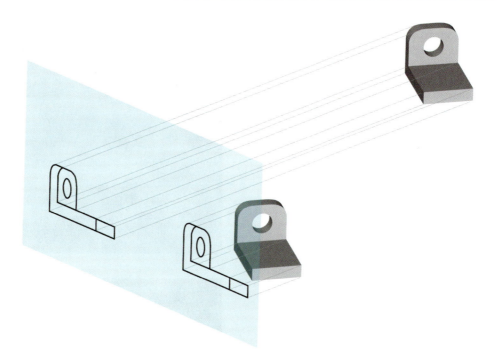

the viewing plane. But as in the case that follows and in most cases, the approximation is close and the convenience and ease of image creation and analysis far outweigh the need to see the image as it really appears.

The effect of an image created by orthogonal projection is similar to a photograph of an object taken at a long distance using a powerful telephoto lens. That type of picture lacks depth; that is, the object appears flat. This lack of depth is attributable to the fact that although the light rays actually extend radially from the surface of an object, the reflected light rays appear less like radial rays and more like parallel rays viewed at a great distances from the object. The greater the distance, the more parallel the light rays. At a long distance, where the light rays compose the image of the object, such as at a camera lens, the light rays are very nearly parallel to each other and very nearly perpendicular to the plane of the lens. This effect is shown in the bottom photograph in Figure 10.06; both photographs are the same object, each taken from a different camera distance. Even though the overall image size of the object is about the same, in the close-up photo, you should be able to see that the parts (for example, the wheels) of the object that are closer appear magnified when compared to the parts that are farther away.

10.02.05 Choice of Viewing Planes

From what was just explained, you should understand that an orthogonal projection of an object is a 2-D drawing of that object as it would appear on a viewing plane. To get a different view of the object, you need to move the object and/or the viewing plane to a different location.

Consider the case of keeping the viewing plane in the same place and rotating the object. One advantage of orthogonal projection is that an object's lines and curves can be seen in their true shape. For example, when the viewing plane is parallel to a circle, the circle actually appears as a circle rather than an ellipse. This may be important, for example, when you want to see how close the edge of a hole in an object actually comes to the edge of the object. It makes sense, therefore, to rotate the object into an orientation where the measurements, such as the diameter of the hole or its distance to the edge, can be seen to represent the true shape, distance, and size. Figure 10.07 shows an object rotated into the best position for this specific analysis versus the same object in

FIGURE 10.06. The top photograph was taken from up close. The bottom photo was taken from a long distance and enlarged so feature sizes could be compared. Can you see the lack of perspective in the long-distance photo?

a poor orientation. In general, in the creation of the first view of an object, it has become common practice to orient the object in a position that shows as many as possible of its lines and curves in their true shape.

However, a single view of an object is usually insufficient to specify all of its features and measurements fully. Figure 10.08 shows how different objects can appear the same using a single view only.

FIGURE 10.07. Good part placement shows most of the part edges in their true length.

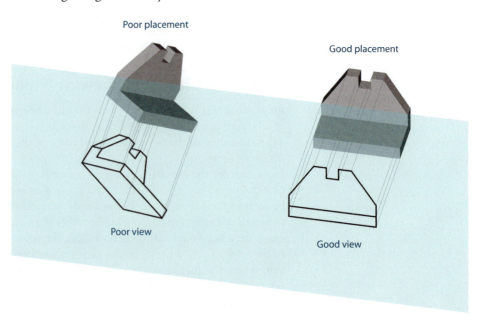

FIGURE 10.08. A single view of a part may have many different interpretations.

This single view...

...can be interpreted in at least six different ways.

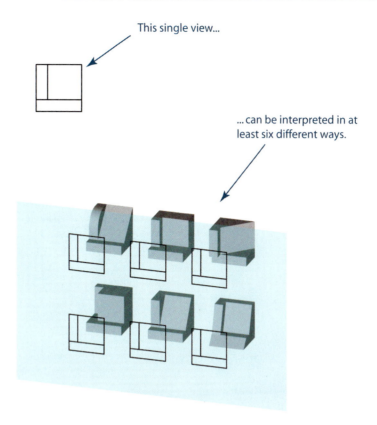

To fully define the 3-D geometry of an object, it is necessary to depict the object in **multiple views**. This means there must be a viewing plane for each of the views. Specifying the location and orientation of each of the additional viewing planes must be done in a standardized way so that 2-D images can be extracted from the object easily. Also, the multiple 2-D images must contain enough information so that the original 3-D image can be re-created from them. One way to do this is to locate and orient the additional viewing planes so that each is orthogonal to the first viewing plane, as shown with a second and third viewing plane in Figure 10.09(a). The images on all of these viewing planes are created using orthogonal projection.

When the location and orientation of the intersection line between the first viewing plane and any one of the additional viewing planes are known, the location and orientation of each of the other additional images can be specified. The intersection line between the first viewing plane and any of the additional viewing planes can be imagined as a hinge between the two planes. By "unfolding" the additional planes at their imaginary hinges, as shown in Figure 10.09(b), the images on all of the viewing planes can be shown on a single plane, or in other words, a 2-D drawing.

Used this way, orthogonal projection and viewing planes offer you the advantage of seeing multiple views of the same object at the same time on a single sheet of paper. Orthogonal projection also can precisely identify the position and orientation of the viewing planes used to create those views by specifying on the single sheet the location of the intersection lines between the viewing planes.

10.02.06 Size and Alignment

When the second and third planes are completely unfolded and are coplanar with the first viewing plane, as shown in Figure 10.09(c), three images can be seen on a single plane. The images from the second and third planes are considered adjacent to (i.e., created immediately next to) the image from the first plane. Note that the size and orientation of the images are not arbitrary. Each image has the same scale (or magnification);

FIGURE 10.09. Two viewing planes that are orthogonal to the first (front) viewing plane (a) can be unfolded (b) to present the images on a single plane (c). The imaginary hinges for the two viewing planes are at the intersections of these planes with the front viewing plane.

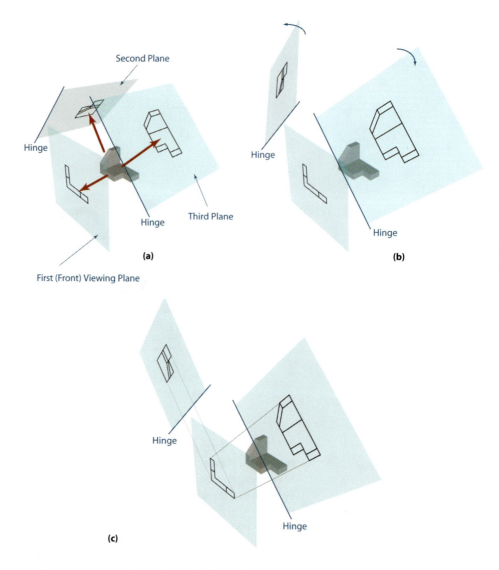

(a)

First (Front) Viewing Plane

Second Plane

Hinge

Hinge

Third Plane

(b)

Hinge

Hinge

(c)

Hinge

Hinge

and the orientation of the image is dependent upon the original location of its viewing plane as defined by the location of the intersection line between the viewing planes, or their hinge. This alignment of the vertices of the object images in **adjacent views** is shown in Figure 10.10, where the three views are presented on a single sheet.

FIGURE 10.10. Viewing planes completely unfolded showing proper size, location, and orientation of the images on a single plane.

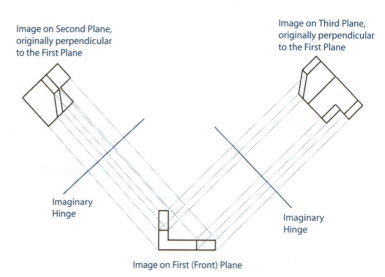

Image on Second Plane, originally perpendicular to the First Plane

Image on Third Plane, originally perpendicular to the First Plane

Imaginary Hinge

Imaginary Hinge

Image on First (Front) Plane

10.03 The Glass Box

Only three or four views are required to fully define most objects. Simple objects may require only one view; complicated objects may require six or more views. Objects such as engineered parts can usually be fully defined when they are viewed through a set of six viewing planes that together form a **glass box**, as shown Figure 10.11(a). The glass box has the unique property that for any viewing plane, all of its adjacent planes are perpendicular to each other and opposite viewing planes are parallel to each other.

When you open (or unfold) the panels of the box, as shown in Figure 10.11(b), you can view all six sides of the object simultaneously on a single plane, as shown in Figure 10.11(c). There is more than one way to unfold the box. Unfolding in the manner shown in Figure 10.11 is the standard way to do it according to accepted drawing practices. The top and bottom and right- and left-side views open about the front view; and the rear view is attached to the left-side view.

Make sure you see and understand that when the viewing planes are completely unfolded, the size and orientation of each image is not arbitrary. The scale in each view is the same. In the case of the complete glass box, each viewing plane is orthogonal to its adjacent viewing planes. When the box is unfolded and presented on a single sheet,

FIGURE 10.11. Viewing an engineered part through a glass box (a) that opens (b) to present the images on a single plane (c).

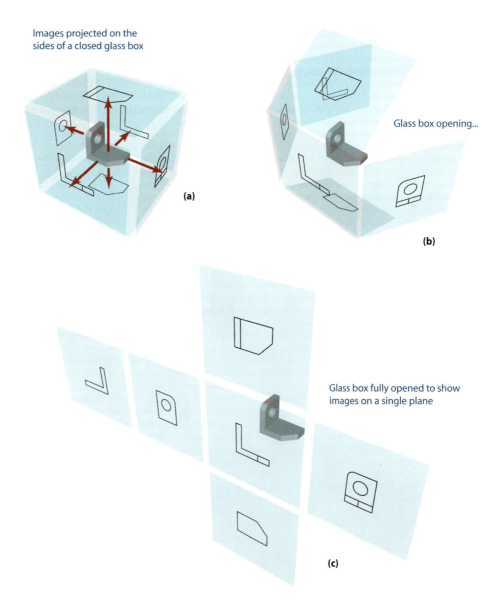

Images projected on the sides of a closed glass box

(a)

Glass box opening...

(b)

Glass box fully opened to show images on a single plane

(c)

as in Figure 10.12, adjacent images are aligned horizontally for horizontally adjacent views or vertically for vertically adjacent views. These alignment properties are very important when the object is analyzed. If you select any point on the object (assume point A on Figure 10.12), the images of that point will be horizontally aligned with each other on horizontally adjacent views and those images will be vertically aligned with each other on vertically adjacent views.

In general, the same point in space seen in adjacent views is aligned along a path that is perpendicular to the intersection line of the viewing planes, as shown in Figure 10.10. What this means for engineering drawing is that features on an object, such as edges or holes, shown in one view can be easily located on the adjacent view because the features are aligned between adjacent views. For complex objects with many features, the ability to identify the same feature on adjacent views is of tremendous utility.

10.03.01 Standard Views

The glass box yields six different views of an object. For a large percentage of engineered parts, six views are more than sufficient. Engineers typically like to design things that are easy and therefore inexpensive to make. Three-axis milling machines and single-axis lathes are common machines in any fabrication shop. These machines easily create surfaces on the workpiece that are parallel, perpendicular, or concentric to each other and that easily cut holes, slots, or other features that are perpendicular to the working surface.

The six views represented by the glass box are the front, top, left side, right side, bottom, and rear views. These views are known as the six standard orthogonal views or the six principal orthogonal views or more simply as the **six standard views** or the **six principal views**, respectively. When a formal drawing is created showing these views, the intersection lines and projection lines between views are not shown because these lines do not add much information to the drawing when it is already understood that adjacent views are orthogonal to each other. Also, each view does not need to be labeled as the front, top, right side, etc., views.

FIGURE 10.12. Alignment of points on adjacent views for all six standard views.

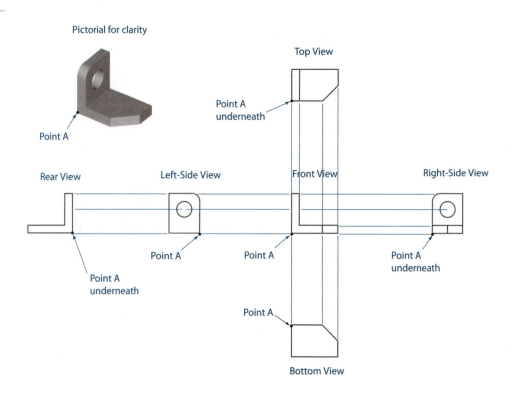

FIGURE 10.13. The preferred presentation configuration showing the front, top, and right-side views of an object. Other views are added only when necessary to show features that cannot be defined in the preferred configuration.

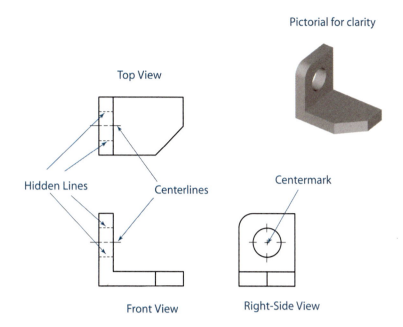

10.03.02 The Preferred Configuration

Are all six views necessary? Usually not. The great percentage of engineered parts can be fully defined geometrically with fewer than all six of the standard views. In fact, most engineered parts can be completely defined for fabrication using only three views.

Although there are no defined rules as to which views must be included or excluded in a formal engineering drawing, there is a **preferred configuration**—the front, top, and right-side views. Additional views are presented only when necessary to reveal and define features that cannot be shown in the preferred views. The preferred configuration for the object in Figure 10.12 is shown in Figure 10.13. Only the front, top, and side views are shown. Make sure every edge of the object can be seen in its true length in at least one view.

It is becoming increasingly popular to include an isometric or trimetric pictorial of the object somewhere on the drawing. When a pictorial is included, it serves only to aid in clarity; it does not need to be properly aligned or scaled with the standard views.

10.04 The Necessary Details

Only the minimum number of views needed to quickly and accurately communicate the geometry of an object should be created. Whenever possible, the preferred configuration of a front, top, and side view should be used unless fewer than three views are needed to see and define all of the features of the object. To minimize the number of required views on complicated objects and to reduce any possible ambiguity, some shorthand notation that describes common geometries such as certain types of holes and screw threads is used in drawing practice. Such notation is detailed in later chapters in this book. There will, however, be cases where additional views become necessary or when the preferred configuration may not be the best.

10.04.01 Hidden Lines and Centerlines

The dashed lines you see on the views shown in Figure 10.13 represent internal features or edges that are obscured by the object. These obscured features or edges are called **hidden lines** in these views. Hidden lines, which are denoted as equally spaced dashed lines on a drawing, represent the edges of an object or its features that cannot be seen on the real object but would be visible if the object were partially transparent. Hidden lines are used to emphasize an object's unseen geometry and thus speed the interpretation of its presentation. Hidden lines also are used to reduce the need for creating additional views. Although hidden edges cannot be seen on an opaque object, they are represented graphically the same way hidden lines are included in a view to emphasize that a feature cannot be seen in that view or to show that a feature cannot be seen from any of the other views. Later in this chapter, hidden lines will be discussed further as you encounter examples of the advantages and problems associated with them.

Looking closely at Figure 10.13, you will see lines located at the center axis of the hole. These are not hidden lines. They are **centerlines**, which are represented graphically by alternating short and long dashes along the length of the center of the circular hole. Centerlines cannot be seen on the real object, but they must be included on the drawing to identify where the center of the circular hole is located on the object. More generally, centerlines are used where there is a cylindrical surface such as a hole or a tube.

The reason for including centerlines is to make it easier for the reader to distinguish between edges, visible or hidden, that are part of a cylindrical surface and edges that result from the intersections of planes. Using centerlines also makes it easier to locate features such as holes, which are commonly defined by their diameters and center locations.

A **centermark**, the end view of a centerline, is identified by a right-angle cross such as that shown in the center of the circular hole in the right-side view of the object in Figure 10.13. Typically, centerlines and centermarks are used where the arc of a cylindrical surface is 180° or greater, although they can be used for lesser arcs as required for clarity in a drawing.

10.04.02 The Necessary Views

How many views should be created to fully define a object? In engineering practice, it is considered poor practice to create more views than are needed. Creating unneeded views means more work for which there is no payoff. However, having too few views can create problems when the fabricator tries to make the part. In the worst-case scenario, the fabricator will try to guess what you want, get it wrong, and deliver a potentially expensive part that cannot be used. In that case, the creator of the drawing would be at fault, not the fabricator. The party responsible for creating the drawing also may be legally responsible for paying for the services of the fabricator.

So how many views are needed to fully define an object? The number depends on how complicated the object is to depict in three dimensions. Start by creating the front, top, and right-side views. Remember, they represent the preferred configuration, which all engineering personnel like to see. Try to orient the object in such a way that these three views reveal as much of the object's features as possible. If you are lucky, these three views will fully define the object; but that is not always the case.

You should ask yourself the following two questions when you finish creating the drawing views:

1. Can the true size of all of the measurements needed to define all of the features of this object be seen in at least one of the views just created?

2. Is it impossible for the geometry of any feature to be misinterpreted as another type of geometry?

Yes to both questions means you have enough views. No to either question means you have more work to do.

FIGURE 10.14. Formal multiview presentation of the Aerotec riser as would be seen in an engineering drawing. The measurements, text, and other specifications are not shown in this example so you can see the views more clearly. (Model and drawing courtesy of Hoyt USA Archery Products)

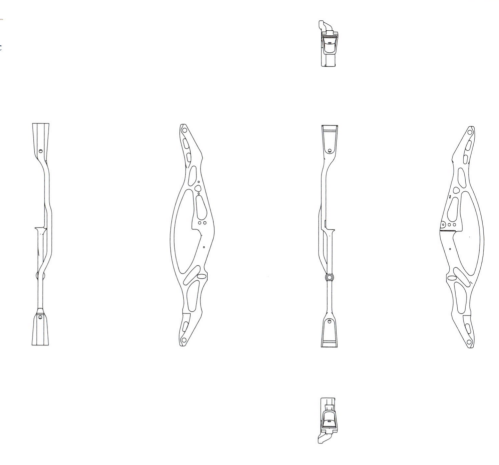

The multiview production drawing for the Hoyt Aerotec riser is shown in Figure 10.14. The complexity of this object requires that all six standard views be used because it has features that can be seen only from each of the six viewing directions.

Objects that are flat can be defined with a single view along with some sort of note specifying the thickness of the object. Flat sheet metal objects and objects that can be cut from a plate of uniform thickness fall into this category. The cuts must be through the entire thickness of the sheet or plate. An example of this type of object is shown in Figure 10.15. Because this object is made of very thin material, the adjacent orthogonal views would appear as lines.

Even when the thickness of the object is constant, a fabricator may find it helpful to see a second view; for example, to emphasize that the thickness of the object is a significant fraction of the object's planar geometry. See how the second view in Figure 10.16 helps depict the relatively large and uniform thickness of the object.

For objects that have 3-D features such as protrusions and cuts, each with a different depth, the problem of finding the proper number of views for a drawing becomes more difficult. Figure 10.17 shows an example of a drawing with two views. In this case, more than one interpretation of the object is possible. The addition of a third view is necessary to completely specify the desired object.

Figure 10.18 shows three original views that, in the absence of hidden lines, could be used to represent two possible objects. A fourth orthogonal view, a bottom view, is required in this case to distinguish between the two possibilities.

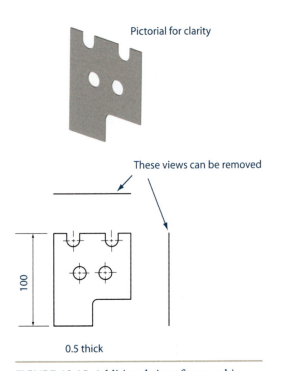

Pictorial for clarity

These views can be removed

100

0.5 thick

FIGURE 10.15. Additional views for very thin parts, such as sheet metal, add little information.

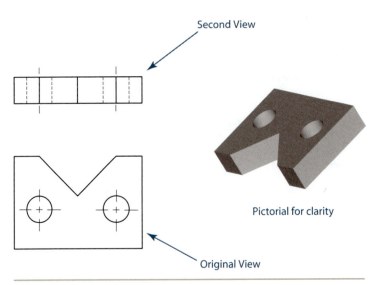

Second View

Pictorial for clarity

Original View

FIGURE 10.16. For a part with a constant but significant thickness, including a second view is a good idea to emphasize the 3-D nature of the part.

FIGURE 10.17. Different interpretations of a drawing with two views. A third view is necessary.

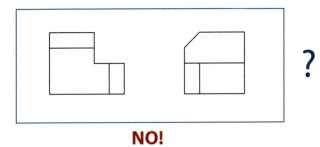

?

NO!

These two views alone are insufficient to define a three-dimensional object

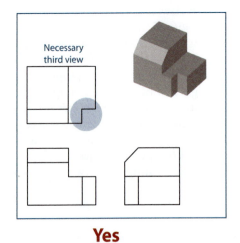

Necessary third view

Yes

Adding the third view uniquely defines the object shown

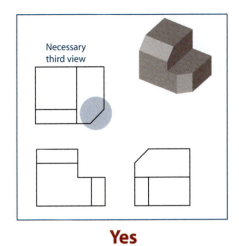

Necessary third view

Yes

Adding the third view uniquely defines the object shown

FIGURE 10.18. In the absence of hidden lines, four views are required to distinguish between these two parts. The fourth view is needed to distinguish the cutout on the underside as being diagonal instead of square.

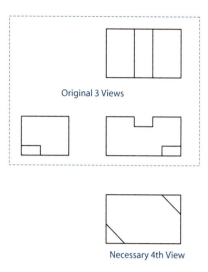

Original 3 Views

Necessary 4th View

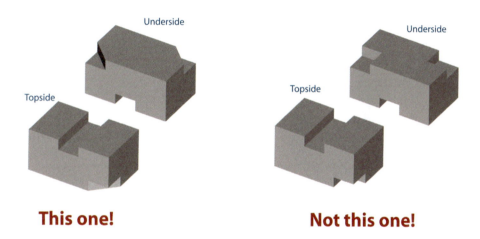

This one!   **Not this one!**

Figure 10.19 shows an example of an object where, in the absence of hidden lines, five views are necessary.

As a rule of thumb, when an object contains inclined surfaces with respect to the standard viewing directions, each of those inclined surfaces must appear inclined in at least one of the orthogonal views representing the object. When the inclined surface is not shown in one of the orthogonal views, a view needs to show the surface as being inclined (i.e., with at least one of its edges at an angle that is not 0° or 90°).

10.04.03 Hidden Lines versus More Views

One way to reduce the number of required views is to use hidden lines. The object shown in Figure 10.20 has some unique features. Try to imagine representing the object without using hidden lines. Without the hidden lines, five views would be required to define all of its features. With only the front, top, and right-side views and no hidden lines, the geometry of the keyway seen on the underside of the object cannot be defined. Moreover, without hidden lines, additional views would be required to show that the hole and slot extend all the way through the object.

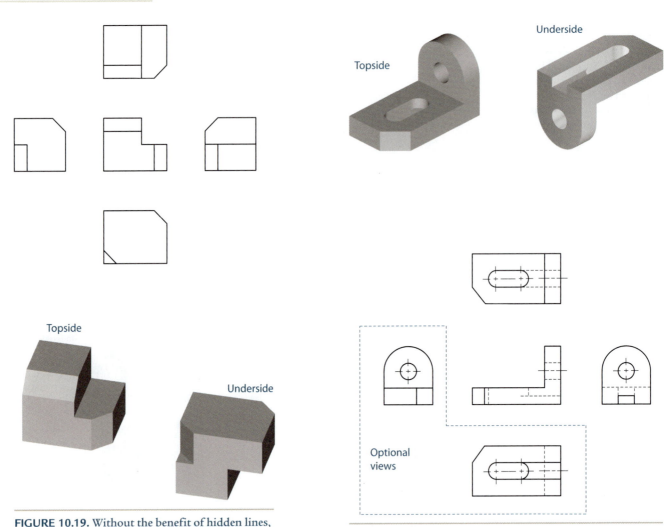

FIGURE 10.19. Without the benefit of hidden lines, five views are required to describe this object.

FIGURE 10.20. Hidden and internal features on a part. Using hidden lines makes the left side and bottom views optional.

By using hidden lines, only three views are required—the preferred configuration of front, top, and side views. Whether you use all five views shown in Figure 10.20 or the preferred three-view presentation depends on your answer to this question: Which presentation would be clearer? You always select the presentation that has, in your opinion, the least ambiguity (and not necessarily the least amount of work to produce). In the case of the five-view presentation, although it would not be an absolute requirement, adding hidden lines would emphasize the internal geometry of the object. For the three-view presentation, adding the hidden lines would be an absolute necessity.

Another use of hidden lines is to reveal the details of internal features that cannot be easily seen in any of the standard orthogonal views. Such details would be, for example, the depth or the profile of holes and slots, as shown in Figure 10.21. Figure 10.20 demonstrates how hidden lines can be used instead of additional views, making the drawing easier to create and more compact without the loss of any information. For the object shown in Figure 10.21, the depth of the slot cannot be seen in any of the standard orthogonal views. If you look carefully at the views for the object shown in Figure 10.21, you see that hidden lines for different features can be separated into different views. But if all of the hidden lines were shown on all of the views, the result would be a jumble of so many hidden lines that it would be difficult to distinguish the different features that they represent.

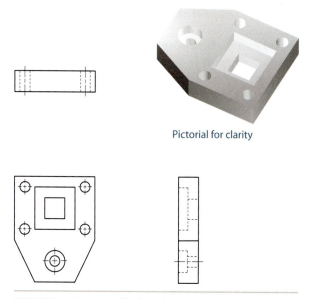

Pictorial for clarity

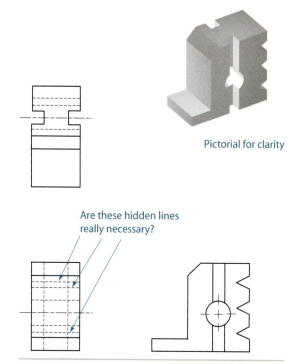

Pictorial for clarity

Are these hidden lines
really necessary?

FIGURE 10.21. Use of hidden lines to reveal internal
features.

FIGURE 10.22. Overuse of hidden lines causes confusion. Exercise judgment. It might be better to create another view, such as a rear view in this case.

A common problem for inexperienced designers is deciding when to use hidden
lines. Hidden lines should be used to add clarity to a drawing. Hidden lines should be
used to emphasize a feature, even if that feature can be seen and defined in the existing
orthogonal views. The goal of the creator or the drawing is to increase the speed at
which the drawing can be interpreted. However, hidden lines must be used to add information when there is no way to obtain this information from the rest of the drawing.

Because hidden lines can be used to avoid creating another view, it is sometimes
tempting to do just that, even when using another view would be better. Figure 10.22
shows that adding too many hidden lines create a complex, confusing drawing. With
this result, it would be better to create extra views. When deciding whether to use hidden lines or to add additional views, simply do whatever will cause less confusion for
the reader of the drawing. However, it is usually not a good idea to create hidden lines
of different features such that the hidden lines cross each other, lie on top of each
other, or even come close to each other.

The purpose of hidden lines is to define or emphasize details that cannot be seen,
which is accepted as standard practice. Deleting unnecessary hidden lines and adding
additional views are considered optional methods of reducing ambiguity when the use
of hidden lines makes the presentation confusing. There must be no confusion as to
which feature a hidden line represents.

10.05 First-Angle Projection versus Third-Angle Projection

The glass box representation of multiviews of an object is formally referred to as **third-
angle projection**. Whenever third-angle projection is specified on a drawing, each
view of the object was created by projecting the image of the object onto the glass box's
transparent viewing plane between you and the object—the object is behind the transparent viewing plane. The viewing planes are then rotated about their intersection lines

until all of the views are shown on a single plane or sheet. This interpretation is the one most commonly used in the United States.

However, in some parts of Asia and Europe, **first-angle projection** is commonly used. In first-angle projection, each viewing plane is behind the object, which means the object is between you and the viewing plane. With first-angle projection, the viewing plane is opaque and the image is projected back and transferred onto the viewing plane. One way to interpret first-angle projection is to imagine the object in front of the opaque panels, as shown in Figure 10.23(a). The image of the object, as seen by a viewer located directly in front of each panel (with the object directly in line between the two panels), is transferred to that panel. Opening the panels, as in Figure 10.23(b), begins to show how the front, top, and right-side views are presented on a single plane, as shown in Figure 10.23(c).

For drawings created using either first-angle or third-angle projection, the primary view is considered to be the front view. The front view in either projection is usually selected as the view containing the most features in their true sizes and shapes, thereby allowing for the most measurement extraction. As you saw earlier, for the six standard views using third-angle projection, the top view appears above the front view and the

FIGURE 10.23. Viewing an object in front of opaque panels for first-angle projection. The images are projected onto the panels (a), which open (b) to present the images on a single plane (c).

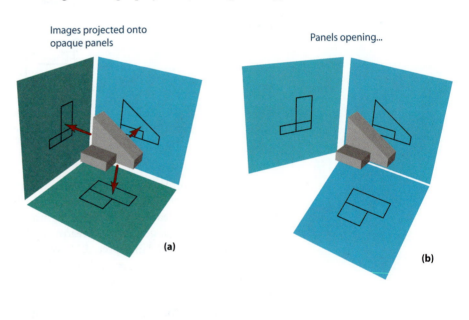

Images projected onto opaque panels

Panels opening...

(a)

(b)

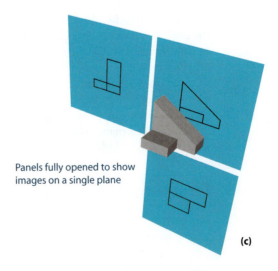

Panels fully opened to show images on a single plane

(c)

bottom view appears below the front view. The right-side view appears to the right of the front view, and the left-side view appears to the left of the front view. The rear view appears, by practice and convention, attached to the left-side view and appears to its left.

Using first-angle projection, the top view of an object appears below the front view and the bottom view appears above the front view. The right-side view appears to the left of the front view, and the left-side view appears to the right of the front view. The rear view appears, by practice and convention, attached to the left-side view and appears on its right. The location of the first-angle projection views is shown in Figure 10.24.

The differences between first-angle projection and third-angle projection are sometimes subtle and confusing, particularly because the front view of the object is the same in both cases. To add further confusion, for a large percentage of engineered parts, the left side and right-side views or the top and bottom views are identical. These reasons explain why drawings need to clearly specify whether first-angle or third-angle projection must be used to interpret the views. Many large companies operate internationally, with engineering and fabrication facilities worldwide. In international business, drawings are often created in one country and the parts fabricated in another country. When a drawing is interpreted incorrectly, the resulting fabricated part may be the mirror image of what was desired. Figure 10.25 shows a multiview drawing of an object and the two different objects that are created when the drawing is interpreted using first-angle projection and third-angle projection.

The symbol added to a drawing to specify first-angle projection or third-angle projection is two views of a truncated cone, shown in Figure 10.26. This symbol depicts how a truncated cone would appear if a drawing of it were made using the projection method used for the entire drawing. The appropriate symbol and/or wording must be added to a formal drawing, usually somewhere in the title block (for which more detail can be found in Chapter 18) to eliminate ambiguities that may arise from misinterpreting which projection was used.

FIGURE 10.24. The six standard views, using first-angle projection, presented on a single sheet.

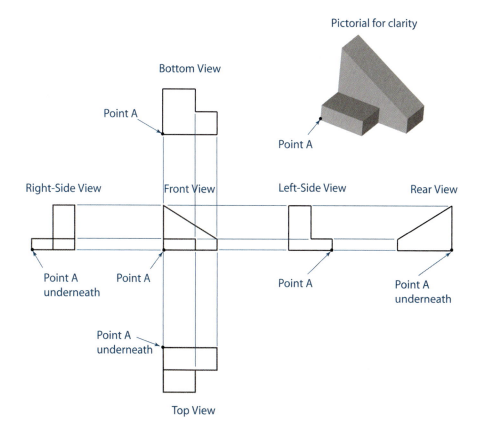

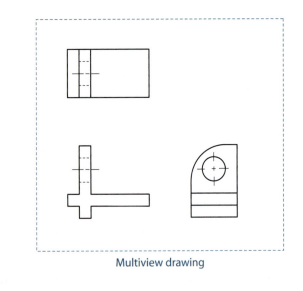

Multiview drawing

Object Represented

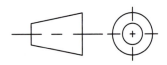

Symbol for First-Angle Projection

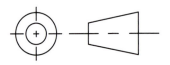

Symbol for Third-Angle Projection

FIGURE 10.26. Drafting symbols for specifying the use of either first-angle or third-angle projection in a drawing.

Interpretation using third-angle projection

Interpretation using first-angle projection

FIGURE 10.25. Drawing interpretation using first-angle or third-angle projection may lead to different parts.

10.06   Strategies for Creating Multiviews from Pictorials

Few people think of an object in terms of its multiview representation. If you are thinking about a pencil, for example, you probably do not imagine it in terms of a front, top, and side view. The image you have is likely to be three-dimensional, perhaps as a pictorial of some sort. Transforming that image into its multiview representation requires some skill. To develop this skill, some rules (and a great deal of practice) are required. And you need to remember that a pictorial image contains 3-D information that must be extracted from the way the pictorial looks in a 2-D medium. A multiview representation is merely a different, more accurate way of presenting this information. Exercises in converting pictorials to multiviews and multiviews to pictorials will help you develop practical skills, as well as improve visualization skills. Engineers should be able to quickly visualize 3-D objects from multiview drawings and quickly create multiview drawings for proposed or existing 3-D objects. You can begin developing these skills using the following step-by-step procedures. For the first few examples that follow, sketching techniques will be used because sketching, as opposed to drawing with instruments or CAD, is an excellent method for developing visualization skills. Later examples in this section will use more formal graphics so the drawings can be more clearly detailed.

Transforming a pictorial image into a multiview drawing usually involves keeping track of the vertices, edges, or surfaces of the object. Regardless of which elements are tracked, the process starts the same way. An eight-step process is used to create the drawing. The first two steps are as follows:

Step 1:   On the pictorial, specify the viewing directions that you intend to create (e.g., front, top, right side, etc.) and create a sheet with areas reserved (and labeled) for the appropriate orthogonal views based on the projection method used.

Step 2:   Find the maximum size of the object in each of the three directions of your coordinate system and in each view, sketch the limits of a rectilinear box that will contain only the entire object in all three directions.

A typical problem of multiview creation is shown in Figure 10.27. An isometric image of an object is presented, and the goal is to create the necessary orthogonal views to specify all of its features completely. Assume that all of the hidden surfaces are flat and that there are no hidden features.

This object is basic, considering all of its surfaces are perpendicular or parallel to each other. When this is the case, the edges of all of the surfaces will appear to be horizontal or vertical in all of the orthogonal views created. The true length of each feature must be shown in at least one view. For convenience in measuring the lengths of the edges on the object, an isometric grid has been placed on the isometric view. Placed on each of the orthogonal views is a corresponding rectangular grid that in each plane direction represents the same grid spacing as the isometric grid, as shown in Figure 10.28. The edge lengths as seen in the isometric view then can be conveniently transferred to the corresponding edges on the orthogonal views.

As an alternative to creating grids, you also can measure the edges in the isometric view using drafting instruments or CAD and transfer these measurements to the corresponding edges on the orthogonal views. When the edge lengths are otherwise specified, such as with notes, the specified edge lengths should be used in the orthogonal views. Carefully note the viewing directions for each view on the orthogonal views and the pictorial and make sure these directions are consistent.

You need to make clear from which point to which other point you are measuring any line in the pictorial —that is, the direction of your measurement—so you can incorporate the same information (direction of point-to-point measurement) in any of the

FIGURE 10.27. How would you create a multiview drawing of this object? The isometric grid is to be used for sizing.

FIGURE 10.28. Defining the foundation space, axes, viewing directions, and anchor point A.

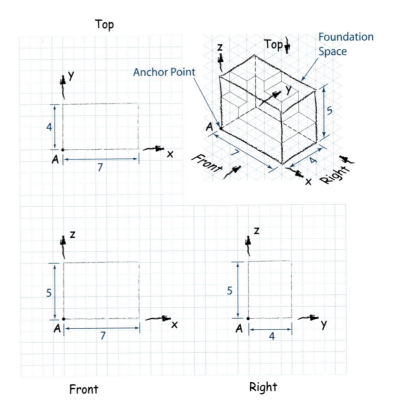

orthogonal views. For example, in Figure 10.28, as you measure from one point to another in the pictorial, you must be able to follow the same direction of measurement in the orthogonal views. For this purpose, it may be convenient to use a set of coordinate axes initially to help you with the directions of measurements until you become more familiar with the directions in the orthogonal views in your drawings. In Figure 10.28, a right-handed Cartesian coordinate system is placed with the origin coincident to one of the corners of the object in the pictorial. This same coordinate system is placed in all three orthogonal views. Make sure you maintain alignment of the origin of the coordinate system in each of the three views (step 1).

For this point to be in the same place in each view, it must be aligned on a vertical line between the top and front views and on a horizontal line between the front and right-side views. The top view looks straight onto the xy plane, so the z-axis points out of the page. The front view looks straight onto the xz plane, so the y-axis points into the page. The right-side view looks straight onto the yz plane, so the x-axis points out of the page.

The next step in creating the multiview drawing is to mark the limit of the size of the object in all three directions (step 2). These limits define a **foundation space**, which represents the rectilinear limits occupied by the object in each view. Although the foundation space is not the outline of the object itself, it helps you visualize the object in each view by delineating the volume that the object can and cannot occupy. If in the process of creating the orthogonal views you start creating lines or points for the object outside its foundation space, you will know you are doing something wrong. The foundation space for the object in Figure 10.27 is shown in Figure 10.28. Examine the foundation space on the pictorial. It extends 7 units in the x-direction, 4 units in the y-direction, and 5 units in the z-direction. Make sure these limits are marked off properly in each of the orthogonal views.

Once the foundation space is defined, there are different ways you can proceed. Students who have practiced and completed many problems in drawing orthogonal views from pictorials are able to proceed intuitively. Most beginners need a little help getting started before intuition kicks in.

10.06.01 Point Tracking

One way of continuing beyond step 2 is to label each vertex on the pictorial as a point, keep track of each point on every orthogonal view, and then connect the points in the views to form an image of the object in these views. (Keep in mind that all of an object's points may not be visible on the pictorial.) This process is called the **point tracking** method. Here is how it works.

After you have established the viewing directions and foundation volume in step 1 and step 2, you are ready to follow the next six steps to complete the drawing. The general procedure is outlined below. Each step is explained in detail in the paragraphs that follow.

Step 3: Define an anchor point.

Step 4: Locate a vertex adjacent to the anchor point and draw that edge.

Step 5: Successively locate other vertices and draw the edges between those vertices.

Step 6: Convert hidden lines.

Step 7: Add internal features.

Step 8: Check model validity.

A point on the object must be selected as an **anchor point** (step 3). There is an anchor point in each of the orthogonal views, and it is the same point in space as is seen from the different views. An anchor point is a point whose location you feel certain you can identify on each of the orthogonal views. Such a point is commonly a vertex located on one of the bottom corners of the object. Call this point A; then locate and label the point on each of the orthogonal views and on the pictorial. Remember, point A in views that are left or right of each other must be aligned horizontally; and in views that are above or below each other each, point A must be aligned vertically. In this case, but not necessarily in all cases, point A also is the origin of the coordinate axes.

Select another point on the pictorial (step 4) near point A, which can be located by traveling along one edge of the object, as shown in Figure 10.29. Call this point B; then locate it with respect to point A in each orthogonal view. You do this by noting on the

FIGURE 10.29. Defining and tracking points on the same surface near the anchor.

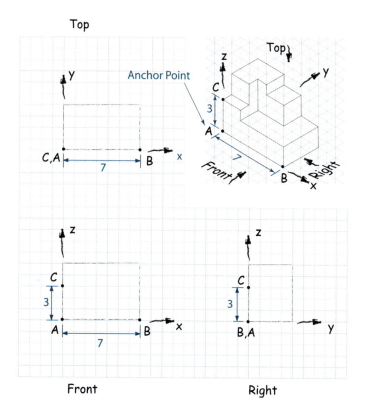

pictorial the direction and distance from point A to point B. From the measurements on the pictorial, you can see that to reach point B from point A, you need to travel 7 units in the positive x-direction. In the top view, the location for point B is 7 units to the right on a horizontal line from point A, which is the x-axis in that viewing plane. In the front view, the location of B also is 7 units to the right on a horizontal line from point A, which is the x-axis in that viewing plane. In the right-side view, the x-axis points out of the page; so point A and point B appear coincident in that view, although point B would actually be closer to you. Finally, connecting point A and point B in each orthogonal view creates an edge in each view.

Next, select another point (step 5) on the object near point A or point B that can be located by traveling along an edge of the object. It does not matter if the point is closer to point A or to point B because eventually all of the points on the object will be selected. Call this point C and locate it on each of the orthogonal views by noting the distance and direction you must travel to get to it from point A or point B. Point C is located 3 units from point A in the positive z-direction. In the front and right-side views, this direction is upward on a vertical line from point A. On the top view, point A and point C appear coincident because the z-axis points out of the page. Once point C is located in each view, its corresponding edge can be created.

The object's other edges are created in the same manner. You should select the points and edges to outline one entire surface of the object before moving to another surface of the object, as shown in Figure 10.30. In this way, you can see the surfaces appear one at a time instead of having a series of connected edges that extend in different directions. This process continues until the entire object is created in all of the views, as shown in Figure 10.31.

Inspecting the object's pictorial for any hidden edges (step 6) reveals that all edges shown are visible; they should be shown with solid lines. There are no internal features (step 7) on the object. As a final check (step 8), make sure each vertex in every view has at least three edges connected to it. Remember, one of these edges may be oriented into the page and thus appear as a point coincident with the vertex. If you determine that no

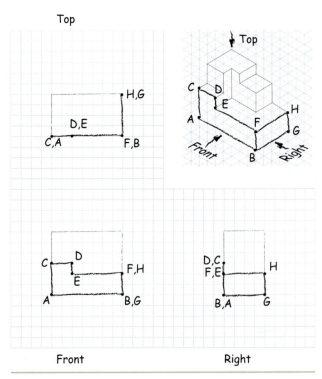

FIGURE 10.30. Connect the points to form edges of a complete surface before proceeding.

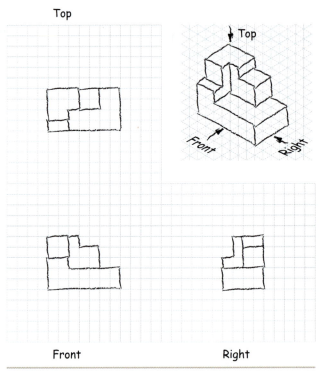

FIGURE 10.31. Continue tracking points, creating edges and surfaces until all points are accounted for.

edges are oriented into the page and only two edges are connected to a vertex, then a line must be missing.

You may have realized that even though three orthogonal views were created, this particular object could have been described using only two orthogonal views—the front and right-side views, shown in Figure 10.32. When only two views are used, it makes a difference as to which two views are used. In the example shown in Figure 10.32, specifying the front and right-side views is correct. If you were to use the front and top views, the possibility exists of either an inclined surface or a step feature, as shown in Figure 10.33.

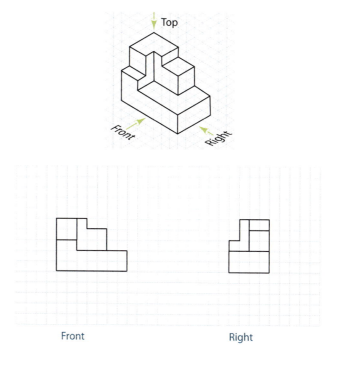

FIGURE 10.32. Two orthogonal views, the front and the right, define all of the features of the object and are an acceptable presentation on an engineering drawing. The use of three views is more common.

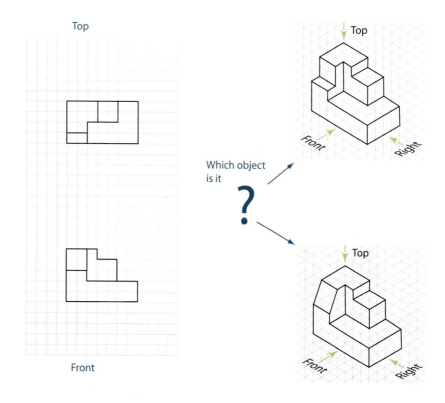

FIGURE 10.33. The wrong two views, the top and the front, lead to ambiguity. These views can represent either of two objects.

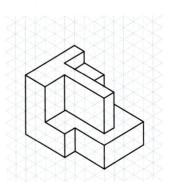

A slightly more complicated object is shown in Figure 10.34. You will try to build the multiview drawing of this object with three views.

The three views and their directions are shown in Figure 10.35. Note that the definitions of the viewing directions for this object are different from the definitions of the directions for the object in the preceding example. This does not matter as long as the definitions are consistent within the same presentation; that is, the right-side view is always on the right of the front view, the top view is always above the front view, etc., on the drawing. Whenever possible, the front, top, and right-side views should be used to represent the object unless one or more of the other standard views offers a better representation of the object's features. This example will proceed using the front, top, and left-side views to show how other views can be created and used.

A set of coordinate axes is defined on the pictorial and then transferred to the orthogonal views (step 1). Note that these directions are different from the coordinate directions in the preceding example. The location and orientation of these axes must be consistent in all views—that is what matters. In fact, as long as you are sure of the travel directions in each view, you can skip the use of coordinate axes altogether.

The foundation space is outlined in each view (step 2); and a convenient anchor point, designated A in the views, is selected (step 3). This time assume you know how to locate points on the pictorial and then transfer the location of each point to its corresponding place in each of the orthogonal views. Points near the anchor are identified on the object and then located on each of the orthogonal views (step 4), as shown in Figure 10.36.

Correctly joining the points creates the edges. The process of successively locating the object's vertices and creating edges on the multiview drawing is then extended to the rest of the object, as shown in Figure 10.37 (step 5). If you carefully examine Figure 10.37, you will notice that although the left-side view shows all of the edges of the object that can be seen in that view, some of the edges would not be seen if the object were solid and opaque. The edges that cannot be seen are hidden lines and should be shown as dashed lines for clarity (step 6).

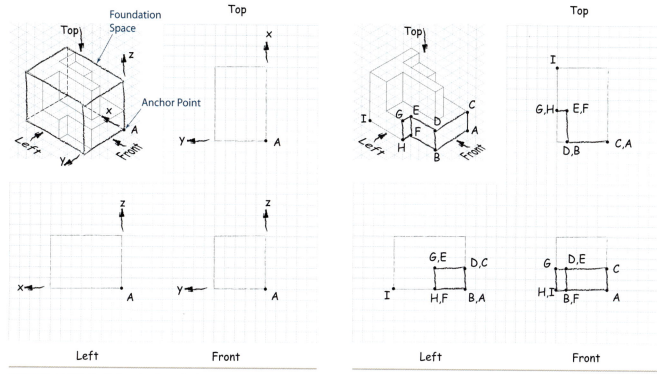

FIGURE 10.35. Defining the foundation space, viewing directions, and anchor point.

FIGURE 10.36. Designate each vertex as a point. Working with one surface at a time, locate the points and then the edges that make up each surface.

An alternative presentation is shown in Figure 10.38, where the right-side view has been added. This additional view allows you to see solid edges that are hidden in the left-side view. Note that the right-side view also has hidden lines, which are edges seen as solid lines in the left-side view. When you add the right-side view in this example, the hidden lines on both side views are no longer mandatory; but it would be wise to keep them since they do not clutter the drawing and would speed its interpretation. With the use of hidden lines, either the left- or right-side view can be deleted without harming the information conveyed about the object's geometry. Or both side views can be presented for additional emphasis to this geometry.

When internal features in this object are examined, none are found (step 7). As a final check, make sure each vertex in every view has at least three edges connected to it (step 8), keeping in mind that one of these edges may be oriented into the page and thus appear as a point coincident with the vertex. When this is not the case, a line is missing.

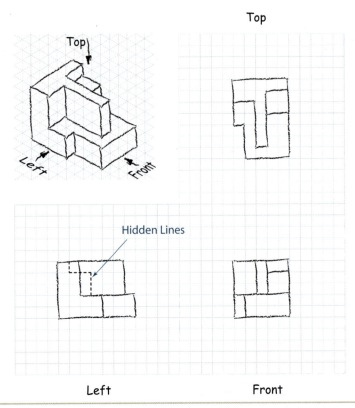

FIGURE 10.37. Continue tracking points, creating edges and surfaces until all points are accounted for. Edges that are obscured are represented with hidden lines.

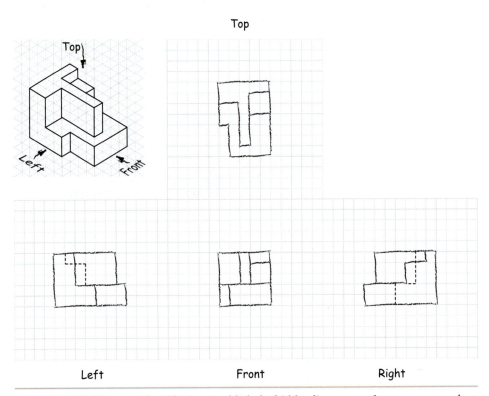

FIGURE 10.38. When a right-side view is added, the hidden lines are no longer necessary but still recommended to show all of the edges of the object.

10.06.02 Edge Tracking

Tracking individual points on an object to create its orthogonal views is a reliable method of creating the views. However, this process is slow and boring. After several trials at using this method, you will be anxious to try something faster. One way is to track an edge instead of tracking two points along the edge. This process is called **edge tracking**. It is like the eight-step process used for point tracking, where steps 1, 2, 6, 7, and 8 are the same, but steps 3, 4, and 5 are modified as follows:

Step 3:   Define an anchor edge.

Step 4:   Locate an edge adjacent to the anchor point, and draw that edge.

Step 5:   Successively locate other adjacent edges.

To create a multiview drawing of the object in Figure 10.39, decide how many orthogonal views you may need; then create the foundation space and directions (steps 1 and 2), as shown in Figure 10.40. Locate and label all of the edges on the object pictorial, keeping in mind that all of its edges may not be visible.

FIGURE 10.39. By keeping track of individual edges, how would you create a multiview drawing of this object?

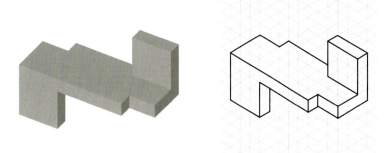

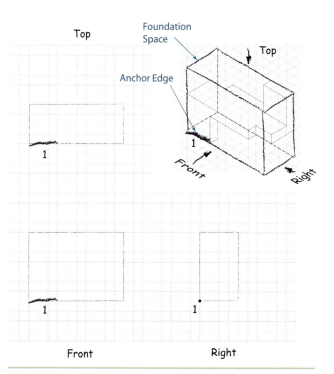

FIGURE 10.40. Defining the foundation space, viewing directions, and anchor edge. The selected anchor edge must be identifiable with confidence in all views.

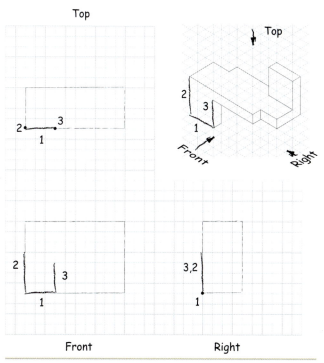

FIGURE 10.41. Defining and tracking edges near the anchor edge on the same surface.

Next, instead of selecting a convenient anchor point, select a convenient edge to use as an **anchor edge** (step 3). A good choice would be an edge whose length, location, and orientation you can easily and confidently find on each of the orthogonal views. For the object in this example, such a choice might be a straightedge on the bottom of the object. Call this edge 1 and locate and label the edge on each of the views, as shown in Figure 10.40.

Next, select an edge on the object that is connected to edge 1 (step 4). Call this edge 2 and note the size, location, and orientation of this edge with respect to edge 1. Find the same location and orientation of edge 2 in each of the orthogonal views and create this edge in those views. Keep in mind that if the edges of the object are parallel to a viewing plane, it will appear as its true length when it is viewed through that plane. If an edge on the object is perpendicular to a viewing plane, that edge will appear with both of its endpoints coincident when viewed through that plane. Note that edge 2 on the object is parallel to the front and side views; so in those views, that edge appears as its true length. Edge 2, however, is orthogonal to the top view. In that view, edge 2 appears as coincident endpoints, which would be drawn as a single point.

Continue to locate each of the object's edges on the respective orthogonal view until the object is completely represented in all orthogonal views (step 5). In the edge tracking method, you should select the edges to outline one entire surface of the object before moving on to outline another surface, as shown in Figure 10.42.

Notice on the completed multiview drawing in Figure 10.43 that for the object in this example, an edge that was originally visible during its creation later became obscured by another surface. The resulting hidden line (step 6) can be removed because the existence of that edge can be easily ascertained from the information already in the drawing. However, in this case, it would be better to leave this hidden line in place because its inclusion would facilitate the interpretation of the drawing.

There are no internal features (step 7) on this object. As a final check, make sure each vertex in every view has at least three edges connected to it (step 8).

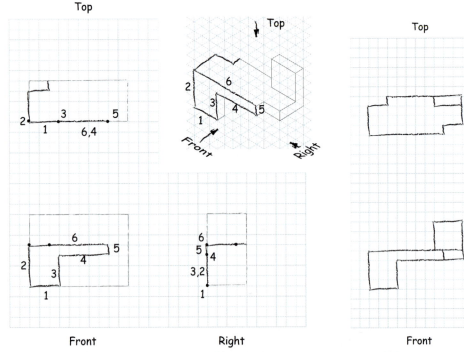

FIGURE 10.42. Define the edges of a complete surface before proceeding to the next surface.

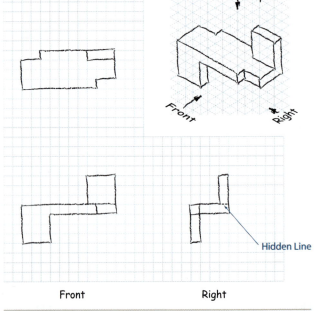

FIGURE 10.43. Continue tracking edges and surfaces until all edges are accounted for.

10.06.03 Surface Tracking

As you gain more experience and become faster at creating orthogonal views using the point tracking or the edge tracking approach, you may want to use an even faster technique. Instead of tracking an object's points or edges, you may want to track its surfaces. This is called **surface tracking**. The initial steps are the same as for point and edge tracking: first, decide how many orthogonal views you may need; then create the foundation space and directions on those views (step 1 and step 2). Steps 6, 7, and 8 are also the same; but steps 3, 4, and 5 are modified as follows:

Step 3: Define an anchor surface.

Step 4: Locate a surface adjacent to the anchor surface and draw its boundary.

Step 5: Successively locate other adjacent surfaces and draw those boundaries.

For the object shown in Figure 10.44, the viewing directions and foundation are shown in Figure 10.45 (steps 1 and 2).

Note that two pictorial views are required to reveal all of the features of this object. This is a clue that more than three orthogonal views (or some use of hidden lines) will be required to specify all the features of this object completely. In addition to front, right side, and top views, the left side and bottom views also will be included. Later the drafter can eliminate views that are unnecessary if she decides they will provide no useful information or function. For surface tracking, you must locate and label all of the surfaces on the object's pictorial, keeping in mind that all of the surfaces may not be visible.

For surface tracking, the first surface selected will be the **anchor surface** (step 3). A good choice would be a surface whose length, location, and orientation you can easily and confidently find on each of the orthogonal views, such as one of the surfaces located on the bottom of the object in this example. Call this surface A and locate and label this surface on each of the orthogonal views. Note that on the top and bottom views, surface A appears as its true shape. On the front and two side views, the surface appears on edge and is represented by a line.

FIGURE 10.44. By keeping track of individual surfaces, how would you create a multiview drawing of this object?

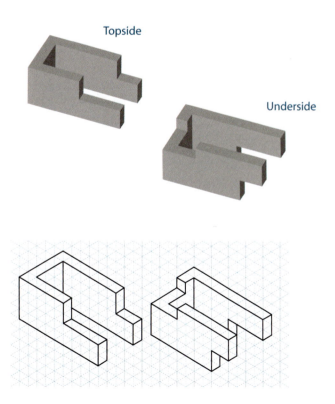

Topside

Underside

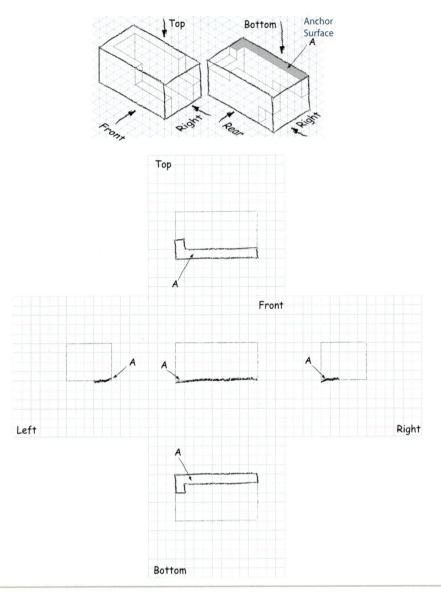

FIGURE 10.45. Create the foundation space, the viewing directions, and an anchor surface. The selected anchor surface must be identifiable with confidence in all views.

Next, select a surface that is adjacent to the anchor (step 4). Call this surface B, locate and orient this surface relative to the anchor surface, and create surface B in each of the orthogonal views. Keep in mind that object surfaces that are parallel to a viewing plane will appear as their true shape when viewed through that plane. If an object surface is perpendicular to a viewing plane, that surface will appear as a single edge when viewed through that plane. Surface B on the object is parallel to the front plane; so in the front view, that surface appears in its entirety. All of its lines will appear as their true length in the front view, and all of its angles will be appear as their true values. Surface B, however, is orthogonal to the top, bottom, and both side views.

In those four views, surface B will appear as an edge, which would be drawn as a line. Creating surface B by surface tracking is shown in Figure 10.46.

Locating each of the object's other surfaces on the orthogonal views is continued until all surfaces have been selected and the drawing is complete (step 5). Be aware that each surface that is created may partially or completely obscure surfaces that have been completed. The edges that have been obscured must be converted to hidden lines (step 6).

Step 7 and step 8 for surface tracking are the same steps used in the point tracking and edge tracking methods. The hidden lines are identified or additional views are created so that hidden lines are not necessary. The completed drawing is shown in Figure 10.47.

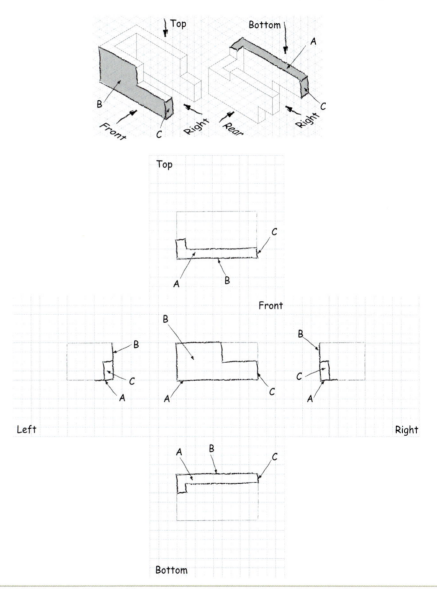

FIGURE 10.46. Locating additional surfaces adjacent to the anchor surface.

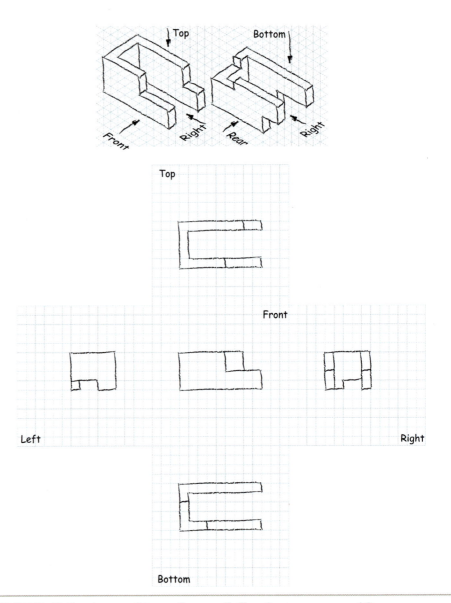

FIGURE 10.47. Continue tracking surfaces until all surfaces are accounted for.

In Figure 10.48, an alternative presentation method is shown. The rear view eliminates the need for a bottom view and a right-side view. The rear view also provides a better description of the rear of the object, which is partially obscured in the front view.

Using hidden lines (step 6), as shown in Figure 10.49, can further reduce the number of views and present the object in the preferred format of the front view, top view, and right-side view.

Whichever presentation method is used, your final step is to assure that each vertex in every view has at least three edges connected to it (step 8). Any vertex not having three edges connected to it means a line is missing. But you need to keep in mind that an edge may be oriented into the page, which means it will appear as a point coincident with the vertex.

In the preceding examples, the problems were simple because all of the surfaces on the object were parallel or perpendicular to each other and, thus, either parallel or perpendicular to the orthogonal views. These orientations are common for many but not all engineered parts.

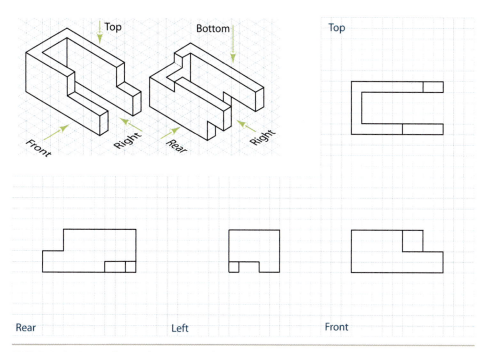

FIGURE 10.48. An alternative presentation using a rear view.

Next, you need to consider more complex objects, starting with an object that has surfaces inclined with respect to one another. That means you have to learn how to represent those surfaces in the orthogonal views of the object. Figure 10.50 shows an object with normal and inclined surfaces. For an object with inclined surfaces, the edges of the surfaces will appear to be inclined in one or more orthogonal views.

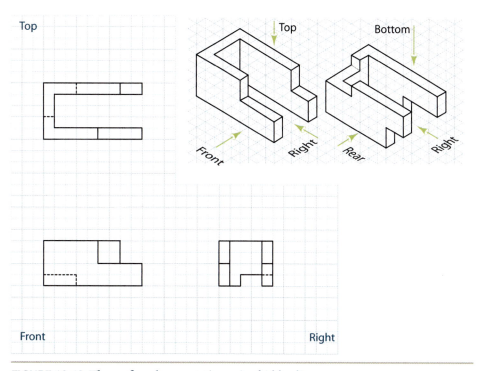

FIGURE 10.49. The preferred presentation using hidden lines.

FIGURE 10.50. Considering the existence of inclined surfaces, how would you create a multiview drawing of this object?

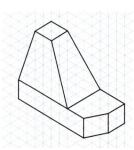

Inclined surfaces

The initial steps to solving this problem are the same as with the preceding problems. Decide how many orthogonal views you think you will need to represent the object completely; then define the space and directions occupied by the object in each of those views, as shown in Figure 10.51. Point tracking, edge tracking, or surface tracking can be used. Assume you have a little experience now and surface tracking can be used for a quicker solution. An anchor surface is necessary. This time, for convenience, surface A will be used, as shown in Figure 10.51.

One way to approach a problem with inclined surfaces is to create the surfaces that are not inclined (i.e., the normal surfaces—parallel and perpendicular surfaces—as was done in the earlier examples). This is done in Figure 10.52. When only one inclined surface is on the object, such as when one edge of the object has been beveled, or chamfered, there is no need to consciously discover the location of the object's edges. Because the other surfaces completely surround the one inclined surface, the edges of the inclined surface are formed by the edges of the other surfaces. The same would be true if there were multiple inclined surfaces on the object, unless the inclined surfaces share a common edge.

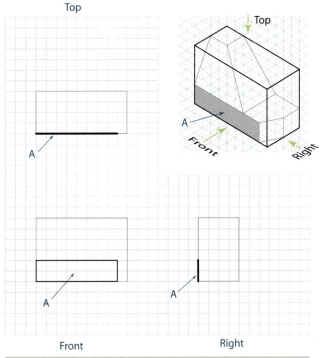

FIGURE 10.51. Define the foundation space, viewing directions, and anchor surface.

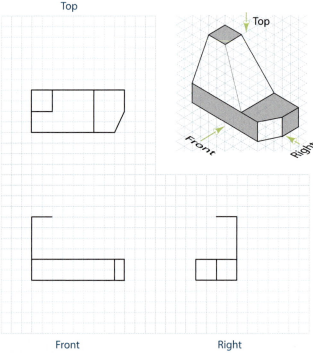

FIGURE 10.52. Continue the process of surface location for the noninclined surfaces.

FIGURE 10.53. By point tracking or edge tracking, add the intersection of the inclined surfaces.

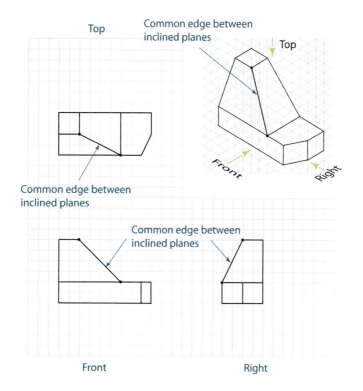

On this object, there are, indeed, two inclined surfaces that share a common edge. Once the parallel and perpendicular surfaces have been created, the shared edge of the inclined surfaces must be located. You can do this by either point tracking or edge tracking. The result is shown in Figure 10.53.

For more complicated objects, the hidden lines are identified or additional views are created so that hidden lines are not necessary.

For any object, no matter which method of tracking you use to create orthogonal views, your final step is to assure that each vertex in every view has at least three edges connected to it and to be aware that one of the edges may be oriented into the page and thus appear as a point coincident with the vertex. When this is not the case, a line is missing.

In the previous example, the inclined surfaces were inclined in a single direction only; that is, each was still perpendicular to one of the three preferred viewing planes. In this case, the entire inclined surface can be seen as an edge from two of the six standard views.

The next example is slightly more complicated. The object in Figure 10.54 has surfaces that are inclined in two directions (i.e., oblique surfaces). Neither of the two oblique surfaces is perpendicular to any of the six standard views; therefore, neither surface appears as an edge view in any of those views. Although creating the multiview drawing of this object may seem daunting, the procedure is the same as that used for creating surfaces inclined in only one direction.

FIGURE 10.54. Considering the existence of oblique surfaces, how would you create a multiview drawing of this object?

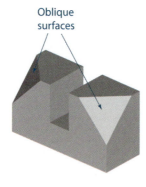
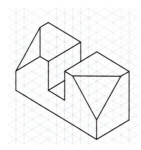

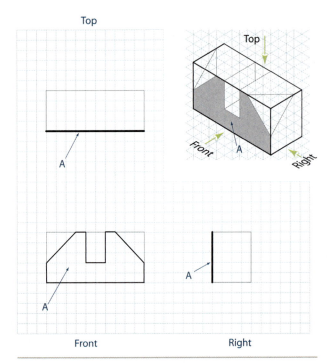

FIGURE 10.55. Define the foundation space, viewing directions, and anchor surface.

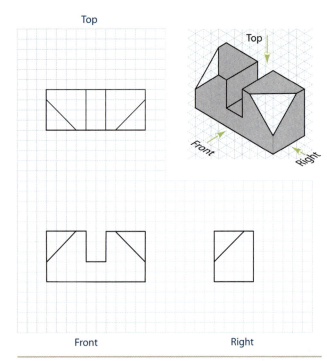

FIGURE 10.56. Continue the process of surface location for the noninclined surfaces. Since the oblique surfaces do no intersect, their boundaries are automatically formed by the normal surfaces.

Create the views for the noninclined surfaces by temporarily ignoring the oblique surfaces and using the surface tracking procedure, which is shown in Figure 10.55.

For the object in this example, the inclined surfaces do not intersect; so their boundaries are formed by the edges of the noninclined surfaces and the drawing is complete, as shown in Figure 10.56.

Next, consider how to represent in orthogonal views an object that has curved surfaces. Although there are an infinite number of types of curved surfaces, the most common are surfaces that are either cylindrical or conical; the most common curved surface is a simple round hole. Drawing a curved surface is unusual because in addition to an edge being shown where it intersects another surface, an artificial edge is drawn where there is an optical limit to the object. Examining Figure 10.57, make sure you understand that the indicated edge is not an intersection of any of the surfaces. This particular edge is the visible limit between the object and the surrounding air when the object is seen in that particular orientation. The location of this limit on the object changes when the orientation of the object changes.

When creating a multiview drawing of an object such as this, you must include one of these limits in each view of the drawing. For the object shown in Figure 10.57, the process of creating the four curved surfaces is made easier by the fact that each surface is orthogonal to one of the standard viewing planes. Therefore, the curved edges will appear in the orthogonal views as their true shape or as an edge view. When viewed as

FIGURE 10.57. Considering the existence of curved surfaces, how would you create a multi-view drawing of this object?

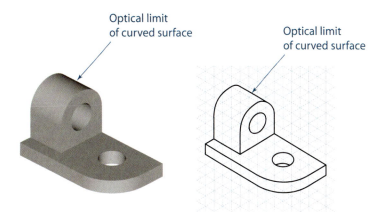

its true shape, a circle will appear as a true circle, with its correct geometry, size, and location. In an edge view, the circle will appear as a straight line. Selecting the necessary number of views, creating the foundation space, defining their directions, and selecting an anchor surface are done as before, which is shown in Figure 10.58.

Holes and rounds appear distorted as ellipses or parts of ellipses in the pictorial. In the process of surface tracking, when a surface contains a circle or a part of a circle, that edge can be drawn as a circle by locating its center point and tangent points to the other edges on the surface. This process is shown in Figure 10.59. In this example, the two holes are internal features; thus, step 7 in the eight-step drawing creation process cannot be dismissed when the views are completed.

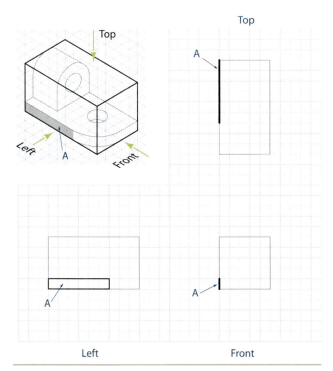

FIGURE 10.58. Define the foundation space, viewing directions, and anchor surface.

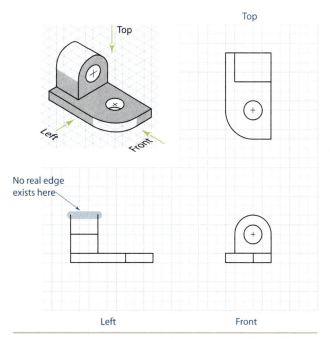

FIGURE 10.59. Locate the remaining planar surfaces of the object in all views.

FIGURE 10.60. Add the optical limits of the curved surface, add hidden lines to show hole depths, remove tangent edges, and add centerlines to the centers of the holes.

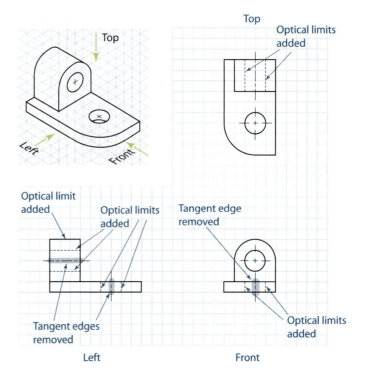

When the orthogonal views are completed, as with the pictorial, additional line segments must be added to delineate the physical limits of the object, as shown in Figure 10.60. Even though an actual edge may not exist there, it is nevertheless what would be seen if the object were real.

The hidden lines are identified or additional views are added to eliminate the need for hidden lines. For this example, it is convenient to use hidden lines to show the depth of the holes. These hidden lines do not represent true edges; rather, the hidden lines delineate the optical limits of curved surfaces internal to the object (i.e., as would be seen if the object were partially transparent).

During the final check, which involves ensuring that each vertex on the object has at least three edges connected to it, it is important to remember that an edge is formed where a curved surface is tangent to a plane. This type of edge is called a **tangent edge** and customarily is not shown on either the pictorial or its multiview drawing, except for objects where not showing the tangent edges deletes key features that may lead to misinterpretation of the drawing.

When an internal feature such as a hole or a round is located on an inclined plane on the object, as shown in Figure 10.61, creating the multiview drawing becomes more difficult. For this problem, the views of the basic object will be created first; then the

FIGURE 10.61. Considering the existence of cutouts on an inclined surface, how would you create a multiview drawing of this object?

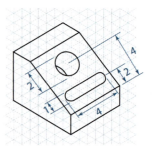

circular features (in this case, a hole and a circular slot) will be added. Because the hole and slot are located on the inclined plane, it is difficult to ascertain their true sizes even with an isometric grid. For convenience, the sizes and locations of these features are given. The center of the hole and slot are otherwise assumed to be symmetrical around the center of the inclined surface.

Determining the necessary number of views, creating the foundation space, defining their directions, and selecting an anchor surface are done as before. The multiview drawing, less the hole and slot, are shown in Figure 10.62.

Because inclined planes are not parallel to any of the six standard orthogonal views, they are not shown in their true shape in any of these six views. This means a circle on an inclined plane will appear not as a circle but as an ellipse and a circular edge will appear as a portion of an ellipse. A circle on an inclined plane will appear as a circle only in an auxiliary viewing plane created to be parallel to the inclined surface, but this is the topic of Chapter 14. The true location of the hole center, as given by the measurements on the pictorial, can be measured only in the front view, where the inclined plane is shown on its edge. The centers of the holes can be found in the other views by point tracking, as shown in Figure 10.63. The fact that points in adjacent views must be aligned vertically or horizontally greatly aids in locating the circle centers on all of the views.

Because the plane containing the circle in Figure 10.63 is inclined in one direction only (i.e., it can be seen as an edge in the front view), you can create the ellipses representing the circle in the other views by realizing that the major axis of the ellipse on the inclined surface will be the same size as the diameter of the circle.

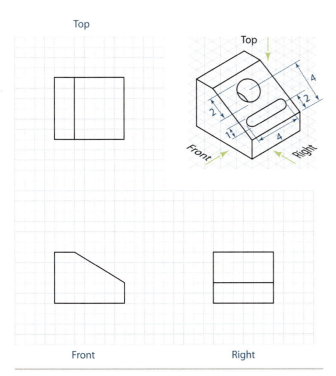

FIGURE 10.62. The basic block without the hole and slot is created.

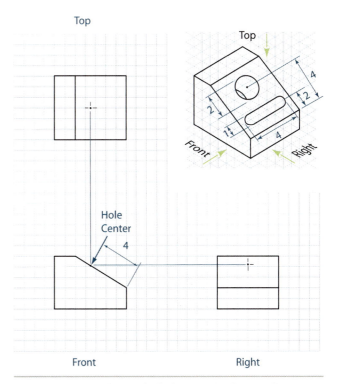

FIGURE 10.63. Locating the hole center by feature alignment in each view.

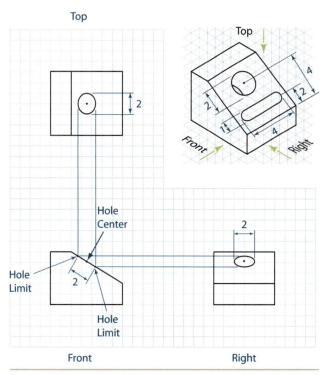

FIGURE 10.64. Construction of the major and minor axes of the ellipse in each view.

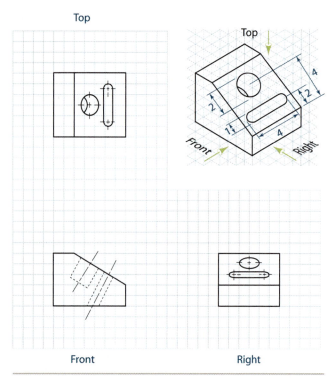

FIGURE 10.65. The addition of the slot by converting the circular edges to elliptical shapes, addition of hidden lines to indicate depth, and addition of centerlines. Note the addition of the hole bottom in the top view.

The measurements of the major and minor axes of the ellipse can be deduced graphically. You do this by marking the limits of the circle on the inclined surface in the front view, as shown in Figure 10.64, and projecting these limits into the right side and top views. Mathematically, the size of the minor axis of the ellipse will be the circle diameter multiplied by the cosine of the inclination angle of the plane from a horizontal plane.

The slot is added to the right-side view, as shown in Figure 10.65, by converting its circular edges to elliptical edges. The depth of the hole and slot are specified using hidden lines in the front view.

10.07 Breaking the Rules—and Why It Is Good to Break Them Sometimes

Creating an engineering drawing using orthogonal views is sometimes a balance between how accurately the drawing can be interpreted and how easily the drawing can be created. Strictly following some of the guidelines presented so far may lead to problems. To avoid those problems, you should consider some generally accepted exceptions to the guidelines, which are usually graphical shortcuts or approximations. These exceptions can reduce the time it takes to create a drawing and/or minimize possible misinterpretation of a drawing. With all of the exceptions that follow, the main question you need to ask yourself before using any of them is whether the approximation or shortcut could lead to misinterpretation of the drawing. If the answer is yes, the exception should not be used.

10.07.01 Threaded Parts

The first shortcut is in the representation of a threaded part, such as the bolt shown in Figure 10.66. A thread is essentially a helical mating surface for a fastener. The thread may be external, such as on the outside of a bolt or screw, as shown in Figure 10.66, or internal, such as on the inside of a nut. An accurate drawing of all surfaces on such an object would result in a very complicated drawing, especially if the drawing had to be created with manual instruments or 2-D CAD software. A much simpler representation of the external thread is shown as the schematic representation in Figure 10.66. For internal threads, the schematic representation is shown in Figure 10.67. These schematic representations are much simpler to construct with very little loss of information, especially since thread sizes are, for the most part, standardized based on the diameter of the part. A note (and arrow) is required to specify the precise thread sizes. Methods for the complete specification of thread sizes are found in Chapter 17 of this book. You can also find thread specifications in most machinists' or engineers' handbooks.

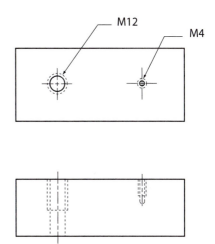

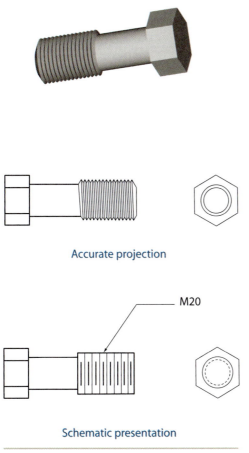

Accurate projection

M20

Schematic presentation

FIGURE 10.67. The schematic presentation of internal threads. The notes specify the metric sizes of the threads.

FIGURE 10.66. The schematic representation of an externally threaded part. The note specifies the metric size of the thread.

10.07.02 Features with Small Radii

An exception to the guidelines is in the representation of edges with small radii. Consider the object shown in Figure 10.68, which has small rounds on some of its edges. Based on the guidelines established in this chapter, a multiview drawing of the object should look like the drawing in Figure 10.68. Recall that cylindrical surfaces have no defined edges and that the tangent lines between curved and planar surfaces are not shown. Following the established guidelines, the top view should look like a featureless plane. Such a presentation, however, would likely cause confusion because upon initial inspection, the front view contains features that are absent in the top view.

A better, albeit not accurate, representation would be a presentation where the small rounds are represented as if they were true edges. The rounded edges are still shown in the front view, where their measurements can be specified. However, the approximation of the small rounds as edges enables the reader of the drawing to grasp the larger shape of the object more quickly. But what exactly is a "small" radius, and when should a small round be approximated as an edge on a drawing? The purpose of the approximation is to clarify the drawing. When the approximation clarifies the drawing, it should be used. As a rule of thumb, when the radius is less than about 5 percent of the overall size of the object, consider using the approximation.

10.07.03 Small Cutouts on Curved Surfaces

An approximation also is allowed when there is a small hole or another cutout on a curved surface. Figure 10.69, for example, shows a small hole and slot on a tube as compared to larger cutouts.

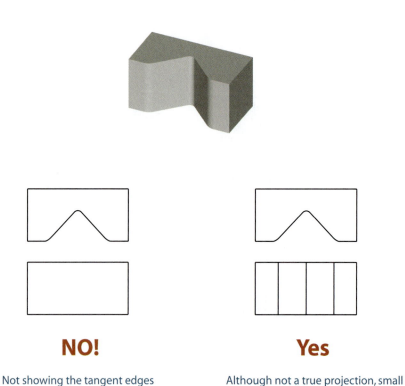

NO!

Not showing the tangent edges on small radii is an accurate projection but creates a deceiving presentation

Yes

Although not a true projection, small radii shown as edges present a clearer representation of the object geometry

FIGURE 10.68. The representation of small radii on a part.

If a true projection were made of these features, the orthogonal views would show a curved depression on the surface of the tube. The shape of this curve is complex and would take time to create. In most applications, the size of the depression on the surface is unimportant; so the depression is not shown on the orthogonal views. The true projection of these features and the accepted shortcut are shown in Figure 10.69. This approximation makes the drawing easier to create, with very little loss of information. However, when the cutouts are large or the size of the depression cannot be ignored in the function of the object, the true projection should be used. Within these guidelines, what is considered "small" is up to whoever is creating the drawing. The question that must be asked is this: Will this approximation possibly lead to misinterpretation of the drawing? If the answer is yes, the shortcut should not be used.

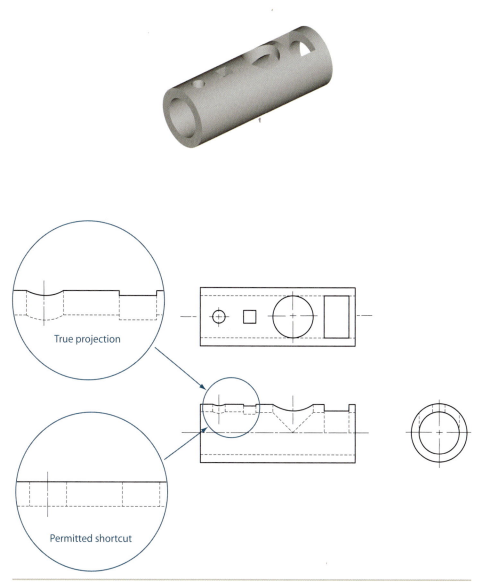

FIGURE 10.69. The true projection and an acceptable shortcut for small holes and slots on a curved surface. The shortcuts should not be used for large holes and slots because the geometric inaccuracies would be too obvious.

10.07.04 Small Intersections with Curved Surfaces

A similar approximation is allowed for small protrusions that extend from a curved surface, as shown in Figure 10.70. As with small cutouts on a curved surface, the appropriate use of this approximation is subjective. When the protrusions are small relative to the arc of the surface, their intersections on the curved surface can be shown as lines without affecting the intended representation of those features. When the protrusions are large relative to the arc of the surface, the approximation cannot be made. Again, the question that must be asked is whether this approximation could lead to misinterpretation of the drawing. If the answer is yes, the shortcut should not be used.

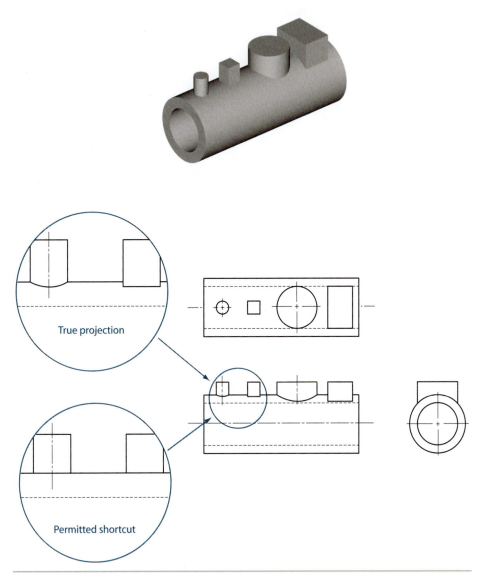

FIGURE 10.70. The true projection and an acceptable shortcut for small protrusions from a curved surface. The shortcuts should not be used for large protrusions because the geometric inaccuracies would be too obvious.

10.07.05 Symmetrical Features

An interesting exception to the rules of true projection occurs in the representation of objects with symmetry, as shown in Figure 10.71. This object has one-third rotational symmetry, which means the object can be divided into three identical sections about its axis of rotation, with three support ribs about the center tube.

An accurate multiview drawing would be the true projection drawing shown in Figure 10.71 using a front and top view. However, using a true projection for the front view in this case has two problems. One problem is that when instruments or 2-D CAD is used, an accurate projection is difficult to create. The other problem is that the true projection of the side view may be incorrectly interpreted as representing a nonsymmetrical object.

A preferred presentation for this drawing is shown in Figure 10.71. This drawing is easier to create and gives the impression that the object is symmetrical. The top view clarifies any possible misinterpretation about the number and locations of the support ribs. This may seem strange, but if the object had one-quarter rotational symmetry, for example, with four equally spaced support ribs instead of three, the front view would be the same as the view for the three support ribs.

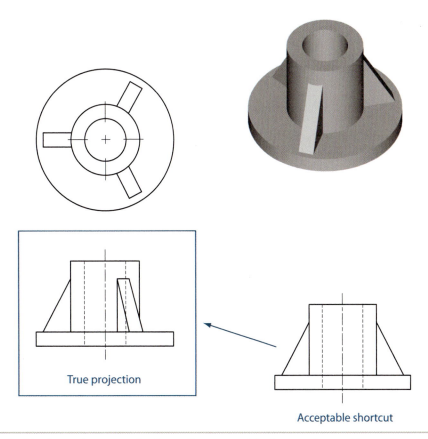

True projection

Acceptable shortcut

FIGURE 10.71. The true projection and an acceptable shortcut for an object with prominent symmetry. This property is emphasized by the use of a projected view that is modified to appear symmetrical.

10.07.06 Representation of Welds

Objects that contain welds, which are very common in civil engineering and some mechanical engineering applications, use special notation to specify the geometry of the weld. The use of this notation increases the speed of drawing creation with little loss of information. A simple object made from individual pieces that are welded together is shown in Figure 10.72. Even though a welded object is composed of two or more smaller pieces, it is common that such an object be fabricated at a single shop and delivered as a single unit. Thus, a single drawing showing the final welded configuration is often desirable.

Drawing the geometry of the welds on the multiview drawing takes time and effort, especially when the object contains many welds. So instead of the weld being drawn, a shorthand symbol is used. The notation specifies the geometry and locations of the weld, as well as any necessary modifications to the individual pieces in preparation for welding.

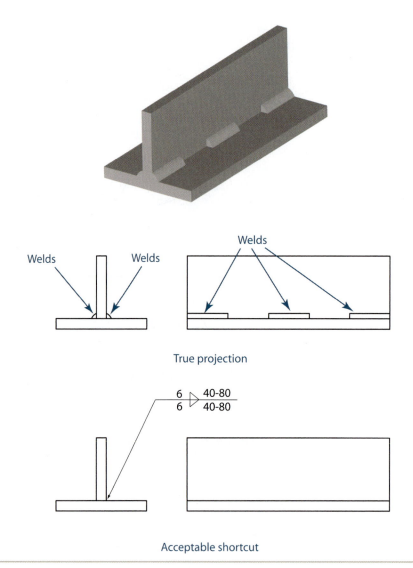

FIGURE 10.72. The acceptable presentation of two parts that are welded together to make a single part. The note specifies the size and location of the welds.

CAUTION

Inexperienced engineers, designers, and drafters can unwittingly introduce errors to their drawings. Despite the errors, the person reading the drawing probably will interpret it as intended because the necessary information is contained on the views not having an error. Nevertheless, errors can cause confusion and slow down interpretation of the drawing. A more serious case of errors can result with an ambiguity that makes the drawing impossible to interpret correctly. In a worst-case scenario, the errors may cause the object to be interpreted as an entirely different object than what was desired. The following sections are a compilation of the most common beginners' errors and ways to fix them.

Missing Lines

A common problem with hand-created or 2D CAD-created drawings is that one or more line segments may be missing from one or more of the orthogonal views. This error is especially difficult to correct when someone else made the drawing. As an example, examine the drawing shown in Figure 10.73, which shows the top, front, and right-side views of an object. Two lines are missing from the side view, and one line is missing from the front view.

The general procedure for locating a missing line is to examine the vertices in the adjacent views, as shown in Figure 10.74. Vertices are formed when surfaces or edges intersect as features on the object. A vertex in one view means there must be a corresponding vertex or edge in its adjacent views. Also, vertices representing the same point or edge on the object must be aligned horizontally or vertically in adjacent views. To discover any missing features, you can start with any view; but eventually you have to examine all of the views. From each vertex in a view, create horizontal or vertical alignment lines into the adjacent view. In the adjacent views along each alignment line, you should see the meeting of two surfaces to form an edge or the meeting of multiple surfaces to form another vertex.

Two vertices in the front view are missing some corresponding feature in the side view. The top view reveals that these vertices are the intersections of perpendicular edges. A horizontal line representing an edge must be added to the side view to keep the

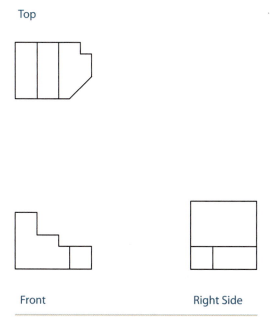

Top

Front   Right Side

FIGURE 10.73. Are any lines missing from this multiview drawing? If so, insert them into the views at their correct locations with their correct visibilities.

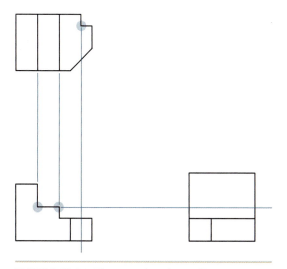

FIGURE 10.74. These vertices do not have corresponding features (i.e., another vertex or edge) in all views.

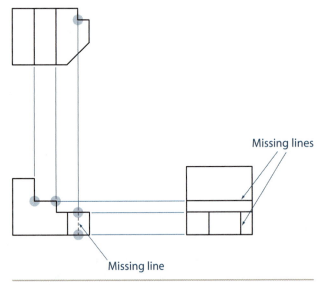

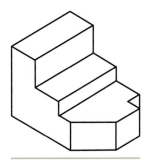

FIGURE 10.76. An isometric presentation of the object shown in the previous figure.

FIGURE 10.75. The missing lines are shown here. The vertex in the top view produces a hidden line in the front view, which, in turn, shows that another line is missing in the side view.

features consistent between the front and top views. Keep in mind that when a line appears to be missing, the next step is to determine whether that line should be visible; that is, the missing line might actually be a hidden line. Because the front view is uncluttered, it would be best to include the hidden line because it would reinforce the presence of the rectangular cutout on the back of the object. This hidden line produces two new vertices, which are missing some corresponding feature in the side view. A vertical line representing an edge must be added to the side view to maintain feature consistency between the front and side views. Figure 10.75 shows the drawing with the missing lines added. Figure 10.76 shows an isometric presentation of the complete object.

As another example, the drawing in Figure 10.77 has two lines missing from the front view. You can find the missing lines by examining the vertices in each view to ensure that a vertex in one view leads to some corresponding feature in an adjacent view, as shown in Figure 10.78. Note that hidden lines are used to show the depth of the

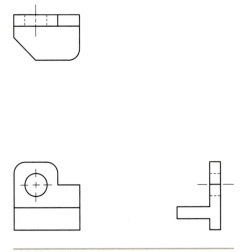

FIGURE 10.77. Are any lines missing from this multiview drawing? If so, insert them into the views at their correct locations with their correct visibilities.

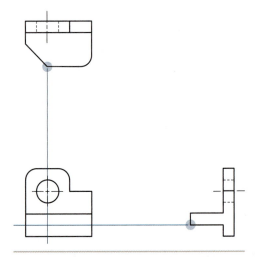

FIGURE 10.78. These vertices do not have corresponding features in all views.

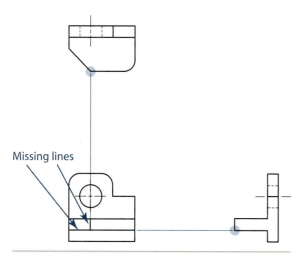

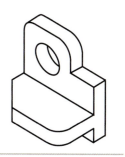

FIGURE 10.80. An isometric presentation of the object in the previous figure.

FIGURE 10.79. The missing lines are shown here. Tangent edges are not shown.

hole. These hidden lines are optical limits, not true edges; therefore, they do not form vertices or intersections as with true edges. Also observe that tangent edges are not shown even though they form intersections and vertices. Figure 10.79 shows the drawing with the missing lines added. Figure 10.80 shows an isometric presentation of the complete object.

Solving problems of missing lines is an excellent way to develop your skills with multiview projection.

Missing Views

When an entire view is not shown, consider it an opportunity to challenge your ability to find missing lines. Figure 10.81 is the drawing of an object where the front view is not shown. Although a missing view of this type rarely occurs in real-world engineering, this is the kind of problem for homework and exams. These types of problems are a test of your ability to recognize and extract 3-D data from 2-D views.

The procedure for finding the missing view in Figure 10.81 is to locate identical vertices in the given views and then transfer the locations of these vertices into the missing view. One way to proceed in this example is to select, one at a time, vertices that

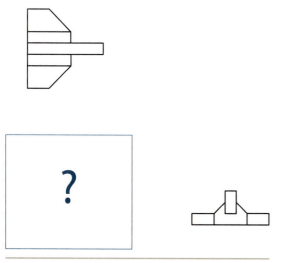

FIGURE 10.81. Create the front view of this object from the top and side views.

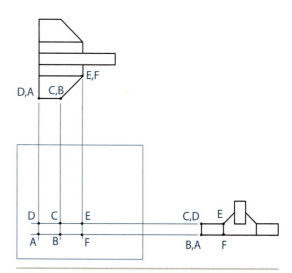

FIGURE 10.82. Locate corresponding vertices by alignment in the given views.

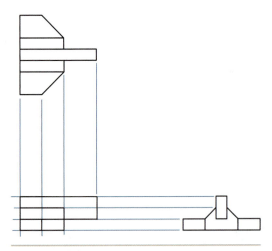

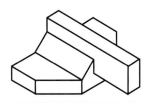

FIGURE 10.84. An isometric presentation of the object in the previous figure.

FIGURE 10.83. Continue locating vertices, adding edges in the front view when they exist in either of the given views.

are closest to the front view and then proceed toward the back of the object. Edges between the vertices in the front view are produced when the existence of the edge is evident for one of the given views. This process is shown in Figure 10.82.

Figure 10.83 shows the drawing with the missing view added. Figure 10.84 shows an isometric presentation of the complete object.

Incorrect Visibility

Sometimes all of the lines are there, but the visibility of one or more lines is incorrect. This means lines that are suppose to be shown dashed (representing hidden lines) are erroneously shown as solid (representing visible lines) or vice versa. Figure 10.85 shows an example of a drawing with an incorrect line visibility in the front view. One of the hidden lines is erroneously shown as a visible edge. Even though a person reading

FIGURE 10.85. Care must be taken to ensure that the visibility of edges is correct.

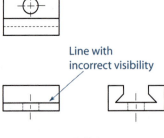

Line with incorrect visibility

NO!

A hidden edge is incorrectly displayed as a solid line

Yes

Hidden edges displayed as dashed lines, visible edges displayed as solid lines

FIGURE 10.86. Overuse of hidden lines causes confusion.

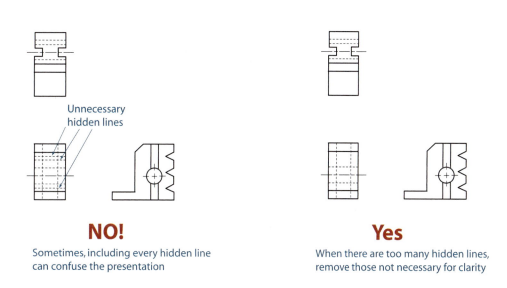

the drawing would probably figure out the error when it is realized that features between the views are not consistent, the error does create some confusion.

Too Many Hidden Lines

Although hidden lines usually aid in the interpretation of a drawing, using more hidden lines than are necessary can lead to confusion. For example, in Figure 10.86, the original front view shows the hidden lines representing the internal features (the hole and slots), as well as the V-grooves on the back of the object, resulting in a large number of hidden lines in that view. There should be no confusion about which feature a hidden line represents. For such cases, consider using hidden lines to emphasize only the most important features or use additional views to characterize the features fully. In this case, limiting the hidden lines in the front view clarifies the presentation. Hidden lines are used for the hole and the two slots in the front view to confirm that these features extend straight and all the way through the object. The fact that the three V-grooves also extend straight and all the way through the object can be confirmed from the hidden lines used in the top view.

Too Few Hidden Lines

Anyone reading a drawing not having any hidden lines, as in Figure 10.87, may have to guess the geometry of features. In this figure, without hidden lines or additional views for clarification, the drawing has two possible interpretations, as shown; the reader will have to guess which interpretation is corect. Never make the reader guess because that guess may be wrong. And if you made the drawing, the wrong guess would be your fault, not the reader's. Legally, a drawing's creator is responsible for ensuring that accepted guidelines for geometry presentation are followed so the drawing's contents cannot be misinterpreted. Just because something is obvious to you does not mean it will be obvious to someone else. Figure 10.88 shows the correct drawing for the desired object, using hidden lines.

FIGURE 10.87. Underuse of hidden lines may delete critical information.

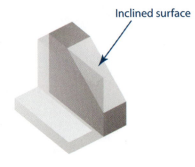

Which one of these objects does the drawing represent

?

FIGURE 10.88. Use hidden lines as often as practical to define and reinforce features.

Squared surfaces

Squared surfaces

Inclined surface

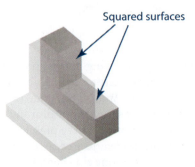

This one

Not this one

FIGURE 10.89. Centerlines and centermarks should be used whenever possible to help define and locate the centers of holes and other axes of rotational symmetry.

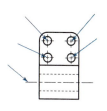

Added centerlines and centermarks

NO!

Centerlines and centermarks are missing from cylindrical surfaces

Yes

Centerlines and centermarks are present on cylindrical surfaces

No Centerlines and/or No Centermarks

To aid in the interpretation of a drawing, centerlines and centermarks mark the location of an axis of rotational symmetry. A centerline on a feature alerts the reader that the edges seen next to the centerline may be optical limits of a surface, not true edges. Without these marks, a drawing cannot be interpreted as quickly, as shown in Figure 10.89. With centermarks and centerlines added, the hidden lines and circles are quickly identified as having been created by holes.

Showing Tangent Edges

A tangent edge is formed when two curved surfaces or a curved surface and a plane are tangent to each other on a line. Tangent edges are normally not shown on drawings because they cannot be seen on a real object. Showing tangent edges, as in Figure 10.90, gives the false impression of the existence of a visible edge. Ideally, a tangent edge on a real object is smooth, without any abrupt changes in surface direction; and the drawing is made to reflect this attribute of real surfaces.

Not Showing Tangent Edges

Some exceptions exist as to when tangent edges should be shown to aid in interpreting the drawing. There are cases, for example, where the precise locations of tangent edges are important for the proper function of the object; and those locations must be emphasized for clarity, as shown in Figure 10.91. In this case, the tangent edges show precisely where the curved surfaces and the flat surfaces intersect. Also, when pictorials of objects that contain many rounded edges are created, the absence of tangent edges can produce relatively featureless presentations. In these cases, it is better to make the tangent edges visible.

FIGURE 10.90. On a real part, tangent edges cannot be seen and thus, in most cases, are not shown on a drawing.

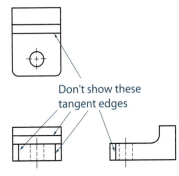

Don't show these tangent edges

Tangent edges deleted

NO!

Tangent edges have incorrectly been included

Yes

Tangent edges generally should not be shown

FIGURE 10.91. In a some cases, tangent edges may be shown to emphasize surface geometry; otherwise, views may appear to be featureless.

Missing tangent edges

Show tangent edges

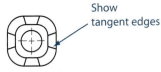

NO!

Removal of tangent edges results in deceptively featureless views

Yes

Including the tangent edges reinforces the existence of curved surface features

FIGURE 10.92. Small radii between surfaces should be shown as edges; otherwise, views may appear featureless.

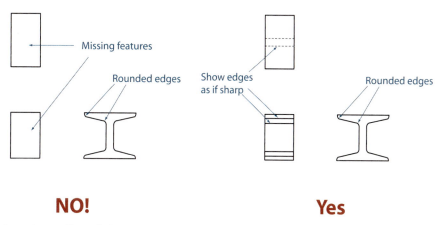

Missing features

Rounded edges

Show edges as if sharp

Rounded edges

NO!

Removing small rounded edges between intersecting surfaces results in deceptively featureless views

Yes

Small rounded edges should be represented as simple visible edges to empahsize their existence

Not Showing Small Radii

The intersection of two surfaces is not usually a sharp edge, but rather a smooth transition with a small radius between the surfaces. The general purpose of this transition is to reduce the number of external sharp edges for safety during use and handling of the object and to eliminate breakage by reducing stress concentrations that exist at sharp internal corners. When the tangent edges of these transitions are removed, the result is a drawing devoid of features in certain views, especially when the object has inclined surfaces, such as the object shown in Figure 10.92. When the radius of a transition is very small relative to the remainder of the object, instead of the object's tangent edges being shown, it is acceptable to show the transition as a single edge, as if the radius is a sharp edge.

Mismatched View Scales

When the scale, or object magnification, in each view of a drawing is different, as shown in Figure 10.93, it becomes very difficult to align features between views. Consequently, when an object has many features, it becomes difficult to identify and characterize those features correctly. All of the orthogonal views on a drawing must have the same scale.

Unaligned Views

When the views are not aligned, as shown in Figure 10.94, it is difficult to align the same features in each view so that each feature can be uniquely identified. The rules of orthogonal projection and multiview presentation mandate that orthogonal views be aligned.

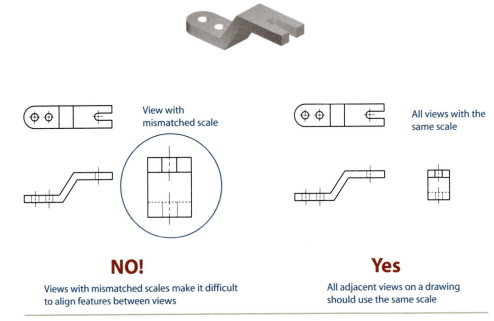

NO!

Views with mismatched scales make it difficult
to align features between views

Yes

All adjacent views on a drawing
should use the same scale

FIGURE 10.93. Orthogonal views must have the same scale.

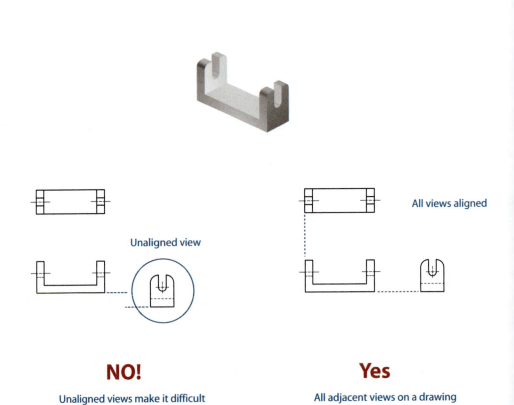

NO!

Unaligned views make it difficult
to align features between views

Yes

All adjacent views on a drawing
should be aligned

FIGURE 10.94. Orthogonal views must be aligned.

Views in Incorrect Relative Locations

The rules of first-angle or third-angle projection dictate that different views on a drawing must be located in certain positions with respect to each other. Consequently, people who read drawings have learned to expect certain views to appear in certain locations, such as a top view located above a front view (using third-angle projection). When the locations of the views are different from what is expected, as shown in Figure 10.95, the reader may become confused because the same features on the object are no longer properly aligned horizontally and vertically between these views.

Poor Choice of Object's Original Orientation

A poor choice in the original orientation of the object leads to a drawing with many hidden lines and/or inclined surfaces, as shown in Figure 10.96. The choice of object orientation should be such that the use of hidden lines and/or inclined surfaces is minimized.

Incorrect Rotational Orientation within a View

Inexperienced drafters may rotate a view by 90 or 180 degrees from its correct orientation with respect to the other views, as shown in Figure 10.97. An indication that this has happened is when the outer edges of the object are not aligned horizontally or vertically in adjacent views or when features on the object do not align. Such a rotation would confuse the person trying to read the drawing. Care should be taken to ensure that the rotational orientation of every view is correct with respect to first-angle or third-angle projection.

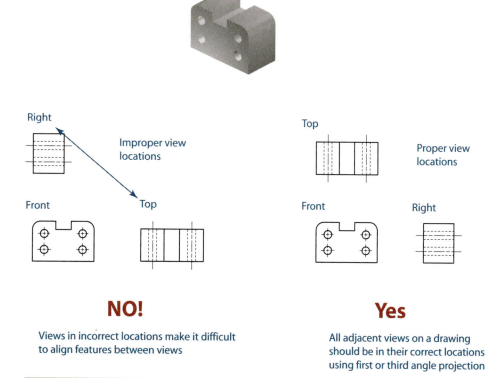

FIGURE 10.95. Orthogonal views must be in their proper locations with respect to one another.

FIGURE 10.96. The part should be oriented to show as many visible lines as possible in their true length.

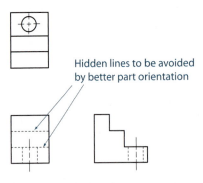

Hidden lines to be avoided by better part orientation

NO!

This object orientation generates many hidden lines which can be avoided by using a differnt orientation

Yes

This object orientation minimizes the number of hidden lines that are generated

FIGURE 10.97. The rotational orientation within each view must be consistent with proper orthogonal projection, as seen in the glass box.

Incorrect rotational orientation of view

NO!

Views in incorrect rotational orientations make it difficult to align features between views

Yes

All adjacent views on a drawing should be in their correct rotational orientation

FIGURE 10.98. An object such as this one cannot be fully described by the six standard views.

FIGURE 10.99. An object with internal features such as this one cannot be fully described by the six standard views.

10.08 When Six Views Are Not Enough

It would seem that the six orthogonal views provided by the glass box would be sufficient to specify the geometry of any object. But the views are not sufficient for every object.

10.08.01 Features at Odd Angles

An example of an object requiring more than six views or nonstandard views is shown in Figure 10.98, where features are located on surfaces that are inclined or oblique. For this object, none of the six standard orthogonal views would show these features in their true shape. A supplementary view, known as an auxiliary view, must be created before measurements can be specified for the feature represented in that view. Auxiliary views are covered in detail in Chapter 14 of this book.

10.08.02 Internal Features

Certain internal features, such as holes, bores, or cutouts with an irregular wall profile and details that are hidden from view, cannot be seen in any of the six standard views. An example of an object with internal features is the wheel shown in Figure 10.99. For this object, the geometry of spokes cannot be seen because the rim of the wheel obscures it. Although hidden lines can sometimes be used to show such features, those features will appear more clearly in a cutaway, or section view. Section views of all sorts are covered in detail in Chapter 13 of this book.

10.09 Considerations for 3-D Modeling

The proliferation of 3-D solids modeling software, especially in mechanical engineering applications, has made the process of creating drawings much easier than in the past. Typically, with solids modeling software, objects are initially modeled as a series of protrusions and cuts to create their 3-D graphical representation. The solids modeling software creates a mathematical model of the geometry from which the projections of the object are used to create drawings. The model can be scaled and rotated for viewing from any orientation direction. Once the solids model is created, it usually is simple to specify the viewing directions needed to for the software to create isometric and other pictorial views. It also is easy to extract a front view, side view, or any of the other orthogonal views from a solids model.

The ease with which pictorials and multiview drawings can be created from a solids model has many advantages, but also some disadvantages. The greatest advantage is the speed and accuracy with which orthogonal views can be created. With most software, additional views can be created by specifying the location of the viewing plane and then picking a location on the drawing where the additional view is to appear. Usually this is done by striking a few keys on a keyboard or making a few clicks with a mouse or another pointing device. The time required to produce the additional view is usually only a few seconds. Hidden lines can be added or removed for individual features or for an entire view. Also, accurate orthogonal projections of features that were previously represented by shortcut practices, such as small cutouts in curved surfaces or thin symmetric features, are easily created. In fact, with most software, it would be difficult to create a view that is *not* an accurate projection.

But there is a disadvantage to having so much ease in creating drawings. Remember, the original process of manually creating projected views from pictorials and mental images and pictorials from projected views depended on the drawer's developed skills of spatial reasoning and mental imaging. When software makes the process of creating drawings too automatic, a person may not be able to apply these skills in the absence of the software because she did not develop adequate drawing skills. In other words, the person may have become too dependent on the software. That person, when faced with a multiview drawing in the shop, may not be able to create a mental image of the object or may not develop the skills necessary to interpret standard drawings. Eventually, the person will develop these skills, but it may require experience with many solids models and their drawings. Whether you are working with instruments, 2-D CAD, or solids modelers, the key to successful development of mental imaging skills is simply to practice—a lot.

10.10 Chapter Summary

Orthogonal projection and the use of the standard views of an object are accepted nationally and internationally as the formal means of creating and presenting images for the purpose of producing the original object. Constructed correctly, these views are used to re-create the same 3-D object, no matter who is viewing the images. Care must be taken to ensure that the rules for view creation, orientation, scale, and alignment are followed. Hidden lines are used for completing the description or for additional emphasis of certain features on the object. Extra views are used as necessary for completing the description of these features. From these formal views, the original 3-D object can be re-created. When done successfully, whether you are the person making the drawing or the person reading the drawing, you will find that the interpretation of the views and the object they represent are the same.

10.11 glossary of key terms

adjacent views: Orthogonal views presented on a single plane that are created immediately next to each other.

anchor edge: The same edge that can be easily and confidently located on multiple views and on a pictorial for an object.

anchor point: The same point, usually a vertex, that can be easily and confidently located on multiple views and on a pictorial for an object.

anchor surface: The same surface that can be easily and confidently located on multiple views and on a pictorial for an object.

centerline: A series of alternating long and short dashed lines used to identify an axis of rotational symmetry.

centermark: A small right-angle cross that is used to identify the end view of an axis of rotational symmetry.

edge tracking: A procedure by which successive edges on an object are simultaneously located on a pictorial image and on a multiview image of that object.

first-angle projection: The process of creating a view of an object by imprinting its image, using orthogonal projection, on an opaque surface behind that object.

foundation space: The rectilinear volume that represents the limits of the volume occupied by an object.

glass box: A visualization aid for understanding the locations and orientations of images of an object produced by third-angle projection on a drawing. The images of an object are projected, using orthogonal projection, on the sides of a hypothetical transparent box that is then unfolded into a single plane.

hidden lines: The representation, using dashed lines, on a drawing of an object of the edges that cannot be seen because the object is opaque.

multiple views: The presentation of an object using more than one image on the same drawing, each image representing a different orientation of the object.

multiview: Refers to a drawing that contains more than one image of an object and whose adjacent images are generated from orthogonal viewing planes.

orthogonal projection: The process by which the image of an object is created on a viewing plane by rays from the object that are perpendicular to that plane.

point tracking: A procedure by which successive vertices on an object are simultaneously located on a pictorial image and a multiview image of that object.

preferred configuration: The drawing presentation of an object using its top, front, and right-side views.

six standard views (or six principal views): The drawing presentation of an object using the views produced by the glass box (i.e., the top, front, bottom, rear, left-side, and right-side views).

surface tracking: A procedure by which successive surfaces on an object are simultaneously located on a pictorial image and a multiview image of that object.

tangent edge: The intersection line between two surfaces that are tangent to each other.

third-angle projection: The process of creating a view of an object by imprinting its image, using orthogonal projection, on translucent surface in front of that object.

viewing plane: A hypothetical plane between an object and its viewer onto which the image of the object, as seen by the viewer, is imprinted.

10.12 questions for review

1. What is orthogonal projection?

2. What are the advantages and disadvantages of using pictorial images, such as isometric images, for the graphical representation of an object?

3. What is a multiview presentation?

4. What are the advantages and disadvantages of using a multiview presentation for the graphical representation of an object?

5. How are different views located with respect to each other on the same drawing?

6. Why should features be aligned between views in a multiview presentation?

7. Why is it important that different views have the same scale?

8. What are the advantages of having features of an object aligned between views?

9. What are the standard (or principal) views?

10. What is the preferred configuration?

11. When should extra orthogonal views be used?

12. When should hidden lines be used?

10.12 questions for review (continued)

13. When should hidden lines not be shown?

14. When should tangent edges be shown?

15. When should tangent edges not be shown?

16. What is the difference between first-angle projection and third-angle projection?

17. When can the rules of orthogonal projection be bent? What are the advantages of doing so?

10.13 problems

The following exercises may be done with instruments, with CAD, or on square and isometric grids. Use third-angle or first-angle projection as specified by your instructor.

1. From the isometric pictorials shown in Figure P10.1, create accurate multiview drawings with a sufficient number of views to specify all details of the object completely. Do not use hidden lines.

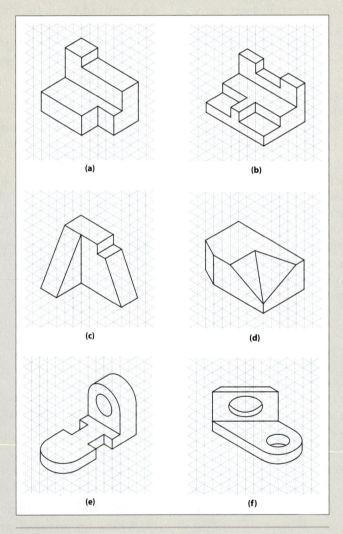

(a) (b)

(c) (d)

(e) (f)

FIGURE P10.1.

10.12 questions for review (continued)

2. From the isometric pictorials shown in Figure P10.2, create accurate multiview drawings in the preferred format of front view, top view, and right-side view. Use hidden lines as necessary to specify all details of the object completely.

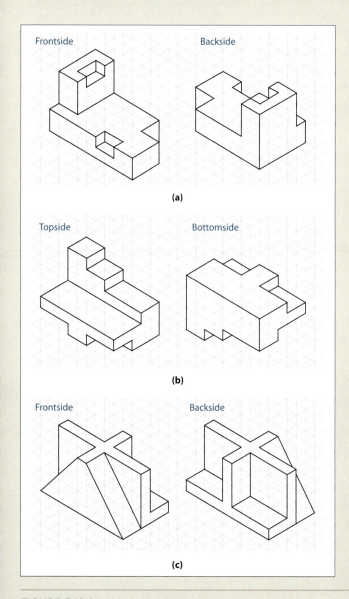

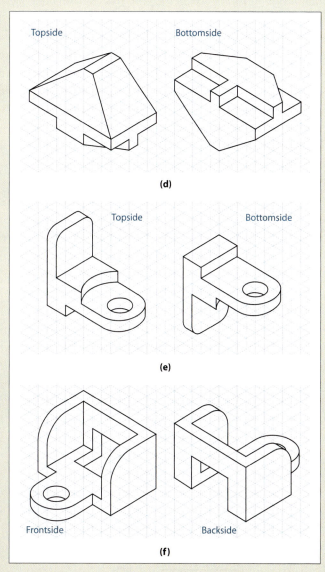

FIGURE P10.2.

10.13 problems (continued)

3. Each set of multiview drawings shown in Figure P10.3 may have visible or hidden lines missing. Add the missing lines to the drawing. An isometric pictorial has been included for clarity.

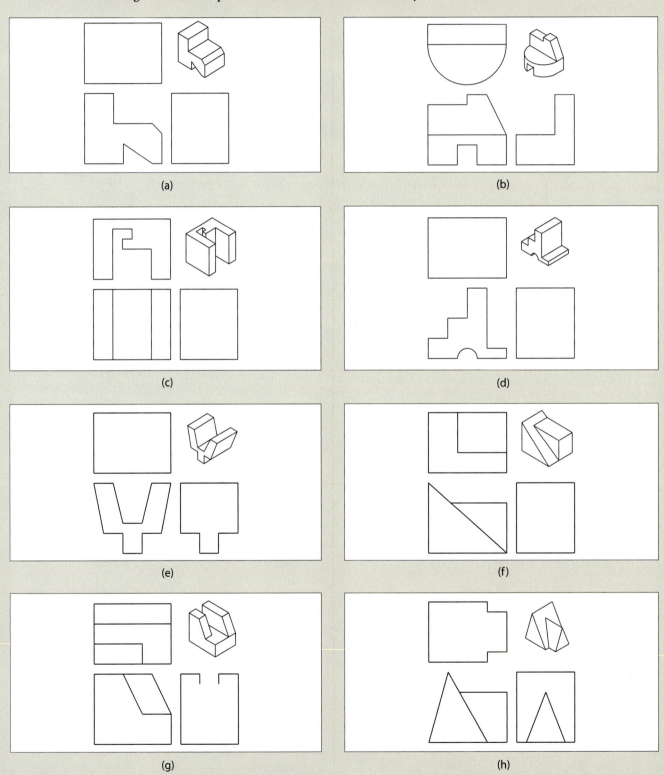

(a) (b)

(c) (d)

(e) (f)

(g) (h)

FIGURE P10.3.

10.13 problems (continued)

4. For each front view shown in Figure P10.4, draw the top view (in the correct scale, location, and orientation) that corresponds to each of the possible side views that are shown.

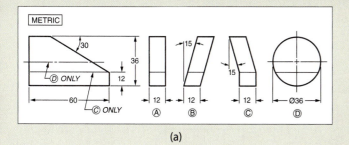

(a)

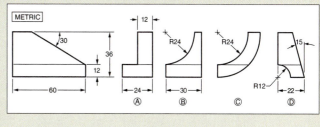

(b)

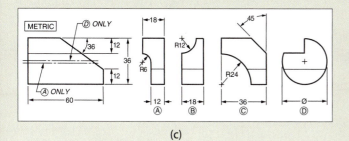

(c)

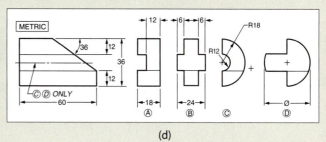

(d)

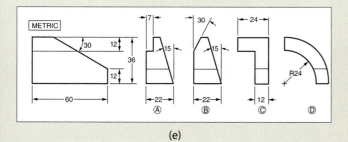

(e)

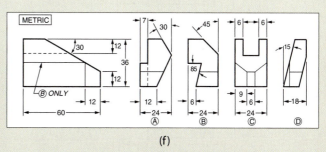

(f)

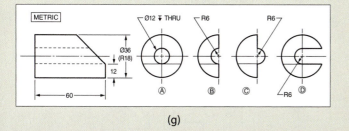

(g)

FIGURE P10.4.

10.13 problems (continued)

5. For each set of multiview drawings shown in Figure P10.5, add the missing view to the drawing in the indicated location. An isometric pictorial has been included for clarity.

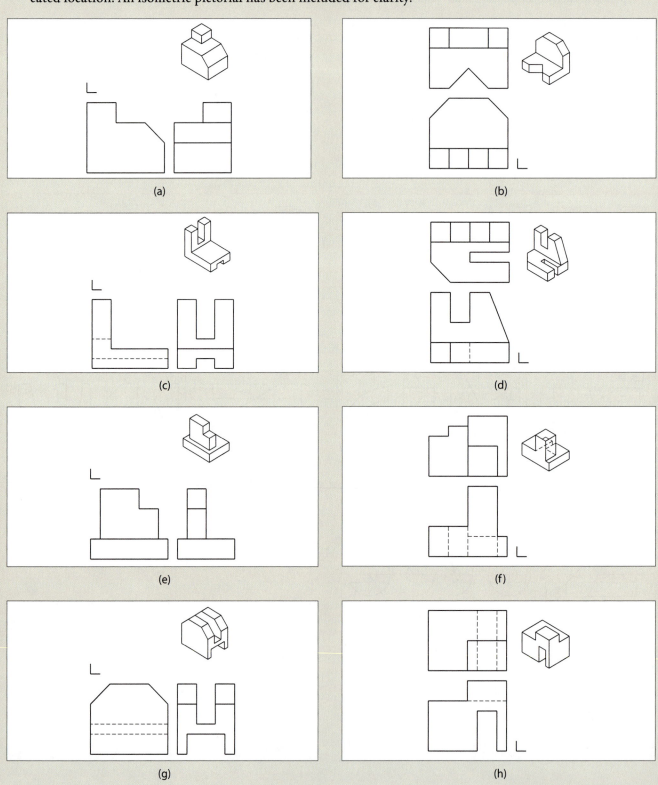

(a)   (b)

(c)   (d)

(e)   (f)

(g)   (h)

FIGURE P10.5.

10.13 problems (continued)

6. Create correctly scaled multiview orthogonal drawings of the objects shown in Figure P10.6. Show at least the front, top, and right-side views. Include hidden lines. Recommend when additional views would be useful to clarify the presentation and add these views to the drawing.

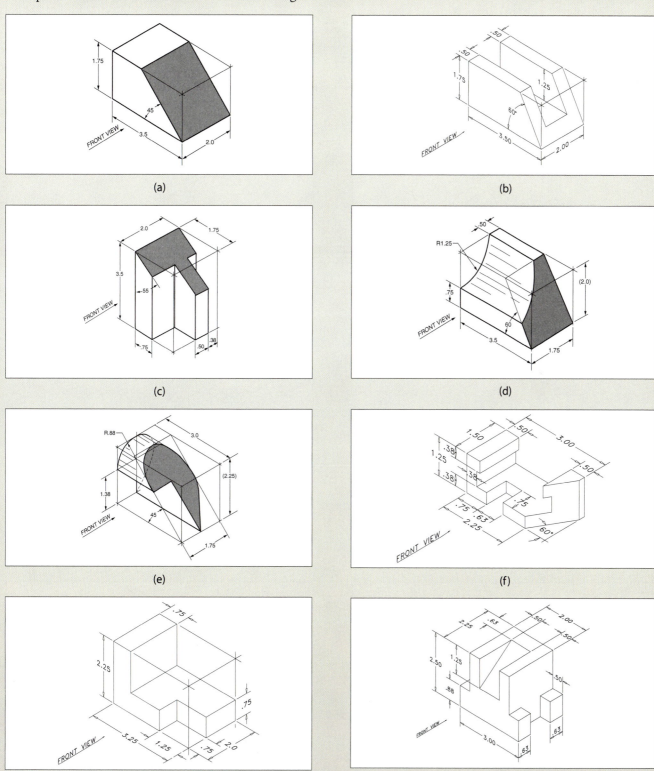

(a)   (b)

(c)   (d)

(e)   (f)

(g)   (h)

10.13 problems (continued)

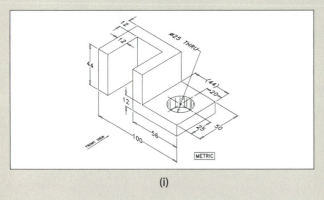

(i)

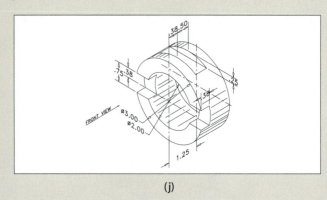

(j)

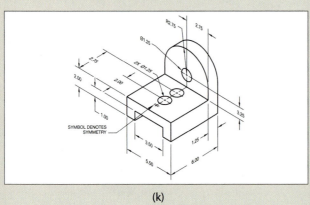

(k)

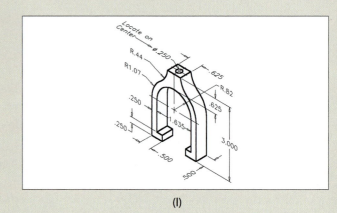

(l)

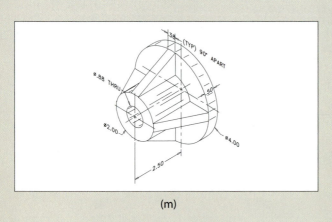

(m)

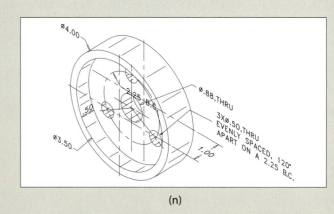

(n)

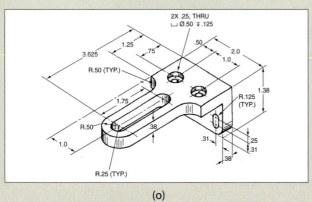

(o)

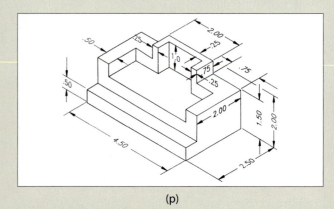

(p)

10.13 problems (continued)

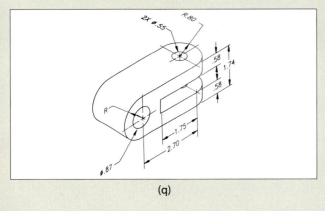

(q)

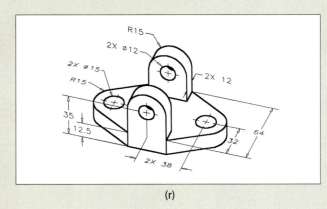

(r)

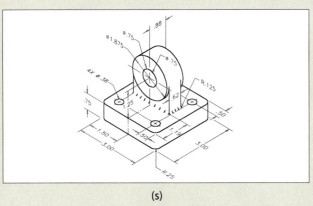

(s)

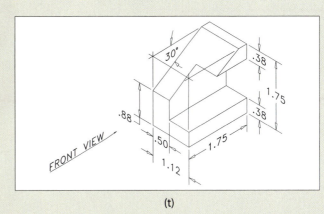

(t)

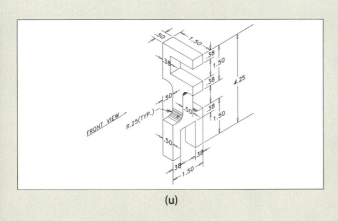

(u)

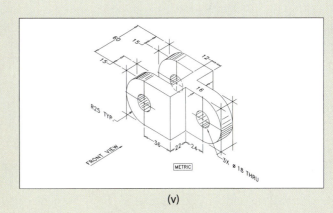

(v)

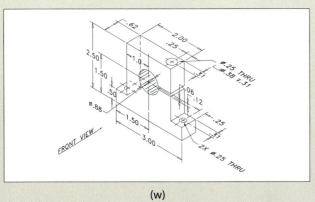

(w)

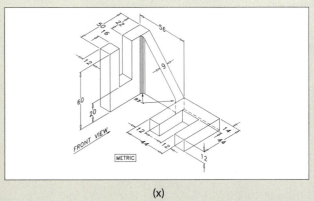

(x)

10.13 problems (continued)

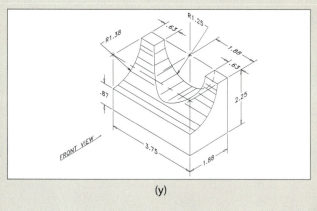

(y)

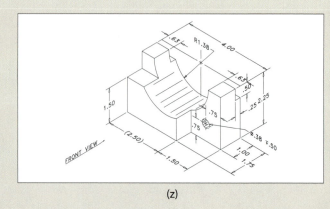

(z)

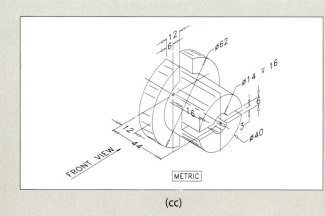

(aa)

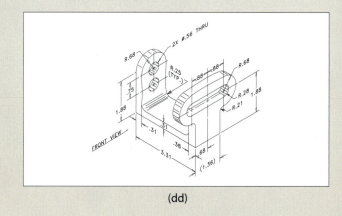

(bb)

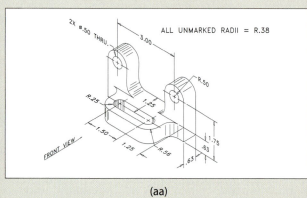

(cc)

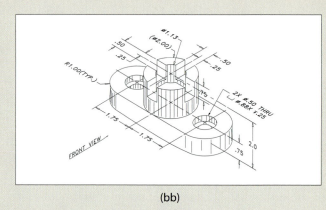

(dd)

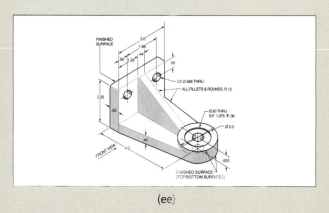

(ee)

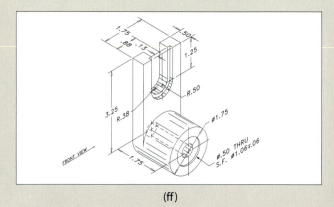

(ff)

10.13 problems (continued)

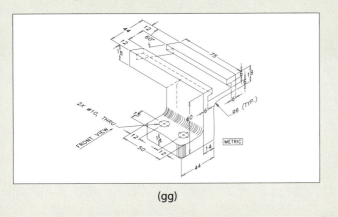

(gg)

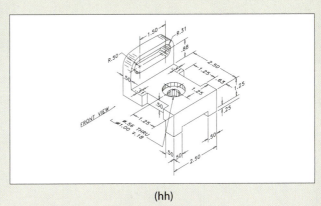

(hh)

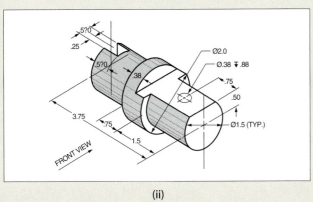

(ii)

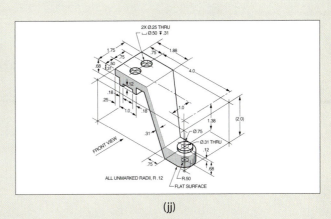

(jj)

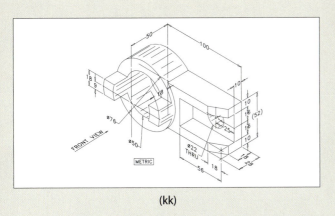

(kk)

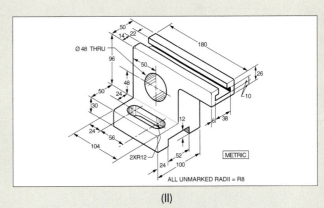

(ll)

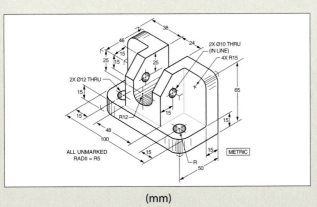

(mm)

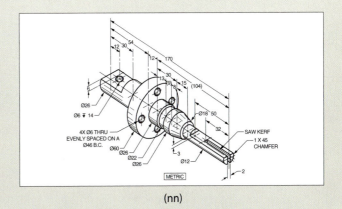

(nn)

10.13 problems (continued)

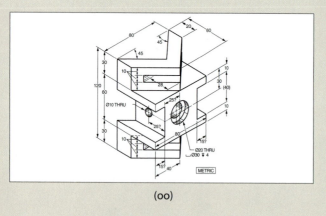

(oo)

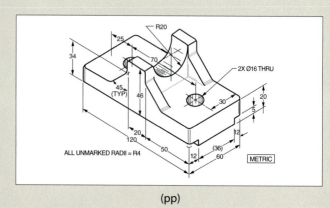

(pp)

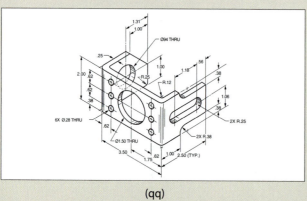

(qq)

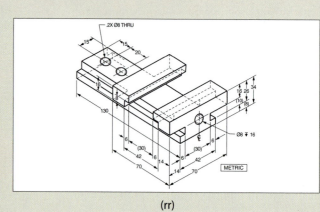

(rr)

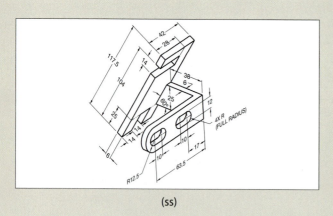

(ss)

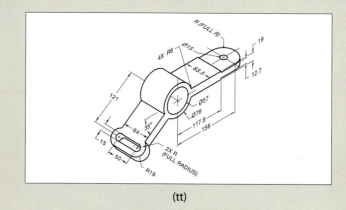

(tt)

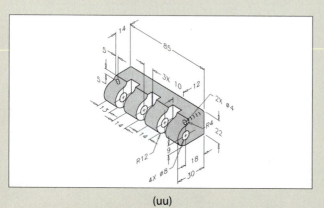

(uu)

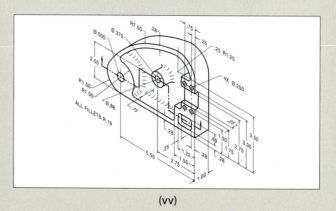

(vv)

10.13 problems (continued)

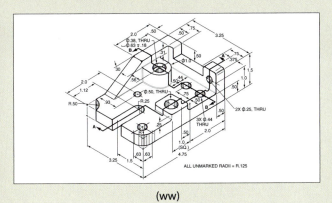

(ww)

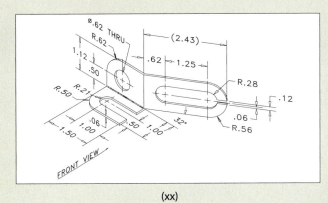

(xx)

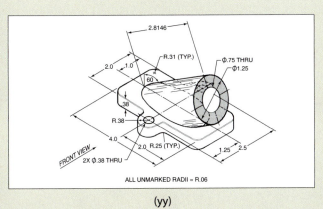

(yy)

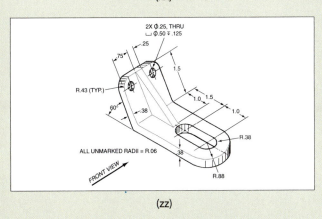

(zz)

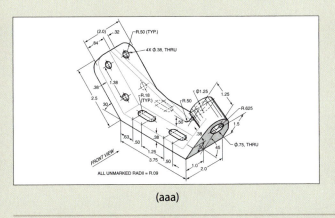

(aaa)

FIGURE P10.6.

11

Advanced Visualization Techniques

objectives

After completing this chapter, you should be able to

- Increase your visualization skills
- Move seamlessly between 2-D and 3-D representations of objects
- Use basic strategies for visualizing complex objects

11.01 introduction

I magine trying to communicate in a foreign county without knowing one word of the native language. You might be able to pantomime adequately to take care of your basic needs—food and drink, for example—but your level of communication would be superficial. You would not be able to exchange views on your philosophy, your feelings, or your thoughts. There would be no chance for a meaningful dialogue. Trying to communicate with other engineers without well-developed graphics and visualization skills is similar to trying to communicate with a French person when you do not know the language. Without graphics skills, you can engage in a superficial conversation about basics; however, you cannot participate in a meaningful discussion about design details. That is why this chapter is included in the text—to help you move beyond the basics in your graphics and visualization skills so you can be a full-fledged member of an engineering design team.

11.02 Basic Concepts and Terminology in Visualization

In previous chapters, you learned about some of the basic building blocks of graphical communication. You learned how to make simple sketches. You developed basic skills for visualizing simple objects, and you learned about pictorial representation of objects. You learned how engineers use multiview drawings to accurately portray 3-D objects on a 2-D sheet of paper. In this chapter, you will bring all of your skills together to tackle increasingly difficult problems in visualization. Mostly through examples, this chapter will illustrate techniques you can use to hone your visualization skills, which will improve your effectiveness in graphical communication.

Before continuing with the chapter, you need to become familiar with some fundamental definitions and concepts. As you learned in a previous chapter, four basic types of surfaces make up 3-D objects. Table 11.01 outlines the basic surface types and explains how they appear in a standard drawing that shows top, front, and right-side views.

The definition of surface types can be further refined based on their characteristics. **Normal surfaces** are defined by the primary view to which they are parallel, and they appear in true shape and size within this primary view. When normal surfaces are parallel to the front view, they are referred to as **frontal surfaces**; when normal surfaces are parallel to the top view, they are referred to as **horizontal surfaces**; and when normal surfaces are parallel to the side view, they are referred to as **profile surfaces**.

SURFACE TYPE	SURFACE CHARACTERISTICS	APPEARANCE
NORMAL	PARALLEL TO ONE OF THE PRIMARY VIEWING PLANES; PERPENDICULAR TO THE OTHER TWO PLANES	SURFACE IN ONE VIEW; EDGE IN TWO REMAINING VIEWS
INCLINED	PERPENDICULAR TO ONE PRIMARY VIEW; ANGLED WITH RESPECT TO THE OTHER TWO VIEWS	SURFACE IN TWO VIEWS; EDGE (ANGLED LINE) IN ONE VIEW
OBLIQUE	NEITHER PARALLEL NOR PERPENDICULAR TO ANY OF THE PRIMARY VIEWS	SURFACE IN ALL THREE VIEWS
CURVED	PERPENDICULAR TO ONE VIEW	CURVED EDGE IN ONE VIEW; EXTENTS OF SURFACE VISIBLE IN TWO REMAINING VIEWS

TABLE 11.01. Basic Surface Types for Objects

Figure 11.01 illustrates the three types of normal surfaces found on objects. Notice that for each normal surface shown, it is viewed as a surface in one view and as edges in the remaining two views.

Inclined surfaces are defined by the primary view to which they are perpendicular—in other words, they are defined by the view in which they appear as an edge. Thus, an inclined frontal plane appears as an edge in the front view and as a surface in the top and side; an inclined horizontal plane appears as an edge in the top view and as a surface in the front and side; and an inclined profile surface appears as an edge in the side view and as a surface in the front and top. Figure 11.02 illustrates the three classifications for inclined surfaces that are found on objects.

Oblique surfaces do not have any special classification since they appear neither as an edge nor in their true shape and size in any of the primary views. Figure 11.03 illustrates the basic characteristic of an oblique surface.

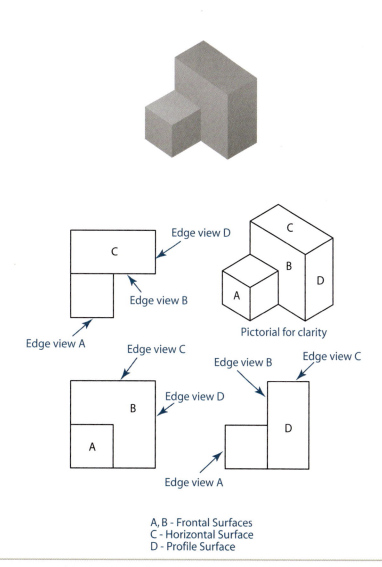

A, B - Frontal Surfaces
C - Horizontal Surface
D - Profile Surface

FIGURE 11.01. An object made up of normal surfaces.

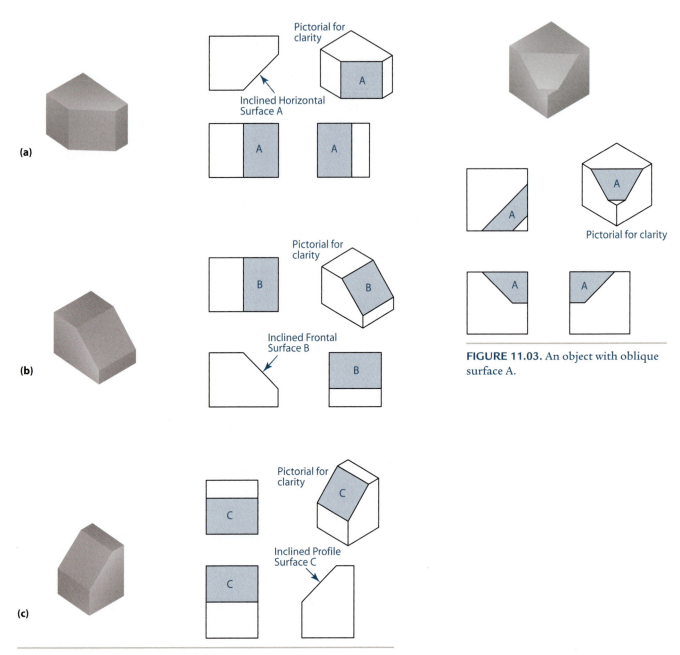

Pictorial for clarity

Inclined Horizontal Surface A

(a)

Pictorial for clarity

FIGURE 11.03. An object with oblique surface A.

Pictorial for clarity

Inclined Frontal Surface B

(b)

Pictorial for clarity

Inclined Profile Surface C

(c)

FIGURE 11.02. Inclined surface definitions.

There is no special classification for **curved surfaces**, although they can be perpendicular to any of the primary views, as shown in Figure 11.04.

Engineering drawings are typically set up in the familiar L-shaped pattern that shows the top, front, and right side views. With this view layout, the top is projected vertically from the front and the right side is projected horizontally from the front with the panels folded out of the glass box and hinged at the edges of the front view. Sometimes you may want to include the top, front, and left side views, resulting in a

FIGURE 11.04. Various orientations for curved surfaces.

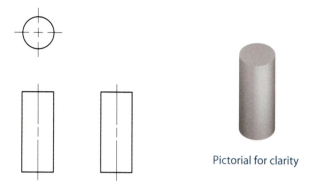

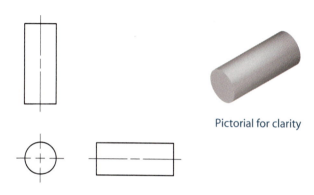

(a) Curved surface perpendicular to top view.

Pictorial for clarity

(b) Curved surface perpendicular to front view.

Pictorial for clarity

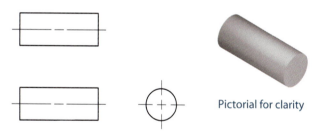

(c) Curved surface perpendicular to side view.

Pictorial for clarity

backward *L* view pattern. In this case, the top hinges on the front view as before, with the left view also hinging on the front, similar to the way the right view was constructed. Other times, you will want the top view as the stationary view and the side and front views hinging from it. This is particularly true when the edge view of a curved surface is visible in the top view. Figure 11.05 shows the alternative ways the glass box can be unfolded to produce varying view layouts. Figure 11.06 shows a comparison of two

FIGURE 11.05. One way to present the drawing of an object: showing the front, top, and right-side views on a hypothetical glass box in (a), the bow panels opening in (b), to present all three views on a single plane in (c), for the multiview drawing in (d). Another way to present the drawing of an object: showing the front, top, and right-side views on a hypothetical glass box in (e), the bow panels opening in (f), to present all three views on a single plane in (g), for the multiview drawing in (h).

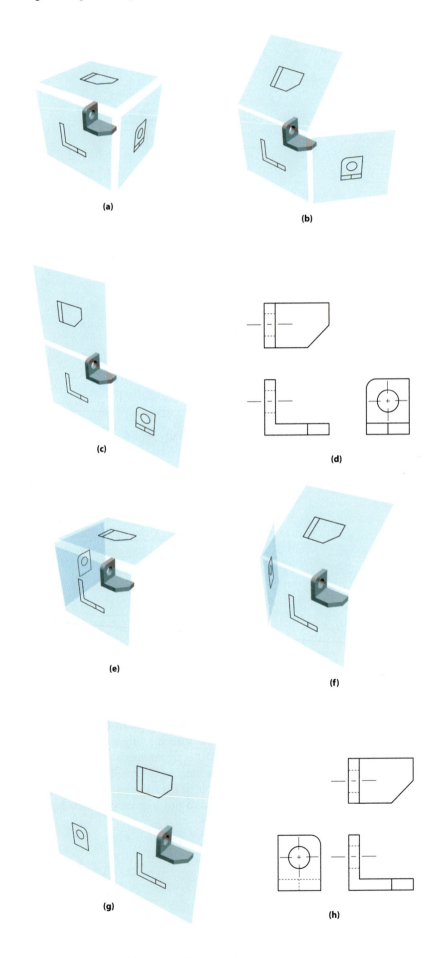

FIGURE 11.05. (continued) Still another way to present the drawing of an object; showing the front, top, and right-side views on a hypothetical glass box in (i), the bow panels opening in (j), to present all three views on a single plane in (k), for the multiview drawing in (l).

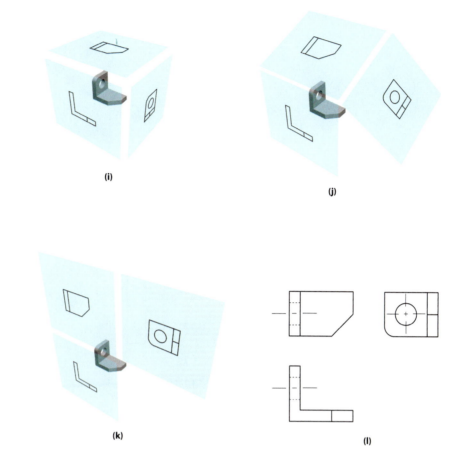

(i)

(j)

(k)

(l)

projection systems used for sketching a cylinder. Notice that the view layout shown in Figure 11.06(b) more clearly defines the curved surface of the cylinder than the layout in Figure 11.06(a). In Figure 11.06(b), any two adjacent views (e.g., either the front and top views or the top and sides views) clearly define the surface as being curved. In Figure 11.06(a), the front and side views together cannot, by themselves, define the curved surface. The addition of a top view is required.

One important visualization concept you must understand is that surfaces retain their basic shapes from view to view. If a surface has four edges and four vertices in one view, it will have four edges and four vertices in the next view. If the surface has a basic L shape in one view, it will have a basic L shape in all views. The exception, of course, is that a normal surface appears as a single edge in at least one of the primary views and merely looks like a line in that view—you will not be able to determine how many edges and vertices it has. Figure 11.07 shows a multiview drawing, including an isometric pictorial, of an object that has an L-shaped surface on it. Notice that this surface (labeled A in all views) maintains its basic shape from view to view, except in the top view where it appears as a single edge.

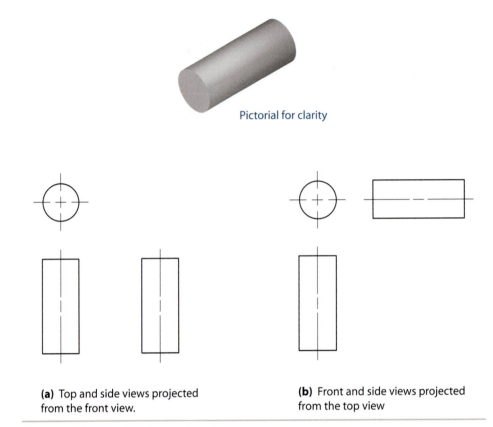

Pictorial for clarity

(a) Top and side views projected from the front view.

(b) Front and side views projected from the top view

FIGURE 11.06. The side view projected from the top view.

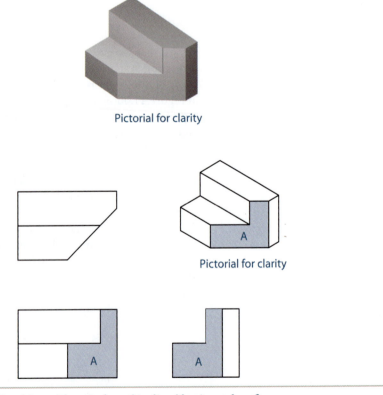

Pictorial for clarity

Pictorial for clarity

FIGURE 11.07. An object with an L-shaped inclined horizontal surface.

Another basic concept to keep in mind when visualizing objects from multiview drawings is that the limits of surfaces should correspond from one view to the next. Because you have a system of orthographic projection, features, and vertices on, the object will be aligned between views. This principle holds true for surfaces as well. If the height of a surface goes from bottom to top in the front view of the object, it also will go from bottom to top in the side view or in an isometric pictorial view. Note that because the top view does not show changes in object height, those changes will not be apparent in that view. Similarly, the width and depth limits of a given surface are maintained from one view to the next. Figure 11.08 shows a multiview drawing of an object along with an isometric pictorial. The height, width, and depth limits of surface B are labeled in each view. Notice that these dimensions are maintained between views.

A final basic concept to keep in mind when looking at multiview drawings is that for normal and inclined surfaces, right angles on surfaces are maintained from view to view. In fact, one way to identify an oblique surface on an object is to observe that an angle appears to change from perpendicular to something other than perpendicular from one view to the next. Figure 11.09 shows an object with a T-shaped inclined surface on it. Notice that in each view of the surface, all angles appear to be right angles (and that the surface has the same basic shape and same number of edges and vertices in each view). Figure 11.10 shows a multiview drawing of an object with an oblique surface (A) on it. Notice that the angle at the vertex labeled X appears to be a right angle in the front view but about 40° in the top view. This change in the relative size of the angle at vertex X is a clear indication that surface A is an oblique surface and, as such, appears as a surface in all views.

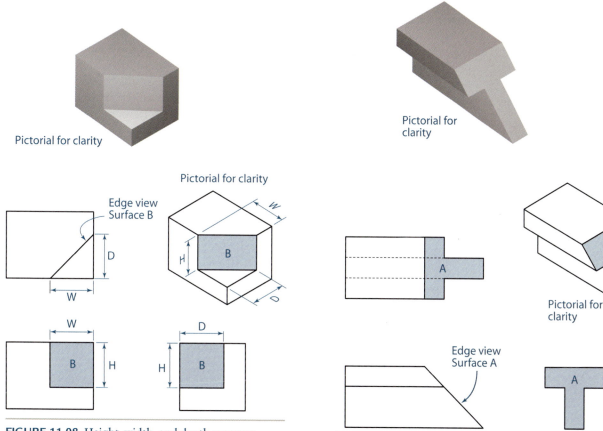

FIGURE 11.08. Height, width, and depth measurements maintained between views.

FIGURE 11.09. An object with inclined T-shaped surface A.

FIGURE 11.10. An apparent change in the angle at vertex X for oblique surface A.

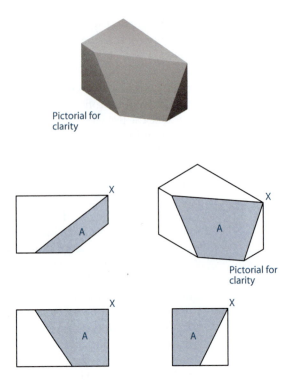

11.03 The Possibilities for a Feature Representation

To understand a drawing or a view completely, you must be able to imagine all of the possibilities inherent to a given feature. Then you can reject the possibilities that do not match the information provided. Figure 11.11 shows the top view of a simple object.

Can you tell what the object looks like from this view alone? Figure 11.12 shows several pictorial sketches of objects that could produce this single top view. Notice that the possible objects include normal, inclined, and curved surfaces as defined in this chapter. Several other possibilities produce this top view, including some that contain oblique surfaces. Can you imagine others? Which one shows the correct object? Without more information, it is impossible to determine the correct object that corresponds to this top view.

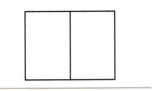

FIGURE 11.11. The top view of an object.

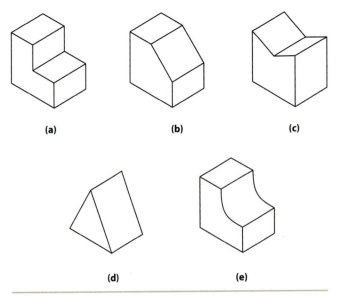

(a)   (b)   (c)

(d)   (e)

FIGURE 11.12. Possible objects from a given top view.

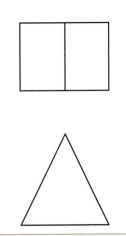

FIGURE 11.13. Top and front views of an object.

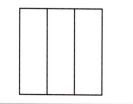

FIGURE 11.14. The top view of an object.

Figure 11.13 shows the top and front views of the object. By examining this additional view, you can rule out all of the possibilities from Figure 11.12 except choice d.

Figure 11.14 shows the front view of an object, and Figure 11.15 shows multiple possibilities for objects that correspond to this view. Thinking about all of the possibilities that exist will help you develop your visualization abilities. Again, which of the objects is correct? It is impossible to tell from this limited information.

Figure 11.16 shows the front and side views of the object. Based on this new information, you can rule out several of the possibilities; only two remain: choices a and e.

To determine which of these is correct, a third view of the object is required. Figure 11.17 shows the top, front, and right-side views of the object. Based on this new information, it is clear that choice a is the correct interpretation of the object.

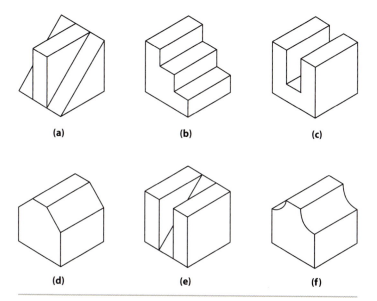

(a)   (b)   (c)

(d)   (e)   (f)

FIGURE 11.15. Possible objects based on the top view given.

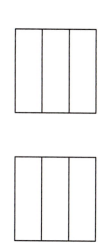

FIGURE 11.16. Top and front views of an object.

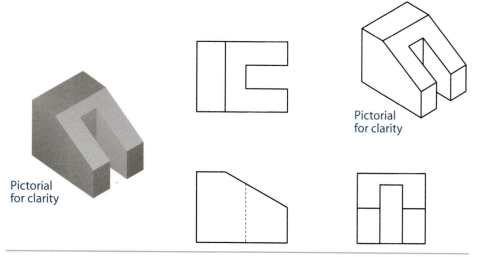

Pictorial for clarity

Pictorial for clarity

FIGURE 11.17. A multiview drawing of an object.

11.04 Other Viewpoints

In a previous chapter, you learned about coded plans and the way they are used to define an object. You also learned how the object can be viewed from any one of the corners defined by the coded plan. Figure 11.18 shows a coded plan and the four corner views defined by it.

In the previous discussion of coded plans and corner views, you learned to look at an object from the indicated corner by imagining that your eye was located *above* the object. Because you are looking at the object from above, this view is sometimes referred to as the bird's-eye view. What if you wanted to look at the object from beneath it? Figure 11.19 shows the same object in Figure 11.18 except that the corner views are defined by locating your eye beneath the object, looking upward toward it. These views are referred to as worm's-eye views. (Think of yourself as a worm burrowing underground and looking up at the object rather than a bird soaring overhead and looking down on the object.)

If you compare the corner views from these two viewpoints, you will notice that with a bird's-eye view, you see all of the top surfaces of the object (except for those that are partially or completely hidden by projections on the object). But with the worm's-eye view, you see the bottom surfaces of the object and none of the top surfaces are shown. Figure 11.20 shows a bird's-eye and worm's-eye view for an object, with the top and bottom surfaces labeled.

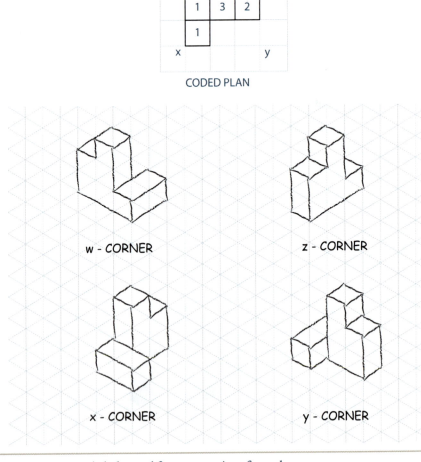

FIGURE 11.18. A coded plan and four corner views from above.

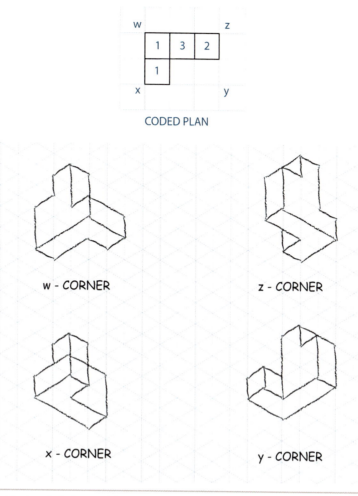

FIGURE 11.19. A coded plan and four corner views from below.

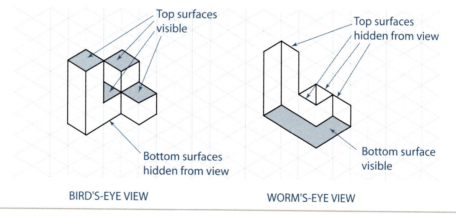

FIGURE 11.20. Bird's-eye view versus worm's-eye view of the same object from the same corner.

Recall that for isometric pictorials, you learned that the axes are set up such that you are looking down the diagonal of a cube. The difference between a bird's-eye view and a worm's-eye view is that you are looking down two different diagonals of the cube. Figure 11.21 shows two cubes with the diagonals for both viewpoints labeled.

FIGURE 11.21. Cube diagonals for bird's-eye (a) and worm's-eye (b) views.

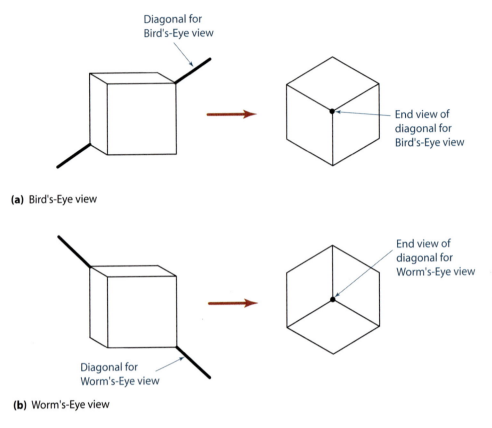

Diagonal for Bird's-Eye view

End view of diagonal for Bird's-Eye view

(a) Bird's-Eye view

End view of diagonal for Worm's-Eye view

Diagonal for Worm's-Eye view

(b) Worm's-Eye view

11.05   Advanced Visualization Techniques

In the next sections, you will learn some systematic strategies for visualizing objects and for seamlessly moving between multiview and pictorial representations. Many of these techniques have been tried and tested with engineering students over the years. Although you may find some of the techniques to be difficult in the beginning, with practice and with improvement in your visualization skills, you will be able to tackle increasingly difficult problems. Some of these techniques may be easier for you than others. You should find the methods that work for you and stick with them as you complete the exercises at the end of the chapter. Good luck!

11.05.01 Visualization with Basic Concepts

Figure 11.22 shows the top and front views of an object, along with three possible right-side views. Using what you know about basic shapes, which of the possible right-side views is correct?

Begin by looking at the angled edge of the object in the front view labeled *A*. The most probable explanation for this angled edge is that it is an inclined frontal surface—seen as an edge in the front and as surfaces in the top and right side. This edge does not extend all the way across the object in the front view, but goes about three-quarters of the way from right to left. You know that this surface must be visible in the top view, and the only surface there that extends from the right side to about three-quarters of the way to the left side is the U-shaped surface labeled *A* in Figure 11.23.

You know that this surface also must appear as a U-shaped surface in the correct side view, since the overall shape of the surface cannot change from view to view. There are two possibilities from the choices given for the correct right-side view—choice a and choice c. Look again at the inclined line that represents the surface in the front view.

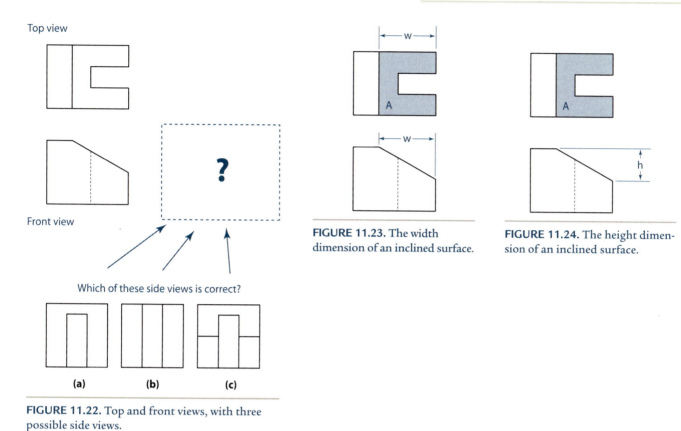

Top view

Front view

?

Which of these side views is correct?

(a)   **(b)**   **(c)**

FIGURE 11.22. Top and front views, with three possible side views.

FIGURE 11.23. The width dimension of an inclined surface.

FIGURE 11.24. The height dimension of an inclined surface.

Notice that it does not extend all the way from the bottom to the top in that view. The inclined line starts at a point about halfway up from the bottom and extends from there to the top of the object, which is illustrated in Figure 11.24.

Of the choices given, only choice c has the correct overall height for the U-shaped surface; the U-shaped surface in choice a goes all the way from the bottom to the top of the object. Figure 11.25 shows the correct multiview sketch of the object. An isometric pictorial sketch also is shown in the figure for clarity.

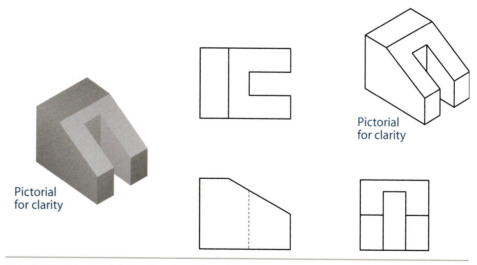

Pictorial for clarity

Pictorial for clarity

FIGURE 11.25. A multiview drawing of an object.

What about the object shown in Figure 11.26? For this particular object, the front and side views are given and you are to select the top from the three choices given.

A U-shaped surface in both views for this object extends all the way from the bottom to the top. Chances are good that this is an inclined surface since it appears as a similarly shaped surface in two views. (It also could be an oblique surface; but for simplicity, assume it is inclined.) If it is an inclined surface, it will be classified as an inclined horizontal surface and, therefore, would be seen as an edge in the top view. Two of the choices, choice b and choice c, show an angled edge in Figure 11.27.

From the front view, the U-shaped surface goes all the way across the object from left to right. Only choice b shows the angled line extending all the way across the object, so it must be the correct top view. Figure 11.28 shows the correct multiview sketch for the object. Once again, a pictorial of the object has been included for clarity.

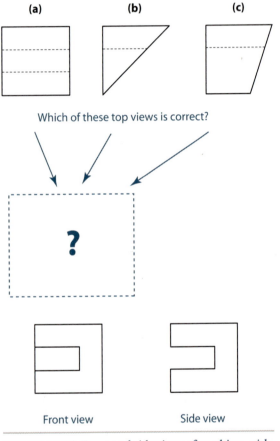

Which of these top views is correct?

?

Front view   Side view

FIGURE 11.26. Front and side views of an object, with three possible top views.

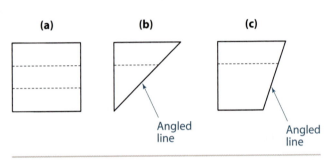

Angled line   Angled line

FIGURE 11.27. Angled lines in choices (b) and (c) for the top view.

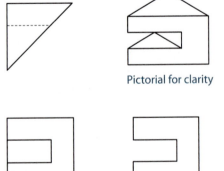

Pictorial for clarity

Pictorial for clarity

FIGURE 11.28. The correct multiview drawing of the object.

11.05.02 Strategy for a Holistic Approach to Constructing Pictorials from Multiview Drawings

The inverse tracking of points, edges, and surfaces covered in the later sections are detailed methods that can be used to create a pictorial from a multiview drawing. A different, more holistic approach also can be employed to achieve the same result, as outlined in the following paragraphs.

A few ground rules must be established before you can begin development of this approach:

- The objects you devise must be solid.

- You cannot have a single, infinitely thin plane on the resulting object.

- You cannot have two objects that are next to each other in the result.

- You should pick one view as the "base" and work from there until all views match the object.

An illustration of this approach will begin with a simple object. Figure 11.29 shows the front and right-side views of an object from which you want to create a pictorial. This example will use the front view as the base view.

The front view shows two surfaces—one is L-shaped; the other, rectangular. When you start with the L-shaped surface, you can assume that it is a normal surface and that it is located all the way forward on the object. You will consider other cases of this surface later; but for this method, it is best to start with the simplest form of the object and then consider increasingly complex solutions. Normal surfaces are the simplest type, so it makes sense to start there. When you also consider the rectangular surface to be normal, it must be parallel to the front view as well. (Otherwise, it would appear as an edge in the front.) The three possibilities for this surface are that it appears all the way at the front of the object, all the way to the back of the object, or somewhere in between. These three possibilities are illustrated in Figure 11.30.

The first of these choices is not a valid solution because if both surfaces were all the way to the front of the object, there would be no need for the edges separating the two areas—lines are not included on a single plane of an object. The second choice produces front and side views that match those that were given initially; however, this choice is not valid since it results in an infinitely thin plane at the back of the object. The third choice also can be eliminated, since it results in an object that does not match the right-side view. It does not matter where the surface is located with respect to the object depth—there is no way to include a normal surface, such as this, on the object to produce a correct right-side view.

FIGURE 11.29. The front and right-side views of an object.

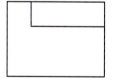

FIGURE 11.30. Three possible solutions when the rectangular area in Figure 11.29 is a normal surface.

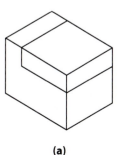
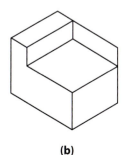
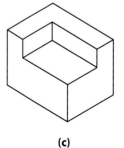

(a)   (b)   (c)

FIGURE 11.31. Two possible inclined horizontal surfaces for an object.

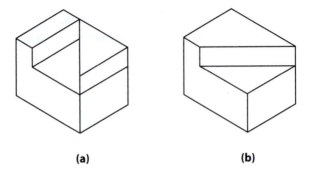

(a)   (b)

When the L-shaped surface is located correctly, there is no possible way the rectangular area represents a normal surface. Now you will assume that the rectangular area represents an inclined surface. In this case, you have to consider only an inclined horizontal or an inclined profile surface, since an inclined frontal would appear as an angled line in the front view. (No such lines are there.) Figure 11.31 shows two possible inclined horizontal surfaces. Notice that the first choice results in an object that does not correspond to the right-side view that is given, while the second choice does. This means that an inclined horizontal surface, such as the one shown, is one possible solution to this problem.

Now consider the possible inclined profile surface shown in Figure 11.32. Notice that this inclined profile surface is not a possible solution, since there are no angled lines in the side view.

Are there other possible solutions? Are curved surfaces a possibility? Figure 11.33 shows two possible solutions where the rectangle seen in the front view represents a curved surface (b and d) and several solutions that include curved surfaces that do not result in valid solutions (a, c, and e). The invalid solutions result in a front or side view that does not match those given.

What about an oblique surface for the rectangular area in the front view? Figure 11.34 shows two possible oblique surfaces that satisfy the front view, but notice that the side view produced by these objects is not valid. You could try any number of different oblique surfaces and not find one that satisfies the two views given.

Now consider choices where the rectangular area in the front view is a normal surface that is located all the way to the front of the object. This time you should again

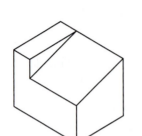

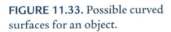

FIGURE 11.32. An inclined profile surface for an object.

FIGURE 11.33. Possible curved surfaces for an object.

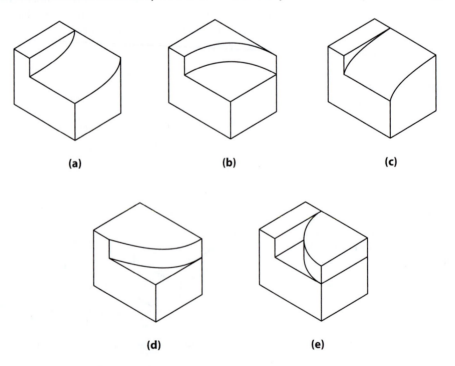

(a)   (b)   (c)

(d)   (e)

FIGURE 11.34. Two possible oblique surfaces.

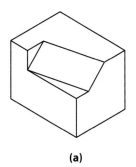
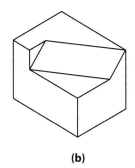

(a)   (b)

consider the L-shaped surface as a normal surface all the way to the front, all the way to the back, or somewhere in between. These three possibilities are shown in Figure 11.35. Once again in this case, the front view of the object is okay for each possibility, but either the side view is incorrect or the result is not permissible.

Next, consider the L-shaped surface to be an inclined profile. The two possibilities are shown in Figure 11.36. Notice that neither of these produces the correct side view.

Figure 11.37 shows the two possibilities for the L-shaped area to represent an inclined horizontal surface. Once again, neither of these produces a side view that corresponds to the one that is given. An inclined frontal surface is not possible, since there are no angled lines in the front view.

FIGURE 11.35. Three possibilities with an L-shaped normal surface.

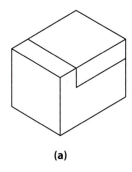
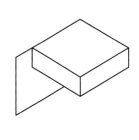
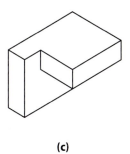

(a)   (b)   (c)

FIGURE 11.36. Two possible solutions with inclined profile L-shaped surfaces.

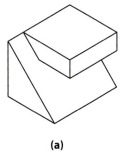
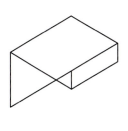

(a)   (b)

FIGURE 11.37. Two possible solutions with inclined horizontal L-shaped surfaces.

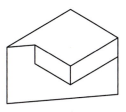
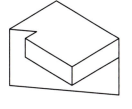

(a)   (b)

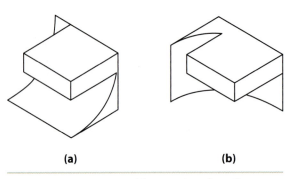

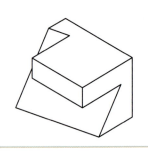

FIGURE 11.38. Two possible curved surfaces for the object.

FIGURE 11.39. A possible L-shaped oblique surface.

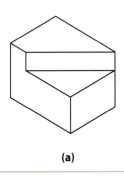
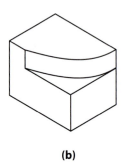
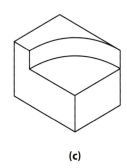

(a)   (b)   (c)

FIGURE 11.40. Possible correct solutions for the object.

You also can consider curved surfaces for the L-shaped area as before, as shown in Figure 11.38. In this case, only one of the possible solutions with a curved surface matches the two views given.

Can the L-shaped surface be an oblique surface? Figure 11.39 shows one possible oblique surface that satisfies the front view. Notice that when the L-shaped area is this oblique surface, once again the side view does not correspond to the one that is given.

After sketching multiple possible pictorials that match the two views that are given, you have identified three possibilities—one that includes an inclined surface and two that contain a curved surface. All of the correct pictorials identified in this exercise are shown in Figure 11.40. You may find this method of creating pictorials from multiview drawings tedious at first; however, as you practice, you will find it increasingly easy. In time, you will probably be able to skip certain pictorials altogether.

11.05.03 Strategy for Constructing Pictorials by Inverse Tracking of Edges and Vertices

The process of creating a multiview drawing from a pictorial (or from an idea or image in your mind) is quickly mastered with practice. However, the inverse process of creating a mental image, or pictorial drawing, of an object from its multiview drawing is considerably more challenging. With some practice, though, you can master this skill as well. People who deal with multiview drawings on a daily basis, such as professional design engineers, drafters, and machinists, usually can create a mental image of a part quickly after they see its multiview drawing.

To develop this skill, it may be best to start with a technique that is well-defined and methodical, as well as an object of fairly simple geometry, such as that shown in Figure 11.41. For this example, sketching techniques will once again be used as a method for developing visualization skills. Later examples will use more formal graphics for presentation clarity.

Inverse Tracking with Edges and Vertices for Normal Surfaces

An eight-step process can be used, but this time the process is inverted compared to the process presented in an earlier chapter where a multiview drawing was created from a pictorial. The steps are as follows:

Step 1: Define the location and directions of a coordinate system consistent in all views.

Step 2: Define an anchor point on the object.

Step 3: Mark the limits of the foundation volume.

Step 4: Locate a vertex or an edge adjacent to the anchor point and draw that edge.

Step 5: Successively locate other vertices and edges and draw those edges.

Step 6: Convert hidden lines.

Step 7: Add internal features.

Step 8: Check model validity.

The first step (Step 1) is to create a set of 3-D axes on an isometric grid and define the viewing directions. Next, select an anchor point for the object (Step 2). The **anchor point** should be a point on the part that you can easily and confidently locate on each of the orthogonal views. A good selection for this example would be the vertex on the lower left front of the part. Called point A, it is identified in each view in Figure 11.41; and a set of axes is drawn at that point in all of the views, including the pictorial grid. Make sure the directions of the coordinate axes are consistent in every view.

The **foundation space** (Step 3) is then outlined with respect to the anchor point on the grid to show the limits of the volume that is occupied by the object in space, as shown in Figure 11.42. To find the foundation space, note that the width and height limits of the part can be seen in the front view. The top view shows the width of the part as well as its depth. The side view shows the height and depth of the part. Thus, the height, width, and depth of the foundation space can be established with only two views as long as those view planes are orthogonal to each other. Any two adjacent views, such as the front and top views or the front and side views, will be orthogonal to each

FIGURE 11.41. To create an isometric pictorial of the object shown in this multiview drawing, define the anchor point and coordinate axes in each view, then the same point and axes in the pictorial grid.

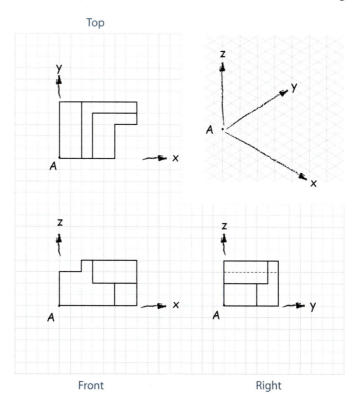

Top

Front   Right

FIGURE 11.42. Coordinate axes and viewing direction are defined on an isometric grid. An anchor point A is selected, and the foundation space is outlined.

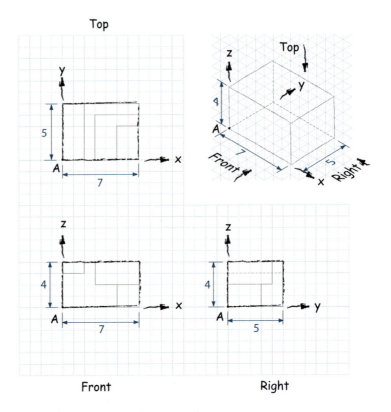

FIGURE 11.42. Coordinate axes and viewing direction are defined on an isometric grid. An anchor point A is selected, and the foundation space is outlined.

other. For this example, the width of the part is 7 units in the x-direction, the height of the part is 4 units in the z-direction, and the depth of the part is 5 units in the y-direction. A rectangular volume with these x, y, and z measurements is created on the isometric grid. Although most objects probably will not be brick-shaped, a foundation shape is easy to draw and easy to visualize. It is typically much easier to locate the points, edges, and surfaces of the object with respect to its foundation than to a set of axes. Also, if you start creating points and lines on the isometric grid that are outside the foundation space, you will know you are doing something wrong.

Next, locate a vertex or an edge adjacent to the anchor point (Step 4). Remember that a vertex can be defined as a point that connects at least three edges and any corner is automatically a vertex. To locate an edge on the pictorial, that edge must be defined in at least two orthogonal views. For parts that contain surfaces that are parallel or perpendicular to the viewing planes, it is likely that two or more edges of the part will appear on top of each other in the orthogonal views. Thus, when you are tracking an edge from a drawing view to the pictorial, the edge needs to be carefully specified in another view. **Point** and **edge tracking** is done by selecting vertices on the part in the orthogonal views and then locating those vertices on the pictorial by moving along the edges of the part. Moving along the part edges in a multiview drawing is more difficult than doing so on a pictorial because you need to track the motion on different orthogonal views simultaneously. Also, on a multiview drawing, many edges and vertices lie directly above or below each other. Identification of the individual vertices may be confusing, but it can be done easily when you exercise some care. The trick is to define the location of the vertices in at least two views simultaneously. Tracking is made easier by the fact that vertices in adjacent orthogonal views must align either horizontally or vertically. For example, as shown in Figure 11.43, as you move 5 units to the right along the x-axis from point A to reach point B in the front view along the edge indicated, you also must identify this motion in the top and side views and on the pictorial simultaneously. To find point B in the top view, draw a vertical alignment line through point B in the front view and see which vertices on the object lie on this line in the top view. The top view shows a few possible vertex locations on the alignment line for point B; but only one of the locations is accessible by moving along a single edge

FIGURE 11.43. Identify and track the edges of the part on the orthogonal views and locate these edges on the pictorial.

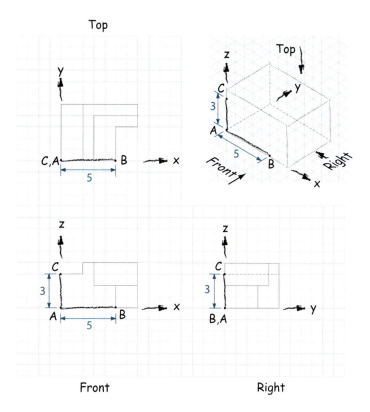

Top

Front

Right

(also in this case, the x-axis from point A). This edge also is parallel to the front viewing plane. In the side view, the location of point B is directly in front of point A, thus appearing coincident with it. With the location of point B identified in each of the orthogonal views, this location and the edge between point A and point B are transferred to the pictorial by moving 5 units along the x-axis.

Next, another vertex on the front view and on the same surface as point A is selected on the orthogonal views. Calling this point C, you can easily see the edge formed by points A and C on the side view. It is located 3 units above point A along the z-axis. Even though the horizontal view alignment line goes through four vertices in the side view, only one of these locations is accessible by moving along an edge from point A. This edge is parallel to the front viewing plane. In the top view, point C is directly above point A; so they are seen as coincident in this view. The location of point C on the pictorial is placed on the z-axis 3 units above point A.

Additional vertices are located by noting their x, y, and z positions on the multiview drawing and then transferring those locations to the pictorial view (Step 5). Before moving to the next surface, you should select the points and edges to outline an entire surface of the object. In this manner, instead of seeing connected edges that appear in a variety of directions on the pictorial, you see that the surfaces of the part appear one at a time. Surfaces or parts of surfaces on the part can be identified easily on the orthogonal views as simple closed loops, as shown in Figure 11.44. Once all of the edges in all of the orthogonal views have been accounted for and placed on the pictorial, the object pictorial should be complete, as shown in Figure 11.45. Note that in the final pictorial, one of the surface loops is partially obscured by some of the other surfaces. In this case, the obscured edges should be removed or shown as hidden lines (Step 6). Since isometric views show only three sides of the glass box, the front, side, and top views are generally sufficient to create the isometric view of most objects. If only two of these views are available, you should create the third view before attempting to construct the pictorial.

The final steps are to add any internal features (Step 7), of which there are none in this example, and then check to ensure that each vertex on the pictorial is connected to at least three edges (Step 8) for the pictorial to be valid. For this pictorial, all vertices are connected to three edges (some of the edges are hidden); so it appears to be valid.

FIGURE 11.44. Surfaces are loops on the orthogonal views. Complete the edges of an entire surface before proceeding to the next surface.

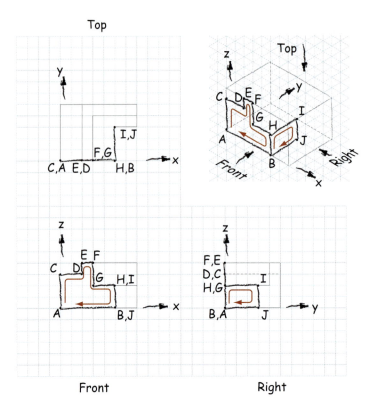

FIGURE 11.44. Surfaces are loops on the orthogonal views. Complete the edges of an entire surface before proceeding to the next surface.

Inverse Tracking with Edges and Vertices for Inclined Surfaces

Anytime there is an edge on a multiview drawing that appears at an angle (i.e., it is not horizontal or vertical), an inclined surface will exist on the part. The drawing of such a part is shown in Figure 11.46. Creating an isometric pictorial of a part with inclined surfaces is only slightly more complicated than creating a part with horizontal or vertical surfaces. The same procedure of vertex and edge tracking is used for both types of parts.

FIGURE 11.45. When all surface loops on the orthogonal views are accounted for, the isometric view should be complete.

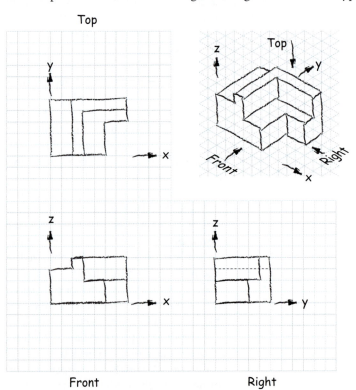

FIGURE 11.46. It is desired to create an isometric pictorial of the object shown in this multi-view drawing.

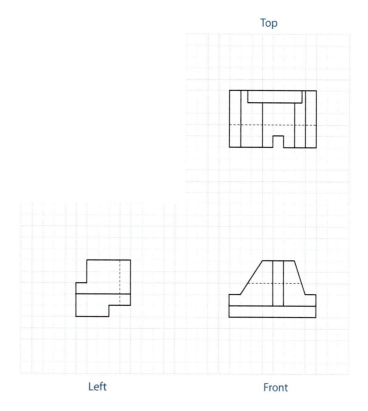

Top

Left   Front

First, an anchor point, called point A, is selected and identified in each view. The anchor point, coordinate axes, and view directions are defined on an isometric grid (Steps 1 and 2). A foundation space is created on the grid (Step 3) based on the limits of the object extracted from the orthogonal views. This space is shown on Figure 11.47. Note that in this particular example, the anchor point is not located at one of the extreme limits of the foundation space. The anchor point can be located just about

FIGURE 11.47. Coordinate axes and viewing directions are defined on an isometric grid. An anchor point, A, is selected, and the foundation space is outlined.

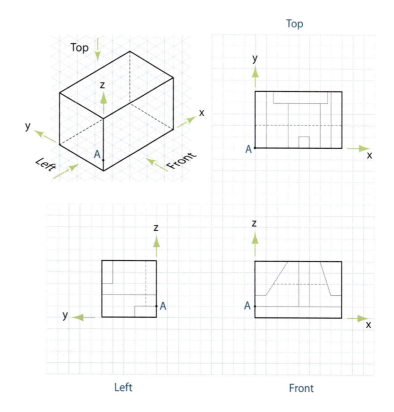

Left   Front

anywhere on the object as long as it is convenient to use and you are confident of its location in all of the orthogonal views and in the pictorial.

Start with vertices near the anchor point (Step 4). By tracking points and identifying the edges of the part in the orthogonal view, build the edges surrounding one of the surfaces on the part. Note that as you build these edges, the direction of travel for any edge at an angle will be two-dimensional; and you must keep track on the pictorial of the distance traveled in each direction. This is shown in Figure 11.48.

Track the edges of one surface at a time (Step 5). Surfaces or parts of surfaces in each view are seen as simple loops, as shown in Figure 11.49. As long as you keep careful track of the location of each vertex on each surface, the inclined surfaces defined by their vertices should appear on the pictorial.

Once all edges in all of the orthogonal views are accounted for, the object should be complete, as shown in Figure 11.50. Note that in this case, one of the surfaces on the vertical slot seen in the pictorial is hidden in the multiview drawing; that is, the surface loop on the multiview drawing contains a hidden line. Surfaces must always be connected on all of their edges by other surfaces. Also note for this example that the hidden edges for a partially obscure surface, such as the ledge on the rear of the object, are not shown in the pictorial (Step 6).

If so desired, the full object with its hidden edges also can be created, as shown in Figure 11.51. For the full model with hidden edges to be created, all of the surfaces of the object need to be created on the pictorial, even the hidden surfaces. The hidden surfaces are identified by creating the remaining three of the six standard views or by carefully extracting the required information from the existing three views. When there are many

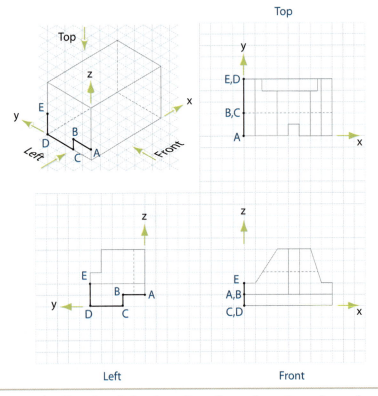

FIGURE 11.48. Identify and track the edges of a surface at the anchor point on the orthogonal views and locate these edges on the pictorial.

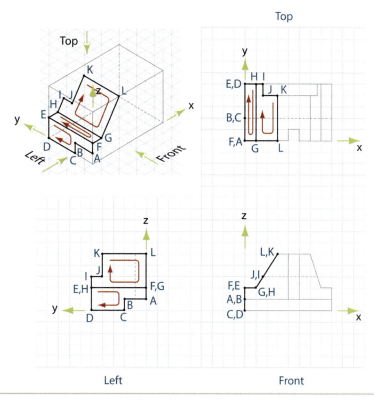

FIGURE 11.49. Surfaces appear as simple loops on the orthogonal views. Complete the entire surface and its edges before proceeding to the next surface.

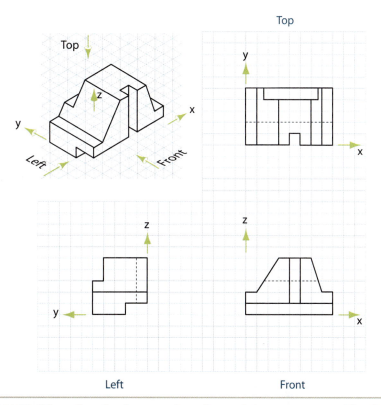

FIGURE 11.50. When all surface loops on the orthogonal views are accounted for, the isometric view should be complete.

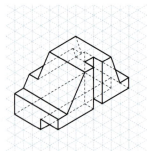

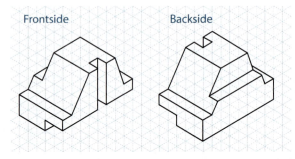

Frontside   Backside

FIGURE 11.51. The full pictorial presentation with visible and hidden edges may be confusing. It is sometimes better to offer a presentation using two separate pictorial views.

hidden edges, it is often better to remove the hidden edges and offer a second pictorial presentation from another viewing position, as shown in Figure 11.51.

Once again, the pictorial is complete after any internal features are added and its model validity is checked (Steps 7 and 8).

Inverse Tracking with Edges and Vertices for Oblique Surfaces and Hidden Features

The next problem is more complicated because of the existence of a hidden feature and an oblique surface. It is desired to create an isometric pictorial of the object shown on the multiview drawing in Figure 11.52. This object is an approximate rectangular solid with one corner that has been cut off obliquely and a cubical cutout at the opposite corner. A five-view presentation would probably have been clearer, but the luxury of additional views is not always offered. Although you can create the additional views yourself, you will proceed with the existing information.

An anchor point, called point A, is selected and identified in each view. The anchor point, coordinate axes, and view directions are defined on an isometric grid. A foundation space is created on the grid based on the limits of the object extracted from the orthogonal views. This space is shown in Figure 11.53.

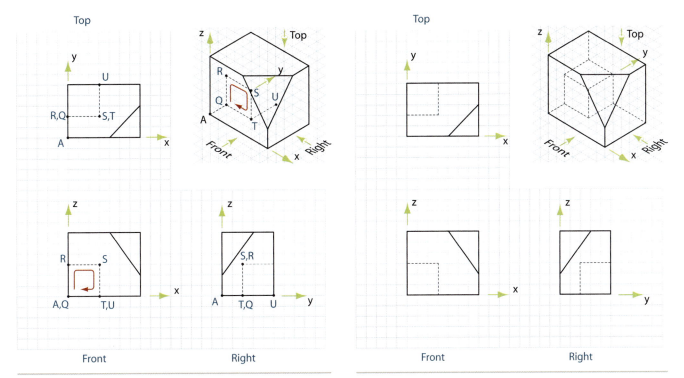

FIGURE 11.55. Track the vertices of the hidden edges. Surfaces are loops formed by the hidden edges. Complete the edges of the entire surface before proceeding to the next surface.

FIGURE 11.56. When all of the edges, visible and hidden, are accounted for, the pictorial should be complete.

have a length of 0.707 units, and the major axis will have a length of 1.225 units. These orientations and sizes for the ellipses are true as long as the circle lies in a plane that is parallel to one of the standard orthogonal views. Also recall that a solid line on the drawing may not be an edge on the part at all, but rather a representation of the optical limit of the part seen along a curved surface. When an isometric pictorial of such an object is constructed, this physical limit of the part must be shown.

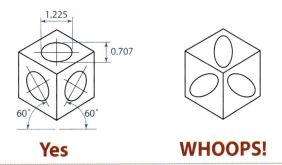

FIGURE 11.57. The size and orientation of a circle of unit diameter on the faces of an isometric cube.

FIGURE 11.58. It is desired to create an isometric pictorial of the object shown in this multiview drawing.

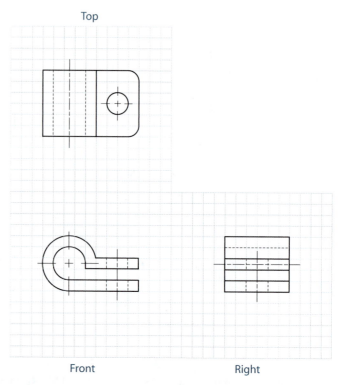

Consider the part shown in Figure 11.58, which has a cylindrical round on its external surface as well as a hole, which is an internal feature.

Start the creation of the isometric view the same way you began the previous cases. An anchor point, point A, is selected and identified in each view. The anchor point, coordinate axes, and view directions are defined on an isometric grid. A foundation space is created on the grid, based on the limits of the object extracted from the orthogonal views. This space is shown in Figure 11.59.

When dealing with curved surfaces, one common strategy you can use is to ignore the curved surfaces temporarily, create the remaining planar surfaces as extended

FIGURE 11.59. Coordinate axes and viewing directions are defined on an isometric grid. An anchor point, A, is selected, and the foundation space is created.

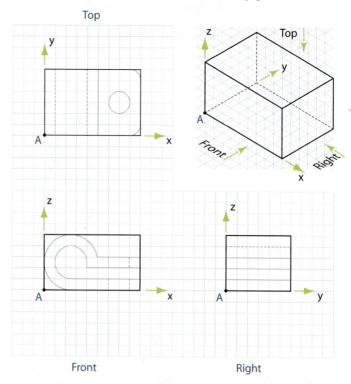

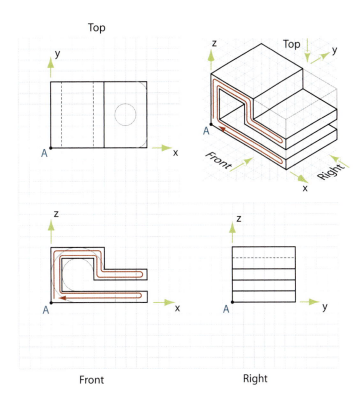

FIGURE 11.60. Create a rough frame by using linear edges instead of curved edges. Use point and edge tracking to create these edges on the pictorial.

surfaces, and then go back to add the curved surfaces. This approach has the advantage of allowing you to quickly draw the rough frame of the part and visualize it. The extended planar surfaces are shown in Figure 11.60.

The circular edges are then added to the appropriate planar surfaces to create the curved surfaces. In the isometric pictorial, the circular edges appear as segments of ellipses. Linear edges that are tangent to circular edges on the orthogonal views also are tangent to their corresponding elliptical edges on the pictorial. This process is shown in Figure 11.61.

FIGURE 11.61. Add the circular edges using elliptical segments in the pictorial.

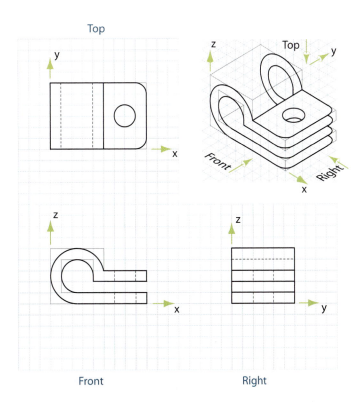

FIGURE 11.62. Add the optical limits of the curved surfaces, seen as lines tangent to the curved edge. Remove portions of edges that are hidden by the curved surfaces.

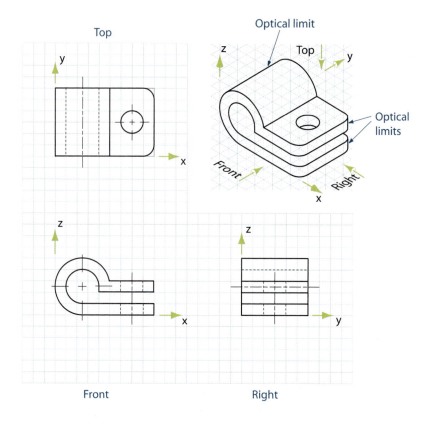

For holes, make sure you include the far edge of the hole when it can be seen through the thickness of the part. For external curves, make sure you include the physical limit of the part as seen on the curved surface, as shown in Figure 11.62, even though a real edge does not exist there. In this example, notice that some surfaces that were hidden or shown on edge in the multiview drawing become partially visible on the pictorial. The continuity of the surfaces must be maintained.

As a final check, make sure each vertex has at least three edges connected to it. Note that tangent edges, created by planes tangent to cylinders or cones, still count as edges, even though they usually are not shown.

11.05.04 Strategy for Constructing Pictorials by Inverse Tracking of Surfaces

The process of tracking vertices and edges to create a mental image, or a pictorial drawing, from a multiview drawing is a slow but reliable method. However, after gaining experience doing this, some people may find it faster and easier to skip this process and go directly to the **surface tracking**. The eight-step model creation process is the same as that used with points and edges except for some slight modifications.

Step 1: Define the location and directions of a coordinate system consistent in all views.

Step 2: Define an anchor surface.

Step 3: Mark the limits of the foundation volume.

Step 4: Locate a surface adjacent to the anchor surface and draw its boundary.

Step 5: Successively locate other adjacent surfaces and draw those boundaries.

Step 6: Convert hidden lines.

Step 7: Add internal features.

Step 8: Check model validity.

FIGURE 11.63. It is desired to create an isometric pictorial of the object shown in this multiview drawing.

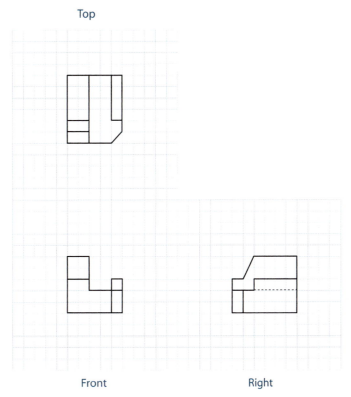

For example, the multiview drawing shown in Figure 11.63 is composed of multiple loops (representing surfaces) in each view. The orientation and viewing directions (Step 1) for the model are shown in Figure 11.64. Also shown in this figure is an **anchor surface**, surface A, which can be easily located in each view and on the pictorial (Step 2). The rectangular extent of the outer loop in each view gives the size of the foundation space as previously described for edges and vertices (Step 3).

FIGURE 11.64. Viewing directions are defined on an isometric grid. The foundation space is created, and an anchor surface is selected.

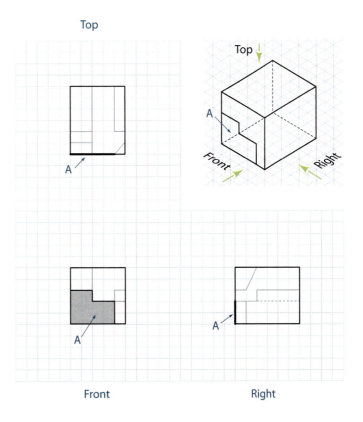

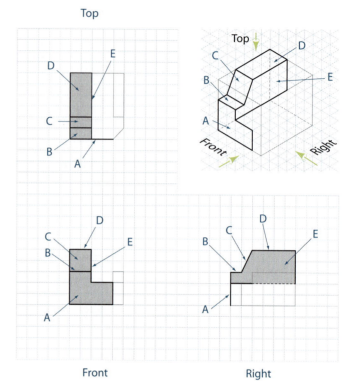

The next step involves identifying a surface adjacent to the anchor surface, locating the adjacent surface in each view, and then adding this surface to the pictorial view (Step 4), as shown in Figure 11.65. Continue to identify surfaces on the object and transfer them to the pictorial, until all of the surfaces are accounted for. The object shown in Figure 11.66 should now be complete (Step 5).

Note on the completed pictorial in Figure 11.66 that some of the surfaces that are fully visible in the multiview drawing may be partially obscured in the pictorial

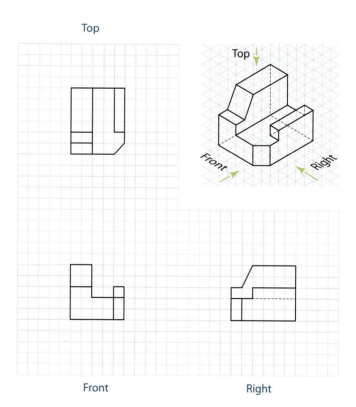

representation. The best way to tell is to look at the view that is adjacent to the view that contains the loop and see if another surface is in "front" of it. If so, you will see one or more hidden lines on the orthogonal view that form the edges of the plane you are trying to create on the pictorial (Step 6). There are no internal features (Step 7) in this example, but remember to check for model validity (Step 8).

Of course, the ultimate goal is to inspect any multiview drawing and create a mental image of the object with very little effort or thought. However, developing this skill takes a great deal of practice. Creating pictorials from the multiview drawing is an exercise that will help you develop this skill. The more you practice, the easier it will become.

11.05.05 Strategy for Improving Spatial Skills through Imagining Successive Cuts to Objects

One last set of exercises to hone your visualization skills is presented next. With this method, you start with a basic shape and remove parts of it through **successive cuts**. With each new cut, you remove different portions of the original object and sketch the result.

You will begin with the basic L-shaped object shown in Figure 11.67. Since this object is made up of only normal surfaces, sketching the missing top view and the isometric pictorial is relatively easy, as shown in Figure 11.68.

FIGURE 11.67. Front and side views of a basic L-shaped object.

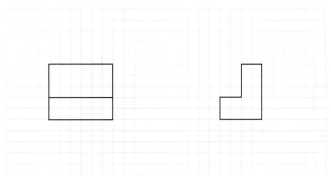

FIGURE 11.68. A multiview drawing of an L-shaped object, with a sketched top view and isometric pictorial.

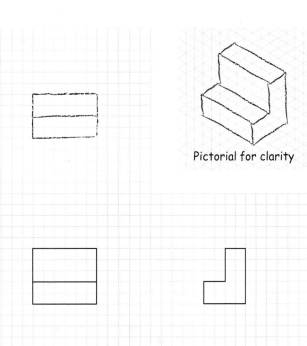

Pictorial for clarity

What happens if you use a block to remove the upper right portion of the original object? Figure 11.69 shows the front and right views of the new object. What do the top and isometric views look like?

The cutting block goes all the way through the object, exposing features as it continues. Since the cutting block cuts through the right side of the object, the left part is undisturbed. You can start by sketching the isometric pictorial of the original object, then superimposing the cutting block on top of this sketch, as shown in Figure 11.70.

If you imagine the intersection between the smaller block and the original object, the result is shown in Figure 11.71a. After the small block has been removed from the original, the object shown in Figure 11.71b results.

Using this isometric pictorial, now you can complete the top view. Recall that the left side of the object is undisturbed, so it will appear exactly as the left side of the top view of the original object. After this cut is made through the object, the multiview drawing shown in Figure 11.72 results.

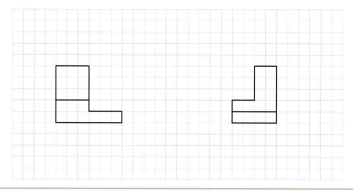

FIGURE 11.69. Front and side views of an object with the upper-right portion removed.

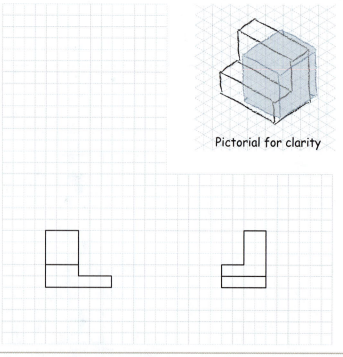

Pictorial for clarity

FIGURE 11.70. A sketched isometric pictorial with a sketched and shaded cutting block shown.

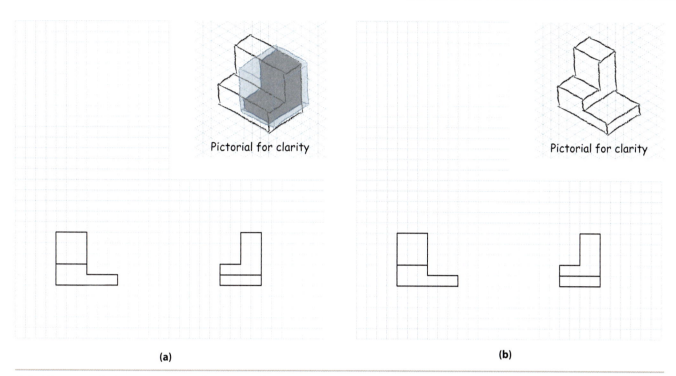

(a)   **(b)**

FIGURE 11.71. The intersections on the cutting block to remove material from an object are shown in (a) on the isometric sketch. The final result (b) on the isometric sketch of using a cutting block to remove material from an object.

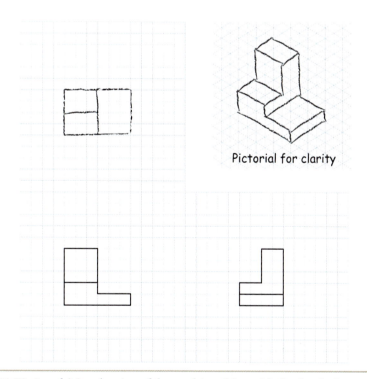

FIGURE 11.72. A multiview drawing of the resulting object with the sketched top view.

What if the cutting block does not go all the way through the original object? Figure 11.73 shows the front- and right-side views of the object in this case. Figure 11.74a shows the isometric with the cutting block superimposed on it, and Figure 11.74b shows the result of using the block to cut the original object.

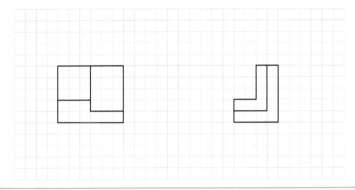

FIGURE 11.73. Front and side views of an object when the cutting block does not go all the way through the object.

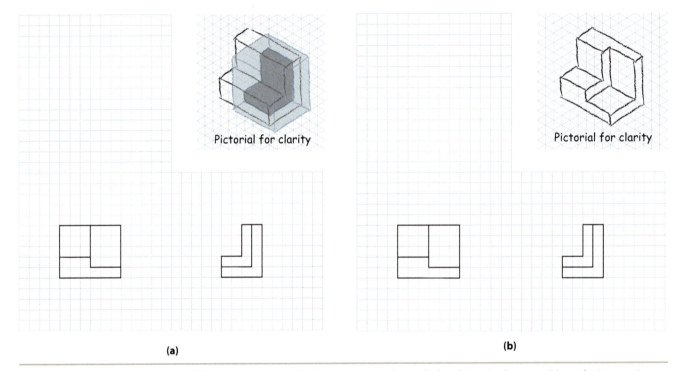

(a)   **(b)**

FIGURE 11.74. The intersections of the cutting block, which cuts partway through the object, is shown in (a) on the isometric sketch. The final result (b) on the isometric sketch of the block cutting partway through an object to remove material from it.

Once again, the new pictorial can be used to create the top view. Remember that the left part of the object was undisturbed by this cut and, therefore, should look identical to the original top view. The correct top view for the object is shown in Figure 11.75.

What if the block had been angled before it was used to cut through the object? Figure 11.76 shows the front- and right-side views of the original L-shaped object after this angled cut has been made. Figure 11.77a shows the original object with the cutting block superimposed on it, and Figure 11.77b shows the result of the cut.

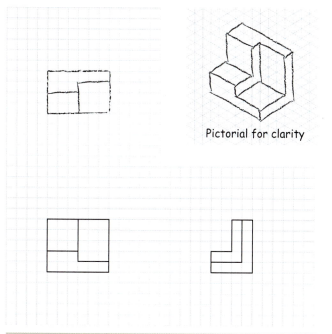

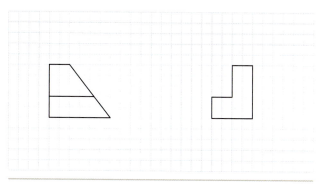

FIGURE 11.76. An L-shaped object with an angled cut.

FIGURE 11.75. A multiview drawing of the resulting object with a sketched top view.

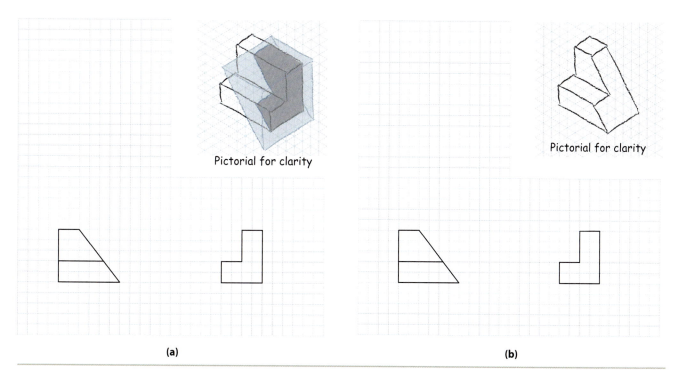

(a)   (b)

FIGURE 11.77. The intersection of the angled cutting block on the original object is shown in (a) in the isometric sketch. The final result (b) on the isometric sketch of the original object cut with the angled cutting block.

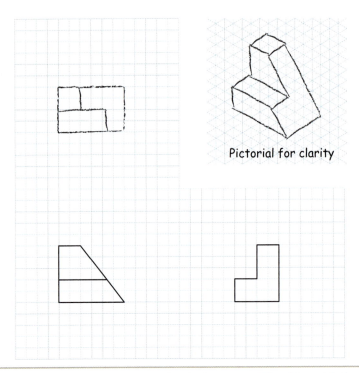

FIGURE 11.78. A multiview drawing with a sketched top view of the object with the angled cut.

What would the top view look like? Notice that the angled cut produced an inclined surface on the object. The surface appears as an L shape in the isometric, an L shape in the side view, and an angled edge in the front view. This inclined surface also must appear as an L-shaped surface in the top view. Figure 11.78 shows a multiview drawing for the object with the correct top view included. Notice how the limits that define the L-shaped surface correspond between the front and top views.

Now what if you take this angled cutting block and move it so that it cuts a *V* out of the middle of the original object? The front and side views that result from this cutting operation are shown in Figure 11.79. Figure 11.80a shows the cutting block superimposed on the original object, and Figure 11.80b shows the result of this cut.

Sketching the correct top view for this object is a bit more difficult since the result is two angled surfaces instead of one. Note that the two angled surface overlap each other in the side view—you see only one outline for the two surfaces. Also note that one of these surfaces is hidden from view in the isometric pictorial because the object is in

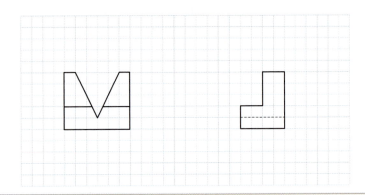

FIGURE 11.79. An object with an angled wedge cut through the center.

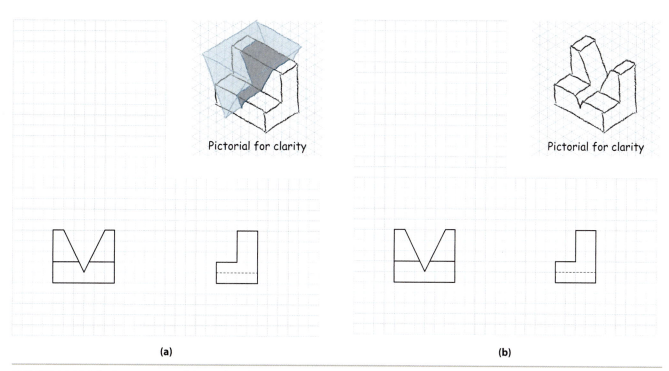

Pictorial for clarity

Pictorial for clarity

(a)

(b)

FIGURE 11.80. The intersection of the wedge-shaped cutting block on the original object is shown in (a) in the isometric sketch. The final result (b) on the isometric sketch of the original object cut with the wedge-shaped cutting block.

the way. However, both inclined surfaces will be visible in the top view. To sketch these surfaces in the top view, use the front view as your guide regarding their limits. The correct multiview drawing, including a top view, is shown in Figure 11.81.

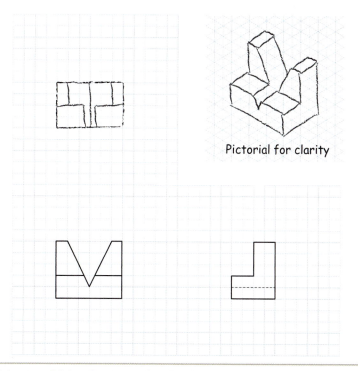

Pictorial for clarity

FIGURE 11.81. A multiview drawing with a sketched top view of the object with the angled wedge removed from it.

FIGURE 11.82. Front and side views of an object with an oblique cut.

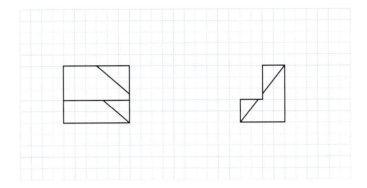

As the last step in this exercise, consider an oblique cut to the object. Figure 11.82 shows the resulting front and side views of the object after it experienced this type of cut. In this case, the cutting block has been oriented in space as shown in Figure 11.83a. The resulting isometric pictorial with this oblique cut is shown in Figure 11.83b.

Recall that oblique surfaces will appear as areas in all views. The triangular surfaces that result from this oblique cut are seen as areas in the front, the side, and the isometric. They also will appear as triangular areas in the top. Use the other views to help you find the limits of the triangular areas in the top view to define the surfaces. The completed multiview drawing for this object is shown in Figure 11.84.

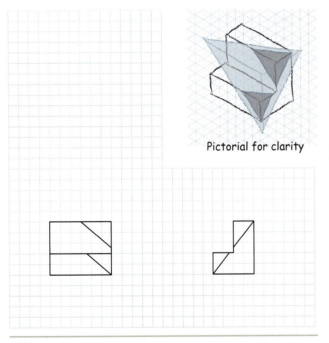

Pictorial for clarity

FIGURE 11.83. The intersection of the oblique surface cutting block on the original object is shown in the isometric sketch.

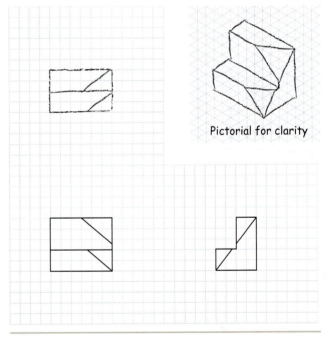

Pictorial for clarity

FIGURE 11.84. A multiview drawing with a sketched top view of the original object with the oblique cut.

11.06 Chapter Summary

The ability to visualize objects in three dimensions is a key skill for engineering design. Physical models for proposed designs often do not exist, and an engineer must be able to envision, manipulate, and modify designs when they are presented in the form of 2-D drawings. You can develop visualization skills only through practice. The more you practice and the more geometrically complex the objects you practice with, the better your skills become. Mental reconstruction of a 3-D object from its 2-D drawing may, at first, seem like a daunting task. However, by keeping in mind some basic rules for the appearance of points, edges, and surfaces between the views on an engineering drawing, you can re-create a pictorial view of the object in a step-by-step process. Eventually, this visualization process becomes easier, quicker, and more natural. The ultimate goal is to be able to work seamlessly with 3-D computer models, pictorial presentations, and 2-D multiview presentations as tools for the development of engineering designs that are in your mind's eye.

11.07 glossary of key terms

anchor point: The same point, usually a vertex, which can be located easily and confidently on multiple views for an object.

anchor surface: The same surface that can be located easily and confidently on multiple views for an object.

curved surface: Any nonflat surface on an object.

edge tracking: A procedure by which successive edges on an object are simultaneously located on a pictorial image and a multiview image of that object.

foundation space: The rectilinear volume that represents the limits of the volume occupied by an object.

frontal surface: A surface on an object being viewed that is parallel to the front viewing plane.

horizontal surface: A surface on an object being viewed that is parallel to the top viewing plane.

inclined surface: A flat surface on an object being viewed that is perpendicular to one primary view and angled with respect to the other two views.

normal surface: A surface on an object being viewed that is parallel to one of the primary viewing planes.

oblique surface: A flat surface on an object being viewed that is neither parallel nor perpendicular to any of the primary views.

point tracking: A procedure by which successive vertices on an object are simultaneously located on a pictorial image and a multiview image of that object.

profile surface: A surface on an object being viewed that is parallel to a side viewing plane.

successive cuts: A method of forming an object with a complex shape by starting with a basic shape and removing parts of it through subtraction of other basic shapes.

surface tracking: A procedure by which successive surfaces are simultaneously located on a pictorial and multiview image of that object.

11.08 questions for review

1. What are normal surfaces on an object? How can they be recognized on a drawing that shows a multiview presentation of the object?

2. What are inclined surfaces on an object? How can they be recognized on a drawing that shows a multiview presentation of the object?

3. What are oblique surfaces on an object? How can they be recognized on a drawing that shows a multiview presentation of the object?

4. Why is it important to be able to visualize an object from its multiview presentation?

5. What are some ways in which the same surface on an object can be recognized in its different orthogonal views?

6. What is an edge view of a surface?

7. In what ways can a curved surface on an object be recognized in its different orthogonal views?

11.08 questions for review (continued)

8. How can inverse tracking of points and edges be used to create a pictorial of an object from a multiview drawing?

9. How can inverse tracking of the surfaces be used to create a pictorial of an object from a multiview drawing?

10. If a pictorial of an object has been incorrectly created from its multiview drawing, what are some indications that the pictorial is incorrect?

11. How are successive cuts on an object used to improve visualization skills?

11.09 problems

1. For the two-view orthographic drawings shown in Figure P11.1, vertices are labeled with numbers to uniquely identify surfaces and edges. Each row of table specifies a geometric element, either a surface or an edge, as seen in one view. Complete each row of the table by identifying the same geometric element seen in the other view.

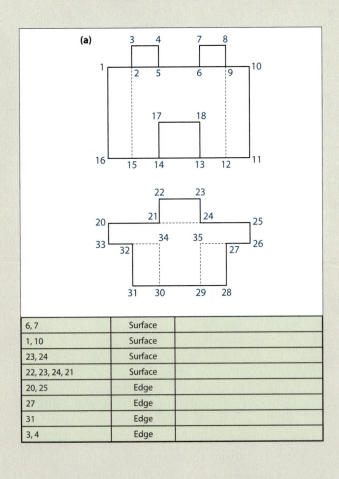

6, 7	Surface	
1, 10	Surface	
23, 24	Surface	
22, 23, 24, 21	Surface	
20, 25	Edge	
27	Edge	
31	Edge	
3, 4	Edge	

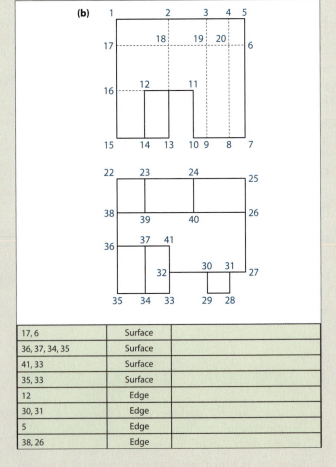

17, 6	Surface	
36, 37, 34, 35	Surface	
41, 33	Surface	
35, 33	Surface	
12	Edge	
30, 31	Edge	
5	Edge	
38, 26	Edge	

11.09 problems (continued)

(c)

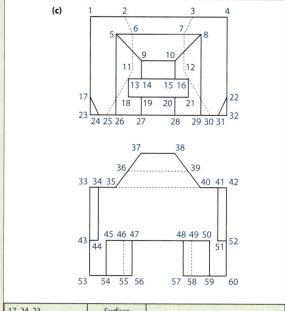

17, 24, 23	Surface	
33, 42	Surface	
35, 37	Surface	
12, 30	Surface	
51, 52	Edge	
35, 37	Edge	
2	Edge	
50, 48	Edge	

(d)

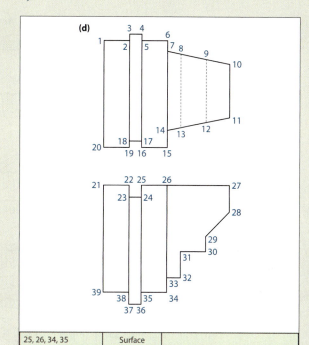

25, 26, 34, 35	Surface	
29, 30	Surface	
14, 11	Surface	
27, 28	Surface	
8, 9	Edge	
33	Edge	
18	Edge	
14, 15	Edge	

(e)

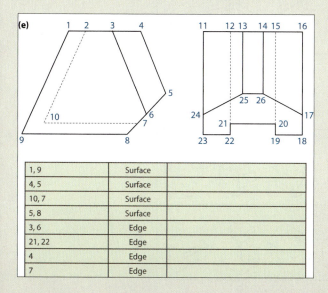

1, 9	Surface	
4, 5	Surface	
10, 7	Surface	
5, 8	Surface	
3, 6	Edge	
21, 22	Edge	
4	Edge	
7	Edge	

(f)

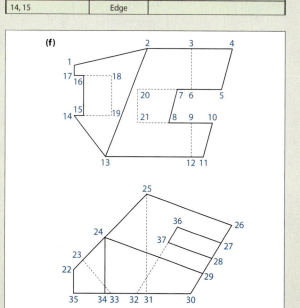

23, 33	Surface	
13, 14	Surface	
11, 13	Surface	
32, 36	Surface	
2, 13	Edge	
36, 27	Edge	
30	Edge	
14	Edge	

FIGURE P11.1.

11.09 problems (continued)

2. For each row shown in Figure P11.2, select the pictorial view of the object that will produce the orthographic views that are given.

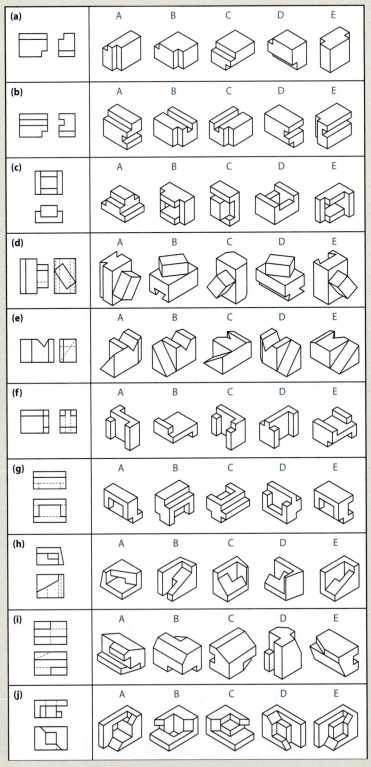

FIGURE P11.2.

11.09 problems (continued)

3. In each lettered cell shown in Figure P11.3, the circle represents the location of a missing view. Select the correct view from the thirty views proposed. A view may be used more than once.

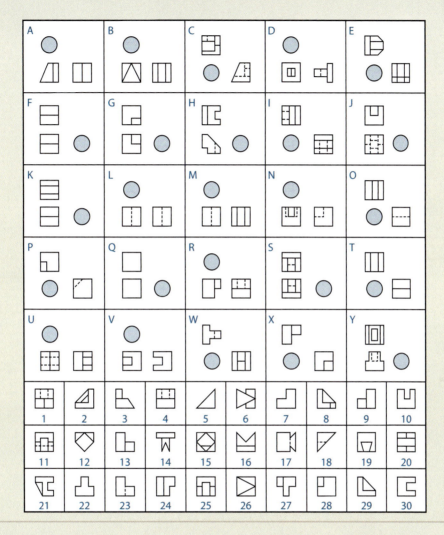

FIGURE P11.3.

11.09 problems (continued)

4. In each lettered cell shown in Figure P11.4, the circle represents the location of a missing view. Select the correct view from the thirty views proposed. A view may be used more than once.

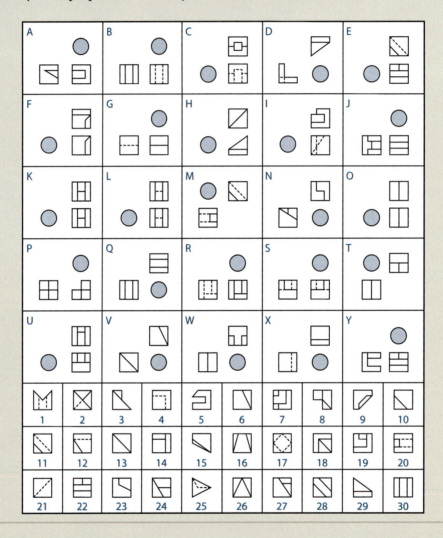

FIGURE P11.4.

11.09 problems (continued)

5. In each lettered cell shown in Figure P11.5, the circle represents the location of a missing view. Select the correct view from the thirty views proposed. A view may be used more than once. Note that all of these problems contain single curved surfaces.

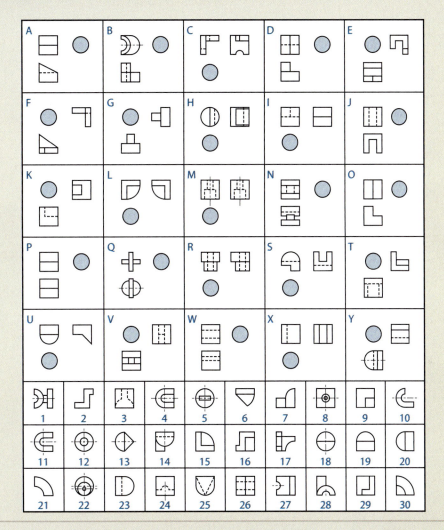

FIGURE P11.5.

11.09 problems (continued)

6. In each lettered cell shown in Figure P11.6, the circle represents the location of a missing view. Select the correct view from the thirty views proposed. A view may be used more than once. Note that all of these problems contain single curved surfaces.

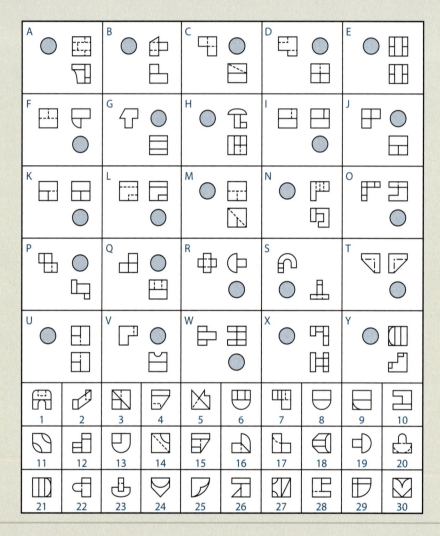

FIGURE P11.6.

11.09 problems (continued)

7. For the object shown in pictorial view in each lettered cell shown in Figure P11.7, select one of the twenty views that correctly shows a top, front, or right-side view of the object having one or more hidden lines. Note that not all of the problems contain single curved surfaces.

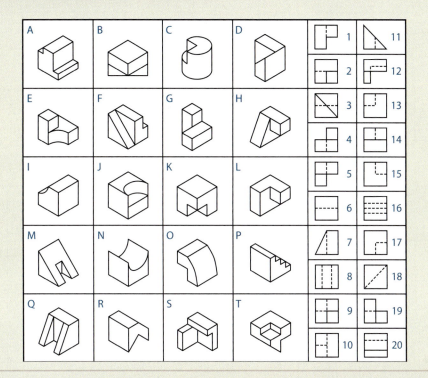

FIGURE P11.7.

11.09 problems (continued)

8. From the multiview drawings shown in Figure P11.8, create an accurate isometric pictorial of each object. Do not show hidden lines on the pictorials.

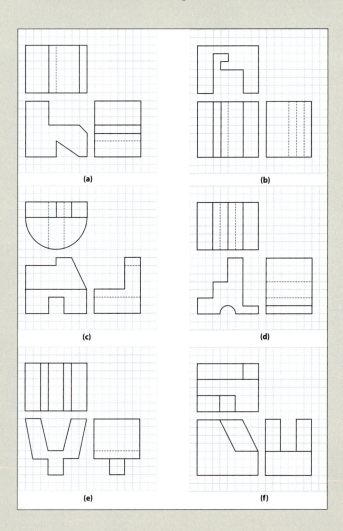

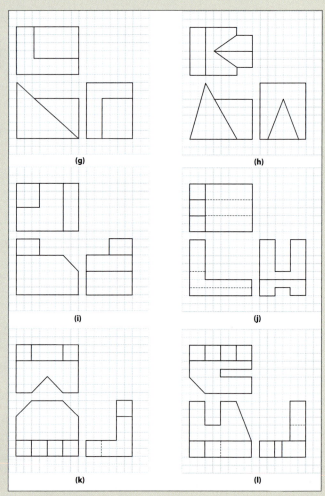

11.09 problems (continued)

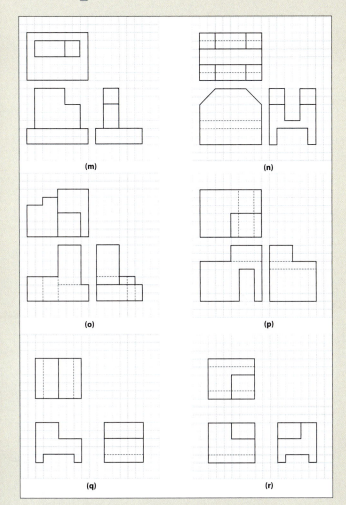

(m)   (n)

(o)   (p)

(q)   (r)

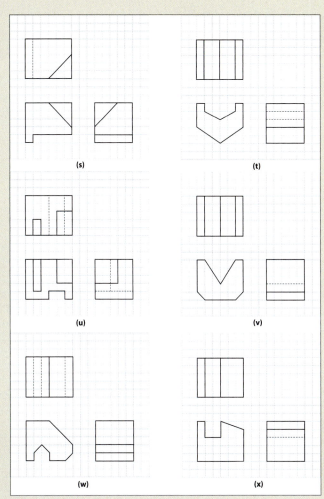

(s)   (t)

(u)   (v)

(w)   (x)

11.09 problems (continued)

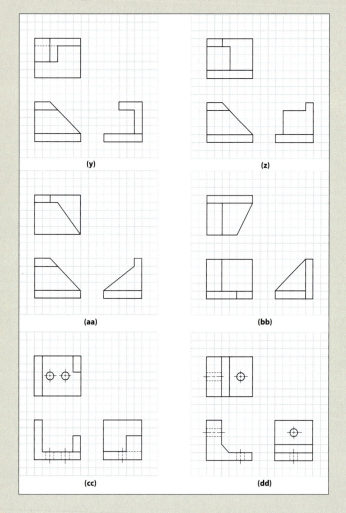

(y)   (z)

(aa)   (bb)

(cc)   (dd)

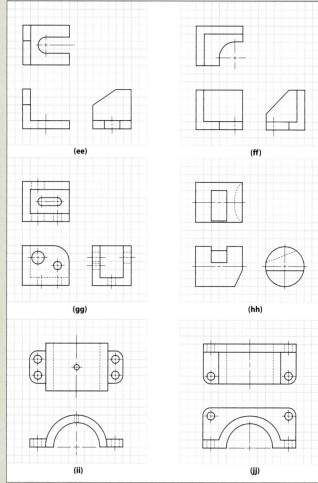

(ee)   (ff)

(gg)   (hh)

(ii)   (jj)

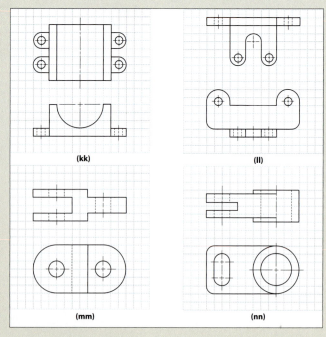

(kk)   (ll)

(mm)   (nn)

FIGURE P11.8.

11.09 problems (continued)

9. From the multiview drawings shown in Figure P11.9, add the missing top- or right-side view and create an accurate isometric pictorial of each object. Do not show hidden lines on the pictorials.

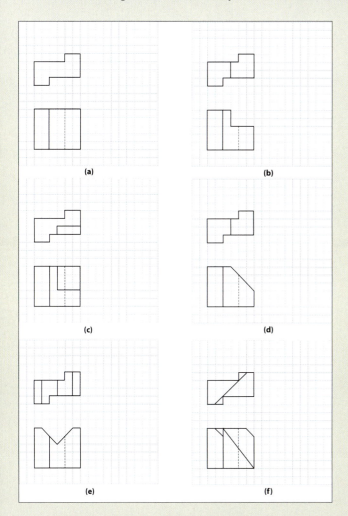

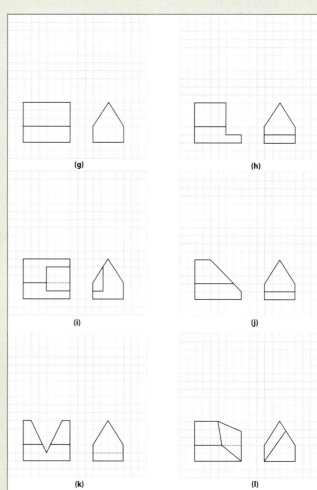

11.09 problems (continued)

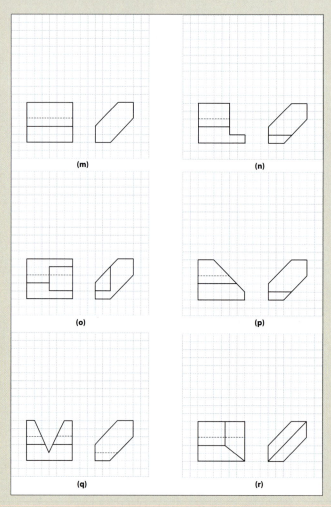

FIGURE P11.9.

11.09 problems (continued)

10. A T-shaped solid object is shown in a multiview and isometric pictorial presentation in Figure P11.10. Subsequent presentations show successive cuts to the original object using a colored solid object whose interfering volume is to be removed from the original object. For each instance shown, draw the multiview and isometric pictorial presentations of the final object after the cut is made.

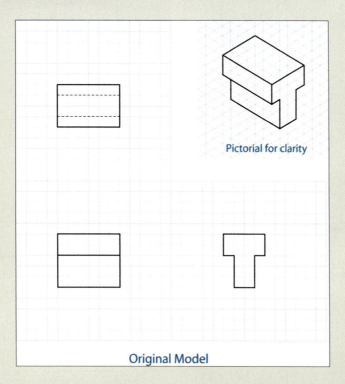

Pictorial for clarity

Original Model

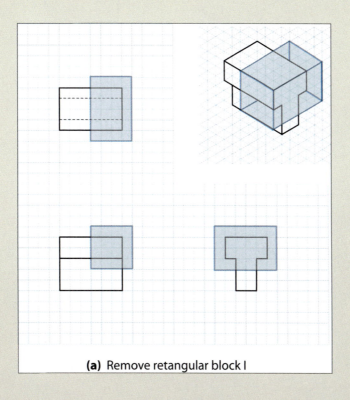

(a) Remove retangular block I

11.09 problems (continued)

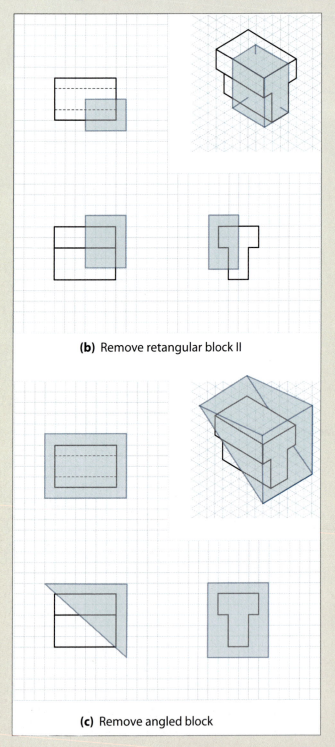

(b) Remove retangular block II

(c) Remove angled block

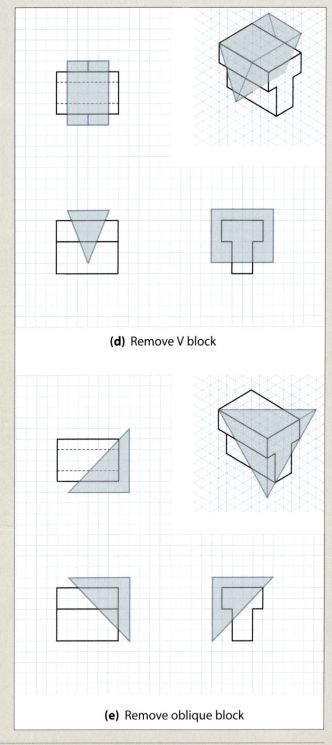

(d) Remove V block

(e) Remove oblique block

FIGURE P11.10.

12

Pictorial Drawings

objectives

After completing this chapter, you should be able to

- Explain the importance of pictorial drawings as an aid in visualization

- Create an isometric drawing of an object composed of principal, inclined, and oblique surfaces

- Draw ellipses on the front, top, and right faces of the isometric to represent cylinders and holes

- Explain the difference between a cavalier and cabinet oblique drawing

- Create an oblique drawing given the orthographic drawing of an object

- Create a two-point perspective drawing given the orientation of the plan view and the location of the elevation view

12.01

introduction

A major problem in engineering graphics is the need to represent a 3-D object using a 2-D sheet of paper. Generally, this is done using orthographic views that show the top, front, and side views of the object as separate entities. Each view represents two dimensions of the object. The top view displays the width and depth dimensions, the front view shows the width and height dimensions, and the side view displays the width and height dimensions. Since each of these views represents only two dimensions, you have to move the views around in your mind's eye to figure out what the 3-D object looks like. Sometimes you will get it right, and sometimes you will get it wrong. A **pictorial** drawing or sketch can be effective in helping you visualize the 3-D shape. The pictorial shows all three dimensions (height, width, and depth) in a single view of the object. You already have been exposed to pictorials in previous chapters of this book, but in a rather informal way. Sketching pictorials helped you develop your visualization skills. The earlier chapters also used pictorials as a visualization aid for making formal multiview drawings. There are cases, however, when pictorial sketches are inappropriate for certain applications; for example, in formal engineering drawings and documents. In these cases, a more formal and more accurate method of creating pictorial drawings is needed.

Pictorials help you visualize, but they also are important when you attempt to assemble parts into mechanisms or need to purchase replacement parts for a tool. For example, when you purchase a lawn mower, the manufacturer will furnish a user's manual. In the user's manual, you are likely to find a series of pictorial drawings that show every lawn mower part with some identifying information. If you need to replace a part, you can use the pictorial to install the part. Pictorial drawings can be an effective way to ensure that a mechanism gets reassembled properly after a part is replaced.

Pictorial drawings are used mainly as visualization aids; they are not used as working drawings from which a part is produced. Historically, before the industrial revolution, craftsmen would sketch a design as a pictorial and use it to produce a product. If part of the design were to fail or break, they would have to custom-make another part in order for the product to be useful again. After the industrial revolution, products were manufactured using mass production and the concept of interchangeable parts came into being. With interchangeable parts, a detail drawing showing lines in true length and angles as true angles was necessary to guarantee that parts could be used interchangeably. Also, proper dimensions and tolerances were required to ensure that the parts would be interchangeable. Because a pictorial drawing shows all three dimensions on a 2-D sheet of paper, lines that define surfaces may not be true length and angles may not be true angles, as shown in Figure 12.01. The figure shows a pictorial representation of a cube. Note that the lines that define the cube are not shown in true length. Also, since a cube is composed of six surfaces that are squares, the angle of line AB and AC should be 90°; but it is not. For the object to be produced by mechanical processes, it is important that the shape be interpreted correctly. An effective way to do that is to use a multiview orthographic drawing (each view contains only two dimensions), which shows the true lengths of lines and the true angles. After the object is produced, the pictorial drawing can be used to compare the 3-D shape of the object with the actual object and the pictorial drawing will verify that the shape is correct.

There are many types and variations of pictorial drawings since they must serve different purposes. Some of the major types of pictorials are **axonometric** drawings, **oblique** drawings, and **perspective** drawings. An example of each is shown in Figure 12.02. The term *axonometric* is a broad category of pictorial drawings that includes **isometric**, **diametric**, and **trimetric** drawings. The term *axonometric* refers to the angle that the coordinate axes make with each other when the three dimensions are defined for the object. Figure 12.02 shows the coordinate axes relative to the type of pictorial drawing represented. An oblique drawing is a pictorial in which the irregular surface (often a circular cylinder) is drawn in the 2-D plane of the paper and the third dimension is drawn at some receding angle to this plane (Figure 12.02). A perspective drawing is one where the receding axis(es) converge at vanishing point(s) located on what is referred to as a horizon. When done correctly, a perspective drawing gives the impression that you are looking at a photograph. Each of these types of pictorial drawings will be discussed in detail later in this chapter.

Most of the following techniques for creating a pictorial drawing are traditional methods based on 2-D CAD and drafting techniques. These techniques can still be used when 3-D modeling tools are not available. For speed, efficiency, and accuracy, pictorial drawings are most easily created when they are extracted from 3-D models.

FIGURE 12.01. An isometric pictorial of a cube.

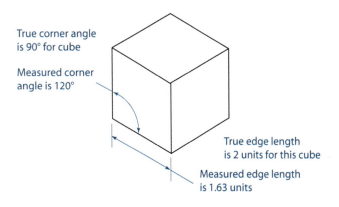

FIGURE 12.02. Samples of pictorial drawing types of the same object.

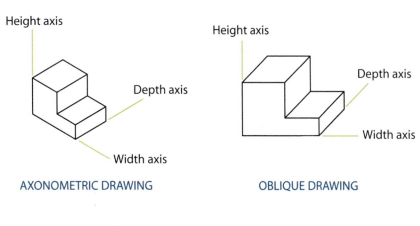

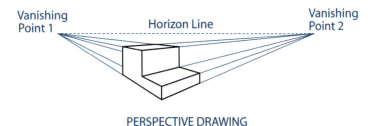

12.02 Axonometric Drawings

The word *axonometric* has its origin in Greek from the word *axon*, which means "axis," and the word *metric*, which means "measure." An axonometric drawing refers to three types of pictorial drawings: isometric, dimetric, and trimetric drawings that are created by measuring along three axes representing width, depth, and height. Isometric drawings are relatively easy to produce and usually can be done quickly. In an isometric drawing, the three dimensions of width, depth, and height are shown along the three isometric axes, as shown in Figure 12.03. When the dimensions of width, height, and depth are plotted in this fashion, the three normal surfaces (frontal, horizontal, and profile) of a rectangular solid object have equal angles between them (120°). In dimetric drawings, the three dimensions of width, depth, and height are shown along the three dimetric axes, as shown in Figure 12.04. When plotted correctly along these axes, two of the normal surfaces (frontal and profile) have equal angles between them, while the third normal surface (horizontal) has a different angle. In a trimetric drawing, the dimensions of width, height, and depth are plotted along axes so the normal surfaces of a rectangular solid (frontal, horizontal, and profile) have none of the three angles equal, as shown in Figure 12.05.

FIGURE 12.03. Three visible normal surfaces (frontal, horizontal, and profile) of a prism have equal angles between them (120°) in an isometric drawing.

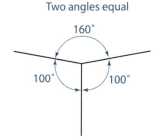

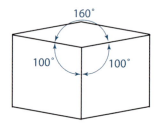

FIGURE 12.04. In a diametric drawing, two of the three visible normal surfaces (frontal and profile) of a prism have edges presented at equal angles, not equal to 120°.

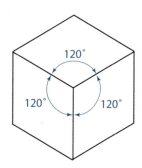

FIGURE 12.05. A trimetric drawing presents the three visible normal surfaces (frontal, horizontal, and profile) of a prism in a position where none of the angles between the surface edges is equal.

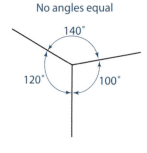

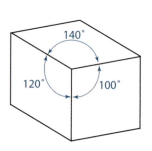

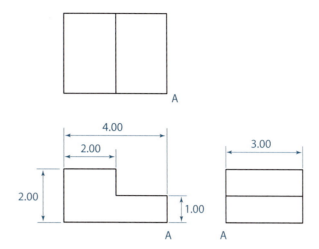

FIGURE 12.06. Orthographic views of a step block.

The drawing of dimetric and trimetic pictorials takes a great deal of time because of the uncommon angles that are used. Therefore, the time it takes to complete this type of pictorial offsets any benefit there may be in producing it. Dimetric and trimetric pictorials are seldom used in engineering work; so they will not be discussed in detail. Instead, the chapter will focus on the easiest and most popular form of axonometric drawing, the isometric drawing.

12.02.01 Isometric Drawing

The best way to learn about formal isometric drawings is to study a simple example. Figure 12.06 shows the orthographic views of a step block. For this object, all of the surfaces are normal surfaces. This means that each surface is viewed in its true size and shape in one of the principal views and will appear as an edge view in the other principal views. A frontal surface appears in true size and shape in the frontal view, while a horizontal surface appears in true size and shape in the horizontal or top view. A profile surface appears in true size and shape in the right side or right profile view. Note that in the front view, you see the width and height dimensions; in the top view, you see the width and depth dimensions; and in the profile view, you see the depth and height dimensions. To draw the isometric pictorial of the step block, you must first set up the isometric axes that will define where to measure the width, height, and depth dimensions.

One way to define the isometric axes is shown in Figure 12.07. Two receding axes intersect at point A and are at 30 degrees to an imaginary horizontal line. The width dimensions will be plotted along the receding axis extending to the left of point A, and the depth dimensions will be plotted along the receding axis extending to the right of point A. The height dimension will be plotted along the vertical axis that extends upward from point A. When you look at the orthographic drawing in Figure 12.06, you see that the maximum width, depth, and height are 4 units, 3 units, and 2 units, respectively.

FIGURE 12.07. The isometric axes.

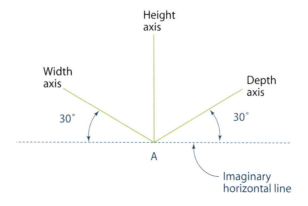

When beginning to draw the isometric, you want to frame the step block on the isometric axes and then take care of the details to finish the isometric. Match the following steps with the steps shown in Figure 12.08 to draw the isometric of the step block.

Step 1: On the isometric axis marked *Width*, measure the maximum width of the step block (4 units) from point A and label this measurement as point B. On the isometric axis marked *Depth*, measure the maximum depth of the step block (3 units) from point A and mark this measurement as point C. On the isometric axis marked *Height*, measure the maximum height of the step block (2 units) from point A and mark this measurement as point D.

Step 2: Draw vertical lines from points B and C. From point D, draw a line parallel with line AC that intersects with the vertical line from C. Label this point as E. In a similar manner, from point D, draw a line parallel with line AB that intersects with the vertical line from B. Label this point as F.

Step 3: Complete the "isometric reference prism" by drawing a line from E that is parallel with line DF. Then draw a line from F that is parallel with DE. The line from E and the line from F intersect at point G. The isometric reference prism contains the step block.

Step 4: Along line FD, measure the width of the upper surface (2 units) and label it point 1. From point 1, draw a line parallel with FG to line GE and label the point on GE as point 2. From point 1 and point 2, draw vertical lines downward parallel with BF.

Step 5: Along the vertical line AD, measure the height of the lower surface (1 inch) and label it on AD as point 3. From point 3, draw a line parallel with AC to line CE. Label the point where this line intersects CE as point 4.

Step 6: From point 3, draw a line parallel with AB that intersects the vertical line from point 1 and label it point 5. In a similar manner, draw a line from point 4 parallel with AB that intersects the vertical line from point 2 and label it point 6. Connect point 5 with point 6.

Step 7: Erase lines 1-D, D-3, 2-E, and E-4. The isometric drawing of the step block is complete. Darken all lines that define the step block.

The step block is relatively easy to draw because all of the lines that define it are called **isometric lines**. All of the lines defining the frontal surface are parallel or perpendicular, as are the lines defining the horizontal and profile surfaces. Isometric drawings can get rather complex when surfaces that need to be drawn are not defined as frontal, horizontal, or profile surfaces because the lines that form the surfaces are not parallel or perpendicular. Therefore, these lines are "nonisometric lines" and need to be plotted using their endpoints.

12.02.02 Inclined Surfaces

Figure 12.09 is an orthographic drawing showing a variation of the step block that has an inclined surface. The procedure to draw the isometric pictorial of this object is similar to drawing the step block in Figure 12.08 except that you have to account for the inclined surface. Use the following steps along with the steps in Figure 12.10.

Step 1: Follow the seven steps to draw the step block shown in Figure 12.08.

Step 2: From point 5, draw a vertical line downward that intersects AB at point 7.

Step 3: From point 7, draw a line to point C; and from point 5, draw a line to point 4. The inclined surface is defined by the points 5-4-C-7.

Step 4: Remove lines 5-3, 3-4, 7-A, and AC. These edges were used for construction only, and are not a part of the object. Darken all visible edges to complete the drawing.

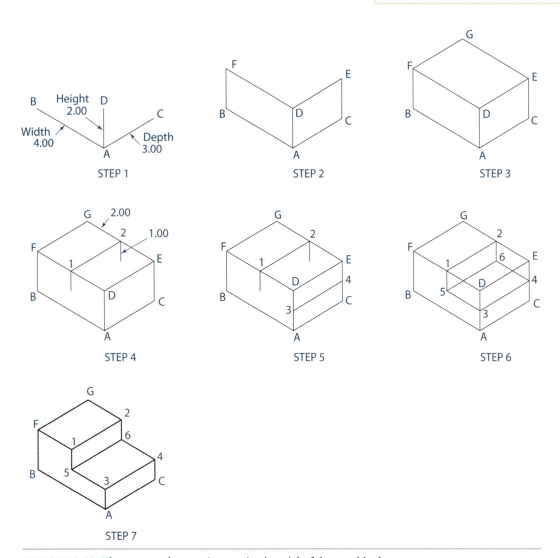

FIGURE 12.08. The steps to draw an isometric pictorial of the step block.

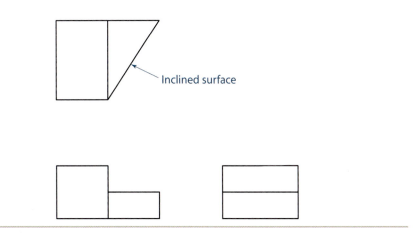

FIGURE 12.09. Orthographic views of a step block with an inclined surface.

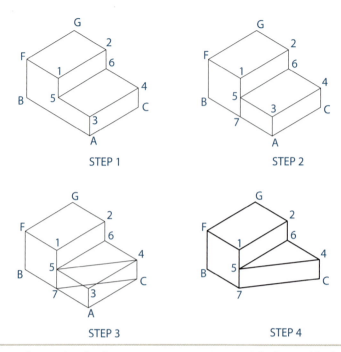

FIGURE 12.10. The step involved to create an isometric pictorial of a step block with an inclined surface.

Lines 5-4 and 7-C in Figure 12.10 are defined as nonisometric lines. They are not parallel to any isometric lines, which is why they are called nonisometric lines. They must be drawn by plotting the endpoints and then connecting them properly, as shown in Figure 12.10.

12.02.03 Oblique Surfaces

An even more complex example involving isometric drawings includes a drawing that has an oblique surface. An oblique surface is a surface that is neither parallel nor perpendicular to the frontal, horizontal, or profile projection plane; and the oblique surface will appear in all three views as its characteristic shape. That is, if the oblique surface is a triangle, it will be a triangle in all three views. It will not appear as an edge in any of the three views. Knowing this, it is reasonable to assume that an isometric drawing showing an oblique surface will show the surface in its characteristic shape. Figure 12.11 shows

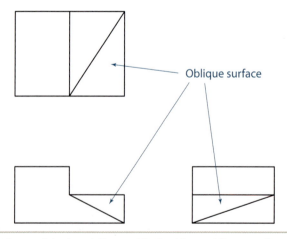

FIGURE 12.11. Orthographic views of a step block with an oblique surface.

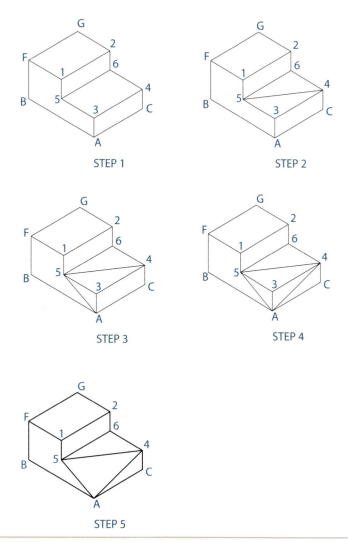

FIGURE 12.12. The steps involved to create an isometric pictorial of a step block with an oblique surface.

three orthographic views of a step block that includes an oblique surface. Use the following steps and the steps shown in Figure 12.12 to draw the isometric of the step block with the oblique surface.

Step 1: Follow the seven steps to draw the step block shown in Figure 12.08.

Step 2: Draw a line from point 5 to point 4.

Step 3: Draw a line from point 5 to point A.

Step 4: Draw a line from point 4 to point A.

Step 5: Erase lines 4-3, 5-3, and 3-A to complete the isometric.

In Figure 12.12, the surface formed by 5-4-A is an oblique surface. The lines 5-4, 4-A, and 5-A are nonisometric lines. All of the remaining lines of the step block are isometric lines.

In an isometric drawing, orientation of the object depends on the placement of the axes, which locates the origin. In Figure 12.08, point A on the lower right-corner of the object was chosen as the location of the origin. From this point, you measured back along the left receding axis to lay off widths and you measured along the right receding axis to measure depth. Then you measured vertically to measure height. For all measurements, you assumed point A to be the origin, or the 0,0,0 point. Selection of the

FIGURE 12.13. The origin used
to locate an object using the
lower-left corner.

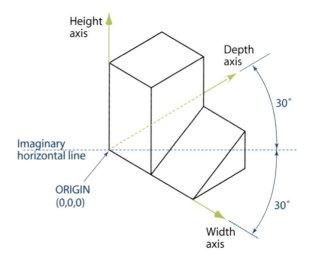

origin (or point 0,0,0) in 3-D coordinates establishes the relative position of each point
that makes up the object. The origin can be placed anywhere on the object, and the isometric can be drawn from this reference. For example, Figure 12.13 shows an origin
that would be located at the lower-left corner of the object. The width, depth, and
height measurements would be measured along the appropriate axes as shown in
Figure 12.13. Notice that the width and depth axes are 30° from an imaginary horizontal line, as explained previously. No matter where the origin is located, all points that
define the object can be plotted in 3-D space; then the points are connected to show the
pictorial.

Refer back to Figure 12.6, the orthographic view of the step block. This set of views
shows a front view, a top or horizontal view, and a right profile view. You also can orient the views for the step block to show a front view, a top or horizontal view, and a left
profile view (Figure 12.14). Each of these orthographic arrangements allows for two
different orientations of the pictorial. The pictorial can open to the right or it can open
to the left. The term *open* refers to how the orthographic views are interpreted. When
the orthographic drawing is oriented to show a front, top, and right-side view, the isometric pictorial opens to the right (Figure 12.15). When the orthographic drawing is
oriented to show a front, top, and left-side view, the isometric pictorial opens to the left
(Figure 12.15).

Orthographic views generally show all kinds of lines that define the object. Lines
may be visible, or they may be hidden. Hidden lines are represented in orthographic
views as dashed lines. In isometric pictorial views, hidden lines are generally not shown

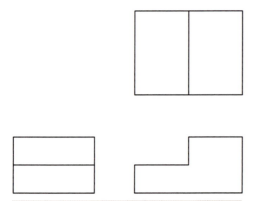

FIGURE 12.14. Orthographic views of the
step block oriented as the front, top, and left
profile views.

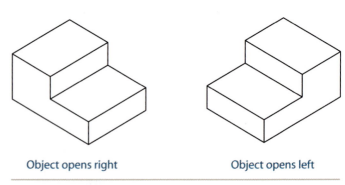

Object opens right   Object opens left

FIGURE 12.15. An isometric drawing has two primary positions
for viewing. The object can be oriented so that it "opens" to the
right or to the left.

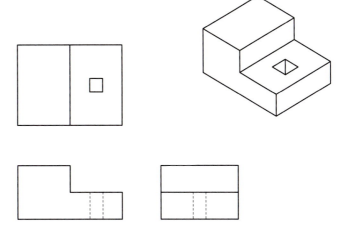

FIGURE 12.16. Orthographic views with a square hole through the base. Hidden lines are shown in the front and right profile views. The isometric pictorial does not show hidden lines.

unless they are necessary for interpretation of the object. For example, Figure 12.16 shows an orthographic drawing of a step block on the left that has a square hole going all the way through it. In the orthographic views, hidden lines in the front view and in the right profile view define the hole. The isometric pictorial of this block is shown on the right. Notice that no hidden lines are shown. The square hole shown is assumed to go through the entire object. This is a correct representation of the step block in isometric. Figure 12.17 shows an orthographic drawing of a step block on the left that has a square hole that does not go all the way through. This is often referred to as a blind hole. As before, hidden lines are shown in the orthographic views; however, since the hole does not go all the way through the object, the isometric pictorial shown on the right must include hidden lines to define how deep the square hole goes into the step block. In this case, hidden lines are necessary to define the depth of the hole.

12.02.04 Cylindrical Surfaces

You have learned that 3-D objects can be composed of normal, inclined, and oblique surfaces, and you have learned how to represent each surface in an isometric pictorial. Another type of surface that is associated with 3-D objects is the cylindrical surface. The cylindrical surface may be positive (such as a post) or negative (such as a hole). When you are looking directly at a normal surface in an orthographic view, the cylinder is represented by a circle. Since the normal surfaces of an isometric pictorial are distorted and you are looking at all three dimensions on one sheet of paper, the cylinder is represented on an isometric drawing by an ellipse. The orientation of the ellipse is dependent upon whether it appears on the top face, the right face, or the left face. An

FIGURE 12.17. Orthographic views showing a square hole that does not go all the way through the base. Hidden lines are shown in the front and right profile views. Hidden lines are shown in the isometric pictorial.

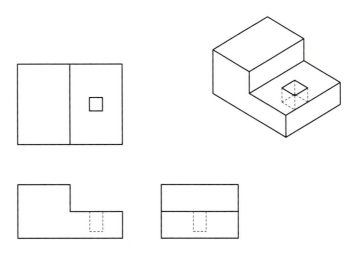

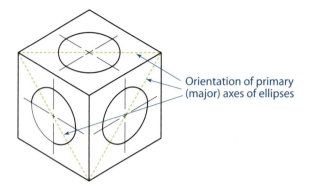

FIGURE 12.18. An isometric cube showing ellipses on each face. Note the orientation of the long (major) axis for the ellipse on each of the three normal faces.

Orientation of primary (major) axes of ellipses

isometric cube that has an ellipse on the right face, the left face, and the top face is shown in Figure 12.18. Note the orientation of the long axis for the ellipses on each of the three normal faces of the cube.

Figure 12.19 illustrates how to construct an ellipse on a horizontal (top) isometric surface. The first step is to establish the limits of the ellipse by drawing a limiting box, which appears as a parallelogram in the pictorial view. For normal surfaces on the object, the sides of the parallelogram are parallel to the isometric lines. The easiest way to create the ellipse is to use an ellipse template (when drawing is done by hand) and selecting the ellipse that is tangent to the sides of the limiting box. When a 2-D CAD tool is used, an ellipse is created by specifying its major and minor diameters in an ellipse creation tool; the ellipse is then rotated into the correct orientation.

The ellipse also can be constructed using the four-center method. The construction locates four centers that can be used for drawing four arcs (two large arcs and two small arcs) that approximate an ellipse. These arcs may be sketched or may be drawn with a compass. The following steps are shown in Figure 12.19.

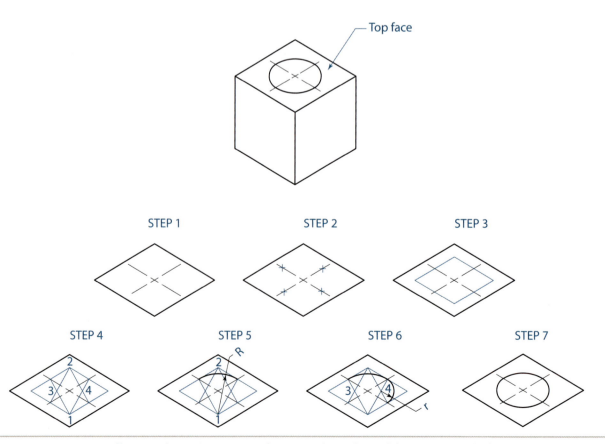

FIGURE 12.19. Drawing an ellipse on the top isometric surface using the traditional four-center method.

Step 1:   Locate the center of the cylinder by showing the centerlines laid out parallel to the isometric sides.

Step 2:   Measure along the centerlines the extreme points of the cylinder.

Step 3:   Frame the limits of the ellipse that will represent the cylinder. The box should look like a diamond. (An isometric square is a diamond-shaped rhombus.)

Step 4:   Note that the near point (point 1) and the far point (point 2) of the diamond are the first two points of the four centers. From these points, draw light construction lines across the diamond where the centerline intersects the box. You should have four light construction lines on the surface. Where these light construction lines cross is the location of point 3 and point 4 of the four centers.

Step 5:   Assuming a long radius (R) as shown in the figure, draw arcs between the centerlines, using point 1 and point 2 as the center.

Step 6:   Assuming a short radius (r) as shown in the figure, draw arcs between the centerlines, using point 3 and point 4 as the center.

Step 7:   The ellipse is now complete.

The steps required to draw an ellipse on the right face are shown in Figure 12.20, and the steps required to draw an ellipse on the left face are shown in Figure 12.21. Essentially, they are the same as drawing the ellipse on the top face.

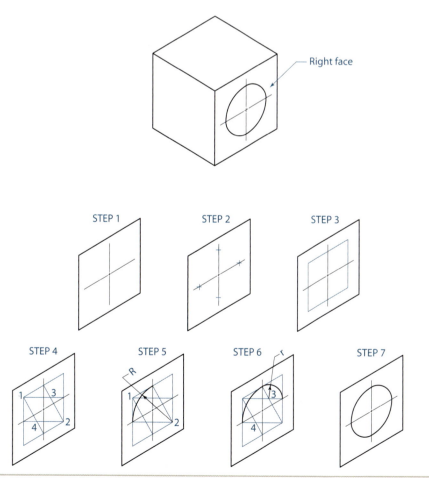

FIGURE 12.20. Drawing an ellipse on the right isometric surface using the traditional four-center method.

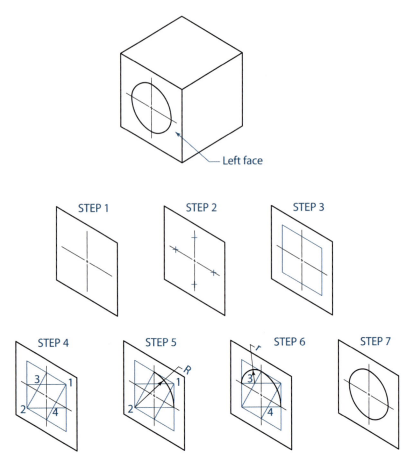

FIGURE 12.21. Drawing an ellipse on the left isometric surface using the traditional four-center method.

Figure 12.22 shows the orthographic views of a step block that has one of the surfaces represented by a semicircle. When drawing the isometric pictorial of this object, you need to incorporate the steps involved in drawing an ellipse; but you are going to draw only half of it. Construct the isometric drawing of the step block shown in Figure 12.08. Figure 12.23 illustrates the steps required to draw the lower surface, which includes a semicircle.

Step 1: Locate the center of the radius (1.5 units) on surface 3-5-6-4. The center can be located by measuring 1.5 units from point 3 along line 3-4 and then measuring 1.5 units from point 3 along 3-5. When you are drawing isometric lines from these two points, the lines will intersect at the center location for the radius.

Step 2: Frame the ellipse by measuring along line 3-5 a distance of 3 units (you will have to extend line 3-5) and marking the point as X. Through X, draw a line parallel with line 3-4 until it intersects line 4-6, which has been extended, and mark this point as Y.

Step 3: From point 3, draw a light line to the midpoint of line 4-Y. From point 3, draw a light line to the midpoint of line X-Y. From point Y, draw a light line to the midpoint of line 3-4. Label this line Z where it intersects the line drawn in from point 3 to the midpoint of 4-Y.

Step 4: Construct a radius (length is the distance from Y to the midpoint of line 3-X) using point Y as the center point and drawing the arc from the midpoint of line 3-X to the midpoint of line 3-4. Construct another radius (length is the distance from Z to the midpoint of line 3-4) using point Z as the center point and drawing the arc from the midpoint of line 3-4 to the midpoint of line 4-Y. The partial ellipse is complete for the top surface.

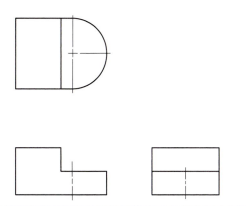

FIGURE 12.22. Orthographic views of a step block with a semicircular base.

Step 5: Repeat the previous steps on the lower surface to create two arcs that are "parallel" to the arcs drawn on the top surface.

Step 6: To complete the pictorial, draw a tangent between the upper and lower surface arcs, as shown in Figure 12.23.

12.02.05 Ellipses on Inclined Surfaces

Sometimes the need arises to draw an ellipse on an inclined surface that appears in an isometric drawing. To do this, you must project some points to ascertain the location of the ellipse on the inclined surface. This task is achieved by using the following steps with steps 1 and 2 shown in Figure 12.24, steps 3 and 4 shown in Figure 12.25, and steps 5 and 6 shown in Figure 12.26. Using these three illustrations, you will create a circular hole in the center of the inclined surface labeled as A-B-C-D.

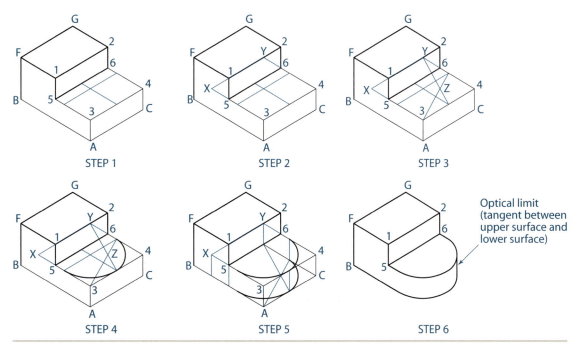

FIGURE 12.23. Drawing an isometric pictorial of a step block with a semicircular base.

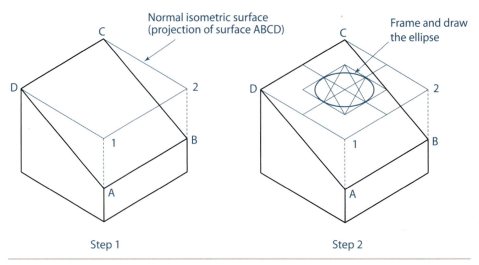

FIGURE 12.24. Steps 1 and 2 to create an isometric pictorial showing a vertical circular hole in an inclined surface.

Step 1: Create a normal isometric surface by locating point 1 using points B and A. Draw a light line from B along the isometric line shown. Draw a light line from A vertically until the line from B is intersected. This intersection is the location of point 1. Repeat step 1 using points C and D as your reference points and locate point 2. The rectangular surface shown as B-1-2-C represents a projection of the inclined surface B-A-D-C. You will use this projection to establish the cylindrical hole.

Step 2: In the center of the horizontal surface shown as B-1-1-C, locate the centerline for the cylindrical hole, which will be shown as an ellipse since this is an isometric surface. Frame the area that will contain the ellipse and draw the ellipse as shown in Figure 12.24.

Step 3: Locate several points on the ellipse that will be used to project to the inclined surface to locate the hole on the inclined surface. Project each point that you located on the ellipse to line C-2. Make sure these projections are parallel to lines BC and 1-2.

Step 4: Where the points projected in step 3 intersect line C-2, project them straight down (parallel to D-2) to line CD, which is on the inclined surface.

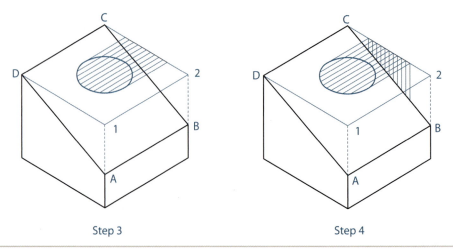

FIGURE 12.25. Steps 3 and 4 to create an isometric pictorial showing a vertical circular hole in an inclined surface.

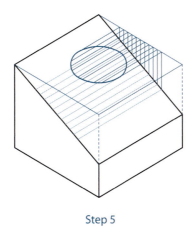
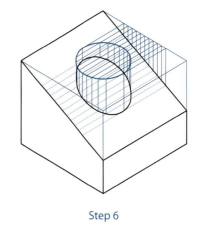

Step 5   Step 6

FIGURE 12.26. Steps 5 and 6 to create an isometric pictorial showing a vertical circular hole in an inclined surface.

Step 5: Where the points projected in step 4 intersect line CD, draw lines parallel to BC and AD on the inclined surface.

Step 6: From the points on the ellipse drawn on the normal isometric surface B-C-2-1, project downward (parallel to A-1 and B-2) to the place where the points intersect their specific projection line on the inclined face (A-B-C-D). Plot a point on this intersection. Using an irregular curve, connect the points on the inclined surface to complete the location of the cylindrical hole on the inclined surface.

Figure 12.27 shows the circular hole in the inclined surface without all of the construction required to create it. This looks as though it would be simple to create; but as you refer back to the steps, the process is very complex.

FIGURE 12.27. The finished construction of the vertical circular hole in an inclined surface.

12.03   Oblique Drawings

Oblique drawings are forms of pictorial drawings that enable the viewer to see the most descriptive view of the object as a front view projected directly onto the plane of the paper. The depth of the object is shown at a receding angle. An advantage of oblique pictorials is that one surface appears as its true size and shape and is not distorted. This means that circular features such as holes and cylinders appear as circles in the plane of the paper and do not appear as ellipses. Objects with cylinders and holes are easier to show in a pictorial representation when the drawing is an oblique pictorial. A disadvantage of oblique pictorials is that they tend to be distorted and appear elongated when viewed because they are not a "true projection" even though they are dimensionally correct. Although any receding angle from 0° to 90° may be used for an oblique drawing, angles between 30° and 60° should be used, which minimizes the distortion and elongation, as shown in Figure 12.28.

12.03.01 Types of Oblique Drawings

There are generally two types of oblique drawings: **cavalier** and **cabinet**. Both types generate a pictorial drawing in a similar manner by showing the most descriptive view of the object in the plane of the paper in true size and shape and show a receding dimension along an axis at some angle between 30° and 60° (45° is preferred) to minimize distortion. The difference between cavalier and cabinet oblique drawings lies in the measurements made along the receding depth axes. A cavalier oblique drawing is generated when the true length of the depth dimension is measured along the receding

FIGURE 12.28. Example of an oblique pictoral.

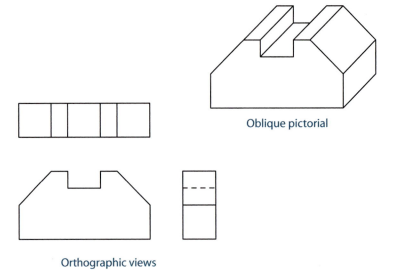

Oblique pictorial

Orthographic views

axes, as shown in Figure 12.29. The cabinet oblique is generated when half the true length of the depth is measured along the receding axes, as shown in Figure 12.30. The cavalier oblique shows the most distortion and elongation, while the cabinet oblique shows the least. For this reason, the cabinet oblique tends to be selected most often.

12.03.02 Construction of Oblique Drawings

An oblique drawing can be constructed using the "framing" technique, which is similar to the technique used when creating an isometric drawing. Essentially, an oblique prism is constructed and then the features are cut away to show the 3-D aspects of the object. Figure 12.31 illustrates the following steps in constructing an oblique drawing.

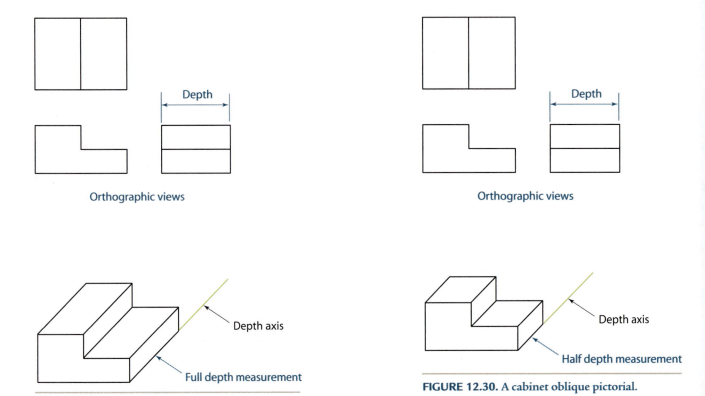

Orthographic views

Orthographic views

Depth axis

Full depth measurement

FIGURE 12.29. A cavalier oblique pictorial.

Depth axis

Half depth measurement

FIGURE 12.30. A cabinet oblique pictorial.

FIGURE 12.31. The steps required to construct an oblique pictorial.

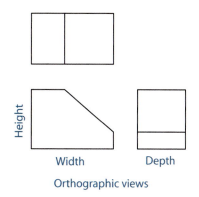

Orthographic views

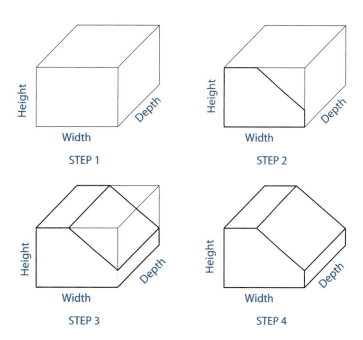

Step 1: Determine the height, width, and depth of the object and frame these dimensions. The width and height should be in the plane of the paper. Draw the receding axis at 45° and measure the depth dimensions along this axis using the true length of the depth.

Step 2: Draw the front view of the object, which matches the orthographic front view.

Step 3: Draw the inclined surface by laying out its location on the front surface and then on the right surface as displayed on the receding axis.

Step 4: Darken all visible lines and erase construction to complete the drawing.

12.03.03 Construction of an Object with Circular Features

The "boxing-in" technique can be used to create an oblique drawing of an object that has circular features such as holes or cylinders. The orthographic views shown in Figure 12.32 indicate that you will have to frame the cylinder with the hole as one step and frame the rectangular prism that also has a hole as another step. This object will need to have two "box-in" steps. Figure 12.32 illustrates the following steps.

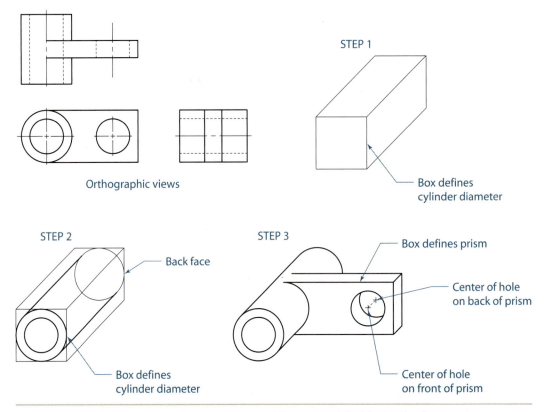

FIGURE 12.32. The steps required to construct an oblique pictorial with circular features.

Step 1: Frame the diameter of the cylinder and draw it in the plane of the paper with a 45° angle for the receding axis.

Step 2: On the front face, locate the center of the hole and cylinder. Frame both the hole and the cylinder. Draw the circles that define the hole and the cylinder on the front face. Frame the cylinder on the back surface and draw the circle (lightly) to define its depth. Connect the front surface cylinder with the back surface cylinder.

Step 3: Frame the rectangular prism and show the circular features that intersect the cylinder. Frame the hole that goes through the prism and show the back edge of the hole as it passes through the far side of the prism. This is done by offsetting the center of the hole by the depth of the prism.

Step 4: Darken all visible lines to complete the drawing.

When a circular feature such as a hole or cylinder appears in a view that is not in the plane of the paper, an ellipse must be drawn using the four-center method discussed in the section describing isometric drawings.

12.04 Perspective Drawings

Perspective drawings and sketches are some of the most lifelike pictorials created. Someone with a great deal of skill can create a perspective drawing that looks as detailed as a photograph. Perspectives incorporate the concept of vanishing points to produce the 3-D shape of an object on the plane of the paper or the computer screen. The concept of a vanishing point is very simple. Suppose you are standing in the middle of a railroad track in flat terrain so you can see where the sky intersects the terrain. The line where the sky intersects the terrain is called the horizon. As you look down the railroad toward the horizon, the outside rails seem to converge to a single

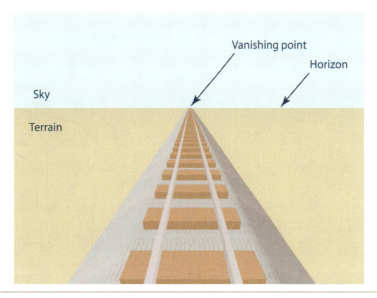

FIGURE 12.33. An illustration showing railroad tracks converging to a vanishing point on the horizon.

point as the railroad intersects the horizon. This point is called the **vanishing point (VP)**, as shown in Figure 12.33. Perspective pictorials employ the use of vanishing points to create a 3-D effect.

12.04.01 Types of Perspective Drawings

There are three types of perspective drawings: one-point perspectives, two-point perspectives, and three-point perspectives. A one-point perspective is illustrated in Figure 12.33. There is one vanishing point at the horizon, and all of the lines converge to this vanishing point. Figure 12.34 shows a two-point perspective. In this type of perspective, there are two vanishing points—one on the right and one on the left. Lines that define the 3-D object converge at the vanishing points. A three-point perspective is shown in Figure 12.35. In this type of perspective, there is a right and left vanishing point and there is a central vanishing point.

Of the three types of perspectives that can be drawn, the two-point perspective is the one most often used to show an object as a pictorial. The one-point and three-point perspectives are limited in their use and do not convey an image that the eye would be likely to perceive. Therefore, the two-point perspective will be the only type discussed in detail.

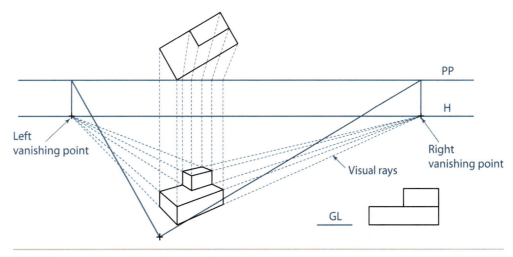

FIGURE 12.34. An example of a two-point perspective drawing.

FIGURE 12.35. An example of a three-point perspective drawing.

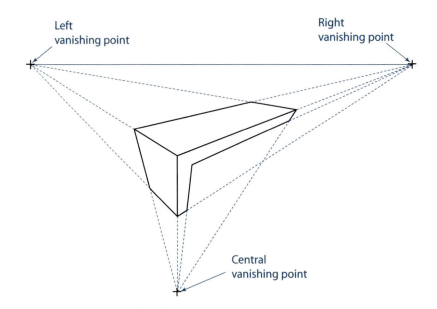

Left vanishing point

Right vanishing point

Central vanishing point

12.04.02 Two-Point Perspective Drawings

Generally, a two-point perspective is generated using the top (plan) orthographic view of the object and an elevation view. The **plan view** is rotated at an appropriate angle to enhance the 3-D aspects of the perspective. The **elevation view** is shown as it normally would be shown in its orthographic position.

An important feature in developing the two-point perspective drawing is the **picture plane (PP)**. The location of this vertical plane (shown as an edge) defines the size and position of the perspective when viewed from the **station point (SP)**. The PP can be located anywhere relative to the object. The simplest position is shown in Figure 12.36, where the PP goes through the corner of the object. When the PP is located in front of the object, the perspective appears farther away from the observer. When the PP is located behind the object, the perspective appears closer to the observer. Selection of the position of the PP controls the size of the perspective. At times, illustrators may choose to place the PP through the middle of the object, giving the appearance that one portion of the object is closer to the observer and one portion is farther away from the observer.

The **horizon line (HL)** is the line that defines where the ground meets the sky. It can be placed anywhere between the PP and the SP. The HL is important because it establishes the position of the left and right vanishing points that locate the position of the perspective drawing.

FIGURE 12.36. The relationship between the PP and the object.

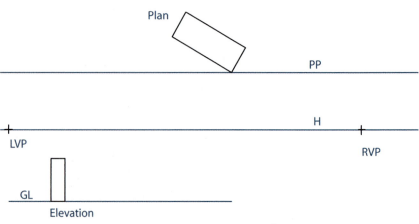

Plan

PP

H

LVP

RVP

GL

Elevation

+ SP

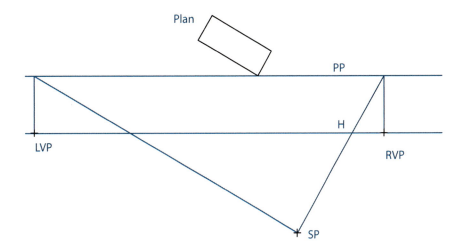

FIGURE 12.37. Establishing and marking the left and right vanishing points.

The **ground line (GL)** defines the position of the elevation view of the object. The GL is important because it determines the vertical location of the perspective drawing and serves as a starting point for the drawing, as shown in Figure 12.36.

12.04.03 Construction of a Two-Point Perspective Drawing

The two-point perspective drawing is constructed using steps similar to those for constructing the isometric pictorial and the oblique pictorial. You will progress through the steps to create a two-point perspective.

Step 1: Establish the vanishing points. Draw lines from the SP to the PP at angles parallel to the left and right visible planes in the plan view. Draw lines from these intersections with the PP vertically downward to the horizon. These intersections become the left vanishing point (LVP) and the right vanishing point (RVP), as shown in Figure 12.37.

Step 2: Establish the **measuring line (ML)**. When the corner of the plan view intersects the picture plane, draw the measuring line downward vertically from the intersection, as shown in Figure 12.38.

When the plan view is behind the picture plane, extend the corner of the plan view to the picture plane by drawing parallel to the line used when establishing the vanishing points (SP to PP). Then draw the measuring line downward vertically from the intersection with the picture plane, as shown in Figure 12.39.

FIGURE 12.38. Establishing the ML.

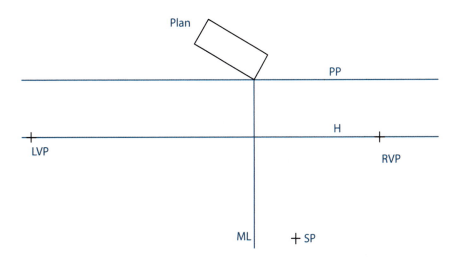

FIGURE 12.39. Establishing the ML when the object is behind the PP.

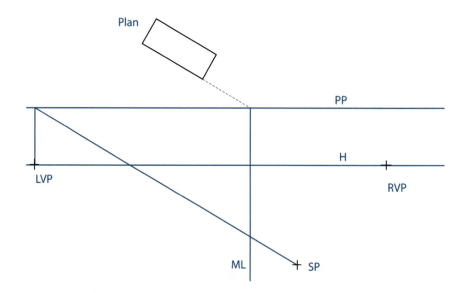

When the plan view is in front of the picture plane, extend the corner of the plan view to the picture plane by drawing parallel to the line used when establishing the vanishing point (SP to PP). Then draw the measuring line down vertically from this intersection with the PP, as shown in Figure 12.40.

Step 3: Project the height measurements to the ML from the elevation view. Project horizontally from the elevation view. Note the position of the GL to the elevation view, as shown in Figure 12.41.

FIGURE 12.40. Establishing the ML when the object is in front of the PP.

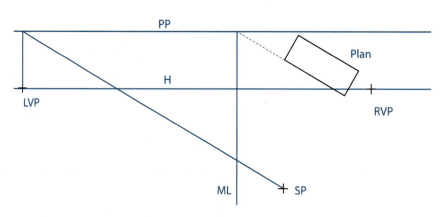

FIGURE 12.41. Establishing the height of the object on the ML.

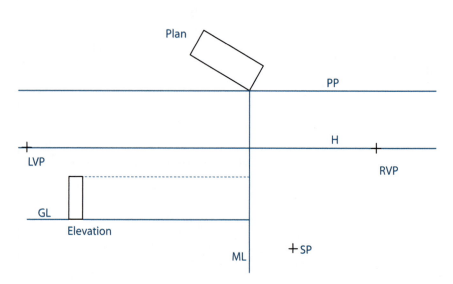

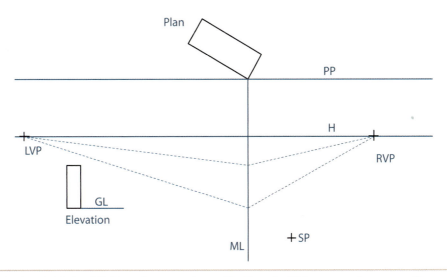

FIGURE 12.42. Establishing the measuring walls when the PP intersects the corner of the object.

Step 4: Establish the **measuring walls**. When the visible corner between the left and right visible planes touches the PP, the measuring walls originate at the ML. The lines may be drawn from the top and bottom of the ML to the left and right vanishing points to establish the left and right measuring walls, as shown in Figure 12.42.

When the corner that will be represented by the measuring line falls behind or in front of the PP, the measuring wall representing the extended visible plane in the plan view must be drawn first, as shown in Figure 12.43.

Special Note: The front-near corner must be projected (steps 5 and 6) before the measuring wall that represents the other visible plane can be drawn, as shown in Figure 12.44.

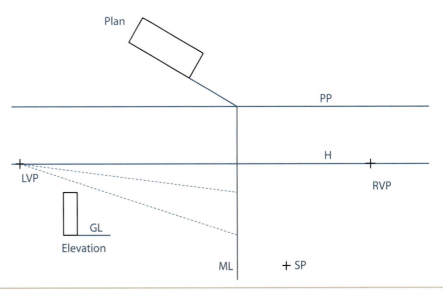

FIGURE 12.43. Establishing the measuring walls when the object is behind or in front of the PP.

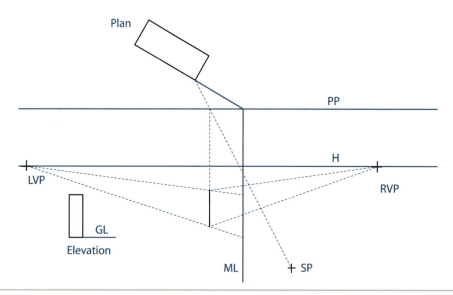

FIGURE 12.44. Projecting the front corner before the measuring wall can be drawn.

Step 5: Draw the visual rays. Align the straight edge from each object intersection in the plan view with the SP, but draw only the plan view intersections to the PP, as shown in Figure 12.45.

Step 6: Project the visual rays. Draw the PP intersections downward perpendicular from the PP to the measuring walls to establish the side details of the object, as shown in Figure 12.46.

Step 7: Lay out details. Repeat steps 3 through 6 to show details. Appropriate points will need to be projected to the vanishing points, as shown in Figure 12.47.

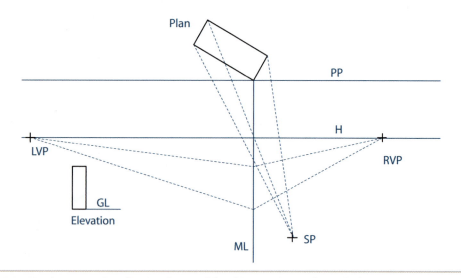

FIGURE 12.45. Drawing the visual rays of the object to the PP through the SP.

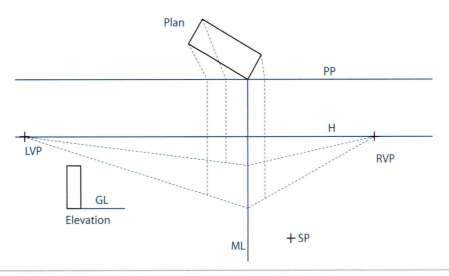

FIGURE 12.46. Projecting the intersection of the visual rays with the PP to the measuring walls.

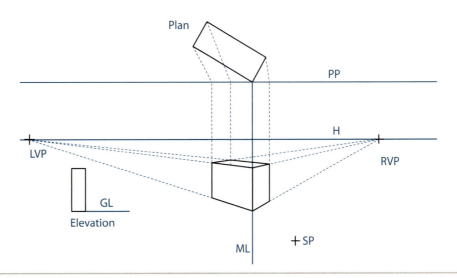

FIGURE 12.47. Completion of the perspective.

12.04.04 Complex Object in Two-Point Perspective

When a complex object is composed of more than one prism, the two-point perspective may require more than one measuring line. Figure 12.48 shows an object composed of two prisms, with one of the prisms behind the PP. When this happens, a second measuring line is required to establish the proper height of the prism located behind the PP. Note in Figure 12.48 that ML-1 establishes the height of the rectangular prism forming the base and ML-2 establishes the height of the prism containing the inclined surface

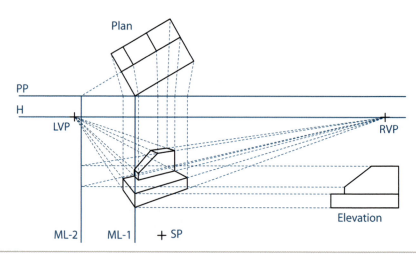

FIGURE 12.48. A two-point perspective drawing of a complex object.

that is behind the PP. The heights projected from the elevation view to ML-2 are transferred along a line from ML-2 to the right vanishing point. The projectors of the prism located behind the PP from the plan view are then drawn to the PP and projected downward to these lines to establish the position of the prism that is behind the PP.

12.05 Considerations for 3-D Modeling

The easiest way to create pictorial drawings is to extract them from solid models. Most solid modeling software has the capability to create engineering drawings from models. These drawings usually include not only traditional orthographic views but also pictorial views to increase the speed with which the parts and assemblies can be visualized. Creating a pictorial from a solid model usually is a matter of specifying the viewing orientation (many times predetermined to give a choice of an isometric or trimetric view) and the amount of perspective when a perspective view is desired. In fact, pictorial drawings are so easily extracted from solid models that it might be foolish not to include them with the orthographic views on working engineering drawings.

12.06 Chapter Summary

Pictorial drawings are designed to enhance your graphic communication skills. Since pictorial drawings describe all three dimensions on the plane of the paper, they are less likely to show the detail that would be expected in the orthographic drawings used in working drawings. Visualization and an understanding of the 3-D relationships of objects are greatly enhanced through the use of pictorial drawings. Different levels of complexity are involved in creating different types of pictorial drawings. For simple communication, isometric drawings usually can be created easily for most objects. Oblique drawings, albeit less realistic in their appearance, are even quicker and simpler to create. For applications that demand the most realistic appearance, especially for large objects such as buildings, perspective drawings can be used. With a solid modeler, pictorial drawings of any type can be created quickly (after the solid model is created) with a few commands. When 2-D CAD or manual drafting instruments are the only tools available, the traditional techniques presented in this chapter may need to be used. Regardless of the graphics tools available, pictorial drawings are now commonly included in formal engineering drawing to add clarity to the traditional orthographic multiview presentation.

12.07 glossary of key terms

axonometric drawing: A drawing in which all three dimensional axes on an object can be seen, with the scaling factor constant in each direction. Usually, one axis is shown as being vertical.

cabinet oblique drawing: An oblique drawing where one half the true length of the depth dimension is measured along the receding axes.

cavalier oblique drawing: An oblique drawing where the true length of the depth dimension is measured along the receding axes.

diametric drawing: An axonometric drawing in which the scaling factor is the same for two of the axes.

elevation view: In the construction of a perspective view, the object as viewed from the front, as if created by orthogonal projection.

ground line (GL): In the construction of a perspective view, a line on the elevation view that represents the height of the ground.

horizon line (HL): In the construction of a perspective view, the line that represent the horizon, which is the separation between the earth and the sky at a long distance. The left and right vanishing points are located on the HL. The PP and the HL are usually parallel to each other.

isometric drawing: An axonometric drawing in which the scaling factor is the same for all three axes.

isometric lines: Lines on an isometric drawing that are parallel or perpendicular to the front, top, or profile viewing planes.

measuring line (ML): In the construction of a perspective view, a vertical line used in conjunction with the elevation view to locate vertical points on the perspective drawing.

measuring wall: In the construction of a perspective view, a line that extends from the object to the vanishing point to help establish the location of horizontal points on the drawing.

oblique pictorial: A sketch of an object that shows one face in the plane of the paper and the third dimension receding off at an angle relative to the face.

perspective drawing: A drawing in which all three-dimensional axes on an object can be seen, with the scaling factor linearly increasing or decreasing in each direction. Usually one axis is shown as being vertical. This type of drawing generally offers the most realistic presentation of an object.

pictorial: A drawing that shows the 3-D aspects and features of an object.

picture plane (PP): In the construction of a perspective view, the viewing plane through which the object is seen. The PP appears as a line (edge view of the viewing plane) in the plan view.

plan view: In the construction of a perspective view, the object as viewed from the top, as if created by orthogonal projection.

station point (SP): In the construction of a perspective view, the theoretical location of the observer who looks at the object through the picture plane.

trimetric drawing: An axonometric drawing in which the scaling factor is different for all three axes.

vanishing point (VP): In the construction of a perspective view, the point on the horizon where all parallel lines in a single direction converge.

12.08 questions for review

1. Why are pictorial drawings useful?

2. When should pictorial drawing be used instead of pictorial sketches?

3. Why should pictorial drawings *not* be used as working drawings to produce parts?

4. What is an axonometric drawing?

5. How do isometric, diametric, and trimetic drawings differ?

6. How do isometric and oblique drawings differ?

7. In what way is an oblique drawing nonrealistic?

8. How do cabinet and cavalier oblique drawings differ?

9. How do isometric and perspective drawings differ?

10. In what way are perspective drawings more realistic than isometric drawings?

11. When should perspective drawings be used in favor of axonometric drawings?

12. Why are two-point perspective drawings more common than one-point or three-point perspective drawings?

12.09 problems

Measure the features shown in the front-, top-, and right- side views of the objects shown in Figure P12.1. Using drafting instruments or CAD, create the following scaled pictorials of each object that is represented.

1. An isometric drawing
2. A cabinet oblique drawing
3. A cavalier oblique drawing
4. A trimetric drawing using your choice of axes angles (one axis must be vertical).
5. A two-point perspective drawing using your choice of plan location and orientation, station point, and vanishing points. The height axis must be vertical, and the two vanishing points must be on the same HL.

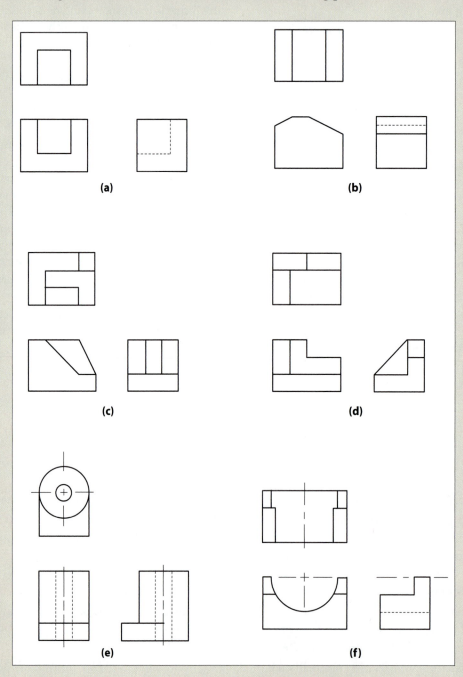

(a) (b) (c) (d) (e) (f)

12.09 problems (continued)

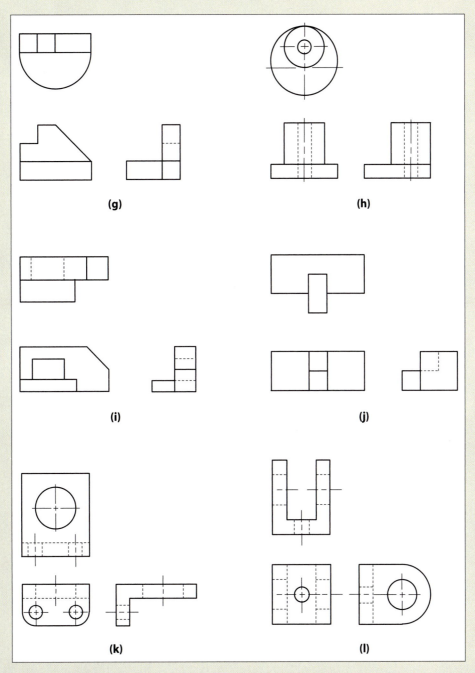

FIGURE P12.1.

13

Section Views

objectives

After completing this chapter, you should be able to

- Use cutaway, or section, views as a method for showing the features of a part that are normally hidden when presented on a multiview drawing

- Decide when a section view is necessary

- Decide what category of section view should be used for particular circumstances

- Create a desired section view such that it adheres to accepted engineering drawing practices

13.01
introduction

The precisely aligned images in a multiview drawing offer an excellent start in defining the exact geometry needed for a part that you want to build. However, this description alone may not be adequate to define all of the features in many types of parts. Some features may be partially or fully obscured in the standard views. The use of hidden lines can alleviate the problem, but too many hidden lines may cause confusion. In these cases, it is useful to have a means of revealing proposed interior detail. This is done by showing cross sections, or section views, at important locations. As with multiview drawings, to minimize ambiguity, you must follow certain guidelines when you want to present a section view.

13.02   A Look Inside

Pick up an everyday object (for example, a coffee mug) and look at it from all directions. If you cannot find a coffee mug, some images have been provided for your convenience in Figure 13.01. You will notice that you cannot view the mug from a direction where the inside depth of the cup or the thickness of the bottom can be directly measured (unless it is made of a clear material). If the mug has a handle on it, look at that as well. Are the edges of the handle rounded? Can you look at the handle from a direction where you get an undistorted view of the radius of the edge? These features are simple examples of measurements that cannot be made from looking at an object in a multiview drawing. Yet there must be some means of showing these types of features so a fabricator will know what to make and what sizes are required. A coffee mug is a very simple example.

Here is an industrial example. Consider the Hoyt AeroTec bow handle again. Its image and multiview engineering drawing are shown in Figure 13.02, with its cross brace highlighted. Note that the edges of the cross brace are rounded. Can these rounded edges be seen on the drawing? Assume the edges of the cross brace are not rounded, (i.e., the surfaces meet at a 90° angle). How would the drawing change? The answer is that in its current state, with all of the complexity and exquisite detail, the drawing cannot show the existence of rounded edges. Clearly, something must be added to the drawing to show that these edges are rounded and to what size they are rounded.

FIGURE 13.01. Two views of an object (a coffee mug) with interior detail.

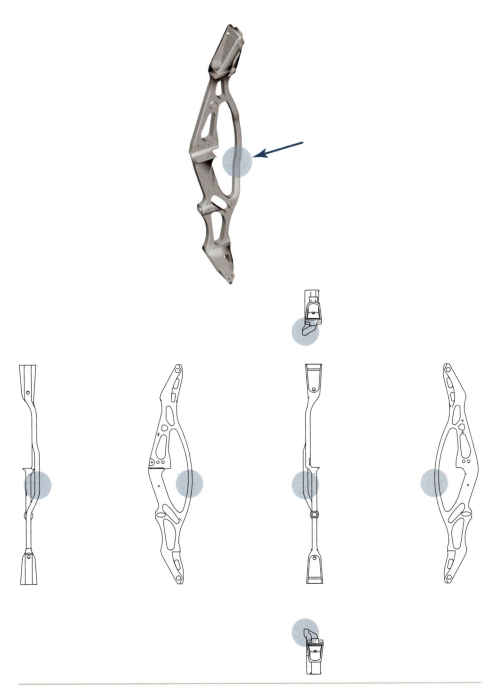

FIGURE 13.02. The geometry of the cross brace on the AeroTec riser cannot be seen in the multiview drawing.

Backing up a bit, look at the mug again to find out what is causing the problem. Figure 13.03 shows the multiview engineering drawing of the mug. Note that the depth of the mug and the radius of the edges of the handle cannot be seen on this drawing. The reason is because the object gets in the way of itself. Portions of the object obscure other portions of the same object. The outside of the mug hides the inside.

A possible solution to this problem is to use hidden lines, as shown in Figure 13.04. The hidden lines show the depth and geometry of the inside of the mug, as well as the geometry of the edges on the handle. However, the use of hidden lines is not always an ideal solution. As objects become more complicated, too many hidden lines make the views confusing, particularly when the images of different features start to fall atop one another.

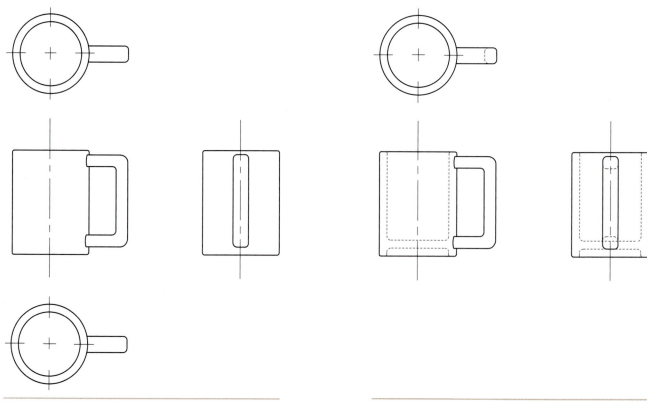

FIGURE 13.03. Orthographic views of the coffee mug fail to define interior detail.

FIGURE 13.04. A multiview drawing of the coffee mug using hidden lines to show interior detail.

In Chapter 3, you learned about cross sections of 3-D objects. If there were a way to cut the mug open, as shown in Figure 13.05, you could take the sliced part and turn it around until you were able to see the desired geometry. This hypothetical slicing is the essence of creating a cross section of the object, to create what is called a **section view**. The slicing, however, must be done following certain rules to ensure that the person who sees a section view on a drawing knows exactly where the slicing has occurred and how it was performed.

A drawing of the mug with three orthogonal views and two types of sections views is shown in Figure 13.06. Do not worry if you have difficulty understanding the extra views in this figure. The following sections discuss in detail how various types of section views are made and how they should be interpreted.

FIGURE 13.05. Hypothetical cutting of the object to reveal interior detail.

FIGURE 13.06. A multiview drawing of the coffee mug using section views to show interior detail.

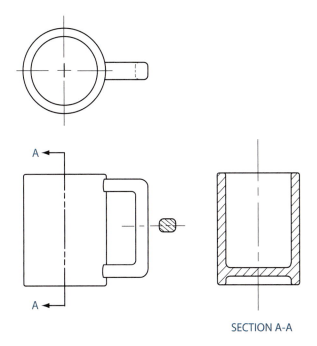

SECTION A-A

13.03 Full Sections

The simplest section view is the **full section**. In a full section, the object is cut completely apart by a **cutting plane** that is perpendicular to one of the standard viewing planes, such as the front, top, or side views. The image of the original whole object is made on the viewing plane using orthogonal projection, and the cutting plane is seen in edge view. A good way to think of a cutting plane is as a very thin knife with the blade held perpendicular to the viewing plane, which hypothetically splits the part into two pieces. This process is shown in Figure 13.07. Note that the cutting plane has an associated **viewing direction**, identified by a set of arrows pointing in the direction of the freshly cut surface that is to be viewed. To create the section view, the image of the split part is imprinted on the cutting plane. The cutting plane and the image are then rotated away from the split part until it is coplanar with the viewing plane. The hinge for this rotation of the cutting plane is its intersection with the viewing plane. With this definition of a section view and its location on the viewing plane, the alignment and orientation of the section view is the same as that used to create an orthogonal view. The section view is the image of the cut object as seen through the cutting plane, and this image is then placed on the viewing plane. In essence, a full section view is just another orthogonal view, but one that reveals the interior of the object.

On an engineering drawing, the original images of the object are not cut apart, as shown in Figure 13.08. A heavy line that extends across the entire part, with alternating short-short-long dashes, represents the edge view of the cutting plane. This line is called a **cutting plane line**. The orientation of the section view relative to the original view of the object is the same as if the viewing plane and the cutting plane were orthogonal viewing plans that had been unfolded. **Section lines**, which are a form of shading, are used to identify areas on the section view that are solid on the original whole object.

There are some important things to note in a full section on an engineering drawing. First, the cutting of the part is imaginary. The part is not to be split into separate pieces. The use of a full section is similar to saying, "If we imagine that the part was cut here, this is what we would see." The cutting plane is flat and goes all the way through the object. Notice the pairs of large, bold capital letters on the cutting plane line next to the arrows. These are used for unique identification of cutting plane lines and their associated section views on a drawing. If the letter *A* is used beside both arrows on a cutting plane line, there must be a note immediately below the corresponding section view that identifies it as "SECTION A-A." The arrows on the cutting plane line point in

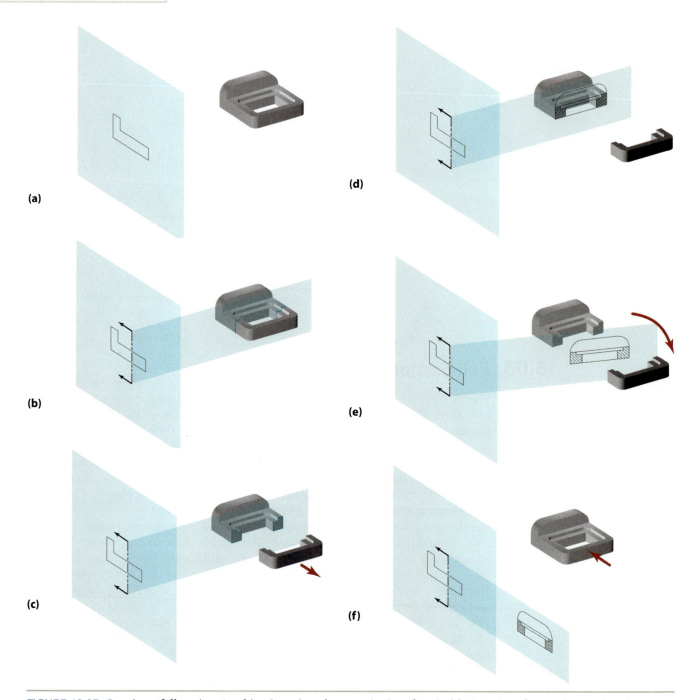

(a)

(b)

(c)

(d)

(e)

(f)

FIGURE 13.07. Creating a full section. An object is projected onto a viewing plane in (a). A cutting plane orthogonal to the viewing plane slices the object in (b). The piece to be viewed remains, while the other piece is removed in (c). The projection of the sliced object is made on the cutting plane in (d). The cutting plane and image are rotated about the section line in (e). The section view is coplanar with the viewing plane in (e).

the direction of viewing. This last point is important because the viewing direction and the orientation of the section view are not arbitrary. An error in either may cause confusion for the reader. Try to visualize the cutting process by comparing Figure 13.07 with Figure 13.08. Correlate the 3-D cutting process in Figure 13.07 with what is shown on the 2-D representation in Figure 13.08. The arrows on the cutting plane point are in the same direction as the arrows on the cutting plane line.

If the arrows on the cutting plane and its corresponding cutting plane line were reversed, as shown in Figure 13.09, the section view would be slightly different. Although the surface that is created by the cutting operation would be the same, the background image of the part would be different. This change is due to the fact that you would be retaining and looking at the other piece that was created when the part was hypothetically split compared to the case in Figure 13.08. When working with multiview drawings, you can remember the proper orientation of the section view by noting that it has the same orientation as the orthogonal view opposite to which the cutting plane line arrows point. For example, if the cutting plane line was located on the front view (and its arrows pointed away from the right-side view), the associated section view would have the same orientation and alignment as the right side view.

FIGURE 13.08. An engineering drawing with a section view to reveal interior detail.

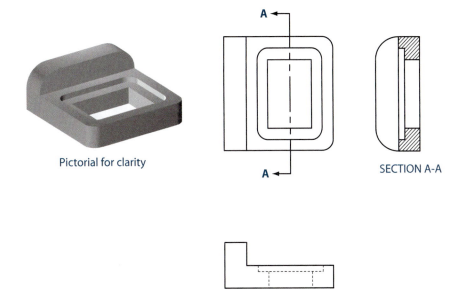

Pictorial for clarity

SECTION A-A

FIGURE 13.09. An engineering drawing with a section view to reveal interior detail.

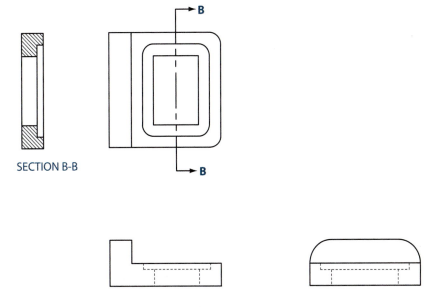

SECTION B-B

The drawing in Figure 13.10 shows a part with multiple section views. If a drawing has multiple section views, a pair of letters must uniquely identify each set of cutting plane lines and corresponding section views. So if there is a second cutting plane line and corresponding section view on the drawing, it may be identified as "SECTION B-B" if "SECTION A-A" already exists. The third set may be called "SECTION C-C," and so on. These identification labels are customarily used even when a drawing has only one section view. The hypothetical interpretation of the multiple sections on the object in Figure 13.10 is shown in Figure 13.11.

One way section views differ from conventional orthogonal views is that, in practice, section views are not required to remain aligned with their adjacent orthogonal views. Although breaking this alignment may violate the rules of orthogonal projection used to create a section view, it is allowed for convenience. Figure 13.12, for example, shows multiple section views of the part. One section is aligned with the view in which it was created. The other section views are nonaligned, but this is permitted in engineering drawing. However, note that even when the section views are nonaligned, they still are required to maintain the same rotational orientation as if they were aligned.

FIGURE 13.10. Multiple section views on a single object.

Pictorial for clarity

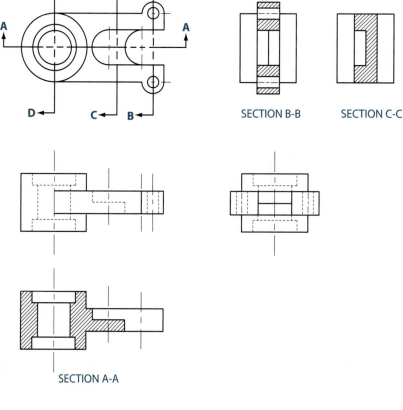

SECTION B-B   SECTION C-C   SECTION D-D

SECTION A-A

FIGURE 13.11. A hypothetical interpretation of the cutting planes for the previous figure.

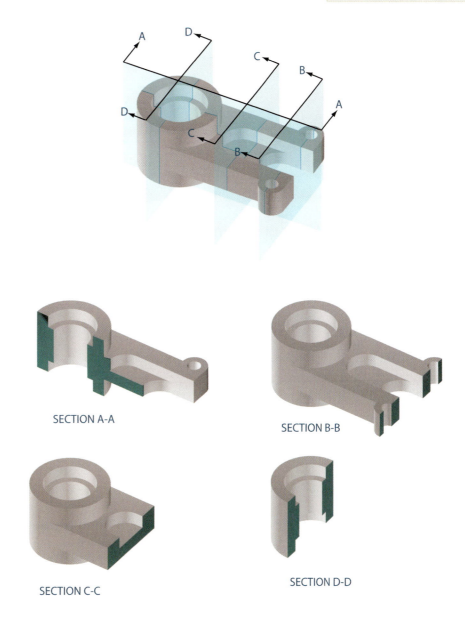

SECTION A-A

SECTION B-B

SECTION C-C

SECTION D-D

Another difference between a section view and a conventional orthogonal view is that a section view is permitted to have a different scale, or magnification, than the view in which it was created. An example of this property is shown in Figure 13.12, where three of the section views are magnified to reveal detail inside the part that would otherwise be difficult to see. When a section view uses a scale that is different from that of the principal views, the new scale must be clearly marked below the note used to identify the section view, as shown in Figure 13.12.

When a section view is created, even though the cutting plane is perpendicular to the viewing plane, there is no requirement that the cutting plane be parallel or perpendicular to any of the other orthogonal views. This property of section views makes it convenient to view features that may be placed at odd angles with respect to the principal views, as with the part shown in Figure 13.13.

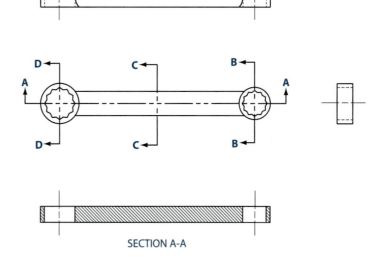

FIGURE 13.12. Multiple sections with nonaligned section views and different scales.

SECTION A-A

SECTION B-B
SCALE 2:1

SECTION C-C
SCALE 2:1

SECTION D-D
SCALE 2:1

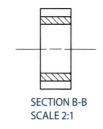

SECTION A-A

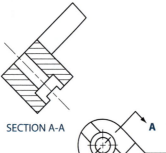

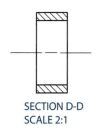

(a)

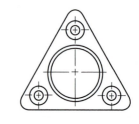

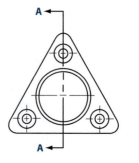

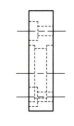

(b)

SECTION A-A

FIGURE 13.13. A full section through a feature placed at an angle.

FIGURE 13.14. The need for many hidden lines in the original drawing (a) is reduced by the use of a section view (b).

13.04 What Happens to the Hidden Lines?

One of the main incentives for using section views is to reduce the use of hidden lines, which until now has been the only method available for revealing the interior and hidden features of many types of objects. When there are too many hidden lines on the view of an object, the drawing becomes confusing. Replacing those hidden lines with one or more section views greatly clarifies the drawing. When section views are used in this manner, there is no longer any need to retain the hidden lines; and they can be removed from the drawing. Hidden lines are typically not shown on the section-line-filled portions of a section view except to indicate the presence of screw threads. Figure 13.14 compares an example of a drawing that originally contained many hidden lines with a revised drawing that replaces one of the orthogonal views with a section view. The improvement in clarity is substantial.

13.05 The Finer Points of Section Lines

Section lines are used to improve the clarity of a section view by indicating the portions of the part that had been solid at the location it was hypothetically cut. However, indiscriminate section line patterns may cause more confusion than clarification. The most basic pattern is a set of lines with a common inclination angle, thickness, and spacing, as shown in Figure 13.15. The line thickness for the pattern is usually no thicker than that used for the part edges. Even with this simple set of variables, the pattern requires some thought. The pattern must be discernible as being section lines when the drawing is read, and the pattern must be reproducible without significant distortion occurring when the drawing is copied. For example, optical copiers, scanners, and fax machines can greatly distort a high-density pattern. A low-density pattern may not appear as section line patterns at all, and the section lines may be misinterpreted as edges on the part.

The pattern should not be parallel or perpendicular to any of the major feature edges of the part; otherwise, there may be some confusion about which lines are part edges and which are section lines. Vertical and horizontal lines are rarely used for section lines.

Different section line patterns can be used to represent different materials. Sample patterns are shown in Figure 13.16. For some materials used in construction, such as concrete or earth, section line patterns are more of a texture than a simple geometric pattern.

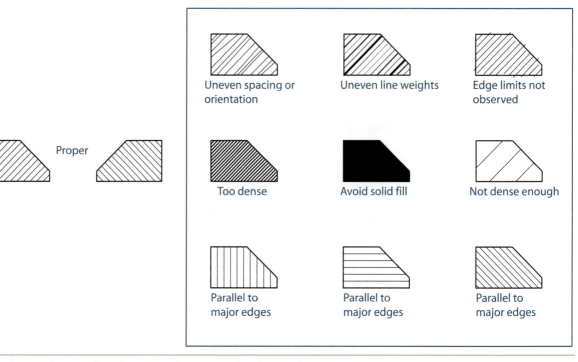

FIGURE 13.15. Examples of proper and poor cross-hatching techniques.

FIGURE 13.16. ANSI standard cross-hatch patterns for various materials.

13.06 Offset Sections

Offset sections can be considered modifications of full sections. An offset section allows multiple features, which normally require multiple section views, to be captured on a single view. As with a full section, an external surface hypothetically cuts through an entire part. However, instead of the part being divided with a single flat cutting plane, the cutting surface is stepped. The size and location of each step is chosen to best capture the features to be displayed. Also, as with a full section, an offset section has its viewing direction indicated by arrows that point at the cut surface to be seen. When the offset cutting surface is rotated onto the viewing plane, the cross sections of multiple features, which could not be shown otherwise with a single cutting plane, can be displayed on a single view. This process is shown in Figure 13.17.

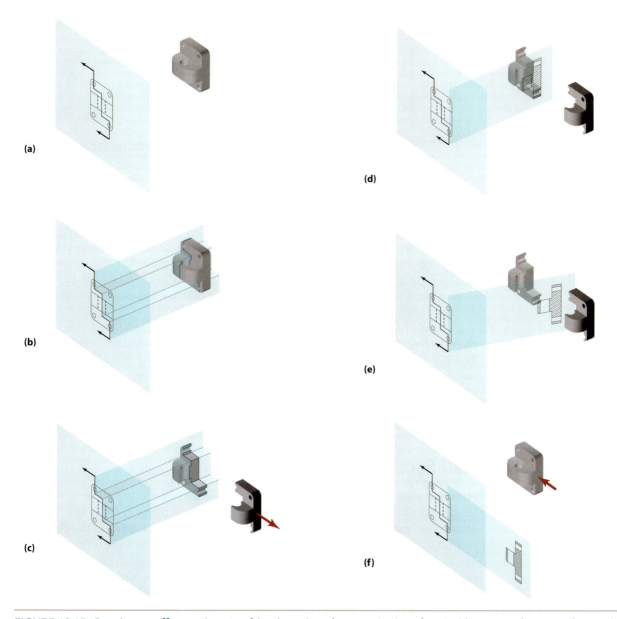

(a)

(b)

(c)

(d)

(e)

(f)

FIGURE 13.17. Creating an offset section. An object is projected onto a viewing plane in (a). A stepped cutting plane orthogonal to the viewing plane slices the object in (b). The piece to be viewed remains, while the other piece is removed in (c). The projection of the sliced part is made on the outermost segment of the stepped cutting plane in (d). The cutting plane and image are rotated about the section line in (e). The section view is coplanar with the viewing plane in (f).

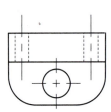

Pictorial for clarity

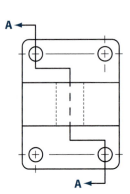

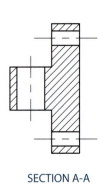

SECTION A-A

On an engineering drawing such as the one shown in Figure 13.18, the edge view of the stepped cutting surface is represented by a heavy stepped line with alternating short-short-long dashes. This is still called a cutting plane line, although technically it is no longer a straight line. The arrows point in the direction of viewing, and the cutting plane line and its associated offset section view are uniquely identified in each drawing with a pair of capital letters, as before.

As with full sections, the rotation orientation of an offset section view must be consistent with the creation of an orthogonal view; but the location and scale of the view is left to the discretion of the person creating the drawing. Note that in an offset section view, it is customary not to show the locations of the steps on the view. The reason is because this information is already available by inspecting the cutting plane line and because adding step lines may cause confusion by showing edges that do not actually exist on the part.

13.07 Half Sections

Half sections are used to save space and labor on an engineering drawing, especially for symmetrical parts. Recall what you learned about symmetry in Chapter 3. When an object is symmetrical about a plane or an axis, it is acceptable to present the object partially in its original state and partially in a sectioned state on the same orthogonal view. The plane of symmetry separates the two states. Another way of visualizing a half section is to imagine a part that is cut such that one-quarter of it is removed to reveal the interior detail. This hypothetical process is shown in Figure 13.19.

In the engineering drawing for this half section, the cutting plane line extends across the object only to the plane of symmetry. The cutting plane line extends partway across the object. A single arrow on the cutting plane line points in the direction of viewing. The absence of a second arrow is an indication that the cutting plane line is for a half section. There is no separate section view. Instead, the orthogonal view and the section view are combined such that the exterior of the part is shown on one half and the interior of the part is shown on the other half. In Figure 13.20, the view types change at the plane of symmetry, which is shown as a centerline. Note that hidden lines are not shown on the unsectioned half of the part.

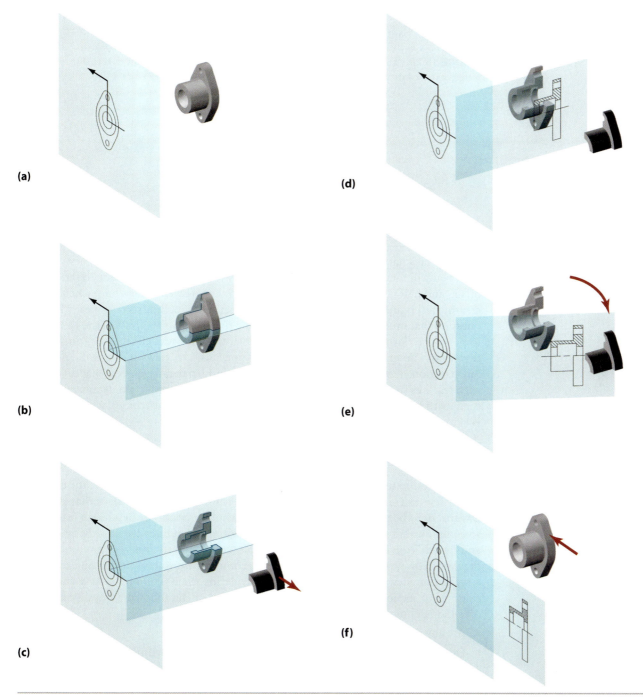

(a)

(b)

(c)

(d)

(e)

(f)

FIGURE 13.19. Creating a half section. An object is projected onto a viewing plane in (a). A stepped cutting plane slices through the object to the plane of symmetry in (b). The piece to be viewed remains, while the other piece is removed in (c). The image of the sliced object is projected onto an orthogonal viewing plane in (d). The viewing plane and image are rotated about the intersection line in (e). The section view is coplanar with the original viewing plane in (f).

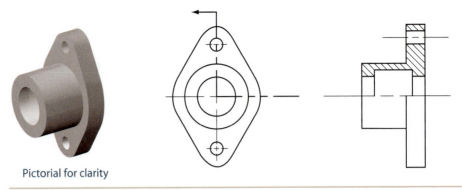

Pictorial for clarity

FIGURE 13.20. An engineering drawing shows the use of a half section to reveal interior as well as exterior detail.

13.08 Procedures for the Creation of Section Views

Later in this chapter, you will learn about some special types of section views; but for now, you will focus on creating the three types you have learned about so far: the full, offset, and half section views. A problem that many new engineers have when they are creating an engineering drawing is deciding whether to include section views. The basic question that must be answered when making this decision is this: Will the section view improve the clarity of the presentation? When the answer is clearly "yes," include section views. When the answer is "probably not," section views may not be necessary. Since section views are used to show interior or hidden details, objects that do not have such features usually fall into the latter category. Objects with such details must reveal them with hidden lines or section views. The question then becomes, which technique is better?

13.08.01 Deciding When to use Section Views

The decision to use section views is somewhat subjective. If the hidden features are relatively few and simple in geometry, such as a few simple holes or slots that go all the way through the part, the use of hidden lines would probably be best. Such features are so common that standard orthogonal views with hidden lines are quickly interpreted, and the addition of section views may actually contribute to unnecessary clutter on the drawing. However, as the hidden features become more numerous or complex in geometry, section views should be used to clearly define their geometries. As a guide, consider using section views when the answer to any of the following questions is yes if only hidden lines will be used.

- Are any hidden lines composed of multiple segments?
- Do any hidden lines of a feature intrude into the area occupied by another hidden feature?

■ Do any hidden lines of a feature share or come close to sharing any hidden lines with another feature unless the lines are exactly common?

Examples of these types of features are shown in Figure 13.21. An overall rule of thumb is that whenever the shear number or the geometric complexity of hidden lines makes the presentation of the object confusing, consider using section views.

FIGURE 13.21. Internal features with hidden lines that have multiple segments (a) or overlap (b and c) are good candidates for using section views.

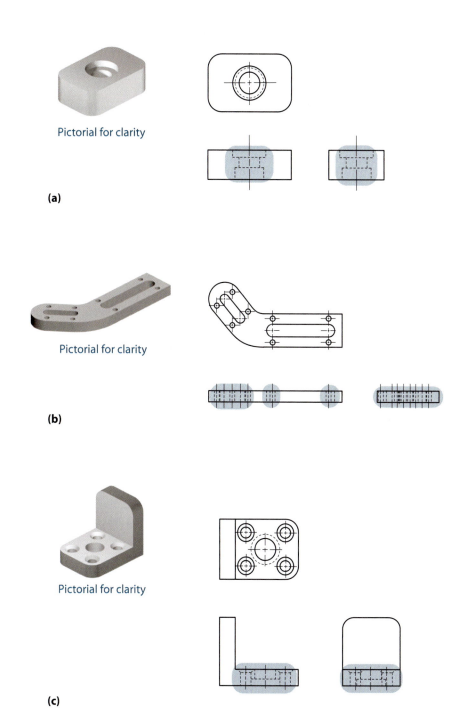

Pictorial for clarity

(a)

Pictorial for clarity

(b)

Pictorial for clarity

(c)

13.08.02 Creating a Full Section View

When the decision has been made to add section views to a drawing, the next step is to create them. When you are the original designer of the part, this task may be simple since you probably already visualize the internal geometry you want to feature. If the part is someone else's design, the process is trickier because you may need to interpret a rather messy drawing that needs the section views to improve the drawing's presentation. If this is the case, you will need to correctly align, identify, and visualize internal features that have been outlined with hidden lines in adjacent views. This task becomes difficult when some hidden lines are missing or have been removed for clarity. The steps to creating a full section view are outlined below and then explained in more detail using an example.

Step 1: Identify the feature(s) to be revealed and the desired viewing direction.

Step 2: Draw the cutting plane line that represents the edge of the cutting plane.

Step 3: Outline the modified part in an adjacent orthogonal view.

Step 4: Identify the intersection points of the part's exterior and interior edges with the cutting plane.

Step 5: Outline the internal features associated with the intersection points on the cutting plane.

Step 6: Find the boundaries between solid and empty space and fill the solid areas with section lines.

Step 7: Add or remove background edges in space and remove edges in solid areas.

When the section view is completed, it may be moved to a more convenient location on the drawing as long as its rotational orientation remains the same.

Consider the object shown in Figure 13.22, which has a pair of grooves inside its main bore. When the geometry of this feature is defined using hidden lines, these lines are composed of multiple segments. Thus, a section view would probably present this feature more clearly.

First (step 1), the bore with its grooves is identified as the feature to be shown in a section view. This feature can be presented best by using a cutting plane that contains the axis of the bore. Slicing the part in this manner would reveal the width and depth of the grooves. In this case, as in most cases, choosing a cutting plane that is parallel to one of the preferred principal views will make section view creation simpler. A cutting plane line (step 2) is drawn in the top view to represent the edge of the cutting plane, as shown in Figure 13.23. The cutting plane is a vertical plane seen as an edge in the top view. Because the cutting plane line is horizontal on the top view of the drawing, the cutting plane also is parallel to the front view; and outlining the part for the next step becomes easier.

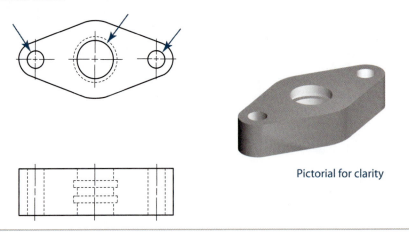

Pictorial for clarity

FIGURE 13.22. Construct a full section view of this part. It is desired to reveal the indicated features (step 1).

FIGURE 13.23. Step 2: Draw the cutting plane and the desired viewing direction.

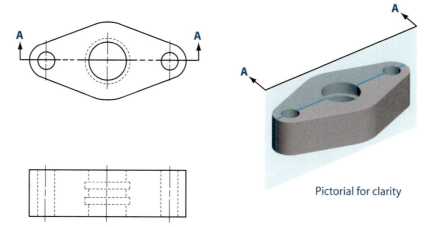

Pictorial for clarity

The direction of the arrows on the cutting plane line shows that the direction of viewing is from front to back. Since the cutting plane is parallel to the front viewing plane, the outline of the part in the section view is likely to be a significant portion of the outline as seen in the front view. This outline (step 3) is shown in Figure 13.24. Note that the size of the outline and its orientation are the same as in the front view from which it was derived. This outline is temporarily placed below the front view for convenience in feature alignment.

Next (step 4), examine the intersections of the cutting plane line with each visible and hidden edge inside the part, as shown in Figure 13.25. Note that some hidden edges may be obscured by visible edges. At every such intersection, a corresponding point will exist in the section view where the edge direction of an internal feature will change. The edges of the internal features appear as hidden lines in the front view but become visible edges in the section view. By tracking the internal edges, the outline of each sectioned internal feature on the cutting plane can be created.

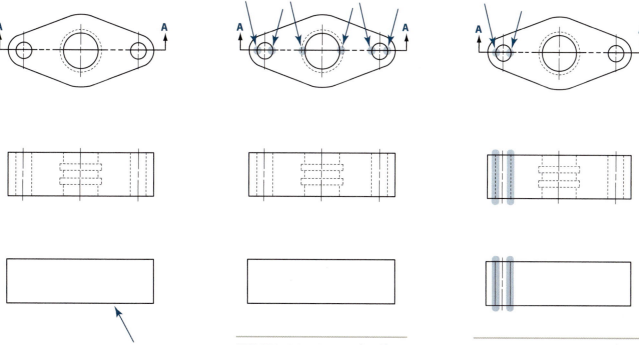

FIGURE 13.24. Step 3: Outline the sectioned part based on the adjacent view.

FIGURE 13.25. Step 4: Identify the intersection points with the cutting plane and see what is happening on the adjacent view.

FIGURE 13.26. Step 5: Outline the internal features associated with the intersection points on the cutting plane.

One way to proceed (step 5) is to examine sets of intersections on the cutting plane line and try to associate and visualize the internal features that created those intersections on the adjacent view. For example, from left to right on the cutting plane line in the top view, the first feature crossed is the visible edge of a circle. When the two intersection points of the cutting plane line on the circle are aligned with their corresponding features on the adjacent view, as shown in Figure 13.26, a pair of hidden lines is seen. The internal feature capable of generating such 2-D geometries on adjacent views is a simple hole that extends all the way through the part. In the section view, the optical limits (or "sides") of the hole become visible edges when the part is hypothetically split at the cutting plane.

The next feature crossed by the cutting plane line is a bit more complicated. It is composed of a visible circular edge inside a hidden circular edge. When these elements are aligned with their corresponding elements in the adjacent view, as shown in Figure 13.27, the feature described is a hole with two internal grooves. In the section view, the hidden edges of this feature seen in the front view become visible edges in the split part.

The final feature crossed by the cutting plane line is another hole. The section view showing all of the edges of the interior features revealed by the cutting plane is shown in Figure 13.28.

The next question that must be asked is, which of the interior areas of the section view were formerly solid, and which were formerly space? The ability to visualize the interior features is important here. The insides of holes, slots, and grooves, for example, contain space. Their edges in the section view separate space from former solid. In Figure 13.29, the portions of the section that are solid have been filled with section lines for easier visualization (step 6). Since the sectioned part also reveals the edges of the slots in the interior surface of the bore, these must be added to the section view (step 7), as shown in Figure 13.30. Note that in the final presentation, the original front

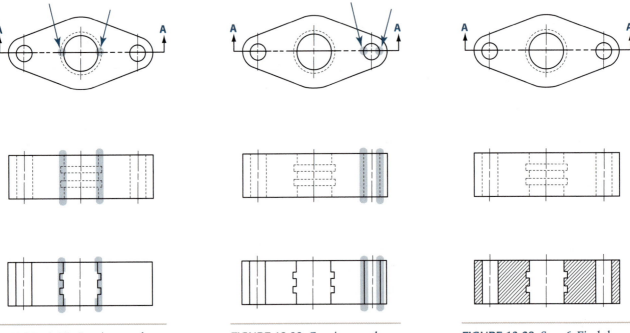

FIGURE 13.27. Continue and outline the next internal feature associated with the intersection points on the cutting plane.

FIGURE 13.28. Continue and outline the last internal feature associated with the intersection points on the cutting plane.

FIGURE 13.29. Step 6: Find the boundaries of air and solid and cross-hatch the solid areas.

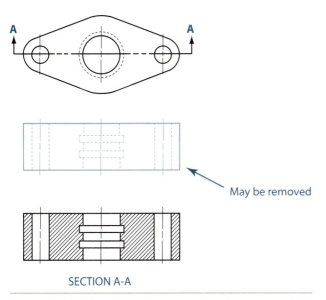

SECTION A-A

FIGURE 13.30. Step 7: Add any new edges that are revealed. Label the view. Note that the former front view may be removed since it adds no additional information.

May be removed

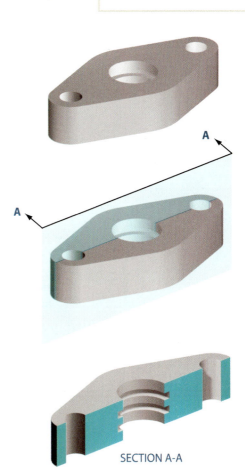

SECTION A-A

FIGURE 13.31. The whole part and the sectioned part.

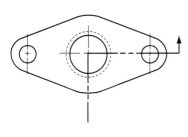

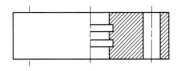

FIGURE 13.32. Presentation as a half section view.

view may be removed because it adds no additional information after the completed section view has been added. A pictorial of the part before and after sectioning is shown in Figure 13.31.

13.08.03 Creating a Half Section

The steps for creating a half section view are identical to those for creating a full section view except that only half the interior of the object needs to be revealed. A drawing with a half section view of the object in Figure 13.31 is shown in Figure 13.32. Note that this object is symmetrical about the centerline of the bore on its left and right sides.

In this case, the cutting plane line is created on the top view and extends only to the plane of symmetry (located at the axis of the bore). The single arrow on the cutting plane line points in the direction that the object is to be viewed. The half section view completely replaces the front view and shows the object in its unsectioned state without hidden lines on one side of the symmetry axis and the interior of the object on the other side of the axis.

13.08.04 Multiple Section Views

When an object is particularly complex, multiple section views should be used to reveal all of its internal details. The object shown in Figure 13.33, for example, has multiple holes and slots that extend from different directions and that intersect. The plethora of hidden lines that result from the internal features make the multiview drawing difficult to interpret.

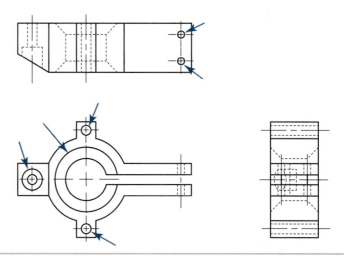

FIGURE 13.33. A complex object requiring multiple sections. It is desired to reveal the interior of the indicated features (step 1).

To reveal the interior details of the part (step 1) and remove the necessity of the hidden lines, three section views are needed. The cutting plane lines are labeled as A-A, B-B, and C-C on Figure 13.34. Note that each section has been chosen to reveal something unique about the interior detail (step 2), although different sections may share the same details. The seven-step process will be used to create Section A-A. The creation of Section B-B and Section C-C are left as a reader exercise, and only the results are shown.

Cutting plane line A-A extends across the length of the part at the keyhole-shaped slot in the front view. Since the arrows of the cutting plane line point away from the top view, an outline of the section view can be begun with a copy of the outline of the top view (step 3), as shown in Figure 13.35. Note that if the arrows of the cutting plane line had been pointing in the other direction (i.e., in the direction away from the (nonexistent) bottom view), the section view would need to have the same alignment and rotational orientation as the bottom view.

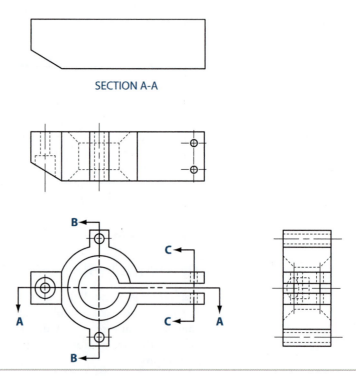

SECTION A-A

FIGURE 13.34. Step 2: Draw the cutting planes with the desired viewing directions.

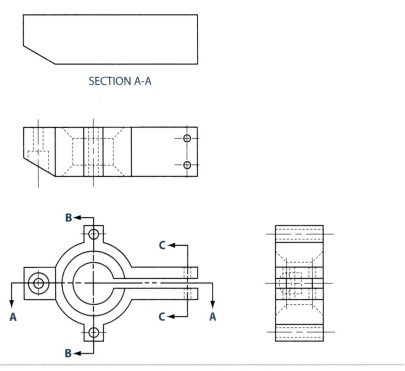

FIGURE 13.35. Step 3: Outline the sectioned part based on the adjacent view.

From left to right, the cutting plane line enters the part; crosses a set of concentric circles, then a set of concentric partial circles; and exits the part. The top view shows that the interior line is created by the edge of the inclined surface. When the intersection between the cutting plane line and concentric circles is aligned to their corresponding hidden edges in the adjacent views (step 4), as shown in Figure 13.36, it can be seen

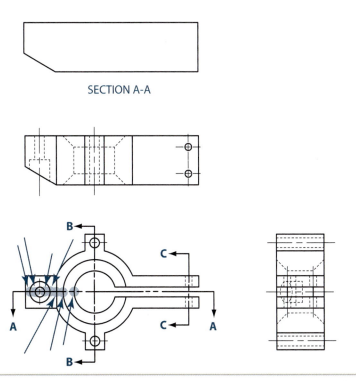

FIGURE 13.36. Step 4: Identify the intersection points with the cutting plane and see what is happening on the adjacent view.

FIGURE 13.37. Step 5: Outline the first internal feature associated with the intersection points on the cutting plane.

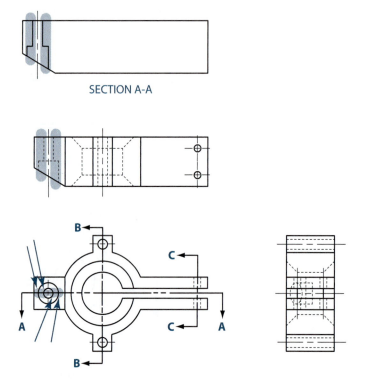

that the concentric circles represent the edges of a counterbored hole. On the section view, the hidden edges of the hole become visible (step 5), as shown in Figure 13.37.

By aligning the intersection between the cutting plane line and circles to their corresponding hidden edges in the adjacent views, it can be seen that the semicircles represent the edges of a countersunk-keyhole-shaped slot. On the section view, the hidden edges of the slot become visible (more of step 5), as shown in Figure 13.38.

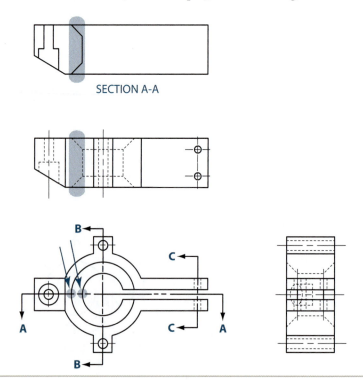

FIGURE 13.38. Step 5 cont: Outline the next internal feature associated with the intersection points on the cutting plane.

FIGURE 13.39. Step 6: Find the boundaries of air and solid and cross-hatch the solid areas.

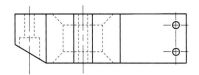

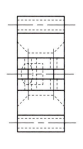

Notice that as you continue to move from left to right along the cutting plane line, it does not intersect any other object feature. This means that the object behind the cutting plane is still visible, but there will not be any section lines to worry about because the cutting plane is not "cutting" through any more solid areas. The insides of the hole, slot, and counterbores contain space. Their edges in the section view separate space from former solid. In Figure 13.39, the portions of the section that were formerly solid have been filled with section lines for visualization (step 6). Note that in this case, the section view reveals the back "uncut" part of the object along with two edges in the keyhole slot that were previously hidden in the regular orthogonal view (step 7), as shown in Figure 13.40.

FIGURE 13.40. Step 7: Add background edges in the air. Label the view.

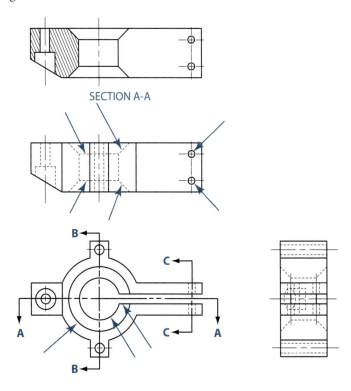

FIGURE 13.41. Add the other two sections using the same method of construction. Note that many of the hidden lines can be removed.

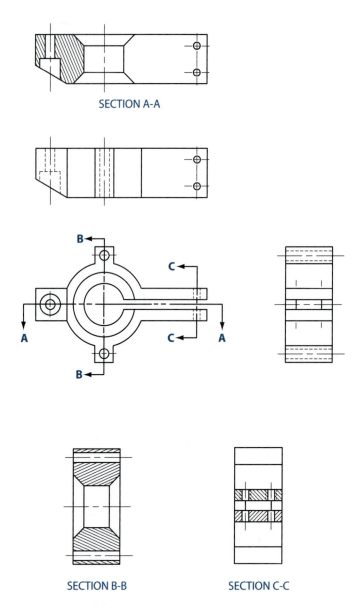

The completed Sections B-B and C-C are shown in Figure 13.41. The addition of Section B-B reveals the interior detail of the two holes beside the countersunk hole, which eliminates the confusion of the multiplicity of hidden lines in the multiview drawing. The addition of Section C-C reveals the interior detail of the cross hole that goes through the keyhole-shaped slot. Note that this view gives the appearance of two disjointed pieces because of the direction in which the section is viewed. A pictorial of the part before and after sectioning is shown in Figure 13.42.

FIGURE 13.42. The whole part and the part split at three different locations.

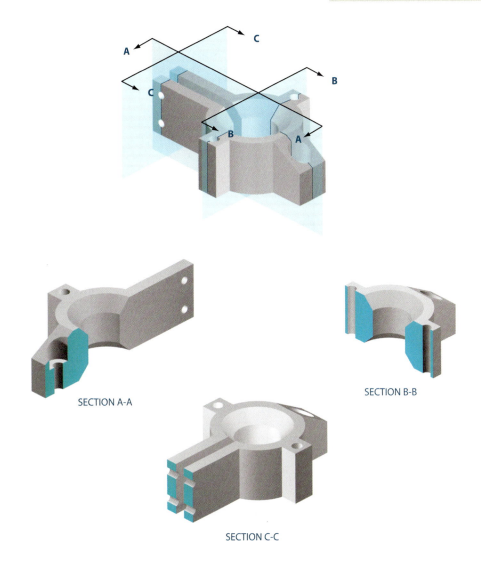

SECTION A-A

SECTION B-B

SECTION C-C

13.08.05 Creating an Offset Section

As mentioned previously, when the multiple internal features do not line up along a single cutting plane, it may be possible to capture most or all of them with an offset cutting plane, which is not a true plane but rather a stepped planar surface. The procedure for creating an offset section is similar to creating a full section, with a few modifications to step 2.

Step 1: Identify the features to be revealed and the desired viewing direction.

Step 2: Draw the stepped cutting plane line to reveal the desired features.

Step 3: Outline the modified part in an adjacent orthogonal view.

Step 4: Identify the intersection points of the part's exterior and interior edges with each segment of the cutting plane.

Step 5: Outline the internal features associated with the intersection points on the cutting plane.

Step 6: Find the boundaries between solid and empty space and fill solid areas with section lines.

Step 7: Add or remove background edges in space and remove edges in solid areas.

In particular, for step 2, it is desired to create a stepped cutting surface that reveals the true dimensions of the features to be revealed with no overlap of these features in the section view.

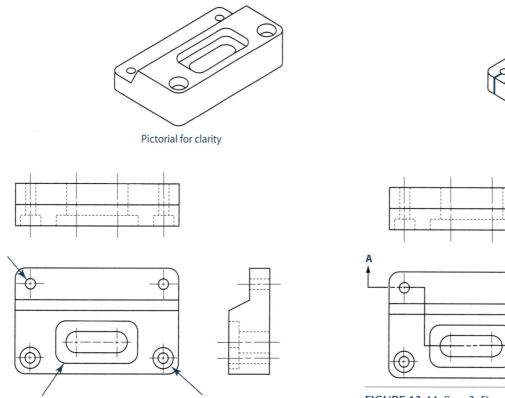

Pictorial for clarity

Pictorial for clarity

FIGURE 13.43. The indicated features (step 1) of this part can be revealed with an offset section.

FIGURE 13.44. Step 2: Draw the cutting plane and select the desired viewing direction.

Consider, for example, the part shown in Figure 13.43, which contains two counterbored holes, a stepped slot, and two simple holes. It is desired to reveal the interior detail of the slot and each type of hole (step 1).

When creating the cutting plane line for an offset, it is important to select a good viewing direction as well as a good line configuration. The hidden lines in the right-side view show that the projections of these features overlap in this view. In the top view, however, the hidden lines show that the projection of these features do not overlap in this view. An offset cutting plane line can be imagined as being composed of two types of segments. The **cutting segments** are the portions that cut through the internal features to be viewed. Each of these segments should pass through entire features if they are to be revealed in their entirety. The **step segments** are transitions between the various cutting segments and are perpendicular to them. The step segments should not pass through any features on the object. A section view that shows all of the desired internal features without interference is desirable; so in this case, the cutting segments are chosen to be parallel to the top view. The precise lengths of the cutting segments are not important as long as they pass through the entire feature. A good offset cutting plane line (step 2) is shown in Figure 13.44.

Since the arrows of the cutting plane line point away from the (nonexistent) bottom view, an outline of the section view can be begun with the outline of the bottom view (step 3). Since a bottom view does not exist, this view or its outline needs to be created with the proper alignment and orientation of a regular orthogonal view, as shown in Figure 13.45. If only the outline of this view is created, the projections of the internal features must then be inferred from the top view.

In this example, each cutting sample passes through only one feature, although in general, a cutting segment can pass through multiple features. From left to right on the cutting plane line in Figure 13.45, the first feature crossed is a circle that represents the edge of a simple hole. The next cutting segment passes through the stepped

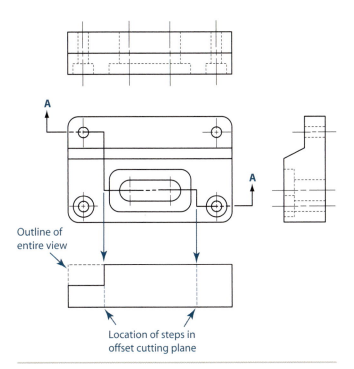

Outline of
entire view

Location of steps in
offset cutting plane

FIGURE 13.45. Step 3: Outline the part based on adjacent
views. Temporarily note locations of steps in the offset line to
use as reference marks.

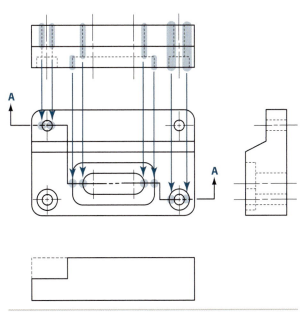

FIGURE 13.46. Step 4: Identify the intersection points
with the cutting plane and see what is happening on the
adjacent view.

slot, and the final cutting segment passes through a set of concentric circles that represent the edges of a counterbored hole. The intersections of the cutting plane line with the edges of these features (step 4), shown in Figure 13.46, indicate the aligned locations on the section view where something should be happening (namely, a change in depth of the feature) to create those edges.

When the part is hypothetically cut, the edges of the counterbored hole, the slot, and the simple hole would be exposed (step 5), as shown in Figure 13.47. Note that if

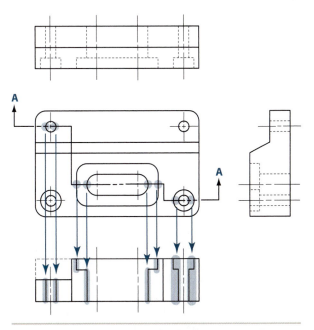

FIGURE 13.47. Step 5: Transfer the features into the section view and form the feature edges.

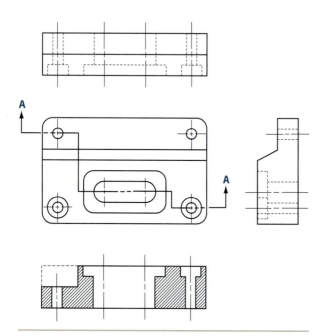

FIGURE 13.48. Step 6: Find the boundaries of air and solid and cross-hatch the solid areas.

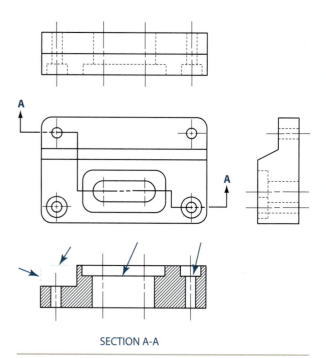

SECTION A-A

FIGURE 13.49. Step 7: Add newly revealed background edges and remove the step edges of the cutting plane. Label the section view.

the cutting implement had actually been a stepped plane, new edges also would have been created at the locations of the step segments. These edges are temporarily shown in Figure 13.47 merely to illustrate the location of the steps. In engineering drawing practice, these edges are removed.

The inside of the hole, slot, and counterbores contain space. Their edges in the section view separate space from solid. In Figure 13.48, the portions of the section that are supposed to be solid have been filled with section lines for visualization (step 6). Note that in this case, the section view reveals edges in the counterbored hole and slot that were previously hidden in the regular orthogonal view (step 7), as shown in Figure 13.49. The boundary edges in the upper left corner of the section view must be deleted because those edges were formed by a portion of the object that would not be seen on the sectioned object. Finally, you must delete any edges formed by the stepped cutting plane that you may have included for reference.

13.08.06 Creating a Sectioned Pictorial

An excellent aid for visualizing sectioned objects is to present pictorials of them. Creating this type of view is also an excellent academic exercise for the development of visualization skills. The process involves removing the portion of the object that is not viewed, creating a multiview drawing of the remaining portion, and creating a pictorial of the remaining portion from the modified drawing. Constructing such a pictorial continues the previous example. The construction of a section pictorial should not be done on the original drawing; rather, it should be done on a copy of the drawing on a separate worksheet to avoid confusion between the real part that is to be built and the model that is for visualization purposes only.

In the drawing in Figure 13.49, the cutting plane line arrows in the front view point toward the portion of the object to be retained and viewed, while the other portion is to be discarded. The modified front view is shown in Figure 13.50.

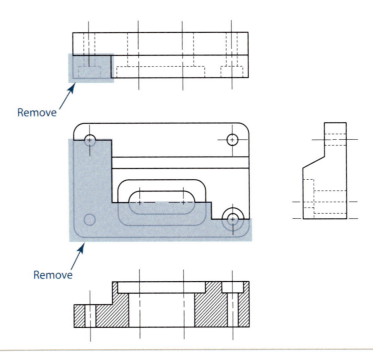

FIGURE 13.50. To visualize the sectioned part, remove the portion that is not seen from the existing orthogonal views.

Once a portion of the part has been removed in the front view, the remaining views also must be modified to be consistent with the modified part. In this case, there will be no changes to the top view. The right-side view must be modified to show that material has been removed. The section view can be used for the bottom view if the section lines are removed and the edges formed by the step segments are once again included. Note that if a left-side view was created, that view also would need to include the new edges formed by the stepped cutting plane. These modified views are shown in Figure 13.51.

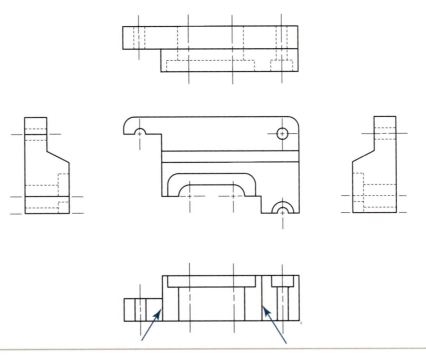

FIGURE 13.51. For ease of visualization, turn the former section view into the bottom view by removing cross-hatching and adding a section step line. Add a left view.

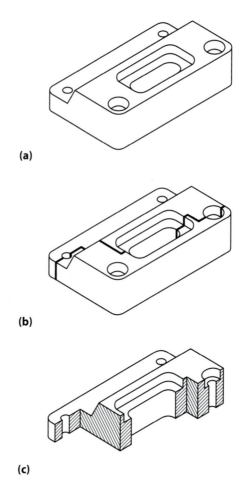

(a)

(b)

(c)

With the completed modified views of the object actually cut, the pictorial can be created by using the point, edge, or surface tracking techniques detailed in Chapter 11. The complete pictorial of the sectioned object is shown in Figure 13.52. Note that the freshly created surfaces from the cutting operation should be shaded or filled with section lines to show that these surfaces were artificially created and do not exist on the real part.

13.09 Removed Sections

In certain cases, it is convenient to use a removed section instead of a full section. A **removed section** offers the convenience of showing only the new surfaces created by a cutting plane, without the complexity of showing the remaining surfaces on an object. A hypothetical procedure for creating a removed section image is shown in Figure 13.53.

In this figure, a cutting plane intersects the object in the area of interest in (a). The cutting plane should be parallel to one of the principal views or perpendicular to the major surfaces of the part where the cut is made in order to reveal its true sizes. In (b), the cutting plane is removed from the part with the image of the intersection imprinted on the plane. The cutting plane and image are then rotated by 90 degrees in (c) such that the arrows point into the page on a drawing (d). The complete removed section view, as would be seen in an engineering drawing, is shown in Figure 13.54.

There is no need for view alignment for the removed section, although its orientation must still follow the rules of multiview presentation and the section view should be located near the cutting plane line. If the scale of the section view is different from that used on the multiview projections, the new scale must be included with the labeling of the section view.

For views where the surfaces created by a hypothetical cut are relatively small compared to the remaining surfaces of the view or where full sections may create unnecessarily large or confusing views, removed sections are a good option for improving clarity while reducing effort and complexity in a drawing. Figure 13.55 shows how removed sections were used for defining various parts of the Hoyt AeroTec bow handle. In this case, full sections or offset sections would have created unnecessary complexity in the drawing.

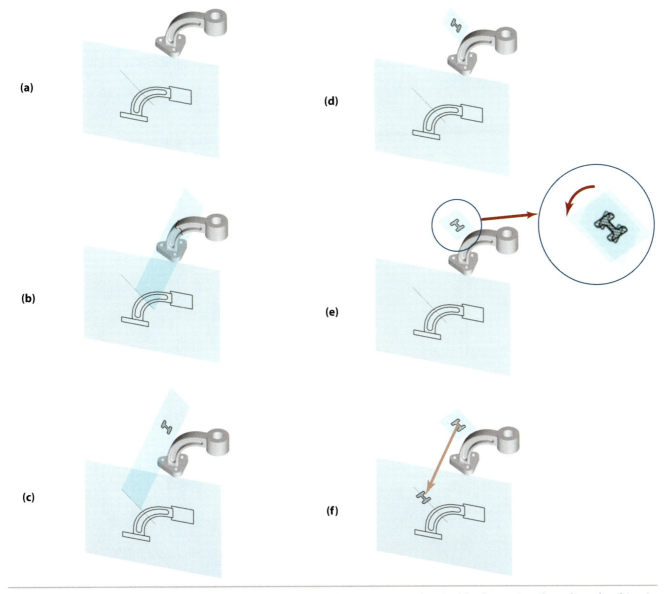

(a)

(b)

(c)

(d)

(e)

(f)

FIGURE 13.53. Creating a removed section. An object is projected into a viewing plane in (a). The cutting plane slices the object in (b). The image of the intersection is removed from the object in (c). The removed image is initially perpendicular to the viewing plane in (d) but is then rotated to be parallel to the viewing plane in (e). The removed image is finally projected onto the viewing plane in (f).

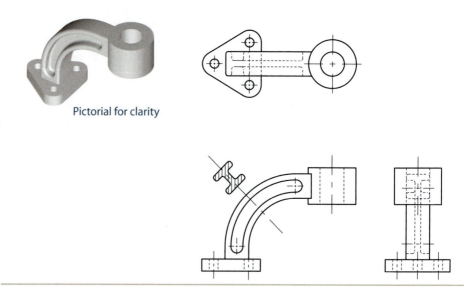

Pictorial for clarity

FIGURE 13.54. A removed section as it would be placed on an engineering drawing.

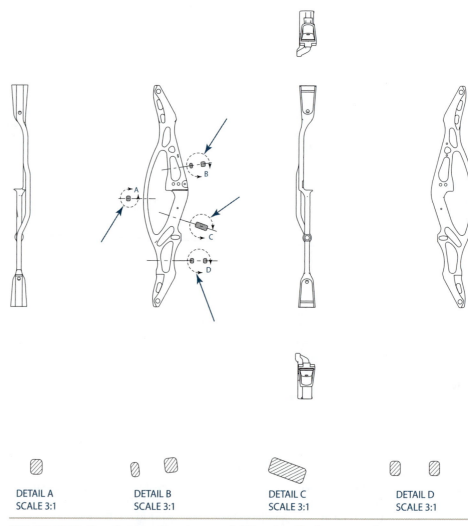

DETAIL A
SCALE 3:1

DETAIL B
SCALE 3:1

DETAIL C
SCALE 3:1

DETAIL D
SCALE 3:1

FIGURE 13.55. Use of removed sections on the Hoyt AeroTec bow example.

13.10 Revolved Sections

Revolved sections are created in a manner similar to that of removed sections. A hypothetical procedure for creating a revolved section image is shown in Figure 13.56.

In this figure, a cutting plane intersects the object in the area of interest and the image of the intersection is imprinted on the plane. The cutting plane should be parallel to one of the principal views or perpendicular to the major surfaces of the part where the cut is made in order to reveal its true sizes. The cutting plane is rotated by 90 degrees such that the arrows point into the page on a drawing. The axis of rotation of the intersection image is on the cutting plane, parallel to the cutting plane line, and through the geometric center of the image. Unlike the removed section, the cutting

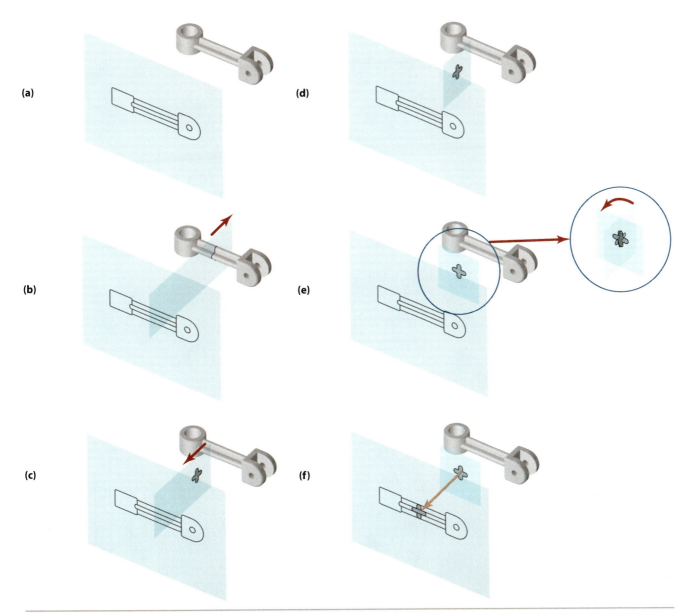

FIGURE 13.56. Creating a revolved section. An object is projected onto a viewing plane in (a). The cutting plane slices the object in (b). An image of the intersection is removed in (c). The intersection image is initially perpendicular to the viewing plane in (d) and then rotated to be parallel to the viewing plane in (e). The image is projected onto the viewing plane in (f).

plane is not removed from the object. Thus, the image of the intersection is superimposed on the orthogonal view. The complete revolved section view, as would be seen in an engineering drawing, is shown in Figure 13.57. Since a revolved section view is constructed at the location of the cutting plane line, there is no need to label the view.

The scale of the revolved section must be the same as for the principal views, and its orientation must follow the rules of multiview presentation. For views where the surfaces created by a hypothetical cut are relatively small compared to the remaining surfaces of the view, revolved sections are another good option for improving clarity while reducing effort and complexity in a drawing. Revolved sections should not be used when the section image interferes significantly with other features in the principal view. Figure 13.58 shows another example where the use of revolved sections is convenient.

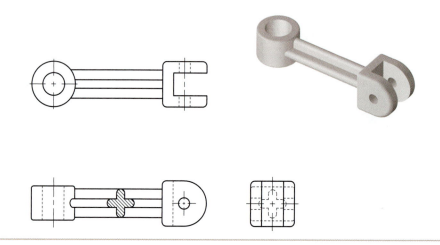

FIGURE 13.57. A revolved section as it would be placed on an engineering drawing.

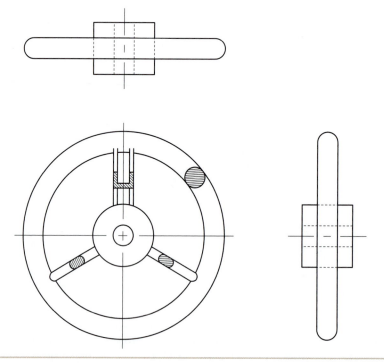

FIGURE 13.58. An example of a part with multiple revolved sections.

13.11 Broken-Out Sections

A **broken-out section** can be used when the internal feature to be revealed is a small portion of the entire object and a full section would not reveal additional details of interest. Use of a broken-out section in this manner would decrease the size and complexity of a drawing, as well as reduce the effort required to make it.

A broken-out section, as with the revolved and removed sections, offers the convenience of slicing only a fraction of the entire object when only a small slice is needed to define an internal geometry. However, a broken-out section offers the added convenience of not requiring the cutting plane to go all the way through the part. With all of the other sections you have studied so far, cutting plane lines start in space, go through the part, and end in space. With a broken-out section, the ends of the cutting plane line can be wholly or partially embedded in the part, as shown in Figure 13.59, where a cutting plane that has its extent limited to the area immediately surrounding a feature is

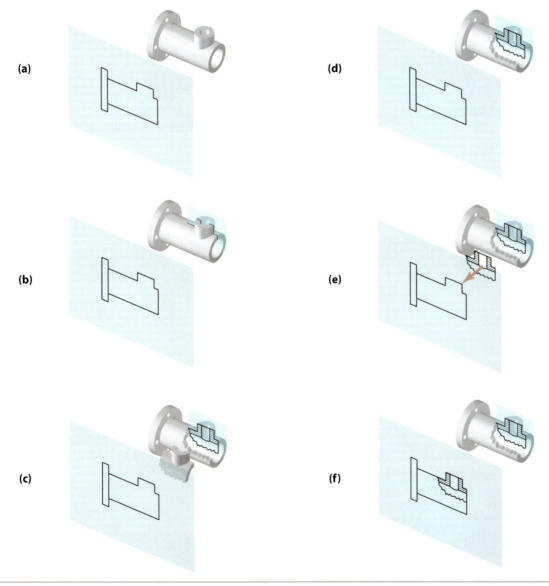

FIGURE 13.59. Creating a broken-out section. The object is projected onto a viewing plane on (a). A cutting plane slices through the feature of interest (but not the entire part) in (b). The portion in front of the cutting is broken out and removed in (c). The interior details of the feature are shown in (d). The image of these features is projected forward in (e) and placed directly on the part image in (f).

imbedded into the part. A piece of the object that is opposite the viewing direction is then hypothetically broken off to reveal the interior details of the feature.

The portion of the cutting plane that is embedded in the part is shown on the section view as an irregular edge to emphasize that the part would hypothetically be broken to reveal the interior details shown at that location. The broken-out-section view may be shown on the corresponding orthogonal view, as shown in Figure 13.60, or in a separate detail view, as shown in another example in Figure 13.61.

FIGURE 13.60. A broken-out section as it would be placed on an engineering drawing.

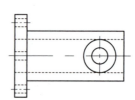
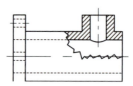

FIGURE 13.61. A broken-out section used to reveal some pocket details of the Hoyt AeroTec bow.

DETAIL E
SCALE 2:1

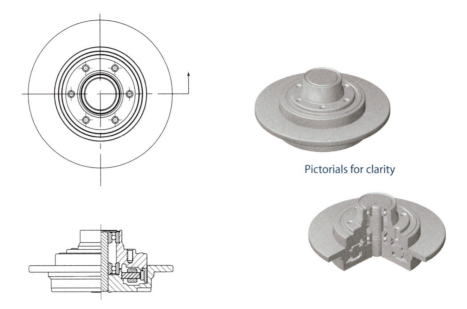

FIGURE 13.62. The method for showing the assembly of many parts.

Pictorials for clarity

13.12 Sections of Assemblies

Section views are commonly used in drawings that show multiple parts in their intended mating configuration to illustrate proper alignment of different features between the parts. When multiple parts are sectioned, as in Figure 13.62, it is advisable to use a different section line pattern for each part in order to distinguish the different parts easily. In everyday practice, assemblies that include pins, keys, shafts, or bolts usually do not show these items sectioned even though the cutting plane line may pass through them. These items usually have standardized geometries and sizes; thus, their sections add little information to a drawing and may even detract from the information presented by parts of greater interest.

13.13 A Few Shortcuts to Simplify Your Life

As with many other engineering drawing practices, acceptable shortcuts for creating section views can be used to reduce the time it takes to create a drawing and/or to minimize possible misinterpretation of a drawing. With all of the shortcuts presented next, the main question you need to ask yourself before using any of them is, "Will this approximation or shortcut increase or decrease the speed and accuracy of interpretation of the drawing?" If the speed or accuracy of interpretation decreases, the shortcuts should not be used.

13.13.01 Small Cutouts on Curved Surfaces

A shortcut is allowed when there is a small hole or another cutout on a curved surface. Figure 13.63, for example, shows a small hole and slot on a tube compared to larger cutouts. If a true projection of these features were made, the orthogonal views would show a curved depression on the surface of the tube. The shape of this curve is complex and would take some time to create. Since in most applications the size of the depression on the surface is unimportant, the depression is not shown on the orthogonal views. The true projection of these features and the accepted shortcut are shown in Figure 13.63. This approximation makes the drawing easier to create, with very little loss of information. However, when the cutouts are large or the size of the depression cannot be ignored in the function of the part, the true projection should be used. What is considered "small" is rather subjective.

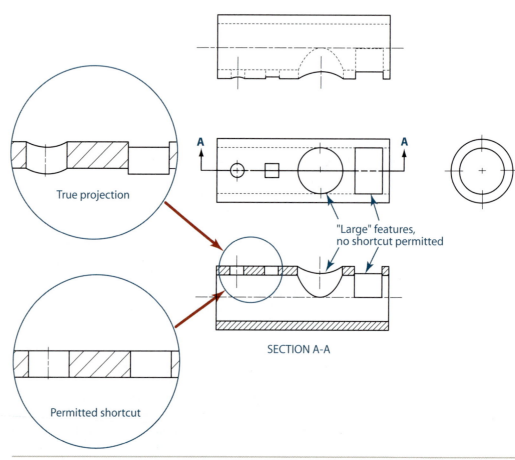

FIGURE 13.63. A permitted shortcut for small holes and slots in curved surfaces.

13.13.02 Threaded Parts

Another shortcut is in the representation of a threaded part, such as the pneumatic fitting shown in Figure 13.64. A thread on the outside of a bolt or screw or the inside of a nut has many complex curved surfaces that would result in a very complicated drawing, especially if it were created with manual instruments or 2-D CAD. Much simpler representations of internal and external threads are included in Figure 13.62. These schematic representations are easier to construct, with very little loss of information, especially since thread sizes are mostly standardized based on the diameter of the part. A note and arrow are required to specify the precise thread sizes. Methods for the complete specification of thread sizes can be found in most machinists' or engineers' handbooks.

FIGURE 13.64. A section of a threaded part.

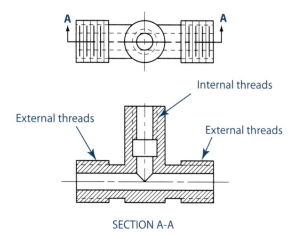

SECTION A-A

13.13.03 Thin Features

For sectioned features that have relatively small thickness when compared to the remainder of a sectioned part, it is acceptable not to fill these features with section lines even when cutting plane lines pass directly through them. As an example, consider the objects in Figure 13.65. These two objects are composed of the same main body but

FIGURE 13.65. The conventional section (a) and recommended variation for a thin feature (b).

(a)

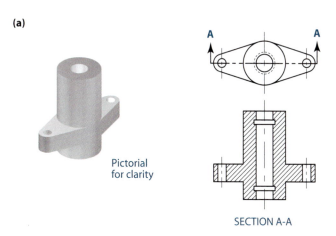

Pictorial for clarity

SECTION A-A

(b)

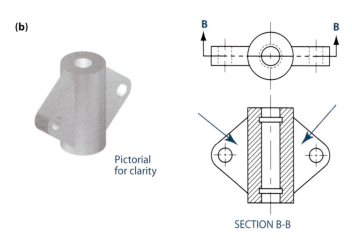

Pictorial for clarity

SECTION B-B

FIGURE 13.66. The
recommended presentation of
thin webs.

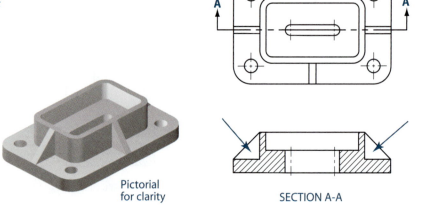

Pictorial
for clarity

SECTION A-A

with mounting flanges turned differently. In both cases, the cutting plane line goes
through both the main body and the flanges. For the part in (a), the thickness of the
two flanges in the section view is about the same as the depth of the main body. In this
case, the flanges are filled with section lines, as normal. For the part in (b), the thickness of the two flanges in the section view is a fraction of the depth of the main body. In
this latter case, it is acceptable not to fill the flanges with section lines because doing so
may give an immediate false impression that the flanges are about the same thickness
as the main body. As an alternative to not filling thin features with section lines, it is
permissible to use a different section line pattern for spokes and vanes than is used for
the main body of the object. Note that for this shortcut, an extra edge must exist to separate the thin feature from the main body in the section view. Webs and fins, such as
those shown in Figure 13.66, are generally treated in this manner.

13.13.04 Vanes, Fins, Spokes, and the Like

Objects with axially symmetric features such as vanes and spokes, as shown in the two
parts in Figure 13.67, also are not filled with section lines even when cutting plane lines
pass directly through them. Filling such features with section lines may give the false
impression that the features are solid throughout the part. It also is permissible to use
a different section line pattern for spokes and vanes than is used for the main body of
the object. Note that for this shortcut, an extra edge must exist to separate spokes and
vanes from the main body in the section view.

13.13.05 Symmetry

An interesting exception to the rules of true projection occurs when parts with rotational
symmetry, which means that the part can be divided into identical wedges along an axis,
are sectioned. Note that rotational symmetry is different from the planar symmetry discussed in Chapter 3, where the image of an entire object can be created by reflecting a portion of it on a plane. For example, examine the part shown in Figure 13.68. This part has
one-third rotational symmetry, with three thin support ribs and three holes about the
center tube.

(a)

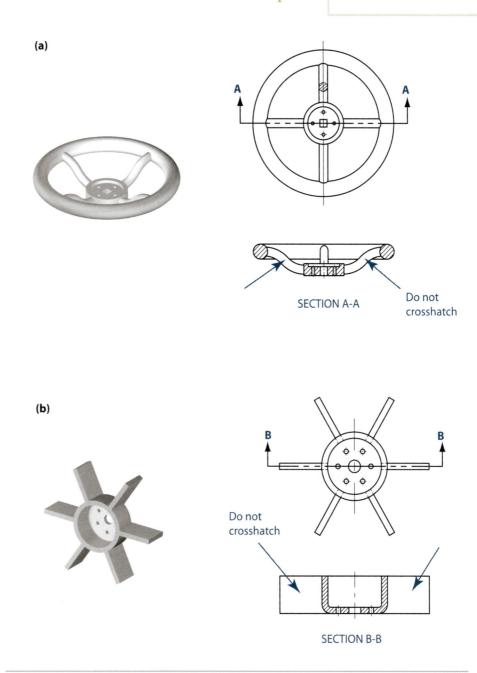

FIGURE 13.67. Two examples of the recommended presentation of spoke, vanes, and fins.

A multiview drawing created using true projection would be like that shown in Figure 13.68. Using a true projection for the front view in this case has some problems. First, using instruments or 2-D CAD, the projection is rather difficult to create. Also, the true projection of the side view may have the negative effect of representing the part as being nonsymmetrical.

An acceptable shortcut for this drawing also is included in Figure 13.68. This drawing is easier to create and gives the impression that the part is symmetrical. The top view clarifies any possible misinterpretation about the number and locations of the support ribs. Interestingly, if the part had one-quarter (or higher) symmetry, for example, four (or more) support ribs instead of three, the front view would be exactly the same as the part with one-third symmetry.

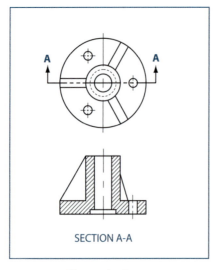

SECTION A-A

True projection

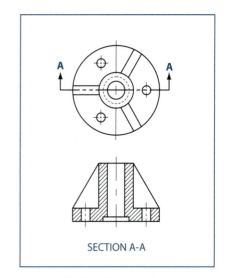

SECTION A-A

Preferred presentation

FIGURE 13.68. The preferred presentation of symmetrical features.

CAUTION

Creating section views is still part science and part art. Even though the rules of orthographic projection generally must be followed (except for the shortcuts mentioned), engineers, designers, and drafters are allowed considerable freedom in choosing when and where section views should be used, what type of section to use, and what presentation method to use. However, experienced people who are required to read and interpret drawings expect certain rules to be followed when the drawings are created; and deviation from these rules may cause confusion. Beginners are sometimes prone to poor choices and errors. In the best case, the person reading the drawing can interpret it because the necessary information is still contained on the remaining views. Still, errors can cause confusion and slow down the process of interpreting the drawing. In a more serious case, errors cause ambiguity that makes the drawing impossible to interpret correctly. In a worst-case scenario, the errors may cause the part to be interpreted as an entirely different part than originally desired. The following sections of this text are a compilation of the most common beginners' errors and ways to avoid them.

Cutting Through Only a Piece of an Internal Feature

Section views should be constructed to reveal true sizes. For example, consider the object shown in Figure 13.67. The full section cuts through the center of the large bore, revealing the true measurable sizes of the diameters inside. The cutting plane line goes through two holes, but not at their centers. The resulting section shows the two holes not at their true diameters. This section may be easily misinterpreted as having hole diameters that are smaller than they really are. A better way to section the object is to use an offset section in which the cutting plane line goes through the center of all of the internal features revealed, also shown in Figure 13.69.

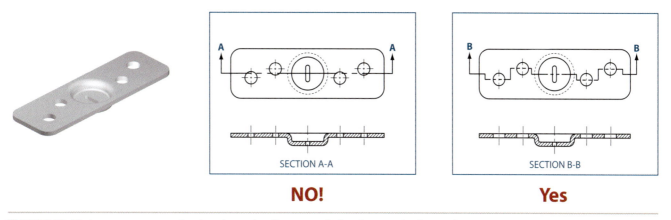

FIGURE 13.69. A common error: Cutting through a feature such that its true size is not shown.

Forgetting the Rest of the Object

A proper full or offset section shows the object as if it had been cut, including any background edges that still may appear outside the cut surfaces. Sometimes it is easy to forget these background edges because the cut surfaces are usually of prime importance. Nevertheless, for a full or offset section to be correct, the background edges must be included, mostly for use as reference locations for the cut surfaces, as shown in Figure 13.70.

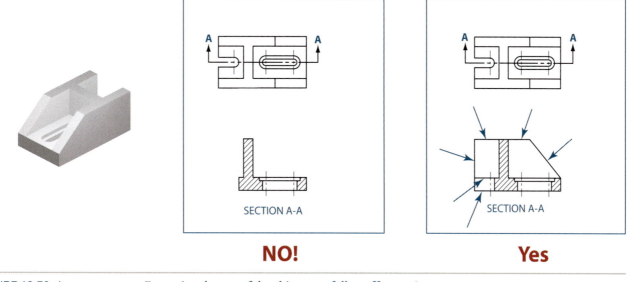

FIGURE 13.70. A common error: Forgetting the rest of the object on a full or offset section.

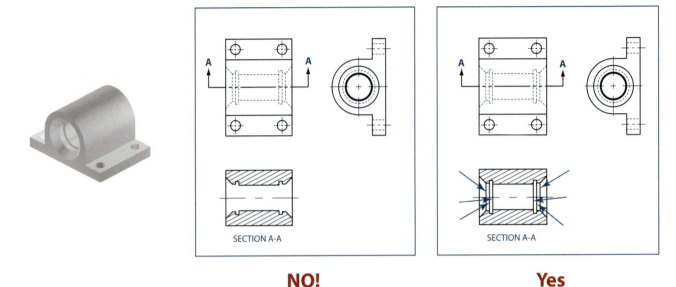

NO!   Yes

FIGURE 13.71. Forgetting internal edges made visible.

Forgetting Back Edges That Are Made Visible

Newly revealed internal edges also are often carelessly omitted. Edges of grooves, counterbores, cross holes, and other similar features have edges that become visible once the object is cut, as shown in Figure 13.71. Neglecting to include these edges may cause confusion about the true geometry of the features.

Incorrect Rotation Orientation

Sometimes in a hastily created section view, the rotational orientation of the view is incorrect, as shown in Figure 13.72. Section views are created and presented using the rules of orthogonal projection and multiview presentation. Even though a section view is forgiven the requirement of proper position alignment with the other orthogonal views (for purposes of drawing convenience), the requirement for proper rotational orientation still exists. Recall that a cutting plane line on an orthogonal view is the edge view of a cutting plane that is perpendicular to that view. Incorrect rotational orientation of the section view may cause confusion with its interpretation.

Viewing the Object from the Wrong Side

The arrows of a cutting plane line point in the viewing direction. A common error, as shown in Figure 13.73, is to show the section view looking at the piece of the object that should have been removed, rather than the piece that is to remain. The areas that have been cut (i.e., the areas that are filled with section lines) are the same for both pieces. However, the background edges and the rotational orientation of the object may be incorrect.

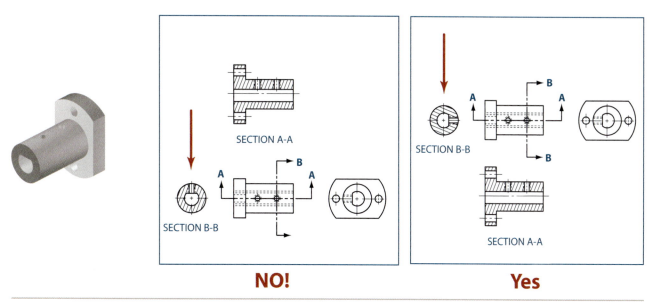

FIGURE 13.72. A common error: Rotational orientation of the section view is incorrect.

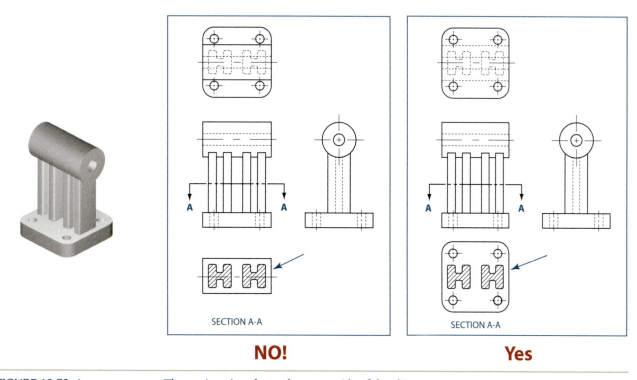

FIGURE 13.73. A common error: The section view shows the wrong side of the object.

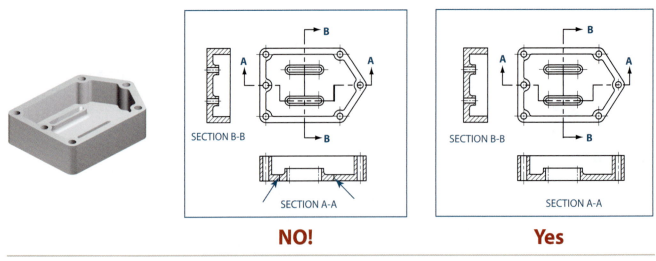

FIGURE 13.74. A common error: Step lines are not removed from the offset section.

Including Step Lines on an Offset Section

The steps in an offset section, as shown in Figure 13.74, create edges; but leaving these edges on a section view is considered improper. Therefore, they should be removed. Actually, if you were to follow the rules of orthogonal projection strictly, the edges created by the changes in direction of the cutting plane should be shown. However, someone in the past thought a nicer-looking drawing would result from their removal; thus, the practice remains today. Until this practice is formally changed, the edges in an offset section should be removed.

No View Label or View Scale

To avoid confusion with parts that have multiple section views, every full or offset section view must be labeled with the same letters used to identify their respective cutting plane line, as shown in Figure 13.75. This labeling is practiced even when only one section view is on the drawing. Half, revolved, and broken-out section views have the same scale as the view on which it was created. Full, offset, and removed section views are allowed to be a different scale than the view on which they were created. However, whenever there is a change in scale, the new scale must be clearly labeled on the view.

Section Lines with Poor Spacing or Angle

Section lines should be created in such a manner that they are easily recognizable as section lines. Section lines should be easily distinguishable from edges on the part. If the section line density is too low or if the angle of the section lines matches the angle of some of the edges of the part, as shown in Figure 13.76, there may be confusion between the section lines and the edges of the part. On the other hand, too dense a section line pattern is difficult to reproduce cleanly when the drawing is printed, copied, or transmitted.

FIGURE 13.75. A common error: No scale or view is missing labels.

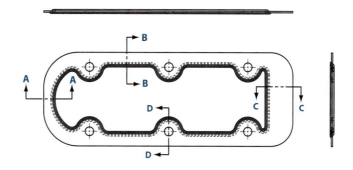

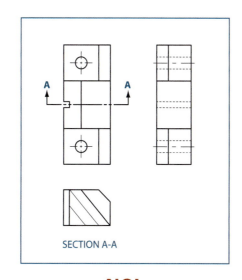

NO!

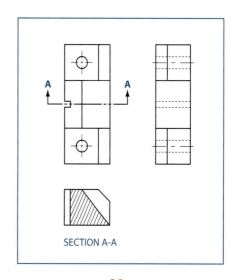

SECTION A-A
SCALE 4:1

SECTION C-C
SCALE 4:1

SECTION B-B
SCALE 4:1

SECTION D-D
SCALE 4:1?

Yes

FIGURE 13.76. A common error: Poor choice of cross-hatching.

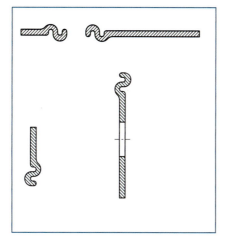

SECTION A-A

NO!

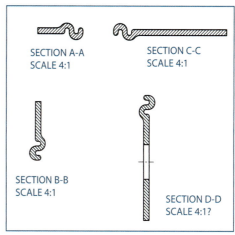

SECTION A-A

Yes

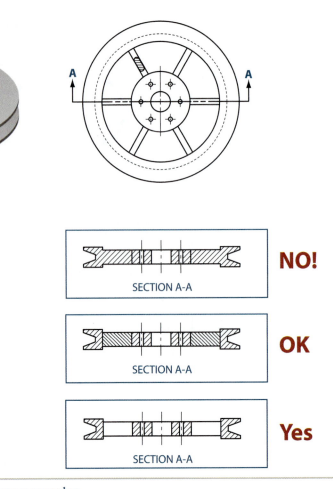

FIGURE 13.77. A common error: Cross-hatching in vanes or spokes.

Filling Vanes or Spokes with Section Lines

Filling spokes or vanes with section lines, as shown in Figure 13.77, gives an immediate but false impression that the object is solid throughout the spoke or vane area. Spokes and vanes should not be filled at all or should be filled with a different section line pattern than is used for the rest of the sectioned object.

Common Section Line Pattern on Different Parts in an Assembly

A common section line pattern used for different parts in a sectioned assembly, as shown in Figure 13.78, gives the immediate but false impression that the separate parts are a single part. Different parts in a sectioned assembly, even when they are made of the same material, should be filled with section lines that are of different patterns.

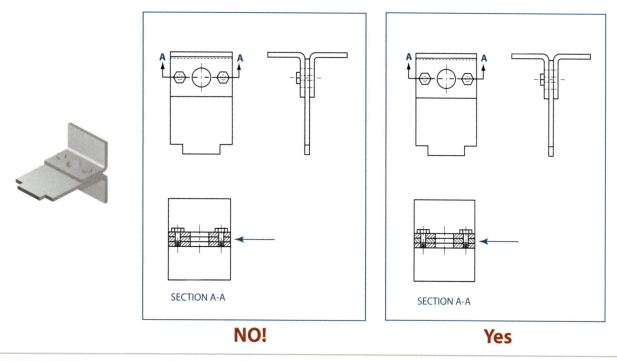

FIGURE 13.78. A common error: The same cross-hatch pattern for different parts in an assembly.

13.14 Considerations for 3-D Modeling

With solids modeling software, parts are initially modeled as a series of protrusions and cuts to create a 3-D graphical model of a part. The solids modeling software creates a mathematical model of the geometry from which the projections of the object are used to create drawings. Once the solids model is created, it is usually a simple matter to extract a front view, side view, or any of the other orthogonal views from the model. A section view is created merely as another orthogonal view, but with a portion of the object removed. The ease with which section views can be created from a solids model has many advantages, but also some disadvantages. The greatest advantage is the speed and accuracy with which section views can be created. With most software, creating additional views is simply a matter of specifying the cutting plane and viewing direction and then picking a location on the drawing where the new view is to appear. Cutting planes can be specified as existing or newly created reference planes. Creating stepped cutting plane lines in the views of interest usually specifies offset cutting planes. The process is often a matter of a few strokes on a keyboard or a few clicks with a mouse or another pointing device. The time required is usually only a few seconds. Also, accurate orthogonal projection of features that were previously represented by shortcut practices, such as small cutouts in curved surfaces or thin symmetrical features, are very easy to create. In fact, with most software, it would be difficult to create a view that is *not* an accurate projection. Using section lines to fill areas that were formerly solid is also a rather simple matter. The software identifies the newly cut surfaces and automatically fills them. All the software user needs to do is specify the section line pattern to be used and modify it if necessary.

The selection of where to section an object to view its interior or the type of section to use is still up to the person making the drawing. One disadvantage of the nearly automatic section creation offered by 3-D modeling is that in some cases, the modeling becomes too accurate. Many of the shortcuts and clarification practices used in traditional drafting are no longer available in some software. For example, a section through a spoke or vane used in a 3-D model would show the spoke or vane filled with section lines, not blank as would be preferred. Also, all projections would be true projections. With an object of odd rotational symmetry, there would be no opportunity to modify the projection to create a symmetrical presentation, as would be preferred. With some software, the step edges of an offset section may be visible in the section view, and not removed as is practiced.

Another disadvantage of 3-D modeling is that manual creation of section views has been a traditional method of developing spatial reasoning and mental imaging skills. When the process is too automatic with software, a person may not adequately develop these skills in the absence of the software and may become too dependent on the software. When faced with multiple section views in a shop drawing, that person may not be able to create a mental image of the part or may not develop the skills necessary to interpret the drawings. Eventually, the person will develop these skills, but it may require exposure to many solids models and their drawings.

13.15 Chapter Summary

With many complex objects, looking only at the exterior may not fully reveal all of their features. The use of section views is a method of looking at the internal details of such objects. The section process involves using a hypothetical cutting plane to hypothetically cut an object into pieces so the interior details one or more features. These features can then be examined more closely, specified in such a manner that the details can be fabricated and inspected to ensure that they meet the desired specifications. On an engineering drawing, the cutting plane appears in an edge view called a cutting plane line. Several types of section views are available for use at the discretion of the drafter, depending on the desired presentation. Whichever type is used, certain rules and practices must be followed to ensure that these views can be interpreted easily and quickly without ambiguity. Of primary importance is that the rules of orthogonal projection and multiview presentation be used.

13.16 glossary of key terms

broken-out section: The section view produced when the cutting plane is partially imbedded into the object, requiring an irregular portion of the object to be removed before the hypothetically cut surface can be seen.

cutting plane: A theoretical plane used to hypothetically cut and remove a portion of an object to reveal its interior details.

cutting plane line: On an orthographic view of an object, the presentation of the edge view of a cutting plane used to hypothetically cut and remove a portion of that object for viewing.

cutting segment: On a stepped cutting plane for an offset section view, that portion of the plane that hypothetically cuts and reveals the interior detail of a feature of interest.

full section: The section view produced when a single cutting plane is used to hypothetically cut an object completely into two pieces.

half section: The section view produced when a single cutting plane is used to hypothetically cut an object up to a plane or axis of symmetry, leaving that portion beyond the plane or axis intact.

offset section: The section view produced by a stepped cutting plane that is used to hypothetically cut an object completely into two pieces. Different portions of the plane are used to reveal the interior details of different features of interest.

removed section: The section view produced when a cutting plane is used to hypothetically remove an infinitesimally thin slice of an object for viewing.

13.16 glossary of key terms

revolved section: The section view produced when a cutting plane is used to hypothetically create an infinitesimally thin slice, which is rotated 90 degrees for viewing, on an object.

section lines: Shading used to indicate newly formed or cut surfaces that result when an object is hypothetically cut.

section view: A general term for any view that presents an object that has been hypothetically cut to reveal the interior details of its features, with the cut surfaces

perpendicular to the viewing direction and filled with section lines for improved presentation.

step segment: On a stepped cutting plane for an offset section view, that portion of the plane that connects the cutting segments and is usually perpendicular to them but does not intersect any interior features.

viewing direction: The direction indicated by arrows on the cutting plane line from the eye to the object of interest that corresponds to the tail and point of the arrow, respectively.

13.17 questions for review

1. When should a section view be used?
2. What does a cutting plane line represent?
3. What does the area filled with section lines on a section view represent?
4. What are some guidelines concerning good drafting practice in creating section line patterns?
5. What is the significance of the direction of the arrow on a cutting plane line?
6. Why is it important that the rotational orientation of a section view, even if it is moved, be maintained as if it were an orthogonal view?
7. When should an offset section be used instead of a full section?
8. When should revolved or removed sections be used instead of full or offset sections?
9. Under what conditions should certain areas on a section view not be filled with section lines even though the cut is through solid material?

13.18 problems

1. In the problem shown in Figure P13.1, the views indicated by the balloons are to be changed to full section views taken along the centerline in the direction indicated by the arrows in the remaining view. For each set of views, select the correct section view from the twenty-four proposed views shown at the right. A section view choice may be used more than once. A correct answer may not be available as a choice.

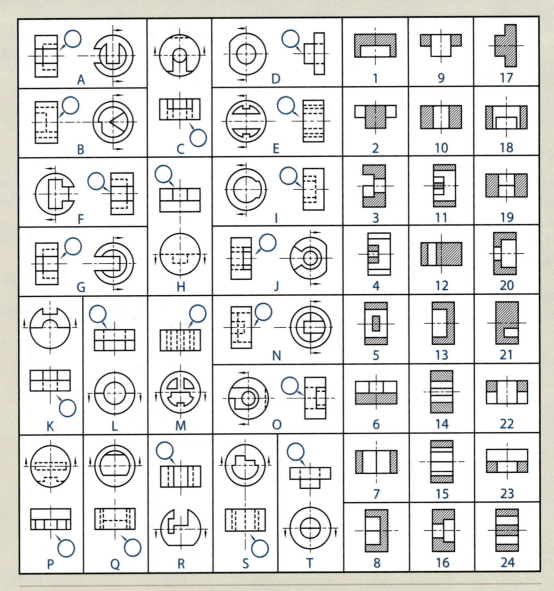

FIGURE P13.1.

13.18 problems (continued)

2. In the problem shown in Figure P13.2, the views indicated by the balloons are to be changed to half section views as indicated by the letters A-A taken along the centerline in the direction in the remaining view. For each set of views, select the correct section view from the twenty-four proposed views shown at the right. A section view choice may be used more than once. A correct answer may not be available as a choice.

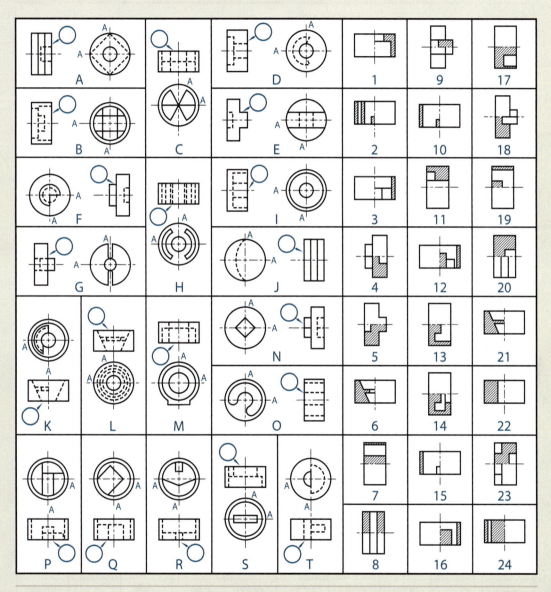

FIGURE P13.2.

13.18 problems (continued)

3. In the problem shown in Figure P13.3, the views indicated by the circles are to be the location of section views. Select the correct section view to complete each problem from the thirty proposed views shown. A section view choice may be used more than once. A correct answer may not be available as a choice.

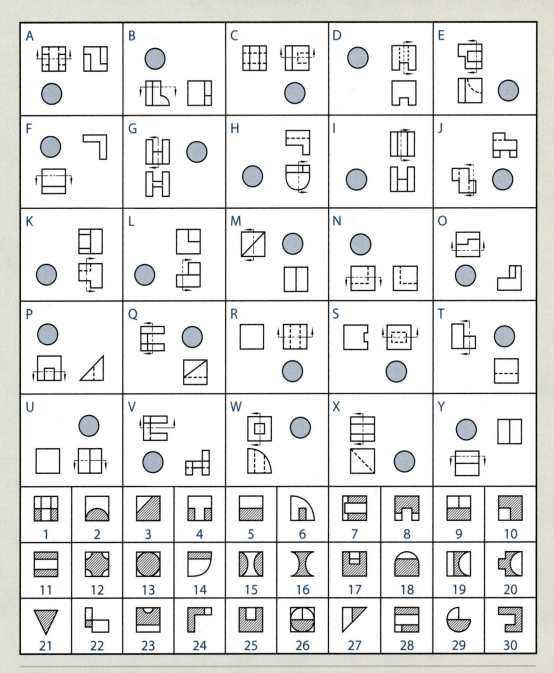

FIGURE P13.3.

13.18 problems (continued)

4. In the problem shown in Figure P13.4, the views indicated by the circles are to be the location of offset section views. Select the correct section view to complete each problem from the thirty proposed views shown. A section view choice may be used more than once. A correct answer may not be available as a choice.

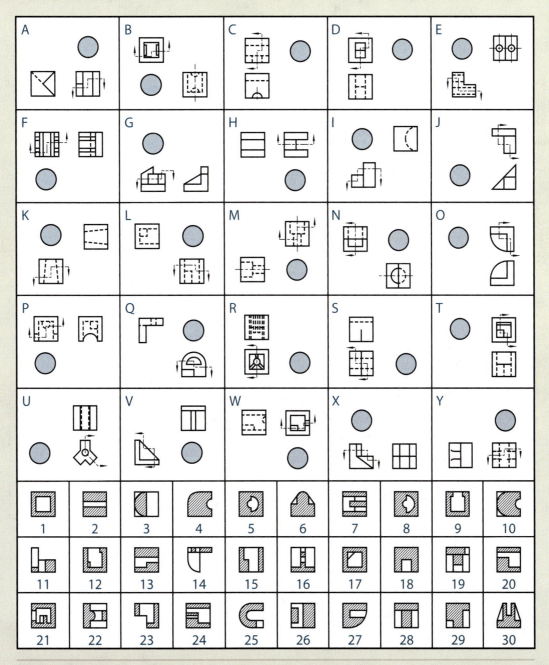

FIGURE P13.4.

13.18 problems (continued)

5. For each object represented in Figure P13.5, create a multiview drawing to fully describe the object, including the indicated full section views to reveal interior detail. When the precise location of the cutting plane line for the full section is not specified, choose the location to best reveal the interior detail.

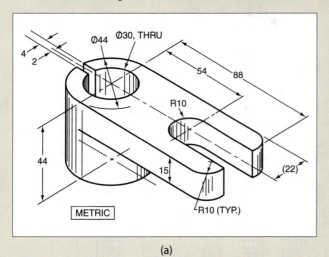

(a)

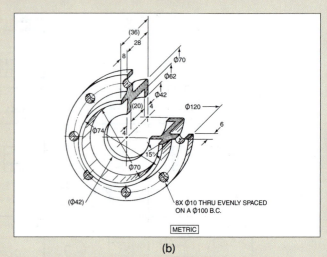

(b)

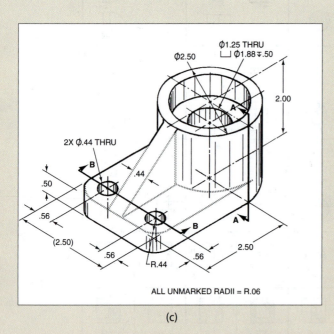

(c)

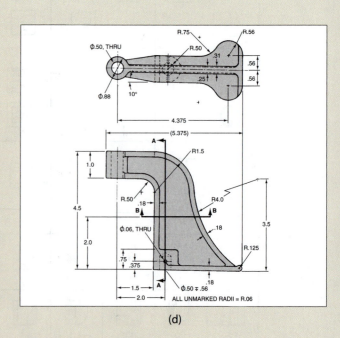

(d)

13.18 problems (continued)

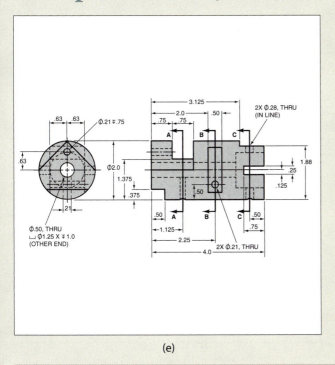

(e)

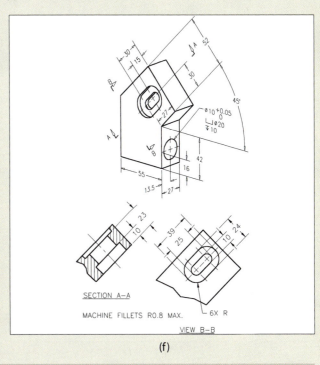

(f)

FIGURE P13.5.

6. For each object represented in Figure P13.6, create a multiview drawing to fully describe the object, including the indicated half section views to reveal interior detail. When the precise location of the cutting plane line for the half section is not specified, choose the location to best reveal the interior detail.

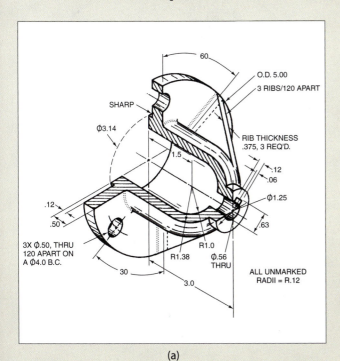

(a)

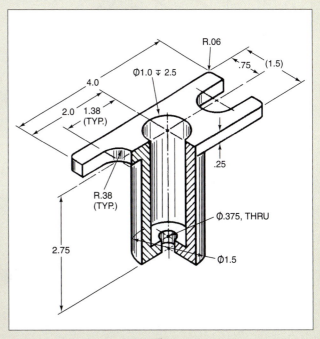

(b)

13.18 problems (continued)

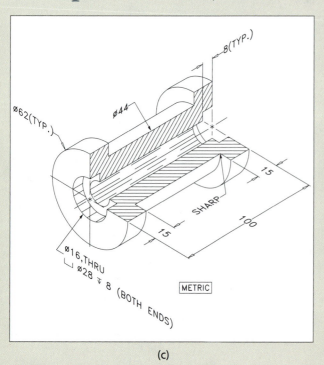

(c)

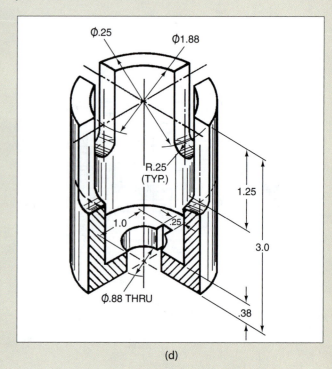

(d)

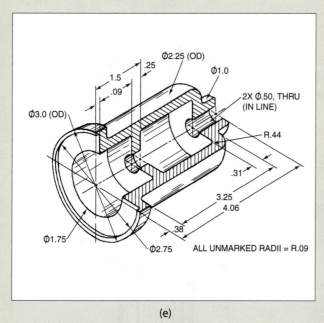

(e)

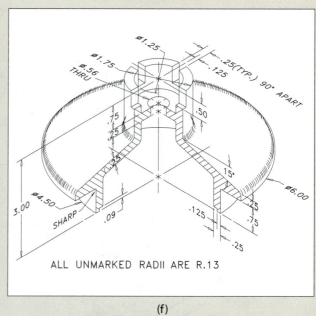

(f)

FIGURE P13.6.

13.18 problems (continued)

7. For each object represented in Figure P13.7, create a multiview drawing to fully describe the object, including the indicated offset section views to reveal interior detail. When the precise location of the cutting plane lines for the offset sections are not specified, choose the locations to best reveal the interior detail.

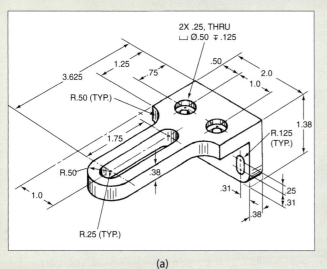

(a)

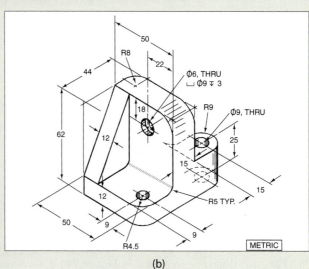

(b)

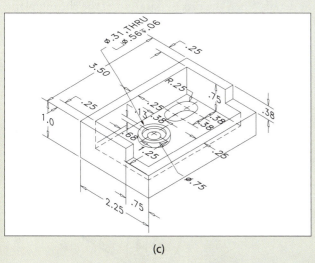

(c)

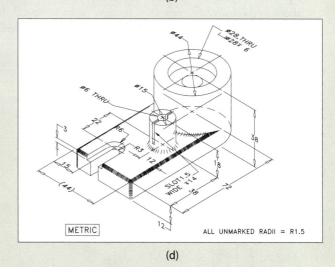

(d)

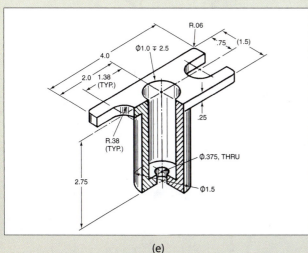

(e)

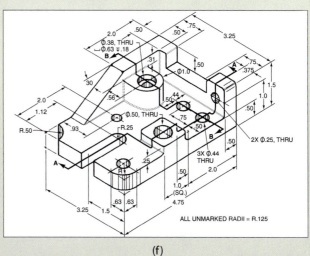

(f)

13.18 problems (continued)

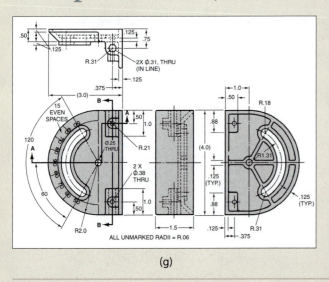

(g)

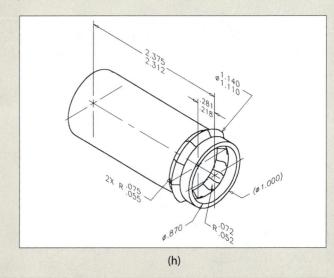

(h)

FIGURE P13.7.

8. For each object represented in Figure P13.8, create a multiview drawing to fully describe the object, including the indicated removed or revolved section views to reveal interior detail. When the precise location of the cutting plane line for the removed or revolved section is not specified, choose the location to best reveal the interior detail.

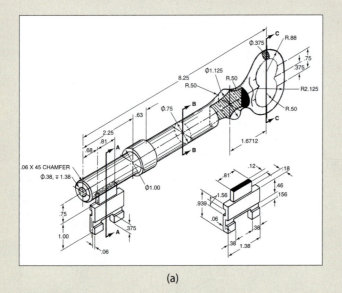

(a)

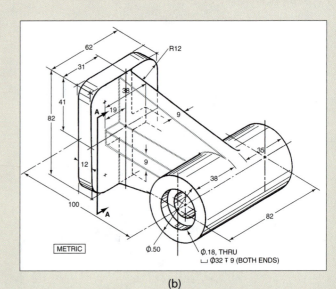

(b)

13.18 problems (continued)

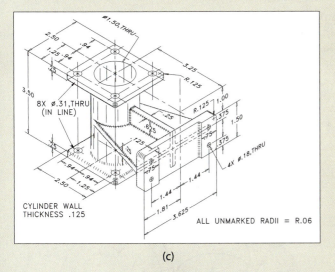

CYLINDER WALL
THICKNESS .125

8X Ø.31,THRU
(IN LINE)

ALL UNMARKED RADII = R.06

(c)

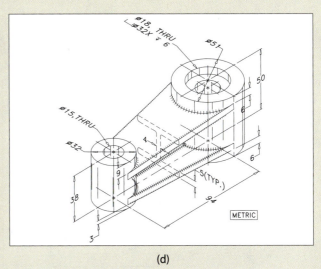

METRIC

(d)

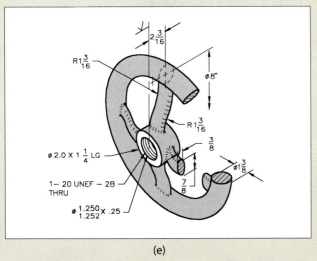

(e)

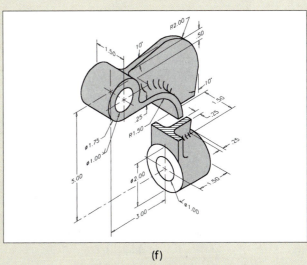

(f)

FIGURE P13.8.

13.18 problems (continued)

9. For each object represented in Figure P13.9, create a multiview drawing to fully describe the object, including the indicated broken-out-section view to reveal interior detail. When the precise location of the broken-out section is not specified, choose the location to best reveal the interior detail.

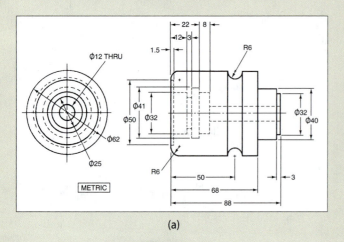

(a)

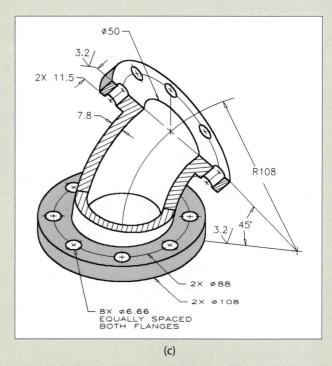

(c)

FIGURE P13.9.

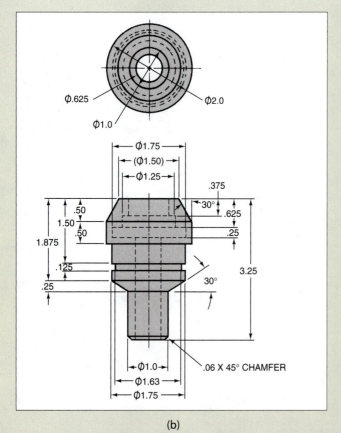

(b)

13.18 problems (continued)

10. For the three objects represented in Figure P13.10, create a multiview drawing to fully describe objects in their assembled state, including a removed section view to reveal the interior detail of how the separate parts are mated.

11. For each object presented in a multiview format in Figures P13.1–P13.10, create an isometric pictorial of the remaining object after it has been sectioned.

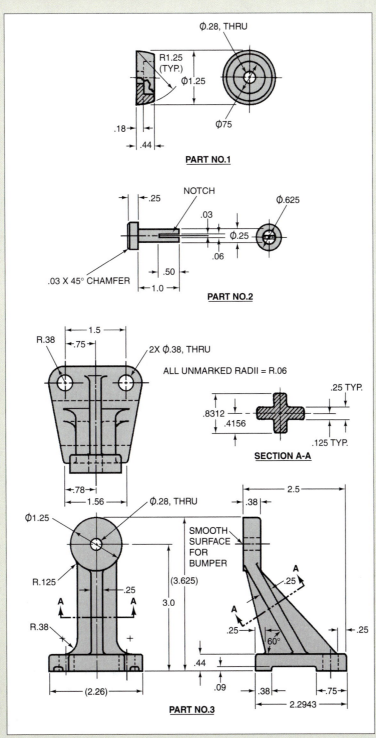

FIGURE P13.10.

14

Auxiliary Views

objectives

After completing this chapter, you should be able to

- Describe an auxiliary view
- Describe situations where an auxiliary view is desired
- Locate top-adjacent, front-adjacent, and side-adjacent auxiliary views constructed from primary views
- Create an auxiliary view of an inclined surface

14.01
introduction

Auxiliary views are most commonly used to determine the true shape of inclined or oblique surfaces. Auxiliary views also are used to locate characteristics and relationships between lines and planes, such as:

- Visibility of lines and planes. For example, at a chemical facility, you will find buildings containing pumps and pipes crossing over and under each other. Auxiliary views and visibility principles are used to determine whether a pipe is on top or in front of another pipe.
- The shortest distance between two lines; for example, designing a brace to separate two pipes at their closest point.
- The shortest distance from a point to a plane. For example, in an ore mine, the entrance tunnel leading to a vein of ore should be short for economic reasons, but it also should have the optimal slope for the transportation of the ore.
- The slope of a line or plane; for example, the downward angle of a feeder that delivers parts to a conveyor belt.
- The angle between two planes, also called a dihedral angle. For example, the angle between the face and flank of a tool bit is ground to 62° to cut steel and 71° to cut cast iron.
- The intersection of two planes; for example, the intersection of one tube with another on a bicycle frame.

In this chapter, you will learn about the basics of using auxiliary views for examining inclined surfaces on solid objects. A supplemental chapter covers fundamentals in descriptive geometry, a graphical technique that was developed to explore characteristics and relationships between lines and planes. In descriptive geometry, auxiliary views are constructed to define relationships between points, lines, and planes in space; however, here you will focus on the fundamentals in creating auxiliary views of object surfaces, leaving the more advanced techniques for your later exploration.

14.02 Auxiliary Views for Solid Objects

In a previous chapter, you learned about multiview drawings of objects. You learned how to construct the top, front, and side views of an object by projecting these views onto the surfaces of a glass box surrounding the object and then unfolding the panes of glass so all of them were in the same plane. You also learned that normal surfaces are parallel to one of the six primary views, inclined surfaces are perpendicular to one of the primary views but are not parallel to any of the primary views, and oblique surfaces are neither parallel nor perpendicular to any of the primary views.

For normal surfaces, a frontal surface is parallel to the front view (or front pane on the glass box). A frontal surface is seen as an edge in the top and side views and is seen in its **true shape** and size in the front view. Likewise, horizontal surfaces are seen in true shape and size in the top view (they are parallel to the top view), and profile surfaces are seen in true shape and size in the side view (they are parallel to the side view). Thus, for a surface to be seen in its true shape and size, it must be parallel to the given view.

What about an inclined or oblique surface? You learned previously that these types of surfaces are not parallel to any of the primary views. Because they are not parallel to any of the primary views, they are seen **foreshortened** in the primary views—they are not seen in their true shape or size in the top, front, or side views. In the remainder of this chapter, you will learn about the creation of auxiliary views that show inclined surfaces in true shape and size. The creation of auxiliary views to determine the true size

FIGURE 14.01. The image of the object is projected onto a horizontal surface to create the top view in (a). The front and top views are shown in (b).

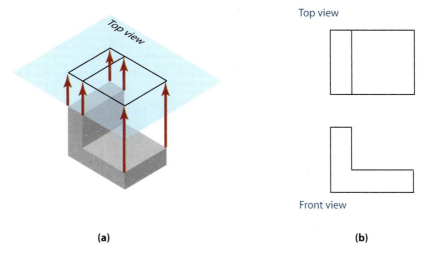

Top view

Front view

(a)
(b)

and shape of oblique surfaces will not be covered here. If you want to create successive auxiliary views to obtain true size and shape of an oblique surface, you can refer to the supplementary chapter on descriptive geometry.

Before you dive into understanding how auxiliary views are created, step back and think about the creation of the primary views. A horizontal surface is parallel to the top view. When you constructed the top view, you projected it onto the top pane of the glass box using **projection rays** that were perpendicular to the pane. In the front view, these projection rays extended perpendicularly from the horizontal **edge view** of the normal surface toward the top viewing plane. When the top pane was unfolded, the projection rays from the front view defined the outer limits of the surface as seen in the top view. The horizontal pane of glass also appears as an edge in the front view—an edge that is parallel to the surface in question. Figure 14.01a shows a simple object with the projection rays used to create the top view indicated, and Figure 14.01b shows the top and front views of the object after the imaginary glass box has been unfolded.

Figure 14.02 shows the top, front, and right-side views of a drill jig; a pictorial view of the jig is also shown for clarity. Surface A on the drill jig is seen as an edge in the front and side views and is seen in its true shape in the top view, signifying that surface A is

FIGURE 14.02. An object with an inclined surface shown in the preferred orthogonal view configuration with top, front, and right-side views. Surface A is parallel to the top viewing plane (and, therefore, will be shown in its true shape in the top view) and in its edge view in the front and right-side views.

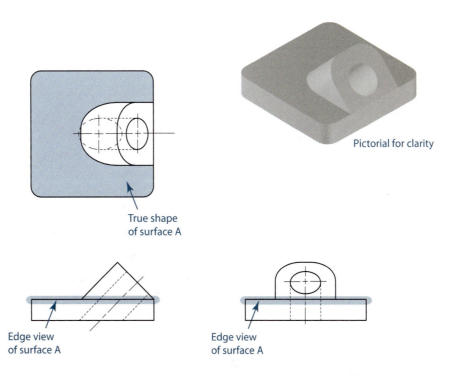

Pictorial for clarity

True shape of surface A

Edge view of surface A

Edge view of surface A

FIGURE 14.03. Surface B is seen in an edge view in the front view but is not seen in its true shape in either the top or right-side views.

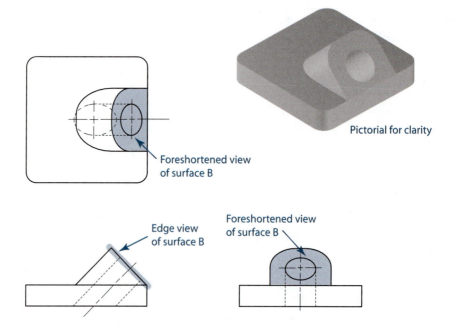

Pictorial for clarity

Foreshortened view of surface B

Edge view of surface B

Foreshortened view of surface B

parallel to the top plane. The edge view of inclined surface B is seen in the front view as indicated in Figure 14.03; however, since neither the top nor right-side views of the object are parallel to the surface, you are not seeing surface B in its true shape and size in either view. Surface B is foreshortened in both the top and side views.

To manufacture this part (you learned about manufacturing processes in an earlier chapter), a machinist will need to know exactly where the hole through the inclined surface B is located. However, since none of the views show the surface in its true shape and size, it would be difficult for the machinist to make this part accurately, based on the given information. If a view that is parallel to this inclined surface could somehow be constructed, surface B would be seen in its true shape and size in this new view, solving the problem of accurately locating the hole for the machinist. Because this new view is not one of the primary views of the object, it is called an auxiliary view—essentially any extra view, other than a pictorial, which has been constructed for overall clarity.

To understand how auxiliary views are created, it is useful once again to think about the object as if it were surrounded by a glass box. This time, however, imagine an extra pane of glass has been added to the glass box that is parallel to the inclined plane as shown in Figure 14.04. The inclined surface can be projected perpendicularly onto

FIGURE 14.04. An additional plane (pane) added to the glass box to show the true shape of the inclined surface.

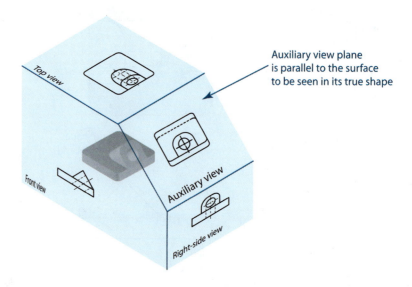

Top view

Front view

Auxiliary view plane is parallel to the surface to be seen in its true shape

Auxiliary view

Right-side view

FIGURE 14.05. The glass box unfolded, showing the top, front, right-side, and auxiliary viewing planes on a single plane.

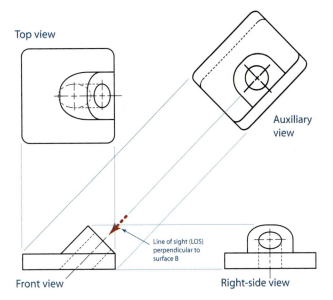

Top view

Auxiliary view

Line of sight (LOS) perpendicular to surface B

Front view

Right-side view

this angled pane of glass, resulting in a view of its true shape and size. The glass box can now be unfolded to show all views, including the auxiliary view, on a single plane. Figure 14.05 shows the glass box with the panes of glass unfolded, including the pane with the auxiliary view on it.

Figure 14.06a shows a full auxiliary view of the drill jig, while Figure 14.06b shows a partial auxiliary view of the drill jig. The difference is that the full auxiliary view shows all surfaces of the object, whether they are true shape or not, whereas the partial auxiliary view shows only the surface for which the true shape and size are required—the inclined surface. Full auxiliary views are usually not necessary to construct because

FIGURE 14.06. A full auxiliary view is shown in (a). A partial auxiliary view, showing only the inclined surface, is shown in (b).

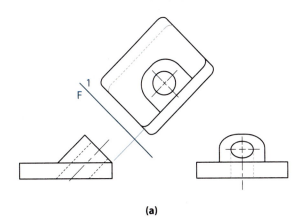

(a)

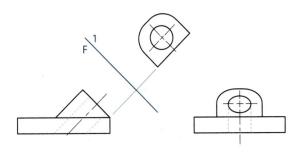

(b)

FIGURE 14.07. The glass box planes labeled.

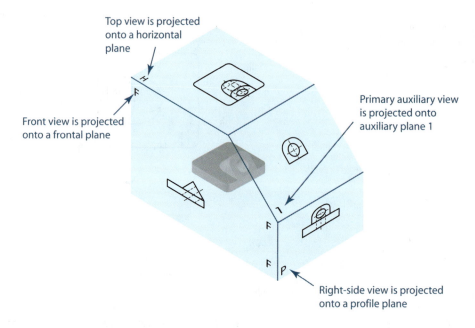

Top view is projected onto a horizontal plane

Primary auxiliary view is projected onto auxiliary plane 1

Front view is projected onto a frontal plane

Right-side view is projected onto a profile plane

you do not gain additional information from including other surfaces , which may even serve to clutter the drawing in such a view—the primary views show you what you need to know about the other surfaces on the object. Since your primary purpose in creating an auxiliary view is for clarity and not for confusion, partial auxiliary views are the most common type of auxiliary views you will create.

Some conventions have been developed to enable you to organize your work when you are working with auxiliary views. One of the conventions is in labeling views. You know the glass box is composed of six principal planes—two horizontal planes (the top and bottom planes), two frontal planes (the front and back planes), and two profile planes (the right and left-side planes). To help identify which views you are working with, when creating auxiliary views, it is standard practice to label the top, front, and side planes with the capital letters *H*, *F*, and *P*, respectively. (In this case, *H* represents the horizontal, or top plane; *F* represents the front plane; and *P* represents the side, or profile plane.) Any auxiliary planes you create should be numbered sequentially.

Figure 14.07 shows the three principal planes of the glass box labeled H, F, and P and the angled glass plane for the auxiliary view labeled 1. Figure 14.08 shows what these planes look like after the glass box has been unfolded. The glass box hinges are referred to as **reference lines** or fold lines and will be used to transfer measurements as you construct your auxiliary views. One of the other conventions that have been developed for work with auxiliary views is that the fold lines are usually shown in the multiview drawings. This is in contrast with standard practice for constructing multiview drawings that you learned about in a previous chapter, where fold lines are not usually shown.

Understanding an auxiliary view is simple if you remember the principles of orthographic projection. Any view that is projected from one view into the next is *adjacent* to the first view. The auxiliary view showing the inclined surface B in true size and shape is adjacent to the front view; the top and side views are also adjacent to the front view. In general, **adjacent views** are aligned side by side and share a common dimension. For example, the right and front views that are adjacent to each other show the height of the object in common. **Related views** are adjacent to the same view and share a common dimension. For the drill jig, the top and auxiliary views are adjacent to the front view. In this case, the top and auxiliary views are related to each other and share a common dimension—the object depth. The depth that point A is away from the fold line is visible in the top view, and this distance is preserved in the related auxiliary view. Figure 14.09 shows how the depth dimension for point A is preserved from the top into the auxiliary view.

FIGURE 14.08. The glass box opened to show proper view alignment.

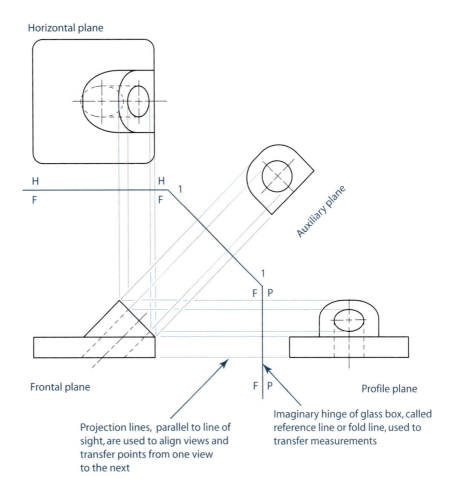

Horizontal plane

Auxiliary plane

Frontal plane

Profile plane

Projection lines, parallel to line of sight, are used to align views and transfer points from one view to the next

Imaginary hinge of glass box, called reference line or fold line, used to transfer measurements

FIGURE 14.09. Comparison between adjacent and related views.

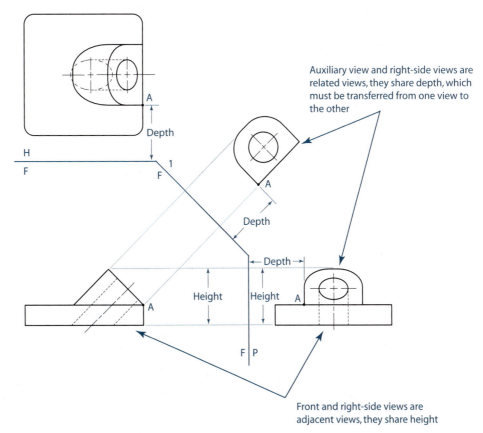

Auxiliary view and right-side views are related views, they share depth, which must be transferred from one view to the other

Depth

Depth

Depth

Height

Height

Front and right-side views are adjacent views, they share height

FIGURE 14.10. A profile-adjacent auxiliary view of surface B.

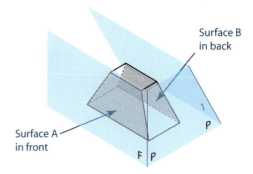

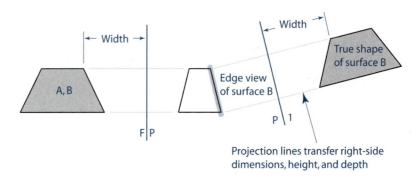

The auxiliary view you have been working with is called a front-adjacent view because it was constructed adjacent to the front view. It also is possible to draw top-adjacent and side-adjacent views, depending on which primary view shows the inclined surface in question as an edge. For example, if the inclined surface is an edge view in the top view, you would project the auxiliary view from the top view to obtain its true size and shape, thus creating a top-adjacent view. Similarly, if the inclined surface is seen as an edge in the side view, a side-adjacent auxiliary view would be used to show the surface in its true size and shape. Figure 14.10 illustrates a side-adjacent auxiliary view that shows the true size and shape of the indicated inclined surface.

14.03 Auxiliary Views of Irregular or Curved Surfaces

Most of the time, you will need to construct an auxiliary view of a surface on an object that shows a curved or irregular feature, like the drill jig from the previous example. Consider the object shown in Figure 14.11. This object contains two holes through an inclined surface. The inclined surface is seen in edge view in the top view, and the holes appear as ellipses in the front view.

When creating an auxiliary view of this surface, you should project several points on the curved edges to define them in the new view. For a circular hole, usually four radial points are sufficient; but, for an irregular curve, you may need to locate several points to obtain an accurate projection. Figure 14.12 shows an auxiliary view of the inclined surface for the object shown in Figure 14.11. In this case, four radial points were transferred into the auxiliary view for each circular hole on the surface.

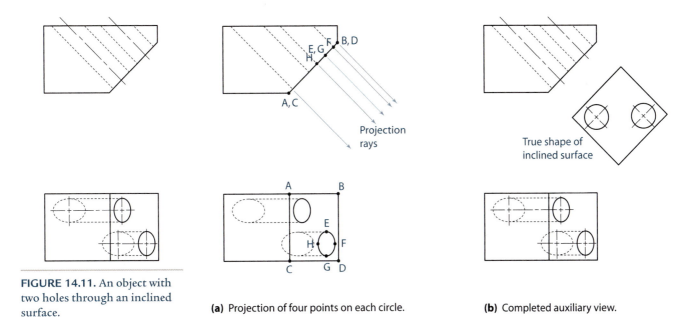

FIGURE 14.11. An object with two holes through an inclined surface.

(a) Projection of four points on each circle.

(b) Completed auxiliary view.

FIGURE 14.12. An auxiliary view of an inclined surface with holes in it.

14.04 Strategies for Auxiliary Views

How do you construct an auxiliary view that shows an inclined surface in true size without the aid of a glass cube? Review what you know so far. To see a surface in true size and shape, you must view it from a plane that is parallel to the surface. Since this viewing plane is parallel to the surface, the edge view of the pane of glass defining the new view also will be parallel to the surface. Finally, the points on the surface will project perpendicularly into this new view, similar to the way points are projected perpendicularly from the front into the top view (or from the front into the side view).

In general, the procedure used to create an auxiliary view that shows an inclined surface in true size and shape is as follows:

1. Identify the edge view in one of the primary views of the surface to be projected.
2. Sketch a "fold" line parallel to the edge view of the surface.
3. Label all of the fold lines, including the one you just created.
4. Project the points that define the surface along rays that are perpendicular to the fold line for the auxiliary view.
5. Obtain the projected dimensions of the surface into the auxiliary view by observing the same dimension in a related view. (Another way to think of this is that the dimensions are found in a view that is two views back from the auxiliary view in which you are working.)

In the case of curved or irregular surfaces, the first step in the procedure is to segment the curved edge into several points and to locate these in the primary views. The points defining the segments also can be projected into the auxiliary view in step 4 of this procedure.

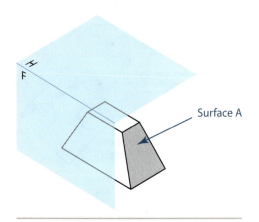

FIGURE 14.13. A truncated pyramid for constructing an auxiliary view.

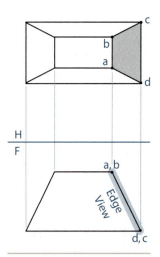

FIGURE 14.14. Top and front views of the truncated pyramid.

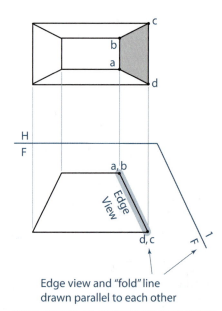

Edge view and "fold" line drawn parallel to each other

FIGURE 14.15. The fold line for the auxiliary view, with labels added.

Figure 14.13 shows a truncated pyramid, and you want to sketch an auxiliary view of the plane A that shows its true size and shape. Figure 14.14 shows the horizontal and front views of the pyramid. Notice that the front view includes the edge view of the surface in question. Also notice for this application the inclusion of the fold line between the top and front views.

To begin the construction of the auxiliary view, sketch a line parallel to the edge view of the surface in the front view. This line represents the fold line for the auxiliary view—the edge view of the pane added to the glass box. After you sketch the fold line, you should label the views appropriately. Figure 14.15 shows the two views of the object with the added fold line and the labels for the fold lines.

The surface will be projected into the auxiliary view along perpendicular projection rays. Only four points define the surface; but in the front view, these points are on top of each other—the endpoints of the line representing the edge view of the plane. Lightly sketch the projection rays from the endpoints of the line into the auxiliary view, keeping in mind that the direction of the projection rays is perpendicular to the fold line. Since the fold line was drawn parallel to the edge view of the plane, the projection rays also are perpendicular to the edge view of the surface. Figure 14.16 shows the two views of the object with the projection rays extending into the auxiliary view.

FIGURE 14.16. Projection rays added, extending from the edge view into the auxiliary view.

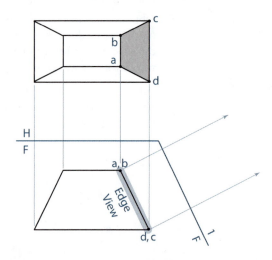

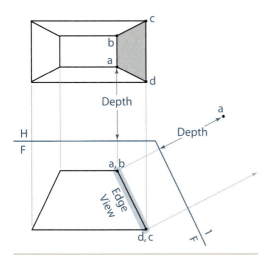

FIGURE 14.17. Point "a" transferred into the auxiliary view.

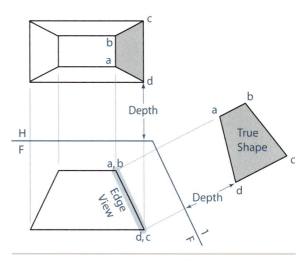

FIGURE 14.18. The auxiliary view showing surface A in true shape.

Transfer the depth dimensions that define the surface in question by looking in the related view (two views back). In this case, the depth is obtained in the top view for transfer into the auxiliary view. Figure 14.17 shows the first point for the surface, point "a," transferred into the auxiliary view.

Continue transferring the remaining points that define the surface into the auxiliary view, each time transferring the depth dimension from the top view. Figure 14.18 illustrates the completed auxiliary view showing the true size and shape of the inclined surface.

Follow this same procedure to obtain the true size and shape of the irregular inclined surface on the object shown in Figure 14.19.

Before you begin, you need to locate several points along the curved edge that you will subsequently project into the auxiliary view. Figure 14.20 shows the same object except that five points along the curved edge have been located and labeled A–E in the top and front views.

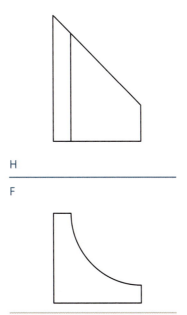

FIGURE 14.19. Top and front views of an object with an irregularly shaped surface for the auxiliary view.

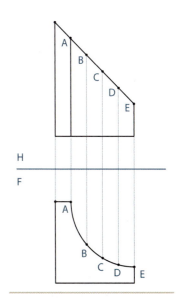

FIGURE 14.20. A curved edge with points A–E located.

FIGURE 14.21. The fold line and projection rays added.

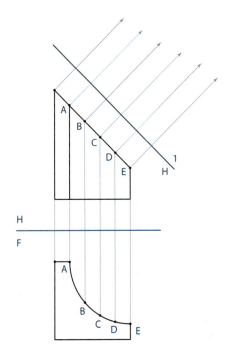

Sketch the fold line parallel to the edge view of the inclined surface and project the points into the auxiliary view using perpendicular projection rays as you did before. Also, make sure you label the fold lines appropriately according to the conventions described earlier. The result is shown in Figure 14.21. When projecting the points, make sure you project each of the points from the segments defining the irregular curved edge.

Locate each of the points in the auxiliary view by measuring the distance each point is from the fold line for the related view (in this case, the front view) and transferring those distances into the view you are creating. Figure 14.22a shows the distance transferred for point A, and Figure 14.22b shows the remainder of the points defining the curved edge transferred into the auxiliary view. Figure 14.22c shows the remaining points defining the surface transferred into the auxiliary view.

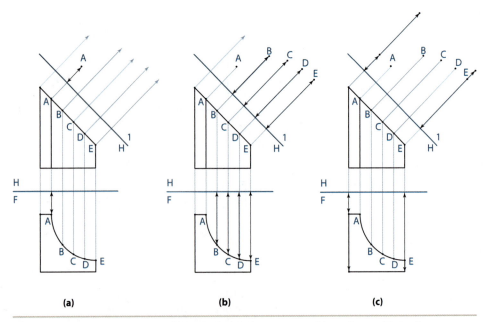

(a)   (b)   (c)

FIGURE 14.22. Points defining an irregular surface projected into the auxiliary view.

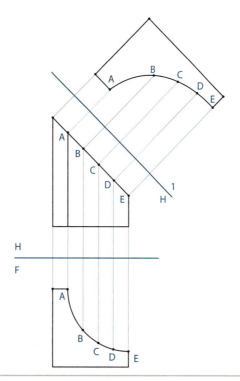

FIGURE 14.23. The completed auxiliary view of the irregular surface.

Finally, connect the dots to create a smooth curved edge and complete the auxiliary view that shows the surface in true shape and size. The completed view is shown in Figure 14.23.

CAUTION   The most common error that students make in creating auxiliary views is when transferring distances into the new view. A typical mistake involves trying to measure the distance from the adjacent view instead of from a related view. For example, Figure 14.24 shows an object with an inclined surface. The parallel fold line is easily created, as are the perpendicular projection rays shown in Figure 14.25.

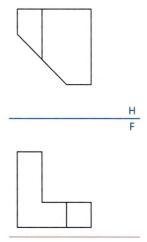

FIGURE 14.24. An object with an inclined surface for auxiliary view.

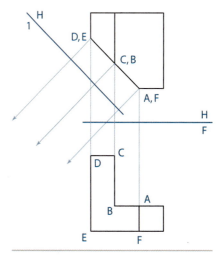

FIGURE 14.25. The fold line and projection rays for auxiliary view creation.

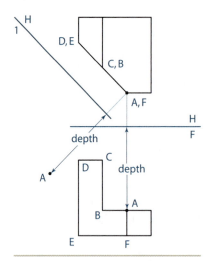

FIGURE 14.26. The depth distance for point A found in the front view.

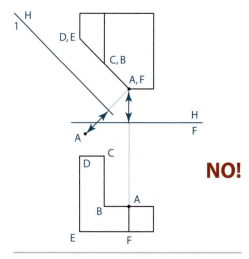

FIGURE 14.27. The incorrect transfer of distance to point A in the auxiliary view.

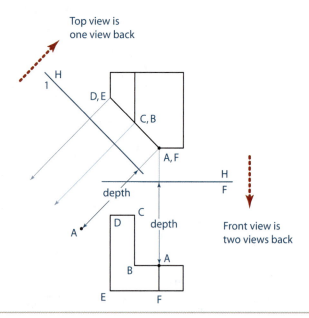

FIGURE 14.28. Choosing the correct view for distance transfer.

To establish point A along the projection ray, you need to determine the distance that it lies from the H/1 fold line. To do this, you need to measure the distance that the point is from the H/F fold line; however, it is the distance from the H/F fold in the *front* view, as shown in Figure 14.26. Many times students incorrectly transfer the distance that the point is from the H/F fold line in the *top* view, as shown in Figure 14.27.

Remember, when transferring distances into the auxiliary view, go back *two* views, not just one, as shown in Figure 14.28.

Solid Modeling Considerations in Creating
14.05 Auxiliary Views

The procedure followed to create an auxiliary view by hand is sometimes tedious and often prone to errors. In the era of 3-D solid modeling, the need for auxiliary views may be somewhat diminished, and the difficulty in creating auxiliary views is greatly reduced. First of all, since the 3-D model is often sent directly to a CAM system for fabrication, it may not be necessary to create an auxiliary view to locate holes and other features accurately on an inclined surface. Because the 3-D model contains all of the necessary information for the creation of the part, an auxiliary view of an inclined surface might not provide any additional information and, therefore, is not needed for manufacturing the part.

Second, when an auxiliary view from a 3-D solid model is needed, creating it is a relatively easy task. Recall that an auxiliary view is created by "looking" perpendicular to the inclined surface. With 3-D modeling software, you usually can select a plane on the object to define the viewing plane. The software will then rotate the object in space so the selected plane is parallel to the computer screen, meaning the plane will appear in true size and shape. With some software, you will be able to show only the plane; with other software, you will be forced to show the entire object from this viewpoint. However, since this auxiliary view is so simple to create, showing the entire object is a small price to pay. Figure 4.29 illustrates a 3-D solid model and an auxiliary view showing the inclined surface in true shape and size.

FIGURE 14.29. The creation of auxiliary views of the AeroTec bow handle from its solid model.

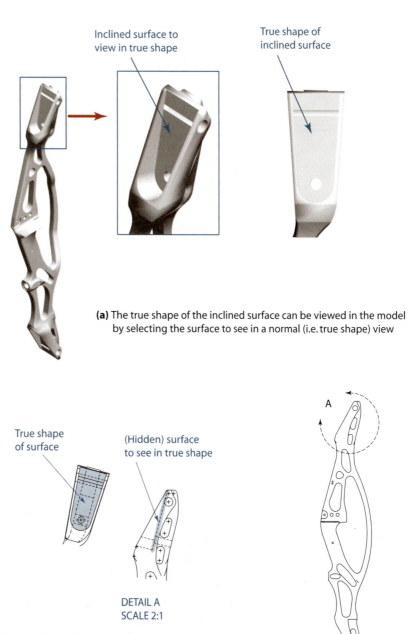

Inclined surface to view in true shape

True shape of inclined surface

(a) The true shape of the inclined surface can be viewed in the model by selecting the surface to see in a normal (i.e. true shape) view

True shape of surface

(Hidden) surface to see in true shape

DETAIL A
SCALE 2:1

(b) The true shape of the inclined surface can be viewed in the drawing by selecting the surface to see in an auxiliary view

FIGURE 14.30. An object with an oblique surface created by a solid model can easily be presented in any view orientation, including a normal (i.e., true shape) view of the oblique surface.

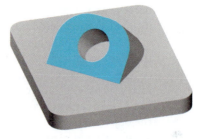

Solid model of object with oblique surface

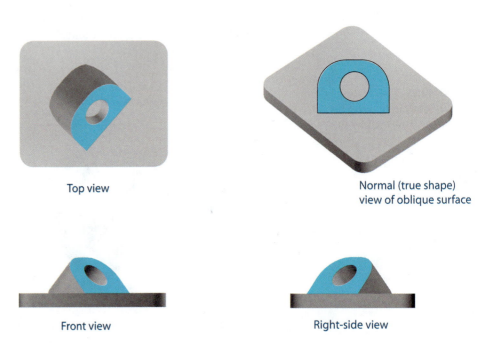

Top view

Normal (true shape)
view of oblique surface

Front view

Right-side view

Also, note that the software can just as easily show the true size and shape of an oblique surface and that successive auxiliary views are not required in this application. Figure 14.30 shows a 3-D model of an object containing an oblique surface and the corresponding auxiliary view in true size and shape.

14.06 Sketching Techniques for Auxiliary Views

In creating auxiliary views as described in this chapter, you should realize by now that you will need to create parallel and perpendicular lines accurately. Many advanced drafting techniques allow you to create these lines; however, you can use a simple set of triangles to create parallel and perpendicular lines. To draw a line parallel or perpendicular to another line, align one leg of one of the triangles with the line. Place the hypotenuse of the second triangle (the base triangle) up against the hypotenuse of the first triangle and hold the base triangle firmly in place. Slide the first triangle up or down the paper, making sure its hypotenuse is always in contact with the hypotenuse of the base triangle. As you slide the triangle, one of its legs will remain parallel to the original line and the other leg will remain perpendicular to the line. You can use either of these straightedge surfaces to draw your parallel or perpendicular lines according to the type of line you need to construct. The method of using triangles to draw parallel or perpendicular lines is illustrated in Figure 14.31.

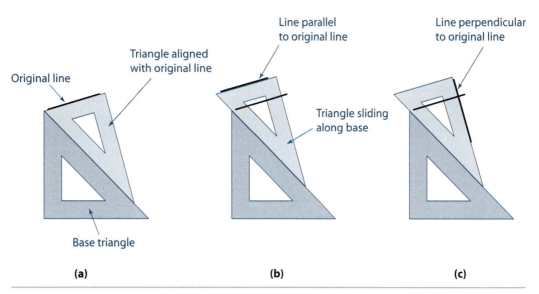

FIGURE 14.31. Sliding triangles used to construct parallel and perpendicular lines.

14.07 Chapter Summary

Auxiliary views allow you to look perpendicular to an inclined surface and see it in its true shape and size. When you are looking perpendicular to the surface, it is important to note that your viewing plane is parallel to the plane in question. Auxiliary views also are used to find information between two lines, a line and a plane, and two planes; however, these applications are left for your exploration in a supplementary chapter on descriptive geometry. The procedure you use to create an auxiliary view is based on the principles from orthographic projection described in detail in previous chapters of this text. In essence, when creating an auxiliary view, you are inserting a pane of glass into the imaginary glass cube that is parallel to the inclined surface. When the glass cube is unfolded, including the extra pane, the auxiliary view shows the inclined surface in true size and shape. In the age of 3-D solid modeling, the need for auxiliary views may be diminished; however, knowing the basics of creating this type of view is important for your understanding of graphic communication. Three-dimensional solid modeling software also enables you to easily create an auxiliary view of an inclined or an oblique surface using a few clicks of the mouse button.

14.08 glossary of key terms

adjacent views: Views that are aligned side by side to share a common dimension.

auxiliary views: Views on any projection plane other than a primary or principal projection plane.

edge view (of a plane): A view in which the given plane appears as a straight line.

foreshortened (line or plane): Appearing shorter than its actual length in one of the primary views.

inclined surface: A plane that appears as an edge view in one primary view but is not parallel to any of the principal views.

oblique surface: A plane that does not appear as an edge view in any of the six principal planes.

projection ray: A line perpendicular to the projection plane. It transfers the 2-D shape from the object to an adjacent view. Projection rays are drawn lightly or are not shown at all on a finished drawing.

14.08 glossary of key terms (continued)

reference line: Edges of the glass box or the intersection of the perpendicular planes. The reference line is drawn only when needed to aid in constructing additional views. The reference line should be labeled in constructing auxiliary views to show its association between the planes it is representing; for example, H/F for the hinged line between the frontal and horizontal planes. A reference line is also referred to as a fold line or a hinged line.

related views: Views adjacent to the same view that share a common dimension that must be transferred in creating auxiliary views.

true shape (of a plane): The actual shape and size of a plane surface as seen in a view that is parallel to the surface in question.

14.09 questions for review

1. What is a primary auxiliary view?

2. What is the purpose of an auxiliary view?

3. How is a full auxiliary view different from a partial auxiliary view?

4. List the five basic steps or procedure for drawing an auxiliary view.

5. Why might the creation of auxiliary views not be necessary with 3-D solid modeling software?

6. How does 3-D solid modeling software make the job of auxiliary view creation easier and less prone to errors?

14.10 problems

1. For each of the pictorials shown in Figure P14.1, create the top, front, and right-side views. Then, create an auxiliary view to present the **true shape** of the inclined surface.

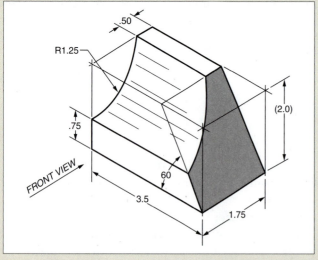

(a)

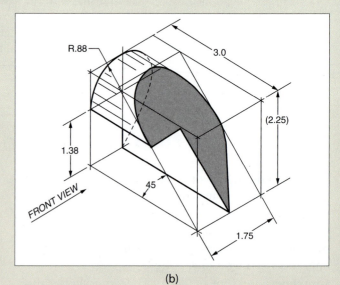

(b)

14.10 problems (continued)

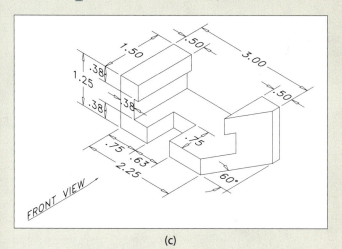

(c)

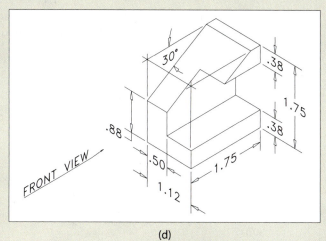

(d)

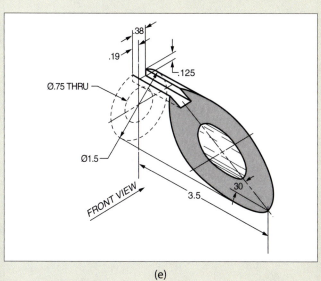

(e)

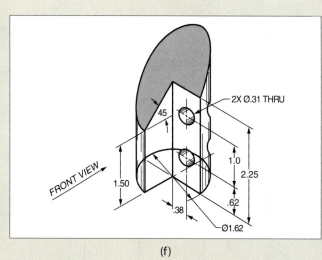

(f)

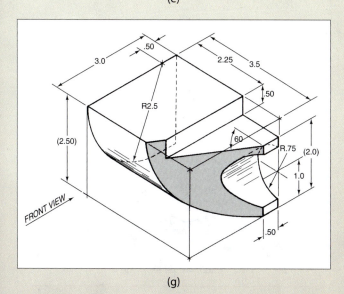

(g)

FIGURE P14.1.

14.10 problems (continued)

2. For each of the pictorials shown in Figure P14.2, create the top, front, and right-side views. Then, create an auxiliary view to present the entire object with the inclined surface shown in its true shape.

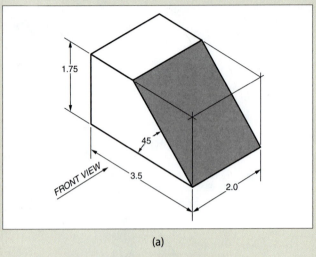

(a)

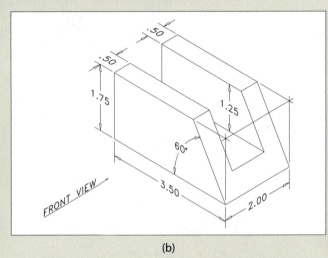

(b)

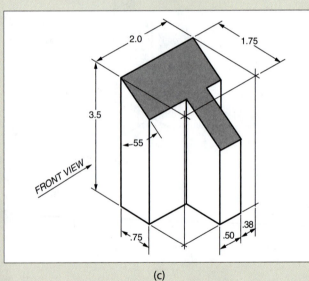

(c)

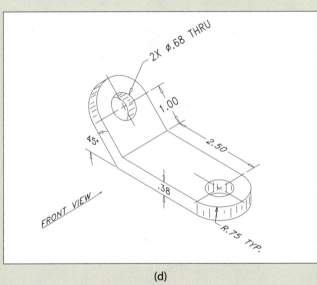

(d)

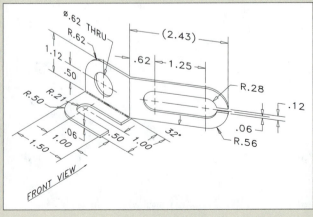

(e)

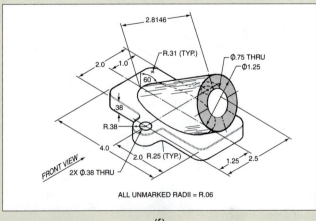

(f)

14.10 problems (continued)

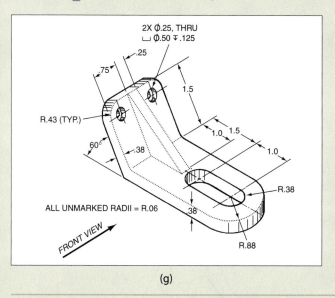

(g)

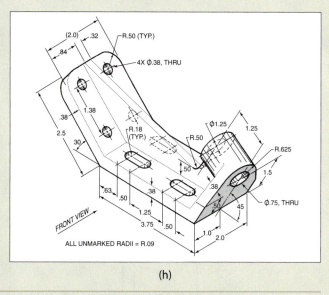

(h)

FIGURE P14.2.

3. For each of the multiview drawings shown in Figure P14.3, create an auxiliary view to present the entire object with the inclined surface shown in its true shape.

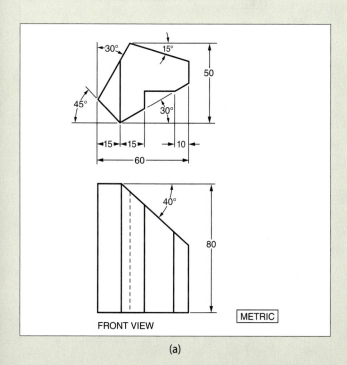

(a)

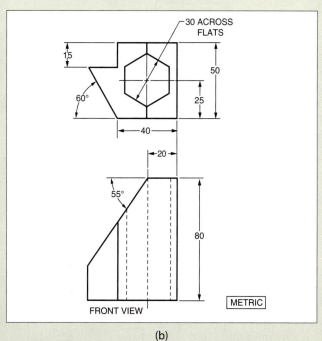

(b)

14.10 problems (continued)

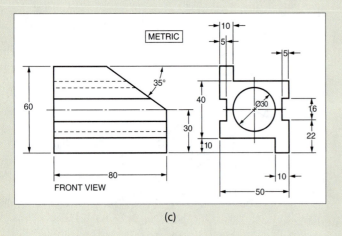

(c)

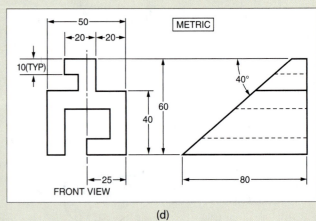

(d)

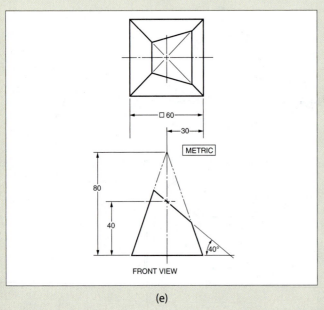

(e)

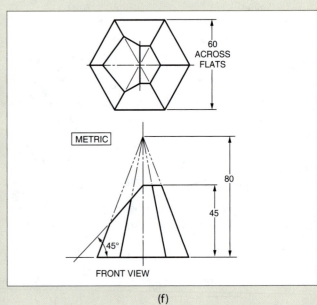

(f)

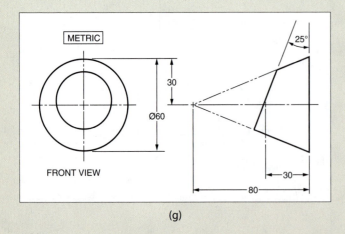

(g)

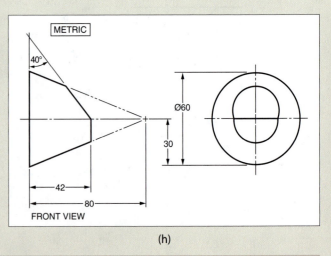

(h)

FIGURE P14.3.

Appendix

casestudy
TiLite Wheelchairs—"The Ultimate Ride"

The freedom to get around is important to everyone, but especially to people with disabilities. Users of manual wheelchairs must be independently mobile to enjoy work, travel, sports, and other social activities. They require comfortable, lightweight wheelchairs with features to suit their active lifestyle. A properly designed and fitted wheelchair not only is comfortable but also minimizes energy expenditure and reduces stresses on the user's body. Today people with disabilities can benefit from breakthroughs in design, materials, and manufacturing methods to obtain a unique, customized wheelchair that fits their personal abilities and active lifestyle.

Founded in 1998, TiLite's goal is to provide a twenty-first-century solution to the age-old problem of mobility. TiLite has successfully combined the unique material properties of titanium; traditional fabrication methods; and modern design tools such as parametric modeling, finite element analysis, and rapid prototyping to provide a unique line of affordable, lightweight, and custom-fit manual wheelchairs. TiLite wheelchairs are fabricated from a titanium alloy that is lightweight and has superior strength. Unlike steel, titanium does not corrode and is very durable. The unique combination of these properties is ideal for wheelchair design.

TiLite designs wheelchairs for a variety of users. Ultralightweight chairs are ideal for sports enthusiasts such as wheelchair basketball players and marathon racers. A TiLite chair has even carried the Olympic Torch. Children require chairs that are lightweight and that adapt to children's growth. Elderly users also benefit from lightweight chairs that are easy to propel. Lightweight folding models are easy to transport. TiLite wheelchairs are designed not only to be functional but also to be stylish, displaying the beautiful patina of polished titanium metal. The ability to custom-design and fabricate a unique chair to fit each individual means that there is a TiLite chair for every user. Modern CAD tools enable designers to modify their designs for custom fabrication.

VARIATIONAL DESIGN OF THE TILITE WHEELCHAIR

Over half of all manual wheelchair users suffer from repetitive motion injuries such as carpal tunnel syndrome, chronic shoulder injury, pressure sores, back pain, postural deformities, and reduced heart and lung function.

Proper fit and low weight are critical to wheelchair design to reduce these injuries. Studies have shown that wheelchair users suffer fewer long-term health problems with a properly fitted wheelchair. The rear axle of the wheelchair should be positioned forward of the user's shoulder, and the center of gravity of the user and chair should be just forward of the rear axle. The seat should be as narrow as possible to keep the rear wheels close to the frame and the arms close to the body during propulsion. The wheelchair frame must be sized to fit the user's body measurements. These requirements translate into specific dimensions on the wheelchair frame.

The TiLite XC Ultralightweight wheelchair.

For the TiLite TX model, the user must specify twelve measurements or parameters, such as seat width, back angle, footrest width, and wheel camber; there is a choice of between three and twenty-five possible values for each measurement. With all of these geometric variables and constraints, the designers of TiLite chairs must utilize parametric models to create solid models and drawings of the custom-fit wheelchair design. Based on the user's desired dimensions, the Design Table function in the solid modeling software allows the designer to create all of the necessary configurations of basic wheelchair frame members. When the user's desired dimensions are inserted into the design tables, all of the remaining dimensions of the parts are automatically adjusted to ensure that the parts fit together and fit the user. From these solid models, all of the necessary drawings for a custom wheelchair can be created in only a few minutes. This saves time for the designer and speeds up delivery of the wheelchair to the user.

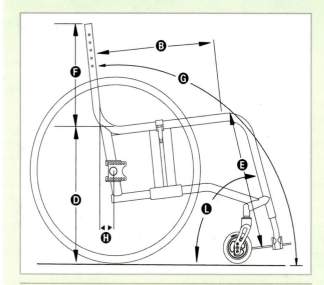

Measurements for ordering a custom wheelchair design.

ASSEMBLY MODELING OF THE TILITE WHEELCHAIR

For a custom-fit wheelchair design, a vast number of different configurations are available depending on the user's body dimensions and seating preferences. Therefore, it is impossible and impractical to build a physical prototype of each wheelchair frame configuration. Nonetheless, the wheelchair must function properly regardless of the size of the frame members. Solid modeling can be used to create virtual prototypes instead of physical models of any wheelchair design.

A folding wheelchair based on a familiar x-frame design is composed of two cross tubes (shown in gray), seat tubes (blue), and hinge members (pink). When designing a folding wheelchair, the designer must check for interference between parts in the open and folded configurations as well as all positions in between. When the sizes of the parts vary for different users, the model must be carefully checked for each configuration. This can be done most efficiently using a solid model. Stick figure models are created to represent the centerlines of each moving part, measured from their attachment points at the pin joints. These skeleton models are manipulated to make sure that the mechanism does not lock up, invert, or toggle. A trial-and-error method is

used to move the locations of the hinge pins until an acceptable design is found that works for all possible sizes of the design.

Although a range of sizes is available for seat tubes and cross tubes, the goal is to have a single design for the hinge members, which may be manufactured by an outside vendor. After the positions of the hinge pins are established with the stick figure models, the hinge members are fleshed out and given a shape that will avoid interference with the cross tubes and seat tubes. A variational solid model allows the designer to check multiple configurations quickly and to ensure that all sizes of the wheelchair frame will function as desired.

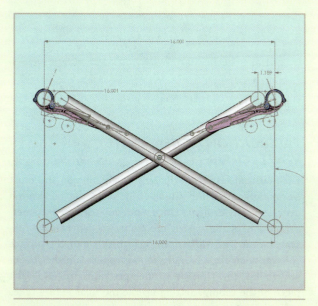

An assembly model of an open wheelchair frame.

Another important design consideration is that the wheelchair can be folded compactly. Solid models can be used to ensure that none of the parts interfere during folding. Changes in the design are easily checked for multiple sizes and configurations of the wheelchair. With the use of solid models, the designer can be assured that the wheelchair will fold to the most compact form possible and function smoothly in the folding operation.

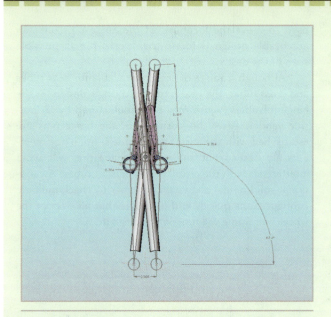

An assembly model of a folded wheelchair frame.

TiLite designer Lindy Anderlini creates full-scale drawings of wheelchair frames.

FABRICATION OF THE TILITE WHEELCHAIR

Each custom-fitted wheelchair frame is unique and must be manufactured individually. To begin the fabrication process, TiLite designer Lindy Anderlini uses the parametric models of the wheelchair frame to generate its full-scale layout. The frame drawing is then laid on the surface of a modular fixture, and the stop blocks on the fixture are bolted down in the proper positions to hold each piece of the tubing. Setting up the modular fixture takes only about ten minutes.

Each tube member is carefully bent to the proper angle using a special bending machine. The tubing is then measured and cut to the correct length based on the scale drawing, and the ends are shaped to fit snugly to the mating parts. Proper sizing and positioning of the frame members is critical to obtain sturdy weld joints. By using the full-scale drawing as a template, the manufacturer can ensure a perfect fit for every customer. After each piece of tubing has been formed and cut, they are laid in the fixture and clamped in the proper position. The parts are tack-welded together, then removed from the fixture and finish-welded. In completion of the fabrication process, the frame is drilled in the necessary locations for assembly of other parts and bead-blasted, hand-buffed, or painted according to the user's preference for surface finish.

DISCUSSION QUESTIONS/ACTIVITIES

1. Explain how a parametric solid modeler can be used to customize the design of wheelchairs to fit individual users. How is the model used in the design process? the manufacturing process?

2. Compare a custom design to a standard wheelchair with adjustable components. What are the advantages and disadvantages of each design?

3. Make a list of the important design and performance specifications for a lightweight wheelchair. How can a solid model be used to check wheelchair design to ensure that the design meets the desired specifications?

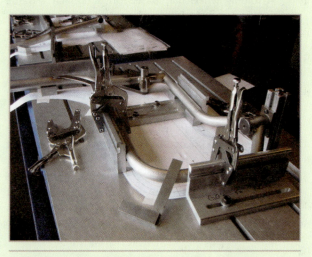

Modular.

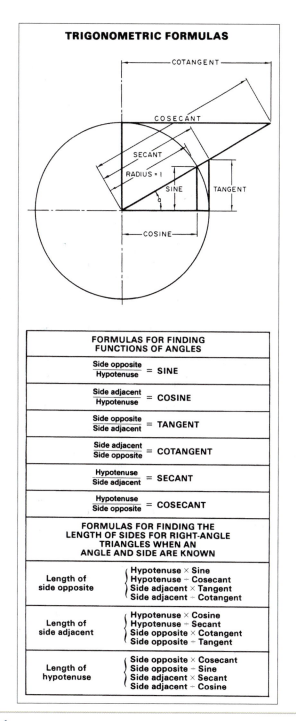

TRIGONOMETRIC FORMULAS

FORMULAS FOR FINDING FUNCTIONS OF ANGLES	
$\dfrac{\text{Side opposite}}{\text{Hypotenuse}}$ = SINE	
$\dfrac{\text{Side adjacent}}{\text{Hypotenuse}}$ = COSINE	
$\dfrac{\text{Side opposite}}{\text{Side adjacent}}$ = TANGENT	
$\dfrac{\text{Side adjacent}}{\text{Side opposite}}$ = COTANGENT	
$\dfrac{\text{Hypotenuse}}{\text{Side adjacent}}$ = SECANT	
$\dfrac{\text{Hypotenuse}}{\text{Side opposite}}$ = COSECANT	

FORMULAS FOR FINDING THE LENGTH OF SIDES FOR RIGHT-ANGLE TRIANGLES WHEN AN ANGLE AND SIDE ARE KNOWN	
Length of side opposite	Hypotenuse × Sine Hypotenuse ÷ Cosecant Side adjacent × Tangent Side adjacent ÷ Cotangent
Length of side adjacent	Hypotenuse × Cosine Hypotenuse ÷ Secant Side opposite × Cotangent Side opposite ÷ Tangent
Length of hypotenuse	Side opposite × Cosecant Side opposite ÷ Sine Side adjacent × Secant Side adjacent ÷ Cosine

Trigonometric Formulas for Triangles.

RIGHT-TRIANGLE FORMULAS

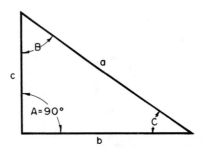

TO FIND ANGLES	FORMULAS	
C	$\frac{c}{a} = $ Sine C	90°−B
C	$\frac{b}{a} = $ Cosine C	90°−B
C	$\frac{c}{b} = $ Tan C	90°−B
C	$\frac{b}{c} = $ Cot C	90°−B
C	$\frac{a}{b} = $ Secant C	90°−B
C	$\frac{a}{c} = $ Csc C	90°−B
B	$\frac{b}{a} = $ Sine B	90°−C
B	$\frac{c}{a} = $ Cosine B	90°−C
B	$\frac{b}{c} = $ Tan B	90°−C
B	$\frac{c}{b} = $ Cot B	90°−C
B	$\frac{a}{c} = $ Secant B	90°−C
B	$\frac{a}{b} = $ Csc B	90°−C

TO FIND SIDES	FORMULAS	
a	$\sqrt{b^2 + c^2}$	
a	$c \times $ Csc C	$\frac{c}{\text{sine C}}$
a	$c \times $ Secant B	$\frac{c}{\text{Cosine B}}$
a	$b \times $ Csc B	$\frac{b}{\text{Sine B}}$
a	$b \times $ Secant C	$\frac{b}{\text{Cosine C}}$
b	$\sqrt{a^2 - c^2}$	
b	$a \times $ Sine B	$\frac{a}{\text{Cosecant B}}$
b	$a \times $ Cos C	$\frac{a}{\text{Secant C}}$
b	$c \times $ Tan B	$\frac{c}{\text{Cotangent B}}$
b	$c \times $ Cot C	$\frac{c}{\text{Tangent C}}$
c	$\sqrt{a^2 - b^2}$	
c	$a \times $ Cos B	$\frac{a}{\text{Secant B}}$
c	$a \times $ Sine C	$\frac{a}{\text{Cosecant C}}$
c	$b \times $ Cot B	$\frac{b}{\text{Tangent B}}$
c	$b \times $ Tan C	$\frac{b}{\text{Cotangent C}}$

Trigonometric Formulas for Triangles. **(CONTINUED)**

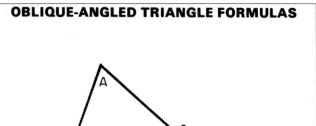

OBLIQUE-ANGLED TRIANGLE FORMULAS

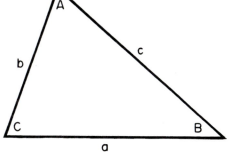

TO FIND	KNOWN	SOLUTION
C	A-B	$180° - (A + B)$
b	a-B-A	$\dfrac{a \times Sin\ B}{Sin\ A}$
c	a-A-C	$\dfrac{a \times Sin\ C}{Sin\ A}$
Tan A	a-C-b	$\dfrac{a \times Sin\ C}{b - (a \times Cos\ C)}$
B	A-C	$180° - (A + C)$
Sin B	b-A-a	$\dfrac{b \times Sin\ A}{a}$
A	B-C	$180° - (B + C)$
Cos A	a-b-c	$\dfrac{b^2 + C^2 - a^2}{2bc}$
Sin C	c-A-a	$\dfrac{c \times Sin\ A}{a}$
Cot B	a-C-b	$\dfrac{a \times Csc\ C}{b} - Cot\ C$
c	b-C-B	$b \times Sin\ C \times Csc\ B$

Trigonometric Formulas for Triangles. (CONTINUED)

RULES RELATIVE TO THE CIRCLE

To Find Circumference—
Multiply diameter by   3.1416 .... Or divide diameter by   0.3183

To Find Diameter—
Multiply circumference by   0.3183 .... Or divide circumference by   3.1416

To Find Radius—
Multiply circumference by   0.15915 .... Or divide circumference by   6.28318

To Find Side of an Inscribed Square—
Multiply diameter by   0.7071
Or multiply circumference by   0.2251 .... Or divide circumference by   4.4428

To Find Side of an Equal Square—
Multiply diameter by   0.8862 .... Or divide diameter by   1.1284
Or multiply circumference by   0.2821 .... Or divide circumference by   3.545

Square—
A side multiplied by   1.4142   equals diameter of its circumscribing circle.
A side multiplied by   4.443   equals circumference of its circumscribing circle.
A side multiplied by   1.128   equals diameter of an equal circle.
A side multiplied by   3.547   equals circumference of an equal circle.

To Find the Area of a Circle—
Multiply circumference by one-quarter of the diameter.
Or multiply the square of diameter by   0.7854
Or multiply the square of circumference by   .07958
Or multiply the square of 1/2 diameter by   3.1416

To Find the Surface of a Sphere or Globe—
Multiply the diameter by the circumference.
Or multiply the square of diameter by   3.1416
Or multiply four times the square of radius by 3.1416

Trigonometric Formulas for Circles.

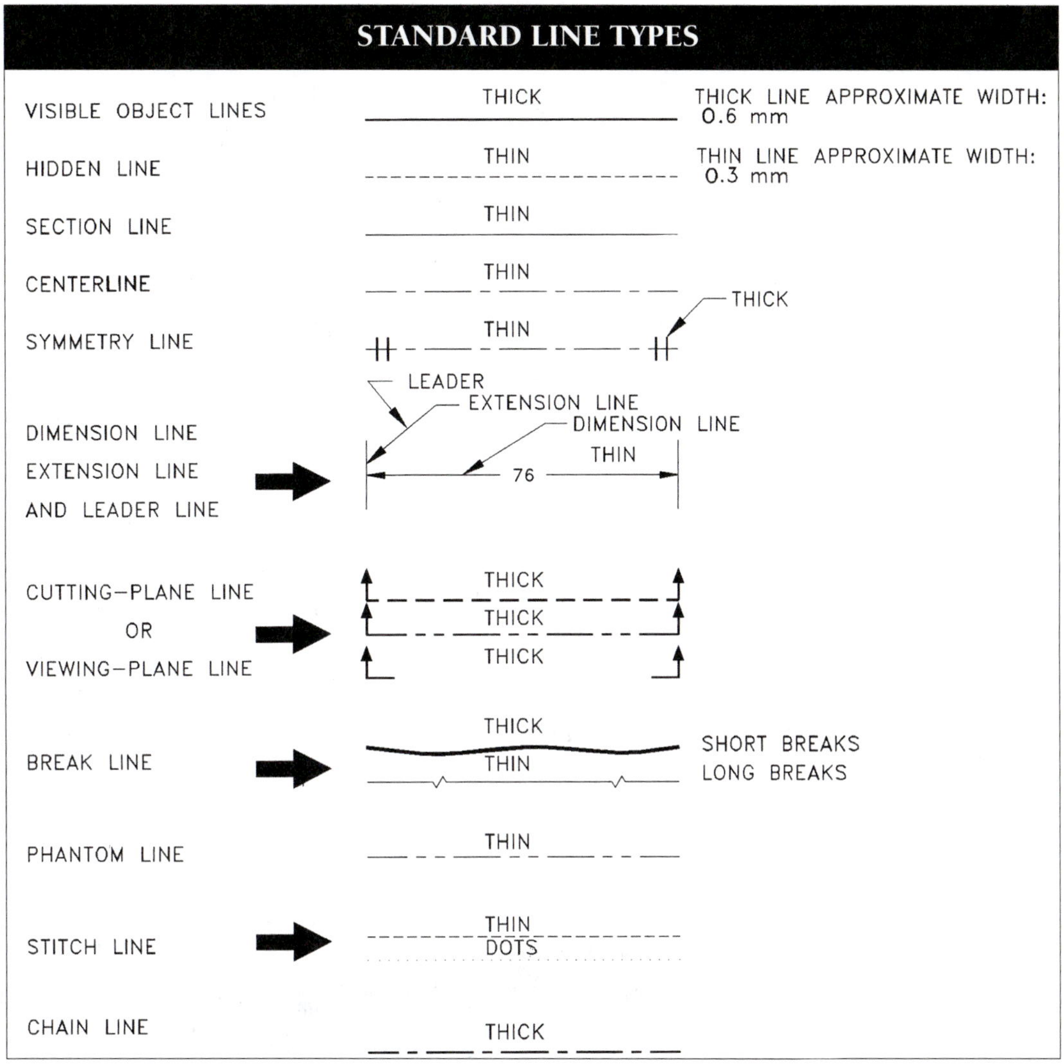

STANDARD LINE TYPES

VISIBLE OBJECT LINES	———————— THICK ————————	THICK LINE APPROXIMATE WIDTH: 0.6 mm
HIDDEN LINE	– – – – – THIN – – – – –	THIN LINE APPROXIMATE WIDTH: 0.3 mm
SECTION LINE	———— THIN ————	
CENTERLINE	— – — THIN — – —	
SYMMETRY LINE	‖ — · — THIN — · — ‖ ← THICK	

DIMENSION LINE
EXTENSION LINE
AND LEADER LINE

LEADER
EXTENSION LINE
DIMENSION LINE
THIN
|← 76 →|

CUTTING—PLANE LINE
OR
VIEWING—PLANE LINE

THICK
THICK
THICK

BREAK LINE

THICK
THIN
SHORT BREAKS
LONG BREAKS

PHANTOM LINE

— – – — THIN — – –

STITCH LINE

THIN
DOTS

CHAIN LINE

— – — THICK — –

Standard Line Types for Drafting.

Inches to Millimeters
Conversion.

INCHES TO MILLIMETERS

in.	mm	in.	mm	in.	mm	in.	mm
1	25.4	26	660.4	51	1295.4	76	1930.4
2	50.8	27	685.8	52	1320.8	77	1955.8
3	76.2	28	711.2	53	1346.2	78	1981.2
4	101.6	29	736.6	54	1371.6	79	2006.6
5	127.0	30	762.0	55	1397.0	80	2032.0
6	152.4	31	787.4	56	1422.4	81	2057.4
7	177.8	32	812.8	57	1447.8	82	2082.8
8	203.2	33	838.2	58	1473.2	83	2108.2
9	228.6	34	863.6	59	1498.6	84	2133.6
10	254.0	35	889.0	60	1524.0	85	2159.0
11	279.4	36	914.4	61	1549.4	86	2184.4
12	304.8	37	939.8	62	1574.8	87	2209.8
13	330.2	38	965.2	63	1600.2	88	2235.2
14	355.6	39	990.6	64	1625.6	89	2260.6
15	381.0	40	1016.0	65	1651.0	90	2286.0
16	406.4	41	1041.4	66	1676.4	91	2311.4
17	431.8	42	1066.8	67	1701.8	92	2336.8
18	457.2	43	1092.2	68	1727.2	93	2362.2
19	482.6	44	1117.6	69	1752.6	94	2387.6
20	508.0	45	1143.0	70	1778.0	95	2413.0
21	533.4	46	1168.4	71	1803.4	96	2438.4
22	558.8	47	1193.8	72	1828.8	97	2463.8
23	584.2	48	1219.2	73	1854.2	98	2489.2
24	609.6	49	1244.6	74	1879.6	99	2514.6
25	635.0	50	1270.0	75	1905.0	100	2540.0

The above table is exact on the basis: 1 in. = 25.4 mm

Millimeters to Inches
Conversion.

MILLIMETERS TO INCHES

in.	mm	in.	mm	in.	mm	in.	mm
1	0.039370	26	1.023622	51	2.007874	76	2.992126
2	0.078740	27	1.062992	52	2.047244	77	3.031496
3	0.118110	28	1.102362	53	2.086614	78	3.070866
4	0.157480	29	1.141732	54	2.125984	79	3.110236
5	0.196850	30	1.181102	55	2.165354	80	3.149606
6	0.236220	31	1.220472	56	2.204724	81	3.188976
7	0.275591	32	1.259843	57	2.244094	82	3.228346
8	0.314961	33	1.299213	58	2.283465	83	3.267717
9	0.354331	34	1.338583	59	2.322835	84	3.307087
10	0.393701	35	1.377953	60	2.362205	85	3.346457
11	0.433071	36	1.417323	61	2.401575	86	3.385827
12	0.472441	37	1.456693	62	2.440945	87	3.425197
13	0.511811	38	1.496063	63	2.480315	88	3.464567
14	0.551181	39	1.535433	64	2.519685	89	3.503937
15	0.590551	40	1.574803	65	2.559055	90	3.543307
16	0.629921	41	1.614173	66	2.598425	91	3.582677
17	0.669291	42	1.653543	67	2.637795	92	3.622047
18	0.708661	43	1.692913	68	2.677165	93	3.661417
19	0.748031	44	1.732283	69	2.716535	94	3.700787
20	0.787402	45	1.771654	70	2.755906	95	3.740157
21	0.826772	46	1.811024	71	2.795276	96	3.779528
22	0.866142	47	1.850394	72	2.834646	97	3.818898
23	0.905512	48	1.889764	73	2.874016	98	3.858268
24	0.944882	49	1.929134	74	2.913386	99	3.897638
25	0.984252	50	1.968504	75	2.952756	100	3.937008

The above table is approximate on the basis: 1 in. = 25.4 mm, 1/25.4 = 0.039370078740+

INCH/METRIC EQUIVALENTS

Fraction	Decimal Equivalent Customary (in.)	Metric (mm)	Fraction	Decimal Equivalent Customary (in.)	Metric (mm)
1/64	.015625	0.3969	33/64	.515625	13.0969
1/32	.03125	0.7938	17/32	.53125	13.4938
3/64	.046875	1.1906	35/64	.546875	13.8906
1/16	.0625	1.5875	9/16	.5625	14.2875
5/64	.078125	1.9844	37/64	.578125	14.6844
3/32	.09375	2.3813	19/32	.59375	15.0813
7/64	.109375	2.7781	39/64	.609375	15.4781
1/8	.1250	3.1750	5/8	.6250	15.8750
9/64	.140625	3.5719	41/64	.640625	16.2719
5/32	.15625	3.9688	21/32	.65625	16.6688
11/64	.171875	4.3656	43/64	.671875	17.0656
3/16	.1875	4.7625	11/16	.6875	17.4625
13/64	.203125	5.1594	45/64	.703125	17.8594
7/32	.21875	5.5563	23/32	.71875	18.2563
15/64	.234375	5.9531	47/64	.734375	18.6531
1/4	.250	6.3500	3/4	.750	19.0500
17/64	.265625	6.7469	49/64	.765625	19.4469
9/32	.28125	7.1438	25/32	.78125	19.8438
19/64	.296875	7.5406	51/64	.796875	20.2406
5/16	.3125	7.9375	13/16	.8125	20.6375
21/64	.328125	8.3384	53/64	.828125	21.0344
11/32	.34375	8.7313	27/32	.84375	21.4313
23/64	.359375	9.1281	55/64	.859375	21.8281
3/8	.3750	9.5250	7/8	.8750	22.2250
25/64	.390625	9.9219	57/64	.890625	22.6219
13/32	.40625	10.3188	29/32	.90625	23.0188
27/64	.421875	10.7156	59/64	.921875	23.4156
7/16	.4375	11.1125	15/16	.9375	23.8125
29/64	.453125	11.5094	61/64	.953125	24.2094
15/32	.46875	11.9063	31/32	.96875	24.6063
31/64	.484375	12.3031	63/64	.984375	25.0031
1/2	.500	12.7000	1	1.000	25.4000

Inch/Metric Equivalents.

Measures of Length

1 millimeter (mm) = 0.03937 inch
1 centimeter (cm) = 0.39370 inch
1 meter (m) = 39.37008 inches
          = 3.2808 feet
          = 1.0936 yards
1 kilometer (km) = 0.6214 mile
1 inch = 25.4 millimeters (mm)
       = 2.54 centimeters (cm)
1 foot = 304.8 millimeters (mm)
       = 0.3048 meter (m)
1 yard = 0.9144 meter (m)
1 mile = 1.609 kilometers (km)

Measures of Area

1 square millimeter = 0.00155 square inch
1 square centimeter = 0.155 square inch
1 square meter = 10.764 square feet
              = 1.196 square yards
1 square kilometer = 0.3861 square mile
1 square inch = 645.2 square millimeters
             = 6.452 square centimeters
1 square foot = 929 square centimeters
             = 0.0929 square meter
1 square yard = 0.836 square meter
1 square mile = 2.5899 square kilometers

Measures of Capacity (Dry)

1 cubic centimeter (cm^3) = 0.061 cubic inch
1 liter = 0.0353 cubic foot
       = 61.023 cubic inches
1 cubic meter (m^3) = 35.315 cubic feet
                     = 1.308 cubic yards
1 cubic inch = 16.38706 cubic centimeters (cm^3)
1 cubic foot = 0.02832 cubic meter (m^3)
            = 28.317 liters
1 cubic yard = 0.7646 cubic meter (m^3)

Measures of Capacity (Liquid)

1 liter = 1.0567 U.S. quarts
       = 0.2642 U.S. gallon
       = 0.2200 Imperial gallon
1 cubic meter (m^3) = 264.2 U.S. gallons
                     = 219.969 Imperial gallons
1 U.S. quart = 0.946 liter
1 Imperial quart = 1.136 liters
1 U.S. gallon = 3.785 liters
1 Imperial gallon = 4.546 liters

Measures of Weight

1 gram (g) = 15.432 grains
          = 0.03215 ounce troy
          = 0.03527 ounce avoirdupois
1 kilogram (kg) = 35.274 ounces avoirdupois
               = 2.2046 pounds
1000 kilograms (kg) = 1 metric ton (t)
                    = 1.1023 tons of 2000 pounds
                    = 0.9842 ton of 2240 pounds
1 ounce avoirdupois = 28.35 grams (g)
1 ounce troy = 31.103 grams (g)
1 pound = 453.6 grams
       = 0.4536 kilogram (kg)
1 ton of 2240 pounds = 1016 kilograms (kg)
                     = 1.016 metric tons
1 grain = 0.0648 gram (g)
1 metric ton = 0.9842 ton of 2240 pounds
            = 2204.6 pounds

Inch/Metric Conversion.

Grade Marking	Specification	Material
NO MARK	SAE—Grade 1	Low or Medium Carbon Steel
	ASTM—A307	Low Carbon Steel
	SAE—Grade 2	Low or Medium Carbon Steel
	SAE—Grade 5	Medium Carbon Steel, Quenched and Tempered
	ASTM—A 449	
	SAE—Grade 5.2	Low Carbon Martensite Steel, Quenched and Tempered
A 325	ASTM—A 325 Type 1	Medium Carbon Steel, Quenched and Tempered Radial dashes optional
A 325	ASTM—A 325 Type 2	Low Carbon Martensite Steel, Quenched and Tempered
A 325	ASTM—A 325 Type 3	Atmospheric Corrosion (Weathering) Steel, Quenched and Tempered
BC	ASTM—A 354 Grade BC	Alloy Steel, Quenched and Tempered
	SAE—Grade 7	Medium Carbon Alloy Steel, Quenched and Tempered, Roll Threaded After Heat Treatment
	SAE—Grade 8	Medium Carbon Alloy Steel, Quenched and Tempered
	ASTM—A 354 Grade BD	Alloy Steel, Quenched and Tempered
	SAE—Grade 8.2	Low Carbon Martensite Steel, Quenched and Tempered
A 490	ASTM—A 490 Type 1	Alloy Steel, Quenched and Tempered
A 490	ASTM—A 490 Type 3	Atmospheric Corrosion (Weathering) Steel, Quenched and Tempered

ASTM and SAE Grade Markings for Steel Bolts and Screws.

Sizes		Basic Major Diameter	Series with graded pitches			Series with constant pitches								Sizes
Primary	Secondary		Coarse UNC	Fine UNF	Extra fine UNEF	4UN	6UN	8UN	12UN	16UN	20UN	28UN	32UN	
0		0.0600	–	80	–	–	–	–	–	–	–	–	–	0
	1	0.0730	64	72	–	–	–	–	–	–	–	–	–	1
2		0.0860	56	64	–	–	–	–	–	–	–	–	–	2
	3	0.0990	48	56	–	–	–	–	–	–	–	–	–	3
4		0.1120	40	48	–	–	–	–	–	–	–	–	–	4
5		0.1250	40	44	–	–	–	–	–	–	–	–	–	5
6		0.1380	32	40	–	–	–	–	–	–	–	–	UNC	6
8		0.1640	32	36	–	–	–	–	–	–	–	–	UNC	8
10		0.1900	24	32	–	–	–	–	–	–	–	–	UNF	10
	12	0.2160	24	28	32	–	–	–	–	–	–	UNF	UNEF	12
¼		0.2500	20	28	32	–	–	–	–	–	UNC	UNF	UNEF	¼
5/16		0.3125	18	24	32	–	–	–	–	–	20	28	UNEF	5/16
3/8		0.3750	16	24	32	–	–	–	–	UNC	20	28	UNEF	3/8
7/16		0.4375	14	20	28	–	–	–	–	16	UNF	UNEF	32	7/16
½		0.5000	13	20	28	–	–	–	–	16	UNF	UNEF	32	½
9/16		0.5625	12	18	24	–	–	–	UNC	16	20	28	32	9/16
5/8		0.6250	11	18	24	–	–	–	12	16	20	28	32	5/8
	11/16	0.6875	–	–	24	–	–	–	12	16	20	28	32	11/16
¾		0.7500	10	16	20	–	–	–	12	UNF	UNEF	28	32	¾
	13/16	0.8125	–	–	20	–	–	–	12	16	UNEF	28	32	13/16
7/8		0.8750	9	14	20	–	–	–	12	16	UNEF	28	32	7/8
	15/16	0.9375	–	–	20	–	–	–	12	16	UNEF	28	32	15/16
1		1.0000	8	12	20	–	–	UNC	UNF	16	UNEF	28	32	1
	1 1/16	1.0625	–	–	18	–	–	8	12	16	20	28	–	1 1/16
1 1/8		1.1250	7	12	18	–	–	8	UNF	16	20	28	–	1 1/8
	1 3/16	1.1875	–	–	18	–	–	8	12	16	20	28	–	1 3/16
1 ¼		1.2500	7	12	18	–	–	8	UNF	16	20	28	–	1 ¼
	1 5/16	1.3125	–	–	18	–	–	8	12	16	20	28	–	1 5/16
1 3/8		1.3750	6	12	18	–	UNC	8	UNF	16	20	28	–	1 3/8
	1 7/16	1.4375	–	–	18	–	6	8	12	16	20	28	–	1 7/16
1 ½		1.5000	6	12	18	–	UNC	8	UNF	16	20	28	–	1 ½
	1 9/16	1.5625	–	–	18	–	6	8	12	16	20	–	–	1 9/16
1 5/8		1.6250	–	–	18	–	6	8	12	16	20	–	–	1 5/8
	1 11/16	1.6875	–	–	18	–	6	8	12	16	20	–	–	1 11/16
1 ¾		1.7500	5	–	–	–	6	8	12	16	20	–	–	1 ¾
	1 13/16	1.8125	–	–	–	–	6	8	12	16	20	–	–	1 13/16
1 7/8		1.8750	–	–	–	–	6	8	12	16	20	–	–	1 7/8
	1 15/16	1.9375	–	–	–	–	6	8	12	16	20	–	–	1 15/16
2		2.0000	4½	–	–	–	6	8	12	16	20	–	–	2
	2 1/8	2.1250	–	–	–	–	6	8	12	16	20	–	–	2 1/8
2 ¼		2.2500	4½	–	–	–	6	8	12	16	20	–	–	2 ¼
	2 3/8	2.3750	–	–	–	–	6	8	12	16	20	–	–	2 3/8
2 ½		2.5000	4	–	–	UNC	6	8	12	16	20	–	–	2 ½
	2 5/8	2.6250	–	–	–	4	6	8	12	16	20	–	–	2 5/8
2 ¾		2.7500	4	–	–	UNC	6	8	12	16	20	–	–	2 ¾
	2 7/8	2.8750	–	–	–	4	6	8	12	16	20	–	–	2 7/8
3		3.0000	4	–	–	UNC	6	8	12	16	20	–	–	3
	3 1/8	3.1250	–	–	–	4	6	8	12	16	–	–	–	3 1/8
3 ¼		3.2500	4	–	–	UNC	6	8	12	16	–	–	–	3 ¼
	3 3/8	3.3750	–	–	–	4	6	8	12	16	–	–	–	3 3/8
3 ½		3.5000	4	–	–	UNC	6	8	12	16	–	–	–	3 ½
	3 5/8	3.6250	–	–	–	4	6	8	12	16	–	–	–	3 5/8
3 ¾		3.7500	4	–	–	UNC	6	8	12	16	–	–	–	3 ¾
	3 7/8	3.8750	–	–	–	4	6	8	12	16	–	–	–	3 7/8
4		4.0000	4	–	–	UNC	6	8	12	16	–	–	–	4
	4 1/8	4.1250	–	–	–	4	6	8	12	16	–	–	–	4 1/8
4 ¼		4.2500	–	–	–	4	6	8	12	16	–	–	–	4 ¼
	4 3/8	4.3750	–	–	–	4	6	8	12	16	–	–	–	4 3/8
4 ½		4.5000	–	–	–	4	6	8	12	16	–	–	–	4 ½
	4 5/8	4.6250	–	–	–	4	6	8	12	16	–	–	–	4 5/8
4 ¾		4.7500	–	–	–	4	6	8	12	16	–	–	–	4 ¾
	4 7/8	4.8750	–	–	–	4	6	8	12	16	–	–	–	4 7/8
5		5.0000	–	–	–	4	6	8	12	16	–	–	–	5
	5 1/8	5.1250	–	–	–	4	6	8	12	16	–	–	–	5 1/8
5 ¼		5.2500	–	–	–	4	6	8	12	16	–	–	–	5 ¼
	5 3/8	5.3750	–	–	–	4	6	8	12	16	–	–	–	5 3/8
5 ½		5.5000	–	–	–	4	6	8	12	16	–	–	–	5 ½
	5 5/8	5.6250	–	–	–	4	6	8	12	16	–	–	–	5 5/8
5 ¾		5.7500	–	–	–	4	6	8	12	16	–	–	–	5 ¾
	5 7/8	5.8750	–	–	–	4	6	8	12	16	–	–	–	5 7/8
6		6.0000	–	–	–	4	6	8	12	16	–	–	–	6

Unified Standard Screw Thread Series.

ISO BASIC METRIC THREAD INFORMATION

Basic Major DIA & Pitch	INTERNAL THREADS			EXTERNAL THREADS		
	Tap Drill DIA	Minor DIA MAX	Minor DIA MIN	Major DIA MAX	Major DIA MIN	Clearance Hole
M1.6 × 0.35	1.25	1.321	1.221	1.576	1.491	1.9
M2 × 0.4	1.60	1.679	1.567	1.976	1.881	2.4
M2.5 × 0.45	2.05	2.138	2.013	2.476	2.013	2.9
M3 × 0.5	2.50	2.599	2.459	2.976	2.870	3.4
M3.5 × 0.6	2.90	3.010	2.850	3.476	3.351	4.0
M4 × 0.7	3.30	3.422	3.242	3.976	3.836	4.5
M5 × 0.8	4.20	4.334	4.134	4.976	4.826	5.5
M6 × 1	5.00	5.153	4.917	5.974	5.794	6.6
M8 × 1.25	6.80	6.912	6.647	7.972	7.760	9.0
M10 × 1.5	8.50	8.676	8.376	9.968	9.732	11.0
M12 × 1.75	10.20	10.441	10.106	11.966	11.701	13.5
M14 × 2	12.00	12.210	11.835	13.962	13.682	15.5
M16 × 2	14.00	14.210	13.835	15.962	15.682	17.5
M20 × 2.5	17.50	17.744	17.294	19.958	19.623	22.0
M24 × 3	21.00	21.252	20.752	23.952	23.577	26.0
M30 × 3.5	26.50	26.771	26.211	29.947	29.522	33.0
M36 × 4	32.00	32.270	31.670	35.940	35.465	39.0
M42 × 4.5	37.50	37.799	37.129	41.937	41.437	45.0
M48 × 5	43.00	43.297	42.587	47.929	47.399	52.0
M56 × 5.5	50.50	50.796	50.046	55.925	55.365	62.0
M64 × 6	58.00	58.305	57.505	63.920	63.320	70.0
M72 × 6	66.00	66.305	65.505	71.920	71.320	78.0
M80 × 6	74.00	74.305	73.505	79.920	79.320	86.0
M90 × 6	84.00	84.305	83.505	89.920	89.320	96.0
M100 × 6	94.00	94.305	93.505	99.920	99.320	107.0

ISO Metric Thread Information.

INCH—METRIC THREAD COMPARISON

INCH SERIES			METRIC			
Size	Dia.(In.)	TPI	Size	Dia. (In.)	Pitch (mm)	TPI (Approx)
			M1.4	.055	.3 .2	85 127
#0	.060	80				
			M1.6	.063	.35 .2	74 127
#1	.073	64 72				
			M2	.079	.4 .25	64 101
#2	.086	56 64				
			M2.5	.098	.45 .35	56 74
#3	.099	48 56				
#4	.112	40 48				
			M3	.118	.5 .35	51 74
#5	.125	40 44				
#6	.138	32 40				
			M4	.157	.7 .5	36 51
#8	.164	32 36				
#10	.190	24 32				
			M5	.196	.8 .5	32 51
			M6	.236	1.0 .75	25 34
¼	.250	20 28				
⁵⁄₁₆	.312	18 24				
			M8	.315	1.25 1.0	20 25
⅜	.375	16 24				
			M10	.393	1.5 1.25	17 20
⁷⁄₁₆	.437	14 20				
			M12	.472	1.75 1.25	14.5 20
½	.500	13 20				
			M14	.551	2 1.5	12.5 17
⅝	.625	11 18				
			M16	.630	2 1.5	12.5 17
			M18	.709	2.5 1.5	10 17
¾	.750	10 16				
			M20	.787	2.5 1.5	10 17
			M22	.866	2.5 1.5	10 17
⅞	.875	9 14				
			M24	.945	3 2	8.5 12.5
1"	1.000	8 12				
			M27	1.063	3 2	8.5 12.5

Inch-Metric Thread Comparison.

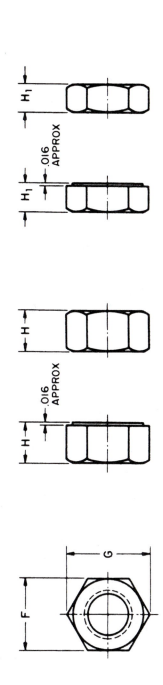

Nominal Size or Basic Major Dia. of Thread		F Width Across Flats			G Width Across Corners		H Thickness Hex Nuts			H₁ Thickness Hex Jam Nuts			Hex Nuts Specified Proof Load Runout of Bearing Face, FIR Max		Jam Nuts All Strength Levels
		Basic	Max	Min	Max	Min	Basic	Max	Min	Basic	Max	Min	Up to 150,000 psi	150,000 psi and Greater	
1/4	0.2500	7/16	0.438	0.428	0.505	0.488	7/32	0.226	0.212	5/32	0.163	0.150	0.015	0.010	0.015
5/16	0.3125	1/2	0.500	0.489	0.577	0.557	17/64	0.273	0.258	3/16	0.195	0.180	0.016	0.011	0.016
3/8	0.3750	9/16	0.562	0.551	0.650	0.628	21/64	0.337	0.320	7/32	0.227	0.210	0.017	0.012	0.017
7/16	0.4375	11/16	0.688	0.675	0.794	0.768	3/8	0.385	0.365	1/4	0.260	0.240	0.018	0.013	0.018
1/2	0.5000	3/4	0.750	0.736	0.866	0.840	7/16	0.448	0.427	5/16	0.323	0.302	0.019	0.014	0.019
9/16	0.5625	7/8	0.875	0.861	1.010	0.982	31/64	0.496	0.473	5/16	0.324	0.301	0.020	0.015	0.020
5/8	0.6250	15/16	0.938	0.922	1.083	1.051	35/64	0.559	0.535	3/8	0.387	0.363	0.021	0.016	0.021
3/4	0.7500	1 1/8	1.125	1.088	1.299	1.240	41/64	0.665	0.617	27/64	0.446	0.398	0.023	0.018	0.023
7/8	0.8750	1 5/16	1.312	1.269	1.516	1.447	3/4	0.776	0.724	31/64	0.510	0.458	0.025	0.020	0.025
1	1.0000	1 1/2	1.500	1.450	1.732	1.653	55/64	0.887	0.831	35/64	0.575	0.519	0.027	0.022	0.027
1 1/8	1.1250	1 11/16	1.688	1.631	1.949	1.859	31/32	0.999	0.939	39/64	0.639	0.579	0.030	0.025	0.030
1 1/4	1.2500	1 7/8	1.875	1.812	2.165	2.066	1 1/16	1.094	1.030	23/32	0.751	0.687	0.033	0.028	0.033
1 3/8	1.3750	2 1/16	2.062	1.994	2.382	2.273	1 11/64	1.206	1.138	25/32	0.815	0.747	0.036	0.031	0.036
1 1/2	1.5000	2 1/4	2.250	2.175	2.598	2.480	1 9/32	1.317	1.245	27/32	0.880	0.808	0.039	0.034	0.039
See Notes	6		3		4								2		

Reprinted from ASME B18.2.1-1981 (R1992), B18.2.2-1987 (R1993), B18.3-1986 (R1993), B18.6.2-1972 (R1993), B18.6.3-1972 (R1991), B18.21.1-1994, B18.22.1-1965 (R1990), B17.1-1967 (R1998), B17.2-1967 (R1990) and B42-1978 (R1994), by permission of The American Society of Mechanical Engineers. All rights reserved.

Inch-Metric Thread Comparison. (continued)

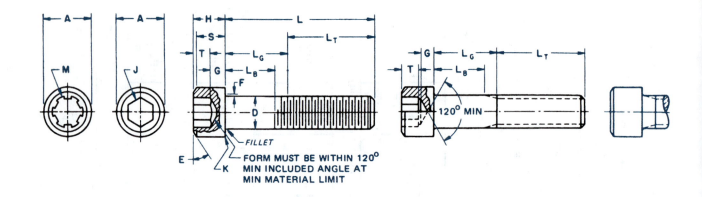

Nominal Size or Basic Screw Diameter		D Body Diameter		A Head Diameter		H Head Height		S Head Side Height	M Spline Socket Size	J Hexagon Socket Size		T Key Engagement	G Wall Thick-ness	K Chamfer or Radius
		Max	Min	Max	Min	Max	Min	Min	Nom	Nom		Min	Min	Max
0	0.0600	0.0600	0.0568	0.096	0.091	0.060	0.057	0.054	0.060		0.050	0.025	0.020	0.003
1	0.0730	0.0730	0.0695	0.118	0.112	0.073	0.070	0.066	0.072	1/16	0.062	0.031	0.025	0.003
2	0.0860	0.0860	0.0822	0.140	0.134	0.086	0.083	0.077	0.096	5/64	0.078	0.038	0.029	0.003
3	0.0990	0.0990	0.0949	0.161	0.154	0.099	0.095	0.089	0.096	5/64	0.078	0.044	0.034	0.003
4	0.1120	0.1120	0.1075	0.183	0.176	0.112	0.108	0.101	0.111	3/32	0.094	0.051	0.038	0.005
5	0.1250	0.1250	0.1202	0.205	0.198	0.125	0.121	0.112	0.111	3/32	0.094	0.057	0.043	0.005
6	0.1380	0.1380	0.1329	0.226	0.218	0.138	0.134	0.124	0.133	7/64	0.109	0.064	0.047	0.005
8	0.1640	0.1640	0.1585	0.270	0.262	0.164	0.159	0.148	0.168	9/64	0.141	0.077	0.056	0.005
10	0.1900	0.1900	0.1840	0.312	0.303	0.190	0.185	0.171	0.183	5/52	0.156	0.090	0.065	0.005
1/4	0.2500	0.2500	0.2435	0.375	0.365	0.250	0.244	0.225	0.216	3/16	0.188	0.120	0.095	0.008
5/16	0.3125	0.3125	0.3053	0.469	0.457	0.312	0.306	0.281	0.291	1/4	0.250	0.151	0.119	0.008
3/8	0.3750	0.3750	0.3678	0.562	0.550	0.375	0.368	0.337	0.372	5/16	0.312	0.182	0.143	0.008
7/16	0.4375	0.4375	0.4294	0.656	0.642	0.438	0.430	0.394	0.454	3/8	0.375	0.213	0.166	0.010
1/2	0.5000	0.5000	0.4919	0.750	0.735	0.500	0.492	0.450	0.454	3/8	0.375	0.245	0.190	0.010
5/8	0.6250	0.6250	0.6163	0.938	0.921	0.625	0.616	0.562	0.595	1/2	0.500	0.307	0.238	0.010
3/4	0.7500	0.7500	0.7406	1.125	1.107	0.750	0.740	0.675	0.620	5/8	0.625	0.370	0.285	0.010
7/8	0.8750	0.8750	0.8647	1.312	1.293	0.875	0.864	0.787	0.698	3/4	0.750	0.432	0.333	0.015
1	1.0000	1.0000	0.9886	1.500	1.479	1.000	0.988	0.900	0.790	3/4	0.750	0.495	0.380	0.015
1 1/8	1.1250	1.1250	1.1086	1.688	1.665	1.125	1.111	1.012	. . . .	7/8	0.875	0.557	0.428	0.015
1 1/4	1.2500	1.2500	1.2336	1.875	1.852	1.250	1.236	1.125	. . . .	7/8	0.875	0.620	0.475	0.015
1 3/8	1.3750	1.3750	1.3568	2.062	2.038	1.375	1.360	1.237	. . . .	1	1.000	0.682	0.523	0.015
1 1/2	1.5000	1.5000	1.4818	2.250	2.224	1.500	1.485	1.350	. . . .	1	1.000	0.745	0.570	0.015
1 3/4	1.7500	1.7500	1.7295	2.625	2.597	1.750	1.734	1.575	. . . .	1 1/4	1.250	0.870	0.665	0.015
2	2.0000	2.0000	1.9780	3.000	2.970	2.000	1.983	1.800	. . . .	1 1/2	1.500	0.995	0.760	0.015
2 1/4	2.2500	2.2500	2.2280	3.375	3.344	2.250	2.232	2.025	. . . .	1 3/4	1.750	1.120	0.855	0.031
2 1/2	2.5000	2.5000	2.4762	3.750	3.717	2.500	2.481	2.250	. . . .	1 3/4	1.750	1.245	0.950	0.031
2 3/4	2.7500	2.7500	2.7262	4.125	4.090	2.750	2.730	2.475	. . . .	2	2.000	1.370	1.045	0.031
3	3.0000	3.0000	2.9762	4.500	4.464	3.000	2.979	2.700	. . . .	2 1/4	2.250	1.495	1.140	0.031
3 1/4	3.2500	3.2500	3.2262	4.875	4.837	3.250	3.228	2.925	. . . .	2 1/4	2.250	1.620	1.235	0.031
3 1/2	3.5000	3.5000	3.4762	5.250	5.211	3.500	3.478	3.150	. . . .	2 3/4	2.750	1.745	1.330	0.031
3 3/4	3.7500	3.7500	3.7262	5.625	5.584	3.750	3.727	3.375	. . . .	2 3/4	2.750	1.870	1.425	0.031
4	4.0000	4.0000	3.9762	6.000	5.958	4.000	3.976	3.600	. . . .	3	3.000	1.995	1.520	0.031

Cap Screw Specifications.

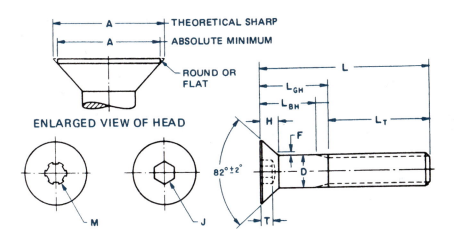

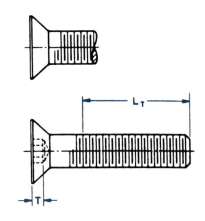

Nominal Size or Basic Screw Diameter		D Body Diameter		A Head Diameter		H Heat Height		M Spline Socket Size	J Hexagon Socket Size		T Key Engagement	F Fillet Extension Above D Max
		Max	Min	Theoretical Sharp Max	Abs. Min	Reference	Flushness Tolerance		Nom		Min	Max
0	0.0600	0.0600	0.0568	0.138	0.117	0.044	0.006	0.048		0.035	0.025	0.006
1	0.0730	0.0730	0.0695	0.168	0.143	0.054	0.007	0.060		0.050	0.031	0.008
2	0.0860	0.0860	0.0822	0.197	0.168	0.064	0.008	0.060		0.050	0.038	0.010
3	0.0990	0.0990	0.0949	0.226	0.193	0.073	0.010	0.072	1/16	0.062	0.044	0.010
4	0.1120	0.1120	0.1075	0.255	0.218	0.083	0.011	0.072	1/16	0.062	0.055	0.012
5	0.1250	0.1250	0.1202	0.281	0.240	0.090	0.012	0.096	5/64	0.078	0.061	0.014
6	0.1380	0.1380	0.1329	0.307	0.263	0.097	0.013	0.096	5/64	0.078	0.066	0.015
8	0.1640	0.1640	0.1585	0.359	0.311	0.112	0.014	0.111	3/32	0.094	0.076	0.015
10	0.1900	0.1900	0.1840	0.411	0.359	0.127	0.015	0.145	1/8	0.125	0.087	0.015
1/4	0.2500	0.2500	0.2435	0.531	0.480	0.161	0.016	0.183	5/32	0.156	0.111	0.015
5/16	0.3125	0.3125	0.3053	0.656	0.600	0.198	0.017	0.216	3/16	0.188	0.135	0.015
3/8	0.3750	0.3750	0.3678	0.781	0.720	0.234	0.018	0.251	7/32	0.219	0.159	0.015
7/16	0.4375	0.4375	0.4294	0.844	0.781	0.234	0.018	0.291	1/4	0.250	0.159	0.015
1/2	0.5000	0.5000	0.4919	0.938	0.872	0.251	0.018	0.372	5/16	0.312	0.172	0.015
5/8	0.6250	0.6250	0.6163	1.188	1.112	0.324	0.022	0.454	3/8	0.375	0.220	0.015
3/4	0.7500	0.7500	0.7406	1.438	1.355	0.396	0.024	0.454	1/2	0.500	0.220	0.015
7/8	0.8750	0.8750	0.8647	1.688	1.604	0.468	0.025	. . .	9/16	0.562	0.248	0.015
1	1.0000	1.0000	0.9886	1.938	1.841	0.540	0.028	. . .	5/8	0.625	0.297	0.015
1 1/8	1.1250	1.1250	1.1086	2.188	2.079	0.611	0.031	. . .	3/4	0.750	0.325	0.031
1 1/4	1.2500	1.2500	1.2336	2.438	2.316	0.683	0.035	. . .	7/8	0.875	0.358	0.031
1 3/8	1.3750	1.3750	1.3568	2.688	2.553	0.755	0.038	. . .	7/8	0.875	0.402	0.031
1 1/2	1.5000	1.5000	1.4818	2.938	2.791	0.827	0.042	. . .	1	1.000	0.435	0.031

Cap Screw Specifications. (CONTINUED)

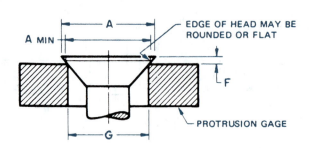

CAP SCREWS
FLAT

Type of Head

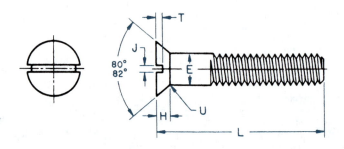

Nominal Size[1] or Basic Screw Diameter		E Body Diameter		A Head Diameter		H[2] Head Height	J Slot Width		T Slot Depth		U Fillet Radius	F[3] Protrusion Above Gaging Diameter		G[3] Gaging Diameter
				Max, Edge Sharp	Min, Edge Rounded or Flat									
		Max	Min			Ref	Max	Min	Max	Min	Max	Max	Min	
1/4	0.2500	0.2500	0.2450	0.500	0.452	0.140	0.075	0.064	0.068	0.045	0.100	0.046	0.030	0.424
5/16	0.3125	0.3125	0.3070	0.625	0.567	0.177	0.084	0.072	0.086	0.057	0.125	0.053	0.035	0.538
3/8	0.3750	0.3750	0.3690	0.750	0.682	0.210	0.094	0.081	0.103	0.068	0.150	0.060	0.040	0.651
7/16	0.4375	0.4375	0.4310	0.812	0.736	0.210	0.094	0.081	0.103	0.068	0.175	0.065	0.044	0.703
1/2	0.5000	0.5000	0.4930	0.875	0.791	0.210	0.106	0.091	0.103	0.068	0.200	0.071	0.049	0.756
9/16	0.5625	0.5625	0.5550	1.000	0.906	0.244	0.118	0.102	0.120	0.080	0.225	0.078	0.054	0.869
5/8	0.6250	0.6250	0.6170	1.125	1.020	0.281	0.133	0.116	0.137	0.091	0.250	0.085	0.058	0.982
3/4	0.7500	0.7500	0.7420	1.375	1.251	0.352	0.149	0.131	0.171	0.115	0.300	0.099	0.068	1.208
7/8	0.8750	0.8750	0.8660	1.625	1.480	0.423	0.167	0.147	0.206	0.138	0.350	0.113	0.077	1.435
1	1.0000	1.0000	0.9900	1.875	1.711	0.494	0.188	0.166	0.240	0.162	0.400	0.127	0.087	1.661
1 1/8	1.1250	1.1250	1.1140	2.062	1.880	0.529	0.196	0.178	0.257	0.173	0.450	0.141	0.096	1.826
1 1/4	1.2500	1.2500	1.2390	2.312	2.110	0.600	0.211	0.193	0.291	0.197	0.500	0.155	0.105	2.052
1 3/8	1.3750	1.3750	1.3630	2.562	2.340	0.665	0.226	0.208	0.326	0.220	0.550	0.169	0.115	2.279
1 1/2	1.5000	1.5000	1.4880	2.812	2.570	0.742	0.258	0.240	0.360	0.244	0.600	0.183	0.124	2.505

[1] Where specifying nominal size in decimals, zeros preceding decimal and in the fourth decimal place shall be omitted.
[2] Tabulated values determined from formula for maximum H, Appendix III.
[3] No tolerance for gaging diameter is given. If the gaging diameter of the gage used differs from tabulated value, the protrusion will be affected accordingly and the proper protrusion values must be recalculated using the formulas shown in Appendix II.

FOOTNOTES REFER TO ANSI B18.6.2–1972 (R1993).

Cap Screw Specifications. (CONTINUED)

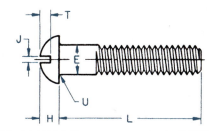

CAP SCREWS
ROUND

Type of Head

Nominal Size[1] or Basic Screw Diameter		E Body Diameter		A Head Diameter		H Head Height		J Slot Width		T Slot Depth		U Fillet Radius	
		Max	Min	Max	Min	Max	Min	Max	Min	Max	Min	Max	Min
1/4	0.2500	0.2500	0.2450	0.437	0.418	0.191	0.175	0.075	0.064	0.117	0.097	0.031	0.016
5/16	0.3125	0.3125	0.3070	0.562	0.540	0.245	0.226	0.084	0.072	0.151	0.126	0.031	0.016
3/8	0.3750	0.3750	0.3690	0.625	0.603	0.273	0.252	0.094	0.081	0.168	0.138	0.031	0.016
7/16	0.4375	0.4375	0.4310	0.750	0.725	0.328	0.302	0.094	0.081	0.202	0.167	0.047	0.016
1/2	0.5000	0.5000	0.4930	0.812	0.786	0.354	0.327	0.106	0.091	0.218	0.178	0.047	0.016
9/16	0.5625	0.5625	0.5550	0.937	0.909	0.409	0.378	0.118	0.102	0.252	0.207	0.047	0.016
5/8	0.6250	0.6250	0.6170	1.000	0.970	0.437	0.405	0.133	0.116	0.270	0.220	0.062	0.031
3/4	0.7500	0.7500	0.7420	1.250	1.215	0.546	0.507	0.149	0.131	0.338	0.278	0.062	0.031

[1] Where specifying nominal size in decimals, zeros preceding decimal and in the fourth decimal place shall be omitted.

Reprinted from ASME B18.2.1-1981 (R1992), B18.2.2-1987 (R1993), B18.3-1986 (R1993), B18.6.2-1972 (R1993), B18.6.3-1972 (R1991), B18.21.1-1994, B18.22.1-1965 (R1990), B17.1-1967 (R1998), B17.2-1967 (R1990) and B4.2-1978 (R1994), by permission of The American Society of Mechanical Engineers. All rights reserved.

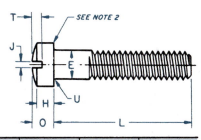

SEE NOTE 2

CAP SCREWS
FILLISTER

Type of Head

Nominal Size[1] or Basic Screw Diameter		E Body Diameter		A Head Diameter		H Head Side Height		O Total Head Height		J Slot Width		T Slot Depth		U Fillet Radius	
		Max	Min	Max	Min	Max	Min	Max	Min	Max	Min	Max	Min	Max	Min
1/4	0.2500	0.2500	0.2450	0.375	0.363	0.172	0.157	0.216	0.194	0.075	0.064	0.097	0.077	0.031	0.016
5/16	0.3125	0.3125	0.3070	0.437	0.424	0.203	0.186	0.253	0.230	0.084	0.072	0.115	0.090	0.031	0.016
3/8	0.3750	0.3750	0.3690	0.562	0.547	0.250	0.229	0.314	0.284	0.094	0.081	0.142	0.112	0.031	0.016
7/16	0.4375	0.4375	0.4310	0.625	0.608	0.297	0.274	0.368	0.336	0.094	0.081	0.168	0.133	0.047	0.016
1/2	0.5000	0.5000	0.4930	0.750	0.731	0.328	0.301	0.413	0.376	0.106	0.091	0.193	0.153	0.047	0.016
9/16	0.5625	0.5625	0.5550	0.812	0.792	0.375	0.346	0.467	0.427	0.118	0.102	0.213	0.168	0.047	0.016
5/8	0.6250	0.6250	0.6170	0.875	0.853	0.422	0.391	0.521	0.478	0.133	0.116	0.239	0.189	0.062	0.031
3/4	0.7500	0.7500	0.7420	1.000	0.976	0.500	0.466	0.612	0.566	0.149	0.131	0.283	0.223	0.062	0.031
7/8	0.8750	0.8750	0.8660	1.125	1.098	0.594	0.556	0.720	0.668	0.167	0.147	0.334	0.264	0.062	0.031
1	1.0000	1.0000	0.9900	1.312	1.282	0.656	0.612	0.803	0.743	0.188	0.166	0.371	0.291	0.062	0.031

[1] Where specifying nominal size in decimals, zeros preceding decimal and in the fourth decimal place shall be omitted.
[2] A slight rounding of the edges at periphery of head shall be permissible provided the diameter of the bearing circle is equal to no less than 90 percent of the specified minimum head diameter.

Reprinted from ASME B18.2.1-1981 (R1992), B18.2.2-1987 (R1993), B18.3-1986 (R1993), B18.6.2-1972 (R1993), B18.6.3-1972 (R1991), B18.21.1-1994, B18.22.1-1965 (R1990), B17.1-1967 (R1998), B17.2-1967 (R1990) and B4.2-1978 (R1994), by permission of The American Society of Mechanical Engineers. All rights reserved.

Cap Screw Specifications. (CONTINUED)

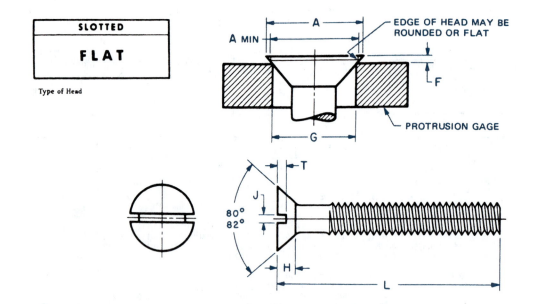

SLOTTED

FLAT

Type of Head

Nominal Size[1] or Basic Screw Diameter		L[2] These Lengths or Shorter are Undercut	A Head Diameter		H[3] Head Height	J Slot Width		T Slot Depth		F[4] Protrusion Above Gaging Diameter		G[4] Gaging Diameter
			Max. Edge Sharp	Min, Edge Rounded or Flat	Ref	Max	Min	Max	Min	Max	Min	
0000	0.0210	—	0.043	0.037	0.011	0.008	0.004	0.007	0.003	*	*	*
000	0.0340	—	0.064	0.058	0.016	0.011	0.007	0.009	0.005	*	*	*
00	0.0470	—	0.093	0.085	0.028	0.017	0.010	0.014	0.009	*	*	*
0	0.0600	1/8	0.119	0.099	0.035	0.023	0.016	0.015	0.010	0.026	0.016	0.078
1	0.0730	1/8	0.146	0.123	0.043	0.026	0.019	0.019	0.012	0.028	0.016	0.101
2	0.0860	1/8	0.172	0.147	0.051	0.031	0.023	0.023	0.015	0.029	0.017	0.124
3	0.0990	1/8	0.199	0.171	0.059	0.035	0.027	0.027	0.017	0.031	0.018	0.148
4	0.1120	3/16	0.225	0.195	0.067	0.039	0.031	0.030	0.020	0.032	0.019	0.172
5	0.1250	3/16	0.252	0.220	0.075	0.043	0.035	0.034	0.022	0.034	0.020	0.196
6	0.1380	3/16	0.279	0.244	0.083	0.048	0.039	0.038	0.024	0.036	0.021	0.220
8	0.1640	1/4	0.332	0.292	0.100	0.054	0.045	0.045	0.029	0.039	0.023	0.267
10	0.1900	5/16	0.385	0.340	0.116	0.060	0.050	0.053	0.034	0.042	0.025	0.313
12	0.2160	3/8	0.438	0.389	0.132	0.067	0.056	0.060	0.039	0.045	0.027	0.362
1/4	0.2500	7/16	0.507	0.452	0.153	0.075	0.064	0.070	0.046	0.050	0.029	0.424
5/16	0.3125	1/2	0.635	0.568	0.191	0.084	0.072	0.088	0.058	0.057	0.034	0.539
3/8	0.3750	9/16	0.762	0.685	0.230	0.094	0.081	0.106	0.070	0.065	0.039	0.653
7/16	0.4375	5/8	0.812	0.723	0.223	0.094	0.081	0.103	0.066	0.073	0.044	0.690
1/2	0.5000	3/4	0.875	0.775	0.223	0.106	0.091	0.103	0.065	0.081	0.049	0.739
9/16	0.5625	—	1.000	0.889	0.260	0.118	0.102	0.120	0.077	0.089	0.053	0.851
5/8	0.6250	—	1.125	1.002	0.298	0.133	0.116	0.137	0.088	0.097	0.058	0.962
3/4	0.7500	—	1.375	1.230	0.372	0.149	0.131	0.171	0.111	0.112	0.067	1.186

[1] Where specifying nominal size in decimals, zeros preceding decimal and in the fourth decimal place shall be omitted.
[2] Screws of these lengths and shorter shall have undercut heads as shown in Table 5.
[3] Tabulated values determined from formula for maximum H, Appendix V.
[4] No tolerance for gaging diameter is given. If the gaging diameter of the gage used differs from tabulated value, the protrusion will be affected accordingly and the proper protrusion values must be recalculated using the formulas shown in Appendix I.
* Not practical to gage.
For additional requirements refer to General Data on Pages 3, 4 and 5.
FOOTNOTES REFER TO ANSI B18.6.3-1972 (R1991).

Machine Screw Specifications.

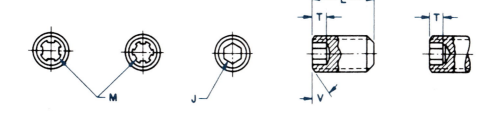

Nominal Size or Basic Screw Diameter		P	Q			B Shortest Optimum Nominal Length To Which Column T_H Applies			B_1 Shortest Optimum Nominal Length To Which Column T_S Applies		
		Half Dog Point									
		Diameter		Length		Cup and Flat	90° Cone and	Half Dog	Cup and Flat	90° Cone and	Half Dog
		Max	Min	Max	Min	Points	Oval Points	Points	Points	Oval Points	Point
0	0.0600	0.040	0.037	0.017	0.013	7/64	1/8	7/64	1/16	1/8	7/64
1	0.0730	0.049	0.045	0.021	0.017	1/8	9/64	1/8	3/32	9/64	1/8
2	0.0860	0.057	0.053	0.024	0.020	1/8	9/64	9/64	3/32	9/64	9/64
3	0.0990	0.066	0.062	0.027	0.023	9/64	5/32	5/32	3/32	5/32	5/32
4	0.1120	0.075	0.070	0.030	0.026	9/64	11/64	5/32	3/32	11/64	5/32
5	0.1250	0.083	0.078	0.033	0.027	3/16	3/16	11/64	1/8	3/16	11/64
6	0.1380	0.092	0.087	0.038	0.032	11/64	13/64	3/16	1/8	13/64	3/16
8	0.1640	0.109	0.103	0.043	0.037	3/16	7/32	13/64	3/16	7/32	13/64
10	0.1900	0.127	0.120	0.049	0.041	3/16	1/4	15/64	3/16	1/4	15/64
1/4	0.2500	0.156	0.149	0.067	0.059	1/4	5/16	19/64	1/4	5/16	19/64
5/16	0.3125	0.203	0.195	0.082	0.074	5/16	25/64	23/64	5/16	25/64	23/64
3/8	0.3750	0.250	0.241	0.099	0.089	3/8	7/16	7/16	3/8	7/16	7/16
7/16	0.4375	0.297	0.287	0.114	0.104	7/16	35/64	31/64	7/16	35/64	31/64
1/2	0.5000	0.344	0.334	0.130	0.120	1/2	39/64	35/64	1/2	39/64	35/64
5/8	0.6250	0.469	0.456	0.164	0.148	5/8	49/64	43/64	5/8	49/64	43/64
3/4	0.7500	0.562	0.549	0.196	0.180	3/4	29/32	51/64	3/4	29/32	51/64
7/8	0.8750	0.656	0.642	0.227	0.211	7/8	1 1/8	63/64	7/8	1 1/8	63/64
1	1.0000	0.750	0.734	0.260	0.240	1	1 17/64	1 1/8	. . .	. . .	. . .
1 1/8	1.1250	0.844	0.826	0.291	0.271	1 1/8	1 25/64	1 3/16	. . .	. . .	. . .
1 1/4	1.2500	0.938	0.920	0.323	0.303	1 1/4	1 1/2	1 5/16	. . .	. . .	. . .
1 3/8	1.3750	1.031	1.011	0.354	0.334	1 3/8	1 21/64	1 7/16	. . .	. . .	. . .
1 1/2	1.5000	1.125	1.105	0.385	0.365	1 1/2	1 51/64	1 9/16	. . .	. . .	. . .
1 3/4	1.7500	1.312	1.289	0.448	0.428	1 3/4	2 7/32	1 61/64	. . .	. . .	. . .
2	2.0000	1.500	1.474	0.510	0.490	2	2 25/64	2 5/64	. . .	. . .	. . .

Reprinted from ASME B18.2.1-1981 (R1992), B18.2.2-1987 (R1993), B18.3-1986
(R1993), B18.6.2-1972 (R1993), B18.6.3-1972 (R1991), B18.21.1-1994, B18.22.1-1965
(R1990), B17.1-1967 (R1998), B17.2-1967 (R1990) and B4.2-1978 (R1994), by
permission of The American Society of Mechanical Engineers. All rights reserved.

Set Screw Specifications.

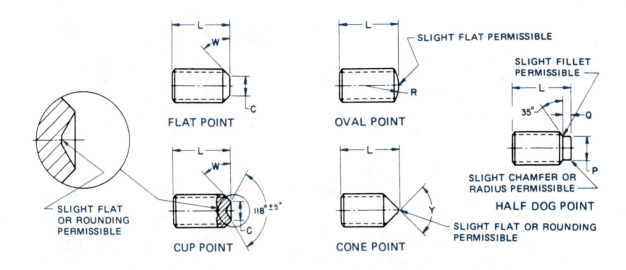

FLAT POINT

OVAL POINT

SLIGHT FLAT PERMISSIBLE

SLIGHT FILLET PERMISSIBLE

35°

SLIGHT CHAMFER OR RADIUS PERMISSIBLE

HALF DOG POINT

SLIGHT FLAT OR ROUNDING PERMISSIBLE

CUP POINT

118° ±5°

CONE POINT

SLIGHT FLAT OR ROUNDING PERMISSIBLE

Y

Nominal Size or Basic Screw Diameter		P		Q		B Shortest Optimum Nominal Length To Which Column T_H Applies			B_1 Shortest Optimum Nominal Length To Which Column T_S Applies		
		Half Dog Point									
		Diameter		Length		Cup and Flat Points	90° Cone and Oval Points	Half Dog Points	Cup and Flat Points	90° Cone and Oval Points	Half Dog Point
		Max	Min	Max	Min						
0	0.0600	0.040	0.037	0.017	0.013	7/64	1/8	7/64	1/16	1/8	7/64
1	0.0730	0.049	0.045	0.021	0.017	1/8	9/64	1/8	3/32	9/64	1/8
2	0.0860	0.057	0.053	0.024	0.020	1/8	9/64	9/64	3/32	9/64	9/64
3	0.0990	0.066	0.062	0.027	0.023	9/64	5/32	5/32	3/32	5/32	5/32
4	0.1120	0.075	0.070	0.030	0.026	9/64	11/64	5/32	3/32	11/64	5/32
5	0.1250	0.083	0.078	0.033	0.027	3/16	3/16	11/64	1/8	3/16	11/64
6	0.1380	0.092	0.087	0.038	0.032	11/64	13/64	3/16	1/8	13/64	3/16
8	0.1640	0.109	0.103	0.043	0.037	3/16	7/32	13/64	3/16	7/32	13/64
10	0.1900	0.127	0.120	0.049	0.041	3/16	1/4	15/64	3/16	1/4	15/64
1/4	0.2500	0.156	0.149	0.067	0.059	1/4	5/16	19/64	1/4	5/16	19/64
5/16	0.3125	0.203	0.195	0.082	0.074	5/16	25/64	23/64	5/16	25/64	23/64
3/8	0.3750	0.250	0.241	0.099	0.089	3/8	7/16	7/16	3/8	7/16	7/16
7/16	0.4375	0.297	0.287	0.114	0.104	7/16	35/64	31/64	7/16	35/64	31/64
1/2	0.5000	0.344	0.334	0.130	0.120	1/2	39/64	35/64	1/2	39/64	35/64
5/8	0.6250	0.469	0.456	0.164	0.148	5/8	49/64	43/64	5/8	49/64	43/64
3/4	0.7500	0.562	0.549	0.196	0.180	3/4	29/32	51/64	3/4	29/32	51/64
7/8	0.8750	0.656	0.642	0.227	0.211	7/8	1 1/8	63/64	7/8	1 1/8	63/64
1	1.0000	0.750	0.734	0.260	0.240	1	1 17/64	1 1/8	...	...	...
1 1/8	1.1250	0.844	0.826	0.291	0.271	1 1/8	1 25/64	1 3/16	...	...	...
1 1/4	1.2500	0.938	0.920	0.323	0.303	1 1/4	1 1/2	1 5/16	...	...	...
1 3/8	1.3750	1.031	1.011	0.354	0.334	1 3/8	1 21/32	1 7/16	...	...	...
1 1/2	1.5000	1.125	1.105	0.385	0.365	1 1/2	1 51/64	1 9/16	...	...	...
1 3/4	1.7500	1.312	1.289	0.448	0.428	1 3/4	2 7/32	1 61/64	...	...	...
2	2.0000	1.500	1.474	0.510	0.490	2	2 25/64	2 5/64	...	...	...

Set Screw Specifications. (**CONTINUED**)

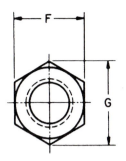
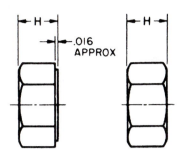
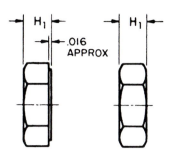

Nominal Size or Basic Major Dia of Thread		F Width Across Flats			G Width Across Corners		H Thickness Hex Nuts			H₁ Thickness Hex Jam Nuts			Hex Nuts Specified Proof Load		Jam Nuts All Strength Levels
													Up to 150,000 psi	150,000 psi and Greater	
		Basic	Max	Min	Max	Min	Basic	Max	Min	Basic	Max	Min	Runout of Bearing Face, FIR Max		
1/4	0.2500	7/16	0.438	0.428	0.505	0.488	7/32	0.226	0.212	5/32	0.163	0.150	0.015	0.010	0.015
5/16	0.3125	1/2	0.500	0.489	0.577	0.557	17/64	0.273	0.258	3/16	0.195	0.180	0.016	0.011	0.016
3/8	0.3750	9/16	0.562	0.551	0.650	0.628	21/64	0.337	0.320	7/32	0.227	0.210	0.017	0.012	0.017
7/16	0.4375	11/16	0.688	0.675	0.794	0.768	3/8	0.385	0.365	1/4	0.260	0.240	0.018	0.013	0.018
1/2	0.5000	3/4	0.750	0.736	0.866	0.840	7/16	0.448	0.427	5/16	0.323	0.302	0.019	0.014	0.019
9/16	0.5625	7/8	0.875	0.861	1.010	0.982	31/64	0.496	0.473	5/16	0.324	0.301	0.020	0.015	0.020
5/8	0.6250	15/16	0.938	0.922	1.083	1.051	35/64	0.559	0.535	3/8	0.387	0.363	0.021	0.016	0.021
3/4	0.7500	1 1/8	1.125	1.088	1.299	1.240	41/64	0.665	0.617	27/64	0.446	0.398	0.023	0.018	0.023
7/8	0.8750	1 5/16	1.312	1.269	1.516	1.447	3/4	0.776	0.724	31/64	0.510	0.458	0.025	0.020	0.025
1	1.0000	1 1/2	1.500	1.450	1.732	1.653	55/64	0.887	0.831	35/64	0.575	0.519	0.027	0.022	0.027
1 1/8	1.1250	1 11/16	1.688	1.631	1.949	1.859	31/32	0.999	0.939	39/64	0.639	0.579	0.030	0.025	0.030
1 1/4	1.2500	1 7/8	1.875	1.812	2.165	2.066	1 1/16	1.094	1.030	23/32	0.751	0.687	0.033	0.028	0.033
1 3/8	1.3750	2 1/16	2.062	1.994	2.382	2.273	1 11/64	1.206	1.138	25/32	0.815	0.747	0.036	0.031	0.036
1 1/2	1.5000	2 1/4	2.250	2.175	2.598	2.480	1 9/32	1.317	1.245	27/32	0.880	0.808	0.039	0.034	0.039

Hex Nut Specifications.

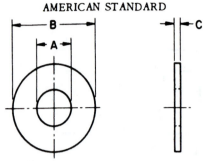

AMERICAN STANDARD

DIMENSIONS OF PREFERRED SIZES OF TYPE A PLAIN WASHERS **

Nominal Washer Size***		Inside Diameter A			Outside Diameter B			Thickness C		
		Basic	Plus (Tolerance)	Minus (Tolerance)	Basic	Plus (Tolerance)	Minus (Tolerance)	Basic	Max	Min
–	–	0.078	0.000	0.005	0.188	0.000	0.005	0.020	0.025	0.016
–	–	0.094	0.000	0.005	0.250	0.000	0.005	0.020	0.025	0.016
–	–	0.125	0.008	0.005	0.312	0.008	0.005	0.032	0.040	0.025
No. 6	0.138	0.156	0.008	0.005	0.375	0.015	0.005	0.049	0.065	0.036
No. 8	0.164	0.188	0.008	0.005	0.438	0.015	0.005	0.049	0.065	0.036
No. 10	0.190	0.219	0.008	0.005	0.500	0.015	0.005	0.049	0.065	0.036
3/16	0.188	0.250	0.015	0.005	0.562	0.015	0.005	0.049	0.065	0.036
No. 12	0.216	0.250	0.015	0.005	0.562	0.015	0.005	0.065	0.080	0.051
1/4 N	0.250	0.281	0.015	0.005	0.625	0.015	0.005	0.065	0.080	0.051
1/4 W	0.250	0.312	0.015	0.005	0.734*	0.015	0.007	0.065	0.080	0.051
5/16 N	0.312	0.344	0.015	0.005	0.688	0.015	0.007	0.065	0.080	0.051
5/16 W	0.312	0.375	0.015	0.005	0.875	0.030	0.007	0.083	0.104	0.064
3/8 N	0.375	0.406	0.015	0.005	0.812	0.015	0.007	0.065	0.080	0.051
3/8 W	0.375	0.438	0.015	0.005	1.000	0.030	0.007	0.083	0.104	0.064
7/16 N	0.438	0.469	0.015	0.005	0.922	0.015	0.007	0.065	0.080	0.051
7/16 W	0.438	0.500	0.015	0.005	1.250	0.030	0.007	0.083	0.104	0.064
1/2 N	0.500	0.531	0.015	0.005	1.062	0.030	0.007	0.095	0.121	0.074
1/2 W	0.500	0.562	0.015	0.005	1.375	0.030	0.007	0.109	0.132	0.086
9/16 N	0.562	0.594	0.015	0.005	1.156*	0.030	0.007	0.095	0.121	0.074
9/16 W	0.562	0.625	0.015	0.005	1.469*	0.030	0.007	0.109	0.132	0.086
5/8 N	0.625	0.656	0.030	0.007	1.312	0.030	0.007	0.095	0.121	0.074
5/8 W	0.625	0.688	0.030	0.007	1.750	0.030	0.007	0.134	0.160	0.108
3/4 N	0.750	0.812	0.030	0.007	1.469	0.030	0.007	0.134	0.160	0.108
3/4 W	0.750	0.812	0.030	0.007	2.000	0.030	0.007	0.148	0.177	0.122
7/8 N	0.875	0.938	0.030	0.007	1.750	0.030	0.007	0.134	0.160	0.108
7/8 W	0.875	0.938	0.030	0.007	2.250	0.030	0.007	0.165	0.192	0.136
1 N	1.000	1.062	0.030	0.007	2.000	0.030	0.007	0.134	0.160	0.108
1 W	1.000	1.062	0.030	0.007	2.500	0.030	0.007	0.165	0.192	0.136
1 1/8 N	1.125	1.250	0.030	0.007	2.250	0.030	0.007	0.134	0.160	0.108
1 1/8 W	1.125	1.250	0.030	0.007	2.750	0.030	0.007	0.165	0.192	0.136
1 1/4 N	1.250	1.375	0.030	0.007	2.500	0.030	0.007	0.165	0.192	0.136
1 1/4 W	1.250	1.375	0.030	0.007	3.000	0.030	0.007	0.165	0.192	0.136
1 3/8 N	1.375	1.500	0.030	0.007	2.750	0.030	0.007	0.165	0.192	0.136
1 3/8 W	1.375	1.500	0.045	0.010	3.250	0.045	0.010	0.180	0.213	0.153
1 1/2 N	1.500	1.625	0.030	0.007	3.000	0.030	0.007	0.165	0.192	0.136
1 1/2 W	1.500	1.625	0.045	0.010	3.500	0.045	0.010	0.180	0.213	0.153
1 5/8	1.625	1.750	0.045	0.010	3.750	0.045	0.010	0.180	0.213	0.153
1 3/4	1.750	1.875	0.045	0.010	4.000	0.045	0.010	0.180	0.213	0.153
1 7/8	1.875	2.000	0.045	0.010	4.250	0.045	0.010	0.180	0.213	0.153
2	2.000	2.125	0.045	0.010	4.500	0.045	0.010	0.180	0.213	0.153
2 1/4	2.250	2.375	0.045	0.010	4.750	0.045	0.010	0.220	0.248	0.193
2 1/2	2.500	2.625	0.045	0.010	5.000	0.045	0.010	0.238	0.280	0.210
2 3/4	2.750	2.875	0.065	0.010	5.250	0.065	0.010	0.259	0.310	0.228
3	3.000	3.125	0.065	0.010	5.500	0.065	0.010	0.284	0.327	0.249

*The 0.734 in., 1.156 in., and 1.469 in. outside diameters avoid washers which could be used in coin operated devices.
**Preferred sizes are for the most part from series previously designated "Standard Plate" and "SAE." Where common sizes existed in the two series, the SAE size is designated "N" (narrow) and the Standard Plate "W" (wide). These sizes as well as all other sizes of Type A Plain Washers are to be ordered by ID, OD, and thickness dimensions.
***Nominal washer sizes are intended for use with comparable nominal screw or bolt sizes.

Reprinted from ASME B18.2.1-1981 (R1992), B18.2.2-1987 (R1993), B18.3-1986 (R1993), B18.6.2-1972 (R1993), B18.6.3-1972 (R1991), B18.21.1-1994, B18.22.1-1965 (R1990), B17.1-1967 (R1998), B17.2-1967 (R1990) and B4.2-1978 (R1994), by permission of The American Society of Mechanical Engineers. All rights reserved.

Dimensions of Preferred Sizes of Type A Plain Washers.

AMERICAN NATIONAL STANDARD
LOCK WASHERS

ANSI B18.21.1-1990

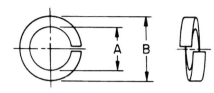

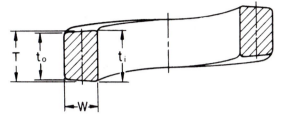

ENLARGED SECTION

DIMENSIONS OF REGULAR HELICAL SPRING LOCK WASHERS[1]

Nominal Washer Size		A Inside Diameter		B Outside Diameter	T Mean Section Thickness $\left(\dfrac{t_i + t_o}{2}\right)$	W Section Width
		Max	Min	Max[2]	Min	Min
No. 4	0.112	0.120	0.114	0.173	0.022	0.022
No. 5	0.125	0.133	0.127	0.202	0.030	0.030
No. 6	0.138	0.148	0.141	0.216	0.030	0.030
No. 8	0.164	0.174	0.167	0.267	0.047	0.042
No. 10	0.190	0.200	0.193	0.294	0.047	0.042
¼	0.250	0.262	0.254	0.365	0.078	0.047
⁵⁄₁₆	0.312	0.326	0.317	0.460	0.093	0.062
³⁄₈	0.375	0.390	0.380	0.553	0.125	0.076
⁷⁄₁₆	0.438	0.455	0.443	0.647	0.140	0.090
½	0.500	0.518	0.506	0.737	0.172	0.103
⁵⁄₈	0.625	0.650	0.635	0.923	0.203	0.125
¾	0.750	0.775	0.760	1.111	0.218	0.154
⁷⁄₈	0.875	0.905	0.887	1.296	0.234	0.182
1	1.000	1.042	1.017	1.483	0.250	0.208
1⅛	1.125	1.172	1.144	1.669	0.313	0.236
1¼	1.250	1.302	1.271	1.799	0.313	0.236
1⅜	1.375	1.432	1.398	2.041	0.375	0.292
1½	1.500	1.561	1.525	2.170	0.375	0.292
1¾	1.750	1.811	1.775	2.602	0.469	0.383
2	2.000	2.061	2.025	2.852	0.469	0.383
2¼	2.250	2.311	2.275	3.352	0.508	0.508
2½	2.500	2.561	2.525	3.602	0.508	0.508
2¾	2.750	2.811	2.775	4.102	0.633	0.633
3	3.000	3.061	3.025	4.352	0.633	0.633

[1]For use with 1960 Series Socket Head Cap Screws specified in American National Standard, ANSI B18.3.
[2]The maximum outside diameters specified allow for the commercial tolerances on cold-drawn wire.

Dimensions of Spring Lock Washers.

AMERICAN NATIONAL STANDARD
LOCK WASHERS

ASME/ANSI B18.21.1-1994

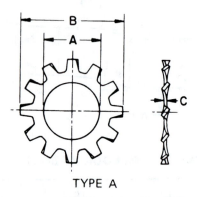
TYPE A

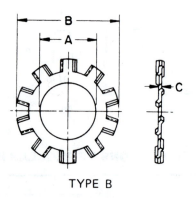
TYPE B

DIMENSIONS OF EXTERNAL TOOTH LOCK WASHERS

Nominal Washer Size		A Inside Diameter		B Outside Diameter		C Thickness	
		Max	Min	Max	Min	Max	Min
No. 3	0.099	0.109	0.102	0.235	0.220	0.015	0.012
No. 4	0.112	0.123	0.115	0.260	0.245	0.019	0.015
No. 5	0.125	0.136	0.129	0.285	0.270	0.019	0.014
No. 6	0.138	0.150	0.141	0.320	0.305	0.022	0.016
No. 8	0.164	0.176	0.168	0.381	0.365	0.023	0.018
No. 10	0.190	0.204	0.195	0.410	0.395	0.025	0.020
No. 12	0.216	0.231	0.221	0.475	0.460	0.028	0.023
¼	0.250	0.267	0.256	0.510	0.494	0.028	0.023
⁵⁄₁₆	0.312	0.332	0.320	0.610	0.588	0.034	0.028
³⁄₈	0.375	0.398	0.384	0.694	0.670	0.040	0.032
⁷⁄₁₆	0.438	0.464	0.448	0.760	0.740	0.040	0.032
½	0.500	0.530	0.513	0.900	0.880	0.045	0.037
⁹⁄₁₆	0.562	0.596	0.576	0.985	0.960	0.045	0.037
⅝	0.625	0.663	0.641	1.070	1.045	0.050	0.042
¹¹⁄₁₆	0.688	0.728	0.704	1.155	1.130	0.050	0.042
¾	0.750	0.795	0.768	1.260	1.220	0.055	0.047
¹³⁄₁₆	0.812	0.861	0.833	1.315	1.290	0.055	0.047
⅞	0.875	0.927	0.897	1.410	1.380	0.060	0.052
1	1.000	1.060	1.025	1.620	1.590	0.067	0.059

Reprinted from ASME B18.2.1-1981 (R1992), B18.2.2-1987 (R1993), B18.3-1986 (R1993), B18.6.2-1972 (R1993), B18.6.3-1972 (R1991), B18.21.1-1994, B18.22.1-1965 (R1990), B17.1-1967 (R1998), B17.2-1967 (R1990) and B4.2-1978 (R1994), by permission of The American Society of Mechanical Engineers. All rights reserved.

Dimensions of Internal and External Tooth Lock Washers.

AMERICAN NATIONAL STANDARD
LOCK WASHERS

ASME/ANSI B18.21.1-1994

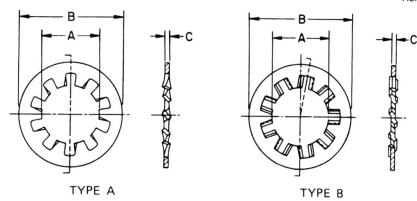

TYPE A   TYPE B

DIMENSIONS OF INTERNAL TOOTH LOCK WASHERS

Nominal Washer Size		A		B		C	
		Inside Diameter		Outside Diameter		Thickness	
		Max	Min	Max	Min	Max	Min
No. 2	0.086	0.095	0.089	0.200	0.175	0.015	0.010
No. 3	0.099	0.109	0.102	0.232	0.215	0.019	0.012
No. 4	0.112	0.123	0.115	0.270	0.255	0.019	0.015
No. 5	0.125	0.136	0.129	0.280	0.245	0.021	0.017
No. 6	0.138	0.150	0.141	0.295	0.275	0.021	0.017
No. 8	0.164	0.176	0.168	0.340	0.325	0.023	0.018
No. 10	0.190	0.204	0.195	0.381	0.365	0.025	0.020
No. 12	0.216	0.231	0.221	0.410	0.394	0.025	0.020
1/4	0.250	0.267	0.256	0.478	0.460	0.028	0.023
5/16	0.312	0.332	0.320	0.610	0.594	0.034	0.028
3/8	0.375	0.398	0.384	0.692	0.670	0.040	0.032
7/16	0.438	0.464	0.448	0.789	0.740	0.040	0.032
1/2	0.500	0.530	0.512	0.900	0.867	0.045	0.037
9/16	0.562	0.596	0.576	0.985	0.957	0.045	0.037
5/8	0.625	0.663	0.640	1.071	1.045	0.050	0.042
11/16	0.688	0.728	0.704	1.166	1.130	0.050	0.042
3/4	0.750	0.795	0.769	1.245	1.220	0.055	0.047
13/16	0.812	0.861	0.832	1.315	1.290	0.055	0.047
7/8	0.875	0.927	0.894	1.410	1.364	0.060	0.052
1	1.000	1.060	1.019	1.637	1.590	0.067	0.059
1 1/8	1.125	1.192	1.144	1.830	1.799	0.067	0.059
1 1/4	1.250	1.325	1.275	1.975	1.921	0.067	0.059

Dimensions of Internal and External Tooth Lock Washers. **(CONTINUED)**

Tap Drill Sizes.

Fraction or Drill Size	Decimal Equivalent	Tap Size
80	.0135	
79	.0145	
1/64	.0156	
78	.0160	
77	.0180	
76	.0200	
75	.0210	
74	.0225	
73	.0240	
72	.0250	
71	.0260	
70	.0280	
69	.0292	
68	.0310	
1/32	.0312	
67	.0320	
66	.0330	
65	.0350	
64	.0360	
63	.0370	
62	.0380	
61	.0390	
60	.0400	
59	.0410	
58	.0420	
57	.0430	0–80
56	.0465	
3/64	.0469	
55	.0520	1–56
54	.0550	1–64, 72
53	.0595	
1/16	.0625	
52	.0635	
51	.0670	
50	.0700	2–56, 64
49	.0730	
48	.0760	
5/64	.0781	3–48
47	.0785	
46	.0810	
45	.0820	3–56, 4–32
44	.0860	4–36
43	.0890	4–40
42	.0935	4–48
3/32	.0938	
41	.0960	
40	.0980	

Fraction or Drill Size	Decimal Equivalent	Tap Size
39	.0995	
38	.1015	5–40
37	.1040	5–44
36	.1065	6–32
7/64	.1094	
35	.1100	
34	.1110	6–36
33	.1130	6–40
32	.1160	
31	.1200	
1/8	.1250	
30	.1285	
29	.1360	8–32, 36
28	.1405	8–40
9/64	.1406	
27	.1440	
26	.1470	10–24
25	.1495	
24	.1520	
23	.1540	
5/32	.1562	
22	.1570	10–30
21	.1590	10–32
20	.1610	
19	.1660	
18	.1695	
11/64	.1719	
17	.1730	
16	.1770	12–24
15	.1800	
14	.1820	12–28
13	.1850	12–32
3/16	.1875	
12	.1890	
11	.1910	
10	.1935	
9	.1960	
8	.1990	1/4–20
7	.2010	
13/64	.2031	
6	.2040	
5	.2055	
4	.2090	
3	.2130	1/4–28
7/32	.2188	
2	.2210	
1	.2280	
A	.2340	

Fraction or Drill Size	Decimal Equivalent	Tap Size
15/64	.2344	
B	.2380	
C	.2420	
D	.2460	
1/4 E	.2500	5/16–18
F	.2570	
G	.2610	
17/64	.2656	
H	.2660	
I	.2720	5/16–24
J	.2770	
K	.2810	
9/32	.2812	
L	.2900	
M	.2950	
19/64	.2969	
N	.3020	
5/16	.3125	3/8–16
O	.3160	
P	.3230	
21/64	.3281	
Q	.3320	3/8–24
R	.3390	
11/32	.3438	
S	.3480	
T	.3580	
23/64	.3594	
U	.3680	7/16–14
3/8	.3750	
V	.3770	
W	.3860	
25/64	.3906	7/16–20
X	.3970	
Y	.4040	
13/32	.4062	
Z	.4130	
27/64	.4219	1/2–13
7/16	.4375	
29/64	.4531	1/2–20
15/32	.4688	
31/64	.4844	9/16–12
1/2	.5000	
33/64	.5156	9/16–18
17/32	.5312	5/8–11
35/64	.5469	
9/16	.5625	
37/64	.5781	5/8–18

Fraction or Drill Size	Decimal Equivalent	Tap Size
19/32	.5938	11/16–11
39/64	.6094	11/16–16
5/8	.6250	
41/64	.6406	3/4–10
21/32	.6562	
43/64	.6719	
11/16	.6875	3/4–16
45/64	.7031	
23/32	.7188	
47/64	.7344	
3/4	.7500	
49/64	.7656	7/8–9
25/32	.7812	
51/64	.7969	7/8–14
13/16	.8125	
53/64	.8281	
27/32	.8438	
55/64	.8594	
7/8	.8750	1–8
57/64	.8906	
29/32	.9062	
59/64	.9219	
15/16	.9375	1–12, 14
61/64	.9531	
31/32	.9688	
63/64	.9844	1 1/8–7
1	1.0000	
1 3/64	1.0469	1 1/8–12
1 17/64	1.1094	1 1/4–7
1 1/8	1.1250	
1 11/64	1.1719	1 1/4–12
1 7/32	1.2188	1 3/8–6
1 1/4	1.2500	
1 19/64	1.2969	1 3/8–12
1 11/32	1.3438	1 1/2–6
1 3/8	1.3750	
1 27/64	1.4219	1 1/2–12
1 1/2	1.5000	

Pipe Thread Sizes

Thread	Drill	Thread	Drill
1/8–27	R	1 1/2–11 1/2	1 47/64
1/4–18	7/16	2–11 1/2	2 7/32
3/8–18	37/64	2 1/2–8	2 5/8
1/2–14	23/32	3–8	3 1/4
3/4–14	59/64	3 1/2–8	3 3/4
1–11 1/2	1 5/32	4–8	4 1/4
1 1/4–11 1/2	1 1/2		

KEY & KEYWAY SIZES

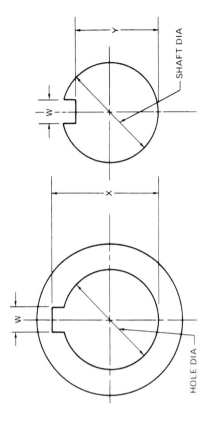

Nom. Size (Inch)	DIA. – (Shaft) Inch	mm	'X' (Collar) Inch	mm	'Y' (Shaft) Inch	mm
1/2	.500	12.700	.560	14.224	.430	10.922
9/16	.562	14.290	.623	15.824	.493	12.522
5/8	.625	15.875	.709	18.008	.517	13.132
11/16	.688	17.470	.773	18.618	.581	14.757
3/4	.750	19.050	.837	21.259	.644	16.357
13/16	.812	20.640	.900	22.860	.708	17.983
7/8	.875	22.225	.964	24.485	.771	19.583
15/16	.938	23.820	1.051	26.695	.791	20.091
1	1.000	25.400	1.114	28.295	.859	21.818
1 1/16	1.062	26.985	1.178	29.921	.923	23.444
1 1/8	1.125	28.575	1.241	31.521	.986	25.044
1 3/16	1.188	30.165	1.304	33.121	1.049	26.644
1 1/4	1.250	31.750	1.367	34.722	1.112	28.244
1 5/16	1.312	33.340	1.455	36.957	1.137	28.879
1 3/8	1.375	34.923	1.518	38.557	1.201	30.505

From DRAFTING FOR TRADES AND INDUSTRY, Mechanical and Electronic (Drafting for Trades & Industry Series) 1st edition by NELSON, 1979. Reprinted with permission of Delmar Learning, a division of Thomson Learning: www.thomsonrights.com. Fax 800 730-2215.

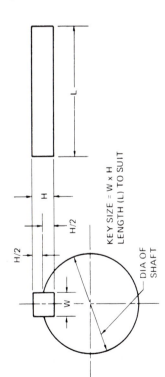

KEY SIZE = W × H
LENGTH (L) TO SUIT

Shaft Nom. Size – DIA. – From	To & Incl.	Square (W = H)	Type	Square Key From	To & Incl.	Tolerance
5/16 (8)	7/16 (11)	3/32 (2.38)	Bar Stock	—	3/4 (19.05)	+.000 – .002 (+.0000 – .0254)
7/16 (11)	9/16 (14)	1/8 (3.175)		—	1 1/2 (38.1)	+.000 – .003 (+.0000 – .0762)
9/16 (14)	7/8 (22)	3/16 (4.76)		3/4 (19.05)	2 1/2 (63.5)	+.000 – .004 (+.0000 – .1016)
7/8 (22)	1 1/4 (32)	1/4 (6.35)		1 1/2 (38.1)	3 1/2 (88.9)	+.000 – .006 (+.0000 – .1524)
1 1/4 (32)	1 3/8 (35)	5/16 (7.94)		2 1/2 (63.5)	—	+.001 – .000 (+.0254 – .0000)
1 3/8 (35)	1 3/4 (44)	3/8 (9.53)	Keystock	—	1 1/4 (31.75)	+.002 – .000 (+.0508 – .0000)
1 3/4 (44)	2 1/4 (57)	1/2 (12.7)		1 1/4 (31.75)	3 (76.2)	+.003 – .000 (+.0762 – .0000)
2 1/4 (57)	2 3/4 (70)	5/8 (15.88)		3 (76.2)	3 1/2 (88.9)	
2 3/4 (70)	3 1/4 (82)	3/4 (19.05)				
3 1/4 (82)	3 3/4 (95)	7/8 (22.23)				

(Figures in parenthesis = mm)

Dimensions of Keys and Slots.

USA STANDARD

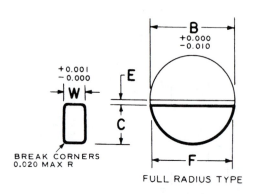

FULL RADIUS TYPE

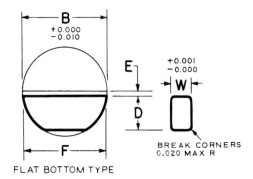

FLAT BOTTOM TYPE

WOODRUFF KEYS

Key No.	Nominal Key Size W × B	Actual Length F +0.000-0.010	Height of Key				Distance Below Center E
			C		D		
			Max	Min	Max	Min	
202	¹⁄₁₆ × ¹⁄₄	0.248	0.109	0.104	0.109	0.104	¹⁄₆₄
202.5	¹⁄₁₆ × ⁵⁄₁₆	0.311	0.140	0.135	0.140	0.135	¹⁄₆₄
302.5	³⁄₃₂ × ⁵⁄₁₆	0.311	0.140	0.135	0.140	0.135	¹⁄₆₄
203	¹⁄₁₆ × ³⁄₈	0.374	0.172	0.167	0.172	0.167	¹⁄₆₄
303	³⁄₃₂ × ³⁄₈	0.374	0.172	0.167	0.172	0.167	¹⁄₆₄
403	¹⁄₈ × ³⁄₈	0.374	0.172	0.167	0.172	0.167	¹⁄₆₄
204	¹⁄₁₆ × ¹⁄₂	0.491	0.203	0.198	0.194	0.188	³⁄₆₄
304	³⁄₃₂ × ¹⁄₂	0.491	0.203	0.198	0.194	0.188	³⁄₆₄
404	¹⁄₈ × ¹⁄₂	0.491	0.203	0.198	0.194	0.188	³⁄₆₄
305	³⁄₃₂ × ⁵⁄₈	0.612	0.250	0.245	0.240	0.234	¹⁄₁₆
405	¹⁄₈ × ⁵⁄₈	0.612	0.250	0.245	0.240	0.234	¹⁄₁₆
505	⁵⁄₃₂ × ⁵⁄₈	0.612	0.250	0.245	0.240	0.234	¹⁄₁₆
605	³⁄₁₆ × ⁵⁄₈	0.612	0.250	0.245	0.240	0.234	¹⁄₁₆
406	¹⁄₈ × ³⁄₄	0.740	0.313	0.308	0.303	0.297	¹⁄₁₆
506	⁵⁄₃₂ × ³⁄₄	0.740	0.313	0.308	0.303	0.297	¹⁄₁₆
606	³⁄₁₆ × ³⁄₄	0.740	0.313	0.308	0.303	0.297	¹⁄₁₆
806	¹⁄₄ × ³⁄₄	0.740	0.313	0.308	0.303	0.297	¹⁄₁₆
507	⁵⁄₃₂ × ⁷⁄₈	0.866	0.375	0.370	0.365	0.359	¹⁄₁₆
607	³⁄₁₆ × ⁷⁄₈	0.866	0.375	0.370	0.365	0.359	¹⁄₁₆
707	⁷⁄₃₂ × ⁷⁄₈	0.866	0.375	0.370	0.365	0.359	¹⁄₁₆
807	¹⁄₄ × ⁷⁄₈	0.866	0.375	0.370	0.365	0.359	¹⁄₁₆
608	³⁄₁₆ × 1	0.992	0.438	0.433	0.428	0.422	¹⁄₁₆
708	⁷⁄₃₂ × 1	0.992	0.438	0.433	0.428	0.422	¹⁄₁₆
808	¹⁄₄ × 1	0.992	0.438	0.433	0.428	0.422	¹⁄₁₆
1008	⁵⁄₁₆ × 1	0.992	0.438	0.433	0.428	0.422	¹⁄₁₆
1208	³⁄₈ × 1	0.992	0.438	0.433	0.428	0.422	¹⁄₁₆
609	³⁄₁₆ × 1¹⁄₈	1.114	0.484	0.479	0.475	0.469	⁵⁄₆₄
709	⁷⁄₃₂ × 1¹⁄₈	1.114	0.484	0.479	0.475	0.469	⁵⁄₆₄
809	¹⁄₄ × 1¹⁄₈	1.114	0.484	0.479	0.475	0.469	⁵⁄₆₄
1009	⁵⁄₁₆ × 1¹⁄₈	1.114	0.484	0.479	0.475	0.469	⁵⁄₆₄

Dimensions of Keys and Slots. **(CONTINUED)**

WOODRUFF KEYS (CONCLUDED)

Key No.	Nominal Key Size W × B	Actual Length F +0.000-0.010	Height of Key				Distance Below Center E
			C		D		
			Max	Min	Max	Min	
610	3/16 × 1¼	1.240	0.547	0.542	0.537	0.531	5/64
710	7/32 × 1¼	1.240	0.547	0.542	0.537	0.531	5/64
810	¼ × 1¼	1.240	0.547	0.542	0.537	0.531	5/64
1010	5/16 × 1¼	1.240	0.547	0.542	0.537	0.531	5/64
1210	3/8 × 1¼	1.240	0.547	0.542	0.537	0.531	5/64
811	¼ × 1⅜	1.362	0.594	0.589	0.584	0.578	3/32
1011	5/16 × 1⅜	1.362	0.594	0.589	0.584	0.578	3/32
1211	3/8 × 1⅜	1.362	0.594	0.589	0.584	0.578	3/32
812	¼ × 1½	1.484	0.641	0.636	0.631	0.625	7/64
1012	5/16 × 1½	1.484	0.641	0.636	0.631	0.625	7/64
1212	3/8 × 1½	1.484	0.641	0.636	0.631	0.625	7/64

All dimensions given are in inches.

The key numbers indicate nominal key dimensions. The last two digits give the nominal diameter B in eighths of an inch and the digits preceding the last two give the nominal width W in thirty-seconds of an inch.

Example:

No. 204 indicates a key 2/32 × 4/8 or 1/16 × ½.
No. 808 indicates a key 8/32 × 8/8 or ¼ × 1.
No. 1212 indicates a key 12/32 × 12/8 or 3/8 × 1½.

Dimensions of Keys and Slots. (CONTINUED)

WOODRUFF KEYS AND KEYSEATS

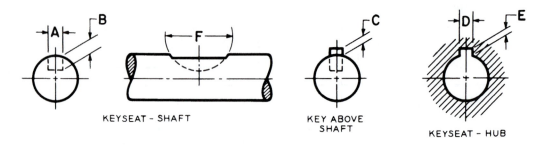

KEYSEAT – SHAFT   KEY ABOVE SHAFT   KEYSEAT – HUB

| Key Number | Nominal Size Key | Keyseat – Shaft | | | | | Key Above Shaft | Keyseat – Hub | |
| | | Width A* | | Depth B | Diameter F | | Height C | Width D | Depth E |
		Min	Max	+0.005 -0.000	Min	Max	+0.005 -0.005	+0.002 -0.000	+0.005 -0.000
202	1/16 × 1/4	0.0615	0.0630	0.0728	0.250	0.268	0.0312	0.0635	0.0372
202.5	1/16 × 5/16	0.0615	0.0630	0.1038	0.312	0.330	0.0312	0.0635	0.0372
302.5	3/32 × 5/16	0.0928	0.0943	0.0882	0.312	0.330	0.0469	0.0948	0.0529
203	1/16 × 3/8	0.0615	0.0630	0.1358	0.375	0.393	0.0312	0.0635	0.0372
303	3/32 × 3/8	0.0928	0.0943	0.1202	0.375	0.393	0.0469	0.0948	0.0529
403	1/8 × 3/8	0.1240	0.1255	0.1045	0.375	0.393	0.0625	0.1260	0.0685
204	1/16 × 1/2	0.0615	0.0630	0.1668	0.500	0.518	0.0312	0.0635	0.0372
304	3/32 × 1/2	0.0928	0.0943	0.1511	0.500	0.518	0.0469	0.0948	0.0529
404	1/8 × 1/2	0.1240	0.1255	0.1355	0.500	0.518	0.0625	0.1260	0.0685
305	3/32 × 5/8	0.0928	0.0943	0.1981	0.625	0.643	0.0469	0.0948	0.0529
405	1/8 × 5/8	0.1240	0.1255	0.1825	0.625	0.643	0.0625	0.1260	0.0685
505	5/32 × 5/8	0.1553	0.1568	0.1669	0.625	0.643	0.0781	0.1573	0.0841
605	3/16 × 5/8	0.1863	0.1880	0.1513	0.625	0.643	0.0937	0.1885	0.0997
406	1/8 × 3/4	0.1240	0.1255	0.2455	0.750	0.768	0.0625	0.1260	0.0685
506	5/32 × 3/4	0.1553	0.1568	0.2299	0.750	0.768	0.0781	0.1573	0.0841
606	3/16 × 3/4	0.1863	0.1880	0.2143	0.750	0.768	0.0937	0.1885	0.0997
806	1/4 × 3/4	0.2487	0.2505	0.1830	0.750	0.768	0.1250	0.2510	0.1310
507	5/32 × 7/8	0.1553	0.1568	0.2919	0.875	0.895	0.0781	0.1573	0.0841
607	3/16 × 7/8	0.1863	0.1880	0.2763	0.875	0.895	0.0937	0.1885	0.0997
707	7/32 × 7/8	0.2175	0.2193	0.2607	0.875	0.895	0.1093	0.2198	0.1153
807	1/4 × 7/8	0.2487	0.2505	0.2450	0.875	0.895	0.1250	0.2510	0.1310
608	3/16 × 1	0.1863	0.1880	0.3393	1.000	1.020	0.0937	0.1885	0.0997
708	7/32 × 1	0.2175	0.2193	0.3237	1.000	1.020	0.1093	0.2198	0.1153
808	1/4 × 1	0.2487	0.2505	0.3080	1.000	1.020	0.1250	0.2510	0.1310
1008	5/16 × 1	0.3111	0.3130	0.2768	1.000	1.020	0.1562	0.3135	0.1622
1208	3/8 × 1	0.3735	0.3755	0.2455	1.000	1.020	0.1875	0.3760	0.1935
609	3/16 × 1 1/8	0.1863	0.1880	0.3853	1.125	1.145	0.0937	0.1885	0.0997
709	7/32 × 1 1/8	0.2175	0.2193	0.3697	1.125	1.145	0.1093	0.2198	0.1153
809	1/4 × 1 1/8	0.2487	0.2505	0.3540	1.125	1.145	0.1250	0.2510	0.1310
1009	5/16 × 1 1/8	0.3111	0.3130	0.3228	1.125	1.145	0.1562	0.3135	0.1622

From The American Society of Mechanical Engineers—ANSI B17.2—1967—R1990

Dimensions of Keys and Slots. (CONTINUED)

KEYS AND KEYSEATS

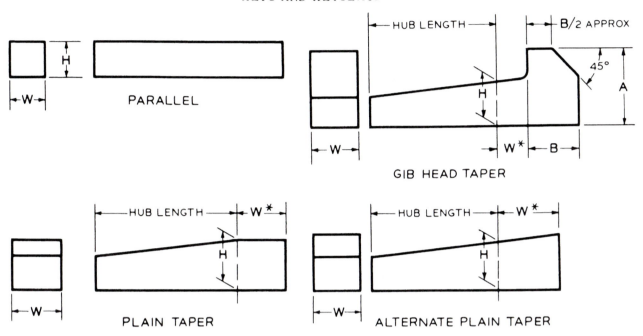

Plain and Gib Head Taper Keys Have a 1/8″ Taper in 12″

KEY DIMENSIONS AND TOLERANCES

KEY			NOMINAL KEY SIZE		TOLERANCE	
			Width, W		Width, W	Height, H
			Over	To (Incl)		
Parallel	Square	Bar Stock	— 3/4 1-1/2 2-1/2	3/4 1-1/2 2-1/2 3-1/2	+0.000 −0.002 +0.000 −0.003 +0.000 −0.004 +0.000 −0.006	+0.000 −0.002 +0.000 −0.003 +0.000 −0.004 +0.000 −0.006
		Keystock	— 1-1/4 3	1-1/4 3 3-1/2	+0.001 −0.000 +0.002 −0.000 +0.003 −0.000	+0.001 −0.000 +0.002 −0.000 +0.003 −0.000
	Rectangular	Bar Stock	— 3/4 1-1/2 3 4 6	3/4 1-1/2 3 4 6 7	+0.000 −0.003 +0.000 −0.004 +0.000 −0.005 +0.000 −0.006 +0.000 −0.008 +0.000 −0.013	+0.000 −0.003 +0.000 −0.004 +0.000 −0.005 +0.000 −0.006 +0.000 −0.008 +0.000 −0.013
		Keystock	— 1-1/4 3	1-1/4 3 7	+0.001 −0.000 +0.002 −0.000 +0.003 −0.000	+0.005 −0.005 +0.005 −0.005 +0.005 −0.005
Taper	Plain or Gib Head Square or Rectangular		— 1-1/4 3	1-1/4 3 7	+0.001 −0.000 +0.002 −0.000 +0.003 −0.000	+0.005 −0.000 +0.005 −0.000 +0.005 −0.000

*For locating position of dimension H. Tolerance does not apply.
See Table 41 for dimensions on gib heads.
All dimensions given in inches.

Dimensions of Keys and Slots. (CONTINUED)

WOODRUFF KEY SIZES FOR DIFFERENT SHAFT DIAMETERS

Shaft Diameter	5/16 to 3/8	7/16 to 1/2	9/16 to 3/4	13/16 to 15/16	1 to 1 3/16	1 1/4 to 1 7/16	1 1/2 to 1 3/4	1 13/16 to 2 1/8	2 3/16 to 2 1/2
Key Numbers	204	304 305	404 405 406	505 506 507	606 607 608 609	807 808 809	810 811 812	1011 1012	1211 1212

Dimensions of Keys and Slots. **(CONTINUED)**

USA STANDARD

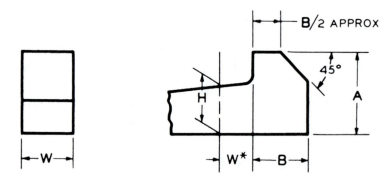

GIB HEAD NOMINAL DIMENSIONS

Nominal Key Size Width, W	SQUARE			RECTANGULAR		
	H	A	B	H	A	B
1/8	1/8	1/4	1/4	3/32	3/16	1/8
3/16	3/16	5/16	5/16	1/8	1/4	1/4
1/4	1/4	7/16	3/8	3/16	5/16	5/16
5/16	5/16	1/2	7/16	1/4	7/16	3/8
3/8	3/8	5/8	1/2	1/4	7/16	3/8
1/2	1/2	7/8	5/8	3/8	5/8	1/2
5/8	5/8	1	3/4	7/16	3/4	9/16
3/4	3/4	1-1/4	7/8	1/2	7/8	5/8
7/8	7/8	1-3/8	1	5/8	1	3/4
1	1	1-5/8	1-1/8	3/4	1-1/4	7/8
1-1/4	1-1/4	2	1-7/16	7/8	1-3/8	1
1-1/2	1-1/2	2-3/8	1-3/4	1	1-5/8	1-1/8
1-3/4	1-3/4	2-3/4	2	1-1/2	2-3/8	1-3/4
2	2	3-1/2	2-1/4	1-1/2	2-3/8	1-3/4
2-1/2	2-1/2	4	3	1-3/4	2-3/4	2
3	3	5	3-1/2	2	3-1/2	2-1/4
3-1/2	3-1/2	6	4	2-1/2	4	3

*For locating position of dimension *H*.

For larger sizes the following relationships are suggested as guides for establishing *A* and *B*.

$$A = 1.8 H \qquad B = 1.2 H$$

All dimensions given in inches.

Reprinted from ASME B18.2.1-1981 (R1992), B18.2.2-1987 (R1993), B18.3-1986 (R1993), B18.6.2-1972 (R1993), B18.6.3-1972 (R1991), B18.21.1-1994, B18.22.1-1965 (R1990), B17.1-1967 (R1998), B17.2-1967 (R1990) and B4.2-1978 (R1994), by permission of The American Society of Mechanical Engineers. All rights reserved.

Dimensions of Keys and Slots. **(CONTINUED)**

Application	SAE No.	Application	SAE No.
Adapters	1145	Chain pins, transmission.......	4320
Agricultural steel	1070	" " "	4815
" "	1080	" " "	4820
Aircraft forgings	4140	Chains, transmission	3135
Axles, front or rear	1040	" "	3140
" " "	4140	Clutch disks	1060
Axle shafts.....................	1045	" "	1070
" "	2340	" "	1085
" "	2345	Clutch springs	1060
" "	3135	Coil springs	4063
" "	3140	Cold-headed bolts..............	4042
" "	3141	Cold-heading steel..............	30905
" "	4063	Cold-heading wire or rod.....	rimmed*
" "	4340	" " "	1035
Ball-bearing races	52100	Cold-rolled steel	1070
Balls for ball bearings........	52100	Connecting-rods	1040
Body stock for cars............	rimmed*	" "	3141
Bolts, anchor	1040	Connecting-rod bolts...........	3130
Bolts and screws	1035	Corrosion resisting..............	51710
Bolts, cold-headed	4042	" "	30805
Bolts, connecting-rod........	3130	Covers, transmission	rimmed*
Bolts, heat-treated............	2330	Crankshafts......................	1045
Bolts, heavy-duty	4815	"	1145
" " "	4820	"	3135
Bolts, steering-arm	3130	"	3140
Brake levers....................	1030	"	3141
" "	1040	Crankshafts, Diesel engine ...	4340
Bumper bars	1085	Cushion, springs.................	1060
Cams, free-wheeling	4615	Cutlery, stainless.................	51335
" " "	4620	Cylinder studs	3130
Camshafts	1020	Deep-drawing steel..............	rimmed*
"	1040	" " "	30905
Carburized parts..............	1020	Differential gears.................	4023
" "	1022	Disks, clutch......................	1070
" "	1024	" "	1060
" "	1320	Ductile steel	30905
" "	2317	Fan blades	1020
" "	2515	Fatigue resisting	4340
" "	3310	" "	4640
" "	3115	Fender stock for cars...........	rimmed*
" "	3120	Forgings, aircraft...............	4140
" "	4023	Forgings, carbon steel.........	1040
" "	4032	" " "	1045
" "	1117	Forgings, heat-treated	3240
" "	1118	" " "	5140

General Applications of SAE Steels.

Application	SAE No.	Application	SAE No.
Forgings, heat-treated	6150	Key stock	1030
Forgings, high-duty	6150	" "	2330
Forgings, small or medium .	1035	" "	3130
Forgings, large .	1036	Leaf springs	1085
Free-cutting carbon steel.....	1111	" "	9260
" " " "	1113	Levers, brake	1030
Free-cutting chro.-ni.steel	30615	" "	1040
Free-cutting mang. steel......	1132	Levers, gear shift	1030
" " " "	1137	Levers, heat-treated	2330
Gears, carburized.............	1320	Lock-washers....................	1060
" "	2317	Mower knives	1085
" "	3115	Mower sections	1070
" "	3120	Music wire......................	1085
" "	3310	Nuts	3130
" "	4119	Nuts, heat-treated	2330
" "	4125	Oil-pans, automobile	rimmed*
" "	4320	Pinions, carburized............	3115
" "	4615	" "	3120
" "	4620	" "	4320
" "	4815	Piston-pins	3115
" "	4820	" "	3120
Gears, heat-treated............	2345	Plow beams	1070
Gears, car and truck	4027	Plow disks	1080
" " " "	4032	Plow shares	1080
Gears, cyanide-hardening ..	5140	Propeller shafts..................	2340
Gears, differential	4023	" "	2345
Gears, high duty.............	4640	" "	4140
" " "	6150	Races, ball-bearing..............	52100
Gears, oil-hardening.........	3145	Ring gears......................	3115
" " "	3150	" "	3120
" " "	4340	" "	4119
" " "	5150	Rings, snap......................	1060
Gears, ring......................	1045	Rivets	rimmed*
" "	3115	Rod and wire......................	killed*
" "	3120	Rod, cold-heading	1035
" "	4119	Roller bearings..................	4815
Gears, transmission	3115	Rollers for bearings.............	52100
" "	3120	Screws and bolts................	1035
" "	4119	Screw stock, Bessemer	1111
Gears, truck and bus	3310	" " "	1112
" " " "	4320	" " "	1113
Gear shift levers..............	1030	Screw stock, open hearth	1115
Harrow disks	1080	Screws, heat-treated............	2330
" "	1095	Seat springs......................	1095
Hay-rake teeth	1095	Shafts, axle......................	1045

General Applications of SAE Steels. **(CONTINUED)**

Application	SAE No.	Application	SAE No.
Shafts, cyanide-hardening ..	5140	Steel, cold-heading	30905
Shafts, heavy-duty	4340	Steel, free-cutting carbon	11111
" " "	6150	" " " "	1113
" " "	4615	Steel, free-cutting chro.-ni.	30615
" " "	4620	Steel, free-cutting mang.	1132
Shafts, oil-hardening	5150	" " " "	0000
Shafts, propeller	2340	Steel, minimum distortion	4615
" "	2345	" " " "	4620
" "	4140	" " " "	4640
Shafts, transmission	4140	Steel, soft ductile	30905
Sheets and strips	rimmed*	Steering arms	4042
Snap rings	1060	Steering-arm bolts	3130
Spline shafts	1045	Steering knuckles	3141
" "	1320	Steering-knuckle pins	4815
" "	2340	" " "	4820
" "	2345	Studs	1040
" "	3115	"	1111
" "	3120	Studs, cold-headed	4042
" "	3135	Studs, cylinder	3130
" "	3140	Studs, heat-treated	2330
" "	4023	Studs, heavy-duty	4815
Spring clips	1060	" " "	4820
Springs, coil	1095	Tacks	rimmed*
" "	4063	Thrust washers	1060
" "	6150	Thrust washers, oil-harden	5150
Springs, clutch	1060	Transmission shafts	4140
Springs, cushion	1060	Tubing	1040
Springs, leaf	1085	Tubing, front axle	4140
" "	1095	Tubing, seamless	1030
" "	4063	Tubing, welded	1020
" "	4068	Universal joints	1145
" "	9260	Valve springs	1060
" "	6150	Washers, lock	1060
Springs, hard-drawn coiled	1066	Welded structures	30705
Springs, oil-hardening	5150	Wire and rod	killed*
Springs, oil-tempered wire ..	1066	Wire, cold-heading	rimmed*
Springs, seat	1095	" " "	1035
Springs, valve	1060	Wire, hard-drawn spring	1045
Spring wire	1045	" " " "	1055
Spring wire, hard-drawn	1055	Wire, music	1085
Spring wire, oil-tempered ...	1055	Wire, oil-tempered spring	1055
Stainless irons	51210	Wrist-pins, automobile	1020
" "	51710	Yokes	1145
Steel, cold-rolled	1070		

General Applications of SAE Steels. (CONTINUED)

Element	Symbol	Melting point, °F	Boiling point, °F	Specific heat,[a] cal/g/°C	Thermal conductivity,[a] Btu/hr/sq ft/°F/ft	Density,[a] g/cm³	Modulus of elasticity in tension, million psi	Coefficient of linear thermal expansion,[a] μ in./in./°F	Electrical resistivity, microhm-cm	Crystal structure
Aluminum	Al	1220	4442	0.215	128.	2.70	9	13.1	2.65	f.c.c
Antimony	Sb	1167	2516	0.049	10.8	6.62	11.3	4.7	39	Rhomb.
Arsenic	As	1503 (28 atm)	1135[b]	0.082	.....	5.72	.....	2.6	33.3	Rhomb.
Barium	Ba	1317	2980	0.068	.....	3.5	.....	....	....	b.c.c.
Beryllium	Be	2332	5020	0.45	84.4	1.85	42	6.4	4	h.c.p.
Bismuth	Bi	520	2840	0.029	4.8	9.80	4.6	7.4	107	Rhomb.
Boron	B	3690	.....	0.309	.....	2.34	.....	4.6	10	Orthorhomb
Cadmium	Cd	610	1409	0.055	53.	8.65	8	16.55	6.83	h.c.p.
Calcium	Ca	1540	2625	0.149	72.3	1.55	3.5	12.4	3.91	f.c.c.
Carbon (graphite)	C	6740[b]	8730	0.165	13.8	2.25	0.7	0.3 to 2.4	1375	Hexag.
Cerium	Ce	1479	6280	0.045	6.6	6.77	6	4.4	75	f.c.c.
Cesium	Cs	84	1273	0.048	.....	1.90	.....	54	20	b.c.c.
Chromium	Cr	3407	4829	0.11	40.3	7.19	36	3.4	12.9	b.c.c.
Cobalt	Co	2723	5250	0.099	41.5	8.85	30	7.66	6.24	h.c.p.
Columbium	Cb	4474	8901	0.065	31.5	8.57	.....	4.06	12.5	b.c.c.
Copper	Cu	1981	4703	0.092	226.	8.96	16	9.2	1.67	f.c.c.
Gallium	Ga	86	4059	0.079	19.4	5.91	.....	10	17.4	Orthorhomb.
Germanium[c]	Ge	1719	5125	0.073	33.7	5.32	.....	3.19	46×10^5	Diam. cubic
Gold	Au	1954	5380	0.031	171.	19.32	11.6	7.9	2.35	f.c.c.
Indium	In	313	3632	0.057	13.8	7.31	1.57	18	8.37	f.c.tetr.
Iridium	Ir	4449	9570	0.031	33.7	22.50	76	3.8	5.3	f.c.c.
Iron	Fe	2798	5430	0.11	43.3	7.87	28.5	6.53	9.71	b.c.c.
Lanthanum	La	1688	6280	0.048	8.	6.19	10.5	2.77	57	Hexag.
Lead	Pb	621	3137	0.031	20.	11.36	2	16.3	20.6	f.c.c.
Lithium	Li	357	2426	0.79	41.	0.534	.....	31	8.55	b.c.c.
Magnesium	Mg	1202	2025	0.245	88.5	1.74	6.35	15.05	4.45	h.c.p.
Manganese	Mn	2273	3900	0.115	.....	7.73	23	12.22	185	Complex cubic
Mercury	Hg	−37	675	0.033	4.7	13.55	.....	....	98.4	Rhomb.
Molybdenum	Mo	4730	10040	0.000	82.	10.22	47	2.7	5.2	b.c.c.
Nickel	Ni	2647	4950	0.105	53.	8.90	30	7.39	6.84	f.c.c.
Osmium	Os	4900	9950	0.031	.....	22.57	81	2.6	9.5	h.c.p.
Palladium	Pd	2826	7200	0.058	40.5	12.02	16.3	6.53	10.8	f.c.c.
Phosphorus (white)	P	112	536	0.177	.....	1.83	.....	70	10	Cubic
Platinum	Pt	3217	8185	0.0314	39.8	21.45	21.3	4.9	10.6	f.c.c.
Plutonium	Pu	1184	6000	0.033	4.8	19.00	14	30.55	141.4	Monoclinic
Potassium	K	147	1400	0.177	58.	0.86	.....	46	6.15	b.c.c.
Rhenium	Re	5755	10650	0.033	41.	21.04	66.7	3.7	19.3	h.c.p.
Rhodium	Rh	3571	8130	0.059	50.6	12.44	42.5	4.6	4.51	f.c.c.
Rubidium	Rb	102	1270	0.080	.....	1.53	.....	50	12.5	b.c.c.
Ruthenium	Ru	4530	8850	0.057	.....	12.20	60	5.1	7.6	h.c.p.
Selenium	Se	423	1265	0.084	.....	4.70	8.4	21	12	Hexag.
Silicon[c]	Si	2570	48660	0.162	48.2	2.33	16.35	1.6 to 1.4	10	Diam. cubic
Silver	Ag	1761	4010	0.056	242.	10.49	11	10.9	1.59	f.c.c.
Sodium	Na	208	1638	0.295	77.2	0.971	.....	39	4.2	b.c.c.
Strontium	Sr	1414	2520	0.176	.....	2.60	.....	.....	23	f.c.c.
Tantalum	Ta	5425	9800	0.034	31.3	16.60	27	3.6	12.45	b.c.c.
Tellurium	Te	841	1814	0.047	3.3	6.24	6	9.3	46×10^5	Hexag.
Thallium	Tl	577	2655	0.031	22.5	11.85	.....	16	18	h.c.p.
Thorium	Th	3182	7000	0.034	21.7	11.66	.....	6.9	13	f.c.c.
Tin	Sn	449	4120	0.054	36.2	7.30	6.3	13	11	Tetrag
Titanium	Ti	3035	5900	0.124	9.8	4.51	16.8	4.67	42	h.c.p.
Tungsten	W	6170	10706	0.033	96.	19.30	50	2.55	5.65	b.c.c.
Uranium	U	2070	6904	0.028	17.1	19.07	24	3.8 to 7.8	30	Orthorhomb.
Vanadium	V	3450	6150	0.119	16.9	6.1	19	4.6	26	h.c.c.
Yttrium	Y	2748	5490	0.071	8.5	4.47	17	....	57	h.c.p.
Zinc	Zn	787	1663	0.092	65.	7.13	.....	22	5.92	h.c.p.
Zirconium	Zr	3366	6470	0.067	9.6	6.49	13.7	3.2	40	h.c.p.

[a] Near 68°F (20°C)
[b] Sublimes—triple point at 2028 atm.
[c] Semiconductor.
Courtesy of "Metals Handbook," vol. 1, 8th ed., American Society for Metals, Cleveland, 1961.

Properties of Common Metals (density, Young's modules, coefficient of thermal expansion).

Metal	Modulus of elasticity E, million psi	Ultimate tensile strength σ_u, thousand psi	Yield strength σ_y, thousand psi	Endurance limit σ_{end}, thousand psi	Hardness Brinell
Gray cast iron, ASTM 20, med. sec.	12	22	....	10	180
Gray cast iron, ASTM 50, med. sec.	19	53	....	25	240
Nodular ductile cast iron:					
Type 60-45-10........................	22–25	60–80	45–60	35	140–190
Type 120-90-02.....................	22–25	120–150	90–125	52	240–325
Austenitic	18.5	58–68	32–38	32	140–200
Malleable cast iron, ferritic 32510..	25	50	33.5	28	110–156
Malleable cast iron, pearlitic 60003	28	80–100	60–80	39–40	197–269
Ingot iron, hot rolled......................	29.8	44	23	28	83
Ingot iron, cold drawn	29.8	73	69	33	142
Wrought iron, hot rolled longit........	29.5	48	27	23	97–105
Cast carbon steel, normalized 70000	30	70	38	31	140
Cast steel, low alloy, 100,000 norm.					
and temp	29–30	100	68	45	209
Cast steel, low alloy, 200,000					
quench. and temp	29–30	200	170	85	400
Wrought plain C steel:					
C1020 hot rolled...................	29–30	66	44	32	143
C1045 hard. and temp. 100°F	29–30	118	88	....	277
C1095 hard. and temp. 700°F	29–30	180	118	....	375
Low-alloy steels:					
Wrought 1330, HT and temp					
1000°F.............................	29–30	122	100	....	248
Wrought 2317, HT and temp					
1000°F.............................	29–30	107	72	....	222
Wrought 4340, HT and temp					
800°F..............................	29–30	220	200	....	445
Wrought 6150, HT and temp					
1000°F.............................	29–30	187	179	....	444
Wrought 2317, HT and temp					
800°F..............................	29–30	214	194	....	423
Ultra high strength steel H11, HT	30	295–311	241–247	132	
300M HT and temp. 500°F......	....	289	242	116	
4340 HT and temp. 400°F.......	30	287	270	107	
25 Ni Maraging.....................	24	319	284		
Austenitic Stainless Steel 302, cold					
worked	28	110	75	34	240
Ferritic Stainless Steel 302, cold					
worked	29	75–90	45–80		
Martensitic Stainless Steel 410, HT	29	90–190	60–145	40	180–390
Martensitic Stainless Steel 440A,					
HT....................................	29	260	240	....	510
Nitriding Steels, 135 Mod, hard					
and temp. (core properties)	29–30	145–159	125–141	45–90	285–320
Nitiding Steels 5Ni-2A, hard. and					
temp...................................	29–30	206	202	90	
Structural Steel............................	30	50–65	30–40	....	120
Aluminum Alloys, cast:					
195 SHT and aged	10.1	36	24	8	75
220 SHT.............................	9.5	48	26	8	75
142 SHT and aged	10.3	28–47	25–42	9.5	75–110
355 SHT and aged	10.2	35–42	25–27	9–10	80–90
A13 as cast...........................	10.3	39	27	19	

ANSI Grades for Steel and Aluminum Alloys (yield strength and ultimate strength for alloys for various conditions including Cold Rolled and Hot Rolled).

Metal	Modulus of elasticity E, million psi	Ultimate tensile strength σ_u, thousand psi	Yield strength σ_y, thousand psi	Endurance limit σ_{end}, thousand psi	Hardness Brinell
Aluminum Alloys, wrought:					
EC ann	10	12	4		
EC H 19, hard......................	10	27	24	7	
3003 H 18 hard.....................	10	29	27	10	55
2024 H T (T3)	10.6	70	50	20	120
5052 H 38, hard....................	10.2	42	37	20	77
7075 HT (T6)	10.4	83	73	23	150
7079 HT (T6)	10.3	78	68	23	145
Copper alloys, cast:					
Leaded red brass BB11-4A........	9–14.8	33–46	17–24	....	55
Leaded tin bronze BB11-2A	12–16	36–48	16–21	....	60–72
Yellow brass BB11-7A	12–14	60–78	25–40	....	80–95
Aluminum bronze BB11-9BHT....	15	90	40	....	180
Copper alloys, wrought:					
Oxygen-free 102 ann..............	17	32–35	10	11	
Hard	....	50–55	45	13	
Beryllium copper, 172 HT........	19	165–183	150–170	35–40	
Cartridge Brass, 260 hard	16	76	63	21	
Muntz metal, 280 ann.............	15	54	21		
Admiralty, 442 ann.................	16	53	22		
Manganese bronze, 675 hard ..	15	84	60		
Phosphor bronze, 521 spring....	16	112	70		
Silicon bronze, 647 HT	18	100	88		
Cupro-Nickel, 715 hard	22	80	73		
Magnesium alloys, cast:					
AZ63A, aged......................	6.5	30–40	14-19	11–15	55–73
AZ92A, aged......................	6.5	36–40	16-21	11–15	80–84
HK31A, T6	6.5	31	16	9–11	55
Magnesium Alloys, wrought:					
AZ61AF, forged......................	6.5	43	26	17–22	55
ZE-10A-H24	6.5	34-38	19-28	20–24	
HM31A-T5	6.5	42	33	12–14	
Nickel-alloy castings 210............	21.5	45–60	20–30	....	80–125
Monel 411 cast	19	65–90	32–45	....	125–150
Inconel 610 cast	23	70–95	30–45	....	190
Nickel alloys, wrought:					
200 Spring	30	90–130	70–115		
Duranickel, 301 Spring	30	155–190			
K Monel, K500 Spring	26	145–165	130–180		
Titanium alloys, wrought, unalloyed	15–16	60–110	40–95	60–70	
5A1-2.5 Sn	16–17	115–140	110–135	95	
13V-11 Cr3A1 HT	14.5–16	190–240	170–220	50–55	
Zinc, wrought, comm. rolled........	....	25–31	....	4.1	
Zinconium, wrought:	14	64	53		
Reactor grade..........................	14	49	29		
Zircaloy 2	13.8	68	61		
Pure metals, wrought:					
Beryllium, ann	44	60–90	45–55		
Hafnium, ann	20	77	32		
Thorium, ann	10	34	26		
Vanadium, ann	20	72	64		
Uranium, ann	30	90	25		

ANSI Grades for Steel and Aluminum Alloys (yield strength and ultimate strength for alloys for various conditions including Cold Rolled and Hot Rolled). **(CONTINUED)**

Metal	Modulus of elasticity E, million psi	Ultimate tensile strength σ_u, thousand psi	Yield strength σ_y, thousand psi	Endurance limit σ_{end}, thousand psi	Hardness Brinell
Precious metals:	12	19		46	25
Gold, ann	11	22	8		25–35
Silver, ann	21	17–26	2–5.5		38–52
Platinum, ann	17	30	5		46
Rhodium, ann....................	42	73			55–156
Osmium, cast	80	....			350
Iridium, ann	74	....			170

Babbitt has a compressive elastic limit of 1.3 to 2.5 ksi and a Brinell hardness of 20.

Compressive yield strength of all metals, except those cold-worked = tensile yield strength.

Poisson's ratio is in the range 0.25 to 0.35 for metals.

Yield strength is determined at 0.2 per cent permanent deformation.

Modulus of elasticity in shear for metals is approximately 0.4 of modulus of elasticity in tension E.

Compressive yield strength of cast iron 80,000 to 150,000 ksi.

From Materials in Design Engineering, Materials Selector Issue, vol. 56, No. 5, 1962.

Courtesy of McGraw-Hill Companies.

ANSI Grades for Steel and Aluminum Alloys (yield strength and ultimate strength for alloys for various conditions including Cold Rolled and Hot Rolled). **(CONTINUED)**

Type	Specific gravity	Coefficient of thermal expansion, 10^{-6} °F	Thermal conductivity, Btu/hr/sq ft/°F/ft	Volume resistivity, ohm-cm	Dialectric strength [a], volts/mil	Modulus of elasticity in tension 10^6 psi	Tensile strength, 10^3 psi
Acrylic, general purpose, type I...............................	1.17–1.19	4.5	0.12	$>10^{15}$	450–530	3.5–4.5	6–9
Cellulose acetnte, type I (med.)	1.24–1.34	4.4–9.0	0.1–0.19	10^{12}	250–600	....	2.7–6.5
Epoxy, general purpose	1.12–2.4	1.7–5.0	0.1–0.8	10^{13}	350–550	....	2–12
Nylon 6.............................	1.13–1.14	4.6–5.4	0.1–0.14	10^{14}	420–485	2.5–3.4	10.2–12
Phenolic, type I (mech.)..........	1.31	3.3–4.4	....	1.7×10^{12}	350–400	4–5	6–9
Polyester, Allyl type	1.30–1.45	2.8–5.6	0.12	$>10^{13}$	330–500	2–3	4.5–7
Silicone, general (mineral)......	1.80–2.0	2.8–3.2	0.09	$>10^{13}$	350–400	....	4.2
Polystyrene, general purpose	1.04–1.07	3.3–4.8	0.00–0.09	10^{18}	>500	4–5	5–8
Polyethylene, low density	0.92	8.9–11	0.19	10^{18}	480	0.22	1.4–2
Polyethylene, medium density............................	0.93	8.3–16.7	0.19	$>10^{15}$	480	....	2
Polyethylene, high density......	0.96	8.3–16.7	0.19	$>10^{15}$	480	....	4.4
Polypropylene......................	0.89–0.91	6.2	0.08	10^{16}	769–820	1.4–1.7	5

[a] Short time.

Harold A. Rothbart, Mechanical Design and Systems Handbook, © 1985, McGraw-Hill.

Reprinted with permission of The McGraw-Hill Companies

Properties of Common Plastics (density, Young's Modulus, coefficient of thermal expansion, strength).

SHEET METAL AND WIRE GAGE DESIGNATION

GAGE NO.	AMERICAN OR BROWN & SHARPE'S A.W.G. OR B. & S.	UNITED STATES STANDARD	MANU-FACTURERS' STANDARD FOR SHEET STEEL	GAGE NO.
0000000	. . . .	.500	. . . .	0000000
000000	.5800	.469	. . . .	000000
00000	.5165	.438	. . . .	00000
0000	.4600	.406	. . . .	0000
000	.4096	.375	. . . .	000
00	.3648	.344	. . . .	00
0	.3249	.312	. . . .	0
1	.2893	.281	. . . .	1
2	.2576	.266	. . . .	2
3	.2294	.250	.2391	3
4	.2043	.234	.2242	4
5	.1819	.219	.2092	5
6	.1620	.203	.1943	6
7	.1443	.188	.1793	7
8	.1285	.172	.1644	8
9	.1144	.156	.1495	9
10	.1019	.141	.1345	10
11	.0907	.125	.1196	11
12	.0808	.109	.1046	12
13	.0720	.0938	.0897	13
14	.0642	.0781	.0747	14
15	.0571	.0703	.0673	15
16	.0508	.0625	.0598	16
17	.0453	.0562	.0538	17
18	.0403	.0500	.0478	18
19	.0359	.0438	.0418	19
20	.0320	.0375	.0359	20
21	.0285	.0344	.0329	21
22	.0253	.0312	.0299	22
23	.0226	.0281	.0269	23
24	.0201	.0250	.0239	24
25	.0179	.0219	.0209	25
26	.0159	.0188	.0179	26
27	.0142	.0172	.0164	27
28	.0126	.0156	.0149	28
29	.0113	.0141	.0135	29
30	.0100	.0125	.0120	30
31	.0089	.0109	.0105	31
32	.0080	.0102	.0097	32
33	.0071	.00938	.0090	33
34	.0063	.00859	.0082	34
35	.0056	.00781	.0075	35
36	.0050	.00703	.0067	36

Sheet Metal and Wire Gage Designation.

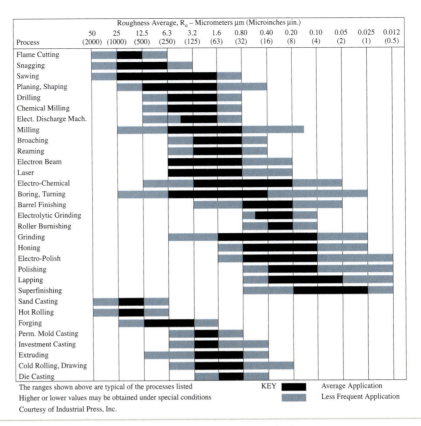

Process	Roughness Average, R_a – Micrometers µm (Microinches µin.)												
	50 (2000)	25 (1000)	12.5 (500)	6.3 (250)	3.2 (125)	1.6 (63)	0.80 (32)	0.40 (16)	0.20 (8)	0.10 (4)	0.05 (2)	0.025 (1)	0.012 (0.5)
Flame Cutting													
Snagging													
Sawing													
Planing, Shaping													
Drilling													
Chemical Milling													
Elect. Discharge Mach.													
Milling													
Broaching													
Reaming													
Electron Beam													
Laser													
Electro-Chemical													
Boring, Turning													
Barrel Finishing													
Electrolytic Grinding													
Roller Burnishing													
Grinding													
Honing													
Electro-Polish													
Polishing													
Lapping													
Superfinishing													
Sand Casting													
Hot Rolling													
Forging													
Perm. Mold Casting													
Investment Casting													
Extruding													
Cold Rolling, Drawing													
Die Casting													

The ranges shown above are typical of the processes listed

Higher or lower values may be obtained under special conditions

Courtesy of Industrial Press, Inc.

KEY ■ Average Application
     ■ Less Frequent Application

Surface Roughness Produced by Common Production Methods.

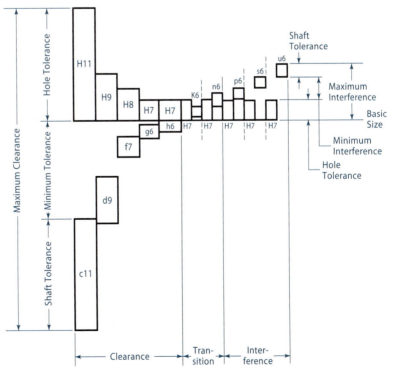

Courtesy of Industrial Press, Inc.

Standard Allowances, Tolerances, and Fits.

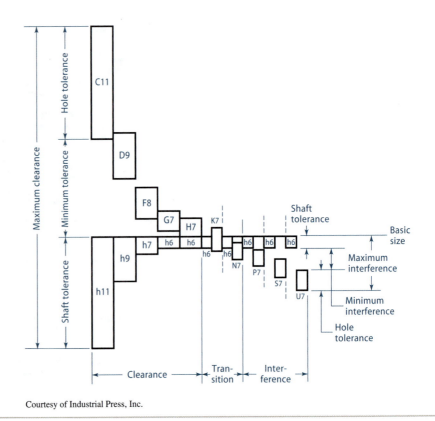

Courtesy of Industrial Press, Inc.

Standard Allowances, Tolerances, and Fits. (CONTINUED)

ISO SYMBOL		DESCRIPTION	
Hole Basis	Shaft Basis		
Clearance Fits H11/c11	C11/h11	*Loose running* fit for wide commercial tolerances or allowances on external members.	**More Clearance** ↑
H9/d9	D9/h9	*Free running* fit not for use where accuracy is essential, but good for large temperature variations, high running speeds, or heavy journal pressures.	
H8/f7	F8/h7	*Close Running* fit for running on accurate machines and for accurate moderate speeds and journal pressures.	
H7/g6	G7/h6	*Sliding fit* not intended to run freely, but to move and turn freely and locate accurately.	
H7/h6	H7/h6	*Locational clearance* fit provides snug fit for locating stationary parts; but can be freely assembled and disassembled.	
Transition Fits H7/k6	K7/h6	*Locational transition* fit for accurate location, a compromise between clearance and interference.	
H7/n6	N7/h6	*Locational transition* fit for more accurate location where greater interferance is permissible.	
Interference Fits H7/p6[a]	P7/h6	*Locational interference* fit for parts requiring rigidity and alignment with prime accuracy of location but without special bore pressure requirements.	**More Interference** ↓
H7/s6	S7/h6	*Medium drive* fit for ordinary steel parts or shrink fits on light sections, the tightest fit usable with cast iron.	
H7/u6	U7/h6	*Force* fit suitable for parts which can be highly stressed or for shrink fits where the heavy pressing forces required are impractical.	

[a] Transition fit for basic sizes in range from 0 through 3 mm.

Courtesy of Industrial Press, Inc.

Standard Allowances, Tolerances, and Fits. (CONTINUED)

AMERICAN NATIONAL STANDARD
PREFERRED METRIC LIMITS AND FITS

ANSI B4.2–1978

Dimensions in mm.

PREFERRED HOLE BASIS CLEARANCE FITS

BASIC SIZE		LOOSE RUNNING Hole H11	LOOSE RUNNING Shaft c11	LOOSE RUNNING Fit	FREE RUNNING Hole H9	FREE RUNNING Shaft d9	FREE RUNNING Fit	CLOSE RUNNING Hole H8	CLOSE RUNNING Shaft f7	CLOSE RUNNING Fit	SLIDING Hole H7	SLIDING Shaft g6	SLIDING Fit	LOCATIONAL CLEARANCE Hole H7	LOCATIONAL CLEARANCE Shaft h6	LOCATIONAL CLEARANCE Fit
1	MAX	1.060	0.940	0.180	1.025	0.980	0.070	1.014	0.994	0.030	1.010	0.998	0.018	1.010	1.000	0.016
	MIN	1.000	0.880	0.060	1.000	0.955	0.020	1.000	0.984	0.006	1.000	0.992	0.002	1.000	0.994	0.000
1.2	MAX	1.260	1.140	0.180	1.225	1.180	0.070	1.214	1.194	0.030	1.210	1.198	0.018	1.210	1.200	0.016
	MIN	1.200	1.080	0.060	1.200	1.155	0.020	1.200	1.184	0.006	1.200	1.192	0.002	1.200	1.194	0.000
1.6	MAX	1.660	1.540	0.180	1.625	1.580	0.070	1.614	1.594	0.030	1.610	1.598	0.018	1.610	1.600	0.016
	MIN	1.600	1.480	0.060	1.600	1.555	0.020	1.600	1.584	0.006	1.600	1.592	0.002	1.600	1.594	0.000
2	MAX	2.060	1.940	0.180	2.025	1.980	0.070	2.014	1.994	0.030	2.010	1.998	0.018	2.010	2.000	0.016
	MIN	2.000	1.880	0.060	2.000	1.955	0.020	2.000	1.984	0.006	2.000	1.992	0.002	2.000	1.994	0.000
2.5	MAX	2.560	2.440	0.180	2.525	2.480	0.070	2.514	2.494	0.030	2.510	2.498	0.018	2.510	2.500	0.016
	MIN	2.500	2.380	0.060	2.500	2.455	0.020	2.500	2.484	0.006	2.500	2.492	0.002	2.500	2.494	0.000
3	MAX	3.060	2.940	0.180	3.025	2.980	0.070	3.014	2.994	0.030	3.010	2.998	0.018	3.010	3.000	0.016
	MIN	3.000	2.880	0.060	3.000	2.955	0.020	3.000	2.984	0.006	3.000	2.992	0.002	3.000	2.994	0.000
4	MAX	4.075	3.930	0.220	4.030	3.970	0.090	4.018	3.990	0.040	4.012	3.996	0.024	4.012	4.000	0.020
	MIN	4.000	3.855	0.070	4.000	3.940	0.030	4.000	3.978	0.010	4.000	3.988	0.004	4.000	3.992	0.000
5	MAX	5.075	4.930	0.220	5.030	4.970	0.090	5.018	4.990	0.040	5.012	4.996	0.024	5.012	5.000	0.020
	MIN	5.000	4.855	0.070	5.000	4.940	0.030	5.000	4.978	0.010	5.000	4.988	0.004	5.000	4.992	0.000
6	MAX	6.075	5.930	0.220	6.030	5.970	0.090	6.018	5.990	0.040	6.012	5.996	0.024	6.012	6.000	0.020
	MIN	6.000	5.855	0.070	6.000	5.940	0.030	6.000	5.978	0.010	6.000	5.988	0.004	6.000	5.992	0.000
8	MAX	8.090	7.920	0.260	8.036	7.960	0.112	8.022	7.987	0.050	8.015	7.995	0.029	8.015	8.000	0.024
	MIN	8.000	7.830	0.080	8.000	7.924	0.040	8.000	7.972	0.013	8.000	7.986	0.005	8.000	7.991	0.000
10	MAX	10.090	9.920	0.260	10.036	9.960	0.112	10.022	9.987	0.050	10.015	9.995	0.029	10.015	10.000	0.024
	MIN	10.000	9.830	0.080	10.000	9.924	0.040	10.000	9.972	0.013	10.000	9.986	0.005	10.000	9.991	0.000
12	MAX	12.110	11.905	0.315	12.043	11.950	0.136	12.027	11.984	0.061	12.018	11.994	0.035	12.018	12.000	0.029
	MIN	12.000	11.795	0.095	12.000	11.907	0.050	12.000	11.966	0.016	12.000	11.983	0.006	12.000	11.989	0.000
16	MAX	16.110	15.905	0.315	16.043	15.950	0.136	16.027	15.984	0.061	16.018	15.994	0.035	16.018	16.000	0.029
	MIN	16.000	15.795	0.095	16.000	15.907	0.050	16.000	15.966	0.016	16.000	15.983	0.006	16.000	15.989	0.000
20	MAX	20.130	19.890	0.370	20.052	19.935	0.169	20.033	19.980	0.074	20.021	19.993	0.041	20.021	20.000	0.034
	MIN	20.000	19.760	0.110	20.000	19.883	0.065	20.000	19.959	0.020	20.000	19.980	0.007	20.000	19.987	0.000
25	MAX	25.130	24.890	0.370	25.052	24.935	0.169	25.033	24.980	0.074	25.021	24.993	0.041	25.021	25.000	0.034
	MIN	25.000	24.760	0.110	25.000	24.883	0.065	25.000	24.959	0.020	25.000	24.980	0.007	25.000	24.987	0.000
30	MAX	30.130	29.890	0.370	30.052	29.935	0.169	30.033	29.980	0.074	30.021	29.993	0.041	30.021	30.000	0.034
	MIN	30.000	29.760	0.110	30.000	29.883	0.065	30.000	29.959	0.020	30.000	29.980	0.007	30.000	29.987	0.000

Reprinted from ASME B18.2.1-1981 (R1992), B18.2.2-1987 (R1993), B18.3-1986 (R1993), B18.6.2-1972 (R1993), B18.6.3-1972 (R1991), B18.21.1-1994, B18.22.1-1965 (R1990), B17.1-1967 (R1998), B17.2-1967 (R1990) and B4.2-1978 (R1994), by permission of The American Society of Mechanical Engineers. All rights reserved. ANSI B4.2-1978

Metric Limits and Fits.

RUNNING AND SLIDING FITS

VALUES IN THOUSANDTHS OF AN INCH

Nominal Size Range Inches		Class RC1 Precision Sliding			Class RC2 Sliding Fit			Class RC3 Precision Running			Class RC4 Close Running			Class RC5 Medium Running		
		Hole Tol. GR5	Minimum Clearance	Shaft Tol. GR4	Hole Tol. GR6	Minimum Clearance	Shaft Tol. GR5	Hole Tol. GR7	Minimum Clearance	Shaft Tol. GR6	Hole Tol. GR8	Minimum Clearance	Shaft Tol. GR7	Hole Tol. GR8	Minimum Clearance	Shaft Tol. GR7
Over	To	-0		+0	-0		+0	-0		+0	-0		+0	-0		+0
0	.12	+0.15	0.10	-0.12	+0.25	0.10	-0.15	+0.40	0.30	-0.25	+0.60	0.30	-0.40	+0.60	0.60	-0.40
.12	.24	+0.20	0.15	-0.15	+0.30	0.15	-0.20	+0.50	0.40	-0.30	+0.70	0.40	-0.50	+0.70	0.80	-0.50
.24	.40	+0.25	0.20	-0.15	+0.40	0.20	-0.25	+0.60	0.50	-0.40	+0.90	0.50	-0.60	+0.90	1.00	-0.60
.40	.71	+0.30	0.25	-0.20	+0.40	0.25	-0.30	+0.70	0.60	-0.40	+1.00	0.60	-0.70	+1.00	1.20	-0.70
.71	1.19	+0.40	0.30	-0.25	+0.50	0.30	-0.40	+0.80	0.80	-0.50	+1.20	0.80	-0.80	+1.20	1.60	-0.50
1.19	1.97	+0.40	0.40	-0.30	+0.60	0.40	-0.40	+1.00	1.00	-0.60	+1.60	1.00	-1.00	+1.60	2.00	-1.00
1.97	3.15	+0.50	0.40	-0.30	+0.70	0.40	-0.50	+1.20	1.20	-0.70	+1.80	1.20	-1.20	+1.80	2.50	-1.20
3.15	4.73	+0.60	0.50	-0.40	+0.90	0.50	-0.60	+1.40	1.40	-0.90	+2.20	1.40	-1.40	+2.20	3.00	-1.40
4.73	7.09	+0.70	0.60	-0.50	+1.00	0.60	-0.70	+1.60	1.60	-1.00	+2.50	1.60	-1.60	+2.50	3.50	-1.60
7.09	9.85	+0.80	0.60	-0.60	+1.20	0.60	-0.80	+1.80	2.00	-1.20	+2.80	2.00	-1.80	+2.80	4.50	-1.80
9.85	12.41	+0.90	0.80	-0.60	+1.20	0.80	-0.90	+2.00	2.50	-1.20	+3.00	2.50	-2.00	+3.00	5.00	-2.00
12.41	15.75	+1.00	1.00	-0.70	+1.40	1.00	-1.00	+2.20	3.00	-1.40	+3.50	3.00	-2.20	+3.50	6.00	-2.20

Nominal Size Range Inches		Class RC6 Medium Running			Class RC7 Free Running			Class RC8 Loose Running			Class RC9 Loose Running		
		Hole Tol. GR9	Minimum Clearance	Shaft Tol. GR8	Hole Tol. GR9	Minimum Clearance	Shaft Tol. GR8	Hole Tol. GR10	Minimum Clearance	Shaft Tol. GR9	Hole Tol. GR11	Minimum Clearance	Shaft Tol. GR10
Over	To	-0		+0	-0		+0	-0		+0	-0		+0
0	.12	+1.00	0.60	-0.60	+1.00	1.00	-0.60	+1.60	2.50	-1.00	+2.50	4.00	-1.60
.12	.24	+1.20	0.80	-0.70	+1.20	1.20	-0.70	+1.80	2.80	-1.20	+3.00	4.50	-1.80
.24	.40	+1.40	1.00	-0.90	+1.40	1.60	-0.90	+2.20	3.00	-1.40	+3.50	6.00	-2.20
.40	.71	+1.60	1.20	-1.00	+1.60	2.00	-1.00	+2.80	3.50	-1.60	+4.00	6.00	-2.80
.71	1.19	+2.00	1.60	-1.20	+2.00	2.50	-1.20	+3.50	4.50	-2.00	+5.00	7.00	-3.50
1.19	1.97	+2.50	2.00	-1.60	+2.50	3.00	-1.60	+4.00	5.00	-2.50	+6.00	8.00	-4.00
1.97	3.15	+3.00	2.50	-1.80	+3.00	4.00	-1.80	+4.50	6.00	-3.00	+7.00	9.00	-4.50
3.15	4.73	+3.50	3.00	-2.20	+3.50	5.00	-2.20	+5.00	7.00	-3.50	+9.00	10.00	-5.00
4.73	7.09	+4.00	3.50	-2.50	+4.00	6.00	-2.50	+6.00	8.00	-4.00	+10.00	12.00	-6.00
7.09	9.85	+4.50	4.00	-2.80	+4.50	7.00	-2.80	+7.00	10.00	-4.50	+12.00	15.00	-7.00
9.85	12.41	+5.00	5.00	-3.00	+5.00	8.00	-3.00	+8.00	12.00	-5.00	+12.00	18.00	-8.00
12.41	15.75	+6.00	6.00	-3.50	+6.00	10.00	-3.50	+9.00	14.00	-6.00	+14.00	22.00	-9.00

VALUES IN MILLIMETERS

Nominal Size Range Millimeters		Class RC1 Precision Sliding			Class RC2 Sliding Fit			Class RC3 Precision Running			Class RC4 Close Running			Class RC5 Medium Running		
		Hole Tol. H5	Minimum Clearance	Shaft Tol. g4	Hole Tol. H6	Minimum Clearance	Shaft Tol. g5	Hole Tol. H7	Minimum Clearance	Shaft Tol. f6	Hole Tol. H8	Minimum Clearance	Shaft Tol. f7	Hole Tol. H8	Minimum Clearance	Shaft Tol. e7
Over	To	-0		+0	-0		+0	-0		+0	-0		+0	-0		+0
0	3	+0.004	0.003	-0.003	+0.006	0.003	-0.004	+0.010	0.008	-0.006	+0.015	0.008	-0.010	+0.015	0.015	-0.010
3	6	+0.005	0.004	-0.004	+0.008	0.004	-0.005	+0.013	0.010	-0.008	+0.018	0.010	-0.013	+0.018	0.020	-0.013
6	10	+0.006	0.005	-0.004	+0.010	0.005	-0.006	+0.015	0.013	-0.010	+0.023	0.013	-0.015	+0.023	0.025	-0.015
10	18	+0.008	0.006	-0.005	+0.010	0.006	-0.008	+0.018	0.015	-0.010	+0.025	0.015	-0.018	+0.025	0.030	-0.018
18	30	+0.010	0.008	-0.006	+0.013	0.008	-0.010	+0.020	0.020	-0.013	+0.030	0.020	-0.020	+0.030	0.040	-0.020
30	50	+0.010	0.010	-0.008	+0.015	0.010	-0.010	+0.030	0.030	-0.015	+0.040	0.030	-0.030	+0.040	0.050	-0.030
50	80	+0.013	0.010	-0.008	+0.018	0.010	-0.013	+0.030	0.030	-0.020	+0.050	0.030	-0.030	+0.050	0.060	-0.030
80	120	+0.015	0.013	-0.010	+0.023	0.013	-0.015	+0.040	0.040	-0.020	+0.060	0.040	-0.040	+0.060	0.080	-0.040
120	180	+0.018	0.015	-0.013	+0.025	0.015	-0.018	+0.040	0.040	-0.030	+0.060	0.040	-0.040	+0.060	0.090	-0.040
180	250	+0.020	0.015	-0.015	+0.030	0.015	-0.020	+0.050	0.050	-0.030	+0.070	0.050	-0.050	+0.070	0.110	-0.050
250	315	+0.023	0.020	-0.015	+0.030	0.020	-0.023	+0.050	0.060	-0.030	+0.080	0.060	-0.050	+0.080	0.130	-0.050
315	400	+0.025	0.025	-0.018	+0.036	0.025	-0.025	+0.060	0.080	-0.040	+0.090	0.080	-0.060	+0.090	0.150	-0.060

Nominal Size Range Millimeters		Class RC6 Medium Running			Class RC7 Free Running			Class RC8 Loose Running			Class RC9 Loose Running		
		Hole Tol. H9	Minimum Clearance	Shaft Tol. e8	Hole Tol. H9	Minimum Clearance	Shaft Tol. d8	Hole Tol. H10	Minimum Clearance	Shaft Tol. e9	Hole Tol. GR11	Minimum Clearance	Shaft Tol. gr10
Over	To	-0		+0	-0		+0	-0		+0	-0		+0
0	3	+0.025	0.015	-0.015	+0.025	0.025	-0.015	+0.041	0.064	-0.025	+0.060	0.100	-0.040
3	6	+0.030	0.015	-0.018	+0.030	0.030	-0.018	+0.046	0.071	-0.030	+0.080	0.110	-0.050
6	10	+0.036	0.025	-0.023	+0.036	0.040	-0.023	+0.056	0.076	-0.036	+0.070	0.130	-0.060
10	18	+0.040	0.030	-0.025	+0.040	0.050	-0.025	+0.070	0.090	-0.040	+0.100	0.150	-0.070
18	30	+0.050	0.040	-0.030	+0.050	0.060	-0.030	+0.090	0.110	-0.050	+0.130	0.180	-0.090
30	50	+0.060	0.050	-0.040	+0.060	0.080	-0.040	+0.100	0.130	-0.060	+0.150	0.200	-0.100
50	80	+0.080	0.060	-0.050	+0.080	0.100	-0.050	+0.110	0.150	-0.080	+0.180	0.230	-0.120
80	120	+0.090	0.080	-0.060	+0.090	0.130	-0.060	+0.130	0.180	-0.090	+0.230	0.250	-0.130
120	180	+0.100	0.090	-0.060	+0.100	0.150	-0.060	+0.150	0.200	-0.100	+0.250	0.300	-0.150
180	250	+0.110	0.100	-0.070	+0.110	0.180	-0.070	+0.180	0.250	-0.110	+0.300	0.380	-0.180
250	315	+0.130	0.130	-0.080	+0.130	0.200	-0.080	+0.200	0.300	-0.130	+0.300	0.460	-0.200
315	400	+0.150	0.150	-0.090	+0.150	0.250	-0.090	+0.230	0.360	-0.150	+0.360	0.560	-0.230

Running and Sliding Fits.

LOCATIONAL CLEARANCE FITS

VALUES IN THOUSANDTHS OF AN INCH

Nominal Size Range Inches		Class LC1			Class LC2			Class LC3			Class LC4			Class LC5			Class LC6		
		Hole Tol. GR6	Minimum Clearance	Shaft Tol. GR5	Hole Tol. GR8	Minimum Clearance	Shaft Tol. GR7	Hole Tol. GR10	Minimum Clearance	Shaft Tol. GR9	Hole Tol. GR7	Minimum Clearance	Shaft Tol. GR6	Hole Tol. GR9	Minimum Clearance	Shaft Tol. GR8	Hole Tol. GR9	Minimum Clearance	Shaft Tol. GR8
Over	To	-0		+0	-0		+0	-0		+0	-0		+0	-0		+0	-0		+0
0	.12	+0.25	0	-0.15	+0.4	0	-0.25	+0.6	0	-0.4	+1.6	0	-1.0	+0.4	0.10	-0.25	+1.0	0.3	-0.6
.12	.24	+0.30	0	-0.20	+0.5	0	-0.30	+0.7	0	-0.5	+1.8	0	-1.2	+0.5	0.15	-0.30	+1.2	0.4	-0.7
.24	.40	+0.40	0	-0.25	+0.6	0	-0.40	+0.9	0	-0.6	+2.2	0	-1.4	+0.6	0.20	-0.40	+1.4	0.5	-0.9
.40	.71	+0.40	0	-0.30	+0.7	0	-0.40	+1.0	0	-0.7	+2.8	0	-1.6	+0.7	0.25	-0.40	+1.6	0.6	-1.0
.71	1.19	+0.50	0	-0.40	+0.8	0	-0.50	+1.2	0	-0.8	+3.5	0	-2.0	+0.8	0.30	-0.50	+2.0	0.8	-1.2
1.19	1.97	+0.60	0	-0.40	+1.0	0	-0.60	+1.6	0	-1.0	+4.0	0	-2.5	+1.0	0.40	-0.60	+2.5	1.0	-1.6
1.97	3.15	+0.70	0	-0.50	+1.2	0	-0.70	+1.8	0	-1.2	+4.5	0	-3.0	+1.2	0.40	-0.70	+3.0	1.2	-1.8
3.15	4.73	+0.90	0	-0.60	+1.4	0	-0.90	+2.7	0	-1.4	+5.0	0	-3.5	+1.4	0.50	-0.90	+3.5	1.4	-2.2
4.73	7.09	+1.00	0	-0.70	+1.6	0	-1.00	+2.5	0	-1.6	+6.0	0	-4.0	+1.6	0.60	-1.00	+4.0	1.6	-2.5
7.09	9.85	+1.20	0	-0.80	+1.8	0	-1.20	+2.8	0	-1.8	+7.0	0	-4.5	+1.8	0.60	-1.20	+4.5	2.0	-2.8
9.85	12.41	+1.20	0	-0.90	+2.0	0	-1.20	+3.0	0	-2.0	+8.0	0	-5.0	+2.0	0.70	-1.20	+5.0	2.2	-3.0
12.41	15.75	+1.40	0	-1.00	+2.2	0	-1.40	+3.5	0	-2.2	+9.0	0	-6.0	+2.2	0.70	-1.40	+6.0	2.5	-3.5

Nominal Size Range Inches		Class LC7			Class LC8			Class LC9			Class LC10			Class LC11		
		Hole Tol. GR10	Minimum Clearance	Shaft Tol. GR9	Hole Tol. GR10	Minimum Clearance	Shaft Tol. GR9	Hole Tol. GR11	Minimum Clearance	Shaft Tol. GR10	Hole Tol. GR12	Minimum Clearance	Shaft Tol. GR11	Hole Tol. GR13	Minimum Clearance	Shaft Tol. GR12
Over	To	-0		+0	-0		+0	-0		+0	-0		+0	-0		+0
0	.12	+1.6	0.6	-1.0	+1.6	1.0	-1.0	+2.5	2.5	-1.6	+1.0	4.0	-2.5	+6.0	5.0	-4.0
.12	.24	+1.8	0.8	-1.2	+1.8	1.2	-1.2	+3.0	2.8	-1.8	+5.0	4.5	-3.0	+7.0	6.0	-5.0
.24	.40	+2.2	1.0	-1.4	+2.2	1.6	-1.4	+3.5	3.0	-2.2	+6.0	5.0	-3.5	+9.0	7.0	-6.0
.40	.71	+2.8	1.2	-1.6	+2.8	2.0	-1.6	+4.0	3.5	-2.8	+7.0	6.0	-4.0	+10.0	8.0	-7.0
.71	1.19	+3.5	1.6	-2.0	+3.5	2.5	-2.0	+5.0	4.5	-3.5	+8.0	7.0	-5.0	+12.0	10.0	-8.0
1.19	1.97	+4.0	2.0	-2.5	+4.0	3.6	-2.5	+6.0	5.0	-4.0	+10.0	8.0	-6.0	+16.0	12.0	-10.0
1.97	3.15	+4.5	2.5	-3.0	+4.5	4.0	-3.0	+7.0	6.0	-4.5	+12.0	10.0	-7.0	+18.0	14.0	-12.0
3.15	4.73	+5.0	3.0	-3.5	+5.0	5.0	-3.5	+9.0	7.0	-5.0	+14.0	11.0	-9.0	+22.0	16.0	-14.0
4.73	7.09	+6.0	3.5	-4.0	+6.0	6.0	-4.0	+10.0	8.0	-6.0	+16.0	12.0	-10.0	+25.0	18.0	-16.0
7.09	9.85	+7.0	4.0	-4.5	+7.0	7.0	-4.5	+12.0	10.0	-7.0	+18.0	16.0	-12.0	+28.0	22.0	-18.0
9.85	12.41	+8.0	4.5	-5.0	+8.0	7.0	-5.0	+12.0	12.0	-8.0	+20.0	20.0	-12.0	+30.0	28.0	-20.0
12.41	15.75	+9.0	5.0	-6.0	+9.0	8.0	-6.0	+14.0	14.0	-9.0	+22.0	22.0	-14.0	+35.0	30.0	-22.0

VALUES IN MILLIMETERS

Nominal Size Range Millimeters		Class LC1			Class LC2			Class LC3			Class LC4			Class LC5			Class LC6		
		Hole Tol. H6	Minimum Clearance	Shaft Tol. h5	Hole Tol. H7	Minimum Clearance	Shaft Tol. h6	Hole Tol. H8	Minimum Clearance	Shaft Tol. h7	Hole Tol. H10	Minimum Clearance	Shaft Tol. h9	Hole Tol. H7	Minimum Clearance	Shaft Tol. g6	Hole Tol. H9	Minimum Clearance	Shaft Tol. f8
Over	To	-0		+0	-0		+0	-0		+0	-0		+0	-0		+0	-0		+0
0	3	+0.006	0	-0.004	+0.010	0	-0.006	+0.015	0	-0.010	+0.041	0	-0.025	+0.010	0.002	-0.006	+0.025	0.008	-0.015
3	6	+0.008	0	-0.005	+0.013	0	-0.008	+0.018	0	-0.013	+0.046	0	-0.030	+0.013	0.004	-0.008	+0.030	0.010	-0.018
6	10	+0.010	0	-0.006	+0.015	0	-0.010	+0.023	0	-0.015	+0.056	0	-0.036	+0.015	0.005	-0.010	+0.036	0.013	-0.023
10	18	+0.010	0	-0.008	+0.018	0	-0.010	+0.025	0	-0.018	+0.070	0	-0.040	+0.018	0.006	-0.010	+0.041	0.015	-0.025
18	30	+0.013	0	-0.010	+0.020	0	-0.013	+0.030	0	-0.020	+0.090	0	-0.050	+0.020	0.008	-0.013	+0.050	0.020	-0.030
30	50	+0.015	0	-0.010	+0.025	0	-0.015	+0.041	0	-0.025	+0.100	0	-0.060	+0.025	0.010	-0.015	+0.060	0.030	-0.040
50	80	+0.018	0	-0.013	+0.030	0	-0.018	+0.046	0	-0.030	+0.110	0	-0.080	+0.030	0.010	-0.018	+0.080	0.030	-0.050
80	120	+0.023	0	-0.015	+0.036	0	-0.023	+0.056	0	-0.036	+0.130	0	-0.080	+0.036	0.013	-0.023	+0.090	0.040	-0.060
120	180	+0.025	0	-0.018	+0.041	0	-0.025	+0.064	0	-0.041	+0.150	0	-0.100	+0.041	0.015	-0.025	+0.100	0.040	-0.060
180	250	+0.030	0	-0.020	+0.046	0	-0.030	+0.071	0	-0.046	+0.180	0	-0.110	+0.046	0.015	-0.030	+0.110	0.050	-0.070
250	315	+0.020	0	-0.023	+0.051	0	-0.030	+0.076	0	-0.051	+0.200	0	-0.130	+0.051	0.018	-0.030	+0.130	0.060	-0.080
315	400	+0.036	0	-0.025	+0.056	0	-0.036	+0.089	0	-0.056	+0.230	0	-0.150	+0.056	0.018	-0.036	+0.150	0.060	-0.090

Nominal Size Range Millimeters		Class LC7			Class LC8			Class LC9			Class LC10			Class LC11		
		Hole Tol. H10	Minimum Clearance	Shaft Tol. e9	Hole Tol. H10	Minimum Clearance	Shaft Tol. d9	Hole Tol. H11	Minimum Clearance	Shaft Tol. c10	Hole Tol. GR12	Minimum Clearance	Shaft Tol. gr11	Hole Tol. GR13	Minimum Clearance	Shaft Tol. gr12
Over	To	-0		+0	-0		+0	-0		+0	-0		+0	-0		+0
0	3	+0.041	0.015	-0.025	+0.041	0.025	-0.025	+0.064	0.06	-0.041	+0.10	0.10	-0.06	+0.15	0.13	-0.10
3	6	+0.046	0.020	-0.030	+0.046	0.030	-0.030	+0.076	0.07	-0.46	+0.13	0.11	-0.08	+0.18	0.15	-0.13
6	10	+0.056	0.025	-0.036	+0.056	0.041	-0.036	+0.089	0.08	-0.56	+0.15	0.13	-0.09	+0.23	0.18	-0.15
10	18	+0.070	0.030	-0.040	+0.070	0.050	-0.040	+0.100	0.09	-0.70	+0.18	0.15	-0.10	+0.25	0.20	-0.18
18	30	+0.090	0.040	-0.050	+0.090	0.060	-0.050	+0.130	0.11	-0.90	+0.20	0.18	-0.13	+0.31	0.25	-0.20
30	50	+0.100	0.050	-0.060	+0.100	0.090	-0.060	+0.150	0.13	-0.100	+0.25	0.20	-0.15	+0.41	0.31	-0.25
50	80	+0.110	0.060	-0.080	+0.110	0.100	-0.080	+0.180	0.15	-0.110	+0.31	0.25	-0.18	+0.46	0.36	-0.31
80	120	+0.130	0.080	-0.090	+0.130	0.130	-0.090	+0.230	0.18	-0.130	+0.36	0.28	-0.23	+0.56	0.41	-0.36
120	180	+0.150	0.090	-0.100	+0.150	0.150	-0.100	+0.250	0.20	-0.150	+0.41	0.31	-0.25	+0.64	0.46	-0.41
180	250	+0.180	0.100	-0.110	+0.180	0.180	-0.110	+0.310	0.25	-0.180	+0.46	0.41	-0.31	+0.71	0.56	-0.46
250	315	+0.200	0.110	-0.130	+0.200	0.180	-0.130	+0.310	0.31	-0.200	+0.51	0.51	-0.31	+0.76	0.71	-0.51
315	400	+0.230	0.130	-0.150	+0.230	0.200	-0.150	+0.360	0.36	-0.230	+0.56	0.56	-0.36	+0.89	0.76	-0.56

Running and Sliding Fits. (CONTINUED)

LOCATIONAL TRANSITION FITS

VALUES IN THOUSANDTHS OF AN INCH

Nominal Size Range Inches		Class LT1			Class LT2			Class LT3			Class LT4			Class LT5			Class LT6		
		Hole Tol. GR7	Maximum Interference	Shaft Tol. GR6	Hole Tol. GR8	Maximum Interference	Shaft Tol. GR7	Hole Tol. GR7	Maximum Interference	Shaft Tol. GR6	Hole Tol. GR8	Maximum Interference	Shaft Tol. GR7	Hole Tol. GR7	Maximum Interference	Shaft Tol. GR6	Hole Tol. GR8	Maximum Interference	Shaft Tol. GR7
Over	To	-0		+0	-0		+0	-0		+0	-0		+0	-0		+0	-0		+0
0	.12	+0.4	0.10	-0.25	+0.6	0.20	-0.4	+0.4	0.25	-0.25	+0.6	0.4	-0.4	+0.4	0.5	-0.25	+0.6	0.65	-0.4
.12	.24	+0.5	0.15	-0.30	+0.7	0.25	-0.5	+0.5	0.40	-0.30	+0.7	0.6	-0.5	+0.5	0.6	-0.30	+0.7	0.80	-0.5
.24	.40	+0.6	0.20	-0.40	+0.9	0.30	-0.6	+0.6	0.50	-0.40	+0.9	0.7	-0.6	+0.6	0.8	-0.40	+0.9	1.00	-0.6
.40	.71	+0.7	0.20	-0.40	+1.0	0.30	-0.7	+0.7	0.50	-0.40	+1.0	0.8	-0.7	+0.7	0.9	-0.40	+1.0	1.20	-0.7
.71	1.19	+0.8	0.25	-0.50	+1.2	0.40	-0.8	+0.8	0.60	-0.50	+1.2	0.9	-0.8	+0.8	1.1	-0.50	+1.2	1.40	-0.8
1.19	1.97	+1.0	0.30	-0.60	+1.6	0.50	-1.0	+1.0	0.70	-0.60	+1.6	1.1	-1.0	+1.0	1.3	-0.60	+1.6	1.70	-1.0
1.97	3.15	+1.2	0.30	-0.70	+1.8	0.60	-1.2	+1.2	0.80	-0.70	+1.8	1.3	-1.2	+1.2	1.5	-0.70	+1.8	2.00	-1.2
3.15	4.73	+1.4	0.40	-0.90	+2.2	0.70	-1.4	+1.4	1.00	-0.90	+2.2	1.5	-1.4	+1.4	1.9	-0.90	+2.2	2.40	-1.4
4.73	7.09	+1.6	0.50	-1.00	+2.5	0.80	-1.6	+1.6	1.10	-1.00	+2.5	1.7	-1.6	+1.6	2.2	-1.00	+2.5	2.80	-1.6
7.09	9.85	+1.8	0.60	-1.20	+2.8	0.90	-1.8	+1.8	1.40	-1.20	+2.8	2.0	-1.8	+1.8	2.6	-1.20	+2.8	3.20	-1.8
9.85	12.41	+2.0	0.60	-1.20	+3.0	1.00	-2.0	+2.0	1.40	-1.20	+3.0	2.2	-2.0	+2.0	2.6	-1.20	+3.0	3.40	-2.0
12.41	15.75	+2.2	0.70	-1.40	+3.5	1.00	-2.2	+2.2	1.60	-1.40	+3.5	2.4	-2.2	+2.2	3.0	-1.40	+3.5	3.80	-2.2

VALUES IN MILLIMETERS

Nominal Size Range Millimeters		Class LT1			Class LT2			Class LT3			Class LT4			Class LT5			Class LT6		
		Hole Tol. H7	Maximum Clearance	Shaft Tol. js6	Hole Tol. H8	Maximum Clearance	Shaft Tol. js7	Hole Tol. H7	Maximum Clearance	Shaft Tol. k6	Hole Tol. H8	Maximum Clearance	Shaft Tol. k7	Hole Tol. H7	Maximum Clearance	Shaft Tol. n6	Hole Tol. H8	Maximum Clearance	Shaft Tol. n7
Over	To	-0		+0	-0		+0	-0		+0	-0		+0	-0		+0	-0		+0
0	3	+0.010	0.002	-0.006	+0.015	0.005	-0.010	+0.010	0.006	-0.006	+0.015	0.010	-0.010	+0.010	0.013	-0.006	+0.015	0.016	-0.010
3	6	+0.013	0.004	-0.008	+0.018	0.006	-0.013	+0.013	0.010	-0.008	+0.018	0.015	-0.013	+0.013	0.015	-0.008	+0.018	0.020	-0.013
6	10	+0.015	0.005	-0.010	+0.023	0.008	-0.015	+0.015	0.013	-0.010	+0.023	0.018	-0.015	+0.015	0.020	-0.010	+0.023	0.025	-0.015
10	18	+0.018	0.005	-0.010	+0.025	0.008	-0.018	+0.018	0.013	-0.010	+0.025	0.020	-0.018	+0.018	0.023	-0.010	+0.025	0.030	-0.018
18	30	+0.020	0.006	-0.013	+0.030	0.010	-0.020	+0.020	0.015	-0.013	+0.030	0.023	-0.020	+0.020	0.028	-0.013	+0.030	0.036	-0.020
30	50	+0.025	0.008	-0.015	+0.041	0.013	-0.025	+0.025	0.018	-0.015	+0.041	0.028	-0.025	+0.025	0.033	-0.015	+0.041	0.044	-0.025
50	80	+0.030	0.008	-0.018	+0.046	0.015	-0.030	+0.030	0.020	-0.018	+0.046	0.033	-0.030	+0.030	0.038	-0.018	+0.046	0.051	-0.030
80	120	+0.036	0.010	-0.023	+0.056	0.018	-0.036	+0.036	0.025	-0.023	+0.056	0.038	-0.036	+0.036	0.048	-0.023	+0.056	0.062	-0.036
120	180	+0.041	0.013	-0.025	+0.064	0.020	-0.041	+0.041	0.028	-0.025	+0.064	0.044	-0.041	+0.041	0.056	-0.025	+0.064	0.071	-0.041
180	250	+0.046	0.015	-0.030	+0.071	0.023	-0.046	+0.046	0.036	-0.030	+0.071	0.051	-0.046	+0.046	0.066	-0.030	+0.071	0.081	-0.046
250	315	+0.051	0.015	-0.030	+0.076	0.025	-0.051	+0.051	0.036	-0.030	+0.076	0.056	-0.051	+0.051	0.066	-0.030	+0.076	0.086	-0.051
315	400	+0.056	0.018	-0.036	+0.089	0.025	-0.056	+0.056	0.041	-0.036	+0.089	0.062	-0.056	+0.056	0.076	-0.036	+0.089	0.096	-0.056

Running and Sliding Fits. (CONTINUED)

LOCATIONAL INTERFERENCE FITS

VALUES IN THOUSANDTHS OF AN INCH

Nominal Size Range Inches		Class LN1 Light Press Fit			Class LN2 Medium Press Fit			Class LN3 Heavy Press Fit			Class LN4			Class LN5			Class LN6		
		Hole Tol. GR6	Maximum Interference	Shaft Tol. GR5	Hole Tol. GR7	Maximum Interference	Shaft Tol. GR6	Hole Tol. GR7	Maximum Interference	Shaft Tol. GR6	Hole Tol. GR8	Maximum Interference	Shaft Tol. GR7	Hole Tol. GR9	Maximum Interference	Shaft Tol. GR8	Hole Tol. GR10	Maximum Interference	Shaft Tol. GR9
Over	To	-0		+0	-0		+0	-0		+0	-0		+0	-0		+0	-0		+0
0	.12	+0.25	0.40	-0.15	+0.4	0.65	-0.25	+0.4	0.75	-0.25	+0.6	1.2	-0.4	+1.0	1.8	-0.6	+1.6	3.0	-1.0
.12	.24	+0.30	0.50	-0.20	+0.5	0.80	-0.30	+0.5	0.90	-0.30	+0.7	1.5	-0.5	+1.2	2.3	-0.7	+1.8	3.6	-1.2
.24	.40	+0.40	0.65	-0.25	+0.6	1.00	-0.40	+0.6	1.20	-0.40	+0.9	1.8	-0.6	+1.4	2.8	-0.9	+2.2	4.4	-1.4
.40	.71	+0.40	0.70	-0.30	+0.7	1.10	-0.40	+0.7	1.40	-0.40	+1.0	2.2	-0.7	+1.6	3.4	-1.0	+2.8	5.6	-1.6
.71	1.19	+0.50	0.90	-0.40	+0.8	1.30	-0.50	+0.8	1.70	-0.50	+1.2	2.6	-0.8	+2.0	4.2	-1.2	+3.5	7.0	-2.0
1.19	1.97	+0.60	1.00	-0.40	+1.0	1.60	-0.60	+1.0	2.00	-0.60	+1.6	3.4	-1.0	+2.5	5.3	-1.6	+4.0	8.5	-2.5
1.97	3.15	+0.70	1.30	-0.50	+1.2	2.10	-0.70	+1.2	2.30	-0.70	+1.8	4.0	-1.2	+3.0	6.3	-1.8	+4.5	10.0	-3.0
3.15	4.73	+0.90	1.60	-0.60	+1.4	2.50	-0.90	+1.4	2.90	-0.90	+2.2	4.8	-1.4	+4.0	7.7	-2.2	+5.0	11.5	-3.5
4.73	7.09	+1.00	1.90	-0.70	+1.6	2.80	-1.00	+1.6	3.50	-1.00	+2.5	5.6	-1.6	+4.5	8.7	-2.5	+6.0	13.5	-4.0
7.09	9.85	+1.20	2.20	-0.80	+1.8	3.20	-1.20	+1.8	4.20	-1.20	+2.8	6.6	-1.8	+5.0	10.3	-2.8	+7.0	16.5	-4.5
9.85	12.41	+1.20	2.30	-0.90	+2.0	3.40	-1.20	+2.0	4.70	-1.20	+3.0	7.5	-2.0	+6.0	12.0	-3.0	+8.0	19.0	-5.0
12.41	15.75	+1.40	2.60	-1.00	+2.2	3.90	-1.40	+2.2	5.90	-1.40	+3.5	8.7	-2.2	+6.0	14.5	-3.5	+9.0	23.0	-6.0

VALUES IN MILLIMETERS

Nominal Size Range Millimeters		Class LN1 Light Press Fit			Class LN2 Medium Press Fit			Class LN3 Heavy Press Fit			Class LN4			Class LN5			Class LN6		
		Hole Tol. GR6	Maximum Interference	Shaft Tol. gr5	Hole Tol. H7	Maximum Interference	Shaft Tol. p6	Hole Tol. H7	Maximum Interference	Shaft Tol. t6	Hole Tol. GR8	Maximum Interference	Shaft Tol. gr7	Hole Tol. GR9	Maximum Interference	Shaft Tol. gr8	Hole Tol. GR10	Maximum Interference	Shaft Tol. gr9
Over	To	-0		+0	-0		+0	-0		+0	-0		+0	-0		+0	-0		+0
0	3	+0.006		-0.004	+0.010	0.016	-0.006	+0.010	0.019	-0.006	+0.015	0.030	-0.010	+0.025	0.046	-0.015	+0.041	0.076	-0.025
3	6	+0.008		-0.005	+0.013	0.020	-0.008	+0.013	0.023	-0.008	+0.018	0.038	-0.013	+0.030	0.059	-0.018	+0.046	0.091	-0.030
6	10	+0.010		-0.006	+0.015	0.025	-0.010	+0.015	0.030	-0.010	+0.023	0.046	-0.015	+0.036	0.071	-0.023	+0.056	0.112	-0.036
10	18	+0.010		-0.008	+0.018	0.028	-0.010	+0.018	0.036	-0.010	+0.025	0.056	-0.018	+0.041	0.086	-0.025	+0.071	0.142	-0.041
18	30	+0.013		-0.010	+0.020	0.033	-0.013	+0.020	0.044	-0.013	+0.030	0.066	-0.020	+0.051	0.107	-0.030	+0.089	0.178	-0.051
30	50	+0.015		-0.010	+0.025	0.041	-0.015	+0.025	0.051	-0.015	+0.041	0.086	-0.025	+0.064	0.135	-0.041	+0.102	0.216	-0.064
50	80	+0.018		-0.013	+0.030	0.054	-0.018	+0.030	0.059	-0.018	+0.046	0.102	-0.030	+0.076	0.160	-0.046	+0.114	0.254	-0.076
80	120	+0.023		-0.015	+0.036	0.064	-0.023	+0.036	0.074	-0.023	+0.056	0.122	-0.036	+0.102	0.196	-0.056	+0.127	0.292	-0.102
120	180	+0.025		-0.018	+0.041	0.071	-0.025	+0.041	0.089	-0.025	+0.064	0.142	-0.041	+0.114	0.221	-0.064	+0.152	0.343	-0.114
180	250	+0.030		-0.020	+0.046	0.081	-0.030	+0.046	0.107	-0.030	+0.071	0.168	-0.046	+0.127	0.262	-0.071	+0.178	0.419	-0.127
250	315	+0.030		-0.023	+0.051	0.086	-0.030	+0.051	0.119	-0.030	+0.076	0.191	-0.051	+0.152	0.305	-0.076	+0.203	0.483	-0.152
315	400	+0.036		-0.025	+0.056	0.099	-0.036	+0.056	0.150	-0.036	+0.089	0.221	-0.056	+0.152	0.368	-0.089	+0.229	0.584	-0.152

Running and Sliding Fits. (CONTINUED)

FORCE AND SHRINK FITS

VALUES IN THOUSANDTHS OF AN INCH

Nominal Size Range Inches		Class FN1 Light Drive Fit			Class FN2 Medium Drive Fit			Class FN3 Heavy Drive Fit			Class FN4 Shrink Fit			Class FN5 Heavy Shrink Fit		
		Hole Tol. GR6	Maximum Interference	Shaft Tol. GR5	Hole Tol. GR7	Maximum Interference	Shaft Tol. GR6	Hole Tol. GR7	Maximum Interference	Shaft Tol. GR6	Hole Tol. GR7	Maximum Interference	Shaft Tol. GR6	Hole Tol. GR8	Maximum Interference	Shaft Tol. GR7
Over	To	-0		+0	-0		+0	-0		+0	-0		+0	-0		+0
0	.12	+0.25	0.50	-0.15	+0.40	0.85	-0.25				+0.40	0.95	-0.25	+0.60	1.30	-0.40
.12	.24	+0.30	0.60	-0.20	+0.50	1.00	-0.30				+0.50	1.20	-0.30	+0.70	1.70	-0.50
.24	.40	+0.40	0.75	-0.25	+0.60	1.40	-0.40				+0.60	1.60	-0.40	+0.90	2.00	-0.60
.40	.56	+0.40	0.80	-0.30	+0.70	1.60	-0.40				+0.70	1.80	-0.40	+1.00	2.30	-0.70
.56	.71	+0.40	0.90	-0.30	+0.70	1.60	-0.40				+0.70	1.80	-0.40	+1.00	2.50	-0.70
.71	.95	+0.50	1.10	-0.40	+0.80	1.90	-0.50				+0.80	2.10	-0.50	+1.20	3.00	-0.80
.95	1.19	+0.50	1.20	-0.40	+0.80	1.90	-0.50	+0.80	2.10	-0.50	+0.80	2.30	-0.50	+1.20	3.30	-0.80
1.19	1.58	+0.60	1.30	-0.40	+1.00	2.40	-0.60	+1.00	2.60	-0.60	+1.00	3.10	-0.60	+1.60	4.00	-1.00
1.58	1.97	+0.60	1.40	-0.40	+1.00	2.40	-0.60	+1.00	2.80	-0.60	+1.00	3.40	-0.60	+1.60	5.00	-1.00
1.97	2.56	+0.70	1.80	-0.50	+1.20	2.70	-0.70	+1.20	3.20	-0.70	+1.20	4.20	-0.70	+1.80	6.20	-1.20
2.56	3.15	+0.70	1.90	-0.50	+1.20	2.90	-0.70	+1.20	3.70	-0.70	+1.20	4.70	-0.70	+1.80	7.20	-1.20
3.15	3.94	+0.90	2.40	-0.60	+1.40	3.70	-0.90	+1.40	4.40	-0.70	+1.40	5.90	-0.90	+2.20	8.40	-1.40

VALUES IN MILLIMETERS

Nominal Size Range Millimeters		Class FN1 Light Drive Fit			Class FN2 Medium Drive Fit			Class FN3 Heavy Drive Fit			Class FN4 Shrink Fit			Class FN5 Heavy Shrink Fit		
		Hole Tol. GR6	Maximum Interference	Shaft Tol. gr5	Hole Tol. H7	Maximum Interference	Shaft Tol. s6	Hole Tol. H7	Maximum Interference	Shaft Tol. t6	Hole Tol. GR8	Maximum Interference	Shaft Tol. gr7	Hole Tol. H8	Maximum Interference	Shaft Tol. t7
Over	To	-0		+0	-0		+0	-0		+0	-0		+0	-0		+0
0	3	+0.006	0.013	-0.004	+0.010	0.216	-0.006				+0.010	0.024	-0.006	+0.015	0.033	-0.010
3	6	+0.007	0.015	-0.005	+0.013	0.025	-0.007				+0.013	0.030	-0.007	+0.018	0.043	-0.013
6	10	+0.010	0.019	-0.006	+0.015	0.036	-0.010				+0.015	0.041	-0.010	+0.023	0.051	-0.015
10	14	+0.010	0.020	-0.008	+0.018	0.041	-0.010				+0.018	0.046	-0.010	+0.025	0.058	-0.018
14	18	+0.010	0.023	-0.008	+0.018	0.041	-0.010				+0.018	0.046	-0.010	+0.025	0.064	-0.018
18	24	+0.013	0.028	-0.010	+0.020	0.048	-0.013				+0.020	0.053	-0.013	+0.030	0.076	-0.020
24	30	+0.013	0.030	-0.010	+0.020	0.048	-0.013	+0.020	0.053	-0.013	+0.020	0.058	-0.013	+0.030	0.084	-0.020
30	40	+0.015	0.033	-0.010	+0.025	0.061	-0.015	+0.025	0.066	-0.015	+0.025	0.079	-0.015	+0.041	0.102	-0.025
40	50	+0.015	0.036	-0.010	+0.025	0.061	-0.015	+0.025	0.071	-0.015	+0.025	0.086	-0.015	+0.041	0.127	-0.025
50	65	+0.018	0.046	-0.013	+0.030	0.069	-0.018	+0.030	0.082	-0.018	+0.030	0.107	-0.018	+0.046	0.157	-0.030
65	80	+0.018	0.048	-0.013	+0.030	0.074	-0.018	+0.030	0.094	-0.018	+0.030	0.119	-0.018	+0.046	0.183	-0.030
80	100	+0.023	0.061	-0.015	+0.035	0.094	-0.023	+0.035	0.112	-0.023	+0.036	0.150	-0.023	+0.056	0.213	-0.036

Running and Sliding Fits. (CONTINUED)

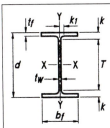

Table 1–1
W Shapes
Dimensions

Shape	Area, A	Depth, d	Web Thickness, t_w	$\dfrac{t_w}{2}$	Flange Width, b_f	Flange Thickness, t_f	k_{des}	k_{det}	k_1	T	Workable Gage
	in.²	in.	in.	in.	in.	in.	in.	in.	in.	in.	in.
W36×800[h]	236	42.6 42½	2.38 2⅜	1 3/16	18.0 18	4.29 4 5/16	5.24	5 9/16	2⅜	31⅜	7½
×652[h]	192	41.1 41	1.97 2	1	17.6 17⅝	3.54 3 9/16	4.49	4 13/16	2 3/16		
×529[h]	156	39.8 39¾	1.61 1⅝	13/16	17.2 17¼	2.91 2 15/16	3.86	4 3/16	2		
×487[h]	143	39.3 39⅜	1.50 1½	¾	17.1 17⅛	2.68 2 11/16	3.63	4	1 15/16		
×441[h]	130	38.9 38⅞	1.36 1⅜	11/16	17.0 17	2.44 2 7/16	3.39	3¾	1⅞		
×395[h]	116	38.4 38⅜	1.22 1¼	⅝	16.8 16⅞	2.20 2 3/16	3.15	3 7/16	1 13/16		
×361[h]	106	38.0 38	1.12 1⅛	9/16	16.7 16¾	2.01 2	2.96	3 5/16	1¾		
×330	97.0	37.7 37⅝	1.02 1	½	16.6 16⅝	1.85 1⅞	2.80	3⅛	1¾		
×302	88.8	37.3 37⅜	0.945 15/16	½	16.7 16⅝	1.68 1 11/16	2.63	3	1 11/16		
×282[c]	82.9	37.1 37⅛	0.885 ⅞	7/16	16.6 16⅝	1.57 1 9/16	2.52	2⅞	1⅝		
×262[c]	77.0	36.9 36⅞	0.840 13/16	7/16	16.6 16½	1.44 1 7/16	2.39	2¾	1⅝		
×247[c]	72.5	36.7 36⅝	0.800 13/16	7/16	16.5 16½	1.35 1⅜	2.30	2⅝	1⅝		
×231[c]	68.1	36.5 36½	0.760 ¾	⅜	16.5 16½	1.26 1¼	2.21	2 9/16	1 9/16		
W36×256	75.4	37.4 37⅜	0.960 15/16	½	12.2 12¼	1.73 1¾	2.48	2⅝	1⅝	32⅛	5½
×232[c]	68.1	37.1 37⅛	0.870 ⅞	7/16	12.1 12⅛	1.57 1 9/16	2.32	2 7/16	1¼		
×210[c]	61.8	36.7 36¾	0.830 13/16	7/16	12.2 12⅛	1.36 1⅜	2.11	2 5/16	1¼		
×194[c]	57.0	36.5 36½	0.765 ¾	⅜	12.1 12⅛	1.26 1¼	2.01	2 3/16	1 3/16		
×182[c]	53.6	36.3 36⅜	0.725 ¾	⅜	12.1 12⅛	1.18 1 3/16	1.93	2⅛	1 3/16		
×170[c]	50.1	36.2 36⅛	0.680 11/16	⅜	12.0 12	1.10 1⅛	1.85	2	1 3/16		
×160[c]	47.0	36.0 36	0.650 ⅝	5/16	12.0 12	1.02 1	1.77	1 15/16	1⅛		
×150[c]	44.2	35.9 35⅞	0.625 ⅝	5/16	12.0 12	0.940 15/16	1.69	1⅞	1⅛		
×135[c,v]	39.7	35.6 35½	0.600 ⅝	5/16	12.0 12	0.790 13/16	1.54	1 11/16	1⅛		
W33×387[h]	114	36.0 36	1.26 1¼	⅝	16.2 16¼	2.28 2¼	3.07	3 3/16	1 7/16	29⅝	5½
×354[h]	104	35.6 35½	1.16 1 3/16	⅝	16.1 16⅛	2.09 2 1/16	2.88	2 15/16	1⅜		
×318	93.6	35.2 35⅛	1.04 1 1/16	9/16	16.0 16	1.89 1⅞	2.68	2¾	1 5/16		
×291	85.7	34.8 34⅞	0.960 15/16	½	15.9 15⅞	1.73 1¾	2.52	2⅝	1 5/16		
×263	77.5	34.5 34½	0.870 ⅞	7/16	15.8 15¾	1.57 1 9/16	2.36	2 7/16	1¼		
×241[c]	71.0	34.2 34⅛	0.830 13/16	7/16	15.9 15⅞	1.40 1⅜	2.19	2¼	1¼		
×221[c]	65.2	33.9 33⅞	0.775 ¾	⅜	15.8 15¾	1.28 1¼	2.06	2⅛	1 3/16		
×201[c]	59.2	33.7 33⅝	0.715 11/16	⅜	15.7 15¾	1.15 1⅛	1.94	2	1 3/16		
W33×169[c]	49.5	33.8 33⅞	0.670 11/16	⅜	11.5 11½	1.22 1¼	1.92	2⅛	1 3/16	29⅝	5½
×152[c]	44.8	33.5 33½	0.635 ⅝	5/16	11.6 11⅝	1.06 1 1/16	1.76	1 15/16	1⅛		
×141[c]	41.6	33.3 33¼	0.605 ⅝	5/16	11.5 11½	0.960 15/16	1.66	1 13/16	1⅛		
×130[c]	38.3	33.1 33⅛	0.580 9/16	5/16	11.5 11½	0.855 ⅞	1.56	1¾	1⅛		
×118[c,v]	34.7	32.9 32⅞	0.550 9/16	5/16	11.5 11½	0.740 ¾	1.44	1⅝	1⅛		

[c] Shape is slender for compression with F_y = 50 ksi.
[h] Flange thickness greater than 2 in. Special requirements may apply per AISC Specification Section A3.1c.
[v] Shape does not meet the h/t_w limit for shear in Specification Section G2.1a with F_y = 50 ksi.

Structural Metal Shape Designations.

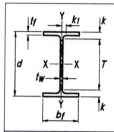

Table 1–1 (continued)
W Shapes
Dimensions

Shape	Area, A	Depth, d		Web Thickness, t_w		$\frac{t_w}{2}$	Flange Width, b_f		Flange Thickness, t_f		k k_{des}	k k_{det}	k_1	T	Workable Gage
	in.²	in.		in.		in.	in.		in.		in.	in.	in.	in.	in.
W30×391[h]	115	33.2	$33\frac{1}{4}$	1.36	$1\frac{3}{8}$	$\frac{11}{16}$	15.6	$15\frac{5}{8}$	2.44	$2\frac{7}{16}$	3.23	$3\frac{3}{8}$	$1\frac{1}{2}$	$26\frac{1}{2}$	$5\frac{1}{2}$
×357[h]	105	32.8	$32\frac{3}{4}$	1.24	$1\frac{1}{4}$	$\frac{5}{8}$	15.5	$15\frac{1}{2}$	2.24	$2\frac{1}{4}$	3.03	$3\frac{1}{8}$	$1\frac{7}{16}$		
×326[h]	95.8	32.4	$32\frac{3}{8}$	1.14	$1\frac{1}{8}$	$\frac{9}{16}$	15.4	$15\frac{3}{8}$	2.05	$2\frac{1}{16}$	2.84	$2\frac{15}{16}$	$1\frac{3}{8}$		
×292	85.9	32.0	32	1.02	1	$\frac{1}{2}$	15.3	$15\frac{1}{4}$	1.85	$1\frac{7}{8}$	2.64	$2\frac{3}{4}$	$1\frac{5}{16}$		
×261	76.9	31.6	$31\frac{5}{8}$	0.930	$\frac{15}{16}$	$\frac{1}{2}$	15.2	$15\frac{1}{8}$	1.65	$1\frac{5}{8}$	2.44	$2\frac{9}{16}$	$1\frac{5}{16}$		
×235	69.2	31.3	$31\frac{1}{4}$	0.830	$\frac{13}{16}$	$\frac{7}{16}$	15.1	15	1.50	$1\frac{1}{2}$	2.29	$2\frac{3}{8}$	$1\frac{1}{4}$		
×211	62.2	30.9	31	0.775	$\frac{3}{4}$	$\frac{3}{8}$	15.1	$15\frac{1}{8}$	1.32	$1\frac{5}{16}$	2.10	$2\frac{1}{4}$	$1\frac{3}{16}$		
×191[c]	56.3	30.7	$30\frac{5}{8}$	0.710	$\frac{11}{16}$	$\frac{3}{8}$	15.0	15	1.19	$1\frac{3}{16}$	1.97	$2\frac{1}{16}$	$1\frac{3}{16}$		
×173[c]	51.0	30.4	$30\frac{1}{2}$	0.655	$\frac{5}{8}$	$\frac{5}{16}$	15.0	15	1.07	$1\frac{1}{16}$	1.85	2	$1\frac{1}{8}$		
W30×148[c]	43.5	30.7	$30\frac{5}{8}$	0.650	$\frac{5}{8}$	$\frac{5}{16}$	10.5	$10\frac{1}{2}$	1.18	$1\frac{3}{16}$	1.83	$2\frac{1}{16}$	$1\frac{1}{8}$	$26\frac{1}{2}$	$5\frac{1}{2}$
×132[c]	38.9	30.3	$30\frac{1}{4}$	0.615	$\frac{5}{8}$	$\frac{5}{16}$	10.5	$10\frac{1}{2}$	1.00	1	1.65	$1\frac{7}{8}$	$1\frac{1}{8}$		
×124[c]	36.5	30.2	$30\frac{1}{8}$	0.585	$\frac{9}{16}$	$\frac{5}{16}$	10.5	$10\frac{1}{2}$	0.930	$\frac{15}{16}$	1.58	$1\frac{13}{16}$	$1\frac{1}{8}$		
×116[c]	34.2	30.0	30	0.565	$\frac{9}{16}$	$\frac{5}{16}$	10.5	$10\frac{1}{2}$	0.850	$\frac{7}{8}$	1.50	$1\frac{3}{4}$	$1\frac{1}{8}$		
×108[c]	31.7	29.8	$29\frac{7}{8}$	0.545	$\frac{9}{16}$	$\frac{5}{16}$	10.5	$10\frac{1}{2}$	0.760	$\frac{3}{4}$	1.41	$1\frac{11}{16}$	$1\frac{1}{8}$		
×99[c]	29.1	29.7	$29\frac{5}{8}$	0.520	$\frac{1}{2}$	$\frac{1}{4}$	10.5	$10\frac{1}{2}$	0.670	$\frac{11}{16}$	1.32	$1\frac{9}{16}$	$1\frac{1}{16}$		
×90[c,v]	26.4	29.5	$29\frac{1}{2}$	0.470	$\frac{1}{2}$	$\frac{1}{4}$	10.4	$10\frac{3}{8}$	0.610	$\frac{5}{8}$	1.26	$1\frac{1}{2}$	$1\frac{1}{16}$		
W27×539[h]	159	32.5	$32\frac{1}{2}$	1.97	2	1	15.3	$15\frac{1}{4}$	3.54	$3\frac{9}{16}$	4.33	$4\frac{7}{16}$	$1\frac{13}{16}$	$23\frac{5}{8}$	$5\frac{1}{2}$[g]
×368[h]	108	30.4	$30\frac{3}{8}$	1.38	$1\frac{3}{8}$	$\frac{11}{16}$	14.7	$14\frac{5}{8}$	2.48	$2\frac{1}{2}$	3.27	$3\frac{3}{8}$	$1\frac{1}{2}$		$5\frac{1}{2}$
×336[h]	98.9	30.0	30	1.26	$1\frac{1}{4}$	$\frac{5}{8}$	14.6	$14\frac{1}{2}$	2.28	$2\frac{1}{4}$	3.07	$3\frac{3}{16}$	$1\frac{7}{16}$		
×307[h]	90.4	29.6	$29\frac{5}{8}$	1.16	$1\frac{3}{16}$	$\frac{5}{8}$	14.4	$14\frac{1}{2}$	2.09	$2\frac{1}{16}$	2.88	3	$1\frac{7}{16}$		
×281	82.9	29.3	$29\frac{1}{4}$	1.06	$1\frac{1}{16}$	$\frac{9}{16}$	14.4	$14\frac{3}{8}$	1.93	$1\frac{15}{16}$	2.72	$2\frac{13}{16}$	$1\frac{3}{8}$		
×258	76.0	29.0	29	0.980	1	$\frac{1}{2}$	14.3	$14\frac{1}{4}$	1.77	$1\frac{3}{4}$	2.56	$2\frac{11}{16}$	$1\frac{5}{16}$		
×235	69.4	28.7	$28\frac{5}{8}$	0.910	$\frac{15}{16}$	$\frac{1}{2}$	14.2	$14\frac{1}{4}$	1.61	$1\frac{5}{8}$	2.40	$2\frac{1}{2}$	$1\frac{5}{16}$		
×217	64.0	28.4	$28\frac{3}{8}$	0.830	$\frac{13}{16}$	$\frac{7}{16}$	14.1	$14\frac{1}{8}$	1.50	$1\frac{1}{2}$	2.29	$2\frac{3}{8}$	$1\frac{1}{4}$		
×194	57.2	28.1	$28\frac{1}{8}$	0.750	$\frac{3}{4}$	$\frac{3}{8}$	14.0	14	1.34	$1\frac{5}{16}$	2.13	$2\frac{1}{4}$	$1\frac{3}{16}$		
×178	52.5	27.8	$27\frac{3}{4}$	0.725	$\frac{3}{4}$	$\frac{3}{8}$	14.1	$14\frac{1}{8}$	1.19	$1\frac{3}{16}$	1.98	$2\frac{1}{16}$	$1\frac{3}{16}$		
×161[c]	47.6	27.6	$27\frac{5}{8}$	0.660	$\frac{11}{16}$	$\frac{3}{8}$	14.0	14	1.08	$1\frac{1}{16}$	1.87	2	$1\frac{3}{16}$		
×146[c]	43.1	27.4	$27\frac{3}{8}$	0.605	$\frac{5}{8}$	$\frac{5}{16}$	14.0	14	0.975	1	1.76	$1\frac{7}{8}$	$1\frac{1}{8}$		
W27×129[c]	37.8	27.6	$27\frac{5}{8}$	0.610	$\frac{5}{8}$	$\frac{5}{16}$	10.0	10	1.10	$1\frac{1}{8}$	1.70	2	$1\frac{1}{8}$	$23\frac{5}{8}$	$5\frac{1}{2}$
×114[c]	33.5	27.3	$27\frac{1}{4}$	0.570	$\frac{9}{16}$	$\frac{5}{16}$	10.1	$10\frac{1}{8}$	0.930	$\frac{15}{16}$	1.53	$1\frac{13}{16}$	$1\frac{1}{8}$		
×102[c]	30.0	27.1	$27\frac{1}{8}$	0.515	$\frac{1}{2}$	$\frac{1}{4}$	10.0	10	0.830	$\frac{13}{16}$	1.43	$1\frac{3}{4}$	$1\frac{1}{16}$		
×94[c]	27.7	26.9	$26\frac{7}{8}$	0.490	$\frac{1}{2}$	$\frac{1}{4}$	10.0	10	0.745	$\frac{3}{4}$	1.34	$1\frac{5}{8}$	$1\frac{1}{16}$		
×84[c]	24.8	26.7	$26\frac{3}{4}$	0.460	$\frac{7}{16}$	$\frac{1}{4}$	10.0	10	0.640	$\frac{5}{8}$	1.24	$1\frac{9}{16}$	$1\frac{1}{16}$		

[c] Shape is slender for compression with F_y = 50 ksi.
[g] The actual size, combination, and orientation of fastener components should be compared with the geometry of the cross-section to ensure compatibility.
[h] Flange thickness greater than 2 in. Special requirements may apply per AISC Specification Section A3.1c.
[v] Shape does not meet the h/t_w limit for shear in Specification Section G2.1a with F_y = 50 ksi.

Structural Metal Shape Designations. (CONTINUED)

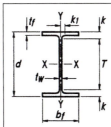

Table 1–1 (continued)
W Shapes
Dimensions

Shape	Area, A	Depth, d		Web Thickness, t_w		$\dfrac{t_w}{2}$	Flange Width, b_f		Flange Thickness, t_f		k_{des}	k_{det}	k_1	T	Workable Gage
	in.²	in.		in.		in.	in.		in.		in.	in.	in.	in.	in.
W24×370[h]	109	28.0	28	1.52	1½	¾	13.7	13⅝	2.72	2¾	3.22	3⅝	1 9/16	20¾	5½
×335[h]	98.4	27.5	27½	1.38	1⅜	11/16	13.5	13½	2.48	2½	2.98	3⅜	1½		
×306[h]	89.8	27.1	27⅛	1.26	1¼	⅝	13.4	13⅜	2.28	2¼	2.78	3 3/16	1 7/16		
x279[h]	82.0	26.7	26¾	1.16	1 3/16	⅝	13.3	13¼	2.09	2 1/16	2.59	3	1 7/16		
×250	73.5	26.3	26⅜	1.04	1 1/16	9/16	13.2	13⅛	1.89	1⅞	2.39	2 13/16	1⅜		
×229	67.2	26.0	26	0.960	15/16	½	13.1	13⅛	1.73	1¾	2.23	2⅝	1 5/16		
×207	60.7	25.7	25¾	0.870	⅞	7/16	13.0	13	1.57	1 9/16	2.07	2½	1¼		
×192	56.3	25.5	25½	0.810	13/16	7/16	13.0	13	1.46	1 7/16	1.96	2⅜	1¼		
×176	51.7	25.2	25¼	0.750	¾	⅜	12.9	12⅞	1.34	1 5/16	1.84	2¼	1 3/16		
×162	47.7	25.0	25	0.705	11/16	⅜	13.0	13	1.22	1¼	1.72	2⅛	1 3/16		
×146	43.0	24.7	24¾	0.650	⅝	5/16	12.9	12⅞	1.09	1 1/16	1.59	2	1⅛		
×131	38.5	24.5	24½	0.605	⅝	5/16	12.9	12⅞	0.960	15/16	1.46	1⅞	1⅛		
×117[c]	34.4	24.3	24¼	0.550	9/16	5/16	12.8	12¾	0.850	⅞	1.35	1¾	1⅛	↓	↓
×104[c]	30.6	24.1	24	0.500	½	¼	12.8	12¾	0.750	¾	1.25	1⅝	1 1/16		
W24×103[c]	30.3	24.5	24½	0.550	9/16	5/16	9.00	9	0.980	1	1.48	1⅞	1⅛	20¾	5½
×94[c]	27.7	24.3	24¼	0.515	½	¼	9.07	9⅛	0.875	⅞	1.38	1¾	1 1/16		
×84[c]	24.7	24.1	24⅛	0.470	½	¼	9.02	9	0.770	¾	1.27	1 11/16	1 1/16		
×76[c]	22.4	23.9	23⅞	0.440	7/16	¼	8.99	9	0.680	11/16	1.18	1 9/16	1 1/16	↓	↓
×68[c]	20.1	23.7	23¾	0.415	7/16	¼	8.97	9	0.585	9/16	1.09	1½	1 1/16		
W24×62[c]	18.2	23.7	23¾	0.430	7/16	¼	7.04	7	0.590	9/16	1.09	1½	1 1/16	20¾	3½[g]
×55[c,v]	16.2	23.6	23⅝	0.395	⅜	3/16	7.01	7	0.505	½	1.01	1 7/16	1	20¾	3½[g]
W21×201	59.2	23.0	23	0.910	15/16	½	12.6	12⅝	1.63	1⅝	2.13	2½	1 5/16	18	5½
×182	53.6	22.7	22¾	0.830	13/16	7/16	12.5	12½	1.48	1½	1.98	2⅜	1¼		
×166	48.8	22.5	22½	0.750	¾	⅜	12.4	12⅜	1.36	1⅜	1.86	2¼	1 3/16		
×147	43.2	22.1	22	0.720	¾	⅜	12.5	12½	1.15	1⅛	1.65	2	1 3/16		
×132	38.8	21.8	21⅞	0.650	⅝	5/16	12.4	12½	1.04	1 1/16	1.54	1 15/16	1⅛		
×122	35.9	21.7	21⅝	0.600	⅝	5/16	12.4	12⅜	0.960	15/16	1.46	1 13/16	1⅛		
×111	32.7	21.5	21½	0.550	9/16	5/16	12.3	12⅜	0.875	⅞	1.38	1¾	1⅛	↓	↓
×101[c]	29.8	21.4	21⅜	0.500	½	¼	12.3	12¼	0.800	13/16	1.30	1 11/16	1 1/16		

[c] Shape is slender for compression with F_y = 50 ksi.
[g] The actual size, combination, and orientation of fastener components should be compared with the geometry of the cross-section to ensure compatibility.
[h] Flange thickness greater than 2 in. Special requirements may apply per AISC Specification Section A3.1c.
[v] Shape does not meet the h/t_w limit for shear in Specification Section G2.1a with F_y = 50 ksi.

Structural Metal Shape Designations. (CONTINUED)

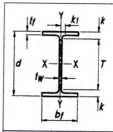

Table 1–1 (continued)
W Shapes
Dimensions

Shape	Area, A	Depth, d		Web Thickness, t_w	$\frac{t_w}{2}$	Flange Width, b_f		Flange Thickness, t_f		k_{des}	k_{det}	k_1	T	Workable Gage
	in.²	in.		in.	in.	in.		in.		in.	in.	in.	in.	in.
W21×93	27.3	21.6	21⁵/₈	0.580	⁹/₁₆	⁵/₁₆	8.42	8³/₈	0.930	¹⁵/₁₆	1.43	1⁵/₈	¹⁵/₁₆	18³/₈	5¹/₂
×83ᶜ	24.3	21.4	21³/₈	0.515	¹/₂	¹/₄	8.36	8³/₈	0.835	¹³/₁₆	1.34	1¹/₂	⁷/₈		
×73ᶜ	21.5	21.2	21¹/₄	0.455	⁷/₁₆	¹/₄	8.30	8¹/₄	0.740	³/₄	1.24	1⁷/₁₆	⁷/₈		
×68ᶜ	20.0	21.1	21¹/₈	0.430	⁷/₁₆	¹/₄	8.27	8¹/₄	0.685	¹¹/₁₆	1.19	1³/₈	⁷/₈		
×62ᶜ	18.3	21.0	21	0.400	³/₈	³/₁₆	8.24	8¹/₄	0.615	⁵/₈	1.12	1⁵/₁₆	¹³/₁₆		
×55ᶜ	16.2	20.8	20³/₄	0.375	³/₈	³/₁₆	8.22	8¹/₄	0.522	¹/₂	1.02	1³/₁₆	¹³/₁₆		
×48ᶜ,ᶠ	14.1	20.6	20⁵/₈	0.350	³/₈	³/₁₆	8.14	8¹/₈	0.430	⁷/₁₆	0.930	1¹/₈	¹³/₁₆	↓	↓
W21×57ᶜ	16.7	21.1	21	0.405	³/₈	³/₁₆	6.56	6¹/₂	0.650	⁵/₈	1.15	1⁵/₁₆	¹³/₁₆	18³/₈	3¹/₂
×50ᶜ	14.7	20.8	20⁷/₈	0.380	³/₈	³/₁₆	6.53	6¹/₂	0.535	⁹/₁₆	1.04	1¹/₄	¹³/₁₆		
×44ᶜ	13.0	20.7	20⁵/₈	0.350	³/₈	³/₁₆	6.50	6¹/₂	0.450	⁷/₁₆	0.950	1¹/₈	¹³/₁₆	↓	↓
W18×311ʰ	91.6	22.3	22³/₈	1.52	1¹/₂	³/₄	12.0	12	2.74	2³/₄	3.24	3⁷/₁₆	1³/₈	15¹/₂	5¹/₂
×283ʰ	83.3	21.9	21⁷/₈	1.40	1³/₈	¹¹/₁₆	11.9	11⁷/₈	2.50	2¹/₂	3.00	3³/₁₆	1⁵/₁₆		
×258ʰ	75.9	21.5	21¹/₂	1.28	1¹/₄	⁵/₈	11.8	11³/₄	2.30	2⁵/₁₆	2.70	3	1¹/₄		
×234ʰ	68.8	21.1	21	1.16	1³/₁₆	⁵/₈	11.7	11⁵/₈	2.11	2¹/₈	2.51	2³/₄	1³/₁₆		
×211	62.1	20.7	20⁵/₈	1.06	1¹/₁₆	⁹/₁₆	11.6	11¹/₂	1.91	1¹⁵/₁₆	2.31	2⁹/₁₆	1³/₁₆		
×192	56.4	20.4	20³/₈	0.960	¹⁵/₁₆	¹/₂	11.5	11¹/₂	1.75	1³/₄	2.15	2⁷/₁₆	1¹/₈	↓	
×175	51.3	20.0	20	0.890	⁷/₈	⁷/₁₆	11.4	11³/₈	1.59	1⁹/₁₆	1.99	2⁷/₁₆	1¹/₄	15¹/₈	
×158	46.3	19.7	19³/₄	0.810	¹³/₁₆	⁷/₁₆	11.3	11¹/₄	1.44	1⁷/₁₆	1.84	2³/₈	1¹/₄		
×143	42.1	19.5	19¹/₂	0.730	³/₄	³/₈	11.2	11¹/₄	1.32	1⁵/₁₆	1.72	2³/₁₆	1³/₁₆		
×130	38.2	19.3	19¹/₄	0.670	¹¹/₁₆	³/₈	11.2	11¹/₄	1.20	1³/₁₆	1.60	2¹/₁₆	1³/₁₆		
×119	35.1	19.0	19	0.655	⁵/₈	⁵/₁₆	11.3	11¹/₄	1.06	1¹/₁₆	1.46	1¹⁵/₁₆	1³/₁₆		
×106	31.1	18.7	18³/₄	0.590	⁹/₁₆	⁵/₁₆	11.2	11¹/₄	0.940	¹⁵/₁₆	1.34	1¹³/₁₆	1¹/₈		
×97	28.5	18.6	18⁵/₈	0.535	⁹/₁₆	⁵/₁₆	11.1	11¹/₈	0.870	⁷/₈	1.27	1³/₄	1¹/₈		
×86	25.3	18.4	18³/₈	0.480	¹/₂	¹/₄	11.1	11¹/₈	0.770	³/₄	1.17	1⁵/₈	1¹/₁₆		
×76ᶜ	22.3	18.2	18¹/₄	0.425	⁷/₁₆	¹/₄	11.0	11	0.680	¹¹/₁₆	1.08	1⁹/₁₆	1¹/₁₆	↓	↓
W18×71	20.8	18.5	18¹/₂	0.495	¹/₂	¹/₄	7.64	7⁵/₈	0.810	¹³/₁₆	1.21	1¹/₂	⁷/₈	15¹/₂	3¹/₂ᵍ
×65	19.1	18.4	18³/₈	0.450	⁷/₁₆	¹/₄	7.59	7⁵/₈	0.750	³/₄	1.15	1⁷/₁₆	⁷/₈		
×60ᶜ	17.6	18.2	18¹/₄	0.415	⁷/₁₆	¹/₄	7.56	7¹/₂	0.695	¹¹/₁₆	1.10	1³/₈	¹³/₁₆		
×55ᶜ	16.2	18.1	18¹/₈	0.390	³/₈	³/₁₆	7.53	7¹/₂	0.630	⁵/₈	1.03	1⁵/₁₆	¹³/₁₆		
×50ᶜ	14.7	18.0	18	0.355	³/₈	³/₁₆	7.50	7¹/₂	0.570	⁹/₁₆	0.972	1¹/₄	¹³/₁₆	↓	↓
W18×46ᶜ	13.5	18.1	18	0.360	³/₈	³/₁₆	6.06	6	0.605	⁵/₈	1.01	1¹/₄	¹³/₁₆	15¹/₂	3¹/₂ᵍ
×40ᶜ	11.8	17.9	17⁷/₈	0.315	⁵/₁₆	³/₁₆	6.02	6	0.525	¹/₂	0.927	1³/₁₆	¹³/₁₆		
×35ᶜ	10.3	17.7	17³/₄	0.300	⁵/₁₆	³/₁₆	6.00	6	0.425	⁷/₁₆	0.827	1¹/₈	³/₄	↓	↓

ᶜ Shape is slender for compression with F_y = 50 ksi.
ᶠ Shape exceeds compact limit for flexure with F_y = 50 ksi.
ᵍ The actual size, combination, and orientation of fastener components should be compared with the geometry of the cross-section to ensure compatibility.
ʰ Flange thickness greater than 2 in. Special requirements may apply per AISC Specification Section A3.1c.

Structural Metal Shape Designations. (CONTINUED)

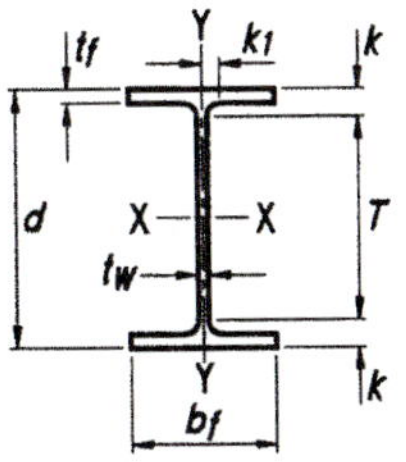

Table 1-1 (continued)
W Shapes
Dimensions

Shape	Area, A	Depth, d		Web			Flange				Distance				
				Thickness, t_w		$\frac{t_w}{2}$	Width, b_f		Thickness, t_f		k		k_1	T	Workable Gage
											k_{des}	k_{det}			
	in.²	in.		in.		in.	in.		in.		in.	in.	in.	in.	in.
W16×100	29.5	17.0	17	0.585	9/16	5/16	10.4	10 3/8	0.985	1	1.39	1 7/8	1 1/8	13 1/4	5 1/2
×89	26.2	16.8	16 3/4	0.525	1/2	1/4	10.4	10 3/8	0.875	7/8	1.28	1 3/4	1 1/16		
×77	22.6	16.5	16 1/2	0.455	7/16	1/4	10.3	10 1/4	0.760	3/4	1.16	1 5/8	1 1/16		
×67[c]	19.7	16.3	16 3/8	0.395	3/8	3/16	10.2	10 1/4	0.665	11/16	1.07	1 9/16	1		
W16×57	16.8	16.4	16 3/8	0.430	7/16	1/4	7.12	7 1/8	0.715	11/16	1.12	1 3/8	7/8	13 5/8	3 1/2[g]
×50[c]	14.7	16.3	16 1/4	0.380	3/8	3/16	7.07	7 1/8	0.630	5/8	1.03	1 5/16	13/16		
×45[c]	13.3	16.1	16 1/8	0.345	3/8	3/16	7.04	7	0.565	9/16	0.967	1 1/4	13/16		
×40[c]	11.8	16.0	16	0.305	5/16	3/16	7.00	7	0.505	1/2	0.907	1 3/16	13/16		
×36[c]	10.6	15.9	15 7/8	0.295	5/16	3/16	6.99	7	0.430	7/16	0.832	1 1/8	3/4		
W16×31[c]	9.13	15.9	15 7/8	0.275	1/4	1/8	5.53	5 1/2	0.440	7/16	0.842	1 1/8	3/4	13 5/8	3 1/2
×26[c,v]	7.68	15.7	15 3/4	0.250	1/4	1/8	5.50	5 1/2	0.345	3/8	0.747	1 1/16	3/4	13 5/8	3 1/2
W14×730[h]	215	22.4	22 3/8	3.07	3 1/16	1 9/16	17.9	17 7/8	4.91	4 15/16	5.51	6 3/16	2 3/4	10	3-7 1/2-3[g]
×665[h]	196	21.6	21 5/8	2.83	2 13/16	1 7/16	17.7	17 5/8	4.52	4 1/2	5.12	5 13/16	2 5/8		3-7 1/2-3[g]
×605[h]	178	20.9	20 7/8	2.60	2 5/8	1 5/16	17.4	17 3/8	4.16	4 3/16	4.76	5 7/16	2 1/2		3-7 1/2-3
×550[h]	162	20.2	20 1/4	2.38	2 3/8	1 3/16	17.2	17 1/4	3.82	3 13/16	4.42	5 1/8	2 3/8		
×500[h]	147	19.6	19 5/8	2.19	2 3/16	1 1/8	17.0	17	3.50	3 1/2	4.10	4 13/16	2 5/16		
×455[h]	134	19.0	19	2.02	2	1	16.8	16 7/8	3.21	3 3/16	3.81	4 1/2	2 1/4		
×426[h]	125	18.7	18 5/8	1.88	1 7/8	15/16	16.7	16 3/4	3.04	3 1/16	3.63	4 5/16	2 1/8		
×398[h]	117	18.3	18 1/4	1.77	1 3/4	7/8	16.6	16 5/8	2.85	2 7/8	3.44	4 1/8	2 1/8		
×370[h]	109	17.9	17 7/8	1.66	1 5/8	13/16	16.5	16 1/2	2.66	2 11/16	3.26	3 15/16	2 1/16		
×342[h]	101	17.5	17 1/2	1.54	1 9/16	13/16	16.4	16 3/8	2.47	2 1/2	3.07	3 3/4	2		
×311[h]	91.4	17.1	17 1/8	1.41	1 7/16	3/4	16.2	16 1/4	2.26	2 1/4	2.86	3 9/16	1 15/16		
×283[h]	83.3	16.7	16 3/4	1.29	1 5/16	11/16	16.1	16 1/8	2.07	2 1/16	2.67	3 3/8	1 7/8		
×257	75.6	16.4	16 3/8	1.18	1 3/16	5/8	16.0	16	1.89	1 7/8	2.49	3 3/16	1 13/16		
×233	68.5	16.0	16	1.07	1 1/16	9/16	15.9	15 7/8	1.72	1 3/4	2.32	3	1 3/4		
×211	62.0	15.7	15 3/4	0.980	1	1/2	15.8	15 3/4	1.56	1 9/16	2.16	2 7/8	1 11/16		
×193	56.8	15.5	15 1/2	0.890	7/8	7/16	15.7	15 3/4	1.44	1 7/16	2.04	2 3/4	1 11/16		
×176	51.8	15.2	15 1/4	0.830	13/16	7/16	15.7	15 5/8	1.31	1 5/16	1.91	2 5/8	1 5/8		
×159	46.7	15.0	15	0.745	3/4	3/8	15.6	15 5/8	1.19	1 3/16	1.79	2 1/2	1 9/16		
×145	42.7	14.8	14 3/4	0.680	11/16	3/8	15.5	15 1/2	1.09	1 1/16	1.69	2 3/8	1 9/16		

[c] Shape is slender for compression with $F_y = 50$ ksi.
[g] The actual size, combination, and orientation of fastener components should be compared with the geometry of the cross-section to ensure compatibility.
[h] Flange thickness greater than 2 in. Special requirements may apply per AISC Specification Section A3.1c.
[v] Shape does not meet the h/t_w limit for shear in Specification Section G2.1a with $F_y = 50$ ksi.

Structural Metal Shape Designations. **(CONTINUED)**

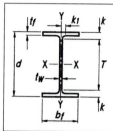

Table 1–1 (continued)
W Shapes
Dimensions

Shape	Area, A (in.²)	Depth, d (in.)	Web Thickness, t_w (in.)	$\frac{t_w}{2}$ (in.)	Flange Width, b_f (in.)	Flange Thickness, t_f (in.)	k_{des} (in.)	k_{det} (in.)	k_1 (in.)	T (in.)	Workable Gage (in.)
W14×132	38.8	14.7 (14 5/8)	0.645 (5/8)	5/16	14.7 (14 3/4)	1.03 (1)	1.63	2 5/16	1 9/16	10	5 1/2
×120	35.3	14.5 (14 1/2)	0.590 (9/16)	5/16	14.7 (14 5/8)	0.940 (15/16)	1.54	2 1/4	1 1/2		
×109	32.0	14.3 (14 3/8)	0.525 (1/2)	1/4	14.6 (14 5/8)	0.860 (7/8)	1.46	2 3/16	1 1/2		
×99[f]	29.1	14.2 (14 1/8)	0.485 (1/2)	1/4	14.6 (14 5/8)	0.780 (3/4)	1.38	2 1/16	1 7/16		
×90[f]	26.5	14.0 (14)	0.440 (7/16)	1/4	14.5 (14 1/2)	0.710 (11/16)	1.31	2	1 7/16	↓	↓
W14×82	24.0	14.3 (14 1/4)	0.510 (1/2)	1/4	10.1 (10 1/8)	0.855 (7/8)	1.45	1 11/16	1 1/16	10 7/8	5 1/2
×74	21.8	14.2 (14 1/8)	0.450 (7/16)	1/4	10.1 (10 1/8)	0.785 (13/16)	1.38	1 5/8	1 1/16		
×68	20.0	14.0 (14)	0.415 (7/16)	1/4	10.0 (10)	0.720 (3/4)	1.31	1 9/16	1 1/16		
×61	17.9	13.9 (13 7/8)	0.375 (3/8)	3/16	10.0 (10)	0.645 (5/8)	1.24	1 1/2	1	↓	↓
W14×53	15.6	13.9 (13 7/8)	0.370 (3/8)	3/16	8.06 (8)	0.660 (11/16)	1.25	1 1/2	1	10 7/8	5 1/2
×48	14.1	13.8 (13 3/4)	0.340 (5/16)	3/16	8.03 (8)	0.595 (5/8)	1.19	1 7/16	1		
×43[c]	12.6	13.7 (13 5/8)	0.305 (5/16)	3/16	8.00 (8)	0.530 (1/2)	1.12	1 3/8	1	↓	↓
W14×38[c]	11.2	14.1 (14 1/8)	0.310 (5/16)	3/16	6.77 (6 3/4)	0.515 (1/2)	0.915	1 1/4	13/16	11 5/8	3 1/2[g]
×34[c]	10.0	14.0 (14)	0.285 (5/16)	3/16	6.75 (6 3/4)	0.455 (7/16)	0.855	1 3/16	3/4		3 1/2
×30[c]	8.85	13.8 (13 7/8)	0.270 (1/4)	1/8	6.73 (6 3/4)	0.385 (3/8)	0.785	1 1/8	3/4	↓	3 1/2
W14×26[c]	7.69	13.9 (13 7/8)	0.255 (1/4)	1/8	5.03 (5)	0.420 (7/16)	0.820	1 1/8	3/4	11 5/8	2 3/4[g]
×22[c]	6.49	13.7 (13 3/4)	0.230 (1/4)	1/8	5.00 (5)	0.335 (5/16)	0.735	1 1/16	3/4	11 5/8	2 3/4[g]
W12×336[h]	98.8	16.8 (16 7/8)	1.78 (1 3/4)	7/8	13.4 (13 3/8)	2.96 (2 15/16)	3.55	3 7/8	1 11/16	9 1/8	5 1/2
×305[h]	89.6	16.3 (16 3/8)	1.63 (1 5/8)	13/16	13.2 (13 1/4)	2.71 (2 11/16)	3.30	3 5/8	1 5/8		
×279[h]	81.9	15.9 (15 7/8)	1.53 (1 1/2)	3/4	13.1 (13 1/8)	2.47 (2 1/2)	3.07	3 3/8	1 5/8		
×252[h]	74.0	15.4 (15 3/8)	1.40 (1 3/8)	11/16	13.0 (13)	2.25 (2 1/4)	2.85	3 1/8	1 1/2		
×230[h]	67.7	15.1 (15)	1.29 (1 5/16)	11/16	12.9 (12 7/8)	2.07 (2 1/16)	2.67	2 15/16	1 1/2		
×210	61.8	14.7 (14 3/4)	1.18 (1 3/16)	5/8	12.8 (12 3/4)	1.90 (1 7/8)	2.50	2 13/16	1 7/16		
×190	55.8	14.4 (14 3/8)	1.06 (1 1/16)	9/16	12.7 (12 5/8)	1.74 (1 3/4)	2.33	2 5/8	1 3/8		
×170	50.0	14.0 (14)	0.960 (15/16)	1/2	12.6 (12 5/8)	1.56 (1 9/16)	2.16	2 7/16	1 5/16		
×152	44.7	13.7 (13 3/4)	0.870 (7/8)	7/16	12.5 (12 1/2)	1.40 (1 3/8)	2.00	2 5/16	1 1/4		
×136	39.9	13.4 (13 3/8)	0.790 (13/16)	7/16	12.4 (12 3/8)	1.25 (1 1/4)	1.85	2 1/8	1 1/4		
×120	35.3	13.1 (13 1/8)	0.710 (11/16)	3/8	12.3 (12 3/8)	1.11 (1 1/8)	1.70	2	1 3/16		
×106	31.2	12.9 (12 7/8)	0.610 (5/8)	5/16	12.2 (12 1/4)	0.990 (1)	1.59	1 7/8	1 1/8		
×96	28.2	12.7 (12 3/4)	0.550 (9/16)	5/16	12.2 (12 1/4)	0.900 (7/8)	1.50	1 13/16	1 1/8		
×87	25.6	12.5 (12 1/2)	0.515 (1/2)	1/4	12.1 (12 1/8)	0.810 (13/16)	1.41	1 11/16	1 1/16		
×79	23.2	12.4 (12 3/8)	0.470 (1/2)	1/4	12.1 (12 1/8)	0.735 (3/4)	1.33	1 5/8	1 1/16		
×72	21.1	12.3 (12 1/4)	0.430 (7/16)	1/4	12.0 (12)	0.670 (11/16)	1.27	1 9/16	1 1/16		
×65[f]	19.1	12.1 (12 1/8)	0.390 (3/8)	3/16	12.0 (12)	0.605 (5/8)	1.20	1 1/2	1	↓	↓

[c] Shape is slender for compression with $F_y = 50$ ksi.
[f] Shape exceeds compact limit for flexure with $F_y = 50$ ksi.
[g] The actual size, combination, and orientation of fastener components should be compared with the geometry of the cross-section to ensure compatibility.
[h] Flange thickness greater than 2 in. Special requirements may apply per AISC Specification Section A3.1c.

Structural Metal Shape Designations. (CONTINUED)

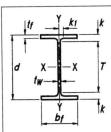

Table 1–1 (continued)
W Shapes
Dimensions

Shape	Area, A (in.²)	Depth, d (in.)		Web Thickness, tw (in.)		tw/2 (in.)	Flange Width, bf (in.)		Flange Thickness, tf (in.)		k kdes (in.)	k kdet (in.)	k1 (in.)	T (in.)	Workable Gage (in.)
W12×58	17.0	12.2	12¼	0.360	3/8	3/16	10.0	10	0.640	5/8	1.24	1½	15/16	9¼	5½
×53	15.6	12.1	12	0.345	3/8	3/16	10.0	10	0.575	9/16	1.18	1⅜	15/16	9¼	5½
W12×50	14.6	12.2	12¼	0.370	3/8	3/16	8.08	8⅛	0.640	5/8	1.14	1½	15/16	9¼	5½
×45	13.1	12.1	12	0.335	5/16	3/16	8.05	8	0.575	9/16	1.08	1⅜	15/16	↓	↓
×40	11.7	11.9	12	0.295	5/16	3/16	8.01	8	0.515	½	1.02	1⅜	7/8	↓	↓
W12×35ᶜ	10.3	12.5	12½	0.300	5/16	3/16	6.56	6½	0.520	½	0.820	1 3/16	3/4	10⅛	3½
×30ᶜ	8.79	12.3	12⅜	0.260	¼	⅛	6.52	6½	0.440	7/16	0.740	1⅛	3/4	↓	↓
×26ᶜ	7.65	12.2	12¼	0.230	¼	⅛	6.49	6½	0.380	3/8	0.680	1 1/16	3/4	↓	↓
W12×22ᶜ	6.48	12.3	12¼	0.260	¼	⅛	4.03	4	0.425	7/16	0.725	15/16	5/8	10⅜	2¼ᵍ
×19ᶜ	5.57	12.2	12⅛	0.235	¼	⅛	4.01	4	0.350	3/8	0.650	7/8	9/16	↓	↓
×16ᶜ	4.71	12.0	12	0.220	¼	⅛	3.99	4	0.265	¼	0.565	13/16	9/16	↓	↓
×14ᶜ,ᵛ	4.16	11.9	11⅞	0.200	3/16	⅛	3.97	4	0.225	¼	0.525	3/4	9/16	↓	↓
W10×112	32.9	11.4	11⅜	0.755	3/4	3/8	10.4	10⅜	1.25	1¼	1.75	1 15/16	1	7½	5½
×100	29.4	11.1	11⅛	0.680	11/16	3/8	10.3	10⅜	1.12	1⅛	1.62	1 13/16	1		
×88	25.9	10.8	10⅞	0.605	5/8	5/16	10.3	10¼	0.990	1	1.49	1 11/16	15/16		
×77	22.6	10.6	10⅝	0.530	½	¼	10.2	10¼	0.870	7/8	1.37	1 9/16	7/8		
×68	20.0	10.4	10⅜	0.470	½	¼	10.1	10⅛	0.770	3/4	1.27	1 7/16	7/8		
×60	17.6	10.2	10¼	0.420	7/16	¼	10.1	10⅛	0.680	11/16	1.18	1⅜	13/16		
×54	15.8	10.1	10⅛	0.370	3/8	3/16	10.0	10	0.615	5/8	1.12	1 5/16	13/16	↓	↓
×49	14.4	10.0	10	0.340	5/16	3/16	10.0	10	0.560	9/16	1.06	1¼	13/16	↓	↓
W10×45	13.3	10.1	10⅛	0.350	3/8	3/16	8.02	8	0.620	5/8	1.12	1 5/16	13/16	7½	5½
×39	11.5	9.92	9⅞	0.315	5/16	3/16	7.99	8	0.530	½	1.03	1 3/16	13/16	↓	↓
×33	9.71	9.73	9¾	0.290	5/16	3/16	7.96	8	0.435	7/16	0.935	1⅛	3/4	↓	↓
W10×30	8.84	10.5	10½	0.300	5/16	3/16	5.81	5¾	0.510	½	0.810	1⅛	11/16	8¼	2¾ᵍ
×26	7.61	10.3	10⅜	0.260	¼	⅛	5.77	5¾	0.440	7/16	0.740	1 1/16	11/16	↓	↓
×22ᶜ	6.49	10.2	10⅛	0.240	¼	⅛	5.75	5¾	0.360	3/8	0.660	15/16	5/8	↓	↓
W10×19	5.62	10.2	10¼	0.250	¼	⅛	4.02	4	0.395	3/8	0.695	15/16	5/8	8⅜	2¼ᵍ
×17ᶜ	4.99	10.1	10⅛	0.240	¼	⅛	4.01	4	0.330	5/16	0.630	7/8	9/16	↓	↓
×15ᶜ	4.41	10.0	10	0.230	¼	⅛	4.00	4	0.270	¼	0.570	13/16	9/16	↓	↓
×12ᶜ,ᶠ	3.54	9.87	9⅞	0.190	3/16	⅛	3.96	4	0.210	3/16	0.510	3/4	9/16	↓	↓

ᶜ Shape is slender for compression with F_y = 50 ksi.
ᶠ Shape exceeds compact limit for flexure with F_y = 50 ksi.
ᵍ The actual size, combination, and orientation of fastener components should be compared with the geometry of the cross-section to ensure compatibility.
ᵛ Shape does not meet the h/t_w limit for shear in Specification Section G2.1a with F_y = 50 ksi.

Structural Metal Shape Designations. (CONTINUED)

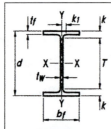

Table 1–1 (continued)
W Shapes
Dimensions

Shape	Area, A	Depth, d		Web				Flange				Distance					
				Thickness, t_w		$\frac{t_w}{2}$		Width, b_f		Thickness, t_f		k		k_1	T	Workable Gage
												k_{des}	k_{det}			
	in.²	in.		in.		in.		in.		in.		in.	in.	in.	in.	in.
W8×67	19.7	9.00	9	0.570	9/16	5/16		8.28	8¼	0.935	15/16	1.33	1⅝	15/16	5¾	5½
×58	17.1	8.75	8¾	0.510	½	¼		8.22	8¼	0.810	13/16	1.20	1½	⅞		
×48	14.1	8.50	8½	0.400	⅜	3/16		8.11	8⅛	0.685	11/16	1.08	1⅜	13/16		
×40	11.7	8.25	8¼	0.360	⅜	3/16		8.07	8⅛	0.560	9/16	0.954	1¼	13/16		
×35	10.3	8.12	8⅛	0.310	5/16	3/16		8.02	8	0.495	½	0.889	13/16	13/16		
×31[f]	9.12	8.00	8	0.285	5/16	3/16		8.00	8	0.435	7/16	0.829	1⅛	¾	↓	↓
W8×28	8.24	8.06	8	0.285	5/16	3/16		6.54	6½	0.465	7/16	0.859	15/16	⅝	6⅛	4
×24	7.08	7.93	7⅞	0.245	¼	⅛		6.50	6½	0.400	⅜	0.794	⅞	9/16	6⅛	4
W8×21	6.16	8.28	8¼	0.250	¼	⅛		5.27	5¼	0.400	⅜	0.700	⅞	9/16	6½	2¾[g]
×18	5.26	8.14	8⅛	0.230	¼	⅛		5.25	5¼	0.330	5/16	0.630	13/16	9/16	6½	2¾[g]
W8×15	4.44	8.11	8⅛	0.245	¼	⅛		4.02	4	0.315	5/16	0.615	13/16	9/16	6½	2¼[g]
×13	3.84	7.99	8	0.230	¼	⅛		4.00	4	0.255	¼	0.555	¾	9/16	↓	↓
×10[c,f]	2.96	7.89	7⅞	0.170	3/16	⅛		3.94	4	0.205	3/16	0.505	11/16	½	↓	↓
W6×25	7.34	6.38	6⅜	0.320	5/16	3/16		6.08	6⅛	0.455	7/16	0.705	15/16	9/16	4½	3½
×20	5.87	6.20	6¼	0.260	¼	⅛		6.02	6	0.365	⅜	0.615	⅞	9/16	↓	↓
×15[f]	4.43	5.99	6	0.230	¼	⅛		5.99	6	0.260	¼	0.510	¾	9/16	↓	↓
W6×16	4.74	6.28	6¼	0.260	¼	⅛		4.03	4	0.405	⅜	0.655	⅞	9/16	4½	2¼[g]
×12	3.55	6.03	6	0.230	¼	⅛		4.00	4	0.280	¼	0.530	¾	9/16	↓	↓
×9[f]	2.68	5.90	5⅞	0.170	3/16	⅛		3.94	4	0.215	3/16	0.465	11/16	½	↓	↓
×8.5[f]	2.52	5.83	5⅞	0.170	3/16	⅛		3.94	4	0.195	3/16	0.445	11/16	½	↓	↓
W5×19	5.56	5.15	5⅛	0.270	¼	⅛		5.03	5	0.430	7/16	0.730	13/16	7/16	3½	2¾[g]
×16	4.71	5.01	5	0.240	¼	⅛		5.00	5	0.360	⅜	0.660	¾	7/16	3½	2¾[g]
W4×13	3.83	4.16	4⅛	0.280	¼	⅛		4.06	4	0.345	⅜	0.595	¾	½	2⅝	2¼[g]

[c] Shape is slender for compression with F_y = 50 ksi.
[f] Shape exceeds compact limit for flexure with F_y = 50 ksi.
[g] The actual size, combination, and orientation of fastener components should be compared with the geometry of the cross-section to ensure compatibility.

Structural Metal Shape Designations. (CONTINUED)

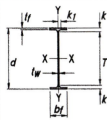

Table 1–2
M Shapes
Dimensions

Shape	Area, A	Depth, d		Web Thickness, t_w		$\frac{t_w}{2}$	Flange Width, b_f		Thickness, t_f		Distance k	k_1	T	Workable Gage
	in.²	in.		in.		in.	in.		in.		in.	in.	in.	in.
M12.5×12.4^{c,v}	3.63	12.5	12½	0.155	⅛	1/16	3.75	3¾	0.228	¼	9/16	⅜	11³⁄₈	—
×11.6^{c,v}	3.40	12.5	12½	0.155	⅛	1/16	3.50	3½	0.211	3/16	9/16	⅜	11³⁄₈	—
M12×11.8^c	3.47	12.0	12	0.177	3/16	⅛	3.07	3⅛	0.225	¼	9/16	⅜	10⁷⁄₈	—
×10.8^c	3.18	12.0	12	0.160	3/16	⅛	3.07	3⅛	0.210	3/16	9/16	⅜	10⁷⁄₈	—
M12×10^{c,v}	2.95	12.0	12	0.149	⅛	1/16	3.25	3¼	0.180	3/16	½	⅜	11	—
M10×9^c	2.65	10.0	10	0.157	3/16	⅛	2.69	2¾	0.206	3/16	9/16	⅜	8⁷⁄₈	—
×8^c	2.37	9.95	10	0.141	⅛	1/16	2.69	2¾	0.182	3/16	9/16	⅜	8⁷⁄₈	—
M10×7.5^{c,v}	2.22	9.99	10	0.130	⅛	1/16	2.69	2¾	0.173	3/16	7/16	5/16	9⅛	—
M8×6.5^c	1.92	8.00	8	0.135	⅛	1/16	2.28	2¼	0.189	3/16	9/16	⅜	6⁷⁄₈	—
×6.2^c	1.82	8.00	8	0.129	⅛	1/16	2.28	2¼	0.177	3/16	7/16	¼	7⅛	—
M6×4.4^c	1.29	6.00	6	0.114	⅛	1/16	1.84	1⅞	0.171	3/16	⅜	¼	5¼	—
×3.7^c	1.09	5.92	5⁷⁄₈	0.0980	⅛	1/16	2.00	2	0.129	⅛	5/16	¼	5¼	—
M5×18.9^t	5.56	5.00	5	0.316	5/16	3/16	5.00	5	0.416	7/16	13/16	½	3⅜	2¾^g
M4×6^f	1.75	3.80	3¾	0.130	⅛	1/16	3.80	3¾	0.160	3/16	½	⅜	2¾	—
×4.08	1.27	4.00	4	0.115	⅛	1/16	2.25	2¼	0.170	3/16	9/16	⅜	2⁷⁄₈	—
×3.45	1.01	4.00	4	0.0920	1/16	1/16	2.25	2¼	0.130	⅛	½	⅜	3	—
×3.2	1.01	4.00	4	0.0920	1/16	1/16	2.25	2¼	0.130	⅛	½	⅜	3	—
M3×2.9	0.914	3.00	3	0.0900	1/16	1/16	2.25	2¼	0.130	⅛	½	⅜	2	—

^{c} Shape is slender for compression with F_y = 36 ksi.
^{f} Shape exceeds compact limit for flexure with F_y = 36 ksi.
^{g} The actual size, combination, and orientation of fastener components should be compared with the geometry of the cross-section to ensure compatibility.
^{t} Shape has tapered flanges while other M-shapes have parallel flange surfaces.
^{v} Shape does not meet the h/t_w limit for shear in Specification Section G2.1b(i) with F_y = 36 ksi.
— Flange is too narrow to establish a workable gage.

Structural Metal Shape Designations. (CONTINUED)

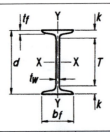

Table 1–3
S Shapes
Dimensions

Shape	Area, A	Depth, d		Web Thickness, t_w		$\frac{t_w}{2}$	Flange Width, b_f		Thickness, t_f		k	T	Workable Gage
	in.²	in.		in.		in.	in.		in.		in.	in.	in.
S24×121	35.5	24.5	24½	0.800	13/16	7/16	8.05	8	1.09	1 1/16	2	20½	4
×106	31.1	24.5	24½	0.620	5/8	5/16	7.87	7⅞	1.09	1 1/16	2	20½	4
S24×100	29.3	24.0	24	0.745	3/4	3/8	7.25	7¼	0.870	7/8	1¾	20½	4
×90	26.5	24.0	24	0.625	5/8	5/16	7.13	7⅛	0.870	7/8	1¾	20½	4
×80	23.5	24.0	24	0.500	1/2	1/4	7.00	7	0.870	7/8	1¾	20½	4
S20×96	28.2	20.3	20¼	0.800	13/16	7/16	7.20	7¼	0.920	15/16	1¾	16¾	4
×86	25.3	20.3	20¼	0.660	11/16	3/8	7.06	7	0.920	15/16	1¾	16¾	4
S20×75	22.0	20.0	20	0.635	5/8	5/16	6.39	6⅜	0.795	13/16	1⅝	16¾	3½g
×66	19.4	20.0	20	0.505	1/2	1/4	6.26	6¼	0.795	13/16	1⅝	16¾	3½g
S18×70	20.5	18.0	18	0.711	11/16	3/8	6.25	6¼	0.691	11/16	1½	15	3½g
×54.7	16.0	18.0	18	0.461	7/16	1/4	6.00	6	0.691	11/16	1½	15	3½g
S15×50	14.7	15.0	15	0.550	9/16	5/16	5.64	5⅝	0.622	5/8	1⅜	12¼	3½g
×42.9	12.6	15.0	15	0.411	7/16	1/4	5.50	5½	0.622	5/8	1⅜	12¼	3½g
S12×50	14.6	12.0	12	0.687	11/16	3/8	5.48	5½	0.659	11/16	1 7/16	9⅛	3g
×40.8	11.9	12.0	12	0.462	7/16	1/4	5.25	5¼	0.659	11/16	1 7/16	9⅛	3g
S12×35	10.2	12.0	12	0.428	7/16	1/4	5.08	5⅛	0.544	9/16	1 3/16	9⅝	3g
×31.8	9.31	12.0	12	0.350	3/8	3/16	5.00	5	0.544	9/16	1 3/16	9⅝	3g
S10×35	10.3	10.0	10	0.594	5/8	5/16	4.94	5	0.491	1/2	1⅛	7¾	2¾g
×25.4	7.45	10.0	10	0.311	5/16	3/16	4.66	4⅝	0.491	1/2	1⅛	7¾	2¾g
S8×23	6.76	8.00	8	0.441	7/16	1/4	4.17	4⅛	0.425	7/16	1	6	2¼g
×18.4	5.40	8.00	8	0.271	1/4	1/8	4.00	4	0.425	7/16	1	6	2¼g
S6×17.2	5.06	6.00	6	0.465	7/16	1/4	3.57	3⅝	0.359	3/8	13/16	4⅜	—
×12.5	3.66	6.00	6	0.232	1/4	1/8	3.33	3⅜	0.359	3/8	13/16	4⅜	—
S5×10	2.93	5.00	5	0.214	3/16	1/8	3.00	3	0.326	5/16	3/4	3½	—
S4×9.5	2.79	4.00	4	0.326	5/16	3/16	2.80	2¾	0.293	5/16	3/4	2½	—
×7.7	2.26	4.00	4	0.193	3/16	1/8	2.66	2⅝	0.293	5/16	3/4	2½	—
S3×7.5	2.20	3.00	3	0.349	3/8	3/16	2.51	2½	0.260	1/4	5/8	1¾	—
×5.7	1.66	3.00	3	0.170	3/16	1/8	2.33	2⅜	0.260	1/4	5/8	1¾	—

g The actual size, combination, and orientation of fastener components should be compared with the geometry of the cross-section to ensure compatibility.
— Flange is too narrow to establish a workable gage.

Structural Metal Shape Designations. (CONTINUED)

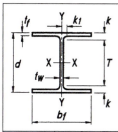

Table 1–4
HP Shapes
Dimensions

Shape	Area, A	Depth, d		Web Thickness, t_w		$\frac{t_w}{2}$	Flange Width, b_f		Thickness, t_f		Distance k	k_1	T	Workable Gage
	in.²	in.		in.		in.	in.		in.		in.	in.	in.	in.
HP14×117[f]	34.4	14.2	14 1/4	0.805	13/16	7/16	14.9	14 7/8	0.805	13/16	1 1/2	1 1/16	11 1/4	5 1/2
×102[f]	30.0	14.0	14	0.705	11/16	3/8	14.8	14 3/4	0.705	11/16	1 3/8	1		
×89[f]	26.1	13.8	13 7/8	0.615	5/8	5/16	14.7	14 3/4	0.615	5/8	15/16	15/16		
×73[c,f]	21.4	13.6	13 5/8	0.505	1/2	1/4	14.6	14 5/8	0.505	1/2	13/16	7/8		
HP12×84	24.6	12.3	12 1/4	0.685	11/16	3/8	12.3	12 1/4	0.685	11/16	1 3/8	1	9 1/2	5 1/2
×74[f]	21.8	12.1	12 1/8	0.605	5/8	5/16	12.2	12 1/4	0.610	5/8	15/16	15/16		
×63[f]	18.4	11.9	12	0.515	1/2	1/4	12.1	12 1/8	0.515	1/2	1 1/4	7/8		
×53[f]	15.5	11.8	11 3/4	0.435	7/16	1/4	12.0	12	0.435	7/16	1 1/8	7/8		
HP10×57[f]	16.8	9.99	10	0.565	9/16	5/16	10.2	10 1/4	0.565	9/16	1 1/4	15/16	7 1/2	5 1/2
×42[f]	12.4	9.70	9 3/4	0.415	7/16	1/4	10.1	10 1/8	0.420	7/16	1 1/8	13/16	7 1/2	5 1/2
HP8×36[f]	10.6	8.02	8	0.445	7/16	1/4	8.16	8 1/8	0.445	7/16	1 1/8	7/8	5 3/4	5 1/2

[c] Shape is slender for compression with F_y = 50 ksi.
[f] Shape exceeds compact limit for flexure with F_y = 50 ksi.

Structural Metal Shape Designations. (CONTINUED)

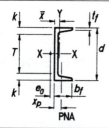

Table 1–5
C Shapes
Dimensions

Shape	Area, A	Depth, d	Web Thickness, t_w		$\frac{t_w}{2}$	Flange Width, b_f		Thickness, t_f		Distance k	T	Workable Gage	r_{ts}	h_o
	in.²	in.	in.		in.	in.		in.		in.	in.	in.	in.	in.
C15×50	14.7	15.0	15	0.716	11/16	3/8	3.72	3³/4	0.650	5/8	1⁷/16	12¹/8	2¹/4	1.17	14.4
×40	11.8	15.0	15	0.520	1/2	1/4	3.52	3¹/2	0.650	5/8	1⁷/16	12¹/8	2	1.15	14.4
×33.9	10.0	15.0	15	0.400	3/8	3/16	3.40	3³/8	0.650	5/8	1⁷/16	12¹/8	2	1.13	14.4
C12×30	8.81	12.0	12	0.510	1/2	1/4	3.17	3¹/8	0.501	1/2	1¹/8	9³/4	1³/4ᵍ	1.01	11.5
×25	7.34	12.0	12	0.387	3/8	3/16	3.05	3	0.501	1/2	1¹/8	9³/4	1³/4ᵍ	1.00	11.5
×20.7	6.08	12.0	12	0.282	5/16	3/16	2.94	3	0.501	1/2	1¹/8	9³/4	1³/4ᵍ	0.983	11.5
C10×30	8.81	10.0	10	0.673	11/16	3/8	3.03	3	0.436	7/16	1	8	1³/4ᵍ	0.925	9.56
×25	7.34	10.0	10	0.526	1/2	1/4	2.89	2⁷/8	0.436	7/16	1	8	1³/4ᵍ	0.911	9.56
×20	5.87	10.0	10	0.379	3/8	3/16	2.74	2³/4	0.436	7/16	1	8	1¹/2ᵍ	0.894	9.56
×15.3	4.48	10.0	10	0.240	1/4	1/8	2.60	2⁵/8	0.436	7/16	1	8	1¹/2ᵍ	0.869	9.56
C9×20	5.87	9.00	9	0.448	7/16	1/4	2.65	2⁵/8	0.413	7/16	1	7	1¹/2ᵍ	0.848	8.59
×15	4.41	9.00	9	0.285	5/16	3/16	2.49	2¹/2	0.413	7/16	1	7	1³/8ᵍ	0.824	8.59
×13.4	3.94	9.00	9	0.233	1/4	1/8	2.43	2³/8	0.413	7/16	1	7	1³/8ᵍ	0.813	8.59
C8×18.7	5.51	8.00	8	0.487	1/2	1/4	2.53	2¹/2	0.390	3/8	15/16	6¹/8	1¹/2ᵍ	0.800	7.61
×13.7	4.04	8.00	8	0.303	5/16	3/16	2.34	2³/8	0.390	3/8	15/16	6¹/8	1³/8ᵍ	0.774	7.61
×11.5	3.37	8.00	8	0.220	1/4	1/8	2.26	2¹/4	0.390	3/8	15/16	6¹/8	1³/8ᵍ	0.756	7.61
C7×14.7	4.33	7.00	7	0.419	7/16	1/4	2.30	2¹/4	0.366	3/8	7/8	5¹/4	1¹/4ᵍ	0.738	6.63
×12.2	3.60	7.00	7	0.314	5/16	3/16	2.19	2¹/4	0.366	3/8	7/8	5¹/4	1¹/4ᵍ	0.721	6.63
×9.8	2.87	7.00	7	0.210	3/16	1/8	2.09	2¹/8	0.366	3/8	7/8	5¹/4	1¹/4ᵍ	0.698	6.63
C6×13	3.81	6.00	6	0.437	7/16	1/4	2.16	2¹/8	0.343	5/16	13/16	4/8	1³/8ᵍ	0.689	5.66
×10.5	3.08	6.00	6	0.314	5/16	3/16	2.03	2	0.343	5/16	13/16	4³/8	1¹/8ᵍ	0.669	5.66
×8.2	2.39	6.00	6	0.200	3/16	1/8	1.92	1⁷/8	0.343	5/16	13/16	4³/8	1¹/8ᵍ	0.643	5.66
C5×9	2.64	5.00	5	0.325	5/16	3/16	1.89	1⁷/8	0.320	5/16	3/4	3¹/2	1¹/8ᵍ	0.617	4.68
×6.7	1.97	5.00	5	0.190	3/16	1/8	1.75	1³/4	0.320	5/16	3/4	3¹/2	—	0.584	4.68
C4×7.2	2.13	4.00	4	0.321	5/16	3/16	1.72	1³/4	0.296	5/16	3/4	2¹/2	1ᵍ	0.563	3.70
×5.4	1.58	4.00	4	0.184	3/16	1/8	1.58	1⁵/8	0.296	5/16	3/4	2¹/2	—	0.528	3.70
×4.5	1.38	4.00	4	0.125	1/8	1/16	1.58	1⁵/8	0.296	5/16	3/4	2¹/2	—	0.524	3.70
C3×6	1.76	3.00	3	0.356	3/8	3/16	1.60	1⁵/8	0.273	1/4	11/16	1⁵/8	—	0.519	2.73
×5	1.47	3.00	3	0.258	1/4	1/8	1.50	1¹/2	0.273	1/4	11/16	1⁵/8	—	0.495	2.73
×4.1	1.20	3.00	3	0.170	3/16	1/8	1.41	1³/8	0.273	1/4	11/16	1⁵/8	—	0.469	2.73
×3.5	1.09	3.00	3	0.132	1/8	1/16	1.37	1³/8	0.273	1/4	11/16	1⁵/8	—	0.455	2.73

ᵍ The actual size, combination, and orientation of fastener components should be compared with the geometry of the cross-section to ensure compatibility.
— Flange is too narrow to establish a workable gage.

Structural Metal Shape Designations. (CONTINUED)

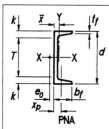

PNA

Table 1–6
MC Shapes
Dimensions

Shape	Area, A	Depth, d		Web				Flange				Distance			r_{ts}	h_o
				Thickness, t_w		$\frac{t_w}{2}$		Width, b_f		Average Thickness, t_f		k	T	Work-able Gage		
	in.²	in.		in.		in.		in.		in.		in.	in.	in.	in.	in.
MC18×58	17.1	18.0	18	0.700	11/16	3/8		4.20	4¼	0.625	5/8	1 7/16	15 1/8	2½	1.35	17.4
×51.9	15.3	18.0	18	0.600	5/8	5/16		4.10	4 1/8	0.625	5/8	1 7/16			1.35	17.4
×45.8	13.5	18.0	18	0.500	½	¼		4.00	4	0.625	5/8	1 7/16			1.34	17.4
×42.7	12.6	18.0	18	0.450	7/16	¼		3.95	4	0.625	5/8	1 7/16	↓	↓	1.34	17.4
MC13×50	14.7	13.0	13	0.787	13/16	7/16		4.41	4 3/8	0.610	5/8	1 7/16	10 1/8	2½	1.41	12.4
×40	11.8	13.0	13	0.560	9/16	5/16		4.19	4 1/8	0.610	5/8	1 7/16			1.38	12.4
×35	10.3	13.0	13	0.447	7/16	¼		4.07	4 1/8	0.610	5/8	1 7/16			1.35	12.4
×31.8	9.35	13.0	13	0.375	3/8	3/16		4.00	4	0.610	5/8	1 7/16	↓	↓	1.34	12.4
MC12×50	14.7	12.0	12	0.835	13/16	7/16		4.14	4 1/8	0.700	11/16	1 5/16	9 3/8	2½	1.37	11.3
×45	13.2	12.0	12	0.710	11/16	3/8		4.01	4	0.700	11/16	1 5/16			1.35	11.3
×40	11.8	12.0	12	0.590	9/16	5/16		3.89	3 7/8	0.700	11/16	1 5/16			1.33	11.3
×35	10.3	12.0	12	0.465	7/16	¼		3.77	3 3/4	0.700	11/16	1 5/16	↓		1.30	11.3
×31	9.12	12.0	12	0.370	3/8	3/16		3.67	3 5/8	0.700	11/16	1 5/16	↓	2¼	1.28	11.3
MC12×10.6[c]	3.10	12.0	12	0.190	3/16	1/8		1.50	1½	0.309	5/16	3/4	10½	—	0.477	11.7
MC10×41.1	12.1	10.0	10	0.796	13/16	7/16		4.32	4 3/8	0.575	9/16	1 5/16	7 3/8	2½[g]	1.44	9.43
×33.6	9.87	10.0	10	0.575	9/16	5/16		4.10	4 1/8	0.575	9/16	1 5/16	7 3/8	2½[g]	1.40	9.43
×28.5	8.37	10.0	10	0.425	7/16	¼		3.95	4	0.575	9/16	1 5/16	7 3/8	2½[g]	1.36	9.43
MC10×25	7.35	10.0	10	0.380	3/8	3/16		3.41	3 3/8	0.575	9/16	1 5/16	7 3/8	2[g]	1.17	9.43
×22	6.45	10.0	10	0.290	5/16	3/16		3.32	3 3/8	0.575	9/16	1 5/16	7 3/8	2[g]	1.14	9.43
MC10×8.4[c]	2.46	10.0	10	0.170	3/16	1/8		1.50	1½	0.280	¼	3/4	8½	—	0.486	9.72
×6.5[c]	1.95	10.0	10	0.152	1/8	1/16		1.17	1 1/8	0.202	3/16	9/16	8 7/8	—	0.364	9.80
MC9×25.4	7.47	9.00	9	0.450	7/16	¼		3.50	3½	0.550	9/16	1¼	6½	2[g]	1.20	8.45
×23.9	7.02	9.00	9	0.400	3/8	3/16		3.45	3½	0.550	9/16	1¼	6½	2[g]	1.18	8.45
MC8×22.8	6.70	8.00	8	0.427	7/16	¼		3.50	3½	0.525	½	1 3/16	5 5/8	2[g]	1.20	7.48
×21.4	6.28	8.00	8	0.375	3/8	3/16		3.45	3½	0.525	½	1 3/16	5 5/8	2[g]	1.18	7.48
MC8×20	5.88	8.00	8	0.400	3/8	3/16		3.03	3	0.500	½	1 1/8	5 3/4	2[g]	1.03	7.50
×18.7	5.50	8.00	8	0.353	3/8	3/16		2.98	3	0.500	½	1 1/8	5 3/4	2[g]	1.02	7.50
MC8×8.5	2.50	8.00	8	0.179	3/16	1/8		1.87	1 7/8	0.311	5/16	13/16	6 3/8	1 1/8[g]	0.624	7.69

[c] Shape is slender for compression with F_y = 36 ksi.
[g] The actual size, combination, and orientation of fastener components should be compared with the geometry of the cross-section to ensure compatibility.
— Flange is too narrow to establish a workable gage.

Structural Metal Shape Designations. (CONTINUED)

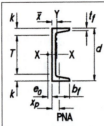

Table 1–6 (continued)
MC Shapes
Dimensions

Shape	Area, A	Depth, d	Web				Flange				Distance			r_{ts}	h_o
			Thickness, t_w		$\frac{t_w}{2}$		Width, b_f		Average Thickness, t_f		k	T	Work-able Gage		
	in.²	in.	in.		in.		in.		in.		in.	in.	in.	in.	in.
MC7×22.7	6.67	7.00	7	0.503	1/2	1/4	3.60	3 5/8	0.500	1/2	1 1/8	4 3/4	2 g	1.23	6.50
×19.1	5.61	7.00	7	0.352	3/8	3/16	3.45	3 1/2	0.500	1/2	1 1/8	4 3/4	2 g	1.18	6.50
MC6×18	5.29	6.00	6	0.379	3/8	3/16	3.50	3 1/2	0.475	1/2	1 1/16	3 7/8	2 g	1.20	5.53
×15.3	4.49	6.00	6	0.340	5/16	3/16	3.50	3 1/2	0.385	3/8	7/8	4 1/4	2 g	1.20	5.62
MC6×16.3	4.79	6.00	6	0.375	3/8	3/16	3.00	3	0.475	1/2	11/16	3 7/8	1 3/4 g	1.03	5.53
×15.1	4.44	6.00	6	0.316	5/16	3/16	2.94	3	0.475	1/2	11/16	3 7/8	1 3/4 g	1.01	5.53
MC6×12	3.53	6.00	6	0.310	5/16	3/16	2.50	2 1/2	0.375	3/8	7/8	4 1/4	1 1/2 g	0.856	5.63
MC6×7	2.09	6.00	6	0.179	3/16	1/8	1.88	1 7/8	0.291	5/16	3/4	4 1/2	—	0.638	5.71
×6.5	1.95	6.00	6	0.155	1/8	1/16	1.85	1 7/8	0.291	5/16	3/4	4 1/2	—	0.630	5.71
MC4×13.8	4.03	4.00	4	0.500	1/2	1/4	2.50	2 1/2	0.500	1/2	1	2	—	0.852	3.50
MC3×7.1	2.11	3.00	3	0.312	5/16	3/16	1.94	2	0.351	3/8	13/16	1 3/8	—	0.657	2.65

g The actual size, combination, and orientation of fastener components should be compared with the geometry of the cross-section to ensure compatibility.
— Flange is too narrow to establish a workable gage.

Structural Metal Shape Designations. (CONTINUED)

glossary of key terms

active animation An animation in which the observer (camera) as well as objects in the scene actively move around and through the scene.

additive color model The RGB color system in which the primary colors of red, green, and blue are added together to create white.

adjacent views Orthogonal views created immediately next to each other, aligned side by side to share a common dimension, and presented on a single plane.

agenda The list of topics for discussion/action at a team meeting.

algebraic constraints Constraint that define the value of a selected variable as the result of an algebraic expression containing other variables from the solid model.

allowance The difference between the maximum material limits of mating parts. It is the minimum clearance or maximum interference between parts.

alpha channel An optional layer of image data containing an additional 8 bits of grayscale data that can be used to control transparency affecting the entire image.

ambient light Indirect light in a scene that does not come directly from a light source, but arrives at a surface by bouncing around or reflecting off other surfaces in the scene.

analysis The study of the behavior of a physical system under certain imposed conditions.

anchor edge The same edge that can be easily and confidently located on multiple views and on a pictorial for an object.

anchor point The same point, usually a vertex, that can be easily and confidently located on multiple views and on a pictorial for an object.

anchor surface The same surface that can be easily and confidently located on multiple views and on a pictorial for an object.

angle of thread The angle between the side of a thread and a line perpendicular to the axis of the thread.

ANSI Y14.5 *(ASME Y14.5M-1994)* Industry standard document that outlines uniform practices for displaying and interpreting dimensions and related information on drawings and other forms of engineering documentation.

approval signatures The dated signatures or initials of the people responsible for certain aspects of a formal drawing, such as the people who did the drafting or the engineer responsible for the function of the part.

arc A curved entity that represents a portion of a circle.

architects Professionals who complete conceptual designs for civil engineering projects.

Architect's scale A device used to measure or draw lines in the English system of units with a base unit of inches and fractions of an inch.

arrowhead A small triangle at the end of dimension lines and leaders to indicate the direction and extent of a dimension.

as-built drawings The marked-up drawings from a civil engineering project that show any modifications implemented in the field during construction.

as-built plans Drawings that show exactly how buildings were constructed, especially when variations exist between the final building and the plans created during the design phase.

aspect A quantitative measure of the direction of a slope face.

assembly A collection of parts and/or subassemblies that have been put together to make a device or structure that performs a specific function.

assembly constraints Used to establish relationships between instances in the development of a flexible assembly model.

assembly dimensions Dimensions that show where parts must be placed relative to other parts when the device is being put together.

associative constraints *See* algebraic constraints.

associativity The situation whereby parts can be modified and the components referenced to the parts will be modified accordingly.

attribute Spatial information that describes the characteristics of spatial features.

auxiliary views Views on any projection plane other than a primary or principal projection plane.

axis The longitudinal centerline that passes through a screw.

axonometric drawing   A drawing in which all three dimensional axes on an object can be seen, with the scaling factor constant in each direction. Usually, one axis is shown as being vertical.

back light   A scene light, usually located behind objects in the scene, which is used to create a defining edge that visually separates foreground objects from the background.

balloons   Closed geometric shapes, usually circles, containing identification numbers and placed beside parts on a layout or assembly drawing to help identify those parts.

bar chart   A chart using bars of varying heights and widths to represent quantitative data.

base feature   The first feature created for a part, usually a protrusion.

base instance   The one fixed instance within an assembly.

baseline dimensioning   A method for specifiying the location of features on a part whereby all the locations are relative to a common feature or edge.

basic dimension   A dimension that is theoretically exact. It is identified by a box around the dimension. It locates the perfect position of features from clearly identified datums.

bearing   The angle that a line makes with a North-South line as seen in a plan view.

benchmarks   Points established by the U.S. Geological Survey that can be used to accurately locate control points on a construction site.

bill of materials (BOM)   A drawing or table in a drawing that lists all of the parts needed to build a device by (at least) the part number, part name, type of material used, and number of times the part is used in the device.

bitmap textures   Texture mapping routines that are based on referencing external image files.

black box diagram   A diagram that shows the major inputs and outputs from a system.

blend   A solid formed by a smooth transition between two or more profiles.

blind extrusion   An extrusion made to a specified length in a selected direction.

blind hole   A hole that does not pass completely through a part.

blueprints   The name sometimes given to construction drawings based on historical blue-on-white drawings that were produced from ink drawings.

bolt   A threaded fastener that passes completely through parts and holds them together using a nut.

Boolean operations   In early versions of 3-D CAD software, commands used to combine solids.

border   A thick line that defines the perimeter of a drawing.

boring   The general process of making a hole in a part by plunging a rotating tool bit into a part, moving a rotating part into a stationary tool bit, or moving a part into a rotating tool bit.

bottom-up modeling   The process of creating individual parts and then creating an assembly from them.

boundary conditions   The constraints and loads added to the boundaries of a finite element model.

boundary representation (b-rep)   A method used to build solid models from their bounding surfaces.

bounding box   A square box used to sketch circles or ellipses.

brainstorming   The process of group creative thinking used to generate as many ideas as possible for consideration.

brainwriting   A process of group creative thinking where sketching is the primary mode of communication between team members.

brazing   A method for joining separate metal parts by heating the parts, flowing a molten metal (solder) between them, and allowing the unit to cool and harden.

brief   A small graphic using word content alone.

broach   A long, shaped cutting tool that moves along the length of a part when placed against it. It is used to create uniquely shaped holes and slots.

broaching   A process of creating uniquely shaped holes and slots using a long, shaped cutting tool that moves along the length of a part in a single stroke when placed against the part.

broken-out section   The section view produced when the cutting plane is partially imbedded into the object, requiring an irregular portion of the object to be removed before the hypothetically cut surface can be seen.

buffer   Measured in units of distance or time, a zone around a map feature. A buffer is useful for proximity analysis.

bump mapping   A technique used to create the illusion of rough or bumpy surface detail through surface normal perturbation.

business diagram   A diagram used in an organization to show organizational hierarchy, task planning or analysis, or relationships between different groups or sets of information.

butt joint   A joint between two parts wherein the parts are butted, or placed next to each other.

CAD   Computer-aided drawing. The use of computer hardware and software for the purpose of creating, modifying, and storing engineering drawings in an electronic format.

CAD designers   Designers who create 3-D computer models for analysis and detailing.

cabinet oblique drawing   An oblique drawing where one half the true length of the depth dimension is measured along the receding axes.

caliper   A handheld device used to measure objects with a fair degree of accuracy.

cap screw   A small threaded fastener that mates with a threaded hole.

casting   A method of creating a part by pouring or injecting molten material into the cavity of a mold, allowing it to harden, and then removing it from the mold.

cavalier oblique drawing   An oblique drawing where the true length of the depth dimension is measured along the receding axes.

center-of-mass (centroid)   The origin of the coordinate axes for which the first moments are zero.

centerline   A series of alternating long and short dashed lines used to identify an axis of rotational symmetry.

centermark   A small right-angle cross that is used to identify the end view of an axis of rotational symmetry.

Central Meridian   The line of longitude that defines the center and is often the x-origin of a projected coordinate system.

Central Parallel   The line of latitude that divides a map into north and south halves and is often the y-origin of a projected coordinate system.

chain dimensioning   A method for specifying the location of features on a part whereby the location of each feature is successively specified relative to the location of the previous feature.

chamfers   Angled cut transitions between two intersecting surfaces.

charts   Charts, graphs, tables, and diagrams of ideas and quantitative data.

child feature   A feature that is dependent upon the existence of a previously created feature.

chief designer   The individual who oversees other members of the design team and manages the overall project.

circle   A closed curved figure where all points on it are equidistant from its center point.

Clarke Ellipsoid of 1866   A reference ellipsoid having a semimajor axis of approximately 6,378,206.4 meters and a flattening of 1/294.9786982. It is the basis for the North American Datum of 1927 (NAD27) and other datums.

clearance   A type of fit where space exists between two mating parts.

clearances   The minimum distances between two instances in an assembly.

clip   A geoprocessing command that extracts the features from a coverage that reside entirely within a boundary defined by features in another coverage.

clipping plane   A 3-D virtual camera technique that allows you to selectively exclude, and not view or render, unnecessary objects in a scene that are either too close or too far away.

CODEC   Video compression-decompression algorithm.

collision detection   A built-in software capability for calculating and graphically animating the results of collisions between multiple objects based on object properties of speed, mass, and gravity.

color mapping   Sometimes called diffuse mapping, color mapping replaces the main surface color of a model with an external image map or texture.

combining solids   The process of cutting, joining, or intersecting two objects to form a third object.

components   References of object geometry used in assembly models.

compositing   The technique and art of rendering in layers or passes, editing the image on each layer as needed, and compiling the edited layers or images into a single unified final image.

computer-aided design (CAD)   The process by which computers are used to model and analyze designed products.

computer-aided manufacturing (CAM)   The process by which parts are manufactured directly from 3-D computer models.

concept mapping   The creative process by which the central idea is placed in the middle of a page and related concepts radiate out from that central idea.

conceptual design   The initial idea for a design before analysis has been performed.

concurrent engineering   The process by which designers, analysts, and manufacturers work together from the start to design a product.

consensus   A process of decision making where an option is chosen that everyone supports.

constraint   A boundary condition applied to a finite element model to prevent it from moving through space.

constraints   Geometric relationships, dimensions, or equations that control the size, shape, and/or orientation of entities in a sketch or solid model.

constructive solid geometry (CSG)   A method used to build solid models from primitive shapes based on Boolean set theory.

construction drawings   Working drawings, often created by civil engineers, that are used to build large-scale, one-of-a-kind structures.

construction line A faint line used in sketching to align items and define shapes.

continuation blocks Header blocks used on the second and subsequent pages of multipage drawings.

contour dimensioning Placing each dimension in the view where the contour or shape of the feature shows up best.

contour interval The vertical distance between contours.

contour rule A drawing practice where each dimension should be placed in the view where the contour shape is best shown.

contours Lines or curves that represent the same elevation across the landscape.

control points Points at a construction site that are referenced to an origin by north, south, east, or west coordinates.

coordinate measuring machine A computer-based tool used to digitize object geometry for direct input to a 3-D CAD system.

corner views An isometric view of an object created from the perspective at a given corner of the object.

cosmetic features Features that modify the appearance of the surface but do not alter the size or shape of the object.

cover sheet The first page in a set of construction drawings showing a map of the location of the project and possibly an index.

crest The top surface or point joining the sides of a thread.

critical path The sequence of activities in a project that have the longest duration.

critical path method (CPM) A tool for determining the least amount of time in which a project can be completed.

cross-section The intersection between a cutting plane and a 3-D object.

curved surface Any nonflat surface on an object.

cut (noun) A feature created by the removal of solid volume from a model.

cut (verb) To remove the volume of interference between two objects from one of the objects.

cutaway diagram A diagram that allows the reader to see a slice of an object.

cutting-plane An imaginary plane that intersects with an object to form a cross section.

cutting plane A theoretical plane used to hypothetically cut and remove a portion of an object to reveal its interior details.

cutting plane line On an orthographic view of an object, the presentation of the edge view of a cutting plane used to hypothetically cut and remove a portion of that object for viewing.

cutting segment On a stepped cutting plane for an offset section view, that portion of the plane that hypothetically cuts and reveals the interior detail of a feature of interest.

data structure The organization of data within a specific computer system that allows the information to be stored and manipulated effectively.

database A collection of information for a computer and a method for interpretation of the information from which the original model can be re-created.

datum A theoretical plane or axis established by real features on an object for the purpose of defining the datum reference frame.

datum geometries Geometric entities such as points, axes, and planes that do not actually exist on real parts, but are used to help locate and define other features.

datum planes The planes used to define the locations of features and entities in the construction of a solid model.

datum reference frame A system of three mutually perpendicular planes used as the coordinate system for geometric dimensioning.

decimal degrees (DD) A measuring system in which values of latitude and longitude are expressed in decimal format rather than in degrees, minutes, and seconds, such as 87.5°.

deep drawing Creating a thin-shelled part by pressing sheet metal into a deep cavity mold.

default tolerances Usually appearing in the drawing header, the tolerances to be assumed for any dimension show on a part when that dimension does not specify any tolerances.

degrees, minutes, seconds (DMS) A measuring system for longitude and latitude values, such as 87° 30' 00", in which 60 seconds equals 1 minute and 60 minutes equals 1 degree.

density The mass per unit volume for a given material.

depiction An illustration describing and simplifying factual information on a real-world system.

depth of thread The distance between the crest and the root of a thread, measured normal to the axis.

depth-mapped shadows Also called shadow-mapped shadows, depth-mapped shadows use a precalculated depth map to determine the location, density, and edge sharpness of shadows.

descriptive geometry A two-dimensional graphical construction technique used for geometric analysis of three-dimensional objects.

design (noun)   An original manifestation of a device or method created for performing one or more useful functions.

design (verb)   The process of creating a design (noun).

design analysts   Individuals who analyze design concepts by computer methods to determine their structural, thermal, or vibration characteristics.

design documentation   The set of drawings and specifications that illustrate and thoroughly describe a designed product.

design process   The multistep, iterative process by which products are conceived and produced.

design table   A table or spreadsheet that lists all of the versions of a family model, the dimensions or features that may change, and the values in any of its versions.

design tree   *See* model tree.

detail designers   The individuals who create engineering drawings, complete with annotation, from 3-D computer models or from engineering sketches.

detail drawing   A formal drawing that shows the geometry, dimensions, tolerances, materials, and any processes needed to fabricate a part.

detail sections   Drawings included in a set of construction plans that show how the various components are assembled.

devil's advocate   The team member who challenges ideas to ensure that all options are considered by the group.

diagram   An illustration that explains information, represents a process, or shows how pieces are put together.

diametric drawing   An axonometric drawing in which the scaling factor is the same for two of the axes.

die   A special tool made specifically to deform raw or stock material into a desired outline of a part or feature in a single operation.

die casting   A method of casting where the mold is formed by cutting a cavity into steel or another hard material. *See* casting.

digital elevation model (DEM)   The representation of continuous elevation values over a topographic surface by a regular array of z-values referenced to a common datum. It is typically used to represent terrain relief.

dimension   A numerical value expressed in appropriate units of measure and used to define the size, location, geometric characteristic, or surface texture of a part or part feature.

dimension line   A thin, dark, solid line that terminates at each end with arrowheads. The value of a dimension typically is shown in the center of the dimension line.

dimension name   The unique alphanumeric designation of a variable dimension.

dimensional constraints   Measurements used to control the size or position of entities in a sketch.

direct dimensioning   Dimensioning between two key points to minimize tolerance accumulation.

directional light   A computer-generated light source designed to simulate the effect of light sources, such as the sun, that are so far away from objects in the scene that lighting and shadow patterns in the scene appear to be parallel.

displacement   A change in the location of points on an object after it has been subjected to external loads.

dissolve   A geoprocessing command that removes boundaries between adjacent polygons that have the same value for a specified attribute.

draft   The slight angling of the walls of a cast, forged, drawn, or stamped part to enable the part to be removed from the mold more easily.

drawing   A collection of images and other detailed graphical specifications intended to represent physical objects or processes for the purpose of accurately re-creating those objects or processes.

drill bit   A long, rotating cutting tool with a sharpened tip used to make holes.

drilling   A process of making a hole by plunging a rotating tool bit into a part.

drill press   A machine that holds, spins, and plunges a rotating tool bit into a part to make holes.

driven dimension   A variable connected to an algebraic constraint that can be modified only by user changes to the driving dimensions.

driving dimension   A variable used in an algebraic constraint to control the values of another (driven) dimension.

double-sided extrusion   A solid formed by the extrusion of a profile in both directions from its sketching plane.

EC Level   A number included in the title block of a drawing indicating that the part has undergone a revision.

edge tracking   A procedure by which successive edges on an object are simultaneously located on a pictorial image and on a multiview image of that object.

edge view (of a plane)   A view in which the given plane appears as a straight line.

EDM   Electric discharge machining; a process by which material is eroded from a part by passing an electric current between the part and an electrode (or a wire) through an electrolytic fluid.

electrical plan   A plan view showing the layout of electrical devices on a floor in a building.

elevation view   In the construction of a perspective view, the object as viewed from the front, as if created by orthogonal projection.

elevation views   Views of a structure that show changes in elevation (side or front views).

ellipse   A closed curve figure where the sum of the distance between any point on the figure and its two foci is constant.

end mill   A rotating cutting tool that, when placed in the spindle of a milling machine, can remove material in directions parallel or perpendicular to its rotation axis.

engineer (noun)   A person who engages in the art of engineering.

engineer (verb)   To plan and build a device that does not occur naturally within the environment.

Engineer's scale   A device used to measure or draw lines in the English system of units with a base unit of inches and tenths of an inch.

engineering   The profession in which knowledge of mathematical and natural sciences gained by study, experience, and practice is applied with judgment to develop and utilize economically the materials and forces of nature for the benefit of humanity.

engineering animation   A dynamic virtual 3-D prototype of a mechanism or system that can be assembled (usually from preexisting or newly created 3-D CAD part models) and/or shown in operation over a period of time.

engineering change (EC) number   A dated number that defines the degree to which the specifications of a part have been updated.

engineering design   The process by which many competing factors of a product are weighed to select the best alternative in terms of cost, sustainability, and function.

engineering scale   A device used to make measurements in much the same way a ruler is used.

explanation diagram   An illustration explaining the way something works; a basic process; or the deconstruction of an object, a plan, or a drawing.

exploded assembly drawing   A formal drawing, usually in pictorial form, that shows the orientation and sequence in which parts are put together to make a device.

exploded configuration   A configuration of an assembly that shows instances separated from one another. An exploded configuration is used as the basis for an assembly drawing.

extension line   A thin, dark, solid line extending from a point on an object, perpendicular to a dimension line used to indicate the extension of a surface or point to a location preferably outside the part outline.

external thread   Threads that are formed on the outside of a cylindrical feature, such as on a bolt or stud.

extrude through all   An extrusion that begins on the sketching plane and protrudes or cuts through all portions of the solid model that it encounters.

extrude to selected surface   An extrusion where the protrusion or cut begins on the sketching plane and stops when it intersects a selected surface.

extrusion (in fabrication)   A process for making long, solid shapes with a constant cross section by squeezing raw material under elevated temperatures and pressure through an orifice shaped with that cross section.

extrusion (in 3D modeling)   A solid that is bounded by the surfaces swept out in space by a planar profile as it is pulled along a path perpendicular to the plane of the profile.

fabricate   To make something from existing materials.

family model   A collection of different versions of a part in a single model that can display any of the versions.

false easting   A value applied to the origin of a coordinate system to modify the x-coordinate readings, usually to make all of the coordinate values positive.

false northing   A value applied to the origin of a coordinate system to modify the y-coordinate readings, usually to make all of the coordinate values positive.

fastener   A manufactured part whose primary function is to join two or more parts.

feature array   A method for making additional features by placing copies of a master feature on the model at a specified equal spacing.

feature generalization   The process of going from the specific to the general in analyzing data.

features   Distinctive geometric shapes on solid parts; 3-D geometric entities that exist to serve some function.

feature-based solid modeling   A solid modeling system that uses features to build models.

feature control frame   The main alphabet of the language of geometric dimensioning and tolerancing. These boxes contain the geometric characteristic symbol, the geometric tolerances, and the relative datums.

feature pattern   *See* feature array.

feature tree   *See* model tree.

feature with size   A cylindrical or spherical surface or a set of two opposed elements or opposed parallel surfaces associated with a size dimension. Typical features with size are holes, cylinders, spheres, and opposite sides of a rectangular block.

feature without size   A planar surface or a feature where the normal vectors point in the same direction.

field   A column in a table that stores the values for a single attribute.

fill light   A light that softens and extends the illumination of the objects provided by the key light.

fillets Smooth transitions of the internal edge created by two intersecting surfaces and tangent to both intersecting surfaces.

Finite Element Analysis An advanced computer-based design analysis technique that involves subdividing an object into several small elements to determine stresses, displacements, pressure fields, thermal distributions, or electromagnetic fields.

first-angle projection The process of creating a view of an object by imprinting its image, using orthogonal projection, on an opaque surface behind that object.

fishbone diagram A diagram that shows the various subsystems in a device and the parts that make up each subsystem.

fixture A mechanical device, such as a clamps or bracket, used for holding a workpiece in place while it is being modified.

flash Bits of material that are left on a part from a casting or molding operation and found along the seams where the mold pieces separate to allow removal of the part.

floor plan A plan view of a single floor in a building that shows the layout of the rooms.

flowchart A quality improvement tools used to document, plan, or analyze a process or series of tasks.

foreshortened (line or plane) Appearing shorter than its actual length in one of the primary views.

forging A process of deforming metal with a common shape at room temperature into a new but similar shape by pressing it into a mold under elevated pressure.

form The shape of the thread cross section when cut through the axis of the thread cylinder.

form feature A recognizable area on a solid model that has a specific function.

forward kinematics In a hierarchical link, total motion in which the motion of the parent is transferred to the motion of the child.

foundation plan A plan view of the foundation of a building showing footings and other support structures.

foundation space The rectilinear volume that represents the limits of the volume occupied by an object.

foundation space The rectilinear volume that represents the limits of the volume occupied by an object.

frame rate The rate of speed, usually in frames per second, in which individual images or frames are played when an animation is viewed.

frontal surface A surface on an object being viewed that is parallel to the front viewing plane.

full section The section view produced when a single cutting plane is used to hypothetically cut an object completely into two pieces.

function curve A graphical method of displaying and controlling an object's transformations.

functional gage An inspection tool built uniquely for the purpose of quickly checking a specific dimension or geometric condition on a part to determine whether or not it fall within tolerance limits.

fused deposition A process where parts are gradually built up by bits of molten plastic that are deposited by a heated tip at selected locations and then solidified by cooling.

Gantt chart A tool for scheduling a project timeline.

general sections Sections through entire structures that show the layout of rooms but provide little detail.

geographic coordinate system A spatial reference system using a grid network of parallels and meridians to locate spatial features on the earth's surface.

geographically referenced data Information that is referenced to a specific geographic location, usually on the earth's surface.

geometric constraints Definitions used to control the shape of a profile sketch through geometric relationships.

geometric dimensioning and tolerancing (GD&T) A 3-D mathematical system that allows a designer to describe the form, orientation, and location of features on a part within precise tolerance zones.

geometric transformation Transformations used to alter the position, size, or orientation of a part, camera, or light over a specified period of time.

georeferenced data *See* geographically referenced data.

glass box A visualization aid for understanding the locations and orientations of images of an object produced by third-angle projection on a drawing. The images of an object are projected, using orthogonal projection, on the sides of a hypothetical transparent box that is then unfolded into a single plane.

global positioning systems (GPS) A system of geosynchronous, radio-emitting and receiving satellites used for determining positions on the earth.

graphical user interface (GUI) The format of information on the visual display of a computer, giving its user control of the input, output, and editing of the information.

green engineering The process by which environmental and life cycle considerations are examined from the outset in design.

grinding A method of removing small amounts of material from a part using a rotation abrasive wheel, thus creating surfaces of very accurate planar or cylindrical geometries.

ground constraint A constraint usually applied to a new sketch to fix the location of the sketch in space.

ground line (GL)   In the construction of a perspective view, a line on the elevation view that represents the height of the ground.

GRS80 spheroid   The satellite-based spheroid for the Geodetic Reference System 1980.

half section   The section view produced when a single cutting plane is used to hypothetically cut an object up to a plane or axis of symmetry, leaving that portion beyond the plane or axis intact.

header   A premade outline on which working drawings are created to ensure that all information required for fabrication and record keeping is entered.

heating and ventilation plan   A plan view of the ventilation systems on a specific floor of a building, including ductwork and devices such as air conditioning units.

hidden lines   The representation, using dashed lines, on a drawing of an object of the edges that cannot be seen because the object is opaque.

hierarchy   The parent-child relationships between instances in an assembly.

hierarchical link   A series of user-defined or linked objects that have a parent-child-grandchild relationship.

history tree   *See* model tree.

holes   A cut feature added to a model that will often receive a fastener for system assembly.

horizon line (HL)   In the construction of a perspective view, the line that represent the horizon, which is the separation between the earth and the sky at a long distance. The left and right vanishing points are located on the HL. The PP and the HL are usually parallel to each other.

horizontal modeling   A strategy for creating solid models that reduces parent-child relationships within the feature tree.

horizontal surface   A surface on an object being viewed that is parallel to the top viewing plane.

identity   A geometric integration of spatial datasets that preserves only the geographic features from the first input layer; the second layer merely adds more information to the dataset.

image   A collection of printed, displayed, or imagined patterns intended to represent real objects, data, or processes.

inclined surface   A flat surface on an object being viewed that is perpendicular to one primary view and angled with respect to the other two views.

index   A list of all sheets of drawings contained in a set of construction plans.

industrial designers   The individuals who use their creative abilities to develop conceptual designs of potential products.

infographic   A shortened form of *informational graphic* or *information graphic*.

information graphics   Often referred to as *infographic*, visual explanations.

injection molding   A process for creating a plastic part by injecting molten plastic into a mold under pressure, allowing the material to solidify, and removing the part from the mold.

instances   Copies of components that are included within an assembly model.

instructional diagram   A diagram showing how a specific action within an object occurs.

instruments   In engineering drawing, mechanical devices used to aid in creating accurate and precise images.

interchangeable manufacturing   A process by which parts are made at different locations and brought together for assembly. For many industries, this process opens the door for third-party companies to produce replacement parts or custom parts.

interference   A fit where two mating parts have intersecting nominal volumes, requiring the deformation of the parts . For example, the diameter of the shaft is larger than the diameter of the hole. When assembled, the intent is that the shaft will not spin in the hole.

interference   The amount of overlap between two instances in an assembly.

internal thread   Threads that are formed on the inside of a hole.

international sheet sizes   The internationally accepted paper dimensions used when drawings are created or printed to their full intended size.

intersect   To create a new object that consists of the volume of interference between two objects.

intersection   A geometric integration of spatial datasets that preserves features or portions of features that fall within areas common to all input datasets.

inverse kinematics   A bidirectional set of constraints that allows motion of a set of linked objects by moving the very end of the hierarchically linked chain and having the rest of the links move in response.

investment casting   A method of casting where the mold is formed by successive dipping of a master form into progressively coarser slurries, allowing each layer to harden between each dipping. *See* casting.

isometric axes   A set of three coordinate axes that are portrayed on the paper at 120 degrees relative to one another.

isometric dot paper   Paper used for sketching purposes that includes dots located along lines that meet at 120 degrees.

isometric drawing   An axonometric drawing in which the scaling factor is the same for all three axes.

isometric grid paper Paper used for sketching purposes that includes grid lines at 120 degrees relative to one another.

isometric lines   Lines on an isometric drawing that are parallel or perpendicular to the front, top, or profile viewing planes.

isometric pictorial A sketch of an object that shows its three dimensions where isometric axes were used as the basis for defining the edges of the object.

item number   A number used to identify a part on a layout or assembly drawing.

join   To absorb the volume of interference between two objects to form a third object.

key   A small removable part similar to a wedge that provides a positive means of transferring torque between a shaft and a hub.

key light   A light that creates an object's main illumination, defines the dominant angle of the lighting, and is responsible for major highlights on objects in a scene.

keyframe   A specific frame located at a specified time within an animation where an object's location, orientation, and scale are defined perfectly.

keyseat   A rectangular groove cut in a shaft to position a key.

keyway   A rectangular groove cut in a hub to position a key.

landscape   The drawing orientation in which the horizontal size is larger than the vertical size.

lap joint   A joint between two parts wherein the parts are overlapped.

laser scanning (three-dimensional)   A process where cameras and lasers are used to digitize an object based on the principle of triangulation.

lathe   A machine used to make axially symmetric parts or features using a material removal process known as turning.

latitude   An imaginary line around the Earth's surface in which all of the points on the line are equidistant from the Equator.

layout drawing   A formal drawing that shows a device in its assembled state with all of its parts identified.

lead   The distance a screw thread advances axially in one full turn.

leader   A thin, dark, solid line terminating with an arrowhead at one end and a dimension, note, or symbol at the other end.

left-handed system Any 3-D coordinate system that is defined by the left-hand rule.

level of detail   The number of polygon mesh triangles used to define the surface shape of a 3-D model. For rendering speed, as a general case, objects close to the camera in a scene require a higher number of polygons to more accurately define their surfaces while more distant objects can be effectively rendered with fewer polygons.

life cycle   The amount of time a product will be used before it is no longer effective.

line   A spatial feature that has location and length but no area and is represented by a series of nodes, points, and arcs.

line chart   A graph showing the relationship between two sets of data, where line segments are used to link the data to show trends in their changes.

list   A boxed series of components, definitions, tips, etc.

location   A dimension associated with the position of a feature on a part.

location grid   An imaginary alphanumeric grid, similar to that of a street map, on a drawing that is used to specify area locations on the drawing.

longitude   An imaginary north-south line on the Earth's surface that extends from the North Pole to the South Pole.

machine screw   A threaded fastener wherein the threads are cut along the entire length of the cylindrical shaft. Machine screws can mate with a threaded hole or nut.

major diameter   The largest diameter on an internal or external thread.

main assembly   A completed device usually composed of multiple smaller parts and/or subassemblies.

main title block   A bordered area of a drawing (and part of the drawing header) that contains important information about the identification, fabrication, history, and ownership of the item shown on the drawing.

manufacturing drawings   Working drawings, often created by mechanical engineers, that are used to mass-produce products for consumers.

map   A diagram of the location of events with geography.

map projection   A systematic arrangement of parallels and meridians on a plane surface representing the geographic coordinate system.

mapping coordinates   Also called UVW coordinates, mapping coordinates are special coordinate systems designed to correctly place and control the shape of external and procedurally generated images on the surfaces of 3-D models.

mass   A property of an object's ability to resist a change in acceleration.

mass properties analysis A computer-generated document that gives the mechanical properties of a 3-D solid model.

master feature A feature or collection of features that is to be copied for placement at other locations in a model.

master model In a collection of similar parts, the model that includes all of the features that may appear in any of the other parts.

matt object An object with a combined material and alpha channel map.

maximum material condition The condition in which a feature of size contains the maximum amount of material within the stated limits of size.

measuring line (ML) In the construction of a perspective view, a vertical line used in conjunction with the elevation view to locate vertical points on the perspective drawing.

measuring wall In the construction of a perspective view, a line that extends from the object to the vanishing point to help establish the location of horizontal points on the drawing.

mechanical dissection The process of taking apart a device to determine the function of each part.

mechanical stress Developed force applied per unit area that tries to deform an object.

mental rotations The ability to mentally turn an object in space.

meridian A line of longitude through the North and South Poles that measure either E or W in a geographic coordinate system.

mesh The series of elements and nodal points on a finite element model.

Metric scale A device used to measure or draw lines in the metric system of units with drawings scales reported as ratios.

metrology The practice of measuring parts.

milling A process of removing material from a part using a rotating tool bit that can remove material in directions parallel or perpendicular to the tool bit's rotation axis.

milling machine A machine used to make parts through a material removal process known as milling.

minor diameter The smallest diameter on an internal or external thread.

mirrored feature A feature that is created as a mirror image of a master feature.

model A mathematical representation of an object or a device from which information about its function, appearance, or physical properties can be extracted.

model builders Engineers who make physical mock-ups of designs using modern rapid prototyping and CAM equipment.

model tree A list of all of the features of a solid model in the order in which they were created, providing a "history" of the sequence of feature creation.

mold A supported cavity shaped like a desired part into which molten material is poured or injected.

moment-of-inertia The measure of an object's ability to resist rotational acceleration about an axis.

morphological chart A chart used to generate ideas about the desirable qualities of a product and all of the possible options for achieving them.

motion blur The amount of movement of a high-speed object recorded as it moves through a single frame.

motion path Spline curves that serve as a trajectory for the motion of objects in animation.

multiple thread A thread made up of two or more continuous ridges side by side.

multiple views The presentation of an object using more than one image on the same drawing, each image representing a different orientation of the object.

multiview Refers to a drawing that contains more than one image of an object and whose adjacent images are generated from orthogonal viewing planes.

node A point at the beginning and end of a line feature or a point that defines a polygon feature.

normal surface A surface on an object being viewed that is parallel to one of the primary viewing planes.

note taker The person who records the actions discussed and taken at team meetings and then prepares the formal written notes for the meeting.

notes Additional information or instructions placed on a drawing that are not contained on the dimensions, tolerances, or header.

nut The threaded mate to a bolt used to hold two or more pieces of material together.

oblate ellipsoid An ellipsoid created by rotating an ellipse around its minor axis.

oblique axes A set of three coordinate axes that are portrayed on the paper as two perpendicular lines, with the third axis meeting them at an angle, typically 45 degrees.

oblique pictorial A sketch of an object that shows one face in the plane of the paper and the third dimension receding off at an angle relative to the face.

oblique surface A flat surface on an object being viewed that is neither parallel nor perpendicular to any of the primary views.

offset section The section view produced by a stepped cutting plane that is used to hypothetically cut an object completely into two pieces. Different portions of the plane are used to reveal the interior details of different features of interest.

one-off A one-of-a-kind engineering project for which no physical prototypes are created.

optimization   Modification of shapes, sizes, and other variables to achieve the best performance based on predefined criteria.

organizational chart   A chart representing the relationships of entities of an organization in terms of responsibility or authority.

orthogonal projection   The process by which the image of an object is created on a viewing plane by rays from the object that are perpendicular to that plane.

outline assembly drawing   *See* layout drawing.

parallel   An imaginary line parallel to the equator that corresponds to a measurement of latitude either N or S in a geographic coordinate system.

parameters   The attributes of features, such as dimensions, that can be modified.

parametric solid modeling   A solid modeling system that allows the user to vary the dimensions and other parameters of the model.

parametric techniques   Modeling techniques where all driven dimensions in algebraic expressions must be known for the value of the dependent variables to be calculated.

parent feature   A feature used in the creation of another feature, which is called its child feature.

part   An object expected to be delivered from a fabricator as a single unit with only its external dimensions and functional requirements specified.

part name   A very short descriptive title given to a part, subassembly, or device.

part number   Within a company, a string of alphanumeric characters used to identify a part, a subassembly, an assembly, or a device.

particle system   Specialized software modules used to generate, control, and animate very large numbers of small objects involved in complex events.

parts list   *See* bill of materials.

path   The specified curve on which a profile is placed to create a swept solid.

passive animation   An animation in which the observer remains still while the action occurs around him or her.

patents   A formal way to protect intellectual property rights for a new product.

pattern   A master part from which molds can be made for casting final parts.

perspective drawing   A drawing in which all three-dimensional axes on an object can be seen, with the scaling factor linearly increasing or decreasing in each direction. Usually one axis is shown as being vertical. This type of drawing generally offers the most realistic presentation of an object.

pictorial   A drawing that shows the 3-D aspects and features of an object.

picture plane (PP)   In the construction of a perspective view, the viewing plane through which the object is seen. The PP appears as a line (edge view of the viewing plane) in the plan view.

pie chart   A circular chart that is divided into wedges like a pie, representing a piece of the whole.

pin   A cylindrical (or slightly tapered) fastener typically used to maintain a desired position or orientation between parts.

pitch   The distance from one point on a thread to the corresponding point on the adjacent thread as measured parallel to its axis.

pitch diameter   The diameter of an imaginary cylinder that is halfway between the major and minor diameters of the screw thread.

pivot point   An independent, movable coordinate system on an object that can be used for location, orientation, and scale transformations.

pixel   The contraction for "picture element"; the smallest unit of information within a grid or raster data set.

plan and profile drawings   Construction drawings typically used for roads or other linear entities that show the road from above as well as from the side, with the profile view usually drawn with an exaggerated vertical scale.

plan view   In the construction of a perspective view, the object as viewed from the top, as if created by orthogonal projection.

plan views   Drawings created from a viewpoint above the structure (top view).

planar coordinate system   A 2-D measurement system that locates features on a plane based on their distance from an origin (0,0) along two perpendicular axes.

point   A spatial feature that has only location, has neither length nor area, and is represented by a pair of xy coordinates.

point light   A computer-generated light source, also called an omni light, that emits light rays and casts shadows uniformly in all directions. Also called an omnidirectional light.

point tracking   A procedure by which successive vertices on an object are simultaneously located on a pictorial image and a multiview image of that object.

polygon   A spatial feature that has location, area, and perimeter and is represented by a series of nodes, points, and arcs that must form a closed boundary.

portrait   The drawing orientation in which the vertical size is larger than the horizontal size.

preferred configuration   The drawing presentation of an object using its top, front, and right-side views.

primary modeling planes   The planes representing the XY-, XZ-, and YZ-planes in a Cartesian coordinate system.

primitives   The set of regular shapes, such as boxes, spheres, or cylinders that are used to build solid models with constructive solid geometry methods (CSG).

principal viewing planes   The planes in space on which the top, bottom, front, back, and right and left side views are projected.

problem identification   The first stage in the design process where the need for a product or a product modification is clearly defined.

procedural textures   Texture mapping routines based on algorithms written into the rendering software that can generate a specialized colored pattern such as wood, water, a checker pattern, a tile pattern, stucco, and many others without reference to external image files.

process check   A method for resolving differences and making adjustments in team performance.

process diagram   An illustration that explains how system elements work and how interactions occur.

professional engineer (PE)   An individual who has received an engineering degree, who has worked under the supervision of a PE for a number of years, and who has passed two examinations certifying knowledge of engineering practice.

profile   A planar sketch that is used to create a solid.

profile surface   A surface on an object being viewed that is parallel to a side viewing plane.

profile views   Views of a structure that show horizontal surfaces in edge view (side or front views).

project   In engineering, a collection of tasks that must be performed to create, operate, or retire a system or device.

projection ray   A line perpendicular to the projection plane. It transfers the 2-D shape from the object to an adjacent view. Projection rays are drawn lightly or are not shown at all on a finished drawing.

prototype   The initial creation of a product for testing and analysis before it is mass-produced.

protrusion   A feature created by the addition of solid volume to a model.

qualitative data   Information collected using words and ideas.

quantitative data   Numeric information.

quantity per machine (Q/M)   The number of times a part is required to build its next highest assembly.

radius-of-gyration   The distance from an axis where all of the mass can be concentrated and still produce the same moment-of-inertia.

rapid prototyping   Various methods for creating a part quickly by selective hardening of a powder or liquid raw material at room temperature.

raster data model   A representation of the geographic location as a surface divided into a regular grid of cells or pixels.

raytraced shadows   Shadows calculated by a process called raytracing, which traces the path that a ray of light would take from the light source to illuminate or shade each point on an object.

raytracing   A method of rendering that builds an image by tracing rays from the observer, bouncing them off the surfaces of objects in the scene, and tracing them back to the light sources that illuminate the scene.

reaming   A process for creating a hole with a very accurate final diameter using an accurately made cylindrical cutting tool similar to a drill bit to remove final bits of material after a smaller initial hole is created.

rebars   Steel bars added to concrete for reinforcement or for temperature control.

receding dimension   The portion of the object that appears to go back from the plane of the paper in an oblique pictorial.

record   A set of related data fields, often a row in a database, containing the attribute values for a single feature.

reference dimensions   Unneeded dimensions shown for the convenience of the reader used to show overall dimensions that could be extracted from other dimensions on the part or from other drawings.

reference line   Edges of the glass box or the intersection of the perpendicular planes. The reference line is drawn only when needed to aid in constructing additional views. The reference line should be labeled in constructing auxiliary views to show its association between the planes it is representing; for example, H/F for the hinged line between the frontal and horizontal planes. A reference line is also referred to as a fold line or a hinged line.

reflection   The process of obtaining a mirror image of an object from a plane of reflection.

reflection mapping   Mapping that allows the use of grayscale values in an image file to create the illusion of a reflection on the surface of a part. White creates reflective highlights, while black is transparent to the underlying color of the surface.

regeneration   The process of updating the profile or part to show its new shape after constraints are added or changed.

related views   Views adjacent to the same view that share a common dimension that must be transferred in creating auxiliary views.

removed section   The section view produced when a cutting plane is used to hypothetically remove an infinitesimally thin slice of an object for viewing.

rendering   The process where a software program uses all of the 3-D geometric object and lighting data to calculate and display a finished image of a 3-D scene in a 2-D viewport.

retaining rings   Precision-engineered fasteners that provide removable shoulders for positioning or limiting movement in an assembly.

reverse engineering   A systematic methodology for analyzing the design of an existing device.

revolved section   The section view produced when a cutting plane is used to hypothetically create an infinitesimally thin slice, which is rotated 90 degrees for viewing, on an object.

revolved solid   A solid formed when a profile curve is rotated about an axis.

ribs   Constant thickness protrusions that extend from the surface of a part and are used to strengthen or stiffen the part.

right-hand rule   Used to define a 3-D coordinate system whereby by pointing the fingers of the right hand down the x-axis and curling them in the direction of the y-axis, the thumb will point down the z-axis.

right-handed system   Any 3-D coordinate system that is defined by the right-hand-rule.

rivet   A cylindrical pin with heads at both ends, one head being formed during the assembly process, forming a permanent fastener often used to hold sheet metal together.

rolling   A process for creating long bars with flat, round, or rectangular cross sections by squeezing solid raw material between large rollers. This can be done when the material is in a hot, soft state (hot rolling) or when the material is near room temperature (cold rolling).

root   The bottom surface or point of a screw thread.

rounds   Smooth radius transitions of external edges created by two intersecting surfaces and tangent to both intersecting surfaces.

sand casting   A casting process where the mold is made of sand and binder material hardened around a master pattern that is subsequently removed to form the cavity. *See* casting.

sawing   A cutting process that uses a multitoothed blade that moves rapidly across and then through the part.

scatter plot   A graph using a pattern of dots showing the relationship between two sets of data.

schedule of materials   A list of the materials, such as doors and windows, necessary for a construction project.

schematic diagram   A diagram explaining how components work together, what the measurements are, how components are set up, or how pieces are connected.

screw thread   A helix or conical spiral formed on the external surface of a shaft or on the internal surface of a cylindrical hole.

scripting   A programming capability that allows a user to access and write code at or near the source code level of the software.

secondary title block   An additional bordered area of a drawing (and part of the drawing header) that contains important information about the identification, fabrication, and history of the item shown on the drawing.

section lines   Shading used to indicate newly formed or cut surfaces that result when an object is hypothetically cut.

section view   A general term for any view that presents an object that has been hypothetically cut to reveal the interior details of its features, with the cut surfaces perpendicular to the viewing direction and filled with section lines for improved presentation.

sectioned assembly drawing   A formal drawing, usually in pictorial form, that shows the device in its assembled form but with sections removed from obscuring parts to reveal formerly hidden parts.

selective laser sintering   A process where a high-powered laser is used to selectively melt together the particles on a bed of powdered metal to form the shape of a desired part.

self-tapping screw   A fastener that creates its own mating thread.

sequence diagram   A group of diagrams that includes process diagrams, timelines, and step-by-step diagrams.

set of construction plans   A collection of drawings, not necessarily all of them plan views, needed to construct a building or infrastructure project.

set screw   A small screw used to prevent parts from moving due to vibration or rotation, such as to hold a hub on a shaft.

shading   Marks added to surfaces and features of a sketch to highlight 3-D effects.

shading algorithms   Algorithms designed to deal with the diffuse and specular light transmission on the surface of an object.

shelling   Removing most of the interior volume of a solid model, leaving a relatively thin wall of material that closely conforms to the outer surfaces of the original model.

sidebar   Small infographics used within a body of text that are subdivided into briefs, lists, and bio profiles.

single thread   A thread that is formed as one continuous ridge.

sintering   A process where a part is formed by placing powdered metal into a mold and then applying heat and pressure to fuse the powder into a single solid shape.

site plan   A plan view showing the construction site for an infrastructure project.

site survey   Data regarding the existing topography and structures gathered during the preliminary design stages by trained surveying crews.

six standard views (or six principal views)   The drawing presentation of an object using the views produced by the glass box (i.e., the top, front, bottom, rear, left-side, and right-side views).

size   The general term for the size of a feature, such as a hole, cylinder, or set of opposed parallel surfaces.

sketches   Collections of 2-D entities.

sketching editor   A software tool used to create and edit sketches.

sketching plane   A plane where 2-D sketches and profiles can be created.

slope   The rate of change of elevation (rise) over a specified distance (run). Measured in percent or degrees.

solid model   A mathematical representation of a physical object that includes the surfaces and the interior material, usually including a computer-based simulation that produces a visual display of an object as if it existed in three dimensions.

solid modeling   Three-dimensional modeling of parts and assemblies originally developed for mechanical engineering use but presently used in all engineering disciplines.

spatial data   A formalized schema for representing data that has both geographic location and descriptive information.

spatial orientation   The ability of a person to mentally determine his own location and orientation within a given environment.

spatial perception   The ability to identify horizontal and vertical directions.

spatial relations   The ability to visualize the relationship between two objects in space, i.e., overlapping or nonoverlapping.

spatial visualization   The ability to mentally transform (rotate, translate, or mirror) or to mentally alter (twist, fold, or invert) 2-D figures and/or 3-D objects.

specifications (specs)   The written instructions that accompany a set of construction plans used to build an infrastructure project.

spheroid   *See* oblate ellipsoid.

spindle   That part of a production cutting machine that spins rapidly, usually holding a cutting tool or a workpiece.

splines   Polynomial curves that pass through multiple data points.

split line   The location where a mold can be disassembled for removal of a part once the molten raw material inside has solidified.

spotlight   A computer-generated light that simulates light being emitted from a point in space through a cone or beam, with the angle and direction of light controlled by the user.

spring pin   A hollow pin that is manufactured by cold-forming strip metal in a progressive roll-forming operation. Spring pins are slightly larger in diameter than the hole into which they are inserted and must be radially compressed for assembly.

sprue   Bits of material that are left on the part from a casting or molding operation and found at the ports where the molten material is injected into the mold or at the ports where air is allowed to escape.

stage   That part of a machine that secures and slowly moves a cutting tool or workpiece in one or more directions.

stamping   A process for cutting and shaping sheet metal by shearing and bending it inside forms with closely fitting cutouts and protrusions.

standard commercial shape   A common shape for raw material as would be delivered from a material manufacturer.

State Plane Coordinate System   The planar coordinate system developed in the 1930s for each state to permanently record the locations of the original land survey monuments in the United States.

station point (SP)   In the construction of a perspective view, the theoretical location of the observer who looks at the object through the picture plane.

statistical tolerancing   A way to assign tolerances based on sound statistical practices rather than conventional tolerancing practices.

step segment   On a stepped cutting plane for an offset section view, that portion of the plane that connects the cutting segments and is usually perpendicular to them but does not intersect any interior features.

step-by-step diagram   An illustration that visually explains a complex process; it is a type of a sequence diagram.

stereolithography   A process for creating solid parts from a liquid resin by selectively focusing heat or ultraviolet light into a pool of the resin, causing it to harden and cure in the selected areas.

storyboard   A sequential set of keyframe sketches or drawings, including brief descriptions, indications of object and camera movement, lighting, proposed frame numbers, and timelines sufficient to produce a complete animation project.

stud   A fastener that is a steel rod with threads at both ends.

subassembly   A logical grouping of assembly instances that is treated as a single entity within the overall assembly model.

subassemblies   Collections of parts that have been put together for the purpose of installing the collections as single units into larger assemblies.

successive cuts   A method of forming an object with a complex shape by starting with a basic shape and removing parts of it through subtraction of other basic shapes.

suppressed   Refers to the option for not displaying a selected feature.

surface area   The total area of the surfaces that bound an object.

surface model   A CAD-generated model created to show a part as a collection of intersecting surfaces that bound a solid.

surface modeling   The technique of creating a 3-D computer model to show a part or an object as a collection of intersecting surfaces that bound the part's solid shape.

surface normal   A vector that is perpendicular to each polygon contained in a polygon mesh model.

surface tracking   A procedure by which successive surfaces on an object are simultaneously located on a pictorial image and a multiview image of that object.

sustainable design   A paradigm for making design decisions based on environmental considerations and life cycle analysis.

swept feature   A solid that is bound by the surfaces swept out in space as a profile is pulled along a path.

symmetry   The characteristic of an object in which one half of the object is a mirror image of the other half.

system   A collection of parts, assemblies, structures, and processes that work together to perform one or more prescribed functions.

table   Data organized in columns and rows.

tangent edge   The intersection line between two surfaces that are tangent to each other.

tap   The machine tool used to form an interior thread. Tapping is the process of making an internal thread.

tap drill   A drill used to make a hole in material before the internal threads are cut.

tapped hole   A hole that has screw threads inside it.

task credit matrix   A table that lists all team members and their efforts on project tasks.

team contract   The rules under which a team agrees to operate (also known as a code of conduct, an agreement to cooperate, or rules of engagement).

team leader   The person who calls the meetings, sets the agenda, and maintains the focus of team meetings.

team roles   The roles that team members fill to ensure maximum effectiveness for a team.

technical diagram   A diagram depicting a technical illustration's measurements, movement, dissection, or relationship of parts.

telephoto   As seen through a camera lens with a focal length longer than 80 degrees, creating a narrow field of view and resulting in a flattened perspective.

texture mapping   The technique of adding variation and detail to a surface that goes beyond the level of detail modeled in the geometry of an object.

thematic layer   Features of one type that are generally placed together in a single georeferenced data layer.

thematic layer overlay   The process of combining spatial information from two thematic layers.

third-angle projection   The process of creating a view of an object by imprinting its image, using orthogonal projection, on translucent surface in front of that object.

thread note   Information on a drawing that clearly and completely identifies a thread.

thread series   The number of threads per inch on a standard thread.

3-D coordinate system   A set of three mutually perpendicular axes used to define 3-D space.

3-D printing   A process for creating solid objects from a powder material by spraying a controlled stream of a binding fluid into a bed of that powder, thus fusing the powder in the selected areas.

three-axis mill   A milling machine whose spindle, which holds the rotating cutting tool, can be oriented along any one of three Cartesian axes.

three-dimensional (3-D) modeling   Mathematical modeling where the appearance, volumetric, and inertial properties of parts, assemblies, or structures are created with the assistance of computers and display devices.

through hole   A hole that extends all the way through a part.

tick mark   A short dash used in sketching to locate points on the paper.

timekeeper   The person who keeps track of the meeting agenda, keeping the team on track to complete all necessary items within the alotted time frame.

timeline   A specific type of sequence diagram used to highlight significant moments in history.

title block   Usually the main title block, which is a bordered area of the drawing (and part of the drawing header) that contains important information about the identification, fabrication, history, and ownership of the item shown on the drawing.

tolerance   The total amount a specific dimension is permitted to vary. It is the difference between the upper and lower limits of the dimension.

tool bit   A fixed or moving replaceable cutting implement with one or more sharpened edges used to remove material from a part.

tool runout   The distance a tool may go beyond the required full thread length.

tooling   Tools and fixtures used to hold, align, create, or transport a part during its production.

top-down modeling   The process of establishing the assembly and hierarchy before individual components are created.

trail   Dashed lines on an assembly drawing that show how various parts or subassemblies are inserted to create a larger assembly.

trajectory   *See* path.

transparency/opacity mapping   A technique used to create areas of differing transparency on a surface or an object.

trimetric drawing   An axonometric drawing in which the scaling factor is different for all three axes.

true shape (of a plane)   The actual shape and size of a plane surface as seen in a view that is parallel to the surface in question.

two-dimensional (2-D) drawing   Mathematical modeling or drawing where the appearance of parts, assemblies, or structures are represented by a collection of two-dimensional geometric shapes.

tumbling   A process for removing sharp external edges and extraneous bits of material from a part by surrounding it in a pool of fine abrasive pellets and then shaking the combination.

turning   A process for making axially symmetric parts or features by rotating the part on a spindle and applying a cutting to the part.

undercut feature   A concave feature in which the removed material expands outward anywhere along its depth.

union   A topological overlay of two or more polygon spatial datasets that preserves the features that fall within the spatial extent of either input dataset; that is, all features from both datasets are retained and extracted into a new polygon dataset.

Universal Transverse Mercator (UTM)   The planar coordinate system that divides the earth's surface between 84° N and 80° S into 60 zones, each 6° longitude wide.

unsuppressed   Refers to the option for displaying a selected feature.

US sheet sizes   The accepted paper dimensions used in the United States when drawings are created or printed to their intended size.

vanishing point (VP)   In the construction of a perspective view, the point on the horizon where all parallel lines in a single direction converge.

variational techniques   Modeling techniques in which algebraic expressions or equations that express relationships between a number of variables and constants, any one of which can be calculated when all of the others are known.

vector data model   A data model that uses nodes and their associated geographic coordinates to construct and define spatial features.

Venn diagram   A type of business diagram that shows the mathematical or logical relationships and overlapping connections between different groups or sets of information.

vertex   A point that is used to define the endpoint of an entity such as a line segment or the intersection of two geometric entities.

video compression   One of a number of algorithms designed to reduce the size and storage requirements of video content.

viewing direction   The direction indicated by arrows on the cutting plane line from the eye to the object of interest that corresponds to the tail and point of the arrow, respectively.

viewing plane   A hypothetical plane between an object and its viewer onto which the image of the object, as seen by the viewer, is imprinted.

visual storytelling diagram   An illustration that displays empirical data or clarification of ideas.

visual thinking   A method for creative thinking, usually through sketching, where visual feedback assists in the development of creative ideas.

visualization   The ability to create and manipulate mental images of devices or processes.

volume   The quantity of space enclosed within an object's boundary surfaces.

volume of interference   The volume that is common between two overlapping objects.

wall sections   Sectional views of walls from foundation to roof for a construction project.

washer   A flat disk with a center hole to allow a fastener to pass through it.

webs   Small, thin protrusions that connect two or more thicker regions on a part.

weighted decision table   A matrix used to weigh design options to determine the best possible design characteristics.

welding   A method for joining two or more separate parts by applying heat to the edges where they meet and melting the edges together along with a filler of essentially the same material composition as the parts.

wide-angle   As seen through a camera lens with a focal length shorter than 30 degrees, creating a wide field of view and resulting in a distorted and exaggerated perspective.

wire drawing   The process of reducing the diameter of a solid wire by pulling it through a nozzle with a reducing aperture.

wireframe models   CAD models created using lines, arcs, and other 2-D entities to represent the edges of the part; surfaces or solid volumes are not defined.

working drawings   A collection of all drawings needed to fabricate and put together a device or structure.

workpiece   A common name for a part while it is still in the fabrication process, that is, before it is a finished part.

z-buffer rendering   A scene-rendering technique that uses visible-surface determination in which each pixel records (in addition to color) its distance from the camera, its angle, light source orientation, and other information defining the visible structure of the scene.

z-value   The value for a given surface location that represents an attribute other than position. In an elevation or terrain model, the z-value represents elevation.

index